PRENTICE-HALL SERIES IN MATHEMATICAL ECONOMICS

Donald V. T. Bear, *Series Editor*

Dennis J. Aigner
BASIC ECONOMETRICS

David A. Bowers and Robert N. Baird
ELEMENTARY MATHEMATICAL MACROECONOMICS

Michael D. Intriligator
MATHEMATICAL OPTIMIZATION AND ECONOMIC THEORY

Ronald C. Read
A MATHEMATICAL BACKGROUND FOR ECONOMISTS
AND SOCIAL SCIENTISTS

Menahem E. Yaari
LINEAR ALGEBRA FOR SOCIAL SCIENCES

A Mathematical
Background
for Economists
and
Social Scientists

RONALD C. READ

Professor of Mathematics
University of Waterloo
Ontario, Canada

A Mathematical Background for Economists and Social Scientists

PRENTICE-HALL, INC., ENGLEWOOD CLIFFS, N. J.

A MATHEMATICAL BACKGROUND
FOR ECONOMISTS AND SOCIAL SCIENTISTS
by Ronald C. Read

© 1972 PRENTICE-HALL, INC., ENGLEWOOD CLIFFS, N. J.

Printed in the United States of America

ISBN: 0-13-560987-9
Library of Congress Catalog Card Number: 70-143436

10 9 8 7 6 5 4 3 2 1

PRENTICE-HALL INTERNATIONAL, INC., *London*
PRENTICE-HALL OF AUSTRALIA, PTY. LTD., *Sydney*
PRENTICE-HALL OF CANADA, LTD., *Toronto*
PRENTICE-HALL OF INDIA PRIVATE LIMITED, *New Delhi*
PRENTICE-HALL OF JAPAN, INC., *Tokyo*

Series Foreword

The Prentice-Hall Series in Mathematical Economics is intended as a vehicle for making mathematical reasoning and quantitative methods available to the main corpus of the undergraduate and graduate economics curricula.

The Series has been undertaken in the belief that the teaching of economics will, in the future, increasingly reflect the discipline's growing reliance upon mathematical and statistical techniques during the past 20 to 35 years and that mathematical economics and econometrics ought not to be "special fields" for undergraduates and graduate students, but that every aspect of economics education can benefit from the application of these techniques.

Accordingly, the Series will contain texts that cover the traditional substantive areas of the curriculum—for example, macroeconomics, microeconomics, public finance, and international trade—thereby offering the instructor the opportunity to expose his students to contemporary methods of analysis as they apply to the subject matter of his course. The composition of the early volumes in the Series will be weighted in favor of texts that offer the student various degrees of mathematical background, with the volumes of more substantive emphasis following shortly thereafter.

As the Series grows, it will contribute to the comprehensibility and quality of economics education at both the undergraduate and graduate levels.

DONALD V. T. BEAR, *Series Editor*

vii

Contents

Preface

It is customary in scientific writing to state the important points just as often as the unimportant ones; exactly once; and to assume thereafter that this vital information is graven in the minds of all readers. This is one of the wonders of science, for it is economical of space in journals and it encourages researchers to write up their findings. It also ensures a certain permanence to the work, for it will be years before it is completely understood.

J. D. WILLIAMS
The Compleat Strategyst

This book is not a textbook of mathematical economics, nor is it meant to be. It is a textbook of mathematics in which the subject matter has been chosen so as to give students of economics and the social sciences a firm grounding in those branches of mathematics that are increasingly being used in these disciplines. Throughout this book, where possible and convenient, I have chosen illustrative examples of economic significance, but I have not done so where this would have entailed too much in the way of non-mathematical preliminaries or, alternatively, the assumption that the reader already had some knowledge of economic theory. Consequently, it is not necessary for the reader to have any prior knowledge of economics or the social sciences before starting this book. A reader who, innocent of any economic theory, reads this book from cover to cover will pick up from it a smattering of theory of a sort as he goes along, but if it is to be of any use this will certainly have to be supplemented by collateral reading of the standard economics textbooks.

Naturally, no amount of printed words can replace a good teacher who knows his subject. Nevertheless, I have attempted to write so that the material will be intelligible to a student working by himself, without the need for interpretation or amplification by an instructor. This has necessitated the use of a discursive style in an attempt to make the book, if not exactly pleasant bedside reading, at least comfortably readable. What is more, I have not hesitated to go over the same ground more than once (though with variations) if the importance of the subject-matter seemed to warrant it, thus showing my agreement with J. D. Williams (in the quotation given above) and his disparagement of the ultra-terse style of much modern scientific writing.

There is one very obvious omission in the topics covered in this book —that of statistics, a subject that is certainly of prime importance for economists and social scientists. This omission is quite deliberate. To have included an adequate treatment of the theory of statistics would have meant increasing quite unconscionably an already long book. This might have been justified were it not for the fact that there are many excellent books on statistics already on the market. Even so, we do encroach on the borders of this subject at several places in this book: in the second half of chapter 4, the theory of probability, and in the treatment of curve-fitting in chapter 9, where the concepts of variance, mean-square deviation and correlation coefficient are defined.

For all this, the student will not get from this book the knowledge of statistics that he needs. This deficiency can be made up by consulting the following books:

Croxton, F. E. and D. J. Cowden, *Applied General Statistics*, 5th ed. Prentice-Hall, Englewood Cliffs, N. J., 1967.
Freund, J. E., *Mathematical Statistics*. Prentice-Hall, Englewood Cliffs, N. J., 1962.
Goldberger, A. S., *Econometric Theory*. Wiley, New York, 1964.
Meroney, M. J., *Facts from Figures*. Penguin Books, London, 1951.
Diegel, S., *Non-parametric Statistics for the Behavioural Sciences*. McGraw-Hill, New York, 1956.

Statistics apart, the student should get from this book exactly what the title suggests, namely a mathematical background against which he can view the various problems that come his way. There are many advanced topics of mathematical economics that are not treated in this book, but the student who has assimilated what is presented here should have little trouble with any further mathematical processes that may turn out to be of interest to him.

A Mathematical
Background
for Economists
and
Social Scientists

1 Elementary

Logical Ideas

Logical consequences are the scarecrows of fools and the beacons of wise men.

T.H. HUXLEY.

1.1 STATEMENTS

The fundamental idea with which we shall be concerned in this chapter is that of a statement. The word *statement* is familiar enough in everyday language; the way that we shall use it mathematically will not differ much from the customary usage. The following are statements:

This book is black.
Cats have nine lives.
$(x + 1)(x - 1) = x^2 - 1$ for all numbers x.
Economics is the dismal science.

Frequently a statement made by a person expresses something that the person believes to be true, for example, "Coffee tastes better than tea." This, however, is not always the case. Apart from the occasions (if any) when we make statements which we do not believe, with deliberate intent to deceive, we often consider statements tentatively, in order to see what would be the consequences of believing them or disbelieving them, or for many other reasons. Sometimes, indeed, it is impossible to believe or to disbelieve a statement. No one, except possibly a very confident astrologer, would profess to believe or to disbelieve the statement "1984 will be a good year for far-mers." Such a statement may nevertheless be considered; in fact, that is what we have just been doing.

A statement is usually expressed as a sentence, but a sentence is not automatically a statement. Sentences such as "Turn left," "What time is it?" and indeed orders and questions in general, are not statements. Excla-mations likewise are not statements. If I drop a hammer on my toe and say "Ouch!" this may possibly convey a certain amount of information, but it is not a statement. It should be remarked that many logicians find the word *statement* unsatisfactory and use the word *proposition* instead. *Statement*, however, will serve our needs.

If we say that a statement is true, we usually mean that the statement corresponds with some occurrence in the world about us (e.g., "The Eiffel Tower is 985 feet high") or within us (e.g., "I am thinking about Brigitte Bardot"). If there were no such correspondence we would say that the statement was false. The question of what is meant exactly by the words *true* and *false* is a difficult one and one over which philosophers are still arguing. Fortunately we shall not need to join in the argument; the reader will be familiar with the customary everyday use of these words, and this, together with some remarks that we shall make later, will provide sufficient understanding for the purposes of this chapter.

1.2 NEGATION

To assert a statement is to say that the statement is true; to deny a statement is to say that it is false. If we deny a statement we automatically assert another statement, called the contradictory of the first statement. Thus if we deny that "This book is black" we are asserting the statement "This book is not black." In the same way if we assert that "This book is black" we deny that "This book is not black."

Thus corresponding to any given statement there is another statement, its *contradictory*, which is false if the statement is true, and true if it is false. The contradictory of a given statement can be formed by putting the statement in place of the dots in the sentence "It is not the case that . . . ." Thus the contradictory of "The Eiffel Tower is 985 feet high" can be written as "It is not the case that the Eiffel Tower is 985 feet high." More often however, we use some grammatical change in the original sentence, generally involving the word *not*, so that it would be usual to say "The Eiffel Tower is not 985 feet high." The process of forming the contradictory of a given statement is known as *negation*.

If two statements are contradictories then they cannot both be true; neither can they both be false. One must be true, the other false. Note that the word *contradictory* as used in logic is more precisely defined than it is in everyday usage. If two men each point to a book and one says "This book is black," and the other says "This book is red" then we might well say that these two men were contradicting each other, or that their statements were contradictory. But this would not agree with the definition that we have given; for although it is not possible for each of these statements to be true, they could both be false. If the book were, in point of fact, green, then the statements would both be false. Hence they are not contradictories. We say they are *contraries*, or that each is contrary to the other. Two statements are contraries when they are so related that they cannot both be true, although they may both be false. Thus the statements

It is hot today.

It is cold today.

are contraries. If the weather were neither hot nor cold, but just warm, then neither statement would be true. Contrast this with the pair of statements,

It is hot today.

It is not hot today.

These are contradictories; they cannot both be true, neither can they both be false. Even if these statements were made by an Eskimo, whose standards of hot and cold weather might well be different from ours, the statements would still be contradictory, since whatever characteristic of the weather may be conveyed by the expression "being hot," the first statement says that the weather has that characteristic, while the second says that it does not.

As a further example take the statements

The number x is even.

The number x is odd.

As they stand, these statements are contraries, for they cannot both be true; but if x is $\frac{1}{2}$ then neither is true. However, if it is known, or understood, that x is a whole number, then they are contradictories, since one must be true and the other false.

We are now in a position to introduce a few symbols. We shall denote statements by the letters $\mathbf{p}, \mathbf{q}, \mathbf{r}, \ldots$ Negation will be denoted by the symbol $\sim$ so that the contradictory of a statement $\mathbf{p}$ will be denoted by $\sim\mathbf{p}$, which we may read as "not $\mathbf{p}$."

Exercise 1

Determine whether the following pairs of statements are contradictories or contraries.

1. (a) The clock on the steeple strikes one.
    (b) The clock on the steeple strikes two.
2. (a) The quality of mercy is not strained.
    (b) The quality of mercy is strained.
3. (a) All Cretans are liars.
    (b) Not all Cretans are liars.
4. (a) Independence day is July 4th.
    (b) Independence day is July 14th.

1.3 CONJUNCTION

Sometimes we make statements to the effect that each of two statements is true:

Yesterday was Sunday and it rained.

$x + y = 3$ and $2x + y = 4$.

Taxes are high and prices are rising.

This method of combining statements is known as *conjunction*, and the *compound statement* thus obtained is called the conjunction of the two constituent statements. The conjunction of two statements **p** and **q** will be written "**p & q**" which may be read as "**p** and **q**".* To assert **p & q** is to assert each of the statements **p** and **q**. Hence **p & q** is true if **p** is true and **q** is true; it is false if either of them is false. Thus if $x = 1$ and $y = 2$ the conjunction

$$x + y = 3 \quad \& \quad 2x + y = 4$$

is true, while if x and y take any other pair of values it is false. It follows that to disprove a conjunction, that is, to demonstrate that it is false, it is sufficient to show that *one* of the two constituent statements is false. It is not necessary to show that both are false, although this might perhaps be the case.

1.4 DISJUNCTION

Sometimes we wish to assert that *at least one* of two statements is true. This is usually done by means of the words *either ... or ...* or frequently by *or* alone.

Either he went to the cinema or he stayed at home.

An integer x is either even or it is odd.

This method of combining two statements is known as *disjunction*. The disjunction of two statements **p** and **q** is the statement "either **p** or **q**," and we shall write this briefly as "**p ∨ q**."

In order to assert the statement **p ∨ q**, we must, according to our definition, be able to assert at least one of the statements **p** and **q**. Thus **p ∨ q** will include the possibility of both **p** and **q** being true. This is sometimes the case with the everyday usage of the words *either ... or ...*, but sometimes not. If a politician were to make the promise "Either wages will be increased or prices will be lowered," and he subsequently managed to increase wages *and* lower prices, then we would hardly accuse him of having broken his pledge. On the other hand if a small boy were told "You may take the lolli-

---

* In many books conjunction is denoted by a period, in others by the symbol ∧. Hence what we have written as **p & q** would appear as **p · q** or **p ∧ q**. This is, of course, purely a matter of convention.

pop or the toffee apple," he would no doubt be considered greedy if he took both! Thus *or* in everyday speech may be used with the idea of "one or the other, and possibly both" (the *inclusive or*), or the idea may be "one or the other, but not both" (the *exclusive or*). If we are to avoid ambiguity we must choose one of these possibilities and stick with it. It is customary to choose the *inclusive or*; hence the definition given above.

We have already seen that to deny the conjunction **p** & **q**, i.e., to assert its contradictory, is the same as showing that at least one of the statements **p** and **q** is false. Now the contradictory of **p** & **q** will be $\sim$(**p** & **q**), the parentheses indicating that the negation is of the whole statement **p** & **q** and not just the statement **p**. The statement that at least one of the statements **p** or **q** is false, i.e., that at least one of the statements $\sim$**p**, $\sim$**q**, is true, will be denoted by $\sim$**p** $\lor$ $\sim$**q**. Thus we see that

$$\sim(\mathbf{p} \ \& \ \mathbf{q}) \text{ means the same as } \sim\mathbf{p} \lor \sim\mathbf{q}.$$

Thus to deny that "Yesterday was Sunday and it rained" is to assert that "Either yesterday was not Sunday or it did not rain"; to deny that "Taxes are high and prices are rising" is to assert that "Either taxes are not high or prices are not rising," and so on.

If a disjunction is false then each of its constituent statements must be false. Thus to deny the statement **p** $\lor$ **q** is to assert the statement $\sim$**p** & $\sim$**q**. The statement $\sim$(**p** $\lor$ **q**) therefore means the same as $\sim$**p** & $\sim$**q**. To deny that "Either he went to the cinema or he stayed at home" is to assert that "He did not go to the cinema and he did not stay at home."

In these last two results we see something analogous to the removal of brackets in algebraic expressions. When we replace $3(x + y)$ by $3x + 3y$, the factor "3" outside the bracket is *distributed* over the two terms inside and the plus sign remains. Similarly when we remove the brackets from the expression $\sim$(**p** & **q**) and $\sim$(**p** $\lor$ **q**), we distribute the symbol $\sim$ over the terms inside the bracket, but we must now change & to $\lor$ and $\lor$ to &.

Using the symbols that we have introduced so far we may construct more complicated statements, such as **p** $\lor$ (**q** & **r**), which is exemplified by the sentence "Either you share the loot or I go to the police and tell them what I know." Note that a statement may occur more than once in a compound statement, for instance the statement

$$(\mathbf{p} \ \& \ \mathbf{q}) \lor (\mathbf{p} \ \& \ \mathbf{r}).$$

Exercise 2

1. The following statements are of the form **p** & **q**. Write the contradictory statements in the form $\sim$**p** $\lor$ $\sim$**q**.

(a)   The boat leaves Dover and sails to Calais.

(b)   The cook could not have committed the murder and the butler has an alibi.

(c)   Little Bo-Peep has lost her sheep and can't tell where to find them.

(d)   Man is born free; and everywhere he is in chains.

2.   The following statements are of the form **p** $\vee$ **q**. Write out their contradictory statements in the form $\sim$**p** & $\sim$**q**.

(a)   Either you leave the room or I do.

(b)   Members of the society either pay $100 for the whole year or $20 for every meeting attended.

(c) $x = \pm 1$. (This is the usual way to write "$x = +1$ or $x = -1$.")

(d)   That's a Chevrolet or I'm a Dutchman!

3.   By assigning suitable letters to the constituent statements, write the following compound statements in symbolic form.

(a)   Either the sun shines and it is hot and humid, or it rains all day and everything is damp.

(b)   The secretary and either the president or the vice-president will attend the meeting; otherwise the meeting will not be held.

(c)   The water gets into the engine and either rusts the pipes or spoils the electrical contacts.

4.   Write out the statements contradictory to those in problem 3 above.

1.5   AXIOMS AND THEOREMS

In all the examples we have analyzed thus far, the truth or falsity of the statement depends on its content and subject matter; it is an empirical matter. The symbol "**p**" may denote any statement and is not, in itself, either true or false; but even if we substitute a definite statement for **p**, the question of whether this statement is true or false will usually lie outside the realm of logic. To determine whether the statement "Cats have nine lives" is true or false we must consult the zoologist, not the logician. There are some statements, however, that are asserted unconditionally i.e., independently of empirical evidence. These are of two kinds: axioms and theorems (or tautologies).

In the days of Euclid, and for a considerable time afterwards, the term *axioms* meant statements that were self-evident and which could therefore be used as material for arriving at other statements that were not self-evident. Modern logicians and mathematicians are much more wary about what is self-evident and consequently when they talk about axioms they mean simply those statements which they have *decided* to regard as true in order that they may use them as fundamental material from which to build up interesting logical or mathematical structures. Just as different kinds of bricks can be used in building houses, so different axioms may be used in building up a logical structure.

An example of a logical axiom is the statement $\mathbf{p} \vee \sim\mathbf{p}$, i.e., either a statement $\mathbf{p}$ is true or its contradictory is true; more briefly, a statement is either true or false. This is one form of what is called in traditional (Aristotelian) logic the *law of the excluded middle*. This axiom essentially expresses what we *mean* by the contradictory of a statement, and by the word *not*. Other axioms indicate that certain expressions may be replaced by other expressions. For example, that $\mathbf{p} \mathbin{\&} \mathbf{p}$ may be replaced by $\mathbf{p}$, i.e., that to assert a statement twice is no more (and no less) than asserting it once. Another such axiom, which we shall use presently, is that $\sim(\sim\mathbf{p})$ may be replaced by $\mathbf{p}$, i.e., that the contradictory of the contradictory of a statement is the same as the statement itself. This is the principle of *double negation*.

Suppose we have two axioms which are embodied in the statements $\mathbf{s}$ and $\mathbf{t}$. Since $\mathbf{s}$ and $\mathbf{t}$ may both be asserted, the statement $\mathbf{s} \mathbin{\&} \mathbf{t}$ may also be asserted. For example, we may assert the statement $(\mathbf{p} \vee \sim\mathbf{p}) \mathbin{\&} (\mathbf{q} \vee \sim\mathbf{q})$. Now this last statement is not an axiom, for it is a consequence of the axiom $\mathbf{p} \vee \sim\mathbf{p}$, but it will be true irrespective of whether the statements $\mathbf{p}$ and $\mathbf{q}$ are true or false. Statements such as this, which are simply consequences of the axioms, and which are therefore true *irrespective* of whether their constituent statements are empirically true or false, are called *tautologies* or *theorems*. We shall discuss later a convenient method for deciding whether a given statement is a tautology or not.

1.6 HYPOTHETICAL STATEMENTS

Consider the statement

If wishes were horses then beggars would ride.

This is called a *hypothetical statement;* it asserts that the two statements "wishes are horses" and "beggars ride" are related in some way. In what way are they related? This question is a little tricky to answer. We might be tempted to say that if the first statement is true then the second is true but this would be merely repeating the hypothetical statement in different words! Let us therefore postpone any attempt at a definition for the moment and examine in a general way, the statement

If $\mathbf{p}$, then $\mathbf{q}$.

Now we have already seen that either $\mathbf{p}$ is true or it is false. If $\mathbf{p}$ is true, then the normal everyday interpretation of sentences of the form "if **p** then **q**" suggests that **q** will be true. But what if **p** is false—can we then say anything about **q**? A moment's reflection will show that we cannot. Suppose we agree that wishes are not horses, does this tell us whether beggars do or do not ride? No, they may or they may not. Consider the statement

If the picture flickers, then the television set is not adjusted properly.

Suppose that the first part of this statement is false, that is, that the picture is not flickering. Does it follow that the set *is* adjusted properly? Not necessarily. There may be things wrong with the adjustment other than those that cause the picture to flicker.

How far do these remarks take us? If we accept the statement **p** $\vee$ $\sim$**p** (since it is an axiom), and take "if **p**, then **q**" in its everyday sense (since we have not yet defined it), we are led to replace **p** by **q** when **p** is true. This gives us the expression **q** $\vee$ $\sim$**p**, or $\sim$**p** $\vee$ **q**, which means the same thing, and this would seem to be as far as we can go since, as we have seen, $\sim$**p** tells us nothing about **q**.

Notice that in the last two paragraphs we have made frequent use of statements of the form "if . . . then . . . ," but this does not matter since we are merely talking around the subject, preparing the way for a proper definition. This we can now state as follows:

"If **p**, then **q**" means "$\sim$**p** $\vee$ **q**"

For brevity we write this as

$$\mathbf{p} \rightarrow \mathbf{q}.$$

Let us see whether this makes sense according to the everyday usage of hypothetical statements. According to our definition the statement

If wishes were horses, then beggars would ride.

should be the same as

Either wishes are not horses, or beggars would ride.

Consider the advertising slogan of the Jamaica Sweep Stakes

If you haven't got a ticket [then] you haven't got a chance.

This should be equivalent to

Either you have a ticket or you haven't a chance.

Perhaps a little more obvious is the fact that the statement

> If you don't leave this house, I'll send for the police.

means, according to our definition, the same as

> Either you leave this house, or I send for the police.

We see that in these examples at any rate, the definition given agrees with everyday usage. There are differences however. The truth of the logical statement $p \rightarrow q$ depends only on the truth or falsity of p and q. Thus if p and q are both true, no matter what their subject matter may be, the statement $p \rightarrow q$ is true. Thus the statement

> If 7 is greater than 5, then Paris is the capital of France.

is a true statement! If this seems surprising, the reason is not difficult to see. It is that in everyday speech we normally use a statement of the form "if p then q" only when there is some sort of *causal* connection between the statements p and q, that is, when the truth of q is, in some sense, the result of p's being true. The statement "If price falls, then the quantity demanded will rise" expresses a connection between the two statements "price falls" and "quantity demanded will rise", a connection deeper than the mere observation that, in point of fact, both are true. We must therefore bear in mind that the definition of "$p \rightarrow q$" is concerned only with the truth or falsity of p and q and not with any causal connections there might be between them. This will explain why the following rather curious statements are true, as the reader may verify using the definition of $p \rightarrow q$.

> If pigs have wings, then the moon is made of green cheese.
> If Julius Caesar was an Eskimo, then Shakespeare wrote *Hamlet*.
> If Julius Caesar was a Roman, then horses have four legs.

Exercise 3

1. Shakespeare wrote in *Henry IV*

"If all the year were playing holidays
To sport would be as tedious as to work."

Assuming that the year is not all playing holidays, can any conclusion be deduced from this statement?

2. Using appropriate symbols, write the following compound statements in symbolic form.

(a) If the present economic trends continue, then either wages must be frozen or the currency must be devalued.

(b) If the wind is from the north and there is a halo round the moon, then we shall have rain.

(c) If you forget your lines or miss your cue then the whole play will be spoiled and I shall be very angry.

3. Given the information that squiths are three-legged animals and that poplids are man-eating fungi, say which of the following statements are true and which are false.

(a) If squiths have four legs then this is not Exercise 3.3(a).

(b) If squiths have three legs then London is in France.

(c) If Greenland is an island in the Pacific then poplids are man-eaters.

(d) If poplids are vegetarians then squiths have no legs.

(e) If poplids are fungi then the present world population is 5 million.

1.7 CONVERSES: REDUCTIO AD ABSURDUM

The statement $\mathbf{p} \to \mathbf{q}$ is the same as $\sim\mathbf{p} \lor \mathbf{q}$; the statement $\mathbf{q} \to \mathbf{p}$ is the same as $\sim\mathbf{q} \lor \mathbf{p}$. The statements $\mathbf{p} \to \mathbf{q}$ and $\mathbf{q} \to \mathbf{p}$ are therefore not the same. If $\mathbf{p}$ were true and $\mathbf{q}$ false, then $\mathbf{p} \to \mathbf{q}$ would be false, but $\mathbf{q} \to \mathbf{p}$ would be true. An example will illustrate:

> If it rains then the crops grow.
>
> If the crops grow then it rains.

These statements clearly do not mean the same thing. Not only that; they are completely unrelated as far as truth and falsity are concerned. If we are told that $\mathbf{p} \to \mathbf{q}$ is true, we cannot say anything about the truth or falsity of $\mathbf{q} \to \mathbf{p}$, at least not without additional information.

Mathematicians call $\mathbf{q} \to \mathbf{p}$ the *converse* of $\mathbf{p} \to \mathbf{q}$.* It appears then that the converse of a true statement may be true, or it may not be true, depending on what the statement is. Thus the two statements

> If two sides of a triangle are equal, then two of its angles are equal.
>
> If two angles of a triangle are equal, then two of its sides are equal.

are converses, each of the other, and it so happens in this particular example that both are true. But the following two statements

---

* Note, however, that logicians use the word *converse* differently. In this book we shall always use it in the mathematical sense.

If two triangles have the same base and the same height then
their areas are equal.

If two triangles have the same area then they have the same
base and the same height.

are also converses of each other, and here the first is true while the second is, in general, false.

Suppose a friend has an invariable rule which he expresses as follows:

If it rains, then I stay at home.

If you met him on the street one day, you could conclude (if you did not know from other evidence) that it was not raining. What is the logic behind this conclusion? You are saying, in effect,

If he is not indoors then it is not raining.

If the original statement is denoted by $\mathbf{p} \to \mathbf{q}$, the latter statement will be $\sim \mathbf{q} \to \sim \mathbf{p}$. It would therefore seem that the statements $\mathbf{p} \to \mathbf{q}$ and $\sim \mathbf{q} \to \sim \mathbf{p}$ are, to say the least, closely related. Now $\mathbf{p} \to \mathbf{q}$ can be written as $\sim \mathbf{p} \vee \mathbf{q}$, while $\sim \mathbf{q} \to \sim \mathbf{p}$ can be written as $\sim(\sim \mathbf{q}) \vee (\sim \mathbf{p})$. But this last expression means the same as $\mathbf{q} \vee \sim \mathbf{p}$, since we have an axiom to the effect that $\sim(\sim \mathbf{q})$ may be replaced by $\mathbf{q}$. In this way we see that $\mathbf{p} \to \mathbf{q}$ and $\sim \mathbf{q} \to \sim \mathbf{p}$ are different ways of writing the same statement. For example, the sweepstake slogan could be rephrased (less effectively) as "If you have a chance, you have a ticket."

This observation is the basis of what is known as proof by *reductio ad absurdum* ("reduction to absurdity"). Suppose we wish to prove a certain theorem $\mathbf{q}$, that is, to show that the statement $\mathbf{q}$ is a consequence of certain other statements (axioms and theorems already proved) which we accept as true. We may imagine these other statements to be lumped together into a single compound statement which we may denote by "$\mathbf{p}$". We then want to show that $\mathbf{p} \to \mathbf{q}$. The proof by *reductio ad absurdum* then proceeds as follows. We assume that $\mathbf{q}$ is false and try to show that we can deduce something which contradicts some statements which we accept as true, that is, which contradicts $\mathbf{p}$. If we succeed in doing this, we have shown that $\sim \mathbf{q} \to \sim \mathbf{p}$. But this means the same as $\mathbf{p} \to \mathbf{q}$, which is what we set out to prove.

This form of argument is perhaps best illustrated by an example, and we take one from Euclid's geometry. It is required to prove that "the line joining the centre of a circle to the point of contact of a tangent is perpendicular to the tangent." This is the statement $\mathbf{q}$. The statement $\mathbf{p}$ may be taken as the combined statement of such of the axioms of Euclidean geometry and the theorems preceding the one we are now proving as we may have occasion

Fig. 1.1

to use. Let t be the tangent to the circle, A the point of contact and O the center of the circle (Fig. 1.1).

We start by assuming $\sim$**q**, i.e., that the line OA is not perpendicular to t. If this is the case then we can draw another line OB through O which *is* perpendicular to t, the point B lying on the tangent t. We then have

$$\text{angle } OBA = 90°.$$

Now the angles of a triangle add up to 180° (this is a known result, i.e., it is included in **p**), and consequently

$$\text{angle } OAB = 180° - \text{angle } OBA - \text{angle } AOB$$
$$= 180° - 90° - \text{angle } AOB$$
$$= 90° - \text{angle } AOB.$$

Thus the angle OAB is less than 90°, and hence less than angle OBA.

But there is a result (also included in **p**) that if two angles of a triangle are unequal then the side opposite the larger angle is larger than that opposite the smaller angle. Hence, since angle $OBA >$ angle OAB we have $OA > OB$. This means that the distance of B from O is less than the radius of the circle, which in turn means that B is *inside* the circle. But this, as Euclid would say, is absurd, for the points of a tangent to a circle (except for point of contact) lie *outside* the circle. The important thing, however, is not so much that the conclusion is absurd, as that it is false; that is, that it contradicts what we have accepted as being true, i.e., it contradicts **p**. Thus if we assume $\sim$**q** we are led to the statement $\sim$**p**, so that $\sim$**q** $\rightarrow \sim$**p**. This is the same as the result **p** $\rightarrow$ **q** that we were aiming for. In our case we conclude that the theorem is true since if it were false, then a point of the tangent would be inside the circle, which we know cannot happen.

This kind of reasoning is commonly used. A man hears a rumor that a prominent statesman has died and says "If that were so it would have been mentioned over the radio; but it was not, so there must be a mistake somewhere." It is also exemplified by the statement

If Mr. X has won the election then my name is Theophilus P. Rosenkranz.

This statement is of the form $\mathbf{p} \rightarrow \mathbf{q}$, and except in the unlikely event that the speaker's name really *is* Theophilus P. Rosenkranz the listener knows that $\mathbf{q}$ is false. Hence the statement becomes merely a roundabout method of denying $\mathbf{p}$. We may note, in passing, that this is a hypothetical statement, the significance of which does not depend on any causal connection between $\mathbf{p}$ and $\mathbf{q}$, and therefore resembles the rather curious examples that we gave earlier.

Exercise 4

1. Which of the following examples is a valid *reductio ad absurdum* argument?

(**a**) I knew he was a bachelor. Bachelors always look untidy, and he looked untidy.

(**b**) If he had caught the train he would have arrived today. Since he missed the train he won't be arriving today.

(**c**) He can't have received the appointment that he went after. He would certainly have let me know by now if he had, and I haven't heard a thing.

2. The following is intended to be a proof, by a *reductio ad absurdum* argument, that all positive integers (whole numbers) are interesting.

PROOF:

Suppose that some positive integers are not interesting. Then there will be one of these "uninteresting" integers that is smaller than any other. This integer has the property of being the "smallest uninteresting integer." But this is an interesting property! Hence this integer is both uninteresting and interesting, which is impossible. Hence all positive integers are interesting.

Comment on this proof. Is it a valid *reductio ad absurdum* argument?

1.8 NEGATION OF A HYPOTHETICAL STATEMENT

What is the negation of the hypothetical statement $\mathbf{p} \rightarrow \mathbf{q}$? It will, of course, be symbolized by $\sim(\mathbf{p} \rightarrow \mathbf{q})$, and this may be written as $\sim(\sim\mathbf{p} \vee \mathbf{q})$ in turn. From this we obtain $\sim(\sim\mathbf{p}) \,\&\, \sim\mathbf{q}$ by the rule for removing brackets, and thence $\mathbf{p} \,\&\, \sim\mathbf{q}$ by the rule for double negation. Thus to deny $\mathbf{p} \rightarrow \mathbf{q}$ we must assert $\mathbf{p} \,\&\, \sim\mathbf{q}$. To deny that "If wishes were horses then beggars would ride," we must assert "Wishes are horses and beggars do not ride." To deny that "If you haven't got a ticket then you haven't got a chance" you must assert that "You haven't got a ticket and you have a chance."

In English there are alternative ways of phrasing hypothetical statements. Instead of "if **p** then **q**" or "if **p**, **q**," we may say "**q**, if **p**." Thus

Beggars would ride if wishes were horses.

Again suppose that **p** stands for the statement "A student is a member of the Debating Society," and **q** stands for "A student may attend the Society's annual dinner." Then **p** → **q** stands for "If a student is a member of the Debating Society, then he may attend the Society's annual dinner." This may be reworded as

1. A student may attend the Society's annual dinner if he is a member of the Debating Society.

Consider now the statement

2. A student may attend the Society's annual dinner *only if* he is a member of the Debating Society.

Statement 1 does not mean the same as statement 2, but what is the connection, if any, between Statements 1 and 2? We may discover it in the following way. Imagine that we are at the Debating Society's annual dinner, and we look around at the students present. If we know that the rule expressed in statement 1 is being adhered to, should we express surprise if we meet students at the dinner who are not members of the Society? No—for the rule states that student members are eligible to attend the dinner, but it does not exclude the possibility of some other students also being eligible. On the other hand, if the rule expressed in statement 2 were to apply, then we *would* be surprised at seeing students present who were not members. In fact, if we ignored any gate-crashers who might be present, we could look around us and say "All students here are members of the Society." Thus statement 2 enables us to state that "If a student is present at the annual dinner then he is a member of the Society." In symbols this is **q** → **p**, so that statement 2 turns out to be the converse of statement 1. Thus "**q** only if **p**" is another way of writing "if **q** then **p**."

1.10 EQUIVALENT STATEMENTS ("IF, AND ONLY IF")

It frequently happens that we want to assert both a hypothetical statement and its converse, that is, we wish to assert $(\mathbf{p} \rightarrow \mathbf{q})$ & $(\mathbf{q} \rightarrow \mathbf{p})$. By what we have said in the previous section this may be concisely rendered in English by the sentence "$\mathbf{p}$ if, and only if, $\mathbf{q}$." Symbolically we write this as "$\mathbf{p} \equiv \mathbf{q}$."* If $\mathbf{p} \equiv \mathbf{q}$ we say that $\mathbf{p}$ and $\mathbf{q}$ are *equivalent statements*. Two equivalent statements are either both true, or both false, since if one is true and the other false it is easily verified that one of the statements $\mathbf{p} \rightarrow \mathbf{q}$, $\mathbf{q} \rightarrow \mathbf{p}$, is false.

It will now be clear that the phrase "means the same as" which we have used from time to time in this chapter, really asserts the equivalence of the two statements that it connects, and is therefore more properly replaced by "if, and only if," or simply $\equiv$. Thus we may write

$$\sim(\sim\mathbf{p}) \equiv \mathbf{p}$$
$$(\mathbf{p} \rightarrow \mathbf{q}) \equiv (\sim\mathbf{p} \vee \mathbf{q})$$

and so on.

NECESSARY AND SUFFICIENT

A statement of the form $\mathbf{p} \rightarrow \mathbf{q}$ is often asserted as "$\mathbf{p}$ is a sufficient condition for $\mathbf{q}$." In other words, to assert the truth of $\mathbf{q}$ it is sufficient to be able to assert the truth of $\mathbf{p}$. For example we might say "A sufficient condition for a student to be eligible to attend the Debating Society's annual dinner is that he be a member of the Society." This is just another way of asserting statement 1 given in the previous section.

A statement of the form $\mathbf{p} \rightarrow \mathbf{q}$ can also be asserted by saying that "$\mathbf{q}$ is a necessary condition for $\mathbf{p}$." In other words, $\mathbf{q}$ is *necessarily* true if $\mathbf{p}$ is true; from the truth of $\mathbf{p}$ the truth of $\mathbf{q}$ necessarily follows. A condition which is sufficient (for the truth of some other statement) may not be necessary. As we have seen, membership in the Debating Society is sufficient to guarantee eligibility for attending the annual dinner, but membership may not be necessary; there might well be regulations which permit some nonmembers also to attend the dinner. In the same way a necessary condition may not also be a sufficient condition. In order to visit a foreign country it is necessary to have an airline ticket (or similar document), but this is not sufficient. One normally needs many other things besides (passports, visa, spending money,

---

* A double arrow is often used instead of $\equiv$. Thus you will meet $\mathbf{p} \longleftrightarrow \mathbf{q}$, and $\mathbf{p} \leftrightarrow \mathbf{q}$. Another common devise is to use "IF" or "iff" to stand for "if, and only if." These latter are usually employed only in personal notes, lectures, and so on, but they occasionally appear in published work.

hotel reservations, etc.), all of which are necessary, but none of which alone is sufficient.

Of particular importance in mathematics are those conditions that are both necessary and sufficient. Now if we say that "**p** is a necessary and sufficient condition for **q**" we are asserting (1) that $q \rightarrow p$ (**p** is necessary) and (2) that $p \rightarrow q$ (**p** is sufficient). Hence we are asserting the statement "**p** if, and only if, **q**."

We have here yet another way of stating the equivalence of two statements. By way of example we see that the following are three different ways of saying the same thing.

1. A student is eligible to attend the annual dinner if, and only if, he is a member of the Debating Society.

2. A necessary and sufficient condition for a student to be eligible to attend the annual dinner is that he be a member of the Society.

3. The statements "A student is eligible to attend the annual dinner" and "The student is a member of the Society" are equivalent statements.

Exercise 5

1. Write down the negations of the hypothetical statements in Exercise 3.2.
2. In everyday language we frequently say "if" when we really mean "if, and only if." In which of the following statements would you judge that this has happened?
(a) If you haven't got a ticket, you haven't got a chance.
(b) You may cross the road if the light is green.
(c) You will be arrested if you are caught gambling.
(d) If the boss agrees, you can have the day off.

1.11 TRUTH TABLES

We have already remarked that although we have a fairly clear intuitive idea of the meanings of the words *true* and *false*, it is not so easy to give a precise definition of them. However, from the point of view of the logician it is not necessary that this be done. Insofar as we are interested only in the purely formal way in which logical statements (like $p \rightarrow q$) are related to each other, we may regard the words *true* and *false* merely as distinctive labels, one, and only one, of which must be attached to any statement. The logician's task then is that of determining which label is to be attached to a given statement. Given a compound statement, and knowing which labels are attached to the constituent statements, he can deduce which of the two labels must be attached to the compound statement. In this way, starting from a handful of statements which he decides to label "true" (the axioms), he can, in theory, determine whether any given statement is true or false.

From this point of view a symbol like **p** appears rather like a variable in algebra. In an expression like "$x^2 + 5x + 6$" the variable x will generally take many different values, say 1, $\frac{1}{2}$, -3, 2, etc.—in fact any number. Sometimes, however, we may want to restrict the range of possible values of x; for example, we frequently restrict x to take only integral values, as when we start the solution to a problem by writing "Let x be the number of cows in the field." If we then arrive at the answer $x = 7\frac{1}{3}$ we know that we have made a mistake somewhere. So it is with symbols for statements, except that the variables are very much more restricted, being allowed to take one of two *values* only, viz., "true" or "false." These are frequently abbreviated to T and F, and referred to as *truth values*.

We may pursue this analogy with algebraic expressions a little further. If we are given an algebraic expression, say $x^2 + y^2$, and we are told the values of the variables x and y in it, then we can calculate the value of the expression itself. In a similar way we may find the truth value of a compound statement if we are told the truth values of the variables (**p, q**, ...) contained in it. Moreover since a statement may take only two values, there is only a finite number of different ways in which truth values may be allocated to these variables; hence we can prepare a list giving the truth value of the whole statement for every combination of truth values of the variables appearing in it. Such a list is usually exhibited in tabular form and called a *truth table*. Thus for the statement **p & q** we have the following truth table.

p	q	p & q
T	T	T
T	F	F
F	T	F
F	F	F

This shows that for only one of the four possible ways of allocating the values T and F to **p** and **q** will the compound statement **p & q** take the value T; for all others it takes the value F. Similarly, for **p** $\lor$ **q** we have

p	q	p $\lor$ q
T	T	T
T	F	T
F	T	T
F	F	F

Thus **p** $\lor$ **q** takes the value *T* unless both **p** and **q** take the value *F*. The final columns of these truth tables simply express the results of applying the definitions of the symbols & and $\lor$.

Again for the statements **p** $\rightarrow$ **q** and $\sim$**p** $\lor$ **q** we have truth tables.

p	**q**	**p $\rightarrow$ q**
T	T	T
T	F	F
F	T	T
F	F	T

p	**$\sim$p**	**q**	**$\sim$p $\lor$ q**
T	F	T	T
T	F	F	F
F	T	T	T
F	T	F	T

The final columns of these two truth tables are identical, showing that for *all* combinations of values of **p** and **q**, the statements **p** $\rightarrow$ **q** and $\sim$**p** $\lor$ **q**, are either both true or both false, i.e., that these two statements are equivalent This, of course, we already knew, but this example illustrates a method of proving two statements to be equivalent—we construct their truth tables and check that the final columns are the same. The truth table for **p** $\lor$ (**p** & **q**) is as follows:

p	**q**	**p & q**	**p $\lor$ (p & q)**
T	T	T	T
T	F	F	T
F	T	F	F
F	F	F	F

and the final column here is the same as the first column, which gives the truth values for **p**. Hence we deduce the equivalence **p** $\lor$ (**p** & **q**) $\equiv$ **p**, which is, perhaps, not immediately obvious.

A tautology, or theorem, is a statement which is true no matter what the truth values of the variables contained in it. Thus a tautology can be recognized by the fact that the final column of its truth table consists entirely of *T*s. This gives a method of proving logical theorems. Thus the truth table of the statement (**p** & **q**) $\rightarrow$ (**p** $\lor$ **q**) is

p	**q**	**p & q**	**p $\lor$ q**	**(p & q) $\rightarrow$ (p $\lor$ q)**
T	T	T	T	T
T	F	F	T	T
F	T	F	T	T
F	F	F	F	T

from which we see at once that the statement is a tautology.

If the preceding examples seem to the reader to be rather obvious, this is no doubt because none of them involves more than two variables, **p** and **q**. More complicated and less obvious statements can be written down using more variables, but the method of constructing their truth tables is essentially the same. The main difference is that the number of possible combinations of truth values of the variables will be greater, being 8 for 3 variables, 16 for 4 variables, and so on. We conclude with one example involving 3 variables, viz., the statement **(p & q) → (q & r)**. The truth table is

p	q	r	p & q	q & r	(p & q → (q & r)
T	T	T	T	T	T
T	T	F	T	F	F
T	F	T	F	F	T
T	F	F	F	F	T
F	T	T	F	T	T
F	T	F	F	F	T
F	F	T	F	F	T
F	F	F	F	F	T

from which we see that the statement is true unless **p** and **q** are true and **r** is false.

Exercise 6

1. Construct truth tables for the following statements:
(a) **(p & q) → r**
(b) **p → (q ∨ r)**
(c) **p & (q → r)**

Exercise 7 (Chapter Review)

1. Determine whether the following pairs of statements are contraries or contradictories:
(a) It always rains on Sunday.
It never rains on Sunday.
(b) The night is young and you're so beautiful.
The night is not young and you're not so beautiful.
(c) The drums go bang and the cymbals clang and the horns they blaze away.
Either the drums do not go bang, or the cymbals do not clang or the horns do not blaze away.
(d) Any two sides of a triangle are together less than the third.
Any two sides of a triangle are together greater than the third.

(e) No one can play Bach fugues on a harmonica.
Albert Finkelstein plays Bach fugues on his harmonica.
(f) If we had some bacon, we could have bacon and eggs if we had some eggs.
We have some bacon, but we can't have bacon and eggs even though we have some eggs.
(g) Stone walls do not a prison make, nor iron bars a cage.
Either stone walls make a prison or iron bars make a cage.
(h) The theory of groups is an important branch of mathematics, and if time permits, it should be included in the high school curriculum provided that there are competent instructors at hand to teach it.
Either the theory of groups is not an important branch of mathematics or it should not be included in the high school curriculum even though time permits and competent instructors are at hand to teach it.
(i) $x + 3y = 17$, and $3x + 2y = 16$.
Either $x \neq 2$ or $y = 6$.

2. The statement $\mathbf{p} \rightarrow \mathbf{q}$ may also be written $\sim\mathbf{p} \vee \mathbf{q}$, $\sim(\mathbf{p}\ \&\ \sim\mathbf{q})$ or $\sim\mathbf{q} \rightarrow \sim\mathbf{p}$. Starting with each of the statements given below, write out the three statements corresponding to it in the above manner.

(a) If I've told you once I've told you a thousand times.
(b) All is well ended if this suit be won
That you express content.
(c) If she sells sea shells on the seashore, then I'm sure she sells seashore shells.
(d) If n is an integer then $\frac{1}{2}n\,(n + 1)$ is an integer.
(e) If food supplies increase then population also increases.
(f) Worms are seldom ferocious if you call them by their first names.
(g) If the sum of the elasticities of demand is greater than unity, then devaluation will improve the balance of payments.
(h) If is it assumed that the floor of the Pacific Ocean has been sinking for a considerable time, then the formation of coral atolls in the Pacific can be explained in terms of the normal growth of corals in comparatively shallow water.

3. Examine the following list and pick out pairs of statements that are equivalent:

(a) Either the moon is not a planet; or 19 is a prime number and jellyfish have teeth.
(b) If a price change takes place then there will be an income effect and a substitution effect.
(c) If you hold the winning ticket and have not previously been successful, you win the prize.
(d) If a price change takes place, then there will be either an income effect or a substitution effect.

(e) Either you do not hold the winning ticket or you have previously been successful; and you win the prize.

(f) Either: if the moon is a planet then 19 is a prime number; or jelly-fish have teeth.

(g) If there is neither an income effect nor a substitution effect then no price change takes place.

(h) You do not hold the winning ticket or you have previously been successful or you win the prize.

(i) It is not the case that: the moon is a planet, and either 19 is not a prime number or jellyfish have no teeth.

(j) It is not the case that: a price change takes place and there is either no income effect or no substitution effect.

(k) You win the prize, and it is not the case that you hold the winning ticket and have not previously been successful.

(l) It is not the case that: the moon is a planet, 19 is not a prime number and jellyfish have no teeth.

4. In one of the exercises to Chapter 4 is a "proof" that all positive integers are equal. Although you have not yet seen it (unless you have been looking ahead) you know at once that this "proof" must contain a fallacy of some sort. By what method of reasoning are you able to make this deduction?

Write out clearly the steps of this deduction.

5. By the method of truth tables. or otherwise, show that the following statements are tautologies:

(a) $p \rightarrow [q \rightarrow (p \ \& \ q)]$

(b) $[(p \rightarrow r) \ \& \ (q \rightarrow s)] \rightarrow [(p \ \& \ q) \rightarrow (r \ \& \ s)]$

(c) $[p \lor (q \ \& \ r)] \equiv [(p \lor q) \ \& \ (p \lor r)]$

(d) $[(p \equiv q) \equiv r] \equiv [p \equiv (q \equiv r)]$

6. Show that

(a) $[(p \ \& \ q) \ \& \ r] \equiv [p \ \& \ (q \ \& \ r)]$

(b) $[(p \lor q) \lor r] \equiv [p \lor (q \lor r)]$

By virtue of (a) we can ignore the parentheses in "$(p \ \& \ q) \ \& \ r$" and in "$p \ \& \ (q \ \& \ r)$," and write either expression as "$p \ \& \ q \ \& \ r$." Similarly (b) shows that either of $(p \lor q) \lor r$ and $p \lor (q \lor r)$ may be written simply $p \lor q \lor r$. These results can clearly be extended to any number of statements $p, q, r, s, \ldots$.

7. Show that the statements

$$(p \rightarrow q) \rightarrow r$$

$$p \rightarrow (q \rightarrow r)$$

are not equivalent.

This shows that we cannot ignore the parentheses in the above expressions without creating doubt as to which expression is intended. The sequence of symbols "$p \rightarrow q \rightarrow r$" thus has no unambiguous meaning.

8. In developing the symbolism of this chapter we defined three basic symbols $\sim$, & and $\vee$, and defined the other two, $\rightarrow$ and $\equiv$ in terms of these. Could we have managed with fewer basic symbols? Yes, we could. We could have defined $\mathbf{p} \vee \mathbf{q}$ to be $\sim(\sim\mathbf{p}\ \&\ \sim\mathbf{q})$ thus reducing the number of basic symbols to two only, viz., $\sim$ and &.

Can we reduce the number of basic symbols still further, namely to one only? The answer is again yes, but it cannot be done with any of the symbols that we have so far introduced. However, we can introduce a symbol $\downarrow$ whose truth table is

p	q	p↓q
T	T	F
T	F	F
F	T	F
F	F	T

Now we can define all the other symbols in terms of this one. The statement $\mathbf{p} \downarrow \mathbf{q}$ may be read as "not **p** and not **q**" or "neither **p** nor **q**."

Verify that

$$\sim\mathbf{p} \equiv (\mathbf{p} \downarrow \mathbf{p})$$
$$\mathbf{p}\ \&\ \mathbf{q} \equiv (\sim\mathbf{p} \downarrow \sim\mathbf{q})$$
$$\equiv [(\mathbf{p} \downarrow \mathbf{p}) \downarrow (\mathbf{q} \downarrow \mathbf{q})]$$

(These show that the symbol $\downarrow$ suffices to define the other symbols.)

$$(\mathbf{p} \vee \mathbf{q}) = \sim(\mathbf{p} \downarrow \mathbf{q})$$
$$= [(\mathbf{p} \downarrow \mathbf{q}) \downarrow (\mathbf{p} \downarrow \mathbf{q})]$$
$$(\mathbf{p} \rightarrow \mathbf{q}) = \sim(\sim\mathbf{p} \downarrow \mathbf{q})$$
$$\equiv \sim[(\mathbf{p}\sim\mathbf{p}) \downarrow \mathbf{q}]$$
$$\equiv \{[(\mathbf{p} \downarrow \mathbf{p}) \downarrow \mathbf{q}] \downarrow [(\mathbf{p} \downarrow \mathbf{p}) \downarrow \mathbf{q}]\}.$$

9. We have seen that the statement $\mathbf{p} \vee \mathbf{q}$ included the possibility of both **p** and **q** being true, that is, that $\vee$ represent the "inclusive or." It is possible to work with the "exclusive or" to which (for a reason which will appear in the next exercises) we shall allot the symbol $+$. Thus $\mathbf{p} + \mathbf{q}$ means "**p** or **q** but not both," and its truth table is

p	q	p+q
T	T	F
T	F	T
F	T	T
F	F	F

Prove that

$$(p + q) \equiv [\sim(p \equiv q)]$$
$$(p \lor q) \equiv [(p + q) + (p \& q)]$$
$$(p \to q) \equiv [\sim p + (p \& q)]$$

10. It is possible to develop an algebraic method of determining whether a statement is a tautology or not, thus avoiding the use of truth tables. To do that we first agree to write 0 and 1 for the truth values in place of F and T respectively; and instead of "**p & q**" we write simply "**pq**." We then see that

$$\text{if } p = 0 \quad \text{and} \quad q = 0, \quad \text{then } pq = 0;$$
$$\text{if } p = 0 \quad \text{and} \quad q = 1, \quad \text{then } pq = 0;$$
$$\text{if } p = 1 \quad \text{and} \quad q = 0, \quad \text{then } pq = 0;$$
$$\text{if } p = 1 \quad \text{and} \quad q = 1, \quad \text{then } pq = 1;$$

where $p = 0$ can be read as **p** has the truth value 0, and so on. This shows that **pq** behaves just like an ordinary algebraic product.

For the symbol $+$ introduced in the last exercise we have that

$$\text{if } p = 0 \quad \text{and} \quad q = 0, \quad \text{then } p + q = 0;$$
$$\text{if } p = 0 \quad \text{and} \quad q = 1, \quad \text{then } p + q = 1;$$
$$\text{if } p = 1 \quad \text{and} \quad q = 0, \quad \text{then } p + q = 1;$$
$$\text{if } p = 1 \quad \text{and} \quad q = 1, \quad \text{then } p + q = 0.$$

The first three of these statements may be written briefly as $0 + 0 = 0$, $0 + 1$, $1 + 0 = 1$, showing that, thus far, the symobl $+$ behaves like the plus sign of arithmetic. However, the fourth statement gives $1 + 1 = 0$, which shows that the resemblance is not complete. This is the only difference, however, so that provided we remember that $1 + 1$ is to be replaced by 0 (and not by 2), we may proceed to manipulate our statements as if they were ordinary algebraic expressions. The fact that the only possible values are 0 and 1 leads to some simplification. Thus $0 + 0 = 1 + 1 = 0$, so that $p + p = 0$ irrespective of the truth value of **p**. Again, $0.0 = 0$ and $1.1 = 1$, so that $p \cdot p$ or p^2 may always be replaced by **p**. The expression $p + 1$ is 1 if $p = 0$ and 0 if $p = 1$, so that $p + 1$ will represent the statement $\sim p$. The reader should verify such of the following results as are new.

(a) $\sim p$ may be written as $1 + p$.
(b) **p & q** may be written as **pq**.
(c) **p $\lor$ q** may be written as $p + q + pq$.
(d) **p $\to$ q** may be written as $1 + p + pq = 1 + p(q + 1)$.
(e) **p $\equiv$ q** may be written as $1 + p + q$.

As an example of the use of this method, consider the statement

$$[(p \to q) \text{ \& } (q \to r)] \to [p \to r].$$

The first part of this statement viz., $(p \to q) \text{ \& } (q \to r)$ may be written

$$(1 + p + pq)(1 + q + qr)$$

which, on expansion, gives

$$1 + q + qr + p + pq + pqr + pq + pq^2 + pq^2r$$

which, by the rules given above, simplifies to

$$1 + p + q + pq + qr.$$

Hence the whole statement can be written as

$$1 + (1 + p + q + pq + qr)[(1 + p + pr) + 1]$$
$$= 1 + (1 + p + q + pq + qr)(p + pr)$$
$$= 1 + p + p^2 + pq + p^2q + pqr + pr + p^2r + pqr + p^2qr + pqr^2$$

which will be found to reduce simply to 1, showing that the statement is always true, i.e., that it is a tautology.

Use this method to prove that the following are tautologies.

(f) $[p \to (q \text{ \& } r)] \to [p \to q]$
(g) $[(p \to q) \lor (r \to s)] \to [(p \lor r) \to (q \lor s)]$
(h) $[(p \text{ \& } q) \lor (r \text{ \& } s)] \equiv [(p \lor r) \text{ \& } (q \lor r) \text{ \& } (p \lor s) \text{ \& } (q \lor s)]$
(i) $[(p \lor q) \text{ \& } (r \lor s)] \equiv [(p \text{ \& } r) \lor (q \text{ \& } r) \lor (p \text{ \& } s) \lor (q \text{ \& } s)]$

11.   Below are four interrelated statements;
Statement (a)   Either (1) statement (b) is false or (2) statement (b) is true and statement (c) is false.
Statement (b)   Either (1) statements (a) and (b) are true, or (2) statement (c) is false.
Statement (c)   Either (1) statement (a) is true, or (2) statement (d) is true and statement (b) is false.
Statement (d)   Either (1) statement (b) is true or (2) statement (a) is true and statement (c) is false.
      Show that these four statements may be symbolised as follows:

$$a \equiv [\sim b \lor (b \text{ \& } \sim c)]$$
$$b \equiv [(a \text{ \& } b) \lor \sim c]$$
$$c \equiv [a \lor (d \text{ \& } \sim b)]$$
$$d \equiv [b \lor (a \text{ \& } \sim c)]$$

Show that if these statements are written in the notation of the previous exercise, and then simplified, the following equations are obtained.

(1) $a = 1 + bc$

(2) $b = 1 + c + abc$

(3) $c = a + d + ad + bd + abd$

(4) $d = a + b + ac + ab + abc$

By replacing **a** by $1 + bc$ in Equation 3, or otherwise, find the values of **a**, **b**, **c** and **d**, and hence determine which of the statements a, b, c, d are true and which false.

Suggestions for Further Reading

So many textbooks of mathematics are on the market these days that the reader should have no difficulty in finding suitable books to supplement his reading of these chapters. The suggestions made throughout the text are intended merely to indicate a few of the many possible reference books; they are by no means the only ones.

There are three books, all highly recommended, that deal very well, and on the proper level, with the subject matter of several chapters of this book. They are:

Kemeny, J. C., Snell, J. L., and Thompson, G. L. *Introduction to Finite Mathematics.* Englewood Cliffs, N.J.: Prentice-Hall, Inc., 1957.

Lipschutz, S. *Theory and Problems of Finite Mathematics.* New York: Schaum Publishing Co., 1966. This is one of the Schaum's "Outline" series, consisting of books that combine brief outlines of theoretical topics with a wealth of both worked and unworked examples.

Mirkil, H., Kemeny, J. C., Snell, J. L., and Thompson, G. L., *Finite Mathematical Structures.* Englewood Cliffs, N.J.: Prentice-Hall, Inc., 1959.

All three of the above books cover the subject matter of Chapter 1 of this text. So does the following:

Sangelosi, V. E. *Compound Statements and Mathematical Logic.* New York: Charles E. Merrill Books, Inc., 1967.

The reader may also consider the following, more advanced, books:

Langer, S. K. *An Introduction to Symbolic Logic.* New York: Dover, 1967.

Rosenbloom, P. *The Elements of Mathematical Logic.* New York: Dover, 1950.

2 Sets

and Relations

Psychologists claim that men are the dreamers and women are the realists,
  But to my mind women are the starriest-eyed of idealists,
  Though I am willing to withdraw this charge and gladly eat it uncomplaineously
  If anyone can explain to me how a person can wear a costume that is different from other people's and the same as other people's, and more expensive than other people's and cheaper than other people's, simultaneously.

OGDEN NASH
"Thoughts Thought on an Avenue"

2.1 SETS

Many of the statements that we make concern single objects: "This book is black," "That cat belongs to the man next door," and so on. But we frequently make statements concerning several objects all at once, for example, "All the books in this room are novels," "All cats are grey at night"; and when we do this we frequently think of these objects as forming a collective whole. This way of thinking of things is very common in mathematics, and the usual term for such a collective whole is a *set*, though other words, such as *aggregate* and *class*, are often used. Thus a set is any collection of objects that, for one reason or another, we choose to regard as a whole. As examples of sets we might take the following

1. The set of all planets of the sun.
2. The set of all one-armed paperhangers.
3. The set of all positive integers.
4. The set of all Acts of Congress dated between 1850 and 1950 inclusive.

The objects that make up a set are called the *members* or *elements* of that set, so that a set is defined once we know which objects are members of the set and which are not. There are essentially two different ways in which we may acquire this information. The person who is defining the set may make a list of the members of the set. Metaphorically speaking, he points to various objects, and says "This, this, this and that and that are the members of the set that I am talking about." Thus if we want to consider the set consisting of a packet of cigarettes, Sophia Loren, the moon, the Taj Mahal and a duck-billed platypus, the simplest way of describing it is to enumerate its several members, as we have just done. This is known as defining a set *by extension*.

It is not often that we have occasion to consider a set as bizarre as the one just mentioned. Usually the reason we think of a collection of objects as forming a set is that they have some property in common—a property that is not shared by objects not in the set. When this is the case we may define the set in a different way, by simply stating what this property is. The set will then consist of those objects, and only those objects, which have the given property. Thus the phrase "The set of all Presidents of the United States of America (past and present)" defines uniquely a certain set of historical persons. George Washington is a member of this set since he possesses (or rather possessed) the property common to its members; the author of this book (for example) does not possess this property and is therefore not a member of the set. This method of defining a set by a common property of its members is known as definition *by intension*.

Many sets can be defined either by extension or intension. The set of U.S. Presidents above could have been defined by extension by compiling a list of the appropriate names. It is clear, however, that many sets can be defined only by intension. This is always the case if the set is an *infinite set*, that is, if it contains an unlimited number of members as, for example, the set of all positive integers.

If we define a set by extension then we can always say exactly how many members it has—we have only to count them as we list them. If we define a set by intension this may not be possible. The phrase "the set of all fish in the Atlantic Ocean at 12 noon G.M.T. on October 1st, 1970" precisely defines a certain set (provided the terms "fish" and "Atlantic Ocean" are precisely defined), but it is not possible to say how many members it has. What is more, not only may we be uncertain as to how many members a set defined by intension may have but, what is more important, we may even be uncertain whether it has any members at all! We may not know whether there are *any* objects having the property that we have used to define the set. The phrase "The set of all Martians" looks like the definition of a certain set, but what if there are no such things as Martians? Does it then define a set? This is an important point, which we must consider.

2.2 THE EMPTY SET

In everyday speech, to define a set by intension is usually regarded as tantamount to asserting that the set has members, i.e., that there are objects having the given property. Consequently, if we are in doubt as to whether there are, in fact, any objects having the given property we customarily add a subordinate phrase like "if there be any such," or just "if any", to indicate our uncertainty. We say, for example,

There will be an examination for all students registered in Course 205 at the end of the term. Those students (if any) who fail this examination will be required to repeat the course.

Here the words "if any" have been added to avoid the suggestion that some students are bound to fail.

This device, though common, is cumbersome to say the least, and in mathematics it would be very awkward if we were frequently uncertain as to whether a definition did or did not define a set. We overcome this difficulty by introducing a special set, called the *empty set*, which is defined as the set having no members (sometimes the phrase "null set" is used). This may seem an artificial, even paradoxical, concept, but it is not really so. The essential thing about the definition of a set is that one should be able to tell, of any given object, whether it is a member of the set or not. For the empty set this is simplicity itself; no matter what the given object may be, it is *not* a member of the set! Paradoxical or not, the empty set is a useful device, since it relieves us of the burden of having to determine, whenever we mention a set, whether it has members. If there are no Martians, then the phrase "The set of all Martians" simply defines the empty set. To say that a set of objects having a given property is the empty set is therefore an alternative, and often convenient, way of saying that there are no objects having that property.

2.3 SYMBOLISM

We can now introduce a few symbols relating to sets. We shall use the small letters $a, b, c, \ldots, x, y, z$, to denote objects, and the capitals $A, B, C, \ldots$, etc., to denote sets. If the object x is a member of the set A we shall write $x \in A$, where the symbol $\in$ may be read as "is a member of" or "belongs to." To denote that an object x does not belong to set A we could write $\sim(x \in A)$, but it is more usual to write $x \notin A$. The empty set will be denoted by $\emptyset$.

2.4 INCLUSION

Let set A be the set of all lions, and let set B be the set of all man-eating lions. There is an obvious connection between these two sets which might well be expressed in everyday speech by saying that set B was a part of set A, or was included in set A. We shall sometimes use the expression "B is included in A," but more often we shall say "B is a *subset* of A." We shall use the notation $B \subset A$ for this statement. A precise definition of the symbol $\subset$, and hence of the term *subset*, is as follows:

> "$B \subset A$" means that for any x (i.e., no matter what the object x may be) $(x \in B) \rightarrow (x \in A)$; that is, if an object belongs to B then it also belongs to A.

Suppose it were the case that *all* lions were man-eaters. Then the sets A and B above would be the same, that is, they would have the same members, a state of affairs which we denote by $A = B$. Do we still write $B \subset A$; do we still call B a subset of A? Yes, we do, for the statement

$$\text{For all } x, (x \in B) \rightarrow (x \in A)$$

still holds. We therefore remark that in saying that B is a subset of A we admit the possibility that B is the *whole* of A. This has an advantage similar to that which we gained by introducing the empty set; we are relieved of the responsibility of determining, for example, whether or not there are any lions that are not man-eaters. There are occasions, however, when it is important to indicate that B is a subset of A but is not the whole of A, and we then say that B is a *proper subset* of A. We shall not introduce any special symbol for this concept.*

There are one or two immediate consequences of these definitions. First we remark that the statement $A \subset A$ is always true, whatever the set A may be, for this simply states that if x belongs to A then x belongs to A, which is clearly true. A useful result is the following:

$$\text{if } A \subset B \text{ and } B \subset A, \text{ then } A = B.$$

This we may prove as follows: from $A \subset B$ and $B \subset A$ we see that,

for all x $\qquad\qquad\qquad (x \in A) \rightarrow (x \in B)$

and $\qquad\qquad\qquad\qquad (x \in B) \rightarrow (x \in A)$

so that we may write

$$(x \in A) \equiv (x \in B).$$

This simply states that the property of being a member of set A and the property of being a member of set B are the same property, hence sets A and B must be the same set. It is, incidentally, the only way of showing that two sets are identical.

---

* Some writers use the symbol $\subset$ differently, taking "$B \subset A$" to mean "B is a proper subset of A," and writing "$B \subseteq A$" for "B is a subset of A" when $B = A$ is possible.

How does the empty set fit in with the concept of inclusion? Is it possible, for example, to have a true statement of the form $A \subset \emptyset$?

The statement $A \subset \emptyset$ means that, for all x

$$(2.4.1) \qquad (x \in A) \rightarrow (x \in \emptyset).$$

Now the statement $x \in \emptyset$ is always false, so the only way in which the statement (2.4.1) can be true is for $x \in A$ to be false (if $\mathbf{p} \rightarrow \mathbf{q}$, then $\sim\mathbf{q} \rightarrow \sim\mathbf{p}$). Therefore from the statement $A \subset \emptyset$ we deduce that, for all x, $x \notin A$. This, by definition, means that A is the empty set, so that the statement $A \subset \emptyset$ is true if, and only if, $A = \emptyset$.

How about the statement $\emptyset \subset A$? This says that, for all x,

$$(x \in \emptyset) \rightarrow (x \in A)$$

which may be rewritten

$$(2.4.2) \qquad (x \notin \emptyset) \vee (x \in A).$$

But the statement $x \notin \emptyset$ is true for any x, and hence the whole statement (2.4.2) is true for any x. In other words, the statement $\emptyset \subset A$ is true, whatever the set A. The empty set is included in any set.

We have referred all along to *the* empty set as if there were only one. Could there not be more than one empty set? No, for if there were, and if we took any two of them, say $\emptyset_1$ and $\emptyset_2$, then, by the result of the last paragraph, $\emptyset_1$ would by included in *any* set and hence in $\emptyset_2$, while $\emptyset_2$ would be included in any set and hence in $\emptyset_1$. Thus we should have $\emptyset_1 \subset \emptyset_2$ and $\emptyset_2 \subset \emptyset_1$, which is the same as $\emptyset_1 = \emptyset_2$. Hence there is only one empty set, and the use of the definite article is justified.*

Exercise 8

1. The following six sets are defined by intension. Which of these would it be *practically* possible to define by extension?

    (a)   The set of all kings of England
    (b)   The set of all inhabitants of the United States on a given day
    (c)   The set of all integers
    (d)   The set of all integers between 365 and 1,965,027 inclusive
    (e)   The set of all human beings alive on a given day
    (f)   The set of all books in the Library of Congress

---

* The idea that there are different empty sets according to what kind of element the set does *not* have, is a common misconception, which is easily seen to be absurd. It is the basis of the joke about the diner in a restaurant who asked for coffee without cream, and was told, "I'm afraid we're out of cream, Sir, would you like it without milk instead?"

2. Of the six sets given below, state which are included in others?

    (a) The set of all film stars

    (b) The set of all female film stars

    (c) The set of all women

    (d) The set consisting of Brigitte Bardot, Sophia Loren and Audrey Hepburn

    (e) The set of all male film stars who are ten feet tall

    (f) The set of all film actresses

2.5 SET UNION

Let set A be the set of all Presidents of the United States, and let set B be the set of all Vice Presidents of the United States; and suppose we wish to refer to the set consisting of all Presidents and all Vice Presidents of the United States, that is to say, we wish to lump these two sets together, so to speak. The new set that we obtain is called the *union* of A and B, and is written $A \cup B$ (sometimes the word *sum* is used). A precise definition for $A \cup B$ is as follows: $A \cup B$ is the set of all objects x such that

$$(x \in A) \lor (x \in B).$$

We can usefully introduce another piece of notation at this point. If we define a set by extension we can write its elements inside braces, as follows, taking the rather odd example of a set that we used earlier in this chapter:

{packet of cigarettes, Sophia Loren, moon, Taj Mahal, duck-billed platypus}

If we define a set by intension we can write "$\{x \mid \ldots \}$" which can be read as "the set of all x such that ...." Using this convenient notation we can write the definition of the union of two sets as

$$A \cup B = \{x \mid (x \in A) \lor (x \in B)\}.$$

What of those men who have been (at different times of course) both President and Vice President? These will certainly be included, for by our definition of $\lor$, the possibility that the statements $x \in A$ and $x \in B$ are both true is allowed. On the other hand the definition of a set determines which objects belong to the set, so that an object either belongs to a given set or it does not—there is no question of its occurring twice in a set, or anything like that.

Examples of the union of two sets are easily constructed. Thus, if

$$A = \text{the set of all mothers}$$
and $$B = \text{the set of all fathers}$$
then $$A \cup B = \text{the set of all parents};$$
if $$A = \text{the set of all smokers}$$
and $$B = \text{the set of all drinkers}$$
then $$A \cup B = \text{the set of all who either smoke or drink (or both).}$$

Certain simple results are easily proved; for example, if $B \subset A$ then $A \cup B = A$; if $C \subset A$ then $C \subset A \cup B$: if $A \subset C$ and $B \subset C$ then $A \cup B \subset C$; for any set A, $A \cup \emptyset = A$. The proof of these results from the original definitions will be left to the reader.

2.6 INTERSECTION

We have just seen that the union of the set of all Presidents of the United States of America and the set of all Vice Presidents of the United States of America will include those people who have been *both* President and Vice President. Suppose we are interested in these people alone; then we are interested in the set of those people who belong to both these sets. We call this set the *intersection* of the two sets (the word *meet* is also used). More generally the intersection of two sets A and B is defined as the set of all objects which belong to *both* A and B. It is denoted by $A \cap B$, and is defined by

$$A \cap B = \{x \,|\, (x \in A) \,\&\, (x \in B)\}.$$

Thus if A is the set of all sword-swallowers, and B is the set of all men with one leg, then $A \cap B$ will be the set of all one-legged sword-swallowers (which may well be $\emptyset$, of course!); if A is the set of all integers divisible by 3, and B is the set of all integers divisible by 5, then $A \cap B$ will be the set of all integers divisible by 15.

As with the union of sets, there are several results which are very easily proved. Thus for any sets A, B we have $A \cap A = A$; $A \cap \emptyset = \emptyset$; $A \cap B \subset A$. We also have the important result

$$A \cap (B \cap C) = (A \cap B) \cap C$$

which serves to define the intersection of three sets, which we may write simply as $A \cap B \cap C$. The extension of this result to any finite number of sets is obvious. The corresponding result for the union of sets is

$$A \cup (B \cup C) = (A \cup B) \cup C$$

which serves to define the union $A \cup B \cup C$ of three sets. Again the extension to any finite number of sets is clear.

If two sets A and B have no elements in common, then $A \cap B = \emptyset$. We then say that the sets A and B are *disjoint*. This is another term that we shall need to extend to more than two sets, but there are two ways in which we might do this. We might say that sets A_1. $A_2, \ldots, A_n$ are disjoint if $A_1 \cap A_2 \cap A_3 \ldots, \cap A_n = \emptyset$. This would mean that there is no object which belongs to *all* of the sets. But this is a rather weak restriction on the sets $A_1, A_2, \ldots, A_n$, and one that we should not need to invoke very frequently. What happens much more often is that we wish to state that no two of the sets $A_1, A_2, \ldots, A_n$ have any objects in common, i.e., that any two of the sets are disjoint, or that given any two of the sets A_i and A_j (i and j standing for any two of the numbers $1, 2, \ldots, n$) we have $A_i \cap A_j = \emptyset$. We shall therefore take this as being what we mean by "disjoint" as applied to several sets.*

Exercise 9

1. If $A =$ the set of all motor mechanics
    $B =$ the set of all men
    $B =$ the set of all housepainters
    $D =$ the set of all inhabitants of France
    $E =$ the set of all astronauts
    $F =$ the set of all women
    describe in words the following sets:
    - (a) $D \cap E$
    - (b) $A \cap D \cap F$
    - (c) $(B \cap C) \cup E$
    - (d) $B \cap (C \cap E)$
    - (e) $(C \cap D) \cup (E \cap F)$
    - (f) $B \cap [E \cup (A \cap C)]$
2. Prove that
    - (a) $A \cap (B \cup C) = (A \cap B) \cup (A \cap C)$
    - (b) $A \cup (B \cap C) = (A \cup B) \cap (A \cup C)$
3. Prove that if A, B and C are any three sets, then
    - (a) $(A \subset B) \to [(A \cap C) \subset (B \cap C)]$
    - (b) $(A \subset B) \to [(A \cup C) \subset (B \cup C)]$

---

* The expression *pairwise disjoint* is sometimes used to obviate any possible misunderstanding.

2.7 COMPLEMENTATION

In the same way that set union and intersection correspond to the disjunction and conjunction of statements, respectively, so there is an operation on sets corresponding to the negation of statements. Suppose that we wish to consider not the set B of all man-eating lions, but the set C of all those lions which do not eat men. Then B is the set of all lions having a given property, and C is the set of all lions *not* having that property; provided that "man-eating" is defined appropriately, B and C will be disjoint, and their union will be A, the set of all lions.

We say that C is the *complement of B in A*, and will write $C = \mathscr{C}(B)$ in A, to denote this situation. A precise definition is as follows: Let A and B be sets such that $B \subset A$. Then the complement of B in A is the set of all elements such that $(x \in A)$ & $(x \notin B)$. Very often writers use a loose terminology and refer just to the complement of B, without explicit mention of the surrounding set A, of which B is a subset. Since the surrounding or *universal* set A is often retained throughout the relevant analysis, this practice does little harm; but occasionally the universal set does change in the course of a problem, so care is necessary.

The reader may easily verify the following properties of the operation of complementation, using the definition given above:

$$\mathscr{C}[\mathscr{C}(B)] = B; \quad (C \subset B) \to [\mathscr{C}(B) \subset \mathscr{C}(C)]$$
$$\mathscr{C}(\emptyset) = A; \quad \mathscr{C}(A) = \emptyset;$$
$$B \cup \mathscr{C}(B) = A; \quad B \cap \mathscr{C}(B) = \emptyset.$$

What about the complement of the union, and of the intersection, of two sets? Suppose that we have two sets B and C such that $B \subset A$ and $C \subset A$. Then their union is defined to mean the set of all x for which $(x \in B) \lor (x \in C)$ is a true statement. Therefore the complement of $B \cup C$ is the set of all x for which $(x \in B) \lor (x \in C)$ is false. By the rules for the negation of a compound statement, this implies that $\mathscr{C}(B \cup C)$ is the set of all x such that $(x \notin B)$ & $(x \notin C)$. But the set of x such that $x \notin B$ is simply $\mathscr{C}(B)$, and the set of x such that $x \notin C$ is $\mathscr{C}(C)$. By definition of the intersection of two sets, therefore, we can say that

$$\mathscr{C}(B \cup C) = \mathscr{C}(B) \cap \mathscr{C}(C).$$

Similarly, we can show that

$$\mathscr{C}(B \cap C) = \mathscr{C}(B) \cup \mathscr{C}(C).$$

Thus we have proved a rule very similar to that for the negation of a compound statement—simply "distribute" the complementation symbol over the sets within the brackets, and change each $\cup$ symbol to a $\cap$ symbol and vice versa; the rule clearly generalizes to unions or intersections of any finite number of sets. The similarity between these rules is scarcely surprising, for complementation of sets is defined in terms of negation of statements, and the *distributive rule* for sets can therefore be derived from the distributive rule for statements.

As a result of this rule, given any result concerning sets, we can always assert a second result concerning their complements. Thus suppose that B, C, D and E are all subsets of a given set A, and are such that

$$B \cup C = D \cap E$$

then $$\mathscr{C}(B \cup C) = \mathscr{C}(D \cap E)$$

so that $$\mathscr{C}(B) \cap \mathscr{C}(C) = \mathscr{C}(D) \cup \mathscr{C}(E)$$

a result which is not immediately obvious. Thus we get two results for the price of one, as it were.

In passing we may note that it is possible to define the *difference* of two sets, with the help of the operation of complementation. Suppose that we were concerned mainly with the set D of all *African* lions, and were particularly interested in those African lions which do not eat *homo sapiens*. The set of such lions clearly consists, in an intuitive sense, of the set D *minus* the set of all African man-eaters. But none of the concepts or symbols we have formulated so far covers this situation.

What we want to distinguish is the set of all lions x which are in Africa, but which do not eat men. It will therefore be perfectly adequate to consider the set of all lions x such that $x \in D$, but whose members are *not* in the set B, of *all* man-eating lions. This set is obviously identical with the set we want. Therefore we may define our required set as the *difference* D minus B (written $D - B$), and give a formal definition as follows:

$$D - B = \{x \,|\, (x \in D) \,\&\, (x \notin B)\}.$$

Now those x such that $x \notin B$ are simply the elements of $\mathscr{C}(B)$—the complementation being with respect to the universal set A (in this case, the set of all lions). So we can rephrase our definition as follows, using the definition of set intersection:

$$D - B = D \cap \mathscr{C}(B)$$

In the same way, we can show that

$$B - D = B \cap \mathscr{C}(D)$$

which indicates that we must not use the word "difference" without specifying the order of the two sets. Notice that if $D \subset B$, then $D \cap \mathscr{C}(B) = \emptyset$, as we would expect from the intuitive idea of "difference." If *all* African lions are man-eaters, then the set of all non-man-eating African lions is empty.

Exercise 10

1. Let us denote by U a set that we are taking as the universal set in a given situation. Show that

(a) $\mathscr{C}(U) = \emptyset$ and $\mathscr{C}(\emptyset) = U$.

(b) $(A \cap B = \emptyset) \to [\mathscr{C}(A) \cup \mathscr{C}(B) = U]$.

(c) $A \cup U = A$ and $A \cap U = A$.

2. Prove that $\mathscr{C}(D - B) = \mathscr{C}(D) \cup B$ and $D - B = D \cap \mathscr{C}(B)$.

3. Taking as the universal set the set of all human beings, express the complements of the sets in examples (a) through (f) of Exercise 9.1 in terms of the complements of the sets A through F. Interpret these sets in everyday language.

2.8 DIAGRAMS

Many results concerning sets are easily comprehended by the use of diagrams, which, in turn, will often indicate how the result can be proved. There are two important methods of constructing these diagrams.

Method 1. In this method we represent the sets by circles (Fig. 2.1). We can think of the points inside the circle as being the elements of the set, or if it should be objected that the set being represented does not have as many elements as the circle has points, we can put some dots inside the circle to represent its elements, though this is not really necessary. We then represent the connections between two or more sets by drawing the circles in the appropriate way. Thus to represent the fact that $B \subset A$, for example, we draw the circle that represents B *inside* the circle representing A (Fig. 2.2). The fact that anything inside circle B is automatically inside circle A corresponds to the statement that for any x, $(x \in B) \to (x \in A)$. If A and B have elements in common, we draw the circles to overlap (Fig. 2.3) and the intersection will be represented by the area common to the two circles. The union will then be represented as in Fig. 2.4.

It is then obvious that $A \cap B \subset B$, or that $A \subset A \cap B$, and so on. Results involving three sets are handled in the same way. Thus from Fig. 2.5 we can see that if $B \subset A$ then $(B \cap C) \subset (A \cap C)$, and $(B \cup C) \subset (A \cup C)$ and so on.

Fig. 2.1

Fig. 2.2

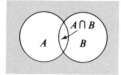

Fig. 2.3

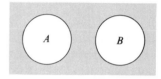

Fig. 2.4

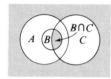

Fig. 2.5

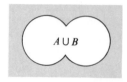

Fig. 2.6

If two sets are disjoint, then they will be represented by circles that do not intersect (Fig. 2.6).

This method is very useful in proving results on the complements and differences of sets. In these cases the universal set is usually represented by a rectangle within which the circles are placed; the various areas are appropriately shaded. Such diagrams are called *Venn diagrams*, named after the 19th century Cambridge logician who invented them.

Method 2. In this method, instead of positioning our circles in a manner which expresses the relationships between the sets, we draw them always in the same way, and such that all possible intersections are shown. Thus for two and three sets we have the basic diagrams of Figs. 2.7 and 2.8.

But what if the sets do not intersect in this particular way? If this is so, then we merely indicate that appropriate subsets are empty. Thus under Method 1, Fig. 2.7 would represent two sets A and B that had elements in common; but in Method 2 we can use Fig. 2.8 to represent disjoint sets,

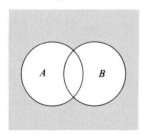

Fig. 2.7

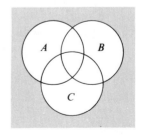

Fig. 2.8

provided that we indicate that the intersection $A \cap B$ is empty. This we can do by putting the symbol "$\emptyset$" in the region which represents $A \cap B$. Thus Fig. 2.9 will represent the fact that the sets A and B are disjoint. If we wish to represent the statement $B \subset A$, we observe that this states that there are no elements that belong to B and do not belong to A. Hence it will be the other portion of circle B that will be empty, so that Fig. 2.10 will represent the statement $B \subset A$.

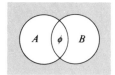

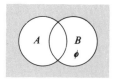

Fig. 2.9   **Fig. 2.10**

Turning to a problem with three sets, let us suppose that we want to show that $(B \subset A) \,\&\, (C \subset B) \rightarrow (C \subset A)$. We start with the basic diagram of Fig. 2.8. We represent $B \subset A$ as above, but now the region that is to be marked as empty is divided into two parts, so we put "$\emptyset$" in each of them, since both are empty. This gives us Fig. 2.11.

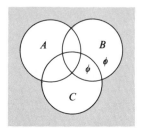

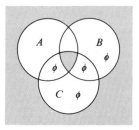

Fig. 2.11   **Fig. 2.12**

Proceeding similarly for $C \subset B$ we obtain Fig. 2.12. We now see that the only portion of the circle C that is not marked as being empty is the shaded portion in the figure, and this lies entirely in A. Hence $C \subset A$. Alternatively the two regions which correspond to elements in C but not in A are marked as being empty. Hence, again, $C \subset A$.

With four or more sets these diagrams become a little more difficult to construct, but the basic ideas are the same.

Exercise 11

1. Revise Exercises 9 and 10 using, where applicable, the two diagrammatic methods of representing set inclusion that have been given in section 2.8. Note: For problems requiring the taking of complements we need to represent the universal set currently being considered. We could do this with just another circle, but it is perhaps better to distinguish it by making it a square or rectangle instead. For example, $\mathscr{C}(A \cup B)$ would be represented by the shaded area in Fig. 2.13.

2. Prove, by any method, that

$$\{[(A \cap C) \subset B] \, \& \, [B \cap C = \emptyset]\} \to [(A \cup B) \cap C = \emptyset].$$

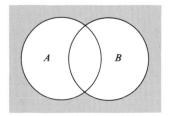

Fig. 2.13

2.9 CARTESIAN PRODUCT OF SETS

The elements of sets are not usually numbers, so there does not seem to be any useful sense in which we can "multiply" two sets together. To take only examples from the beginning of this chapter, what meaning could we possibly ascribe to the multiplication of the planet Mars (an element of the first set mentioned) by the Sherman Act (an element of the set defined in the fourth example)? Yet it happens there is at least one sense in which we can form the *product* of two (or more) sets, and that this concept is of great use in the development of set theory.

Suppose we have two sets A and B. We consider the set of all ordered pairs of elements of A and B, that is, all pairs (x, y) where $x \in A$ and $y \in B$. The significance of the term *ordered pairs* is that the order in which we take the elements, the first from A, the second from B, is important. The set of pairs we have now obtained is a new set, in general larger than either A or B, which is called the *Cartesian product* (or just the "product") of the two sets. This Cartesian product set is denoted by $A \times B$. In symbols the definition is:

$$A \times B = \{(x, y) \,|\, x \in A \, \& \, y \in B\}.$$

As an example, let A and B each have three elements, and be defined by

$$A = \{\text{Antony, Samson, Napoleon}\}.$$
$$B = \{\text{Cleopatra, Delilah, Josephine}\}.$$

Then the Cartesian product of these two sets consists of nine elements, namely the pairs:

(Antony, Cleopatra),	(Antony, Delilah),	(Antony, Josephine),
(Samson, Cleopatra),	(Samson, Delilah),	(Samson, Josephine),
(Napoleon, Cleopatra),	(Napoleon, Delilah),	(Napoleon, Josephine).

Since the sets A and B are disjoint it is not possible for the pairs (x, y) and (y, x) both to occur, for if $x \in A$ then $x \notin B$. But suppose we form the Cartesian product of A with itself. We obtain a set, denoted by $A \times A$, containing the following nine elements (pairs):

(Antony, Antony),	(Antony, Samson),	(Antony, Napoleon),
(Samson, Antony),	(Samson, Samson),	(Samson, Napoleon),
(Napoleon, Antony),	(Napoleon, Samson),	(Napoleon, Napoleon).

Here we see that, for example, (Antony, Samson) and (Samson, Antony) are both included, since they are different *ordered* pairs.

Product sets such as this are far from being mere playthings of the logician or the mathematician. They crop up time and time again in all branches of science, in economics, in business, and in many other walks of life. A Cartesian product of common occurrence in business is one in which A is a set of periods of time (say the months of the year), and B is the set of possible production figures for a factory. An element of $A \times B$ is then a pair consisting of a month and a production figure. Of special significance are those pairs in which the production figure is that which was actually achieved in the stated month. These pairs will form a subset of the product set, a subset of particular importance to the management of the factory since it summarizes the production over the year.

To take an example in which the number of pairs is comparatively small, let us consider an aircraft factory which turns out between one and five aircraft per month.

If $A = \{$January, February, March, April, May, June, July,

August, September, October, November, December$\}$

and

$$B = \{1, 2, 3, 4, 5\}$$

then the product set $A \times B$ consists of pairs like

(February, 3),　(July, 2),　(September, 5),

and so on.

There is a convenient method for displaying in diagrammatical form the elements of a Cartesian product of this kind. We draw two straight lines —one horizontal, one vertical. These lines are called *axes*. On the horizontal axis we make marks denoting the elements of A (we shall assume for the moment that there is only a finite number of them), and we make marks on the vertical axis to represent the elements of B. An element of $A \times B$, say (x, y), is then represented by a dot or point lying immediately above the mark on the horizontal axis which represents x, and lying on a level with the mark on the vertical axis which represents y. (The terms "horizontal," "vertical," "above" and "on a level with" should be self-explanatory, though if the diagram is drawn on a horizontal sheet of paper the "vertical" axis is not *really* vertical.)

The set of months and production figures that we have just considered will then be represented by the following diagram in which each dot denotes an element of the set $A \times B$.

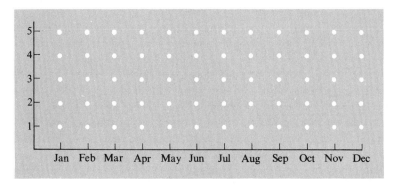

Fig. 2.14

If we now eliminate from the diagram all dots except those representing pairs that correspond to an actual production, that is, pairs (x, y) where y is the production in month x, we obtain the diagram of Fig. 2.15. This is, of course, the well-known business graph or chart.

It often happens that a subset of a Cartesian product set has the property (as this one has) that an element of the first set can occur in only one pair. Here, for example, there can only be one pair whose first item is "January," since there is only one production figure for each month. This is not always so. Suppose we consider the same product set as represented in Fig. 2.14, but let the digits "1," "2," "3," "4," "5" represent the individual cars owned by a small car hire firm. Consider that subset of $A \times B$ consisting of those pairs (x, y) for which car y was available for hire during month x. Such a subset might well look like Fig. 2.16.

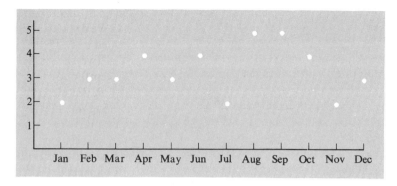

Fig. 2.15

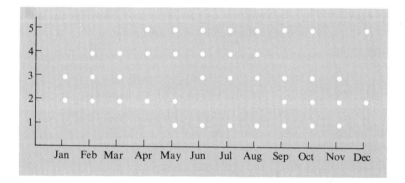

Fig. 2.16

In these examples the sets have all been finite sets, but this need not be so; a similar diagrammatical representation is possible for Cartesian products of infinite sets. One such product, of fundamental importance in mathematics, is the product with itself of the set of all numbers (strictly speaking, the set of all *real* numbers—but until we have discussed what a real number is we shall not use this term). The diagrammatical representation of subsets of this product is then a graph in the mathematical, as opposed to the business, sense of the word. Since we shall be studying these concepts at length in the next chapter, we shall not pursue them further here.

If the sets A and B are not the same, then either there is an element of A which is not an element of B, or else there is an element of B which is not in A. In the first case let x belong to A but not to B, and let y be any element of B. Then $(x, y) \in A \times B$, but since $x \notin B$ it follows that $(x, y) \notin B \times A$. In the second case we get a similar result using y in place of x. Hence if A and B are *different* sets, then

$$A \times B \neq B \times A.$$

Thus the two product sets are not the same. We say that the operation of forming a Cartesian product is a *noncommutative* operation, unlike the familiar multiplication in ordinary arithmetic which is a *commutative* operation.

Naturally if $A = B$ then the two products are the same, being the set $A \times A$ of all ordered pairs of elements of A. The converse is also true; if $A \times B = B \times A$ then $A = B$. For we have just shown that if $A \neq B$ then $A \times B \neq B \times A$. Hence $A \times B = B \times A$ if, and only if, $A = B$.

The idea of the Cartesian product of two sets is readily extended to three or more sets. If we have k sets $A_1, A_2 \ldots, A_k$, we consider the collection of all possible ordered k-*tuples** of the form $(x_1, x_2, \ldots, x_k)$ where $x_1 \in A_1$, $x_2 \in A_2 \ldots, x_k \in A_k$. This set of k-tuples is denoted by

$$A_1 \times A_2 \times \ldots \times A_k$$

and is called the Cartesian product of the k sets. Thus the Cartesian product of two sets is a set of ordered pairs, of three sets it is a set of ordered triples, and so on.

Exercise 12

1. Draw a graph corresponding to the following set of data:

Month	Jan	Feb	Mar	Apr	May	Jun	Jul	Aug	Sep	Oct	Nov	Dec
Profits	10.2	10.4	10.8	9.8	9.7	9.7	9.9	10.3	10.4	10.7	10.9	10.5

2. (a) Prove that $(A \cup B) \times C = (A \times C) \cup (B \times C)$
and that $\qquad (A \cap B) \times C = (A \times C) \cap (B \times C)$.
   (b) Deduce that
$$(A \cap B) \times (C \cup D)$$
$$= [(A \times C) \cup (A \times D)] \cap [(B \times C) \cup (B \times D)].$$

2.10 RELATIONS

We frequently make statements to the effect that something has a certain relationship to something else, for example,

Mr. Smith is the father of Tom Smith.
The set A is a subset of the set B.

---

* This ungainly word is about the best we can do to generalize the terms *triple*, *quadruple*, etc.

7 is the square root of 49.

29 is less than 35.

The statement **p** is equivalent to the statement **q**.

Statements of this form are known as *relations*; they are of the form "*x* has such and such a relationship to *y*." If we use the symbol *R* to denote the relationship, we may abbreviate this statement to *x R y*, which we may read as "*x* has the relation *R* to *y*." Thus in the first of the above examples *x* stands for Mr. Smith, *y* stands for Tom Smith, and *R* stands for the relation "father of" and so on.

What has just been said will (I hope) convey to the reader what is meant by a relation, but how are we to give a strict definition of this term? The essential thing about a relation *R* is that if we take two particular objects *x* and *y*, we should be able to say whether or not the relation *R* holds between them, that is, whether the statement *x R y* is true or false. Thus if we know exactly what is meant by a particular relation *R* we are able to distinguish certain pairs of objects, *x* and *y*, (namely, those for which *x R y* is true), from other pairs of objects (those for which *x R y* is false).

Suppose that we consider the set *U* consisting of all possible objects (a very large set indeed).* and take the Cartesian product $U \times U$ of this set with itself. This product set will consist of all possible ordered pairs of objects (x, y). Then, to any given relation *R*, there corresponds a subset *V* of $U \times U$, namely the set of all pairs (x, y) for which *x R y* is true. Thus if *R* is the relation "father of," the subset *V* will be the set of all pairs of animals between which the relationship of paternity holds, with the father being the first element of the ordered pair, and the offspring the second element.

Does this work the other way? If we are given a subset *V* of $U \times U$, does this correspond to some relation? Yes, because we can *define* the relation, *S* say, by asserting that *x S y* if and only if $(x, y) \in V$.

We now see that corresponding to a relation there is a subset of $U \times U$, and corresponding to a subset of $U \times U$ there is a relation. We can therefore define a relation as being determined in the above way by a subset *V* of $U \times U$, the set of all ordered pairs (x, y). We could, in fact, go the whole hog and define a relation as *being* a subset of $U \times U$; but the reader will perhaps find it clearer to preserve a distinction between a relation and the corresponding subset of $U \times U$.

One further property of our definition may be noted before we proceed. As we have defined it, a relation can hold only between *pairs* of objects. But what about such relationship as the following? "Chicago is between New York and Los Angeles"; "I am between the devil and the deep blue sea"

---

* There are certain logical difficulties which are liable to arise from the use of phrases like "the set of *all* objects." In practice, as we shall see, we shall always restrict the choice of *x* and *y*, in which case these difficulties disappear. (See section 2.13.)

"An unstable equilibrium point must lie between two stable equilibria." In each of these cases, the relationship concerned has reference to *triples* of objects, not pairs. Such *ternary relations*, as they are called, define subsets of the product set $U \times U \times U$ of ordered triples of objects. Thus the first element of the triple could refer to the object between the other pair; or one might choose the second element for such a role. In any case, the relation cannot be defined by a subset of $U \times U$, and hence does not fit our earlier definition.

Relations which are equivalent to a subset of $U \times U$ are referred to as *binary relations*; it is only those to which our definition refers. Ternary, quaternary, . . . , *n*-ary relations may be defined similarly as subsets of $U \times U \times \ldots \times U$ where the number of "multiplication" signs is one less than the degree of the relation. Such relations may occasionally be important. For example, the assertion that the utility difference between two batches of goods a and b is greater than the utility difference between two other batches c and d, expresses a relation involving four objects a, b, c and d; hence it is a quaternary relation. But binary relations are by far the "leading species of the large genus" of relations, and our discussion will be restricted to them entirely.

Our definition of binary relations (as a subset of $U \times U$) is quite satisfactory for logical and mathematical purposes, but it should be borne in mind that this way of looking at relations does entail a certain departure from the everyday ideas from which the concept originally came. This is especially the case when the subset of $U \times U$ is defined by intension, for in everyday speech there is usually something more to a relation than the mere fact that certain pairs of objects are connected by it while others are not. If a visitor from another planet wished to know what was meant by the expression "father of," we might satisfy his curiosity by parading before him all pairs of fathers and offspring. He would then be able to use properly statements of the form "x is the father of y," and in this sense he would know what was meant by the term "father of"; but he might well be quite unaware of many aspects of the relation of paternity which, in everyday speech, would be considered essential to a full understanding of the relationship. This discrepancy between ordinary language and the language of logic is similar to the discrepancy mentioned in the last chapter, between statements of the form $\mathbf{p} \rightarrow \mathbf{q}$ and the causal connection which usually underlies such statements in the everyday usage. In mathematics this difference is not likely to give rise to any confusion, so it is quite in order to regard a relation as being, or being determined by, a subset of the set $U \times U$ defined above.

2.11 RELATIONS ASSOCIATED WITH A GIVEN RELATION *R*

NEGATION OF *R*.

If a certain relation holds between x and y, then there is another relation which *automatically* does not hold between x and y. It is denoted by $\tilde{R}$ and is defined as follows:

$$x \ \tilde{R} \ y \text{ if, and only if, } \sim(x \ R \ y).$$

The relation $\tilde{R}$ is called the *negation* of R. Thus if $x \ R \ y$ stands for "x is the father of y," then $x \ \tilde{R} \ y$ stands for "x is not the father of y"; $\tilde{R}$ is the relation of nonpaternity.

CONVERSE OF *R*.

If x has a certain relation R to y, then y will automatically have a certain relation to x. This relation is called the *converse* of R, is denoted by R', and is defined as follows:

$$y \ R' \ x \text{ if, and only if, } x \ R \ y.$$

Thus the converse of the relation "father of" is the relation "offspring of"; the converse of "less than" is "greater than or equal to," and so on.

2.12 PROPERTIES OF RELATIONS

Sometimes it is possible for an object to have a certain relation to itself. Thus if $x \ R \ y$ stands for "x is an admirer of y," then it is usually true to say that "x is an admirer of x," i.e., $x \ R \ x$ although there are possibly a few individuals for whom this is not true! There are some relations, however, for which the statement $x \ R \ x$ is *always* true (i.e., true whatever the object x may be) as for example when $x \ R \ y$ stands for

$$x = y$$

or x is the same age as y

or x is a contemporary of y

and so on. Such relations are said to be *reflexive*. If $x \ \tilde{R} \ x$ is always true, i.e., no object has the relation R to itself, then R is said to be *irreflexive*. A relation that is neither reflexive nor irreflexive will be said to be *nonreflexive*.

If we say "John loves Mary," then although it may be true that "Mary loves John," this does not necessarily follow; there are many unrequited lovers. Thus in general the converse of a relation does not follow from the relation itself. For some relations, however, the converse always follows; that is to say whenever $x \, R \, y$ then $y \, R \, x$ or briefly, $(x \, R \, y) \equiv (y \, R \, x)$ for all x and y. This means that the relations R and R' are the same, which we may write as $R = R'$. Such relations are said to be *symmetric*. Thus the relations contained in the sentences

x is a relative of y.
x is 10 feet away from y.
x lives in a different country from y.

and so on, are symmetric relations.

Sometimes if $x \, R \, y$ is true then $x \, \tilde{R}' \, y$, i.e., $y \, R \, x$ cannot be true. R is then said to be *asymmetric*. Examples are given by the statements

x is the father of y.
x is better than y.
$x < y$.

A relation which is neither symmetric nor asymmetric is called *nonsymmetric*, for example

x loves y
$x \leq y$
x is a brother of y

and so on.

If $x \, R \, y$ and $y \, R \, z$ then, in general, it will not be the case that $x \, R \, z$; thus if x is the father of y and y is the father of z, then x *cannot* be the father of z. But if we take R to be the relation "ancestor of" then from $x \, R \, y$ and $y \, R \, z$ it *always* follows that $x \, R \, z$. Similarly if R is the relation $<$, and x, y and z are numbers, then grade school arithmetic teaches us that $x < y$ and $y < z$ automatically imply $x < z$. Relations such as these latter are said to be *transitive*. Thus a relation R is transitive if

$$(x \, R \, y) \, \& \, (y \, R \, z) \rightarrow (x \, R \, z) \text{ for all } x, y, z.$$

Other examples are "x is to the north of y," "x is prettier than y," "x costs more than y," and so on.

If on the other hand we have

$$(x \, R \, y) \, \& \, (y \, R \, z) \rightarrow x \, \tilde{R} \, z \text{ for all } x, y, z$$

as, for example, with the relation "father of," we say that R is *intransitive*. A relation which is neither transitive nor intransitive is said to be *nontransitive*.

Exercise 13

1. Decide whether the relations specified below are reflexive, irreflexive or nonreflexive; symmetric, asymmetric or nonsymmetric; transitive, intransitive or nontransitive.

    (a) x was built by y.
    (b) x is a compatriot of y.
    (c) x earns more than y.
    (d) $x < \frac{1}{2}y$.
    (e)   the set x has non-empty intersection with the set y.
    (f) x is at least as good a trombone player as y.
    (g) x is two inches taller than y.
    (h)   the universal relation, defined as the relation V for which $x \, V \, y$ is true no matter what x and y are.
    (i)   the empty relation. This is the relation N for which the statement $x \, N \, y$ is never true.

These last two relations are the analogues in the theory of relations to the universal set and the empty set.

2. Form the negations of the relations in **(a)** through **(i)** of problem 1 and classify them according to reflexivity, symmetry, and transitivity.

3. Form the converse relations to those in problem 1 and classify them according to reflexivity, symmetry and transitivity.

2.13 DOMAIN AND CONVERSE DOMAIN

In the expression $x \, R \, y$ we have taken the symbols x and y to stand for any objects whatever. In general, most of the statements that we get by replacing x and y by specific objects will be false, and what is more, false in a rather obvious way. For example, if R is the relation "father of," and we replace x by "Brooklyn Bridge" and y by "Beethoven's Ninth Symphony" we obtain the statement "Brooklyn Bridge is the father of Beethoven's Ninth Symphony." This statement is false, but its falsity is of a trivial kind; for the very natures of the objects x and y preclude any possibility of the statement $x \, R \, y$ being true. This is not so with a statement like "George V was the father of Edward VII." This also is false, but its falsity is, so to speak, informative. There is nothing in the nature of either George V or Edward

VII *considered separately* which would automatically prevent the statement from being true; the falsity of this statement is a historical fact and not anything trivial.

Thus in dealing with a given relation R it is natural to ignore those objects which if put in place of x would, by their very nature, make the statement $x\ R\ y$ false (i.e., those objects which we know cannot have the relation R to anything) and to concentrate on the set of those objects which have the relation R to something. This set is known as the *domain* of R and is defined as the set of all objects x such that the statement $x\ R\ y$ is true for at least one object y. These objects x are sometimes called the *predecessors* with respect to the relation R, and thus the domain of R may be called the set of all its predecessors. If R is the relation "father of," the domain of R is the set of all fathers; the domain of the relation "pupil of" is the set of all pupils, and so on.

In the same way we may consider the set of all objects to which something has the relation R, i.e., the set of objects y such that $x\ R\ y$ is true for at least one object x. These objects are often called the *successors* with respect to the relation R, and the set of all successors of R is called the *converse (or counter) domain* of R, since it is the domain of the converse R' of R. The converse domain is also known as the *range*. The union of the domain and the converse domain of R is called the field of R*. We note in passing that if R is reflexive *or* symmetric then the domain and the converse domain will coincide.

2.14 EQUIVALENCE RELATIONS

Using different combinations of the various properties that relations can have, we may isolate certain special types of relations of particular importance in mathematics and its applications. The first such type we shall deal with are those known as *equivalence relations*. These are relations which are *reflexive, symmetric* and *transitive*. For example,

1. x was born on the same day of the week as y. (For this relation the domain is, shall we say, the set of all people.)
2. x is parallel to y. (The domain is the set of all lines, planes, etc.)
3. x is the same price as y. (The domain is the set of all goods.)
4. $x = y$. (The domain is the set of all real numbers, say.)
5. $\mathbf{p} \equiv \mathbf{q}$. (The domain is the set of all statements.)

---

* This use of the word *field* is unfortunate since the same word is used for an important algebraic concept. Nevertheless the usage is fairly common.

Such relations are called *equivalence relations* and we shall in general denote such a relation by the symbol E. The importance of equivalence relations resides in a property that they all possess and that is stated in the following theorem:

THEOREM 2.1

IF E IS AN EQUIVALENCE RELATION, AND D IS ITS DOMAIN,* THEN D CAN BE EXPRESSED AS THE UNION OF A NUMBER OF *disjoint* SUBSETS WITH THE PROP-ERTY THAT IF x AND y ARE ELEMENTS OF D, THEN $x \, E \, y$ IF, AND ONLY IF, x AND y BELONG TO THE SAME ONE OF THESE SUBSETS.

The meaning of this theorem becomes clear if we consider a specific example, say example 1 above. Here D is the set of all people. Let A_1 be the set of all people born on a Sunday, A_2 the set of all people born on a Monday, and so on up to A_7, the set of all people born on a Saturday. Then clearly the union of the seven sets A_1, A_2, A_3, A_4, A_5, A_6, and A_7 is the set D, and these seven sets are *disjoint*—the intersection of any two is empty. Moreover it is easy to see that in this case $x \, R \, y$ is true if, and only if, x and y belong to the same subset.

We now proceed to the proof of the theorem. Let a be an element of D. Corresponding to a there is a subset A of D, namely the set of elements y of D to which a has the relation E, i.e.,

$$A = \{y \mid a \, E \, y\}.$$

There is one such subset corresponding to each element of D, but, in general, it will often happen that a subset will correspond to two or more elements of D. Consider now the collection of all *distinct* subsets which can be obtained in this way (so that each subset is counted only once in the collection). These subsets are the ones required by the theorem.

To prove this we must first show that their union is D. This is easy, for every element of D belongs to at least one of the subsets, namely the one that corresponds to it, since E is reflexive. Next we must show that no two subsets intersect. Suppose two of the subsets, A and B, corresponding to elements a, b of D, have a nonempty intersection, and let $c \in A \cap B$. Then $a \, E \, c$, (from the definition of A, since $a \in A$) and similarly $b \, E \, c$, or, since E is symmetric, $c \, E \, b$. But from $a \, E \, c$ and $c \, E \, b$ it follows that $a \, E \, b$, since E is transitive. If y is *any* element of B, we have also $b \, E \, y$, which with $a \, E \, b$, shows that $a \, E \, y$. In other words y belongs to A. Thus we have shown that if $y \in B$, then $y \in A$, or, in short, that $B \subset A$.

If we now repeat the argument with A and B interchanged we find that

---

* Since E is reflexive and symmetric, its converse domain is also D.

$A \subset B$, from which it follows that $A = B$. But this contradicts the fact that the subsets of the collection are distinct. Thus our original assumption (that A and B intersect) is shown to be false, by a *reductio ad absurdum* argument.

Finally we must show that $x E y$ if, and only if, x and y belong to the same subset. Because of the "if, and only if" we have two things to prove—a statement and its converse. Suppose first that $x E y$, and that A, corresponding to a, is the subset to which x belongs. Then $a E x$ and $x E y$, so that $a E y$ (by the transitivity of E). But $a E y$ means that $y \in A$, i.e., y belongs to the same subset as x. Hence if $x E y$, then x and y belong to the same subset.

Conversely, if x and y belong to the same subset, then $a E x$ (from which follows $x E a$) and $a E y$. But if $x E a$ and $a E y$, then $x E y$ by transitivity. Hence if x and y belong to the same subset, then $x E y$. This completes the proof of the theorem.

A division of a set D into a number of disjoint subsets whose union is D, is often called a *partition* of D. The subsets which are determined by an equivalence relation in the manner we have just described are usually called *equivalence classes*. We may therefore restate Theorem 2.1 a little more succinctly as follows:

AN EQUIVALENCE RELATION E PARTITIONS ITS DOMAIN D INTO SUBSETS (EQUIVALENCE CLASSES) SUCH THAT IF x, $y \in D$ THEN $x E y$ IF, AND ONLY IF, x AND y BELONG TO THE SAME EQUIVALENCE CLASS.

This basic theorem regarding equivalence relations makes them of great importance. Mathematics abounds with equivalence relations (e.g., equality, congruence, similarity), and so does the world around us. In particular, they are often used—explicitly or implicitly—in much modern theoretical work in the social sciences.

Exercise 14

1. Determine the domain and converse domain for each of the relations in Exercise 13.1.

2. Determine whether or not the following relations are equivalence relations:

    (a) x hates y.
    (b) x is a brother of y.
    (c) x pays exactly as much income tax as y.
    (d) x has the same number of legs as y.
    (e) x is a pupil of y.
    (f) x likes the same kind of music as y.

3. Show that the following relations are equivalence relations, and describe in words the equivalence classes determined by them.

(a) x is in the same profession as y.

(b) x is in the same college class as y.

(c) x is the same age as y.

(d) x has the same I.Q. as y.

4. Consider the set of all pairs of positive integers (x, y). Over this set we define a relation R by

$$(x, y) \, R \, (u, v) \equiv (xv = yu).$$

Prove that R is an equivalence relation.

2.15 ORDERING RELATIONS

By combining, in different ways, the various properties of relations that we have described, we can generate whole classes of relations, one of which—the class of equivalence relations—has just been discussed. Before introducing other important classes, however, we will first describe another property which is of immediate relevance here.

Suppose we consider the relation "at least as pretty as" (denoted by ALP), and the set X of all female college students in the world (with a suitable definition of *college*). It seems reasonable to suppose that if we took *any* pair of elements (x_i, x_j) from the product set of $X \times X$, we would be able to assert the statement $(x_i \text{ ALP } x_j) \lor (x_j \text{ ALP } x_i)$; that is, that any two female college students are comparable (one way or the other) as regards prettiness. If we now adjoin to X the set Y, consisting of all faculty members in the aforesaid colleges, then it is highly unlikely that we would be willing to assert, with any degree of comfort, that for every pair (z_i, z_j) in $Z \times Z$ (where $Z \equiv X \cup Y$) the statement $(z_i \text{ ALP } z_j) \lor (z_j \text{ ALP } z_i)$ is true. We might well object that the implied concept of relative pulchritude was inapplicable to many members (the author, for one!) of the new set Y.

If the statement $(x_i \, R \, x_j) \lor (x_j \, R \, x_i)$ holds for *all* elements (x_i, x_j) of the Cartesian product $X \times X$ of a set X, then the relation is said to be *connected* in the set X. If the statement holds for *no* such pair, R is said to be *vacuous* (in X). Notice that this property of connectedness involves both R *and* X. It is false, as our example shows, to say that a relation R is connected, without specifying its field; while it is *meaningless* to say that a set X is connected. Connectedness refers to the interplay, as it were, of the relation R and the set X. Thus if we take the relation R to be "less than," and X to be the set of all numbers, then R is connected in X; but if R were the relation "less than by at least 2," then R would not be connected in X. Alternatively, if R were "less than," and X not the set of numbers but the set of all possible ordered pairs of numbers, then R would not be connected in X.

This brings out the importance of the distinction between a *set* and an *ordered set*. A set is simply a collection of certain elements without regard to the way in which those elements are arranged; the set $\{x, y\}$ is the same set as $\{y, x\}$. But for many purposes the way in which those elements are arranged, or *ordered*, is of great importance. We have seen, in our discussion of Cartesian products, the importance of distinguishing the ordered pair (x, y) from the ordered pair (y, x). Similarly, there is only one *set* consisting of the elements $\{x, y, z\}$, but it provides the basis for six different ordered triples: (x, y, z), (x, z, y), (y, x, z), (y, z, x), (z, x, y) and (z, y, x). The reader should check that, in addition, although $\{x, y, z\}$ has eight distinct subsets, we can make sixteen different ordered subsets from it.

In general there are many possible distinct ways of arranging (i.e., ordering) the elements of a set, but in applications we are not usually interested in the number of possible orderings of the relevant set, if only because that set often contains an unlimited number of elements (e.g., the set of all the positive integers). We are usually much more concerned with some particular *criterion* with respect to which the elements may be brought into order. As Bertrand Russell succinctly expressed it: "The only thing that is arbitrary about the various orders of a set of terms (elements) is our attention, for the terms have always all the orders of which they are capable."*

A leading criterion of this kind—indeed the original example in the theory of ordering—is the quality of being "less than" $(<)$, applied to the set J of the positive integers. The elements of this set may be paraded by order of size, with "1" first, "2" second, and so on; arranged in this way it is *ordered* by the relation $<$. We are so accustomed to this ordering that it is commonly called the *natural* ordering of the positive integers; so "natural" is it, indeed, that the reader probably instinctively thought of the set of numbers under consideration as already being arranged "naturally," though of course there is no reason why it should; it could just as well be 2, 1, 4, 3, 6, 5, ...

Now this criterion, applied to the set J, has certain of the properties that we have already introduced. First, it is *connected*, since either a positive integer i is less than another integer j, or $j > i$. Secondly, grade school arithmetic taught us that if $(i < j)$, then $\sim (j < i)$; in words, that $<$ is *asymmetric* in J. But even if we had not learned this, we could now deduce it from two other properties of this ordered set, probably gleaned from the same source. For clearly $\sim (i < i)$ for all $i \in J$, so that $<$ is *irreflexive* in J, while we know that $\{(i < j)$ & $(j < k)\} \to (i < k)$, for all i, j, k in J, so that $<$ is *transitive* in J. Now suppose that $<$ were not asymmetric, so that there was a pair of elements in J such that $(i < j)$ & $(j < i)$. Then, by transitivity (since its definition does not assume that i, j, k are distinct elements), we can assert $i < i$, which contradicts reflexivity. Hence $<$ *is* asymmetric in J.

* Bertrand Russell, *Introduction to Mathematical Philosophy* (N.Y.: The Macmillan Co., 1924, Reprinted 1948), p. 30.

It is appropriate here to note in passing an error into which the brighter beginning students often fall. By using an argument similar to that just given, they attempt to show that reflexivity is unnecessary in the definition of an equivalence relation, since it follows from symmetry and transivity. But such a proof would fail for equivalence classes containing only one member (since symmetry could not be defined), and so reflexivity is an essential part of a careful definition.

To return to our ordering ($<$) suppose we had called it "precedes," instead of "less than." This does not, of course, change any of its properties but does bring out more clearly the basic idea of an ordering or *ranking* of the integers. Moreover, this change of terminology suggests other, similar relations in the world around us that imply the idea of *precedence*, such as the quality of being "heavier," or "prettier," or "cleverer." For all such relations, it will usually be found that everyday usage assigns them the properties of irreflexivity and transitivity (and, therefore, of asymmetry); but it is not usually the case that the relations are connected in every relevant set.

These considerations suggest, to the mathematical mind, the idea of *abstracting* the properties of these precedence relations, and so defining a whole class of relations as those that possess these properties. Thus we define the class of *ordering relations* (or, for brevity, of *orderings*) as those binary relations which are *transitive* and *irreflexive*.* Any set X on which an ordering relation S is defined is called an *ordered set*, which we will denote by $(X ; S)$. The process of ordering imposes a certain *structure* upon that set, namely the ordering itself.

The set of all ordered sets may be partitioned into two subsets, the class of ordered sets for which the ordering is *connected* in the relevant set, and the class of sets for which this is not the case. Those ordered sets belonging to the first class are called *totally*, or *completely ordered sets*, and those belonging to the second are termed *partially ordered sets*. The set $(X ; \text{ALP})$ of our first example was totally ordered, the set $(Z ; \text{ALP})$ was partially ordered. Sometimes, when the underlying set is taken for granted, mathematicians refer loosely to complete orderings and partial orderings; but, strictly speaking, this terminology is incorrect. It is the (ordered) *sets* which have these properties, not the ordering relations themselves. Mathematics abounds with ordered sets (e.g., the set of all numbers, or the positive integers, both ordered "naturally") and these concepts also play an important role in the social sciences.

Suppose that we consider not $<$ but the relation "less than or equal to," denoted by $\leq$, and apply it to the set of positive integers. The result will not be an ordered set, since although transitive in this set, $\leq$ is not irreflexive but reflexive. Nevertheless it is an interesting kind of relation,

* Unfortunately, textbooks in mathematics and in logic are not standard in their terminology concerning orderings. Orderings are sometimes called simple orderings or (in economics) strong orderings.

and following the same procedure as before, we abstract its properties and define relations which are *transitive* and *reflexive* to be *weak ordering relations* (or weak orderings).* Again, the set of all weakly ordered sets may be partitioned into two subsets, those for which the ordering is connected (the completely weakly ordered sets), and those for which it is not (the partially weakly ordered sets).

Notice that the class of equivalence relations is included in the class of weak orderings, since all equivalence relations are transitive and reflexive. But they possess a property—symmetry—not shared by weak orderings in general. There is, in fact, a close connection between weak orderings, equivalence relations and orderings, which is expressed in the following simple theorem.

THEOREM 2.2

GIVEN ANY COMPLETELY WEAKLY ORDERED SET $(X; R)$, IT IS POSSIBLE TO PARTITION X INTO EQUIVALENCE CLASSES BY MEANS OF AN EQUIVALENCE RELATION DERIVED FROM R. FURTHER, THESE EQUIVALENCE CLASSES ARE THEMSELVES COMPLETELY ORDERED BY AN ORDERING DERIVED FROM R.

PROOF:

Define a relation E by $(x_i \ E \ x_j) \equiv (x_i \ R \ x_j) \ \& \ (x_j \ R \ x_i)$. Then it is easy to check that E is an equivalence relation.
1. *Reflexivity.* Since R is a weak ordering, $x_i \ R \ x_i$ for all x_i which, together with the definition of E, implies that E is reflexive.
2. *Symmetry.* Obvious from the definition of E.
3. *Transitivity.* Suppose that x_i, x_j, and x_k are such that $x_i \ E \ x_j$ and $x_j \ E \ x_k$. Then, from the definition of E, we have $(x_i \ R \ x_j) \ \& \ (x_j \ R \ x_k)$ which, by the transivity of R, yields $x_i \ R \ x_k$. Similarly we may assert $(x_k \ R \ x_j) \ \& \ (x_j \ R \ x_i)$ which gives $x_k \ R \ x_i$. Hence $(x_i \ R \ x_k) \ \& \ (x_k \ R \ x_i)$, which is equivalent to $x_i \ E \ x_k$.

Therefore, by Theorem 2.1, the set X can be partitioned into equivalence classes. Note that since $(X; R)$ is connected and reflexive, it follows that each equivalence class is nonempty, though we cannot be sure that any of them has more than one member.

It remains to prove the second part of the theorem. Define a relation S by $(x_i \ S \ x_j) \equiv (x_i \ R \ x_j) \ \& \ (x_j \ \tilde{R} \ x_i)$. We now show that S is a complete ordering of the equivalence classes obtained above. First observe that if

---

* The terminology on weak orderings is even less standard than that for orderings; that used here follows common usage in economics.

x_i, x_i' belong to the same equivalence class, and some other element x_k does not, but is such that x_i S x_k, then x_i' S x_k. By the definition of S, we have to show that $(x_i'$ R $x_k)$ & $(x_k$ $\tilde{R}$ $x_i')$. Now $(x_i'$ E $x_i)$ → $(x_i'$ R $x_i)$ which, together with x_i S x_k and the transivity of R, implies x_i' R x_k. To prove the second part of the conjunction, suppose it false; that is, suppose that x_k R x_i'. Then, by x_i' E x_i and the transitivity of E, we have x_k R x_i, which contradicts x_i S x_k. Hence x_k $\tilde{R}$ x_i, and so we have x_i' S x_k.

A similar proof shows that if x_j S x_i, then x_j S x_i'. These two results together demonstrate that if two elements x_i, x_k are such that x_i S x_k, then all the elements in the same equivalence class as x_i have the relation S to all the elements in the same equivalence class as x_k; so that S *operates* on the equivalence classes of E, as it were.

To show that S is in fact a complete ordering of these classes we prove that it is:

1. *Connected* (i.e., for all x_i, x_k such that x_i $\tilde{E}$ x_k, $(x_i$ S $x_k)$ $\vee$ $(x_k$ S $x_i)$. Since R is a complete weak ordering of X, $(x_i$ R $x_k)$ $\vee$ $(x_k$ R $x_i)$. If x_i R x_k, then since x_i $\tilde{E}$ x_k, we have $(x_i$ R $x_k)$ & $(x_k$ $\tilde{R}$ $x_i)$, or x_i S x_k. Similarly if x_k R x_i, then x_i S x_k. This proves the result.

2. *Irreflexive.* Suppose this is false, so that x_i S x_i for some x_i. But then, by definition of S, $(x_i$ R $x_i)$ & $(x_i$ $\tilde{R}$ $x_i)$ which is obviously contradictory; hence x_i $\tilde{S}$ x_i for all x_i.

3. *Transitive* [i.e., for all x_i, x_j, x_k in pairwise different equivalence classes, $(x_i$ S $x_j)$ & $(x_j$ S $x_k)$ → $(x_i$ S $x_k)$]. By the definition of S, and the transitivity of R, $(x_i$ S $x_j)$ & $(x_j$ S $x_k)$ → x_i R x_k. Suppose x_k R x_i. Then $(x_k$ R $x_i)$ & $(x_i$ S $x_j)$ → $(x_k$ R $x_j)$, contrary to the hypothesis that x_j S x_k. Hence $(x_i$ R $x_k)$ & $(x_k$ $\tilde{R}$ $x_i)$, which is equivalent to x_i S x_k. This completes the proof of the theorem.

Exercise 15

1. Determine which of the following relations are ordering relations, and whether they are weak orderings or strong orderings
   (a) x has a higher cost-of-living than y. (x, y are countries.)
   (b)   The rate of industrial development in country x is no less than in country y.
   (c) x and y have the same mother.
   (d)   Airport x can be reached by direct flight from airport y.
   (e) x is downstream from y.
Does the above list include any equivalence relations?
2. Below are given some sets together with an ordering relation. Determine in each case whether the set is totally ordered, or partially ordered by the corresponding relation.

(a) The set of all human beings. x is not older than y.
(b) The set of all books. x has more pages than y.
(c) All human beings (past and present). x is an ancestor of y.
(d) The set of all circles in a plane. x is included in y.
(e) The set of all integers. x divides y.
In each case state whether the ordering relation is strong or weak.

Exercise 16 (Chapter Review)

1. If A is the set of all nonnegative integers (i.e., positive and zero);
    if B is the set of all positive nonzero integers;
    if C is the set of all positive even numbers;
    if D is the set of all positive prime numbers;
    if E is the set of all positive integers divisible by 3;
and if F is the set of all nonpositive integers;
prove the following results:
    (a) $B \subset A$
    (b) $C \subset B$
    (c) $D \subset B$
    (d) $B \cap F = \emptyset$
    (e) $C \cap F = \emptyset$
    (f) $(B \cup E) \subset (A \cup E)$
Identify the following sets.
    $A \cap F$; $E \cap F$; $(C \cup E) \cap D$; $C \cap E$; $A \cup F$; $C \cup F$.
2. Prove by means of diagrams, or otherwise, that
    (a) if $D \subset A$, $D \subset B$ and $D \subset C$, then $D \subset A \cap (B \cap C)$
    (b) $A \cap (B \cap C) \subset A \cap B$
    (c) $A \cap (B \cap C) = (A \cap B) \cap C$
    (d) $A \cap (B \cup C) = (A \cap B) \cup (A \cap C)$
    (e) $A \cup (B \cap C) = (A \cup B) \cap (A \cup C)$
    (f) $A \cup (B \cap C) \neq (A \cup B) \cap C$
3. Let B, C, D all be subsets of A, and denote by $\mathscr{C}$ the complement with respect to A. Show that
    (i) $\mathscr{C}[B \cup (C \cap D)] = \mathscr{C}(B) \cap [\mathscr{C}(C) \cup \mathscr{C}(D)]$
    (ii) $\mathscr{C}\{(B - D) \cap D\} = \mathscr{C}(C - D) \cup \mathscr{C}(B - C)$
4. For each of the following statements of the form $x \, R \, y$ state which of the properties: *reflexive, irreflexive, symmetric, asymmetric, transitive, intransitive*, apply to the relation R.
    (a) x is the father of y.
    (b) x is the same height as y.
    (c) x is greater than y.
    (d) x lives in the same country as y.
    (e) x is at least as insufferable as y.

(**f**) x loves y.

(**g**) x is an ancestor of y.

Repeat this exercise with the negations $\tilde{R}$, and converses R' of the above relations.

Pick out those relations that are equivalence relations, and identify the corresponding equivalence sets.

Suggestions for Further Reading

Good discussions of many of the topics treated in this chapter may be found in:

Russell, Bertrand, *Introduction to Mathematical Philosophy*. New York: The Macmillan Co., 1924, Reprinted 1948.

Sigler, L. E., *Exercises in Set Theory*. Princeton, New Jersey: D. Van Nostrand Co., Inc., 1966.

Suppes, P., *Introduction to Logic*. Princeton, New Jersey: D. Van Nostrand Co., Inc., 1957.

Tarski, A., *Introduction to Logic*. New York: Oxford University Press, 1941.

Rather more advanced are the following:

Kamke, E., *Theory of Sets*. New York: Dover, 1950.

Rosenbloom, P., *The Elements of Mathematical Logic*. New York: Dover, 1950.

Wilder, R. L., *Introduction to the Foundations of Mathematics*. New York: John Wiley and Sons, Inc., 1952.

The three books on finite mathematics mentioned on page 27 also cover the material of this chapter. There is, in addition, a Schaum "Outline" publication specifically on the theory of sets:

Lipschutz, S., *Set Theory and Related Topics*. New York: Schaum Publishing Co., 1964.

3.1 FUNCTIONAL RELATIONS AND FUNCTIONS

The concept of a *relation* was defined very generally in the last chapter; we shall now take a look at some rather special types of relations. Let us consider a relation R and the corresponding statement $x\,R\,y$. The relation R will then be said to be a *functional relation* if, whatever element x of its domain is given, there is exactly one object y such that the statement $x\,R\,y$ is true. As a simple example of a functional relation we may take the converse of the relation "father of." If $x\,R\,y$ means "y is the father of x" and if we specify an object x belonging to the domain of R (which in this case will be the set of all offspring), then there is one and only one object y for which $x\,R\,y$ is true, namely x's father. Of course, if we were to choose for x an object which did not belong to the domain of R, then we would not have a corresponding object y; if we take x to be the Panama Canal, then there is no object which is the father of x.

If we wish to give a more formal definition of a functional relation, we may do it as follows:

A relation R is a functional relation if for all x in the domain of R

$$(3.1.1) \qquad (x\,R\,y_1\ \&\ x\,R\,y_2) \to (y_1 = y_2),$$

(where $y_1 = y_2$ means that y_1 and y_2 are the same object).

What this says is that there cannot be two *different* objects to which a given x has the relation R. From the definition of the domain of R there will be no object to which x has the relation R if x is not a member of the domain of R, whereas if x does belong to the domain, there will be at least one object (and hence, in this case, exactly one object) to which x has the relation. This brings us around to the original definition.

As further examples of functional relations we may cite the following:

y is the cost of x (in a certain shop at a certain time, shall we say).
y is the area of x.
y is the weight of x.
y is the positive square root of x.
y is the reciprocal of x.

The last two examples will indicate that these functional relations are of common occurrence in mathematics. Functional relations which occur in mathematics are usually known as *functions*, and we say that y is a function of x. Moreover, it is customary to use a different notation. Instead of $x \, R \, y$, $x \, S \, y$, etc., we write $y = f(x)$, $y = \phi(x)$, and so on. Of course, symbols other than x and y can be used, but these are perhaps the commonest, and it is for this reason that the examples just given are all of the form $y \, R' \, x$, rather than $x \, R \, y$. The symbols x and y stand for elements of the domain and converse domain of R respectively, and are frequently referred to as *variables*, x being called the *independent variable* and y the *dependent variable*. Alternatively (and preferably) we may call x the *argument* and y the *value* of the function. The converse domain of R, i.e., the set of values of the function, is often called the *range* of the function.

The notation $y = f(x)$ has an advantage over the notation $x \, R \, y$, which is that the symbol $f(x)$, which represents the same object as the symbol y, contains within itself an indication of the object which has the relation in question to it. In other words we have, in $f(x)$, a convenient way of indicating the value of the function corresponding to the argument x. If we wish, we can replace the general symbol x by a symbol representing some particular element of the domain of the function, and the resulting symbol will represent the unique object having the given relation to it. Thus if $y = f(x)$ is taken to mean "y is the father of x," then

$$f(\text{George Washington})$$

will denote a unique individual, viz., George Washington's father. Similarly, if $y = f(x)$ stands for "y is greater than x by 1.32," then

$$f(2.57)$$

denotes a unique number, viz., 3.89.

If the symbol "$f(x)$" represents the value of the function corresponding to an argument x, what symbol shall we use to denote the function itself: that is to say, what symbol in our new notation corresponds to the symbol R in the old notation? The obvious choice would be to write "$f(\ \ )$" or preferably just "f," and this is certainly good enough so long as we are talking about functions in general, for which arbitrary letters of an alphabet, like f, g, ϕ, etc., are used. But as soon as we start talking about particular mathematical functions, for which special symbols have been reserved, we immediately run into a grave difficulty. Thus, for example, there are two functions that are specified in the following way:

$$y = \log x \quad \text{and} \quad y = \sin x.$$

(The reader need not worry if he has not yet met these functions—he will do so soon enough. Their definitions are not needed for the understanding of the point being made.) These functions may be conveniently referred to as the functions *log* and *sin*, or even by their full names of *logarithm* and *sine*, and this is analogous to the use of "$f(\ \ )$" or "f" to which we have already alluded. But what are we to do about statements like

$$y = x^2 \quad \text{or} \quad y = \frac{1}{x}?$$

Here the simple omission of the x on the right hand side of the $=$ sign will hardly do; while for functions like

$$y = x - 3; \quad y = -3x; \quad y = 2x; \quad y = 2^x$$

it clearly will not do at all. How then are we to refer to these functions?

The method that has in actual fact been devised is to make no attempt to omit the x, but to leave it where it is. This means that, for example, the function *square of* is referred to as the function "x^2"; the function *reciprocal of* is referred to as the function "$1/x$," and so on. Logically, this usage is quite indefensible, for it means that these symbols "x^2," "$1/x$," etc., are ambiguous. They may stand for the value of the function corresponding to the argument x, or they may stand for the function itself. What is more, this is not a matter of academic hair-splitting; this ambiguous notation can, and does, give rise to quite a few difficulties. Mathematicians are well aware of this problem and suggestions have been made from time to time for solving it.* Unfortunately, none of these suggestions has received universal approval. Accordingly we shall accept the situation as it is, and agree to allow symbols like "log x," "x^2," "$2x$," etc., to stand for the functions "logarithm of," "square of," "twice times," etc., as well as for "logarithm of x," "square of x," "twice x," etc. The purist may throw up his hands in horror at this, but after all, the purpose of this book is to prepare the reader to understand the mathematics that he is likely to encounter in economics and social sciences, and the convention described above concerning the notation for functions is what he will, in fact, encounter.**

Exercise 17

1. Determine which of the following relations are functional relations.
   (a) y is a son of x.
   (b) y is a great-grandfather of x.

---

* See, for example, the amusingly written papers by K. Menger, "Gulliver in the Land without One, Two, Three," *Mathematical Gazette*, 53 (1959), 241–50; "Gulliver's Return to the Land without One, Two, Three," *American Mathematical Monthly*, 67 (1960), 641–48.

** For the same reason, this is no place for the author to air his own pet theories on this topic. Suffice it to say that while admitting that the customary notation does give rise to difficulties, I feel that the extent of these difficulties is often exaggerated, and that a student who has been made aware of this source of possible confusion is not likely to have much trouble because of it.

(c) y is the midpoint of the line-segment x.

(d) y differs from x by 2.

(e) y is the annual salary of x.

(f) y is the length of x.

2. Show that if y is a function of x, and x is a function of t, then y is a function of t.

3.2 SOME SIMPLE FUNCTIONS

Having dealt with this question of notation, let us now consider the kinds of functions that are of most common occurrence in mathematics. For most of these the domains and converse domains are sets of numbers of some kind, such as *integers, rational numbers* (i.e., fractions, both greater than and less than 1; or in decimal notation, numbers represented by either terminating or recurring decimals), or *real numbers* (i.e., including also *irrational numbers* such as $\sqrt{2}$, which are represented by nonterminating, nonrecurring decimals).* Here are some examples.

1. y = the number of positive integers which divide the integer x exactly. Here the domain and converse domain are the set of positive integers. If we denote this function by $f(x)$, then we will have, for example

$$f(24) = 8$$

since the integers 1, 2, 3, 4, 6, 8, 12 and 24 are the only integers which divide 24.

2. $y = [x]$, where $[x]$ stands for the "integral part" of x, that is to say the largest integer which is less than or equal to x. Thus we have $[5] = 5$; $[6] = 6$; $[17.513] = 17$, and so on. Note that $[-4\frac{3}{4}] = -5$, and not -4. For this function the domain will usually be the set of all real numbers, whereas since y is necessarily an integer, the range will be the set of all integers.

3. $y = x^2$. Here the domain is the set of real numbers; the range will be the set of positive real numbers.

4. $y = |x|$. Here $|x|$ stands for the "absolute value of x," that is to say it is just x if x is a positive number, and is $-x$ if x is negative. Clearly $|x|$ is necessarily a non-negative number.

---

* This sentence conveys some idea of what a real number is, though it is not a satisfactory definition. We shall therefore make use of the term "real number" from now on, anticipating its eventual definition in Chapter 12.

Note. The notations [*x*] and |*x*| are standard, and will be used later on.

It should not be thought from the foregoing that the domains and ranges of functions that occur in mathematics are necessarily sets of numbers. This is very often so, but not always, as the following examples will show.

5. N = the number of elements in the set A.
Here the arguments are sets.
6. S = the area of K.
Here the arguments are, for example, portions of a plane, or of some other surface.
7. l = the length of C, where C is, shall we say, a curve in the plane.
8. d = the distance between the points P and Q.
Here the argument of the function is a pair of points.
    It is not necessary, either, that the value of a function should be any sort of number.
9. L = the set of points of a plane equidistant from the points P and Q.
Here the arguments are pairs of points, as in example 8, while the values of the function are lines, as the reader will know from his school geometry. The relation in question is that of "being the perpendicular bisector of the segment PQ."

3.3 FUNCTIONS OF TWO OR MORE VARIABLES

Consider the equation $z = u^2 + v^2$. If we give values to u and v we arrive at a value for z. We therefore have a function in the sense in which we have defined the term; the arguments are pairs of numbers, denoted by u and v, and the values are numbers, denoted by z. For convenience a function like this is generally called a *function of two variables,* and u and v are regarded as two (independent) variables rather than as component parts of the single argument that is required by our definition of a function. This is purely a matter of convenience, and in no way prevents us from including such functions under the heading of functions in general. A typical function of two variables can be written as

$$z = f(u, v).$$

In the same way we can talk of functions of three, four or any number of variables. We write

$$z = f(u, v, w), \text{ etc.}$$

A function of two variables is thus a set of ordered pairs (as is any function) and of the two elements in a typical pair the first is itself an ordered pair (u, v). Similarly a function of three, four, or more variables is specified by ordered pairs whose first elements are ordered triples, quadruples, etc.

It is of interest to note briefly a few functions which occur in economics. As examples of functions of a single variable we can cite

1. The *consumption function*, which is aggregate real consumption demand as a function of aggregate real disposable income.
2. The level of personal income tax payments as a function of the level of income.
3. The level of demand for investment goods as a function of the change in the level of output. This is the *accelerator relation*.

We can also cite some functions of several variables, for example,

4. Aggregate real consumption demand as a function of aggregate real disposable income, the rate of interest, and the real value of the stock of liquid assets held by households.
5. Quantity demanded of corn as a function of the price of corn, the price of wheat and the level of real income.
6. Demand for money as a function of the interest rate and total wealth.

Many other examples of functions of common occurrence in economics will appear from place to place in this book.

3.4 GRAPHS

We have already seen that a relation given by a statement of the form $x \, R \, y$ is defined once we can specify for which pairs of values of x and y the statement is true. Hence, in particular, a function whose arguments and values are numbers will give rise to a certain subset of the set of all pairs of numbers (the Cartesian product of the set of all real numbers with itself), namely, the set of all pairs (x, y) for which $y = f(x)$ (or $x \, R \, y$) is true, where (x, y) is the usual notation for a typical pair.

Now there is a very effective method of representing pairs of real numbers diagramatically, similar to that for products of finite sets discussed in section 2.9. We take two perpendicular lines, which we shall call the *axes*. One of them (usually drawn horizontally across the page) we shall call the *axis of x*; the other the *axis of y*. The point of intersection of the two axes is called the *origin* and is usually denoted by "O". The axes are to be thought of as extending as far as may be necessary on either side of the origin.

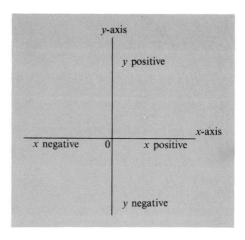

Fig. 3.1

The portion of the x-axis which lies on the right of the origin (see Fig. 3.1) is called the positive portion of the x-axis; the other half being the negative portion. Similarly the positive and negative portions of the y-axis are those which lie "above" and "below" the origin respectively.

Let us now consider a point of the plane in which the axes lie—the point P of Fig. 3.2, say. If we draw perpendiculars from P to the two axes, and call the feet of these perpendiculars M and N, then we can associate two numbers with the point P, namely the two lengths OM and ON measured in suitable units. Thus if the unit of measurement is the centimeter, and $OM = 3$ cm and $ON = 2$ cm then the numbers associated with the point P would

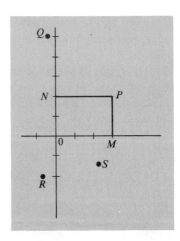

Fig. 3.2

be 3 and 2 respectively. These numbers are called the *coordinates* of P; that corresponding to OM being called the x-coordinate, and the other the y-coordinate. If M happens to lie in the negative portion of the x-axis, so that OM will be measured in the opposite direction, we shall associate a negative sign with the length OM, and the x-coordinate will be the resulting negative number. Similarly if N lies in the negative portion of the y-axis, the y-coordinate will be a negative number.

Alternatively we may define the x-coordinate of P in terms of the distance of P from the y-axis. This amounts to using the length PN in place of the length OM, but since $ONPM$ is a rectangle and $OM = NP$ the final result is the same. In the same way we may define the y-coordinate in terms of the distance PM. Thus to every point of the plane we can allot a pair of numbers, its coordinates. A typical pair of coordinates can be denoted, as before, by (x, y). The reader may care to verify that the coordinates of the point P, Q, R and S of Fig. 3.2 are $(3, 2)$, $(-1, 5)$, $(-1, -2)$ and $(2.5, -1.3)$, respectively.

It is clear that we can reverse the above process. If we are given a pair (x, y) of numbers we can measure off x units along the x-axis, obtaining a point M for which $OM = x$ units, and likewise find the point on the y-axis for which $ON = y$ units, and by completing the rectangle $ONPM$ construct the point P which has (x, y) as its coordinates. In this way pairs of real numbers can be represented by points of a plane.

We now return to our functions. Certain pairs of values (x, y) will make the functional relation $y = f(x)$ true, while the others will not. Let us choose a suitable *coordinate system*, i.e., x- and y-axes and a unit of measurement, and consider the set of all points of the plane whose coordinates *satisfy* the equation $y = f(x)$, i.e., make this a true statement. We obtain a set of points which, so to speak, represents geometrically the function in question. Thus the graph of a function $f(x)$ is the set of all points $(x, f(x))$ of the plane, for all x in the domain of f. This set of points is called the *graph* of the function.

The graph of a function will, in general, consist of an infinite number of points, so that there is no possibility of ever being able to draw a complete graph of any of the usual run of functions. What frequently happens however is that we construct (plot) a certain number of points belonging to the graph, and notice that these lie (or appear to lie) on a smooth curve, which we then proceed to sketch in. In this way graphs (or rather diagrams of graphs) can be drawn that may be very helpful in visualizing and discovering properties of functions, but it should be clear that a graph, by itself, cannot be used to prove anything about a function. In this respect graphs are like figures in geometry; although you cannot prove a geometrical theorem simply by drawing a figure, it would make life very difficult if the drawing of figures were prohibited.

3.5 EXAMPLES OF GRAPHS—SOME COMMON FUNCTIONS

LINEAR FUNCTIONS

A linear function is a function of the form $y = mx + d$, where m and d are constants. For example, the following are linear functions:

$$y = 2x + 3; \quad y = 3x - 2; \quad y = -\tfrac{3}{4}x + \tfrac{1}{2}.$$

Let us consider the first of these, $y = 2x + 3$, and make a table showing a few pairs of values of x and y which satisfy this equation, as follows.

x	-3	-2	-1	0	1	2	3
y	-3	-1	1	3	5	7	9

If we now plot these points we find that they lie on a straight line, shown in Fig. 3.3. This is no coincidence; we shall see in the next chapter that the graph of a linear funcion will always be a straight line. This is, of course, the reason why these functions have been called "linear."

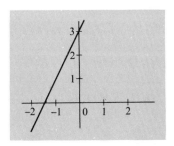

Fig. 3.3

QUADRATIC FUNCTIONS

A quadratic function is a function given by an equation of the form $y = ax^2 + bx + c$, where a, b and c are constants. As examples we may take

$$y = x^2 - 5x; \quad y = x^2 - 8x + 15; \quad y = 3 - x^2; \quad y = -x^2 + x - 1.$$

Let us look at the equation $y = x^2 - 8x + 15$. We first find a few pairs of values of x and y, as in the following table

x	-1	0	1	2	3	4	5	6	7	8
y	24	15	8	3	0	-1	0	3	8	15

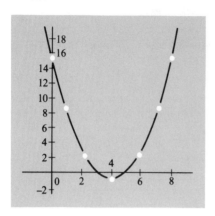

Fig. 3.4

The points corresponding to these pairs of values seem, when plotted, to lie on a smooth curve, which is shown in Fig. 3.4.

Note that for convenience, and to obtain a diagram of manageable proportions, it is permissible to use a different scale for the two axes. This has been done in Fig. 3.4, as can be seen from the gradations on the two axes. Provided the scale for each axis is properly indicated, no ambiguity will arise from such a difference in scale.

The graph of this function, and indeed of any quadratic function, is called a *parabola*. It will be seen that the curve in Fig. 3.4 crosses the x-axis twice; in other words there are two values of x for which $y = 0$. These are the two roots of the quadratic equation $x^2 - 8x + 15 = 0$, viz., $x = 3$ and $x = 5$. We recall from high school algebra that the roots of the quadratic equation $ax^2 + bx + c = 0$ are given by the formula

$$(3.5.1) \qquad\qquad x = \frac{-b \pm \sqrt{b^2 - 4ac}}{2a}$$

In plotting the graph of a quadratic function it is worthwhile to find these roots, at least roughly, in order to get a clear idea of how the parabola is situated with respect to the axes.

Now if it so happens that $b^2 < 4ac$, then on attempting to apply formula (3.5.1) we find that we have an expression containing the square root of a negative number. This simply means that there is no real value of x which makes $y = 0$, in other words that the graph of the function does not cross the x-axis. Such is the case with the function given by $y = -x^2 + x - 1$. A table of values and the graph of this function are given below.

x	-4	-3	-2	-1	0	1	2	3	4	5
y	-21	-13	-7	-3	-1	-1	-3	-7	-13	-21

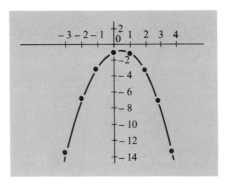

Fig. 3.5

CUBIC AND OTHER POLYNOMIALS

If we add a term in x^3 to a quadratic function we obtain a cubic function, which is therefore given by an equation of the form

$$y = ax^3 + bx^2 + cx + d$$

where a, b, c, d are constants. By way of example, a table of values and the graph of the function given by

$$y = \tfrac{1}{3}x^3 - 3x^2 + \tfrac{2}{3}x + 16$$

are given below.

x	-3	-2	-1	0	1	2	3	4	5	6	7	8	9
y	-22	0	12	16	14	8	0	-8	-14	-16	-12	0	22

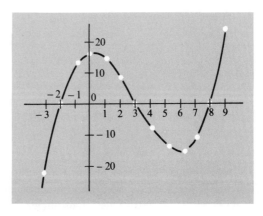

Fig. 3.6

The functions that have been given so far are all examples of what are known as polynomials. A *polynomial* is a function which is the sum of a number of terms, each of which is a constant multiple of an integral power of x (or whatever letter denotes the argument of the function). The highest power occurring in any term of the polynomial is called the *degree* of the polynomial. Thus a general polynomial of degree n may be written

$$a_0x^n + a_1x^{n-1} + a_2x^{n-2} + \ldots + a_{n-1}x + a_n$$

where $a_0, a_1, a_2, \ldots, a_{n-1}, a_n$ are constants. We see also that linear functions, quadratic functions, and cubic functions are the polynomials of degrees 1, 2, and 3, respectively.

Exercise 18

1. Draw graphs of the following functions for x between -2 and $+2$.

(a) $\dfrac{x}{x^2 + 1}$

(c) $3x^3 + 5x^2 - 2x - 7$

(b) $x^2 - 2x + 4$

(d) $\dfrac{x^2 + 3}{x^2 + 4}$

2. Using the same axes draw the graphs of

$$y = x^2 - 6x + 8 \qquad \text{and} \qquad y = \frac{1}{x}$$

between $x = 1$ and $x = 5$.
Hence find approximately two values of x which satisfy the equation

$$x^3 - 6x^2 + 8x - 1 = 0.$$

3. Sketch the graph of the polynomial

$$y = x^4 + 2x^3 - 13x^2 - 14x + 24$$

from $x = -5$ to $x = 4$.

3.6 TRIGONOMETRIC FUNCTIONS

We now define a set of functions which are of great importance and which will be much used in the following chapters.

Let OAB be a triangle, right-angled at A, for which the angle AOB

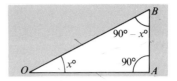

Fig. 3.7

is $x°$ (Fig. 3.7). The ratio AB/OB is called the *sine* of the angle AOB. This ratio depends only on the magnitude of the angle, for any right-angled triangle with an angle of $x°$ will be similar to the one just considered, and the ratio will the same as before. Hence we may talk about the *sine* of $x°$ or, as it is usually written, sin $x°$. We have now defined a new function of x, namely the sine of an angle of x degrees.

In a similar way, using the same figure, we define the function *cosine* by means of the ratio OA/OB. We write cos $x° = OA/OB$.

The functions sine and cosine are what are known as trigonometric functions, though they are not the only ones. Another trigonometric function is the *tangent*, abbreviated to *tan*. This is defined by tan $x° = AB/OA$. There are other trigonometric functions but we shall not consider them at the moment.

The three functions that we have just defined are related to each other. For since

$$\frac{AB}{OA} = \frac{AB/OB}{OA/OB}$$

it follows that

(3.6.1) $\tan x° = \sin x°/\cos x°.$

Furthermore, the theorem of Pythagoras shows that

$$OA^2 + AB^2 = OB^2$$

or

$$\left(\frac{OA}{OB}\right)^2 + \left(\frac{AB}{OB}\right)^2 = 1$$

so that

(3.6.2) $(\sin x°)^2 + (\cos x°)^2 = 1.$

The notation $(\sin x°)^2$ is rather cumbersome and is usually simplified. The simple omission of the brackets would make the "2" denoting "square of" clash with the degree sign. Even if the degree sign does not appear (and we shall see shortly how this can happen), an expression like sin x^2 is ambiguous; it might mean "the sine of the square of x" or "the square of the sine of x." To avoid ambiguity, the index 2 is written immediately after the "sin," so that in the usual notation (3.6.2) would be written

(3.6.3) $$\sin^2 x° + \cos^2 x° = 1.$$

By virtue of equations (3.6.1) and (3.6.3), we could dispense with the expressions cos $x°$ and tan $x°$ altogether, replacing them by

$$\sqrt{1 - \sin^2 x°} \quad \text{and} \quad \frac{\sin x°}{\sqrt{1 - \sin^2 x°}}$$

respectively; but these functions are of such common occurrence that the retention of the separate names and symbols is well worth while.

By the use of Figs. 3.8 and 3.9, which represent an equilateral triangle

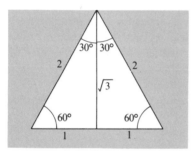

Fig. 3.8   **Fig. 3.9**

of side 2 units with an altitude drawn in, and an isosceles right-angled triangle, respectively, the reader should verify the following standard results:

(3.6.4)
$$
\begin{cases}
\sin 30° = \dfrac{1}{2} & \cos 30° = \dfrac{1}{2}\sqrt{3} & \tan 30° = \dfrac{1}{\sqrt{3}} \\[2ex]
\sin 45° = \dfrac{1}{\sqrt{2}} & \cos 45° = \dfrac{1}{\sqrt{2}} & \tan 45° = 1 \\[2ex]
\sin 60° = \dfrac{1}{2}\sqrt{3} & \cos 60° = \dfrac{1}{2} & \tan 60° = \sqrt{3}
\end{cases}
$$

Since the angles of a triangle sum to 180°, the angle OBA in Fig. 3.7 is $90° - x°$. It follows that

(3.6.5)
$$
\begin{cases}
\sin (90° - x°) = \cos x° \\
\cos (90° - x°) = \sin x°
\end{cases}
$$

If we consider a triangle with a very small angle (Fig. 3.10) we see that the sine of this small angle will be a very small number; and indeed the smaller the angle the nearer its sine will be to zero. We therefore say that sin 0° = 0, though our original definition of the sine function is difficult to

Fig. 3.10

apply in this case since a triangle with one angle zero is, in a sense, no triangle at all.

By similar considerations or by use of (3.6.3) and (3.6.5), we have

(3.6.6)
$$\begin{cases} \cos\ 0° = 1 \\ \sin 90° = 1 \\ \cos 90° = 0 \end{cases}$$

The value of tan 0° presents no difficulty—it is 0 by (3.6.1). But tan 90° is another matter. Equation (3.6.1) gives $\tan 90° = \frac{1}{0}$, which has no meaning (since division by zero is not allowed in the rules of arithmetic). We therefore do not attempt to define the "tangent" function for this value of the argument. It is true that in tables of trigonometric functions, where it may be necessary to have something occupying the space reserved for the tangent of 90°, and sometimes elsewhere, the symbol ∞ (standing for "infinity," and suggested by the fact that as x gets closer and closer to 90 so tan $x°$ gets larger and larger) is frequently used; but the reader should avoid the temptation to read into this symbol any more than what has just been said about it.

Since we now have the values of these three trigonometric functions for $x = 0$, 30, 45, 60, 90 we can sketch the graphs of these functions for $0 \le x \le 90$. These are shown in Fig. 3.11.

The question now arises whether we can define the trigonometric functions for angles larger than 90°. This might seem impossible at first, since we

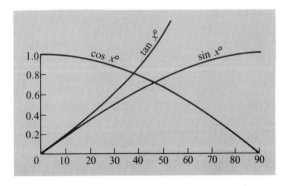

Fig. 3.11

cannot draw a right-angled triangle with an angle greater than 90°, but a slight change in our way of looking at our original definition of, for example, the sine function, will enable us to extend the application of these functions without much difficulty.

Figure 3.7 can be thought of as having been constructed by taking two lines making an angle of $x°$, taking any point B on one line and drawing a perpendicular from B to the other line. This construction will give us Fig. 3.7 precisely if $x°$ is an acute angle, but can be carried out just as well whatever the size of the angle. What have to be watched are the signs associated with the lengths involved. To facilitate this we shall call the two lines OA and OB and treat O and OA as if they were the origin and x-axis, respectively, of a coordinate system. The perpendicular from B will be denoted by NB always. The signs of ON and NB will then be determined as were the signs of x and y in section 3.4.

For an angle $x°$ between 90° and 180° we then have a figure like Fig. 3.12.

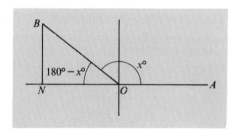

Fig. 3.12

The effect of our modification to the definition of the functions is merely that the angle in question is now not an angle of the triangle but an exterior angle. As before, we have sin $x° = NB/OB$ and NB is positive. By comparing Fig. 3.12 and Fig. 3.7 with an angle of $(180 - x)°$, it is readily seen that

(3.6.7) $\sin x° = \sin (180 - x)°.$

We also have cos $x° = ON/OB$, but here ON will be counted as negative. Hence the cosines of angles between 90° and 180° are negative. The reader should verify, again by comparison with Fig. 3.7, that

(3.6.8) $\cos x° = -\cos (180 - x)°$

From this, (3.6.7) and (3.6.1), or directly from Fig. 3.12, it follows that

(3.6.9) $\tan x° = -\tan (180 - x)°$

for angles within the range considered.

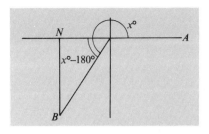

Fig. 3.13

Let us now look at angles between 180° and 270°. For these our construction gives a figure like that of Fig. 3.13. Here both ON and NB will be counted as negative. Hence $\sin x° = NB/OB$ will be negative, and so will $\cos x° = ON/OB$. Comparison with Fig. 3.7 with an angle of $(x - 180)°$ shows that

(3.6.10) $$\sin x° = -\sin(x - 180)°$$

(3.6.11) $$\cos x° = -\cos(x - 180)°.$$

On the other hand

(3.6.12) $$\tan x° = \tan(x - 180)°$$

by virtue of (3.6.1).

The figure for angles between 270° and 360° is like Fig. 3.14.* ON is

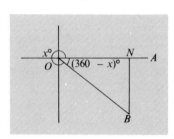

Fig. 3.14

now positive again, but NB is negative. The corresponding results for these angles are

$$\sin x° = -\sin(360 - x)°$$

(3.6.13) $$\cos x° = \cos(360 - x)°$$

$$\tan x° = -\tan(360 - x)°.$$

---

* Note that in this figure the "$(360 - x)°$" denotes the positive number of degrees in the angle NOB. This is consistent with the notation in the two previous figures, but here one could be led to expect $(360 - x)$ to be negative.

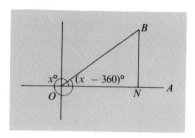

Fig. 3.15

An angle $x°$ which is greater than $360°$ will give values of ON and NB which are the same as for an angle of $(x - 360)°$. (See Fig. 3.15.) Hence the addition of $360°$ to an angle gives us an angle with the same sine, cosine and tangent. Thus we may write

(3.6.14)
$$\begin{cases} \sin (360 + x)° = \sin x° \\ \cos (360 + x)° = \cos x° \\ \tan (360 + x)° = \tan x°. \end{cases}$$

Equations (3.6.14) enable us to extend the definition of sine, cosine and tangent to angles having any positive number of degrees, since it shows that addition of multiples of $360°$ is, so to speak, immaterial. Because of this property the trigonometric functions are what is known as *periodic functions*; their values repeat themselves at intervals (periods) of $360°$.

It remains only to define these functions for negative values of the argument. What do we mean by a negative angle? If we think of the angle BOA in, for example, Fig. 3.12 as having been obtained by rotating a line, initially coinciding with OA, about O in a counter-clockwise direction until it coincides with the line OB, thus "sweeping out" an angle of $x°$, then it is reasonable to define an angle of $-x°$ to be the result of a similar rotation but in a clockwise direction. We thus arrive at Fig. 3.16 for an angle between $-90°$ and $0°$. By comparison with the figure for the corresponding positive angle, we see that

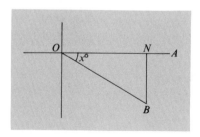

Fig. 3.16

$$(3.6.15) \qquad \begin{cases} \sin(-x)^\circ = -\sin x^\circ \\ \cos(-x)^\circ = \cos x^\circ \\ \tan(-x)^\circ = -\tan x^\circ. \end{cases}$$

These last results will, in fact, apply whatever the angle x° may be. We leave the reader to verify this for himself by drawing the appropriate figures.

Using the foregoing results we can now draw the graphs of the trigonometric functions more fully than in Fig. 3.11. This we do in Fig. 3.17.

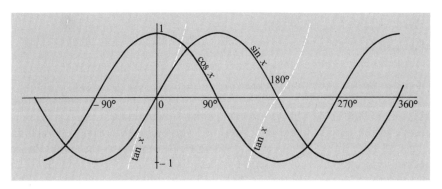

Fig. 3.17

3.7 RADIAN MEASURE

Exercise 19

1. Write down the values of the sine, cosine and tangent of the following angles:

(a) 30°   (c) 225°   (e) −60°
(b) 135°   (d) 300°   (f) 330°

2. Use a table of trigonometric functions and the rules given in this chapter to find the values of

(a) $\sin(-52°30')$ (c) $\cos(186°20')$ (e) $\sin(320°4')$
(b) $\tan(121°5')$ (d) $\cos(256°14')$ (f) $\tan(97°53')$

People who use angles for practical purposes (such as surveyors, astronomers, draughtsmen, and the like) usually measure their angles in degrees, as we have done in the previous section. This unit of measurement, though very good for these purposes, has some disadvantages when used for theoretical work. For this reason another unit of angular measurement has come into use, known as the *radian*. A *radian* is defined as the measure of the angle which is subtended at the center of a circle by an arc of length equal to the radius of the circle. It is the angle *AOB* shown in Fig. 3.18, where the length

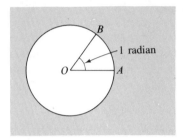

Fig. 3.18

of the arc *AB* is equal to the radius *OA*. This angle will be the same whatever the size of the circle. How large is an angle of one radian? We can easily find this out, for the angle which an arc of a circle subtends at the center is proportional to the length of the arc. Now the whole circumference subtends an angle of 360°, and its length is $2\pi r$, where r is the radius of the circle. Hence if 360° corresponds to an arc length of $2\pi r$, then $(360/2\pi)°$ corresponds to an arc length of r. In other words, one radian is $(360/2\pi)°$, or $(180/\pi)°$. This works out at a little over 59°.

An angle of θ radians* will therefore be an angle of $\dfrac{180}{\pi}\theta$ degrees. Similarly an angle of $x°$ will be $\pi x/180$ radians. Conversion from degrees to radians and vice versa is easily effected by remembering that $180° = \pi$ radians. Now this may seem a very ungainly unit to use—it does not even make a right angle come out to a round number. Nevertheless it has certain advantages, as we shall see later, which make its use worthwhile. So much so, in fact, that in future we shall always assume that an angle is measured in radians unless it is expressly stated that it is not. In particular, we shall not use any symbol (such as the ° for "degree") to indicate that an angle is measured in radians, so that in an expression like "sin θ" it will be understood that θ means "θ radians."

Just to indicate the use of this unit we shall write out some of the important results that we have already obtained, this time using radian measure. The proofs of these results are, of course, exactly the same as before, except for the change in the unit.

(3.7.1) $$\tan \theta = \frac{\sin \theta}{\cos \theta}$$

(3.7.2) $$\sin^2 \theta + \cos^2 \theta = 1$$

---

* The Greek letter θ (theta) is commonly used for angles measured in radians.

$$
(3.7.3) \quad
\begin{cases}
\sin\left(\tfrac{1}{2}\pi - \theta\right) = \cos\theta \\
\cos\left(\tfrac{1}{2}\pi - \theta\right) = \sin\theta \\
\sin\left(\pi - \theta\right) \;= \sin\theta \\
\cos\left(\pi - \theta\right) \;= -\cos\theta \\
\sin\left(\pi + \theta\right) \;= -\sin\theta \\
\cos\left(\pi + \theta\right) \;= -\cos\theta \\
\sin\left(-\theta\right) \quad\;\; = -\sin\theta \\
\cos\left(-\theta\right) \quad\;\; = \cos\theta \\
\sin\left(2\pi + \theta\right) = \sin\theta \\
\cos\left(2\pi + \theta\right) = \cos\theta
\end{cases}
$$

Similar results to those of (3.7.3) for $\tan\theta$ can be deduced from (3.7.3) by means of (3.7.1). Fig. 3.19 gives the graphs of the three functions, using

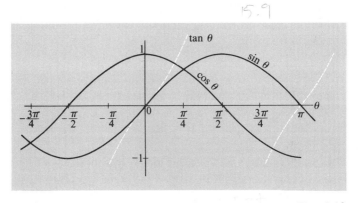

Fig. 3.19

the new unit. In Fig. 3.19, unlike Fig. 3.17, the same scale has been used for both axes.

The results (3.6.4) and (3.6.6) when translated into radian measure, become

$$
\begin{aligned}
&\sin 0 = \; 0 \;\; ; \;\; \cos 0 \;= \;\; 1 \;\; ; \;\; \tan 0 \;= 0 \\
&\sin \tfrac{1}{6}\pi = \;\; \tfrac{1}{2} \;\; ; \;\; \cos \tfrac{1}{6}\pi = \tfrac{1}{2}\sqrt{3} \;; \;\; \tan \tfrac{1}{6}\pi = \tfrac{1}{3}\sqrt{3} \\
(3.7.4) \quad &\sin \tfrac{1}{4}\pi = \tfrac{1}{\sqrt{2}} \;\; ; \;\; \cos \tfrac{1}{4}\pi = \tfrac{1}{\sqrt{2}} \;\; ; \;\; \tan \tfrac{1}{4}\pi = 1 \\
&\sin \tfrac{1}{3}\pi = \tfrac{1}{2}\sqrt{3} \;; \;\; \cos \tfrac{1}{3}\pi = \;\; \tfrac{1}{2} \;\; ; \;\; \tan \tfrac{1}{3}\pi = \sqrt{3} \\
&\sin \tfrac{1}{2}\pi = \;\; 1 \;\; ; \;\; \cos \tfrac{1}{2}\pi = \;\; 0.
\end{aligned}
$$

Values of trigonometric functions for angles other than these rather special ones are, of course, found by referring to tables of values of these functions.

Finally we must stress that the functions sin $x°$ and sin θ are *different* functions. The difference is most easily seen if we review just what is meant by the function sin $x°$. The first thing to note is that, strictly speaking, the argument of this function is a real number (not an angle). To find the value of the function corresponding to a given real number x we construct an angle of x degrees, and from this angle we construct the appropriate right-angled triangle from which the ratio AB/OB is calculated. For the function sin θ we construct an angle of θ radians, and again read off the ratio AB/OB. Now if x and θ take the same value, then clearly the angle that we construct will be different in the two cases; hence, in general, the value of the two functions will be different, showing that they are different functions. Of course, they are, in a sense, not *very* different, but the distinction between them should be understood, since it will be of importance. Similar remarks apply to the functions cos $x°$ and cos θ, and so on.

It is helpful to be able to remember easily the signs that the three trigonometric functions sine, cosine and tangent have in the four quadrants. They are as follows:

		sin	cos	tan
First quadrant	$0° - 90°$	positive	positive	positive
Second quadrant	$90° - 180°$	positive	negative	negative
Third quadrant	$180° - 270°$	negative	negative	positive
Fourth quadrant	$270° - 360°$	negative	positive	negative

Note that all three functions are positive in the first quadrant, and exactly one of them is positive in each of the other quadrants. This can be remembered by the mnemonic

$$\text{"ALL} - \text{SIN} - \text{TAN} - \text{COS"}$$

or, diagrammatically, by Fig. 3.20.

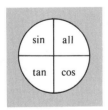

Fig. 3.20

Exercise 20

1. Convert the following angles (in degrees and minutes) into radian measure
   (a) 27°34'   (c) 98°7'   (e) 126°13'
   (b) −51°8'   (d) 54°41'   (f) 120°51'
2. Convert the following angles, in radians, into degrees and minutes.
   (a) 0.873   (c) 1.902   (e) −3.046
   (b) −0.017   (d) 2.103   (f) 0.894
3. Use tables of trigonometric functions to find the sine, cosine and tangent of the angles given in problem 2.

3.8 EXPONENTIAL FUNCTIONS

In an expression like a^b (a to the power of b) the number a is called the *base*, and the number b is called the *exponent* or *power*. We have already met functions of this form in the polynomials x^2, x^3, etc., and in these the base was the argument of the function and the exponent was a constant. If we make the exponent the argument and let the base be a constant, we obtain another kind of function called an exponential function. Examples are 2^x, 10^x, $(5.72)^x$ and so on. The graphs of these functions are all very much alike, so we shall consider only one of them—the function $y = 2^x$.

Now if x is a positive integer there is no difficulty in finding the corresponding value of y; we simply raise 2 to the appropriate power. The following table therefore gives us some points of the graph.

x	1	2	3	4	5
y	2	4	8	16	32

What is the interpretation of 2^x if x is not a positive integer? The reader may know already, but we shall answer the question anyway, in case he does not.

The most important thing about exponential functions is that they obey the *laws of indices*, viz., that

$$2^m \times 2^n = 2^{m+n} \quad \text{and} \quad (2^m)^n = 2^{mn}.$$

If we want to extend the definition of 2^x to include values of x other than positive integers, then certainly we should strive to do it in such a way that the laws of indices continue to hold. It turns out that this can be done, and in only one way.

Let us first look at $2^{1/2}$. If the laws of indices are to hold then we must have

$$(2^{1/2})^2 = 2^1 = 2$$

so that $2^{1/2}$ is the number whose square is 2, i.e., $2^{1/2} = \sqrt{2}$. In the same way, if q is any positive integer, then since we want to have

$$(2^{1/q})^q = 2$$

it follows that we must interpret $2^{1/q}$ as the qth root of 2, i.e., $\sqrt[q]{2}$.

If p is also a positive integer, then

$$2^{p/q} = (2^{1/q})^p$$

i.e., $2^{p/q}$ is the pth power of the qth root of 2. This defines 2^x when x is a *positive rational* number.

We next ask, "What is 2^0"? The first law of indices shows that

$$2^0 \times 2^1 = 2^1, \text{ i.e., } 2^0 \times 2 = 2,$$

whence $2^0 = 1$, which settles that question.

Finally we ask, "If x is a positive rational number, what is 2^{-x}?" Appealing to the first law of indices we see that

$$2^{-x} \times 2^x = 2^{x-x} = 2^0 = 1$$

Thus $2^{-x} = (\frac{1}{2})^x$, i.e., it is the reciprocal of 2^x. In particular $2^{-1} = \frac{1}{2}$. We have now extended the definition of 2^x to cover all rational values of x, both positive and negative.

The same observations will apply to whatever the base may be (provided it is positive). We may therefore summarize the results for a general positive base a as follows

(3.8.1)
$$a^{p/q} = (\sqrt[q]{a})^p$$
$$a^0 = 1$$
$$a^{-x} = \frac{1}{a^x}$$

Using these results we can find many more points on the graph of $y = 2^x$. Some of these are given in the following table.

x	-2	-1.5	-1	-0.5	0	0.25	0.5	0.75	1.25	1.5	1.75
y	0.25	0.354	0.5	0.707	1.0	1.189	1.414	1.683	2.378	2.828	3.366

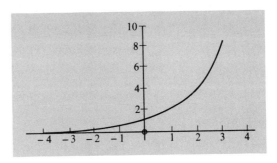

Fig. 3.21

These points have been plotted in Fig. 3.21 and joined by a smooth curve. It will be noticed that we have still not defined 2^x for irrational values of x. This is a much more difficult matter, which we shall have to postpone to a later chapter.

Exercise 21

1. Calculate the following

(a) $\dfrac{(64)^{1/4}}{(8)^{1/2}}$

(b) $\dfrac{6^{5/4}3^{1/4}}{2^{3/4}}$

(c) $2^{1/3} \cdot 12^{1/3} \cdot \left(\dfrac{1}{9}\right)^{-1/3}$

(d) $2^{5/7}3^{-2/7}\left(\dfrac{4}{243}\right)^{1/7}$

2. Sketch the graphs of $y = 2^x$, $y = (\frac{5}{2})^x$ and $y = 3^x$ between $x = 1$ and 3 using the same set of axes. Note that these curves do not intersect within this range. Could they do so for any value of x?

3. Sketch the graphs of $y = x^n$ for $n = -1, 2, \frac{5}{2}, 3$ and x between 0 and 2. What special difficulty is encountered in drawing the graph for $n = -1$?

3.9 SOME UNCOMMON FUNCTIONS

The graphs of the functions that we have so far considered have all been what we would describe as "smooth curves," though it is not so easy to say exactly what one means by this term. It is not always the case that the graph of a function is a smooth curve, or indeed any sort of curve at all. To emphasize this we now give some functions, some of them not without importance, whose graphs are not curves.

Some functions are defined by extension, that is to say we actually enumerate all the pairs (x, y) for which the functional relation $y = f(x)$ holds. Suppose a schoolteacher gives a class of 12 pupils a test and grades them in order of merit. Suppose he also measures the height of each pupil to the nearest inch, and draws a graph of the function $y = f(x)$ where y is the height of the pupil whose grade is x. This graph will consist of 12 points only, since a point (x, y) will belong to the graph if, and only if, there is a pupil whose height is y inches and whose grade is x. The graph might look something like that in Fig. 3.22.

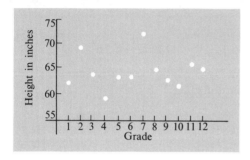

Fig. 3.22

One would not expect that there would be much regularity in such a graph since one does not expect that there would be much connection between a pupil's height and his grading in a test. This example is, indeed, highly artificial, and we turn quickly to one that is more reasonable though, at first glance, not so very different.

The heights of 1,000 students in a graduating class are measured to the nearest inch, and the graph of the function $y = f(x)$ is constructed, where y is the number of students whose height is x inches. (x is an integer, by the conditions of the measurement.) The graph will consist of a set of isolated points as did the previous one, but this time there is some sort of pattern in the way that they are situated in the plane. The sort of graph that one gets with functions of this sort is shown in Fig. 3.23. Functions like this (known

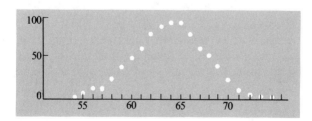

Fig. 3.23

as *distributions*) are part of the subject matter of statistics; but since we shall not be concerned with statistics in this book, we shall not consider them further. We merely note their existence.

It may happen that a single expression, such as $x^3 + 3x - 5$, will suffice to specify a function over the whole range of values that its argument assumes. At the other extreme we have functions like those just considered where for every value of the argument the value of the function has to be individually specified. In between these two extremes we have functions for which no single expression will suffice but where several expressions are needed, one or another of them being used according to the particular value of the argument. This method of defining a function often strikes people as being rather artificial, but this is only because so many of the functions which occur in elementary mathematics can, in fact, be specified by a single expression. The important thing to remember about a function is that for each value of the argument there should be a corresponding value of the function. This relation between the values of function and argument can be expressed in any convenient way, no matter how arbitrary or artificial it may appear to be. A few examples of functions of this kind should make these remarks clear.

1. Consider the function defined by

$$y = -1 \quad \text{if} \quad x < 0$$
$$y = 0 \quad \text{if} \quad x = 0$$
$$y = 1 \quad \text{of} \quad x > 0.$$

Here we have three possible values -1, 0 and 1; which of them is correct depends on the value of x, as indicated. The graph of this function is shown in Fig. 3.24.

Note that the line AB "approaches," but does not include the point $(0, -1)$, and that the line DC "approaches" but does not include the point $(0, 1)$. The origin is an isolated point of the graph.

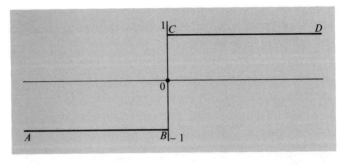

Fig. 3.24

2. The function given by

$$y = 1, \quad \text{if } x \text{ is an integer}$$
$$y = 0, \quad \text{if } x \text{ is not an integer.}$$

The graph of this function consists of the x-axis from which the points $(n, 0)$ (where n is an integer) have been removed, together with the points $(n, 1)$, as shown in Fig. 3.25.

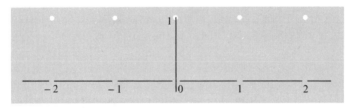

Fig 3.25

3. The function given by

$$y = 1, \quad \text{if } x \text{ is a rational number,}$$
$$y = 0, \quad \text{if } x \text{ is not a rational number.}$$

The graph of this function appears to consist of two lines, the x-axis and the line consisting of all the points (x, y) for which $y = 1$ (briefly "the line $y = 1$"). This is not the case however, since the part of the graph which appears to be the x-axis does not contain all the points of the x-axis, but only those points of the form $(x, 0)$ where x is irrational. Similarly the part of the graph that looks like the line $y = 1$ does not, in fact, consist of all the points of that line, but only those of the form $(x, 1)$ where x is rational. Graphs drawn on paper are only diagrams at most, but the attempt made in Fig. 3.26 to depict the graph of this function is even more diagrammatic than usual. It has been drawn as two dotted lines, but it should be remembered that, to speak loosely for a moment, the dots are really infinitely numerous and infinitely close together!

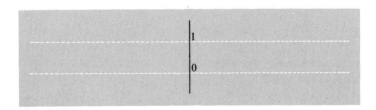

Fig. 3.26

4. This example has been included mainly in order to use a notation that has already been introduced and that will be used very frequently later on. This notation consists in the enclosing of a symbol for a number between two vertical lines, viz., $|x|$. This is read "the absolute value of x" and by that is meant the numerical value of x, counted positive if x is not already positive. Alternatively, we can define it by

$$|x| = \quad x, \text{ if } x \geq 0$$
$$|x| = -x, \text{ if } x < 0.$$

Thus $|6| = 6$, but $|-5| = 5$ and $|-100| = 100$, and so on. The graph of this function is given in Fig. 3.27.

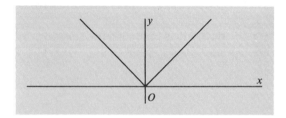

Fig. 3.27

It is perhaps worthwhile at this stage to indicate one of the ways in which this notation is useful. Consider the expression $|a - b|$. If $a \geq b$ then $a - b \geq 0$ and $|a - b|$ is the same as $a - b$. If, however, $a < b$ then $a - b < 0$ and $|a - b| = -(a - b) = b - a$. Thus $|a - b|$ simply means the (positive) difference between a and b.* Now the (positive) difference between two numbers is something that we often want to talk about and write about, and the notation just described enables us to do this concisely. The notation "$a \sim b$" is sometimes met, especially in older textbooks, to denote the difference between a and b, but the reader should avoid this usage; the symbol $\sim$ is much too convenient to be wasted in this way.

Exercise 22

1. Sketch the graphs of the following functions.
    (**a**) $y = [x]^2$ (**b**) $y = \sqrt{|x|}$ (**c**) $y = x - [x]$
In (*b*) take the positive square root.

---

* The English language does not offer much to distinguish between "$a - b$" and "$|a - b|$." We may call the first the "difference *of a* and b" (this may be a negative number), and the second can be called the "difference *between a* and b" (this is necessarily positive). To avoid possible ambiguity we may talk about the "positive difference between a and b." A better plan is perhaps to translate the mathematical symbols literally and say "a minus b" and "the absolute value of a minus b," respectively.

2. Using the usual system of coordinates, draw a graph of the set of points which satisfy

  (a) $x^2 + y^2 = 1$ **(b)** $|x + y| = 1$

Prove that these graphs are in fact what they seem to be.

3.10 SOME FURTHER RESULTS CONCERNING TRIGONOMETRIC FUNCTIONS

In section 3.6 we introduced the three trigonometric functions—sine, cosine and tangent—and some basic results concerning these functions were summarized in (3.7.1), (3.7.2) and (3.7.3). We shall need to introduce three other trigonometric functions and to derive some further results, and this is as good a place as any to do so.

The three other trigonometric functions are called *cotangent*, *secant* and *cosecant*, and when applied to an argument θ they are written

$$\cot \theta, \quad \sec \theta \quad \text{and} \quad \text{cosec } \theta$$

respectively. Their definitions are as follows:

$$\cot \theta \; = \frac{\cos \theta}{\sin \theta} = \frac{1}{\tan \theta}$$

(3.10.1) $\sec \theta \; = \dfrac{1}{\cos \theta}$

$$\text{cosec } \theta = \frac{1}{\sin \theta}$$

Thus these functions represent nothing very new, being all expressible in terms of the functions we have had already. For this reason they tend to be used rather less than the others, though many formulae are rendered neater by their use. From the values given in (3.7.4) we obtain the following values of these functions

$$(3.10.2) \quad \begin{cases} & \sec \; 0 = 1 \\ \cot \tfrac{1}{6}\pi = \sqrt{3} & \sec \tfrac{1}{6}\pi = \tfrac{2}{3}\sqrt{3} & \text{cosec } \tfrac{1}{6}\pi = 2 \\ \cot \tfrac{1}{4}\pi = 1 & \sec \tfrac{1}{4}\pi = \sqrt{2} & \text{cosec } \tfrac{1}{4}\pi = \sqrt{2} \\ \cot \tfrac{1}{3}\pi = \tfrac{1}{3}\sqrt{3} & \sec \tfrac{1}{3}\pi = 2 & \text{cosec } \tfrac{1}{3}\pi = \tfrac{2}{3}\sqrt{3} \\ \cot \tfrac{1}{2}\pi = 0 & & \text{cosec } \tfrac{1}{2}\pi = 1 \end{cases}$$

As before we have omitted the values for those arguments for which the definition gives a zero denominator. We forbear from writing $\cot 0 = \infty$, for example, though this is often done.

The theorem of Pythagoras gives us

$$\cos^2 \theta + \sin^2 \theta = 1$$

as we saw in section 3.6. It also gives us

$$(3.10.3) \quad \begin{cases} \sec^2 \theta = 1 + \tan^2 \theta \\ \operatorname{cosec}^2 \theta = 1 + \cot^2 \theta \end{cases}$$

as the reader should verify.

We now derive expressions for the sine and cosine of the sum of two numbers. For this we shall need a figure, and Fig. 3.28 will serve for both results.

Figure 3.28 is obtained by constructing the lines OQ, OP making angles

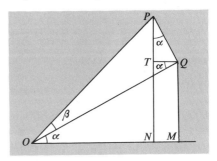

Fig. 3.28

of α and $\alpha + \beta$ with a line ON. P is any point on the second of these lines, and Q the foot of the perpendicular from P to the first line. PN and QM are perpendicular to ON, and QT is perpendicular to PN. The reader should check that this makes both the angles TQO and TPQ equal to α. We want to find expressions for $\sin(\alpha + \beta)$ and $\cos(\alpha + \beta)$.

First we have

$$PN = PT + TN$$
$$= PT + QM$$
$$= PQ \cos \alpha + OQ \sin \alpha$$

since $PT/PQ = \cos \alpha$ and $QM/OQ = \sin \alpha$. Hence

$$\sin(\alpha + \beta) = \frac{PN}{OP}$$
$$= \frac{PQ}{OP} \cos \alpha + \frac{OQ}{OP} \sin \alpha$$
$$= \sin \beta \cos \alpha + \cos \beta \sin \alpha$$

since $PQ/OP = \sin \beta$ and $OQ/OP = \cos \beta$. Hence the first result is

(3.10.4) $\qquad \sin (\alpha + \beta) = \cos \alpha \sin \beta + \sin \alpha \cos \beta.$

To find $\cos (\alpha + \beta)$ we observe that

$$ON = OM - NM$$
$$= OQ \cos \alpha - PQ \sin \alpha$$

since $\qquad\qquad NM = TQ = PQ \sin \alpha.$ Hence

$$\cos (\alpha + \beta) = \frac{ON}{OP}$$
$$= \frac{OQ}{OP} \cos \alpha - \frac{PQ}{OP} \sin \alpha$$
$$= \cos \beta \cos \alpha - \sin \beta \sin \alpha.$$

Hence the second result is

(3.10.5) $\qquad \cos (\alpha + \beta) = \cos \alpha \cos \beta - \sin \alpha \sin \beta.$

From the last two results we obtain at once

$$\tan (\alpha + \beta) = \frac{\sin (\alpha + \beta)}{\cos (\alpha + \beta)}$$
$$= \frac{\cos \alpha \sin \beta + \sin \alpha \cos \beta}{\cos \alpha \cos \beta - \sin \alpha \sin \beta}.$$

If we divide top and bottom by $\cos \alpha \cos \beta$ we find that

(3.10.6) $\qquad \tan (\alpha + \beta) = \frac{\tan \alpha + \tan \beta}{1 - \tan \alpha \tan \beta}.$

Further, if we put $\alpha = \beta$ we obtain the following useful results for double angles.

$$\sin 2\alpha = 2 \sin \alpha \cos \alpha$$
(3.10.7) $\qquad \cos 2\alpha = \cos^2 \alpha - \sin^2 \alpha$
$$\tan 2\alpha = \frac{2 \tan \alpha}{1 - \tan^2 \alpha}$$

Exercise 23

1. Prove the following statements:

   (a) $\sin (\alpha - \beta) = \sin \alpha \cos \beta - \cos \alpha \sin \beta$

(b) $\cos(\alpha - \beta) = \cos\alpha\cos\beta + \sin\alpha\sin\beta$

(c) $\tan(\alpha - \beta) = \dfrac{\tan\alpha - \tan\beta}{1 + \tan\alpha\tan\beta}$

2. By adding (3.10.4) and the result of problem 1(a) above, show that

(a) $\sin(\alpha + \beta) + \sin(\alpha - \beta) = 2\sin\alpha\cos\beta$

Show similarly that

(b) $\sin(\alpha + \beta) - \sin(\alpha - \beta) = 2\cos\alpha\sin\beta$

(c) $\cos(\alpha + \beta) + \cos(\alpha - \beta) = 2\cos\alpha\cos\beta$

(d) $\cos(\alpha - \beta) - \cos(\alpha + \beta) = 2\sin\alpha\sin\beta$

3. By putting $A = \alpha + \beta$ and $B = \alpha - \beta$ in the results of problem 2 above, show that

(a) $\sin A + \sin B = 2\sin\dfrac{A+B}{2}\cos\dfrac{A-B}{2}$

(b) $\sin A - \sin B = 2\cos\dfrac{A+B}{2}\sin\dfrac{A-B}{2}$

(c) $\cos A + \cos B = 2\cos\dfrac{A+B}{2}\cos\dfrac{A-B}{2}$

(d) $\cos A - \cos B = 2\sin\dfrac{A+B}{2}\sin\dfrac{B-A}{2}$

4. Prove from (3.10.7) that

(a) $\cos 2\alpha = 1 - 2\sin^2\alpha$

(b) $\cos 2\alpha = 2\cos^2\alpha - 1$

and hence that

(c) $\sin\dfrac{A}{2} = \sqrt{\tfrac{1}{2}(1 - \cos A)}$

(d) $\cos\dfrac{A}{2} = \sqrt{\tfrac{1}{2}(1 + \cos A)}$

5. Find exactly (i.e., in terms of $\sqrt{3}$, $\sqrt{2}$, etc.) the values of

(a) $\sin 15°$ **(d)** $\cos 22\tfrac{1}{2}°$

(b) $\cos 15°$ **(e)** $\sin 37\tfrac{1}{2}°$

(c) $\sin 22\tfrac{1}{2}°$ **(f)** $\cos 7\tfrac{1}{2}°$

Exercise 24 (Chapter Review)

1. Let xRy stand for "x was written by y" where x belongs to the set of all books, and y to the set of all authors. Is R a functional relation? If not, why not? Is R' a functional relation?

2. Give some examples from everyday life of functional relations R such that R' is also a functional relation.

3. Let the domain and converse domain of a relation R be a finite set of objects. We may represent these objects diagrammatically by dots on a sheet of paper and indicate that xRy by drawing an arrow from the dot representing x to that representing y. Which of the following diagrams (obtained in this way) represents a functional relation?

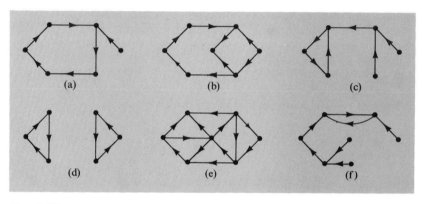

Fig. 3.29

What simple property distinguishes the diagrams which correspond to functional relations?

4. Draw the graph of the function f defined as follows

$$f(x) = 9 - x^2 \qquad \text{for} \quad -3 \leq x \leq 0.$$
$$f(x) = 9 - x \qquad \text{for} \quad 0 \leq x \leq 2.$$
$$f(x) = x^2 - 4x + 11 \quad \text{for} \quad 2 \leq x \leq 5.$$
$$f(x) = \frac{64}{x-1} \qquad \text{for} \quad 5 \leq x \leq 9.$$

5. Draw the graph of the function

$$y = 2x^2 + 2^x + 3^{-x} - 6$$

for x between -2 and 2.

6. Draw the graph of the function

$$\cos A + \sin 3A$$

for A between $-90°$ and $270°$.

7. Draw the graph of the function

$$\sin A \cos 2A$$

for A between $-180°$ and $180°$.

8. By using the formulae

$$\cos 2\theta = \cos^2 \theta - \sin^2 \theta$$

and $$\sin 2\theta = 2\sin \theta \cos \theta$$

in conjunction with the result $\cos^2 \theta + \sin^2 \theta = 1$, show that

$$\cos \alpha = \frac{1 - t^2}{1 + t^2} \quad \text{and} \quad \sin \alpha = \frac{2t}{1 + t^2}$$

where $t = \tan \tfrac{1}{2}\alpha$.

9. Prove the following results.

(a) $\cos 3\theta = \cos^3 \theta - 3 \sin^2 \theta \cos \theta$
$$= 4\cos^3 \theta - 3 \cos \theta.$$

(b) $\sin 3\theta = 3\sin \theta - 4\sin^3 \theta.$

(c) $\dfrac{\cos 3\theta - \cos \theta}{\sin 3\theta - \sin \theta} = \dfrac{2\tan \theta}{\tan^2 \theta - 1}$

(d) $\dfrac{\sin 2\theta + \sin \theta}{1 + \cos 2\theta + \cos \theta} = \tan \theta.$

10. Prove that

(a) $\dfrac{\sin A + \sin B}{\cos A + \cos B} = \tan \tfrac{1}{2}(A + B).$

(b) $\cos^2 A - \sin^2 B = \cos (A + B) \cos (A - B).$

(c) $\cot A + \cot B = \dfrac{\sin (A + B)}{\sin A \sin B}.$

Suggestions for Further Reading.

Insofar as the bulk of Chapter 3 is concerned with High School or first-year College algebra and trigonometry, the reader should have no difficulty in finding plenty of supplementary reading. For the trigonometry, the following book is recommended:

Ayres, F. *Trigonometry*. New York: Schaum Pub. Co., 1954.

By the same author is another book which is useful in a slightly different way:

Ayres, F. *First Year College Mathematics*. New York: Schaum Pub. Co., 1958. This book not only covers the topics of this, and other, chapters; it also deals with many elementary topics with which the readers of this book will be assumed to be already familiar, such as the solution of quadratic equations, to mention but one.

4 Combinatorics

and Probability

Dr. Beattie observed, as something remarkable which had happened to him, that he had chanced to see both No. 1, and No. 1000, of the hackney-coaches, the first and the last; "Why, Sir, (said Johnson,) there is an equal chance for one's seeing those two numbers as any other two".

BOSWELL

Life of Johnson

4.1 INTRODUCTION

The Japanese are renowned for their flower arranging; the zealous housewife is continually rearranging her furniture; a general deploys his army; in all walks of life we find people whose occupation consists in arranging or rearranging things. Usually the purpose of this rearranging is to achieve some sort of optimum result, but in this chapter we shall not be concerned with arrangements made with this object in mind, though we shall come to this later in the book. Right now we shall be concerned only with the number of ways in which it is possible to arrange things subject to certain given conditions. The study of such problems is a branch of mathematics known as *Combinatorial Analysis* or, more briefly, *Combinatorics*, and the first part of this chapter will be an elementary introduction to this subject.

Many results of combinatorial analysis, though interesting in themselves, have little relevance to economic problems, and we shall therefore concentrate on those problems whose solutions will be needed later in the book. In particular, many results will be required for the discussion of probability which makes up the second half of this chapter.

4.2 PERMUTATIONS

Let us consider first a very straightforward problem. A Dean of students at a university has the task of allocating 6 new students to 6 different rooms; in how many ways can this be done?

This is not a difficult problem. Let us suppose that the Dean first decides where to put student number 1; clearly there are 6 choices of room that he can make. Next he decides where to put student number 2; for this there are only 5 ways, since only 5 rooms are now available. Let us review the situation so far. There are 6 ways of housing the first student, and *whichever* of these ways is in fact chosen there are 5 ways of housing the second student. Thus if we were interested only in the number of ways of housing the first two students, the answer required would be $6 \times 5 = 30$. But this is not what is wanted, and we must continue the process. No matter which of the 6×5 ways of housing the first two students is decided upon, there will be 4 ways of housing the third student, since only 4 rooms are now available. Hence the first three students can be allocated rooms in $6 \times 5 \times 4$ different ways.

The general trend of the argument should now be apparent. The fourth and fifth students can be given rooms in 3 and 2 ways; but there is only one way of housing the last student—he must have the one remaining room. Hence the number of different ways of allocating 6 students to 6 rooms is

$$6 \times 5 \times 4 \times 3 \times 2 \times 1 = 720.$$

Let us just make sure that we have not overlooked anything. It might seem as if the order in which the Dean allocates the students to rooms might be important; in particular it might seem a little hard on the last student that he should *have* to go in the one and only remaining room. But this would be to misunderstand the problem. What we are counting is the possible ways of allocating rooms, and a moment's reflection will show that *any* allocation of students to rooms can be arrived at by the procedure just outlined. Thus the arrangement

(4.2.1)

Student number	1	2	3	4	5	6
Room number	5	1	6	4	3	2

will certainly be among the possibilities considered. For one of the ways of housing student number 1 is that of putting him in room 5; one way of then housing student number 2 is to put him in room 1, and so on. The same allocation would appear among the others if the students were allocated rooms in a different order, say in reverse order. One method of allocation would be to put the first student (number 6) in room 2, the next student in room 3, and so on. This would give the same allocation of rooms to students as before. Thus we see that since we are asking only for the *number* of ways of allocating the students to rooms, the order in which we take the students is immaterial. The important thing (and we shall have more to say about this in a moment) is that the students should be *different*, that is, that one should be able to distinguish one from another. This is usually possible. Likewise the formulation of the problem implies that the rooms are different—that it is possible to distinguish any one room from any other.

This problem is readily generalized. Suppose we have n different "objects" and we have to place them in n different "boxes," one object in each box. Then the first object can be placed in any of the n boxes, and the second must be placed in one of the remaining $n - 1$ boxes. There are $n - 2$ boxes in which the next object can be placed, $n - 3$ for the next, and so on. Carrying on in this way we see that the total number of possible ways is

$$n(n - 1)(n - 2) \ldots 3 \times 2 \times 1.$$

Expressions like this—the product of all the integers from a given integer n down to 1—are of such common occurrence in mathematics that a special notation has been devised for them. We write "$n!$" (pronounced *factorial n* or *n factorial*) for such an expression. That is,

$$n! = n(n - 1)(n - 2) \ldots 3 \times 2 \times 1.$$

by definition.

The terms *objects* and *boxes* that we used above are convenient general terms to use for thinking about combinatorial problems and can be interpreted in many ways according to the subject matter of the problem being considered. Thus in the problem with which we opened the chapter, the "objects" are the students, and the "boxes" are the rooms. The number of ways of allocating the students to rooms is therefore

$$6! = 6 \times 5 \times 4 \times 3 \times 2 \times 1 = 720$$

as we saw. In this problem we could just as correctly take the "objects" to be the rooms and the "boxes" to be students, since it makes no difference whether we allocate rooms to students or students to rooms. Notice that if we label the objects with the numbers $1, 2, \ldots, n$, and likewise label the boxes $1, 2, \ldots, n$ we can specify any allocation of objects to boxes by writing the number of the object under the number of the box in which it is put. Thus

Box	2	3	1	4	6	5
Object	1	3	6	2	5	4

means that object 1 goes in box 2, and so on. We can also write this same allocation as

(4.2.2)
$$\begin{pmatrix} 1 & 2 & 3 & 4 & 5 & 6 \\ 6 & 1 & 3 & 2 & 4 & 5 \end{pmatrix}$$

where, following convention, we have omitted references to boxes and objects, arranged the upper line in ascending order and enclosed the two lines by brackets to denote that they go together. A symbol like (4.2.2) is called a *permutation*, since the lower line is a rearrangement, or permutation, of the symbols on the upper line. Thus the number of permutations of n objects is $n!$.

Exercise 25

1. Calculate the values of $n!$ for $n = 1, 2, \ldots, 10$.

2. Write down the $4! = 24$ ways of arranging the letters A, B, C, D in order on a line, e.g., $ABCD, ACBD, DACB$, etc. In how many of these arrangements does B occur in the second place? In how many does C occur in the first place and A in the second place?

3. How many of the $10!$ arrangements of the digits $1, 2, 3, 4, 5, 6, 7, 8, 9, 0$ begin with $246 \ldots$? How many end with $\ldots 2715$?

4.3 COMBINATIONS

Let us now turn to a different problem. Suppose the Dean of students has to allocate to 3 available rooms 3 out of 6 new students. (What he does with the other 3 is not our concern.) This is not a difficult problem. There are 6 ways in which he can allocate a student to the first room. Then, for each such choice, there are 5 ways of assigning one of the remaining students to the second room, and then 4 ways of allocating to the third room. Thus we easily see that the allocation to rooms can be made in

$$6 \times 5 \times 4 = 120$$

different ways. If there had been 7 students the number would have been $7 \times 6 \times 5 = 210$, and so on. The general result is easy to see. If there are n objects from which r different boxes are to be filled ($r < n$), then the first box can be filled in n ways; the second in $n - 1$ ways, and so on until each box is occupied. The required number will be the product $n(n - 1)(n - 2) \ldots$, there being r factors in all. It is easily verified that the last factor is $(n - r + 1)$. Hence this number—which is called the *number of permutations of n objects r at a time*—will be

$$n(n - 1)(n - 2) \ldots (n - r + 1).$$

It will be seen that this expression reduces to $n!$ when $r = n$, as we would expect.

Let us now suppose that the Dean of students wants to choose 3 of the 6 students. We may suppose that he wants to decide *which* students are to have the available rooms, without actually deciding which student is to have which room. The difference from the previous problem is clear. In each case the task is that of choosing a subset of 3 objects (students) from a set of 6 objects; but in the previous problem this subset was an *ordered* subset—the students being ordered by reason of the rooms to which they were assigned. In the present problem the choice is of a subset in the usual sense, with no ordering attached to it. Such a subset is often called a *combination* of 3 of the 6 objects. How many such combinations are there? Since we do not know, let us call the number X.

Now if the Dean's ultimate aim is to allocate the students to individual rooms, he can achieve this aim by first choosing which three students are to be given rooms, and then allocating rooms to the chosen students. Once the three students have been chosen they can be allocated to rooms in $3! = 3 \times 2 \times 1 = 6$ ways; and this can be done for every combination of 3 out of 6 students. Hence he can fill the 3 rooms in $6X$ ways. But this number is already known; it is the number of permutations of 6 objects 3 at a time and is $6 \times 5 \times 4 = 120$. Hence $6X = 120$ or $X = 20$. Thus the number of combinations of 6 objects 3 at a time is 20.

Again, this is easily generalized. Let X denote the number of combinations of n objects r at a time, i.e., the number of different subsets of r objects that can be chosen from a set of n objects. Let us reconsider the problem of finding the number of permutations of n objects r at a time, the answer to which we know to be

$$n(n-1)(n-2) \ldots (n-r+1).$$

We can obtain this number by first choosing an *unordered* set of r objects (possible in X ways) and then ordering this set. This ordering can be done in $r!$ ways. Thus we have the equation

$$r!X = n(n-1)(n-2)\ldots(n-r+1)$$

from which it follows that

(4.3.1) $$X = \frac{n(n-1)(n-2) \ldots (n-r+1)}{r!}$$

Expression (4.3.1) is of such great importance that a special notation is usually given to it. Unfortunately there is not yet universal agreement on what notation to use. In older books this expression is denoted by

$$_nC_r \quad \text{or} \quad {}^nC_r \quad \text{or} \quad C_{n,r}$$

or something of the sort, C standing for *combination*. These notations are going out of fashion, and should be avoided. The modern tendency is to use the notation

$$\binom{n}{r}$$

Note: no horizontal line between the *n* and the *r*. Thus we have the definition

(4.3.2)
$$\binom{n}{r} = \frac{n(n-1)(n-2)\dots(n-r+1)}{r!}.$$

This can be written in a more compact form. Let us multiply the numerator of (4.3.2) so as to make it into *n*!, i.e., let us multiply the numerator by

$$(n-r)(n-r-1)\dots 3 \times 2 \times 1 = (n-r)!$$

To keep the expression the same we must multiply the denominator by the same amount. Thus we have

$$\binom{n}{r} = \frac{n(n-1)(n-2)\dots(n-r+1)(n-r)(n-r-1)\dots 3 \times 2 \times 1}{r!\,(n-r)!}$$

or

(4.3.3)
$$\binom{n}{r} = \frac{n!}{r!\,(n-r)!}$$

Thus (4.3.3) gives the notation and the formula for the number of combinations of *n* objects *r* at a time.

Exercise 26

1. Find the values of the following expressions.

(a) $\binom{5}{3}$ (c) $\binom{4}{2}$ (e) $\binom{6}{2}$ (g) $\binom{8}{4}$

(b) $\binom{6}{4}$ (d) $\binom{5}{4}$ (f) $\binom{7}{4}$ (h) $\binom{3}{2}$

2. Find the number of ways of choosing 4 cards from a deck consisting of 13 cards.

Express in terms of factorials the number of ways of choosing a hand of 13 cards from a standard deck of 52 playing cards. By what proportion is this number increased if an extra card (the joker) is included in the deck.

3. A man wishes to make a selection of *r* objects from a set of 10 objects. For what value of *r* is the number of possible selections a maximum? Make a conjecture concerning the corresponding general problem of 2*n* objects. Can you prove your conjecture?

4. In how many ways can a selection of *at least* 6 books be made from a set of 9 books?

4.4 PERMUTATIONS OF OBJECTS NOT ALL DISTINCT

So far we have considered permutations and combinations of objects that are all distinct. A new kind of combinatorial problem arises if we consider arrangements of objects that are not all distinguishable from each other. As an example of such a problem we may take the following: A gardener wishes to plant 9 plants in a flower bed in which there are 9 empty places. He has 4 dahlia seeds, 3 nasturtium seeds and 2 chrysanthemum seeds. In how many ways can he do this?

We shall assume that seeds of the same plant are indistinguishable from each other; that is, that although one *could* detect small differences between one seed and another, nevertheless *for the purposes of this problem* we shall imagine that no such differences exist. Thus all that matters is what *kind* of seed goes into a given place, not what *particular* seed goes there.

Shorn of all its trimmings, this problem is that of finding the number of permutations of 9 objects, comprising 4 of one kind, 3 of another kind and 2 of a third kind. Since we do not yet know what this number is, let us call it X, provisionally.

Let us consider one of these X permutations, which specifies that the 4 dahlia seeds go in such and such places, the nasturtium seeds in certain other places, and the chrysanthemum seeds in the remaining places. We could denote a typical permutation of this kind by

$$D \quad C \quad D \quad N \quad D \quad N \quad D \quad C \quad N$$

where D denotes a dahlia seed and so on. Let us now imagine that we have temporarily regained our ability to distinguish individual seeds. Then we might observe that the above arrangement was in fact

$$D_3 \quad C_1 \quad D_1 \quad N_2 \quad D_4 \quad N_3 \quad D_1 \quad C_2 \quad N_1$$

where D_1, D_2, D_3, D_4 stand for the individual dahlia seeds, and so on. There are many other possibilities, however, for example,

$$D_4 \quad C_1 \quad D_1 \quad N_3 \quad D_2 \quad N_2 \quad D_3 \quad C_2 \quad N_1$$
$$D_4 \quad C_2 \quad D_1 \quad N_1 \quad D_2 \quad N_2 \quad D_3 \quad C_1 \quad N_3.$$

All these would be regarded as identical if we did not distinguish the seeds. Let us now see how many such arrangements there are corresponding to a particular one of the X permutations we are interested in. The 4 dahlia seeds must occupy the 4 specified places, but within those spaces may be rearranged in 4! ways. Similarly the 3 nasturtium seeds can be arranged in 3! ways and the chrysanthemum seeds in 2! ways. Thus, if we distinguish the seeds each from the other, we have $4! \times 3! \times 2!$ arrangements for each arrangement that we are interested in, i.e., X times $4! \times 3! \times 2!$ in all. But this number must be 9!, since the 9 seeds are now 9 distinguishable objects, and all the 9! possible permutations of these objects will be included in the above enumeration.

Hence

(4.4.1)
$$X = \frac{9!}{4!\,3!\,2!}$$
$$= 1260.$$

This result is easily generalized. Suppose we have n objects, of which r_1 are of one kind, r_2 of another kind, r_3 of yet another kind, and so on. If k is the number of different *kinds* of objects then we must have

(4.4.2)
$$r_1 + r_2 + r_3 + \ldots + r_k = n$$

In how many ways can we permute these objects, i.e., place them in n distinguishable boxes, if objects of the same kind are not distinguishable? To answer this question, let us imagine that we temporarily "label" the r_i objects of the ith kind ($i = 1, 2, \ldots, k$), i.e., we render them distinguishable in some way. Any one of the X (say) permutations that we want to count will give rise to

$$r_1!\,r_2! \ldots r_k!$$

permutations of the labelled objects. Thus as in the specific problem considered above, the required number X of permutations is

(4.4.3)
$$\frac{n!}{r_1!\,r_2! \ldots r_k!}$$

This then is the number of permutatons of n objects, r_1 of one kind, r_2 of another kind, etc.

It is easily shown that the formula for the number of combinations of n objects r at a time which we derived in (4.3.2) is a special case of (4.4.3). Imagine that we have n labels, r of which bear the word "chosen," $n - r$ of which bear the word "rejected." We shall attach one label to each of our n objects. Since there are two kinds of labels and we shall not distinguish between labels of the same kind, the number of ways of disposing these on the objects is, by (4.4.3),

$$(4.4.4) \qquad\qquad \frac{n!}{r!(n-r)!} = \binom{n}{r}$$

But any such disposal of the labels will give us a choice of r of the n objects; we choose exactly those to which have been attached the labels bearing the word "chosen." Thus (4.4.4) gives us the number of combinations of r of the n objects. In this way we see the connection between (4.3.3) and (4.4.3).

Exercise 27

1. A car park contains spaces for 10 cars. In how many ways can 4 Buicks, 3 Chevrolets and 3 Cadillacs be parked, assuming that no distinction is made between cars of the same make?

2. In the same car park, how many ways are there of parking 2 Buicks, 3 Chevrolets and 1 Cadillac?

3. How many different sequences of letters can be made from the letters of the word *SELFLESSNESS*?

4. In how many distinct ways can the letters of the word *INDEFINITE* be arranged?

4.5 SELECTIONS WITH REPETITIONS

Let us now consider another problem of a somewhat different sort. We shall suppose that there are available a certain number of kinds of objects, and that there are unlimited supplies of each kind. We want to know in how many ways it is possible to make a selection of a certain given number, call it n, of objects. There is no restriction on how many objects of each kind we take in our selection, and since there are unlimited supplies of each kind of object, i.e., sufficient to meet any need that may arise, we could, if we wished, choose all n objects to be the same kind. On the other hand we could choose a few of each kind, and so on. If there are k different kinds of objects we can describe this problem as that of choosing n objects from a set of k objects *with repetitions allowed*. Another way of describing this would be as the number of ways of choosing n objects from k *sets* of objects, where each set consists of a sufficiently large number of identical objects. Note that there is no question of ordering the objects; all that matters is how many of each kind of object occur in the selection.

To fix our ideas, let us think of the following particular example. A fruit shop is stocked with apples, grapefruits, oranges and peaches, and from these stocks we want to make a selection of 8 pieces of fruit. In how many ways can this be done?

To specify any particular selection we need only say what number of each kind of object has been selected. If we use the initial letters to stand for the kinds of fruit, then the following are possible selections—

1.	O	A	G	A	P	P	G	A
2.	P	P	P	P	P	G	P	P
3.	O	A	P	A	A	O	O	O
4.	G	A	G	O	O	A	G	O

etc.

We shall now show that this problem can be restated in such a way that its solution follows readily from the results of the previous sections. To do this we write our selections in a standard way. We agree to take the k kinds of objects in some particular order. This order can be arbitrary, but in our example a natural order to choose would be alphabetical order. We could therefore write the above four examples as,

A	A	A	G	G	O	P	P
G	P	P	P	P	P	P	P
A	A	A	O	O	O	O	P
A	A	G	G	G	O	O	O

As a further simplification we note that we can drop the letters altogether, provided we indicate in some way whereabouts we change from one letter to the next. Let us therefore replace all the letters by xs and put a vertical line wherever there is a change from one letter to the next. Thus the first of the above selections will assume the form

1. $\qquad xxx\,|\,xx\,|\,x\,|\,xx$

indicating that there are 3 of the first kind of object, 2 of the second, and so on. The remaining selections become

2. $\qquad |\,x\,\|\,xxxxxxx$
3. $\qquad xxx\,\|\,xxxx\,|\,x$
4. $\qquad xx\,|\,xxx\,|\,xxx\,|.$

Notice that the nonappearance of one or more objects in a selection does not raise any difficulties. If there is no object of the first kind, as in (2), we simply start with a vertical line indicating that we change at once to the second kind of object. A final vertical line denotes the absence of the last kind of object, as in (4), while the absence of the others will be signalled by the appearance of consecutive vertical lines within the specifications. Thus for every selection we can write down in a unique manner a sequence of xs and vertical lines (there will be 8 xs and 3 vertical lines for our example). Moreover, to every sequence of 8 xs and 3 vertical lines there corresponds a selection; we simply start writing As for xs until we come to a vertical line, then we change to the next letter and carry on. In this way, we establish what is known as a *one-to-one correspondence* between the set of selections on the one hand and the sequence of xs and vertical lines on the other. Two sets M and N are in one-to-one correspondence when to every element of M there corresponds a unique element of N, and for every element of N there is exactly one element of M to which it corresponds. Now when a one-to-one correspondence can be set up between the elements of two sets we can conclude that the two sets have the same number of elements.* Hence if we can find the number of sequences of xs and $|$s, this will also be the required number of selections. This problem, however, is easily solved: for it is the number of arrangements of 8 xs and 3 $|$s, i.e., combinations of 8 objects of one kind and 3 of another, and by the result (4.4.3) this is

$$\frac{11!}{8!3!} = \binom{11}{3}$$
$$= \frac{11.10.9}{3.2.1}$$
$$= 165.$$

The general formula is readily obtained. In each sequence the number of xs is the number of objects n. It is easy to see that the number of $|$s is one less than the number of *kinds* of objects, and is therefore $k - 1$. Hence the number we need to find is the number of arrangements of n objects of one kind and $k - 1$ objects of a second kind, making $n + k - 1$ objects altogether. This number, by (4.4.3), is

(4.5.1)
$$\frac{(n + k - 1)!}{n!(k - 1)!} = \binom{n + k - 1}{n}.$$

This result has several important applications, but it so happens that it will not play any great part in what follows in this chapter. The reason it

---

* This is not, as might be thought, so much a deduction or a theorem as a matter of definition. The statement "A has the same number of elements as B" is defined by the existence of some one-to-one correspondence between the elements.

has been included here is that it provides a very good illustration of a dodge that is frequently used in combinatorial analysis: that is, that in order to count the number of permutations or combinations of a particular kind—something which may be difficult to tackle directly—an indirect approach is used. Instead of counting the permutations or combinations required by the problem, we count instead some other set of permutations or combinations which we have previously shown to be *equinumerous* with the required set. The proof that the set we count and the set we were supposed to count are equinumerous, i.e., have the same number of elements, is effected by exhibiting a one-to-one correspondence between them. This is exactly the procedure that we adopted in the above example.

Exercise 28

1. A snack bar offers 5 possible breakfast menus. A man takes breakfast every morning for a week. In how many ways can he choose the various menus, taking no account of which days he chooses them?

2. **(a)** In a gambling casino are 6 one-armed bandits, each requiring a quarter to operate it. A man wishes to spend \$3 on operating the machines but does not use any money that he may win. In how many ways can he divide his money between the various machines?

  (b) In how many ways can he play the machines, if we take into account the order in which he plays them?

3. **(a)** The organizer of an international conference wishes to invite a total of 11 delegates from 5 different countries. In how many ways can he do this? (Do not assume that all five countries are necessarily represented.)

  (b) How many of these possible conferences will be "international" in the sense that at least two countries will be represented?

4.6 THE BINOMIAL THEOREM

Result (4.3.3) for the number of combinations of n objects taken r at a time can be used to prove a most important theorem in mathematics, the famous *binomial theorem*. A binomial is an expression containing two terms, that is, of the form $x + y$. To see what the binomial theorem is about, let us first see what happens if we form the square of this expression. By the ordinary rules of algebra we see that

$$(x + y)^2 = (x + y)(x + y)$$
$$= x^2 + yx + xy + y^2$$
$$= x^2 + 2xy + y^2.$$

(4.6.1)

The second line above shows that the final answer is obtained from four terms, which are formed by taking every product of a term (either x or y) from the first factor with a term from the second factor. We can use this result to find the cube $(x + y)^3$. We have

(4.6.2)
$$(x + y)^3 = (x + y)(x + y)^2$$
$$= (x + y)(x^2 + 2xy + y^2)$$
$$= x^3 + 2x^2y + xy^2 + yx^2 + 2xy^2 + y^3$$
$$= x^3 + 3x^2y + 3xy^2 + y^3.$$

We could, however, obtain this same result without making use of (4.6.1). For we could write

$$(x + y)^3 = (x + y)(x + y)(x + y)$$

and then observe that every term in this threefold product must be obtained by choosing either x or y from the first factor, either x or y from the second factor, and either x or y from the third. There are eight ways in which this can be done. One way is to choose x from each factor—this gives the term x^3. Further terms arise if we choose a y from one factor, and x from the other two factors, giving the term x^2y. The number of ways of doing this is 3, since there are three factors from which the y can be chosen. In the same way, there are three ways of obtaining the term xy^2, since there are three factors from which we can choose the x, and having made this decision we must take y from the other two. Finally we could choose y from each factor. Taking all these terms together we obtain the right-hand side of (4.6.2).

Let us now look at the general power of $x + y$, namely $(x + y)^n$, where n is some whole number. What is the expansion in terms of powers of x and y? For $n = 1, 2, 3$, we already know the answers, and this will provide a check on what we subsequently discover. We first write $(x + y)^n$ as the product of n factors, as follows:

(4.6.3) $$(x + y)^n = (x + y)(x + y) \ldots (x + y)$$

where the number of binomial factors in the product on the left-hand side is n. To expand this expression by the usual rules of algebra we must choose either an x or a y from each factor in the product, and the product of the xs and ys that are chosen will be a term in the expansion. This observation tells us at once what the terms will look like. If we choose a y from each of r of the factors, and from the remaining $n - r$ factors choose the x, then the term we shall obtain will be the product of $n - r$ xs and r ys; i.e., it will be $x^{n-r}y^r$. If $r = 0$ we have the term x^n (an x from each factor) and if $r = n$ we have y^n. Most of these terms will be obtained more than once, and our main task is to determine how many times each term occurs. In other words, we want to determine the coefficient of $x^{n-r} y^r$ in the expansion.

This is quite easy to do. We get the term $x^{n-r}y^r$ when we choose a y from each of exactly r of the factors on the right-hand side of (4.6.3). The number of ways in which we can make this kind of choice is simply the number of ways of choosing r objects (factors) from a set of n objects; the objects chosen are the factors from which a y is to be picked. Hence the required number of ways is $\binom{n}{r}$, and the term in $x^{n-r}y^r$ is

$$(4.6.4) \qquad \binom{n}{r} x^{n-r}y^r.$$

To get all possible terms we must take all values of r from 0 to n, since the extreme cases are those of choosing a y from *no* factors, and choosing it from every factor. In each of these extreme cases there is exactly one way of making the choice. Hence

$$(4.6.5) \qquad \binom{n}{0} = \binom{n}{n} = 1$$

There is nothing startling about (4.6.5)—after all, the number of ways of choosing *no* objects from a set is clearly one, as is that of choosing all objects. However, taken in conjunction with (4.3.3), these results necessitate a slight extension of our definition of the factorial, for they imply that

$$(4.6.6) \qquad \frac{n!}{0!\,n!} = 1$$

in which the symbol 0! occurs. This defies interpretation in terms of our original definition of *factorial*, which applied only to positive (and therefore nonzero) integers, but we can easily extend the definition of *factorial* so as to apply to the integer zero as well. The definition of 0! is at our disposal, but if we want (4.6.6) to be true (and we do) then we have no choice but to define 0! to be 1. This done, (4.3.3) is then valid for $k = 0$ and n, and the statements (4.6.5) are true, as common sense dictates that they must be.

We can now write out the final result of all this. It is that

$$(4.6.7) \qquad \begin{aligned} (x+y)^n = \binom{n}{0} x^n + \binom{n}{1} x^{n-1}y + \binom{n}{2} x^{n-2}y^2 + \cdots \\ + \binom{n}{r} x^{n-r}y^r + \cdots + \binom{n}{n-1} xy^{n-1} + \binom{n}{n} y^n. \end{aligned}$$

This result, (4.6.7), is known as the *binomial theorem*. The right-hand side contains $n + 1$ terms of which we have given the first three and the last two. We have also specified the general term. For the first few values of n we have

$$(x + y)^1 = x + y$$
$$(x + y)^2 = x^2 + 2xy + y^2$$
$$(x + y^3) = x^3 + 3x^2y + 3xy^2 + y^3$$
$$(x + y^4) = x^4 + 4x^3y + 6x^2y^2 + 4xy^3 + y^4$$
$$(x + y)^5 = x^5 + 5x^4y + 10x^3y^2 + 10x^2y^3 + 5xy^4 + y^5.$$

Notice that the coefficients read the same from either end. This is a consequence of the fact that

$$\binom{n}{n-r} = \frac{n!}{(n-r)!\,r!} = \binom{n}{r}.$$

It can also be deduced from the fact that the expansion of $(x + y)^n$ is clearly the same as that of $(y + x)^n$ except that the terms will appear in the reverse order.

The statement of the binomial theorem can be expressed more succinctly by using a notational device which is very convenient in many applications—the use of the Greek capital letter Σ (sigma). The right-hand side of (4.6.7) is a common type of expression in mathematics, namely the sum of a number of terms all of which have the same general form, individual terms being specified by the value of an integer occurring in the general form (in this case it is r, which denotes an integer). When we wish to denote an expression of this type, we can use the symbol Σ and interpret it as meaning "take the sum of all terms of the following form." Naturally this symbol must be followed by the general form of the terms that are added together. Thus we get an expression like

$$\Sigma \binom{n}{r} x^{n-r}y^r$$

which almost represents the right-hand side of (4.6.7). One thing is missing; there is no indication of what values r is to take. We need to add something which shows that r takes all integer values between 0 and n. This is done by writing these *limits* below and above the sigma sign. In this way we obtain the expression

(4.6.8)
$$\sum_{r=0}^{n} \binom{n}{r} x^{n-r}y^r$$

which *does* represent the right-hand side of (4.6.7).

With this $\sum$ notation we can now state the theorem in the following form.

THE BINOMIAL THEOREM

If n is any integer, then

$$(x + y)^n = \sum_{r=0}^{n} \binom{n}{r} x^{n-r} y^r.$$

Because of their appearance as coefficients in the binomial theorem, the numbers $\binom{n}{r}$, which we first met as numbers of combinations, are called the *binomial coefficients*.

PASCAL'S TRIANGLE

If we look at the expressions of the first few powers of $x + y$ that we gave above, and concentrate on the coefficients only we find that they can be arranged very symmetrically in the form of a triangle, as follows

```
                    1
                1       1
            1       2       1
        1       3       3       1
    1       4       6       4       1
1       5      10      10       5       1
                  etc.
```

The "1" which heads this triangle corresponds to the expansion $(x + y)^0 = 1$, which is a special case of the binomial theorem. This arrangement of the binomial coefficients is known as *Pascal's triangle*. An interesting property of this triangular array of numbers—one which is easily found by inspection, and which will be proved in a moment—is that each number is the sum of the two numbers next above it. The particular example

```
        1     4  +  6   4       1
    1       5     = 10  10  5       1
```

should make this clear. This property does not quite apply to the 1s at the ends of each row, but will do so if we interpret an empty space as zero.

This property of Pascal's triangle must represent some property of the binomial coefficients and it is easy to see what it is. If we call the rows in the triangle row 0, row 1, row 2, and so on, and number the entries in each row from 0 upwards, then the rth entry in row $n + 1$ is $\binom{n+1}{r}$. According to the property of the triangle we would expect this to be the sum of the entries above it. These will be entries $r - 1$ and r in row n. Hence we conjecture that

(4.6.9)
$$\binom{n+1}{r} = \binom{n}{r-1} + \binom{n}{r}.$$

Let us see if we can prove this conjecture. The right-hand side of (4.6.9) can be written as

$$\frac{n!}{(r-1)!(n-r+1)!} + \frac{n!}{r!(n-r)!}$$

or, with a bit of algebraic juggling, as

$$\frac{r}{r!} \cdot \frac{n!}{(n-r+1)!} + \frac{n!}{r!} \cdot \frac{(n-r+1)}{(n-r+1)!}$$

Taking out common factors, we can write this as

$$\frac{n!}{r!(n-r+1)!} \{r + n - r + 1\}$$

which reduces to

$$\frac{n!}{r!(n-r+1)!} \{n+1\} = \frac{(n+1)!}{r!(n-r+1)!}$$

$$= \binom{n+1}{r}.$$

This completes the proof of (4.6.9).

Exercise 29

1. Extend Pascal's Triangle as far as the 14th row.
2. Use the binomial theorem to expand the following expressions:

(a) $(2x + 3y)^4$ (e) $(x - 4y)^5$

(b) $(4x + 1)^3$ (f) $(\frac{1}{3}x + \frac{2}{3}y)^4$

(c) $(1 - x)^7$ (g) $(2x - 5y^2)^3$

(d) $(1 + 2x^2)^4$ (h) $\left(x + \dfrac{1}{x}\right)^6$

3. By considering the ratio of the coefficient of $x^r y^{n-r}$ to that of $x^{r+1} y^{n-r-1}$, show that the largest coefficient in the expansion of $(x + y)^n$ is the middle one (if n is an even number) and that if n is odd the two middle coefficients are equal and larger than all the others.

4. Using the method of problem 3, or otherwise, find which term in each of the following expansions has the largest coefficient.

(a) $(x + 2y)^7$ (d) $(\frac{2}{5}x + \frac{3}{5}y)^6$

(b) $(3x + 2y)^5$ (e) $(\frac{4}{7}x + \frac{3}{7}y)^{10}$

(c) $(x + 5y)^{14}$ (f) $(3x + 7y)^9$

4.7 MATHEMATICAL INDUCTION

We have all of us in our childhood, played the game of setting up a row of toy soldiers, wooden blocks, or what have you, and then, by pushing over the first one, making the whole row topple over one after the other, each toy soldier knocking down the next in turn. It may seem unlikely that such a childish game should be an example of a powerful mathematical method, but it is. The method is called *mathematical induction*, and is a very useful tool for proving certain kinds of mathematical theorems.

To be amenable to proof by mathematical induction a theorem must assert some statement about the set of all integers, that is, that something is true for all values of an integer variable, n say. Examples of such statements are the following:

(4.7.1) $$1 + 2 + 3 + \ldots + n = \tfrac{1}{2}n(n + 1)$$

(4.7.2) $$1 + 3 + 5 + \ldots + (2n + 1) = n^2$$

(4.7.3) $$1 + r + r^2 + \ldots + r^n = \frac{1 - r^{n+1}}{1 - r}$$

Another example is the binomial theorem itself. To be quite general, let us denote such a statement by $P(n)$. $P(n)$ is therefore some assertion involving a positive integer n. It may contain other variables as well as n, for example, the r of (4.7.3) or the x and y of the binomial theorem, but at present only its dependence on n is of interest to us.

Now every infant knows that in order to knock over a row of tin soldiers, two things are essential:

1. You have to give the first one a push.

2. The soldiers must be set out so that the fall of each one initiates the fall of the next.

In exactly the same way, if we are to achieve our object, which is to prove that $P(n)$ is a true statement for *every* positive integer value of n, it suffices to prove two things, viz.,

(4.7.4) $P(1)$ is true

(4.7.5)   For any positive integer m, $P(m) \rightarrow P(m + 1)$

If we can prove these two assertions concerning the statement $P(n)$, then we can assert the following conclusion:

(4.7.6) $P(n)$ is true for every positive integer n.

A proof which proceeds in this way is called a proof by *mathematical induction.* *

It may not be immediately obvious that (4.7.6) follows from (4.7.4) and (4.7.5), but the argument is quite simple. $P(1)$ is known to be true, by (4.7.4). Since (4.7.5) is true for any positive integer it is true for $m = 1$, and then reads $P(1) \rightarrow P(2)$. Since $P(1)$ is true $P(2)$ is true. Putting now $m = 2$ in (4.7.5) we have $P(2) \rightarrow P(3)$, and since $P(2)$ is true it follows that $P(3)$ is true. Similarly $P(3) \rightarrow P(4)$ so that $P(4)$ is true, and so on.

Perhaps that "and so on" is a little glib. Let us phrase the proof slightly differently. Suppose we are given an integer N. Then we can write out a chain of implications, obtained by giving different values to m in (4.7.5) as follows

$$P(1) \rightarrow P(2)$$
$$P(2) \rightarrow P(3)$$
$$P(3) \rightarrow P(4)$$
$$\cdots$$
$$P(N - 2) \rightarrow P(N - 1)$$
$$P(N - 1) \rightarrow P(N)$$

the final outcome of which is seen to be $P(1) \rightarrow P(N)$, by the results of the first chapter. But $P(1)$ is known to be true. Hence $P(N)$ is true. In this way we

---

* The process is often called simply *induction*, which in a sense is incorrect, since *induction* is the name of an entirely different process of reasoning used in the natural sciences. Insofar as *induction* in this latter sense has no place at all in mathematics, there is no real ambiguity in using the term as an abbreviation for *mathematical induction*, and the reader may do this if he wishes. It is a pity that no-one has come up with a shorter term—but there it is.

could, in a finite number of steps, write out the proof of $P(N)$ for any positive integer N. The analogy with the game of toppling the toy soldiers should be clear from these remarks.

Exercise 30

1. What two statements would suffice to imply that $P(n)$ was true
    (a) for all positive even integer values of n?
    (b) for all positive odd integer values of n?
2. Suppose the statement

$$P(m) \longrightarrow P(m + 3) \text{ for all positive integers } m$$

had been proved. Indicate what further statements would suffice to prove the statement

$$P(n) \text{ is true for all positive integers } n.$$

3. Write down a set of statements that would suffice to prove that $P(n)$ was true for all integral values of n, positive, negative and zero.

4.8 EXAMPLES OF MATHEMATICAL INDUCTION

Let us now see how all this works out in a specific example. We take the example (4.7.1) above, viz.,

$$1 + 2 + 3 + \ldots + n = \tfrac{1}{2}n(n + 1) \text{ for all } n$$

(it will be assumed from now on that n is a positive integer, unless the contrary is stated).

We must first prove $P(1)$ in the notation of the previous section. Putting $n = 1$ in the given equation we have

$$1 = \tfrac{1}{2} . 1 . 2$$

which is clearly true.

Now we must show that $P(m) \longrightarrow P(m + 1)$, i.e., that if m is an integer for which P is true, so is $m + 1$. If m is such an integer then

$$1 + 2 + 3 + \ldots + m = \tfrac{1}{2}m(m + 1)$$

Hence

$$(4.8.1) \qquad 1 + 2 + 3 + \ldots + m + (m + 1) = \tfrac{1}{2}m(m + 1) + (m + 1)$$
$$= \tfrac{1}{2}(m + 1)(m + 2).$$

Now (4.8.1) is simply the statement $P(m + 1)$—the original statement but with $m + 1$ in place of m—so we have reached our objective.

A proof by mathematical induction is easier to follow if it is properly set out, and it is a help to display proofs in a manner which exhibits the logic behind the method. Too many students are led astray by being told to start a proof by mathematical induction by a statement something like "Assume that the theorem is true for the integer n", which gives a strong impression that the method starts by assuming what is to be proved. This impression is erroneous, but it is worthwhile to avoid giving it. The following three rules show how to set out the argument clearly and concisely.

RULE 1:
Start by proving that the statement or theorem is true when $n = 1$.

RULE 2:
Say "Let m be any value of n for which the theorem is true." Then prove that it is also true for $n = m + 1$.

This is the step $P(m) \rightarrow P(m + 1)$. What we are saying is that if A is the set of n for which the theorem is true then

$$(4.8.2) \qquad (m \in A) \rightarrow (m + 1 \in A).$$

At this stage we already know that A has at least one member, namely the integer 1. In fact it would not matter if we did not know this, since if A were the null set, i.e., if the theorem were not true for any value of n, then (4.8.2) would still hold. A false statement implies any statement, as we saw in Chapter 1. However, it seems to inspire confidence to know that one is not talking about the empty set—if nothing else it gets rid of any suggestion that one is making any sort of assumption about the validity of the theorem.

RULE 3:
Say "Hence, by mathematical induction the theorem is true for all positive integer values of n." Just that. It is not necessary to go into the long rigmarole, so often advocated, of saying, for example, "We have proved that the theorem is true for $n = 1$; therefore it is true for $n = 2$, and for $n = 3$, and so on. Hence it is true for all value of n." This merely reiterates

the proof of the method being used. Unless you are particularly asked to do this, it is reasonable to take the attitude that if you use the phrase "mathematical induction" you should do your reader the honor of assuming that he is acquainted with this method of proof.

Exercise 31

1. The left-hand sides of (4.7.1) and (4.7.2) are examples of what is known as an *arithmetic progression*, namely a series of numbers such that the difference between successive numbers is constant. The general arithmetic progression, having n terms is

$$a + (a + d) + (a + 2d) + \ldots + (a + (n - 1)d)$$

where d is the common difference. Prove by mathematical induction that the sum of this progression is

$$\tfrac{1}{2}n(2a + (n - 1)d).$$

2. The left-hand side of (4.7.3) is an example of a *geometric progression* in which each term is a constant multiple (r) of the preceding. Prove (4.7.3) by mathematical induction.

3. Prove the following results by mathematical induction

(a) $\dfrac{1}{1.2} + \dfrac{1}{2.3} + \dfrac{1}{3.4} + \ldots + \dfrac{1}{(n-1)n} = 1 - \dfrac{1}{n}$

(b) $1^2 + 2^2 + 3^2 + \ldots + n^2 = \tfrac{1}{6}n(n + 1)(2n + 1)$

4. Deduce from problem 2 above, or prove by mathematical induction that

$$a^{n-1} + a^{n-2}b + a^{n-3}b^2 + \ldots + a^2b^{n-3} + ab^{n-2} + b^{n-1}$$

$$= \dfrac{a^n - b^n}{a - b}$$

5. The following purports to be a proof by mathematical induction that all positive integers are equal (!).

PROOF:

Let max (p, q) denote the larger of the positive integers p and q. We first show that if $0 < p \le q$, then max $(p, q) = p$.

If $q = 1$ then p must be 1 and this result is true.

Suppose k is an integer such that the result is true for $q = k$. Consider any pair p and $k + 1$, and the integer max $(p, k + 1)$. If we subtract

1 from each of p and $k+1$, then $\max(p, k+1)$ will also be reduced by 1. Hence $\max(p, k+1) = 1 + \max(p-1, k)$. Since the result is true for $q = k$ we have $\max(p-1, k) = p - 1$, and hence

$$\max(p, k+1) = 1 + (p-1) = p.$$

Hence the theorem is true for $q = k + 1$. Hence by mathematical induction, the result is true for all positive integers q.

But if $\max(p, q) = p \leq q$, then $p = q$. This proves the theorem that any two positive integers are equal.

Find the fallacy in the above argument.

4.9 THE BINOMIAL THEOREM BY MATHEMATICAL INDUCTION

Since we have already proved the binomial theorem it might seem pointless to do so again. But the binomial theorem is of such importance that it is to be hoped that the reader will not reprove the author if he reproves the theorem. We shall set out the proof in the manner suggested above.

The theorem to be proved is that

$$(x+y)^n = \binom{n}{0} x^n + \binom{n}{1} x^{n-1}y + \binom{n}{2} x^{n-2}y + \dots$$
$$+ \binom{n}{r} x^{n-r}y^r + \dots + \binom{n}{n-1} xy^{n-1} + \binom{n}{n} y^n.$$

We first prove that the theorem is true when $n = 1$. This is obvious because

$$(x+y)^1 = \binom{1}{0} x + \binom{1}{1} y$$

since

$$\binom{1}{0} = \binom{1}{1} = 1.$$

Now let m be any value of n for which the theorem is true. Then we have

(4.9.1)
$$(x+y)^m = \binom{m}{0} x^m + \binom{m}{1} x^{m-1}y + \dots + \binom{m}{r} x^{m-r}y^r + \dots$$
$$+ \binom{m}{m-1} xy^{m-1} + \binom{m}{m} y^m.$$

Multiply both sides of (4.9.1) by $x + y$. On the left-hand side we obtain

(4.9.2) $$(x + y)^m(x + y) = (x + y)^{m+1}$$

and on the right-hand side

(4.9.3)
$$(x + y)\left\{\binom{m}{0}x^m + \binom{m}{1}x^{m-1}y + \ldots + \binom{m}{r}x^{m-r}y^r + \ldots + \binom{m}{m}y^m\right\}.$$

If we multiply out the two bracketed expressions in (4.9.3) we shall get a collection of terms, each of which will be of the form

$$x^{m+1-s}y^s \qquad (0 \le s \le m + 1)$$

with some appropriate coefficient. Let us see what this coefficient will be.

In order to get a term in $x^{m+1-s}y^s$ we can either (1) multiply the x from the $(x + y)$ by the term in $x^{m-s}y^s$ inside the braces, thus obtaining a coefficient $\binom{m}{s}$, or (2) multiply the y from $(x + y)$ by the term $x^{m+1-s}y^{s-1}$ inside the braces. This will give us $\binom{m}{s-1}$. These are the only two ways of obtaining this term, and the required coefficient is therefore

$$\binom{m}{s} + \binom{m}{s-1}.$$

But we showed earlier in the chapter that $\binom{m}{s} + \binom{m}{s-1} = \binom{m+1}{s}$. Hence (4.9.3) becomes

(4.9.4)
$$\binom{m+1}{0}x^{m+1} + \binom{m+1}{1}x^m y + \ldots + \binom{m+1}{s}x^{m+1-s}y^s + \ldots$$
$$+ \binom{m+1}{m+1}y^{m+1}.$$

(The case $s = 0$ does not quite come under the general heading because of the occurrence of $\binom{m}{s-1}$, but the coefficient of x^{m+1} is clearly 1.)

The statement, now proved, that (4.9.2) and (4.9.4) are equal is the binomial theorem stated for $n = m + 1$. Hence $m + 1$ is also an integer for which the theorem is true.

Hence by mathematical induction, the theorem is true for all positive integral values of n.

This completes the proof of the binomial theorem by mathematical induction.

Exercise 32

1. If, instead of the binomial expression $(x + y)$, we consider the trinomial $(x + y + z)$ and its nth power, we are led to a generalization of the binomial theorem. A general term in the expansion of $(x + y + z)^n$ will be a term in $x^p y^q z^r$ where $p + q + r = n$. Show that the coefficient of this term is

$$\frac{n!}{p!q!r!}.$$

2. If we continue the generalization started in problem 1, and consider a general multinomial $x_1 + x_2 + \ldots + x_k$, then we arrive at the *multinomial theorem*, which states that the coefficient of

$$x_1^{p_1} x_2^{p_2} x_3^{p_3} \ldots x_k^{p_k}$$

in the expansion of $(x_1 + x_2 + \ldots + x_k)^n$ is

$$\frac{n!}{p_1! p_2! p_3! \ldots p_k!}$$

where, of course, $p_1 + p_2 + \ldots + p_k = n$. Prove this result.

3. Find the following coefficients
   (a) the coefficient of x^2yz^3 in $(2x - y + z)^6$
   (b) the coefficient of xy^3z^4 in $(x + 2y - 4z)^8$
   (c) the coefficient of x^2y^2 in $(2 - x + 3y)^7$
   (d) the coefficient of x^7 in $(1 + 3x + 4x^2)^3$
   (e) the coefficient of x^4 in $(1 + x - 3x^2 + 4x^3)^4$

4.10 PROBABILITY

Words or phrases denoting degrees of probability—that is, which convey in some manner the amount of confidence that we have in a statement that we are making—are of common occurrence in everyday language. We say that it is "very likely" that the weather will be fine tomorrow, or that "the chances are slight" that it will snow. We may say that a certain horse is "certain" to win a particular race (and the "chances" are "quite high" that we shall be wrong!). People talk of the "likelihood" of a candidate winning an election, and so on. By these phrases, and many others, we give a qualitative indication of our assessment of what we think is going to happen, and while admitting our uncertainty about the future we convey our feelings, hopes or expectations of what will come to pass. These phrases are in common use and are generally understood.

Behind these colloquial phrases is a vague concept of "likelihood," "chance" or, as we shall call it, "*probability*." If this concept is to be of use in mathematics it should be made precise and quantitative; but as soon as we try to do this we immediately run into difficulties. These difficulties are the subject matter of many philosophical arguments, and are still unresolved, so the reader must not expect to see them cleared up in this book. We must of necessity gloss over the finer points that will arise when we define what we mean by probability and set up a quantitative measure for it.

In setting up a measure for probability, that is, a way of assigning a number to the "probability" of a given event, there is no argument about the extreme values. If an event *cannot* occur under any circumstances, then we shall say that the probability of its happening is 0. If an event *must* happen we say that the probability that it will occur is 1. Thus if we take a conjurer's deck of cards in which every card is the ace of spades, and choose a card from it, the probability is 1 that we will choose the ace of spades, and 0 that we will choose, say, the queen of hearts. The probability of obtaining a "head" with a two-headed penny is 1, and so on. Events which are neither certain nor impossible will be assigned probabilities which are numbers between 0 and 1, since nothing can be more unlikely than an impossible event, or more likely than something that is bound to occur. But how are we to assign these numbers? What are we to mean, for example, by a probability of 0.65? This is where things start to get difficult.

Take a simple example, the tossing of an ordinary coin. Judging by our everyday experience we would be tempted to say that there is an equal chance of the coin landing head-up or tail-up. We might be tempted to ascribe the probability $\frac{1}{2}$ to the event of obtaining a head. Again, if we throw a die there are six possible ways in which it can land, all equally likely if the die is "true," and we would be tempted to say that the probability of its showing a particular face is $\frac{1}{6}$. This all sounds very reasonable and is leading up to a definition of probability which we shall give in a moment, but there is one serious snag. In describing this reasoning we have used the phrases "equally likely" and "equal chances" and these phrases are not defined. What is meant by saying that two events are equally likely? We cannot say that they have the same probability, since probability is what we are striving to define. We might say that we have no reason to expect a head to be more likely to turn up than a tail; but this would seemingly make the definition of "equally likely" (and hence, ultimately, of probability) depend on our opinions, which, to say the least, would be extremely odd. There might seem to be a way out in a statistical approach, saying that heads and tails are equally likely because in a long series of tosses the number of heads attained and the number of tails tend to be equal. But even this would not help much. To say that the chances of getting a head when a brand-new penny is tossed depends on what will happen if (in the future) the penny is tossed a large number of times, is far-fetched, and even if true, would not help much in discovering what the probability was.

It is all very confusing, and there is no way out of the difficulty unless we are willing to take something on trust. We shall cut the Gordian knot by assuming that we know what is meant by saying that two events are "equally likely" or "equiprobable." This will enable us to give a definition of probability which is good enough for our purposes.

Let us set up a somewhat idealistic situation, as follows. We have a set or collection U of objects of some kind, and we are going to choose one element from this set. Let A be some subset of U. Then if the elements of U are all equally likely to be chosen (this is where the assumption comes in) then we *define* the probability that the chosen element (call it x) will belong to the subset A as follows:

$$(4.10.1) \qquad Pr(x \in A) = \frac{\text{Number of elements in } A}{\text{Number of elements in } U}$$

Interpreting this less abstractly we can say that U is the set of all possible outcomes of a particular action; A is subset consisting of those outcomes which are of particular interest to us. We can call them *favorable* outcomes. The probability defined in (4.10.1) is then the probability of a favorable outcome.

Some actual examples should make this idea clear.

1. A box contains 7 black balls and 5 red balls. One ball is drawn at random. What is the probability that the ball chosen is red?

The phrase "at random" means, effectively, that the balls are equally likely to be chosen. Our set U consists of all possible choices, 12 in all. A favorable choice will be the drawing of a red ball, and there are 5 of these. Hence the required probability is $\frac{5}{12}$. Similarly, the probability of choosing a black ball is easily seen to be $\frac{7}{12}$.

2. A balanced die is thrown. What is the the probability that point showing is a 1 or a 6?
   Here there are 6 possible outcomes, all equally probable (this is the

sense of the word "balanced"). Of these two are favorable. Hence the probability is $\frac{2}{6} = \frac{1}{3}$.

3. Two dice are thrown. What is the probability of getting a total of 7?

Here the number of possible outcomes is 36, viz., any of the numbers 1 to 6 on one die, with any numbers from 1 to 6 on the second. The combinations which will produce a total of 7 are as follows:

Die 1		Die 2
1	+	6
2	+	5
3	+	4
4	+	3
5	+	2
6	+	1

Thus there are 6 favorable cases out of a total of 36. Hence the probability of throwing a 7 with two dice is $\frac{6}{36} = \frac{1}{6}$.

4. Four cards are chosen at random from a standard deck of cards. What is the probability that the 4 cards will be the 4 aces?

Here the number of possibilities is the number of ways of choosing 4 cards from a total of 52, i.e., the number of combinations of 52 objects 4 at a time. This is

$$\binom{52}{4} = \frac{52.51.50.49}{4.\ 3.\ 2.\ 1} = 270725$$

There is only one favorable outcome—the drawing of the four aces. Hence the required probability is $\frac{1}{270725}$.

5. If 5 cards are chosen from a standard deck, what is the probability that they are all of the same suit?

The number of possible choices of 5 cards is $\binom{52}{5}$. To see how many of these are favorable let us consider how many choices there are of 5 cards all of which are spades. Since there are 13 spades in the deck the number of such choices is $\binom{13}{5}$. To this we must add the equal number of choices for the other three suits. Hence the total number of favorable outcomes is $4 \times \binom{13}{5}$. Thus the required probability is

$$\frac{4 \times \binom{13}{5}}{\binom{52}{5}} = 4 \times \frac{13.12.11.10.9}{5!} \times \frac{5!}{52.51.50.49.48}$$

which reduces to $\frac{33}{16660}$.

In this chapter we shall express our estimate of the likelihood of an event as a probability, defined in the manner just given. The man in the street however often talks, not of probability, but of "chances," or of *odds*. Thus he will say " the chances of a win are 3 to 2," or that "the odds are 3 to 2 in favor of a win." These statements mean the same as "the probability of a win is $\frac{3}{5}$."

In expressing chances or odds we compare the probability of success with that of failure, multiplying these probabilities by a suitable constant to give convenient integers. In general, odds of p to q means a probability of $p/(p + q)$. A probability of $\frac{2}{27}$ could be described as odds of 2 to 25, or 4 to 50 and so on. Often the speaker avoids possible confusion by saying whether the odds are *in favor* or *against*. Thus if the probability of an event were $\frac{2}{27}$ one could say that the odds or chances were 8 to 100 in favor, or 100 to 8 against, for example. A probability of $\frac{1}{2}$ is often described as a 50–50 chance, for the same reason.

At this stage we can formulate the first of several "rules" that will help us to calculate probabilities. If, of N possible outcomes, there are n favorable ones, then there are $N - n$ unfavorable ones. The probability of a favorable outcome is

$$p = \frac{n}{N}$$

as we have seen. The probability of an unfavorable outcome is likewise

$$q = \frac{N - n}{N}.$$

Hence we have

(4.10.2) $p + q = 1.$

We express this as

RULE 1.
If the probability of an event is p, the probability of the event *not* occurring is $q = 1 - p$.

Exercise 33

1. Twelve cards are numbered from 1 to 12. Two cards are chosen at random. What is the probability that the sum of the numbers on them is 7?
2. A man writes 4 letters and addresses 4 envelopes, one for each letter. If he puts the letters into envelopes at random what are the probabilities that (a) every letter is put into the correct envelope, and (b) no letter is put into the correct envelope? Do these two probabilities sum to 1? Should they?
3. Four different pairs of shoes are mixed up in a closet. A man takes two shoes at random. What are the probabilities of his getting
    (a) two left shoes
    (b) a left shoe and a right shoe
    (c) a matching pair of shoes
4. Determine the numbers of ways of throwing totals of 2, 3, 4, . . . , 12 with two dice, and hence calculate the probabilities of getting these scores.

4.11 FURTHER RULES FOR CALCULATING PROBABILITIES

When the outcome of some event or trial is one of a finite set of equiprobable possibilities, the probability of an outcome being one of a certain subset of the possibilities can be found from the definition that was given in the last section. This will not always suffice for all situations in which we will wish to assess the probability of such and such an event, but it will do for the present. In theory one can always calculate probabilities of this sort by careful counting of the numbers of the favorable and unfavorable outcomes; in practice certain useful results often enable us to calculate probabilities without thus having to go back to first principles. We now look at some of these rules.

PROBABILITY OF OCCURRENCE OF EACH OF TWO INDEPENDENT EVENTS.

The previous examples have been concerned with the conducting of some sort of trial or experiment for which the result can be "favorable" or "unfavorable." Suppose now that we conduct *two* experiments simultaneously. What is the probability of getting a favorable result in each experiment? Let us suppose that in the first experiment there are N_1 possible outcomes, of which n_1 are favorable, and that N_2 and n_2 are the corresponding numbers for the second experiment. If p_1 and p_2 are the probabilities of a favorable outcome in the two experiments respectively, then

(4.11.1) $$p_1 = \frac{n_1}{N_1}; \quad p_2 = \frac{n_2}{N_2}.$$

We shall assume that the experiments are *independent*, that is that the result of one does not affect the outcome of the other in any way. That being the case any of the equiprobable outcomes of the first experiment can occur simultaneously with any of the equiprobable outcomes of the second, giving a total of $N_1 N_2$ possible pairs of outcomes. The "favorable outcomes" of this composite experiment will be pairs of favorable outcomes, whose number is $n_1 n_2$. Hence the probability of a favorable outcome to both experiments is, as before, the ratio of favorable to total possibilities, i.e., it is

$$\frac{n_1 n_2}{N_1 N_2} = \frac{n_1}{N_1} \cdot \frac{n_2}{N_2}$$

(4.11.2)

$$= p_1 p_2.$$

We can state this in the form of a rule, as follows:

RULE 2.

If the probability of an event A_1 is p_1, and the probability of another *independent* event A_2 is p_2 then the probability of both events occurring together is $p_1 p_2$.
A few examples should make this idea clear.

EXAMPLE 1.

Two dice are thrown. What is the probability of getting two sixes?

We could, of course, obtain the answer from first principles, by enumerating all possible pairs of faces of the dice, but it is easier to use the rule just given. The throwing of one die clearly does not affect what happens to the other, so these events are independent. The probability of throwing 6 with one die is $\frac{1}{6}$. Hence

$$p_1 = p_2 = \tfrac{1}{6}$$

and the required probability is $\frac{1}{6} \times \frac{1}{6} = \frac{1}{36}$.

EXAMPLE 2.

A card is drawn from a shuffled deck, noted, and replaced. The deck is shuffled again and another card drawn. What is the probability that both cards are aces?

Since the card first chosen is replaced, both choices are made from a complete shuffled deck, and are independent of each other. The situation is essentially the same as if the cards were drawn from different decks. The probability of drawing an ace from a deck is $\frac{4}{52} = \frac{1}{13}$. Hence the probability of drawing two aces under the above conditions is

$$\tfrac{1}{13} \times \tfrac{1}{13} = \tfrac{1}{169}.$$

Note that the situation would be quite different if the card first chosen were not replaced, for the outcome of the second draw would then depend on the outcome of the first. If the first card drawn were an ace, then there would be less chance of drawing an ace the second time, since there would be then only three of them in the pack; if the first card were not an ace, the chances of drawing an ace the second time would be slightly enhanced, since one unfavorable possibility would have been removed. We consider problems of this kind later on.

Exercise 34

1. Box A contains 3 red balls, 2 white balls and 5 green balls; box B contains 1 red ball, 4 white balls and 3 green balls. A ball is drawn at random from each box. What is the probability
   (a) that both are red
   (b) that both are white
   (c) that both are green
2. A card is drawn at random from a standard playing card deck, noted and replaced. This is repeated two more times. What is the probability
   (a) that all three cards are aces
   (b) the three cards are all hearts
   (c) the three cards are all court cards (Jacks, Queens or Kings)
3. The standard Monte Carlo roulette wheel has 37 compartments, labelled $0, 1, 2, \ldots, 36$, all equally likely to occur. What is the probability of the 12 occurring three times in succession? What is the probability of an even number occurring in each of 10 successive spins of the wheel?

4.12 PROBABILITY OF OCCURRENCE OF EITHER OF TWO EVENTS

Let us suppose that we again perform two independent experiments, but this time we ask for the probability that one or the other experiments is successful (or possibly that both are successful). The rule for finding this probability is easily derived. Let a success in the first experiment be the occurrence of one of n_1 possible outcomes from a total of N_1 outcomes so that the probability is $p_1 = n_1/N_1$, and let n_2, N_2, p_2 be similarly defined for the second experiment. Then there are clearly $N_1 N_2$ possible pairs of outcomes, exactly as before. The number of possibilities for the first experiment succeeding and the second failing is

$$n_1(N_2 - n_2)$$

since each of the n_1 successes in the first experiment may occur with each of the $N_2 - n_2$ failures in the second, the two experiments being independent. Similarly, the number of ways in which the first experiment may fail and the second succeed is $n_2(N_1 - n_1)$. Finally the number of ways in which both experiments succeed is $n_1 n_2$.

All these possibilities will now be regarded as a success for the composite experiment, thus giving us

$$n_1(N_2 - n_2) + n_2(N_1 - n_1) + n_1 n_2 = n_1 N_2 + n_2 N_1 - n_1 n_2$$

successes in all. Hence from first principles the probability of a success with one or the other, or possibly both of the experiments is

(4.12.1)
$$\frac{n_1 N_2 + n_2 N_1 - n_1 n_2}{N_1 N_2} = \frac{n_1}{N_1} + \frac{n_2}{N_2} - \frac{n_1 n_2}{N_1 N_2}$$
$$= p_1 + p_2 - p_1 p_2.$$

Thus we have the following rule:

RULE 3.

The probability of one or other (or both) of two independent events occurring is

$$p_1 + p_2 - p_1 p_2$$

where p_1, p_2 are the separate probabilities of the events.

EXAMPLE 1:

A card is drawn from each of two decks. What is the probability that at least one is a spade? The separate probabilities are

$$p_1 = p_2 = \tfrac{1}{4}.$$

Hence the required probability is $\tfrac{1}{4} + \tfrac{1}{4} - \tfrac{1}{16} = \tfrac{7}{16}$.

EXAMPLE 2:

Two dice are thrown and the faces noted. They are then thrown again. What is the probability that on at least one of the two throws the two faces are both aces? The probability of getting aces with both dice is $\tfrac{1}{6} \times \tfrac{1}{6} = \tfrac{1}{36}$, by rule 2. Hence the probability required is, by rule 3,

$$\tfrac{1}{36} + \tfrac{1}{36} - (\tfrac{1}{36})^2 = \tfrac{71}{1296}.$$

There is another way of proving the result (4.12.1). The probability of failure in experiment 1 is $1 - p_1$, by rule 1; that of failure in experiment 2 is $1 - p_2$. Hence, by rule 2, the probability of failing in *both* experiments is $(1 - p_1)(1 - p_2)$. Applying rule 1 again we see that the probability of *not* failing in both experiments, i.e., of succeeding in at least one, is

$$1 - (1 - p_1)(1 - p_2) = p_1 + p_2 - p_1 p_2$$

which is the desired result.

The reader may notice here a resemblance to the sort of manipulations we performed with statements in Chapter 1. We have just asserted that getting a success with one or other of two experiments is the same as not getting a failure with both. This is the logical assertion.

$$\mathbf{p} \vee \mathbf{q} \equiv \sim(\sim\mathbf{p} \mathbin{\&} \sim\mathbf{q})$$

which we have met before.

This resemblance can be brought out by means of a useful notation. Suppose we let $Pr(\mathbf{p})$ mean "the probability that the statement $\mathbf{p}$ is true". Then the three rules we have had so far can be stated as

Rule 1. $Pr(\sim\mathbf{p}) = 1 - Pr(\mathbf{p})$
Rule 2. $Pr(\mathbf{p}_1 \mathbin{\&} \mathbf{p}_2) = Pr(\mathbf{p}_1) \cdot Pr(\mathbf{p}_2)$
Rule 3. $Pr(\mathbf{p}_1 \vee \mathbf{p}_2) = Pr(\mathbf{p}_1) + Pr(\mathbf{p}_2) - Pr(\mathbf{p}_1) \cdot Pr(\mathbf{p}_2)$

provided the events referred to by $\mathbf{p}_1$ and $\mathbf{p}_2$ are independent.

Exercise 35

1.   Obtain an expression for the probability

$$Pr(\mathbf{p}_1 \vee \mathbf{p}_2 \vee \cdots \vee \mathbf{p}_k)$$

in terms of the probabilities

$$Pr(\mathbf{p}_1), Pr(\mathbf{p}_2), \ldots, Pr(\mathbf{p}_k)$$

given that the events referred to are independent.

2.   Two dice are thrown. What is the probability that at least one is a six? What is the probability of getting at least one six when n dice are thrown?

3. Assuming that children are born in equal numbers of boys and girls, what is the probability of a family of four children consisting of

    (a) all boys

    (b) at least one boy

4.13 INCOMPATIBLE EVENTS

It frequently happens that we wish to consider two events that are incompatible, that is, they cannot occur at the same time. When this is the case the probability of one or the other occurring is given by a different rule. We can divide the possible outcomes of the double experiment into three sets

    1. Those which give a success for experiment 1,

    2. Those which give a success for experiment 2,

    3. Those for which both experiments fail.

The set of outcomes where both experiments succeed is empty, by our supposition of incompatability. Let the numbers in the three sets be n_1, n_2 and n_3 so that $n_1 + n_2 + n_3 = N$ is the total number of possible outcomes. Then we have

$$p_1 = \frac{n_1}{N} \quad \text{and} \quad p_2 = \frac{n_2}{N}$$

as the probabilities of success of the first and second experiments. Hence the probability of one or other experiment succeeding is

(4.13.1) $$\frac{n_1 + n_2}{N} = p_1 + p_2$$

This gives us

RULE 4:

If two incompatible events have probabilities p_1 and p_2 of occurring, then the probability that one or other of them will occur is $p_1 + p_2$.

In terms of statements this rule can be expressed as

$$Pr(\mathbf{p}_1 \vee \mathbf{p}_2) = Pr(\mathbf{p}_1) + Pr(\mathbf{p}_2)$$

if $\mathbf{p}_1$ and $\mathbf{p}_2$ are incompatible.

EXAMPLE 1:

A die is loaded so that the probabilities of the various faces turning up are as follows

Face	1	2	3	4	5	6
Prob.	0.20	0.16	0.18	0.19	0.17	0.10

What is the probability of getting (1) a 5 or a 6, (2) an odd number?

1. By rule 4 the probability is $0.17 + 0.10 = 0.27$. (The events are incompatible—we cannot throw 5 *and* 6.)
2. Here we want the probability of a 1 or a 3 or a 5. The probability of 1 or 3 is

$$0.20 + 0.18 = 0.38$$

by rule 4. The probability of getting this or the (incompatible) alternative of a 5 is

$$0.38 + 0.17 = 0.55$$

also by rule 4. Thus the required probability is 0.55.

This example shows that we can extend rule 4 to cover any number of mutually incompatible events. Symbolically we can write

$$Pr(\mathbf{p}_1 \lor \mathbf{p}_2 \lor \mathbf{p}_3) = Pr[(\mathbf{p}_1 \lor \mathbf{p}_2) \lor \mathbf{p}_3]$$
$$= Pr(\mathbf{p}_1 \lor \mathbf{p}_2) + Pr(\mathbf{p}_3) \text{ by rule 4}$$
$$= Pr(\mathbf{p}_1) + Pr(\mathbf{p}_2) + Pr(\mathbf{p}_3) \text{ by rule 4.}$$

The extension to larger numbers of statements (or of events) is obvious.

EXAMPLE 2:

A box (box *A*) contains 5 red counters and 3 black counters; another box (box *B*) contains 4 red counters and 2 black counters. If three counters are drawn at random from each box, what is the probability that 4 red counters and 2 black counters are drawn altogether? The possible outcomes are as follows:

	Box A	*Box B*
1.	3 red, 0 black	1 red, 2 black
2.	2 red, 1 black	2 red, 1 black
3.	1 red, 2 black	3 red, 0 black

Now there are $\binom{5}{3} = 10$ ways of drawing 3 red counters from box A. Hence

the probability of this is $10 \Big/ \binom{8}{3} = \frac{5}{28}$. The probability of 1 red and 2 black

from box B is the number of possibilities (which is 4) divided by $\binom{6}{3} = 20$,

i.e., it is $\frac{1}{5}$.

The number of ways of drawing 2 red and 1 black from box A is

$\binom{5}{2}\binom{3}{1} = 30$. Hence the probability of this drawing is $30 \Big/ \binom{8}{3} = \frac{15}{28}$. The

probability of 2 red and 1 black from box B is $\binom{4}{2}\binom{2}{1} = 12$ divided by $\binom{6}{3}$

$= 20$, i.e., it is $\frac{3}{5}$.

The probability of 1 red and 2 black from box A is $\binom{5}{1}\binom{3}{2}$ divided by

$\binom{8}{3}$, i.e., $\frac{15}{56}$. The probability of 3 red and no black counters from box

B is $\binom{4}{3}\binom{2}{0}$ divided by 20, i.e., it is $\frac{1}{5}$. Thus

the probability of outcome (1) is $\frac{5}{28} \cdot \frac{1}{5} = \frac{1}{28}$;

the probability of outcome (2) is $\frac{15}{28} \cdot \frac{3}{5} = \frac{9}{28}$;

and   the probability of outcome (3) is $\frac{15}{56} \cdot \frac{1}{5} = \frac{3}{56}$.

The required probability is the sum of these numbers, viz., $\frac{1}{28} + \frac{9}{28} + \frac{3}{56} = \frac{23}{56}$.

Exercise 36

1. Under the conditions of Exercise 34.1, viz.,

	Red balls	White balls	Green balls
Box A	3	2	5
Box B	1	4	3

what is the probability that the two balls (drawn one from each box) will be of different colors?

2. One card is drawn from each of two standard decks. What is the probability that at least one of the cards is a spade? What is the probability that one is a spade and the other a heart?

3. A pair of dice is thrown twice. What is the probability that at least one of the throws will be a 7?

4. What is the probability that a family of 5 children will consist of 3 boys and 2 girls?

4.14 PROBABILITY IN TERMS OF SETS—CONDITIONAL PROBABILITY

We defined probability in terms of the ratio of favorable outcomes to the total number of outcomes. The set A of favorable outcomes is a subset of the set U of all outcomes. If $v(A)$ denotes the number of elements in the set A, then we can write

$$Pr(A) = \frac{v(A)}{v(U)}$$

for the probability that an outcome belongs to A, i.e., is favorable. As before we assume all outcomes equally probable.

By use of this set-theoretical notation we can formulate our rules in yet another way. Rule 1 becomes

(4.14.1) $Pr(U - A) = 1 - Pr(A).$

Rule 2 becomes

(4.14.2) $Pr(A_1 \cap A_2) = Pr(A_1) \cdot Pr(A_2).$

Rule 3 becomes

(4.14.3) $Pr(A_1 \cup A_2) = Pr(A_1) + Pr(A_2) - Pr(A_1) \cdot Pr(A_2).$

Rule 4 becomes

(4.14.4) $Pr(A_1 \cup A_2) = Pr(A_1) + Pr(A_2)$ if $A_1 \cap A_2 = \emptyset.$

Note that we can combine rules 3 and 4 into a more general rule. If A_1 and A_2 are two general sets then

(4.14.5) $v(A_1 \cup A_2) = v(A_1) + v(A_2) - v(A_1 \cap A_2)$

since in adding $v(A_1)$ and $v(A_2)$ we count twice all elements of $A_1 \cap A_2$. Thus by subtracting $v(A_1 \cap A_2)$ we get the number of elements of $A_1 \cup A_2$ each counted once (see Fig. 4.1). From (4.14.3) we have

$$\frac{v(A_1 \cup A_2)}{v(U)} = \frac{v(A_1)}{v(U)} + \frac{v(A_2)}{v(U)} - \frac{v(A_1 \cap A_2)}{v(U)}$$

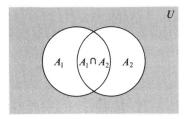

Fig. 4.1

or

(4.14.6) $Pr(A_1 \cup A_2) = Pr(A_1) + Pr(A_2) - Pr(A_1 \cap A_2).$

This equation together with rule 2—in the form (4.14.2)—gives rule 3 (4.14.3); while if $A_1 \cap A_2 = \emptyset$ we obtain rule 4 (4.14.4), since clearly $Pr(\emptyset) = 0$.

Suppose U is, as above, a universal set of equiprobable possibilities. Let A be a subset of U, and B a subset of A (see Fig. 4.2).

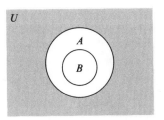

Fig. 4.2

The probability of an outcome in B (i.e., one of the possibilities represented by the elements of B) is clearly

(4.14.7) $Pr(B) = \dfrac{v(B)}{v(U)}$

Suppose we now restrict ourselves to the set A, treating this now as the universal set. Then the probability of an outcome in B is now

$$\dfrac{v(B)}{v(A)}$$

This is the probability of an outcome in B when we already know that the choice is restricted to A. We can call this briefly the "probability of B given A" and write it

(4.14.8)
$$Pr(B \mid A) = \frac{v(B)}{v(A)}.$$

From (4.14.7) and (4.14.8) it follows that, since

$$\frac{v(B)}{v(U)} = \frac{v(A)}{v(U)} \cdot \frac{v(B)}{v(A)}$$

therefore

(4.14.9)
$$Pr(B) = Pr(A) \cdot Pr(B \mid A).$$

If B is not a subset of A, and if we are told that the choice is restricted to A then we can choose an element of B only by choosing an element of $A \cap B$. Equation (4.14.9) then becomes

(4.14.10)
$$Pr(A \cap B) = Pr(A) \cdot Pr(B \mid A).$$

The probability $Pr(B \mid A)$ is known as a conditional probability. It is the probability of a certain event when it is already known that some other event has already happened. Conditional probabilities can be found from (4.14.10) in the form

(4.14.11)
$$Pr(B \mid A) = \frac{Pr(A \cap B)}{Pr(A)}.$$

In terms of statements this reads

(4.14.12)
$$Pr(\mathbf{p} \mid \mathbf{q}) = \frac{Pr(\mathbf{p} \ \& \ \mathbf{q})}{Pr(\mathbf{q})}.$$

We shall take the statement analogue of (4.14.10) as our next rule but interchange $\mathbf{p}$ and $\mathbf{q}$. Hence

RULE 5.

$$Pr(\mathbf{p} \ \& \ \mathbf{q}) = Pr(\mathbf{p}) \cdot Pr(\mathbf{q} \mid \mathbf{p}).$$

Although the above equations define what conditional probability is, they may not readily convey to the reader the "feel" of this concept. The following examples and exercises, however, should suffice to get the idea across.

EXAMPLE 1:

Two cards are drawn at random from a deck, the first card being retained and not replaced. What is the probability that at least one of the cards is a spade?

There are two mutually exclusive possibilities for the first card. Either it is a spade or it is not. If it is a spade (the probability of which is $\frac{1}{4}$) then it does not matter what the next card is, the experiment is a success. If the first card is not a spade (probability $\frac{3}{4}$) then we need to know the chances of getting a spade on the second draw. This is a conditional probability, since it depends on the fact that a card (not a spade) has been removed from the deck. The deck now contains 51 cards of which 13 are spades. and the probability of a spade is thus $\frac{13}{51}$.

The probability we are looking for is that of drawing a nonspade and then a spade. From rule 5 we have

$$Pr(\mathbf{p}) = \text{probability of a nonspade} = \tfrac{3}{4}$$
$$Pr(\mathbf{q}\,|\,\mathbf{p}) = \text{probability of a spade, given that a}$$
$$\text{nonspade has been drawn} = \tfrac{13}{51},$$

and hence $Pr(\mathbf{p}\ \&\ \mathbf{q}) = $ probability of a nonspade followed by a spade $= \frac{3}{4} \times \frac{13}{51} = \frac{13}{68}$.

Thus the probability of succeeding by getting a spade first is $\frac{1}{4}$; that of succeeding on the second draw is $\frac{13}{68}$. These events are incompatible. Hence by rule 4, the total probability of success is

$$\tfrac{1}{4} + \tfrac{13}{68} = \tfrac{15}{34}.$$

Compare this with the probability of at least one spade when the first card is replaced. This was found above to be $\frac{7}{16}$, which is slightly less than $\frac{15}{34}$. This is to be expected since replacing the first nonspade decreases slightly the chances of a spade on the second draw.

We could, of course, have obtained this result from first principles. There are 52×51 ways of drawing two cards from a deck without replacement, and of these 13×39 are of the form (spade, nonspade), an equal number are (nonspade, spade), and 13×12 are of the form (spade, spade). The total number of favorable outcomes is $13(39 + 39 + 12) = 13 \times 90$ and the probability of a favorable outcome is thus

$$\frac{13 \times 90}{52 \times 51} = \frac{15}{34}.$$

As problems become more complex, the appeal to first principles becomes less and less attractive, while the use of the rules continues to be a practical method. This is illustrated by the next example, where it would be extremely tedious to work from first principles.

EXAMPLE 2:

A card is drawn from a deck. If it is a spade the experiment is a success. If it is a heart it is put back in the deck; if it is a diamond or a club it is left out. This is repeated for two more draws. Only if none of the three draws yields a spade does the experiment fail. What is the probability of success?

On the first draw there are three mutually exclusive possibilities—a spade, a heart, or neither of these. The probabilities are $\frac{1}{4}, \frac{1}{4}, \frac{1}{2}$. If the card is a heart, then two more draws are made starting with a complete pack. If the card is neither a spade nor a heart then two more draws are made with a deck having 13 spades, 13 hearts and 25 other cards.

Let x be the probability of a spade in two draws starting with a complete deck, and y the similar probability starting with the shortened deck. These are the conditional probabilities. The probabilities of success under the mutually exclusive possibilities for the first draw are thus

$$\tfrac{1}{4}, \tfrac{1}{4}x \quad \text{and} \quad \tfrac{1}{2}y$$

respectively. Hence the required probability is

$$p = \tfrac{1}{4} + \tfrac{1}{4}x + \tfrac{1}{2}y.$$

We now have to find x and y.

To find x we proceed exactly as above; we find that

$$x = \tfrac{1}{4} + \tfrac{1}{4}u + \tfrac{1}{2}v$$

where u is the probability of a spade in 1 draw from a full deck, and v is the similar probability starting with a deck from which either a diamond or a club has been removed. Clearly $u = \tfrac{1}{4}$ and $v = \tfrac{13}{51}$. Hence

$$x = \tfrac{1}{4} + \tfrac{1}{16} + \tfrac{13}{102} = \tfrac{359}{816}$$

The calculation for y is similar but not quite the same. The probabilities for the first card are now

$$\text{Spade} = \tfrac{13}{51}, \qquad \text{Heart} = \tfrac{13}{51}, \qquad \text{Other} = \tfrac{25}{51}.$$

Proceeding as before we find that

$$y = \tfrac{13}{51} + \left(\tfrac{13}{51}\right)^2 + \tfrac{25}{51} \cdot \tfrac{13}{50}$$
$$= \tfrac{2327}{5202}.$$

We now have all that is necessary to find p. It is

$$p = \tfrac{1}{4} + \tfrac{1}{4} \cdot \tfrac{359}{816} + \tfrac{1}{2} \cdot \tfrac{2327}{5202}$$
$$= \tfrac{97157}{166464} = 0.585 \text{ approximately.}$$

In problems of this type it is often convenient to draw a diagram showing how the various choices depend on those that have gone before. A suitable diagram for the problem is shown in Fig. 4.3.

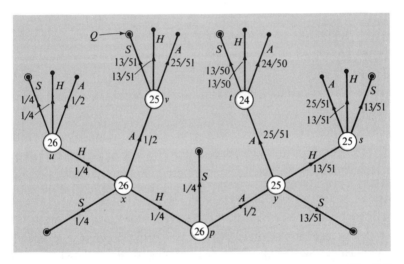

Fig. 4.3

The various possible stages are denoted by circles "○." For convenience a number has been written in each circle, namely the number of diamonds and clubs in the deck at that stage (the spades and hearts are always all present since a heart, if drawn, is replaced, and the drawing of a spade concludes the experiment). From the lowest circle, which we can call the *root*, and which represents the initial deck, there are three arrows, representing the various ways in which the experiment may proceed. The drawing of a spade (*S*) leads to a successful termination of the experiment (denoted by "⊙"). Drawing a heart (*H*) or another suit (*A*) leads us to another circle, the one labelled "*x*" or the one labelled "*y*," respectively. From these we move on to other circles, according to the outcomes of the successive draws. Each arrow leading from one circle to another is allotted a letter (*S, H* or *A*) according to the suit to whose drawing it corresponds, and a number which is the probability of drawing that suit. Unsuccessful terminations of the experiment are denoted by dots.

A diagram such as Fig. 4.3 contains all the information we need to solve the problem. The circles have been labelled with letters denoting the probability of final success assuming that you have reached as far as that circle. They are therefore conditional probabilities and the letters have been chosen to correspond to those used when we calculated the probability earlier. If we have reached as far as the circle marked "u" the present situation can be represented by Fig. 4.4.

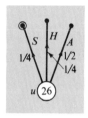

Clearly, from this the probability of a successful outcome is $\frac{1}{4}$. Hence $u = \frac{1}{4}$. Similarly we find

$$v = \tfrac{13}{51}; \qquad s = \tfrac{13}{51}; \qquad t = \tfrac{13}{50}$$

as before.

Now look at the circle marked "x." From it there is probability $\frac{1}{4}$ of an immediate success; probability $\frac{1}{4}$ of going to "u," at which stage the probability of success is again $\frac{1}{4}$; and probability $\frac{1}{2}$ of going to "v" at which stage the probability of success is $\frac{13}{51}$. This gives us the equation

$$x = \tfrac{1}{4} + \tfrac{1}{4} \cdot \tfrac{1}{4} + \tfrac{1}{2} \cdot \tfrac{13}{51}$$

exactly as before. In the same way the equations for y and p can be written down just by looking at the diagram.

The diagram can be used in other ways. In Fig. 4.3 one of the successful outcomes has been labelled Q. What is the probability of a success in the particular manner specified by this outcome? There is only one way of getting from the root to Q, and it corresponds to the following sequence of events.

1. A heart is drawn and replaced.
2. A club or diamond is drawn and not replaced.
3. A spade is drawn.

We are therefore asking for the probability of this particular sequence of draws. To find this we have only to multiply the probabilities associated with the three arrows leading from the root to Q. Hence the probability is

$$\tfrac{1}{4} \cdot \tfrac{1}{2} \cdot \tfrac{13}{51} = \tfrac{13}{408}.$$

Note that the events of reaching the various successful terminations of the experiment are mutually incompatible; hence the probability of reaching *some* successful termination is the sum of the probabilities of the separate terminations. Hence another way of finding the value of p is to take each successful termination in turn, multiply the numbers associated with the arrows leading from the root to that termination (as for the termination Q given above) and add the results. In this way we obtain

$$p = \tfrac{1}{4} \cdot \tfrac{1}{4} + \tfrac{1}{4} \cdot \tfrac{1}{4} \cdot \tfrac{1}{4} + \tfrac{1}{4} \cdot \tfrac{1}{2} \cdot \tfrac{13}{51} + \tfrac{1}{4} + \tfrac{1}{2} \cdot \tfrac{25}{51} \cdot \tfrac{13}{50} + \tfrac{1}{2} \cdot \tfrac{13}{51} \cdot \tfrac{13}{51} + \tfrac{1}{2} \cdot \tfrac{13}{51}$$

which reduces to the same number as before—$\tfrac{97157}{166464}$.

A diagram like Fig. 4.3 is called (for obvious reasons) a *tree*.

EXAMPLE 3:

For our next example we take the well known dice game of "craps." In this game one player, the banker, throws the two dice. If the faces showing total to 7 or 11 he wins; if they total 2, 3 or 12 he loses. If he throws some other total (called his "point") he must then keep throwing the two dice until he again throws his point (in which case he wins) or until he throws a 7 (in which case he loses, and must hand over the bank to another player).

Our ultimate question will be "what is the probability that the banker will win?" but we shall approach it in easy stages.

Let us first ask "What is the probability that in a succession of throws of two dice we shall throw a total of 4 before throwing a total of 7?" There are two ways of approaching this question.

The first method is to consider the various ways that we can get a 4 before a 7. If x means a number between 2 and 12, inclusive, that is not 4 or 7, then the possibilities are

1. Throw 4
2. Throw x then throw 4
3. Throw x, throw x, then throw 4
    etc.

The probability of throwing 4 is $\tfrac{3}{36} = \tfrac{1}{12}$ since (as we saw in Exercise 33.4) there are 3 ways of obtaining this total. There are 27 ways of getting a total which is not 4 or 7 so that the probability of getting x is $\tfrac{3}{4}$. Hence the probabilities of the above events (1), (2), (3), etc., are the terms in

$$\tfrac{1}{12} + \tfrac{3}{4} \cdot \tfrac{1}{12} + (\tfrac{3}{4})^2 \cdot \tfrac{1}{12} + (\tfrac{3}{4})^3 \cdot \tfrac{1}{12} + \cdots$$

We have ended up with what is known as an *infinite series*. The reader may well know already how to handle such series, but we shall not assume that he does. Accordingly we abandon this line of attack, though we shall hark back to it in Chapter 8.

The preceding method was a brute force, hammer-and-tongs approach to the problem. By being a little more subtle we can solve the problem with ease. The crucial point is that the throwing of any number other than 4 or 7 is in fact irrelevant. Such a throw is not a possible outcome of the experiment because a further throw is mandatory if a number other than 4 or 7 is thrown. Hence the only possible outcomes of the experiment are the 3 ways of throwing a 4 and the 6 ways of throwing a 7 — 9 possibilities in all. Of these, 3 are favorable. Hence the required probability is

$$p_4 = \tfrac{3}{9} = \tfrac{1}{3}.$$

In the same way we see that if the banker's point is 5, then there are 10 possible outcomes to this part of the game, 4 of which are favorable. The probability of the banker making his 5 is thus

$$p_5 = \tfrac{2}{5}.$$

Similarly, if the point is 6 the probability is

$$p_6 = \tfrac{5}{11}.$$

The reader should verify this, and also the fact that, using an obvious notation

$$p_8 = \tfrac{5}{11}, \quad p_9 = \tfrac{2}{5}, \quad \text{and} \quad p_{10} = \tfrac{1}{3}.$$

We can now answer our original question "What is the probability of the banker winning?" To do this we can draw a tree diagram as for the last example. It will look something like Fig. 4.5.

If we now calculate the probabilities of getting from the root (*START*) to each of the winning circles Ⓦ and add the results, we obtain the required probability. It is

$$p = \tfrac{1}{12} \cdot \tfrac{1}{3} + \tfrac{1}{9} \cdot \tfrac{2}{5} + \tfrac{5}{36} \cdot \tfrac{5}{11} + \tfrac{5}{36} \cdot \tfrac{5}{11} + \tfrac{1}{9} \cdot \tfrac{2}{5} + \tfrac{1}{12} \cdot \tfrac{1}{3} + \tfrac{2}{9}$$

$$= \tfrac{976}{1980} = 0.493 \text{ approximately.}$$

Thus the banker has somewhat less than an even chance of winning—a result worth remembering if you play this particular game!

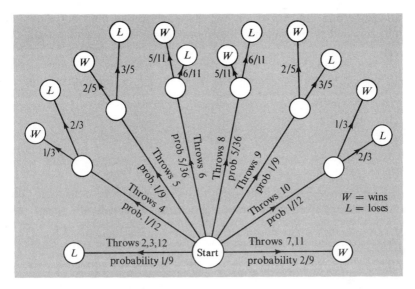

Fig. 4.5

Exercise 37

1. In a certain country the total supply of cigarette lighters is provided by four factories A, B, C and D, which account for 40 percent, 30 percent, 25 percent, and 5 percent of the total production respectively. Factory A's output contains 5 percent of defective lighters, and the corresponding figures for factories, B, C and D are 10 percent, 5 percent and 20 percent.

Find the probability that a lighter bought at random will be defective

(a) by drawing a tree diagram and using conditional probabilities,

(b) from first principles.

2. Box A contains 2 red balls and 3 green balls; box B contains 3 red and 5 green balls. A ball is chosen from box A and replaced. Next a ball is chosen from box B, and is replaced if it is the same color as that originally chosen from box A, and is put into box A if it is different. Finally a ball is drawn from box A, and is replaced in box A or box B according as it is the same color or the opposite color to the second ball chosen.

What is the probability that after these three operations box A contains exactly 3 green balls?

3. Two integers are chosen at random from the set

$$0, \ 1, \ 2, \ 3, \ 4, \ 5, \ 6, \ 7$$

these integers being equally likely to be chosen. What is the probability that the two integers differ by 3 or less?

4. Three dice are thrown on a table. Any die that does not show a six is taken up and thrown again. Once more any die not showing a six is thrown again. Draw a tree diagram that will enable you to calculate the probability that the above procedure will, at some stage, give 3 sixes. Find an expression for this probability, but do not evaluate it (unless you want to).

4.16 EXPECTATION

Let us imagine that you are fortunate enough to have the opportunity of playing games of coin tossing with a generous opponent who insists on paying you $8 every time you throw a head and collecting only $2 from you if you throw a tail. Under these circumstances you will clearly stand to gain money—but how much?

If you toss a coin N times then, since the probability of a head is $\frac{1}{2}$, you could reasonably expect to get a head approximately $\frac{1}{2}N$ times. Each time you do so you receive $8 so that you will collect approximately $\frac{1}{2}N \times 8$ dollars altogether. On the other hand you will throw approximately $\frac{1}{2}N$ tails, and each time lose $2—a total loss of about $\frac{1}{2}N \times 2$ dollars. Hence after N throws you will end up approximately

$$\tfrac{1}{2}N \times 8 - \tfrac{1}{2}N \times 2 = 3N$$

dollars to the good. Now if you gain approximately $3N$ dollars over N throws, this is much the same as if you had gained about $3 with each throw. Of course, the rules of this game do not permit you to win exactly $3 in any one throw—you either gain $8 or lose $2. The "$3 per throw" represents a sort of average —on the average you expect to be about $3 to the good on every throw.

Suppose now that you are throwing a die and that your opponent pays you $3 if you throw a 1 or a 6, but you pay him $3 if you throw any other number. What do you stand to gain or lose on the average? In a large number N of throws you will win roughly $\frac{1}{3}N$ times for a total gain of $3 \times \frac{1}{3}N = N$ dollars; on the other hand you will lose roughly $\frac{2}{3}N$ times for a loss of $3 \times \frac{2}{3}N = 2N$ dollars. Hence you expect to win $N - 2N = -N$ dollars, that is, you expect to lose N dollars. This is an average *loss* of $1 per throw.

These two examples serve to introduce an important concept, that of *expectation*, for which we now give a general definition. Suppose we conduct an experiment (such as tossing a coin, choosing a card, or anything which has a finite number k of possible outcomes) and that the probabilities of the k outcomes are

$$p_1, p_2, p_3, \ldots, p_k.$$

Naturally $p_1 + p_2 + \ldots + p_k = 1$. Suppose further that with each outcome there is associated a "payoff," a number indicating an amount of money (or other items of value) that you receive if the experiment yields this particular outcome. These numbers may be negative, meaning that you have to pay out if the experiment gives the appropriate outcome. These payoffs will be denoted by

$$Q_1, Q_2, Q_3, \ldots, Q_k.$$

Then the expected gain from the experiment, or the *expectation*, is defined to be

(4.16.1) $$E = p_1 Q_1 + p_2 Q_2 + \ldots + p_k Q_k.$$

It is easy to verify that this expectation E is the average amount that one could expect to win per trial if the experiment is repeated over and over again for many trials. For in N trials the ith outcome will occur approximately $p_i N$ times giving a gain of $p_i N Q_i$. Thus the total gain is

(4.16.2) $$\sum_{i=1}^{k} p_i N Q_i$$

over the N trials, or

(4.16.3) $$E = \sum_{i=1}^{k} p_i Q_i$$

per trial.

To illustrate the concept of expectation let us consider the game of roulette. For a reason that will appear later we shall first consider a version of roulette that is not actually played anywhere. In this hypothetical game the roulette wheel has 36 compartments numbered 1 through 36. The wheel is spun and a small ball eventually settles in one of the numbered compartments as the wheel comes to a stop. The players bet sums of money that a particular number will thus "come up" or that one of a certain set of numbers will come up. For example, a player may bet that an even number will come up, or an odd number, or a red number, or a black number (half the numbers are red, half are black). These are all even chances—the probability of the player being right is $\frac{1}{2}$. On the other hand, bets can be placed on the number being 1 through 12, or 13 through 24, or 25 through 36, for each of which the probability is $\frac{1}{3}$; or a player may bet on a set of four numbers, in which case he has a $\frac{4}{36} = \frac{1}{9}$ probability of winning. Finally he may bet on a single number, which gives him a probability $\frac{1}{36}$ of winning. Other combinations of numbers are possible, but the general pattern is the same; if the player bets on m numbers his probability of winning is $m/36$.

Let us postulate a philanthropic casino whose aim is that every player of this game should "break even" in the end; that is, that although there may be gains and losses, in the long run a player will go away with about as much money as he had when he came in. What payoffs would the management have to pay for the various bets to achieve this?

Suppose a player consistently backs single numbers. In N plays of the game he will win $\frac{1}{36}N$ times and receive the payoff Q, whatever it is. This gives a gain of $\frac{1}{36}NQ$. On $\frac{35}{36}N$ plays he will lose his stake, of amount say x. Thus his total losses are $\frac{35}{36}Nx$. We wish his net gains to be zero, and hence we have

(4.16.4)
$$\frac{1}{36}NQ - \frac{35}{36}Nx = 0$$

or

$$Q = 35x.$$

Thus for every dollar the player gives up if he loses, he receives 35 dollars if he wins. What would actually happen is that he would place a chip for a dollar on the table when play begins; if he loses, this is taken up by the croupier; if he wins it is returned to him together with 35 similar chips.

By comparing (4.16.3) with (4.16.4) we see that the aim of the philanthropic casino is to make the player's expectation zero. Thus if a player bets x dollars on a set of m numbers, the number Q of dollars he should win is given by

$$\frac{m}{36}Q - \frac{36 - m}{36}x = 0$$

or
$$Q = \frac{36 - m}{m}x$$

$$= \frac{36}{m}x - x.$$

Thus if he wins he is given back $36x/m$ dollars, x dollars of which was his own money which he put on the table to indicate his bet. If he bet on red, by putting say $10 on the appropriate section of the table, he would receive back on winning a total of $20 ($m = 18$ so $36/m = 2$)—his own $10 stake and another $10 winning ($36x/m - x = 10$ here).

One point that arises here is that a player who stakes x dollars, shall we say, on 12 numbers simultaneously, as is permitted by the rules of the game, could just as well stake $\frac{1}{12}x$ dollars on each of the 12 numbers separately. The reader should verify that the expectation is zero just the same. Such a procedure would make life difficult for the croupier, but makes no difference to the analysis. Hence we do not lose anything by assuming that the players bet only on single numbers. We can even assume that they play always on the same number; it really makes no difference!

Clearly a casino operating the game just described would soon go bankrupt. A casino must make enough money to pay its staff and cover expenses, and *its* expectation at the game must therefore be positive, which means that the player's expectation must be negative. This is achieved at Monte Carlo, for example, by using a roulette wheel having 37 compartments, the additional compartment (colored green) being numbered "0". Payoffs on the nonzero compartments are made *as if* the wheel had only 36 compartments. (There are some variations when a zero comes up, which we will not go into.) Thus a player betting \$1 on a number is paid \$35 (in addition to receiving back his stake) if he wins. Since the probability of winning is now only $\frac{1}{37}$, his expectation is therefore

$$\frac{1}{37} \cdot 35 - \frac{36}{37} \cdot 1 = -\frac{1}{37}.$$

Thus on the average the bank wins $\frac{1}{37}$ of the money staked. It is this which makes the running of casinos a profitable concern.

In the western hemisphere where everything (including the extraction of money from gamblers) tends to be carried on at a faster rate, roulette wheels have *two* extra compartments (0 and 00). Some even have three! A player's expectation per dollar staked is then

$$\frac{1}{38} \cdot 35 - \frac{37}{38} \cdot 1 = -\frac{1}{19} \text{ dollars}$$

and he is thus relieved of his money in a shorter time.

As we observed earlier, this expectation does not depend on how the bets are placed, and it follows that no "system" of placing of bets, however plausible, can alter in any way the fact that the European roulette player loses, on the average, $\frac{1}{37}$ of every stake placed (the American player loses $\frac{1}{19}$, or more). The only sure-fire way of winning at roulette is to own the casino!*

We have treated this example at some length, but further examples can now be dealt with quite quickly. Suppose you are the banker at a game of craps, and you bet even money that you will pass, that is, win in the manner described in section 4.15. The probability of this is 0.493, near enough. Hence your expectation is

$$0.493Q - 0.507x = -0.014$$

* Provided you have a sufficiently large bankroll. (See the next section.)

since $Q = x = 1$ for an even money bet. Clearly you are not going to make a fortune this way, though you are better off than at roulette (even European style).

We have tended to regard an expectation as if it were a sum of money. This is often the case, but is not necessarily so, as the following example shows.

Suppose we toss a coin, and keep tossing it until we get a head. How many times are we likely to have to toss it? More mathematically, what is the expected number of tosses we shall need to make?

Let us ask first what the probability is that we shall have to make exactly k tosses to get a head. This is the probability of getting $k - 1$ tails followed by a head, and is clearly $(\frac{1}{2})^k$. Hence we have probability

$(\frac{1}{2})$ that the number of tosses is 1
$(\frac{1}{2})^2$ that it is 2
$(\frac{1}{2})^3$ that it is 3

and so on. Hence by (4.14.1) the expected number of tosses is

$$E = \tfrac{1}{2} \times 1 + (\tfrac{1}{2})^2 \times 2 + (\tfrac{1}{2})^3 \times 3 + \ldots$$

with the difference that these terms continue indefinitely. As happened earlier in this chapter, we have ended up with an infinite series, and for the same reason as before we shall fight shy of tackling the problem in this form but look for a subterfuge for avoiding it. This is not difficult.

Let E be the expected number of tosses. There is probability $\frac{1}{2}$ of success on the first toss. If we get a tail on the first toss we are effectively back where we started except that we have clocked up 1 toss already. Looking ahead we expect to have to make E more tosses before getting a head. Hence the situation is that there is

probability $\frac{1}{2}$ that the number of tosses is 1
and   probability $\frac{1}{2}$ that it is $1 + E$.

Hence by (4.14.1) the expected number of tosses is

$$\tfrac{1}{2} \times 1 + \tfrac{1}{2} \times (1 + E)$$

which gives us the equation

(4.16.5) $E = \tfrac{1}{2} + \tfrac{1}{2}(1 + E).$

Solving this for E we find that $E = 2$.

Notice how we have made equation (4.16.5) turn round and bite its own tail, so to speak, by using the (as yet) unknown expectation E to obtain an expression for E, and solving the resulting equation. This is a very useful dodge for finding expectations relating to series of events such as this, especially when the individual probabilities are difficult to obtain.

A similar example is to ask for the expected number of throws of a die before a six is obtained. If this expected number is E, then on the first throw there is a probability of $\frac{1}{6}$ that one throw is required, and a probability $\frac{5}{6}$ that, the first throw being unsuccessful, $1 + E$ throws are required. Hence as with (4.16.5) we have

$$E = \tfrac{1}{6} + \tfrac{5}{6}(1 + E)$$

whence $E = 6$. On the average we would expect to have to throw the die 6 times to make a six.

A more elaborate example is the following.* In the game of craps what is the expected number of throws that the banker will make before he loses by failing to make his point? (When this happens the banker relinquishes the dice. Under all other circumstances he retains them.)

Let us call the number u. The possibilities for the first throw are

a. 4 or 10. Let the expected number of subsequent throws be v_1.
b. 5 or 9. Let the expected number of subsequent throws be v_2.
c. 6 or 8. Let the expected number of subsequent throws be v_3.
d. 2, 3, 7, 11 or 12. In this case the game starts again with the same banker, so that the expected number of subsequent throws is u.

Note that we are not interested here in whether the banker wins or loses, merely in how many throws he is likely to make before the dice pass to another player.

Hence the expected number of throws after the first is

$$\tfrac{1}{6}v_1 + \tfrac{2}{9}v_2 + \tfrac{5}{18}v_3 + \tfrac{1}{3}u$$

since $\frac{1}{6}, \frac{2}{9}, \frac{5}{18}, \frac{1}{3}$ are the probabilities of the four occurrences. Thus we have

(4.16.6) $$u = 1 + \tfrac{1}{6}v_1 + \tfrac{2}{9}v_2 + \tfrac{5}{18}v_3 + \tfrac{1}{3}u.$$

Now consider v_1. The player is trying to throw a 4 (or a 10). On the next throw (his second in all) he may

a. throw a 4 (probability $\frac{1}{12}$) and start again
b. throw a 7 (probability $\frac{1}{6}$) which ends the sequence or
c. throw some other number (probability $\frac{3}{4}$) leaving an expected v_1 further throws.

* See J. C. Barton, "Mathematical Notes No. 2,638," *Mathematical Gazette*, 40, No. 334 (1956), pp 276–77.

Thus

(4.16.7) $$v_1 = 1 + \tfrac{1}{12}u + \tfrac{1}{6} \cdot 0 + \tfrac{3}{4}v_1.$$

A precisely similar analysis gives

(4.16.8) $$v_2 = 1 + \tfrac{1}{9}u + \tfrac{1}{6} \cdot 0 + \tfrac{13}{18}v_2$$

(4.16.9) $$v_3 = 1 + \tfrac{5}{36}u + \tfrac{1}{6} \cdot 0 + \tfrac{25}{36}v_3.$$

We now have four equations (4.16.6–9) for four unknowns, and can expect to be able to solve them. From (4.16.7) we have

$$\tfrac{1}{4}v_1 = 1 + \tfrac{1}{12}u \quad \text{or} \quad v_1 = \tfrac{1}{3}u + 4.$$

From (4.16.8) we have $v_2 = \tfrac{2}{5}u + \tfrac{18}{5}$ and from (4.16.9) we have $v_3 = \tfrac{5}{11}u + \tfrac{36}{11}$. We then have, from (4.16.6)

$$u = 1 + \tfrac{1}{6}(\tfrac{1}{3}u + 4) + \tfrac{2}{9}(\tfrac{2}{5}u + \tfrac{18}{5}) + \tfrac{5}{18}(\tfrac{5}{11}u + \tfrac{36}{11}) + \tfrac{1}{3}u$$

or

$$(1 - \tfrac{1}{18} - \tfrac{4}{45} - \tfrac{25}{198} - \tfrac{1}{3})u = 1 + \tfrac{2}{3} + \tfrac{4}{5} + \tfrac{10}{11}$$

from which we find that the expected number of throws is $8\tfrac{103}{196}$.

Exercise 38

1. A player throws two dice, and receives the number of dollars given by the sum of the faces showing, except that if he throws 2 sixes he pays out $200. What is his expectation per throw?

2. As in Exercise 37.4 three dice are thrown, and those not showing a six are taken up and thrown again. This process is repeated. What is the expected number of times that this process must be performed before 3 sixes are obtained?

Hint: Let E_3 denote the expected number of times, let E_2 be the corresponding expectation when 2 dice are thrown, and E_1 that when one die only is thrown. Obtain and solve the equations

$$E_3 = \left(\frac{5}{6}\right)^3 (1 + E_3) + 3 \cdot \frac{5^2}{6^3}(1 + E_2) + 3 \cdot \frac{5}{6^3}(1 + E_1) + \frac{1}{6^3}$$

$$E_2 = \left(\frac{5}{6}\right)^2 (1 + E_2) + 2 \cdot \frac{5}{6^2}(1 + E_1) + \frac{1}{6^2}$$

$$E_1 = \left(\frac{5}{6}\right)(1 + E_1) + \frac{1}{6}$$

3. In a certain town there are 5 toyshops, *A*, *B*, *C*, *D* and *E*. A man wishes to buy a chemistry set which is sold only in shop *E*. He starts by inquiring for the set in shop *A*.

In shop *A*, 2 of the 6 salesmen will redirect him to shop *B*, another 2 will redirect him to shop *C*, while the other 2 will suggest shops *D* and *E*, respectively.

At shop *B* there are 5 salesmen of which 1 will redirect him to shop *A*, 1 will redirect him to shop *C*, 2 will redirect him to shop *D*, and 1 will redirect him to shop *E*.

At shop *C*, 2 of the 4 salesmen will suggest trying shop *B*, while the other two will suggest shop *E*. At shop *D* the 4 salesmen will each suggest a different shop—*A*, *B*, *C* and *E*.

At each shop the man inquires of a salesman chosen at random and follows his advice; except that if told to go to a shop he has already visited he will pick another salesman and inquire again.

(a) What is the probability that he will have visited all the other shops (in some order) before reaching shop *E*?

(b) What is the expected number of shops he will have to visit?

4. A carton of 12 eggs contains 3 bad ones. An egg is chosen at random. If it is bad it is thrown away and another is chosen. What is the expected number of eggs that will have to be chosen before a good one is found?

4.17 GAMBLER'S RUIN PROBLEMS

Suppose two players *A* and *B* play a coin tossing game as follows. *A* tosses a coin. If it is a head, *A* pays *B* \$1; if it is a tail, *B* pays *A* \$1. *A*'s expectation is

$$\tfrac{1}{2} \times 1 - \tfrac{1}{2} \times 1 = 0$$

since there is probability $\tfrac{1}{2}$ that he gains 1 and probability $\tfrac{1}{2}$ that he loses 1. Similarly *B*'s expectation is also 0. Thus over a long run of plays both players can expect to break even. But this presupposes that the players are equipped for a long run of plays, and this may not be the case. If *A* has only \$5 on him he would not be able to continue playing if he had the misfortune to lose in each of the first five plays. Provided we make the realistic assumption that each player has only a finite sum of money to gamble with, there is always a possibility at least that sooner or later one of the players will lose all his money and be unable to continue playing. He is then said to be *ruined*, or forced out of the game.

Similar considerations obtain if there are several players, and it makes no essential difference whether the "players" are individual people or business firms, industries or nations; or if the "games" are simple coin tossing, or the more serious "games" of business rivalry, industrial competition or warfare. Problems of the type "what is the expected number of plays before player *A* is forced out of the game"—usually called *gambler's ruin problems*—are therefore of importance in economics. Every gambler knows that one needs a good sum of money to tide over losing runs and to stay in the game. In the same way a business firm must have adequate financial backing even though its organization is such that on the average it will prosper; otherwise it will not be able to survive the occasional economic crises.

In this section we give a brief introduction to problems of this type by considering in some detail the coin-tossing game between *A* and *B* as described above.

Suppose *A* starts with *m* dollars and *B* with *n* dollars. Let $E(m, n)$ be the expected number of plays before one of the players (we shall not require it to be a particular one) is ruined. Naturally $E(m,n)$ depends on *m* and *n*, as the functional notation indicates. After the first play there is probability $\frac{1}{2}$ that *A* wins, in which case 1 play has already gone by and the expected number of further plays is $E(m + 1, n - 1)$ since *B* has paid *A* \$1; there is probability $\frac{1}{2}$ that *B* wins and that $E(m - 1, n + 1)$ further plays will be required. Hence we have

$$E(m, n) = \tfrac{1}{2}[1 + E(m + 1, n - 1)] + \tfrac{1}{2}[1 + E(m - 1, n + 1)]$$

or

(4.17.1) $E(m, n) = 1 + \tfrac{1}{2}E(m + 1, n - 1) + \tfrac{1}{2}E(m - 1, n + 1).$

We have now reduced the problem to mathematical terms. We want to find a function $E(m, n)$ which satisfies equation (4.17.1). There are in fact many functions which will do this, but the particular one we want must satisfy a further requirement, namely that

(4.17.2) $E(0, m + n) = E(m + n, 0) = 0$

which simply says that when one player has no money, no more plays are needed to force him out of the game. Equation (4.17.1) is not any very standard sort of equation, and we shall "solve" it by exhibiting a function which satisfies it and (4.17.2), and by showing that it is the only one. The function is very simple; it is

$$E(m, n) = mn.$$

Clearly this satisfies (4.17.2), and if we substitute in (4.17.1) the right-hand side becomes

$$1 + \tfrac{1}{2}(m + 1)(n - 1) + \tfrac{1}{2}(m - 1)(n + 1)$$
$$= 1 + \tfrac{1}{2}mn + \tfrac{1}{2}n - \tfrac{1}{2}m - 1 + \tfrac{1}{2}mn - \tfrac{1}{2}n + \tfrac{1}{2}m - \tfrac{1}{2}$$
$$= mn$$

which is the left-hand side. The function mn therefore satisfies the requirements, but is it really the expectation we want? Could the required expectation perhaps be some other function which *also* satisfies (4.17.1) and (4.17.2)? Let us see. Suppose $F(m, n)$ also satisfies (4.17.1) and (4.17.2). Then

(4.17.3) $F(m, n) = 1 + \tfrac{1}{2}F(m + 1, n - 1) + \tfrac{1}{2}F(m - 1, n + 1).$

Now subtract (4.17.1) from (4.17.3). We get

$$F(m, n) - E(m, n) = \tfrac{1}{2}F(m + 1, n - 1) - \tfrac{1}{2}E(m + 1, n - 1)$$
$$= \tfrac{1}{2}F(m - 1, n + 1) - \tfrac{1}{2}E(m - 1, n + 1).$$

If we write $G_k = F(m + n - k, k) - E(m + n - k, k)$, this becomes

(4.17.4) $G_n = \tfrac{1}{2}G_{n-1} + \tfrac{1}{2}G_{n+1}.$

Let us write this as

(4.17.5) $G_{n+1} = 2G_n - G_{n-1};$

then with $n = 1$ we have

(4.17.6)
$$G_2 = 2G_1 - G_0$$
$$= 2G_1$$

since $G_0 = 0$ by (4.17.2). Also from (4.17.5) we have

(4.17.7)
$$G_3 = 2G_2 - G_1$$
$$= 4G_1 - G_1$$
$$= 3G_1.$$

Again we have

$$G_4 = 2G_3 - G_2$$
$$= 6G_1 - 2G_1$$
$$= 4G_1.$$

Proceeding in this way we see that G_2, G_3, G_4 and so on are all positive multiples of G_1. In particular G_{m+n} is also a multiple of G_1. But

$G_{m+n} = F(0, m+n) - E(0, m+n) = 0$, hence $G_1 = 0$. From this it follows that all the functions G_k are zero. Thus $F(m, n) = E(m, n)$ for all m and n. In other words, if we can find a solution of (4.17.1) satisfying also (4.17.2) then it is the only solution. Hence the solution $E(m, n) = mn$ is the one we want, and is the required expectation.

Another type of gambler's ruin problem is that of assessing a player's chance of forcing his opponent out of the game. Again we can illustrate this by reference to the coin-tossing game.

Let p_k denote the probability that player A will eventually win all of player B's money (thus forcing him out of the game) when player A starts with k dollars and B starts with $m + n - k$ dollars. Clearly $p_0 = 0$, since if $k = 0$, A is already out of the game himself, and $p_{m+n} = 1$ since then A has already forced B out.

If A wins the first toss (probability $\frac{1}{2}$) his probability of success thereafter is p_{k+1}, since he now has $k + 1$ dollars. If he loses, his probability is p_{k-1}. Hence our basic equation is

$$p_k = \tfrac{1}{2}p_{k+1} + \tfrac{1}{2}p_{k-1}$$

or

$$p_{k+1} = 2p_k - p_{k-1}.$$

With $k = 1$ we have

$$p_2 = 2p_1 \text{ since } p_0 = 0.$$

With $k = 2$ we have

$$p_3 = 2p_2 - p_1$$
$$= 3p_1.$$

Similarly $p_4 = 4p_1$; $p_5 = 5p_1$ and so on. In general we have $p_k = kp_1$, exactly as with G_k in the previous example. But now, since $p_{m+n} = (m+n)p_1 = 1$, it follows that $p_1 = \dfrac{1}{m+n}$ and hence

$$p_k = \frac{k}{m+n}.$$

Hence if A starts with m dollars and B with n dollars, and the game continues until one player is ruined, then A's probability of success is $\dfrac{m}{m+n}$.

Thus success tends to go to the player with the largest capital—a well-known fact both at the gaming table and in the economic field.

Exercise 39

1. Two players A and B have \$7 and \$3 respectively. They toss a coin and according to the way it lands either A pays B \$1 or B pays A \$1, each with probability $\frac{1}{2}$. They continue to do this until either (a) B has lost all his money, or (b) each player has \$5. What is the expected number of tosses of the coin before the game finishes?

   Hint: Draw a tree diagram and show that, if E_n is the expected number of further tosses when player A has n dollars, then

$$E_7 = \tfrac{1}{2}(1 + E_8) + \tfrac{1}{2}(1 + E_6)$$
$$E_8 = \tfrac{1}{2}(1 + E_9) + \tfrac{1}{2}(1 + E_7)$$
$$E_6 = 1 + \tfrac{1}{2}E_7$$
$$E_9 = 1 + \tfrac{1}{2}E_8.$$

Hence show that $E_7 = 6$.
2. In the game described in problem 1, what is the probability that the game will end by both players holding \$5?
3. Answer the previous two questions for the modified game in which, at each play, the probability of A paying B is p, while B pays A with probability $q = 1 - p$.

4.18 THE BINOMIAL DISTRIBUTION

If the probability of a successful outcome of an experiment is p, and we perform the experiment n times, what is the probability that we shall get k successes?

Let us first consider the probability that k *specified* experiments will succeed and the rest fail. The probability of the k successes is p^k (by rule 2) and the probability of the $n - k$ failures is $(1 - p)^{n-k}$ (by rules 1 and 2). Hence the probability in this case is

(4.18.1) $p^k(1 - p)^{n-k}.$

It will be convenient to put $q = 1 - p$ and write (4.18.1) as

(4.18.2) $p^k q^{n-k}.$

To find the probability that some unspecified set of k experiments will succeed we invoke rule 4, and add up the probabilities (4.18.2) for every possible set of k out of n experiments. The probabilities are all the same, and the number of sets is $\binom{n}{k}$. Hence the required probability is

$$(4.18.3) \qquad\qquad P(k) = \binom{n}{k} p^k q^{n-k}.$$

Now (4.18.3) gives a whole set of probabilities, one for each of the values $0, 1, 2, \ldots, n$ of k. $P(k)$ is a function of k, and its graph (for $n = 8$, $p = \frac{1}{3}$) is given in Fig. 4.6. Such a set of probabilities is known in statistics as a *distribution*. This particular distribution is called the *binomial distribution*, since the general expression contains the binomial coefficients $\binom{n}{k}$. In fact (4.18.3) can be directly related to the binomial theorem for, by that theorem

$$(4.18.4) \qquad\qquad (pt + q)^n = \sum_{k=0}^{n} \binom{n}{k} p^k q^{n-k} t^k$$

where t is a variable. Thus (4.18.3) is the coefficient of t^k in the binomial expansion of $(pt + q)^n$.

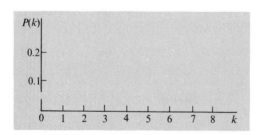

Fig. 4.6

This device, just used, of expressing a function of k (an integer) as the coefficient of t^k in a polynomial in t, is a very useful one, and one which is frequently used in combinatorial analysis and probability theory. The polynomial in question is called the *generating function* for the function of k; in more advanced applications the generating function may be a more general function than a polynomial.

We would expect that the sum of the probabilities (4.18.3) would be 1, since they relate to all possible outcomes of the n experiments. By using the generating function (4.18.4) we can readily see that this is so. If we put $t = 1$ in (4.18.4) we obtain

$$(4.18.5) \qquad\qquad (p + q)^n = \sum_{k=0}^{n} \binom{n}{k} p^k q^{n-k}$$

which shows that the sum of the probabilities [the right-hand side of (4.18.5)] is $(p + q)^n$, which is 1 since $p + q = 1$.

Let us now ask for what value of k is $P(k)$ the largest; more loosely, what is the most likely number of successful experiments?

We shall answer this question by comparing $P(k)$ with $P(k + 1)$. We have

(4.18.6)

$$
\begin{aligned}
\frac{P(k + 1)}{P(k)} &= \frac{\binom{n}{k + 1} p^{k+1} q^{n-k-1}}{\binom{n}{k} p^k q^{n-k}} \\
&= \frac{n! p^{k+1} q^{n-k-1}(n - k)! k!}{(n - k - 1)!(k + 1)! n! p^k q^{n-k}} \\
&= \frac{n - k}{k + 1} \cdot \frac{p}{q}.
\end{aligned}
$$

Now (4.18.6) is certainly positive, and the larger k is the smaller this expression becomes, since increasing k decreases the numerator and increases the denominator. Hence $P(k + 1)/P(k)$ will in general be > 1 for small values of k, but will ultimately become < 1 (even if only for $k = n$, when it is zero). Thus the values $P(k)$ increase at first, but at some stage they begin to decrease. The exception is if (4.18.6) is < 1 even when $k = 0$, i.e., if $np < q$. In this case $P(k)$ gets smaller all the time, and $P(0)$ is the largest value.

To find the largest value of $P(k)$ we therefore look for a value k such that

$$
P(k - 1) \le P(k) \quad \text{and} \quad P(k) \ge P(k + 1).
$$

From (4.18.6) we have

$$
\frac{n - (k - 1)}{k} \cdot \frac{p}{q} \ge 1
$$

and

$$
\frac{n - k}{k + 1} \cdot \frac{p}{q} \le 1.
$$

These inequalities can be written as

$$
[n - (k - 1)]p \ge kq
$$

and

$$
(n - k)p \le (k + 1)q,
$$

or as

$$(n + 1)p \geq k(p + q) = k$$

and

$$np - q \leq k(p + q) = k$$

since $p + q = 1$.
Hence

(4.18.7) $$np - q \leq k \leq (n + 1)p.$$

Dividing both sides of (4.18.7) by np we find that

(4.18.8) $$1 - \frac{q}{pn} \leq \frac{k}{np} \leq 1 + \frac{1}{n}.$$

The answer to the question is given by (4.18.7). For example, if $n = 8$ and $p = \frac{1}{3}$ (as in Fig. 4.6) then k must satisfy

$$\tfrac{8}{3} - \tfrac{2}{3} \leq k \leq 9 \cdot \tfrac{1}{3}.$$

In this case $k = 2$ and $k = 3$ both satisfy the criterion and we have two (equal) maximum values. If $n = 16$, $p = \frac{1}{4}$, then

$$3\tfrac{1}{4} \leq k \leq 4\tfrac{1}{4}$$

whence $k = 4$ gives the maximum value.

From (4.18.8) we see that if n is large, k/np is close to the value 1, the approximation becoming better as n becomes larger. Thus the most likely number of successes is an integer approximately equal to np.

Exercise 40

1. If an experiment is performed n times, and the probability of success is p, then the probability that exactly r experiments succeed is

$$\binom{n}{r} p^r q^{n-r}$$

where $q = 1 - p$, as we have seen. Deduce that the expected number of successes is

$$E = \sum_{r=0}^{n} \binom{n}{r} r p^r q^{n-r}.$$

Show that

$$E = np \sum_{r=1}^{n} \binom{n-1}{r-1} p^{r-1} q^{n-r}$$

and deduce that the expected number of successes is np.

2. Six dice are thrown. What is the probability of throwing four or more sixes?

3. An experiment which has a probability $\frac{1}{3}$ of success is performed six times. Draw a graph of the function $f(n)$ where $f(n)$ is the probability of obtaining exactly n successes. (The graph will contain 7 points, corresponding to $n = 0, 1, 2, \ldots, 6$).

4. A balanced coin is tossed 10 times. What is the probability that the numbers of heads and tails differ by more than 4?

4.19 SOME GAMBLING FALLACIES

Very many people have peculiar ideas concerning probability. Even habitual gamblers, who might be expected to have acquired at least some rule-of-thumb knowledge of the subject, are frequently found to entertain wildly fallacious notions, as a result of which they may lose large sums of money. It is not the aim of this chapter to teach the reader how to become a successful gambler, but it is instructive to take a look at some of these mistaken ideas, since in finding out just why they are mistaken we acquire a greater familiarity with the subject matter of the theory of probability. We shall therefore wind up this chapter by finding the flaws in one or two of the many "systems" which have done so much to make the casinos of the world into going concerns.

To show that it is not the complexity of the game which occasions these wrong ideas we shall take the simplest gamble of all—betting on a 50–50 chance. Guessing the result of tossing a coin will do as well as any; if the player guesses wrong he loses his stake, if he guesses right he retrieves his stake together with an equal sum. We first consider a gambler's attitude towards unusual runs. Suppose, for example, that the coin has come up heads 10 times in a row; how should we act if we are required to bet on the eleventh toss? There are three possible attitudes we could take, as follows:

1. prefer to bet on getting a tail
2. prefer to bet on getting another head
3. have no preference one way or the other.

Let us examine these three attitudes in some detail.

In attitude (1) the assumption is that since the coin has fallen heads so many times already, and since we expect roughly equal numbers of heads and tails on the average, therefore the initial preponderance of heads will now be balanced by an excess of tails in the tosses to come. This attitude is the most obviously absurd, yet it is perhaps the commonest. The person who considers that there is an advantage in betting on a tail under these circumstances is saying, in effect, that the fact that the coin has turned up heads so often in the past will make the coin turn up tails more often in the future. This idea is often referred to as the *maturing of chances*. Clearly if this were to happen it would mean that the coin somehow "remembered" what had happened to it in the past—that the previous history of the coin had an effect on the result of the next toss—which is absurd. Nevertheless the argument that the numbers of heads and tails should "even up" in a long run does seem to have some sort of relevance, and we ought to discover precisely where the fallacy lies.

From the result of the last section we see that for n tosses of the coin the most likely number of heads is $\frac{1}{2}n$. It is also true that the *expected* number of heads is $\frac{1}{2}n$, (see exercise 40.1). The interpretation of this is that in a large number of tosses the ratio

$$\frac{\text{Number of correct guesses}}{\text{Number of incorrect guesses}}$$

should be approximately 1. Here we have the root of the fallacy. For what determines how much we gain or lose over a run of tosses is the *difference* between the numbers of correct and incorrect guesses, not their ratio. There is no reason why this difference, having once become positive, shall we say, should have any particular tendency to decrease—it may or it may not. A specific example will make this clear. Suppose the player bets always on heads, and that after 57 tosses of the coin he has won 27 times and lost 30 times, at a dollar a time. He is then $3 down, and the ratio of wins to losses is

$$\frac{27}{30} = 0.90.$$

The player might now expect that if he continues playing, his "luck will turn," that (to use a common analogy) the "pendulum" will swing the other way, so that he will recoup his losses. But it may well happen that after a total of 195 tosses the numbers are as follows:

$$\text{Heads 95,} \qquad \text{Tails 100,}$$

in which case the player is worse off than before ($5 down) even though the ratio of the numbers of heads and tails, viz.

$$\frac{95}{100} = 0.95$$

is closer to 1 than it was before. Thus the fact that the ratio of heads to tails becomes approximately 1 for a large number of tosses does not mean that the difference between these numbers must oscillate between positive and negative values. The so-called "swing of the pendulum," which forms the basis of so many get-rich-quick gambling systems has no existence outside the brains of the inventors of these systems.

Similar considerations show that there is no advantage to be gained from changing from betting heads to betting tails at any stage.

Turning now to attitude (2) towards a long run of heads we first observe that if we accept the possibility of the coin being biased in favor of showing heads (or even of being two-headed) then there is something to be said for betting that the coin will continue to show a head. To take another example, suppose that Mrs. Smith has had four children, all boys, and is now expecting a fifth. Is it more or less likely to be another boy? In the population as a whole the probability of a child being a boy is near enough $\frac{1}{2}$, and one would tend to say that this is the probability of the next Smith offspring being a boy. There is biological evidence however that some individual parents are more likely to produce children of one sex than the other, and it could be that Mrs. Smith is more likely to produce a boy than a girl. Even so, Mrs. Smith's all-male family does not give much indication that she is in any way unusual. The probability of getting four boys in a row is $\frac{1}{16}$, which represents a fairly common occurrence.

Returning to the problem of the ten consecutive heads and assuming that the coin is a normal one, we still find that many gamblers will nevertheless take a run of heads, or in general a run of successes, as indicating that they have hit a "lucky break," and for this reason will continue to bet in the same way. A common piece of gambling advice is "If you hit a lucky break, make the most you can of it." This advice is not accompanied by any indication of when one should *stop* making the most of it, though no gambler would be sanguine enough to expect a lucky break to last forever! Needless to say that provided the game is one of pure chance these runs of luck are purely imaginary, since they would imply, just as surely as does the idea of the maturing of chances, that the outcome of tossing a coin is affected in some way by what has happened to the coin in the past.

We therefore come to the correct, but uninteresting conclusion that attitude (3) is the only reasonable attitude to take. It really does not matter whether you bet heads or tails. In fact there is really no point in playing the game at all!

Much of the confusion evident above stems from an imperfect appreciation of the concept of conditional probability. The person who expects the eleventh toss of a coin to give tails because the last 10 tosses were heads, very likely has at the back of his mind the awareness that a run of 11 heads is a very improbable event. So it is; the probability is $\frac{1}{2048}$. But this is the probability of getting 11 heads from scratch, so to speak, and is not what we want here. What we want is the probability of getting 11 heads *given that 10 heads have already been obtained*. This is a conditional probability, and by (4.14.12) its value is

$$\frac{Pr(11 \text{ heads})}{Pr(10 \text{ heads})} = \frac{\frac{1}{2048}}{\frac{1}{1024}} = \frac{1}{2}$$

as we would expect.

We can also look upon these fallacious ideas as arising from an unwillingness to accept that successive tosses of a coin are independent. For the belief that a run of heads makes the outcome of the next toss more likely (or perhaps, less likely) to be a head amounts to a denial of the independence of the successive events.

Another gambling system which is worth looking at as an illustration of some ideas connected with expectation is the well known *Martingale* or *double-up* system. This system is very straightforward and at first sight the logic behind it seems unassailable. Betting on a 50–50 chance we bet $1 at first. If we win, fine—we start again. If we lose we now bet $2 on the second toss of the coin (or whatever it is). If we win now we get back $4—$1 more than the $1 + 2 = 3$ dollars so far staked. If we lose we bet $4 on the third toss, and so on. Whenever we lose we double the amount staked for that toss. When we win (as we must do sooner or later) our winnings amount to $1 more than the total amount so far staked. Hence, it is claimed, our winnings will gradually mount up.

This reasoning is fallacious, and the fault lies in the fact that to work this system exactly as specified we would need to have unlimited resources, for otherwise we could not be sure of continuing to double our stake until such time as we achieve a win. Suppose, for example, that we had only a limited capital, say $1,023. This is just enough to enable us to continue to bet on 10 tosses without winning, since

$$1 + 2 + 4 + 8 + 16 + 32 + 64 + 128 + 256 + 512 = 1023.$$

If the tenth toss is also a loss then our whole capital is gone and we cannot continue the system. Promoters of this and other similar systems usually say that the chances of this happening are so slight that the possibility can be ignored for practical purposes; but can it? True, the probability of 10 consecutive heads is small $(\frac{1}{1024})$ but the amount that is at stake is correspondingly large. Let us see what the expectation is. There is a probability of $\frac{1}{1024}$ of getting 10 heads in a row and losing $1,023; there is a probability $\frac{1023}{1024}$ of this not happening, in which case we gain $1. Thus our expectation is

$$-\frac{1}{1024} \cdot 1023 + \frac{1023}{1024} \cdot 1 = 0.$$

Thus the probability of 10 heads in a row, small though it is, is just large enough that, when this event occurs, it will (on the average) exactly wipe out our gains!

There are many variations on the Martingale system, but they all fail for a similar reason—the existence of a certain sequence of successes or failures which, while improbable, occurs sufficiently often to annul the gains that accrue under all other circumstances.

Exercise 41 (Chapter Review)

1. In a certain factory 8 employees are in Administration, and 6 are in Research and Development. A committee is to be set up consisting of 5 representatives from Administration and 3 from Research and Development. In how many ways can such a committee be chosen?

2. In England, up to a few years ago, car license plates could be of the following two forms

  (a) One or two letters followed by a number of up to 4 digits,

  (b) Three letters followed by a number of up to 3 digits.

  (Other possibilities have since been introduced.)

  In both cases the number portion of the plate did not start with one or more zeros. What was the total number of cars that could be registered using this system?

3. In how many different ways can the letters of the word *TEMPESTUOUS* be arranged?

4. A greengrocer has large stocks of apples, pears, oranges and tangerines. He wishes to make up packages, each of which will contain 10 fruit in all. How many different kinds of packages can he make up?

5. What is the constant term in the expansion of each of the following expressions?

  (a) $\left(x + \dfrac{1}{x}\right)^4$ **(c)** $\left(x^2 + \dfrac{3}{x}\right)^3$ **(e)** $\left(\dfrac{x}{3} - \dfrac{1}{4x^2}\right)^6$

  (b) $\left(\dfrac{x^3}{2} - \dfrac{1}{x^2}\right)^5$ **(d)** $\left(\dfrac{2}{5}x - \dfrac{1}{4x^3}\right)^8$.

6. By considering the coefficient of t^k on both sides of

$$(1 + t)^n (1 + t)^n = (1 + t)^{2n},$$

or otherwise, show that

$$\sum_{r=1}^{n}\binom{n}{r}\binom{n}{k-r}=\binom{2n}{k}.$$

Deduce that

$$\sum_{r=1}^{n}\binom{n}{r}^{2}=\binom{2n}{n}.$$

7. Prove by mathematical induction, or otherwise, that

$$\cos\theta+\cos 2\theta+\cdots+\cos n\theta=\frac{\sin\left(n+\frac{1}{2}\right)\theta-\sin\frac{\theta}{2}}{2\sin\frac{\theta}{2}}.$$

8. A circle is drawn on a piece of paper. A chord of the circle is drawn; this divides the circle into 2 regions. Another chord is drawn intersecting the first; the circle is now divided into 4 regions. A third chord is drawn intersecting the first two, and the circle is now divided into 7 regions. (See Fig. 4.7.) If R_n denotes the number of regions into which the circle is

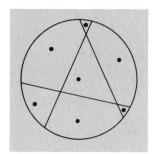

Fig. 4.7

divided by n chords, assuming that each chord intersects every other chord inside the circle, prove by mathematical induction that

$$R_n=\tfrac{1}{2}n(n+1)+1.$$

9. A defect in a piece of manufacturing equipment causes it to produce 3 defective parts in each batch of 20 parts that it manufactures. What is the probability that none of 10 parts, chosen at random from such a batch, will be defective? What is the probability that not more than one will be defective?

10. A man at a fairground bowls 6 balls towards 6 receptacles. Under the (rather unlikely) assumption that each ball is equally likely to go into any receptacle, what is the probability

(a) that he gets one ball in each receptacle,

(b) that one receptacle only has no ball in it?

11. Two cards are drawn from each of two standard decks of cards. What is the probability that they will be four different aces? What is the probability of their being four aces, not necessarily all different?

12. A man wishes to choose a number by means of a spinning arrow device as shown in Fig. 4.8. In this device the radii which separate the 6 regions of

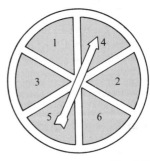

Fig. 4.8

the disk are printed in thick lines, and there is a probability $\frac{1}{60}$ that the arrow will come to rest sufficiently close to a *given* radius that the outcome of the spin is in doubt. When this happens, the man deems the arrow to have stopped in the region bearing the larger number. The probability of the arrow stopping unequivocably in a given region is the same for all six regions.

(a) Find the probabilities of obtaining the numbers 1 to 6 with this device, used in this way.

(b) If two spins are made, what is the probability that the two numbers obtained add up to 7?

13. A card is drawn from a deck of playing cards. If the card is a red one, the experiment is a success. If the card is a spade it is replaced in the deck; if a club it is not replaced. This procedure is repeated two more times. What is the probability of success?

14. A man tosses 4 pennies on a table. He picks up any that have come down tails and tosses them again. He continues to do this until all 4 coins on the table show a head. What is the expected number of tosses required to do this?

15. In a multiple-choice examination, consisting of 25 questions, the candidates must choose one of five possible answers to each question. The examination is marked by giving 4 points for a correct answer, and deducting one point for a wrong answer. Show that if a candidate answers the questions at random, his expected mark is zero.

What is the candidate's expected mark if, in each question, one of the possible answers is obviously wrong, so that he makes his choice at random from the remaining four possibilities?

16. A backer gives the following instructions to a bookmaker:

(a) A sum of x dollars is to be placed on a certain horse in the first race.

(b) If, and only if, this first horse loses, y dollars are to be placed on a horse in the second race.

(c) If, and only if, the second horse also loses, z dollars are to be placed on a horse in the third race.

Let the probabilities that these horses will win their respective races be p_1, p_2, p_3, respectively. Assume that if the probability of a horse's winning is p, the odds will be "q to p" (where $q = 1 - p$), i.e., that a stake of amount x will be converted to x/p if the horse wins. (This is equivalent to assuming a benevolent bookie who strives to make the backer's expectation, and his own, zero).

Calculate the amount that the backer can expect to win, using this system. (The answer to this should be clear, without any calculation, but prove it anyway.)

Suggestions for Further Reading

Most textbooks on probability go considerably beyond the modest limits of this chapter; but the first two chapters of:

Parzen, E. *Modern Probability Theory and its Applications.* New York: John Wiley & Sons, Inc., 1960,

can profitably be read, and will give the reader an idea of how probability can be defined in a more general way than the "equiprobable" method used in this chapter.

For a more light-hearted treatment of the subject, see:

Huff, D., *How to Take a Chance.* London: Victor Gollancz, 1960;

and while you are about it you might as well read its companion volume

_____. *How to Lie with Statistics.* London: Victor Gollancz, 1954,

even though it deals only indirectly with probability.

For an encyclopedic coverage of everything to do with betting, games of chance, systems, gambling fallacies, and so on, the reader cannot do better than consult:

Scarne, J., *Scarne's Complete Guide to Gambling.* New York: Simon & Schuster, Inc., 1961.

The three "finite mathematics" books (see page 27) give a full treatment of all the topics covered in this chapter, and the reader should certainly consult them.

5 Plane

Coordinate Geometry

5.1 INTRODUCTION

In this chapter we shall be concerned with the geometry of straight lines in a plane. If we think of a line as a set of points, i.e., as a graph of a particular kind, and of these points as pairs of numbers (their coordinates) which satisfy a certain equation (the equation of a line), then we can study the geometrical properties of these lines, not by constructions based on the lines themselves as in elementary geometry in the style of Euclid, but by algebraic manipulations of the appropriate equations and coordinates. The branch of mathematics which studies the properties of geometrical figures in this algebraic fashion is known as *coordinate geometry*.

We first need a coordinate system, in order that every point of the plane can be allotted its coordinates. We choose perpendicular axes OX and OY through a suitable point O as origin, as we did in Chapter 3. With a suitable scale of measurement along the axes (we need not specify what it is) our coordinate system is now set up and we can proceed.

5.2 DISTANCES

We shall first look at the problem of finding the distance between two points. Let the two points be denoted by P_1 and P_2, and let their coordinates be (x_1, y_1) and (x_2, y_2) respectively. We want to find the length of the line segment P_1P_2.

If N_1 and N_2 are the feet of the perpendiculars from P_1 and P_2 respectively to the x-axis (as in Fig. 5.1), then $ON_1 = x_1$, $N_1P_1 = y_1$, $ON_2 = x_2$, $N_2P_2 = y_2$; and if M is the foot of the perpendicular from P_1 to P_2N_2 then

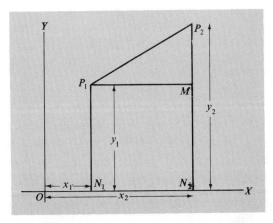

Fig. 5.1

173

$$P_1M = N_1N_2 \quad \text{(since } P_1MN_2N_1 \text{ is a rectangle)}$$
$$= ON_2 - ON_1$$
$$= x_2 - x_1;$$

and
$$P_2M = N_2P_2 - N_2M$$
$$= N_2P_2 - N_1P_1 \quad \text{(since } N_1P_1 = N_2M)$$
$$= y_2 - y_1.$$

Now by the familiar theorem of Pythagoras, since the angle P_2MP_1 is a right angle, we have

$$P_1P_2{}^2 = P_1M^2 + MP_2{}^2$$
$$= (x_2 - x_1)^2 + (y_2 - y_1)^2.$$

Hence we have the result that the distance D between the points (x_1, y_1) and (x_2, y_2) is given by the equation

(5.2.1) $$D^2 = (x_2 - x_1)^2 + (y_2 - y_1)^2$$

i.e., the square of the required distance is the square of the difference of the x-coordinates plus the square of the difference of the y-coordinates.

Exercise 42

1. Figure 5.1 has been drawn for $x_2 > x_1$ and $y_2 > y_1$. By drawing other diagrams show that the result (5.1) is true whatever the relative positions of the points P_1 and P_2.
2. Find the distances between the following pairs of points.
    (a) (1, 3) and (4, 7)   (f) $(0, -6)$ and $(-3, -2)$
    (b) (9, 8) and (30, 28)   (g) $(-7, 4)$ and $(-12, 16)$
    (c) $(-2, 9)$ and (13, 17)   (h) $(1, -\frac{9}{5})$ and $(\frac{2}{5}, -2)$
    (d) $(3, -2)$ and $(5, -4)$ (i) $(\frac{3}{4}, -\frac{1}{3})$ and $(\frac{5}{2}, \frac{4}{3})$
    (e) $(\frac{2}{5}, -\frac{7}{3})$ and $(-\frac{6}{5}, -\frac{2}{3})$ (j) $(-\frac{1}{2}, \frac{1}{3})$ and $(\frac{4}{5}, \frac{3}{8})$
3. Show that the triangle which has its vertices at the points (1, 1), $(1 + \sqrt{3}, 4)$ and $(1 - \sqrt{3}, 4)$ is equilateral.
4. Find the length of the line AB in each of the following examples:
    (a) $A(0, 0)$, $B(3, 4)$ (d) $A(-2, -3)$, $B(1, 1)$
    (b) $A(-4, -6)$, $B(-1, -2)$ (e) $A(2, 4)$, $B(-3, 16)$
    (c) $A(-1, 3)$, $B(11, -2)$ (f) $A(0, 1)$, $B(6, -7)$
5. Prove that two sides of the triangle $A(-6, -7)$, $B(6, -2)$, $C(-1, 5)$ are equal.

5.3 VARIOUS FORMS OF THE EQUATION OF A LINE

If we draw a line* on a given coordinate system (as in Fig. 5.2) then there will be some relation between the coordinates x and y of those points which are on the line—a relation which is not shared by points which are not on the line. These coordinates will, in fact, satisfy a certain equation— the equation of the line. If we are given enough information about the line to specify it uniquely, i.e., so that we know exactly which points are on it and which are not, then we can determine its equation.

The equation of a line may be written in several different forms, each form bringing into prominence some particular feature of the line. We shall now derive several standard forms which the equation of a line may take.

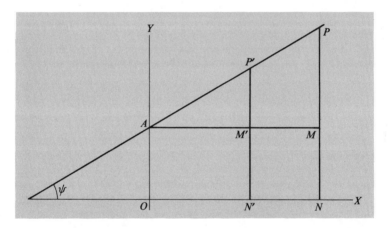

Fig. 5.2

Let us suppose that the line meets the y-axis in the point A, where $OA = d$. The length d is called the *intercept* of the line on the y-axis, and is reckoned negative if A is on the negative half of the y-axis. Let P be any point of the line, and let us draw PN perpendicular to OX and AM perpendicular to PN (Fig. 5.2). We shall call the ratio PM/AM the *slope* of the line, and denote it by m. If we had taken some other point P' of the line, and carried out the same construction, we would have obtained the same value for m; for since the triangles $AP'M'$ and APM in Fig. 5.2 are similar, it follows that $P'M'/AM' = PM/AM$. We are therefore quite in order to call this ratio the slope of the *line* since it depends only on the line, and not on which point

* Henceforth we shall use *line* instead of *straight line* (since there aren't really any other kinds!).

of the line we choose as the point P. It is this fact, indeed, which distinguishes a line from some other kind of curve. We may notice, in passing, that

$$m = \frac{PM}{AM} = \tan \psi$$

where ψ is the angle that the line makes with the x-axis.

If the coordinates of P are (x, y) then

$$PM = PN - MN$$
$$= y - d \quad \text{(since } NM = OA = d\text{)}$$

and $\qquad\qquad AM = x \quad \text{(since } AM = ON = x\text{)}.$

Hence, $PM/AM = (y - d)/x = m$, from which we get the result

(5.3.1) $\qquad\qquad\qquad y = mx + d$

which is the simplest form of the equation of a line.

Any point of the line will do as the point P, so that the coordinates (x, y) of any point on the line will satisfy the equation $y = mx + d$; moreover, it is easy to see that the coordinates of any point *not* on the line will *not* satisfy the equation. Hence the functional relation $y = mx + d$ holds if, and only if, the point (x, y) is on the line.

Exercise 43

1. Prove that the result just mentioned, viz., that if a point is not on the line $y = mx + d$ then its coordinates do not satisfy this equation.
2. Find the equations of the lines whose intercepts on the y-axis are of lengths

   (a) 4,   **(b)** −1,   **(c)** 17.813   **(d)** −50

and whose slopes are

   (a) −5   **(b)** −2   **(c)** 4.016   **(d)** 29

respectively.

5.4 LINE WITH GIVEN SLOPE CONTAINING A GIVEN POINT

To specify the intercept of a line on the y-axis is to say that the line contains a particular point on the y-axis. A more general way of specifying a line is to give its slope, m, say, and to say that it contains some given point, for example the point A with coordinates (α, β). This information specifies the line, and Fig. 5.3 shows us how to obtain its equation.

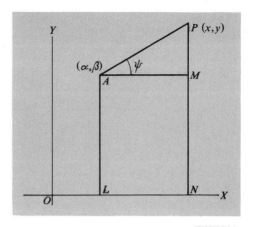

Fig. 5.3

For
$$PM = PN - MN$$
$$= PN - AL$$
$$= y - \beta;$$
and
$$AM = LN$$
$$= ON - OL$$
$$= x - \alpha.$$

Hence, since the slope of this line is PM/AM, the equation of the line containing the point (α, β) and having slope m is

$$\frac{y - \beta}{x - \alpha} = m, \quad \text{or}$$

(5.4.1)
$$y - \beta = m(x - \alpha).$$

Thus the equation of the line containing the point $(1, 5)$ and having slope $-\frac{2}{3}$ is

$$y - 5 = -\tfrac{2}{3}(x - 1).$$

Note that this may be simplified, either to

$$2x + 3y - 17 = 0$$

or to
$$y = -\tfrac{2}{3}x + \tfrac{17}{3}$$

which is of the form $y = mx + d$.

Exercise 44

1. Find the equations of the lines defined below, and simplify them to the form $y = mx + d$.

(a) The line with slope $\frac{1}{2}$ containing the point $(2, -3)$

(b) the line with slope 10 containing the point $(7, 5)$

(c) the line with slope -3 containing the point $(-6, -4)$

(d) the line with slope $-\frac{1}{3}$ containing the point $(3, 0)$

(e) the line with slope 2 containing the point $(2, 4)$

2. What must the slope of a line through the point $(2, -3)$ be if the line also contains the point $(3, 7)$?

3. What must the slope of a line through (x_1, y_1) be if the line also contains the point (x_2, y_2)?

5.5 LINE CONTAINING TWO GIVEN POINTS

Yet another way of fixing a line is to specify two points on it. Suppose the coordinates of two points P_1 and P_2 are given, say (x_1, y_1) and (x_2, y_2), respectively. How do we set about finding the equation of the line which joins them? This is a problem which we can reduce to the previous one by first finding the slope of this line. Referring back to Fig. 5.1 we see that

$$m = \frac{P_2 M}{P_1 M} = \frac{y_2 - y_1}{x_2 - x_1}.$$

Now that we have the slope m, and the coordinates of a point of the line (we can take either P_1 or P_2 for this—it doesn't matter) we use the previous result (5.4.1) to write down the required equation. It is

$$y - y_1 = \frac{y_2 - y_1}{x_2 - x_1}(x - x_1)$$

or

(5.5.1)
$$\frac{y - y_1}{x - x_1} = \frac{y_2 - y_1}{x_2 - x_1}.$$

The form of (5.5.1) suggests another way of proving this result; for $(y - y_1)/(x - x_1)$ is the slope of the line joining P_1 to a general point, while $(y_2 - y_1)/(x_2 - x_1)$ is the slope of the line $P_1 P_2$, and if P lies on $P_1 P_2$ then these slopes must be the same. Another form for this result is

(5.5.2)
$$\frac{y - y_1}{y_2 - y_1} = \frac{x - x_1}{x_2 - x_1}$$

which offers some advantage in that the variables x and y are on opposite sides of the equation.

A special case of this result, worthy of mention, applies when a line is specified by its intercepts on the two axes. If the intercepts of a line on the x- and y-axes are A and B, respectively, then the line joins the two points $(A, 0)$ and $(0, B)$. By the result just obtained, its equation is therefore

$$\frac{y - 0}{b - 0} = \frac{x - A}{0 - A}$$

which simplifies to

(5.5.3) $$\frac{x}{A} + \frac{y}{B} = 1$$

a neat result, which is occasionally useful.

Exercise 45

1. Write down the equations of the lines joining the pairs of points given in Exercise 42.2.

2. Find the equations of the lines whose intercepts on the x- and y-axes respectively are

(a) 5 and 6   (c) -1 and 3

(b) $-\frac{1}{2}$ and $\frac{7}{3}$ (d) -2 and $-\frac{5}{3}$.

3. Find the equation of a straight line subject to the two data in each of the following examples:

(a) through the origin and $(-5, 6)$

(b) through $(7, 0)$ and $(0, 8)$

(c) through $(1, 2)$ and $(3, -4)$

(d) through $(-2, 3)$ and with slope 2

(e) through $(0, 0)$ and with slope -1

(f) through $(-1, 0)$ and with slope $-\frac{2}{5}$.

5.6 GENERAL EQUATION OF A LINE

Although we have derived several different forms for the equations of lines in the preceding paragraphs, it is clear that the differences of form are not of any significance—an equation of one form can always be rearranged into any other. In particular, if we simplify any of the equations as far as is possible, by removing brackets and collecting up like terms, we shall always end up with an equation containing at most a term in x, a term in y and a constant term, i.e., an equation of the form

(5.6.1) $$ax + by + c = 0.$$

Conversely, any equation of this form will be the equation of some line. For if $b \neq 0$ we can write it as

$$y = -\frac{a}{b}x - \frac{c}{b}$$

which is of the form $y = mx + d$, with $m = -a/b$ and $d = -c/b$; and this, we know, is the equation of the line whose slope is $-a/b$ and whose intercept on the y-axis is $-c/b$. If $b = 0$, then (5.6.1) becomes $x = -c/a$, which gives a line parallel to the y-axis. In consequence of this the expression $ax + by + c$ is called a *linear expression*, or a *linear function* of the variables. Algebraically speaking, a linear function of two variables (or indeed of any number of variables) is defined as a polynomial, each term of which is either constant or a constant multiple of one of the variables.

Exercise 46

1. Simplify the equations of the lines in Exercises 44 and 45 to the form $ax + by + c = 0$.
2. Write down the equations of the lines parallel to $4x - 2y + 7 = 0$ through the points
    (a)   (0, 0)
    (b)   (−1, 3)
    (c)   (1, 6)

5.7 PARAMETRIC EQUATIONS OF A LINE

Before going on to a different topic we should consider briefly one further method of specifying a line algebraically—a method somewhat different from those already dealt with. Let us return to the line containing a given point (α, β) and with slope m. In Fig. 5.3 the angle $PAM = \psi$, where $\tan \psi = m$, and hence we may write

$$x = OL + LN$$
$$= OL + AP \cos \psi$$
$$= \alpha + r \cos \psi$$

where $r = AP$; and

$$y = NM + MP$$
$$= \beta + r \sin \psi.$$

Now if we choose any value for r, and substitute it in the equations

(5.7.1)
$$x = \alpha + r \cos \psi$$
$$y = \beta + r \sin \psi$$

the values x and y thus obtained will be the coordinates of some point of the line. Hence we can use the equations (5.7.1) to construct as many points of the line as we wish, just as we can do with equation (5.4.1). What we have done is that, instead of linking x and y by a single equation as in (5.4.1), we have linked them via a third variable r, and expressed each of them as a function of r. Such a representation of the relationship between x and y is known as a *parametric representation*. The equations (5.7.1) are called *parametric equations*, and the "intermediate" variable r is known as the *parameter*.

This parametric representation of the line is often more useful than the usual single equation, but in any case we can always change it to a single equation if we want to. We need only eliminate the parameter from the parametric equations. Thus from equations (5.7.1), provided that $\cos \psi \neq 0$ and $\sin \psi \neq 0$,

$$r = \frac{x - \alpha}{\cos \psi} \quad \text{and} \quad r = \frac{y - \beta}{\sin \psi},$$

and hence

$$\frac{x - \alpha}{\cos \psi} = \frac{y - \beta}{\sin \psi}$$

or
$$y - \beta = \tan \psi \, (x - \alpha)$$
$$= m \, (x - \alpha),$$

which brings us back to equation (5.4.1).

Exercise 47

1. Find the parametric equations of the following lines.
   (a) the line containing $(-3, 5)$ making an angle of $60°$ with the x-axis
   (b) the line containing $(\frac{1}{2}, -1)$ making an angle of $45°$ with the x-axis
   (c) the line containing $(\frac{1}{6}, -\frac{5}{6})$ making an angle of $120°$ with the x-axis.

2. Obtain the equations, in the form $y = mx + c$, of the lines whose parametric equations are

(a) $x = -17 + r/\sqrt{2}$, $y = 9 - r/\sqrt{2}$

(b) $x = \dfrac{2}{3} - \dfrac{r\sqrt{3}}{2}$, $y = -\tfrac{1}{3} + \tfrac{1}{2}r.$

3. A pair of parametric equations of the form

$$x = \alpha + \lambda t$$
$$y = \beta + \mu t$$

where t is the parameter and λ and μ are constants, will also represent a line, even though the numbers λ and μ are not the cosine and sine of the angle ψ. The only difference that this makes will be that the parameter t is now no longer the distance of the point (x, y) from (α, β), but is proportional to it. Find the equations, in the form, $y = mx + c$, of the lines given by the parametric equations

(a) $x = 3 + 5t$, $y = -2 + 7t$

(b) $x = -5 - 17t$, $y = 7 + 8t$

(c) $x = \tfrac{2}{3} - \tfrac{1}{5}t$, $y = -\tfrac{1}{5} + \tfrac{2}{3}t.$

5.8 INTERSECTIONS OF LINES

Suppose we have two lines whose equations are, let us say,

(5.8.1) $3x + 4y - 7 = 0$

and

(5.8.2) $2x - 5y + 3 = 0.$

The first line is the set of all points whose coordinates satisfy (5.8.1), and the second is the set of points whose coordinates satisfy (5.8.2). What is the intersection (in the sense of Chapter 2) of these two sets? Clearly it will be the set of points common to the two lines. Now we know from elementary geometry that two lines are either parallel or else they meet (intersect) in a single point. There are three possibilities. If the lines coincide (this is a special case of parallelism) then their intersection will be the set of all points of the (common) line. If the lines are distinct and parallel then their intersection is the empty set. If the lines are not parallel the intersection set will consist of one point only. In the latter case the coordinates of the point will satisfy both equations (5.8.1) and (5.8.2), and we can find these coordinates by solving these two equations for x and y.

This we may do by the usual process of eliminating one of the variables. From (5.8.2) we can write

(5.8.3) $2x = 5y - 3,$

while (5.8.1) can be written as

$$6x + 8y = 14.$$

Substitute for x from (5.8.3) and we have

$$3(5y - 3) + 8y = 14$$

i.e., $17y = 17$

or $y = 1.$

Going back to (5.8.3) we have

$$2x = 5 - 3$$

or $x = 1.$

Hence the point of intersection of the two lines is the point $(1, 1)$.

In this same way we can find the point of intersection of any two lines, of the form

(5.8.4)
$$ax + by + c = 0$$
$$px + qy + r = 0$$

provided that the lines are not parallel. Now parallel lines will make the same angles with the x-axis, and hence two lines are parallel if, and only if, their slopes are equal. The slopes of the lines in (5.8.4) are $-a/b$ and $-p/q$, so that these lines are parallel if, and only if, $-a/b = -p/q$, i.e., $aq - bp = 0$.

In what we have just done we assumed the geometrical result that two lines are either parallel or meet in a single point, and deduced that equations (5.8.4) have either an unlimited number of solutions or no solutions, when $aq - bp = 0$, and have a unique solution if $aq - bp \neq 0$. We could have gone the other way about. On attempting to solve equations (5.8.4) we easily find that if $aq - bp \neq 0$ then x and y are uniquely determined, giving a unique point of intersection of the lines, whereas if $aq - bp = 0$ the method fails to give a unique solution. From this the geometrical statements can be deduced.

Exercise 48

1. Find which of the following pairs of lines are parallel, and for those that are not, find the point of intersection.
    - (a) $3x - y + 12 = 0$; $3x - 7y - 6 = 0$
    - (b) $3x - 7y - 6 = 0$; $-9x + 21y - 10 = 0$
    - (c) $2x + 7y - 38 = 0$; $4x - 3y - 8 = 0$
    - (d) $3x + 5y + 4 = 0$; $-7x + 2y - 23 = 0$
    - (e) $x + 2y + 3 = 0$; $5x + 10y - 4 = 0$
    - (f) $3x - 4y = 71$; $5x + y = 57$

2. On a single sheet of graph paper draw the following four lines; (a) $20x - 9y + 42 = 0$ (b) $x + 7y - 23 = 0$ (c) $10x - y - 17 = 0$ (d) $x - 3y - 3 = 0$. These four lines will intersect in pairs in 6 points. Find the coordinates of these six points.

3. Find the point of intersection of the two given lines in each of the following examples:
    - (a) $2x + 3y = 0$; $3x - 2y = 0$
    - (b) $x - 2y = 0$; $x + y - 6 = 0$
    - (c) $x - y + 2 = 0$; $3x + 2y + 6 = 0$
    - (d) $4x - y - 3 = 0$; $x + 2y - 3 = 0$
    - (e) $3x + 2y - 13 = 0$; $2x + 3y - 12 = 0$

5.9 POINT DIVIDING A SEGMENT IN A GIVEN RATIO

Let P and Q be the points (x_1, y_1) and (x_2, y_2), and let λ, μ be *any* two numbers. What are the coordinates of the point R which divides the line segment PQ in the ratio $\lambda : \mu$ i.e., the point R on PQ, such that

$$\frac{PR}{RQ} = \frac{\lambda}{\mu}?$$

Suppose R is the point (x, y). Then from Fig. 5.4 we have

$$\frac{PR}{RQ} = \frac{PU}{RV} = \frac{PU}{UM}$$

since the triangles PRU and RQV are similar. Moreover

$$PU = x - x_1$$

and $$UM = x_2 - x.$$

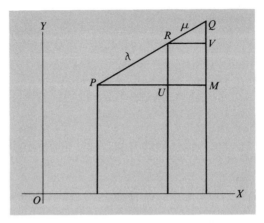

Fig. 5.4

Hence

$$\frac{x - x_1}{x_2 - x} = \frac{\lambda}{\mu}$$

i.e.,
$$\mu x - \mu x_1 = \lambda x_2 - \lambda x$$

or
$$x = \frac{\mu x_1 + \lambda x_2}{\lambda + \mu}.$$

In the same way, from $PR/RQ = RU/QV = VM/QV$ it follows that

$$y = \frac{\mu y_1 + \lambda y_2}{\lambda + \mu}.$$

Thus the required point R is the point

(5.9.1)
$$\left(\frac{\mu x_1 + \lambda x_2}{\lambda + \mu}, \ \frac{\mu y_1 + \lambda y_2}{\lambda + \mu}\right).$$

This result is not so difficult to remember as it might seem at first glance. The coordinates have the same denominator, $\lambda + \mu$, and the numerators are similar combinations of the corresponding coordinates of P and Q. The main thing to notice is the form of the numerator, i.e., that λ goes with x_2 and μ with x_1—not the other way about. This is what we would expect. For suppose $\lambda > \mu$. Then R will be nearer to Q than to P. Hence it is to be expected that the coordinates of Q will carry more weight, so to speak, than those of P in determining the coordinates of R, and will therefore be multiplied by the larger number λ.

Exercise 49

1. For each of the following pairs of points find the points which divide the segments joining them in the ratios (i) $3:2$, (ii) $5:1$, (iii) $3:7$.
- **(a)** $(2, -1)$ and $(6, 3)$
- **(b)** $(7, 5)$ and $(-2, -9)$
- **(c)** $(8, 0)$ and $(-5, 1)$
- **(d)** $(-3, -7)$ and $(2, 5)$.

2. The vertices of a triangle are the points $A(3, 4)$, $B(-1, 2)$ and $C(2, -1)$. Find the coordinates of the midpoint D of AB. Find the coordinates of the point G which divides CD in the ratio $2:1$. (See Fig. 5.5).

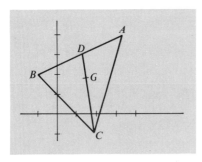

Fig. 5.5

Prove further that if these same three points are labelled A, B, C in a different way, the same point G is obtained.

3. Generalize the result of problem 2. That is, show that if A, B, C are three general points (x_1, y_1), (x_2, y_2) and (x_3, y_3) then the point G obtained by the above construction does not depend on the way the points are labelled.

Find the coordinates of G.

(The point G defined in this way is called the *centroid* of the triangle ABC).

Other forms of this result

The result (5.9.1) may be written in other forms. A given ratio may be written in many ways. Thus the ratio $3:5$ may also be written as $6:10$ or $-30:-50$, etc. In particular, by multiplying the two parts of a ratio by a suitable constant, we can arrange that the first of them is 1. Thus the ratio $3:5$ can be written as $1:\frac{5}{3}$ (multiplying by $\frac{1}{3}$), and in general any ratio can be written in the form $1:k$. Clearly $k = \mu/\lambda$. If we assume that this has been done, then (5.9.1) becomes

(5.9.2) $$\left(\frac{kx_1 + x_2}{k + 1},\ \frac{ky_1 + y_2}{k + 1}\right).$$

Alternatively, by multiplying the two parts by a suitable constant we can arrange that in the new form of the ratio the sum of the parts is 1. Thus $3:5$ can be written as $\frac{3}{8}:\frac{5}{8}$ where $\frac{3}{8} + \frac{5}{8} = 1$, and in general any ratio may be written as $\lambda:\mu$ where $\lambda + \mu = 1$. If we assume this to have been done, then (5.9.1) becomes

(5.9.3) $$(\mu x_1 + \lambda x_2,\ \mu y_1 + \lambda y_2).$$

To carry this one stage further, put $\mu = t$, thus making $\lambda = 1 - t$. Then R is given by

(5.9.4) $$(tx_1 + (1 - t)x_2,\ ty_1 + (1 - t)y_2)$$

i.e., if R is the point (x, y) then

(5.9.5) $$\begin{aligned} x &= tx_1 + (1 - t)x_2 \\ y &= ty_1 + (1 - t)y_2. \end{aligned}$$

If we regard t not as a given constant, but as a parameter, then the equations (5.9.5) can be interpreted as parametric equations for the line joining (x_1, y_1) and (x_2, y_2).

The ratio $\lambda:\mu$ in (5.9.1) may be negative. If it is, then the point R is on the line containing P and Q but outside the segment PQ. This makes no difference to the above results. It is worth noting that, in (5.9.5), R is inside the segment if $0 < t < 1$; but is outside if $t < 0$ (in which case the three points occur on the line in the order P, Q, R) or $t > 1$ (in which case the order of the points is R, P, Q). (See Fig. 5.6)

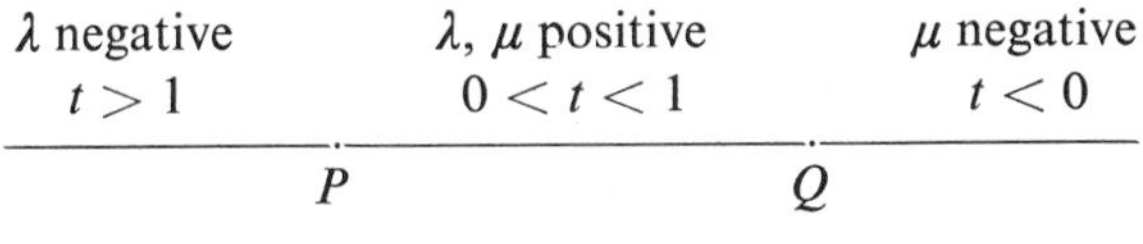

λ negative	λ, μ positive	μ negative
$t > 1$	$0 < t < 1$	$t < 0$

$$\overline{\hspace{2cm} P \hspace{4cm} Q \hspace{2cm}}$$

Fig. 5-6

We can usefully anticipate here a term that we shall make much use of later on (Chapter 7 and elsewhere). When λ and μ are nonnegative and $\lambda + \mu = 1$ then the point R, which can be written as $\mu P + \lambda Q$, is called a *convex combination* of P and Q.

Exercise 50

1. On a piece of graph paper plot the points $(1, 8)$ and $(7, 3)$. Label them P and Q respectively. Using (5.9.5.) plot the points which divide PQ in the ratio $1 - t : t$ where $t = \frac{1}{5}, \frac{2}{3}, \frac{1}{2}, 2, -2$.

2. Repeat problem 1 with

$$P \equiv (-3, -2) \quad \text{and} \quad Q \equiv (7, 9).*$$

3. Repeat problem 1 with

$$P \equiv (6, 5) \quad \text{and} \quad Q \equiv (-1, 1).$$

4. If P_1, P_2 are the points $(1, 2)$, $(8, 9)$, find the coordinates of the points which divide $P_1 P_2$ in the following ratios:

    (a) 1:6   **(c)** 3:4   **(e)** 5:2

    (b) −2:5   **(d)** −4:3   **(f)** −6:1

Check your results graphically.

5.10 ANGLE OF INTERSECTION OF TWO LINES

We have already seen that two lines will, in general, intersect in a point. When they do they form two angles at the points of intersection—one acute and one obtuse, as is illustrated in Fig. 5.7 by the angles marked θ and $\pi - \theta$. If the equations of the lines are given, these angles can be determined.

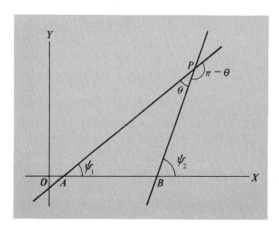

Fig. 5.7

Let the lines be given by

$$y = m_1 x + d_1$$

and

$$y = m_2 x + d_2,$$

---

  * $P \equiv (-3, -2)$, etc., is a brief, and standard, way of writing "P is the point whose coordinates are $(-3, -2)$."

then, referring to Fig. 5.7 we see that

$$m_1 = \tan \psi_1 \quad \text{and} \quad m_2 = \tan \psi_2.$$

From elementary geometry of the triangle we know that

(5.10.1) $\qquad\qquad \theta = \text{angle } APB = \psi_2 - \psi_1$

and since ψ_1 and ψ_2 can be found (their tangents m_1 and m_2 are known), equation (5.10.1) is sufficient to give the value of θ, so that the problem is in a sense, already solved. But the solution can be put in a better form. We write

$$\tan \theta = \tan (\psi_2 - \psi_1)$$
$$= \frac{\tan \psi_2 - \tan \psi_1}{1 + \tan \psi_1 \tan \psi_2}$$

by virtue of (3.10.6). Hence

(5.10.2) $\qquad\qquad \tan \theta = \dfrac{m_2 - m_1}{1 + m_1 m_2}$

—a result which enables us to find θ without finding ψ_1 and ψ_2 separately.

One might be tempted to ask "Why is the numerator $m_2 - m_1$, why not $m_1 - m_2$?" The answer is that two lines determine *two* angles at their point of intersection. These angles are supplementary (their sum is $180°$ or π radians) and therefore their tangents are equal and opposite. Hence

(5.10.3) $\qquad\qquad \tan \theta' = \dfrac{m_1 - m_2}{1 + m_1 m_2}$

will give the supplement θ' of the angle given by (5.10.2). In short, either of (5.10.2) or (5.10.3) will determine *one* of the angles in which the lines intersect, and the magnitude of the other is found immediately by taking the supplement. Moreover there can be no confusion over which angle is which since $\tan \theta$ is positive if θ is acute, and negative if it is obtuse.

Exercise 51

1. Find the angles in which the following pairs of lines intersect.
   (a) $x + y + 1 = 0, \qquad x - 2y - 3 = 0$
   (b) $5x - 3y + 1 = 0, \qquad 2x + 7y - 5 = 0$
   (c) $7x + 4y - 3 = 0, \qquad x - y + 7 = 0$
2. By finding the angles of intersection, or otherwise, prove that the angle of intersection of the lines

(a) $x - 3y + 12 = 0$

and

(b) $x + 2y + 5 = 0$

is the same as the angle of intersection of the lines

(c) $3x + y - 2 = 0$

and

(d) $2x - y - 1 = 0.$

5.11 PERPENDICULAR LINES

Two special cases of (5.10.2) are worthy of notice. First, if the lines are parallel then $\theta = 0$ and hence $m_1 = m_2$. This is something we already know. Secondly, if the lines are perpendicular, then $\theta = \pi/2$, and $\cot \theta = 0$. But

$$\cot \theta = \frac{1 + m_1 m_2}{m_2 - m_1},$$

so that we must have

(5.11.1) $\qquad 1 + m_1 m_2 = 0$ or $m_1 m_2 = -1.$

Conversely, if $m_1 m_2 = -1$ then $\cot \theta = 0$ and the lines are perpendicular. This then is the condition for two lines to be perpendicular—an important result.

It is to be expected that (5.10.2) would depend only on the slopes m_1 and m_2, and not on the constant terms d_1 and d_2; for altering the values of d_1 and d_2 simply replaces the lines by parallel lines, and this does not affect the angles at which they intersect.

Exercise 52

1. Show that the following pairs of lines are perpendicular.
    (a) $3x - y + 5 = 0,\quad 3x + y + 7 = 0$
    (b) $\frac{2}{7}x + \frac{1}{3}y + 1 = 0,\quad \frac{1}{6}x - \frac{1}{7}y - 3 = 0$
2. Show that any line perpendicular to the line $ax + by + c = 0$ has an equation of the form $bx - ay = $ constant. Hence find the equation of the lines which are perpendicular to the line $2x + 3y - 7 = 0$ and which contain the points $(4, -5)$, $(7, 9)$ and $(-8, -11)$ respectively.
3. Find the equations of the lines perpendicular to

$$2x + 3y + 4 = 0$$

and passing through

(a) (0, 1)

(b) (1, 2)

(c) (−1, −2)

(d) (0, 7)

4. Find the equations of the lines passing through the point (4, −2) and respectively

(a) parallel

(b) perpendicular to the line $2x - 3y = 4$.

Find also the coordinates of the foot of the perpendicular from (4, −2) to $2x - 3y = 4$.

5.12 LINES THROUGH THE INTERSECTION OF TWO LINES

Consider the following problem. What line contains the point of intersection of the lines

(5.12.1) $$2x - 5y - 61 = 0$$

and

(5.12.2) $$7x - 2y - 43 = 0$$

and has slope $-5/3$? We could solve this problem by first finding the point of intersection, and then using (5.4.1) to write down the equation of the line. We would find the point of intersection to be the point (3, −11), and the required equation of the line to be $5x + 3y + 18 = 0$. If the reader cares to check this result he will find that the calculation is quite straightforward and simple, so that it would seem pointless to look for an alternative method. Nevertheless we *shall* consider another method, because it will turn out to have certain advantages over the one just described.

If k is a constant, the equation

(5.12.3) $$(2x - 5y - 61) + k(7x - 2y - 43) = 0$$

which is a sort of combination of (5.12.1) and (5.12.2), is certainly linear and hence is the equation of a line. If we knew the coordinates of the point of intersection of the given lines, and substituted them into (5.12.3), then each of the two expressions $2x - 5y - 61$ and $7x - 2y - 43$ would become zero, because the point lies on both lines, and hence equation (5.12.3) would be satisfied. Hence the line given by (5.12.3) contains the point of intersection of the given lines. What is more, we have not yet given any particular value to k. We therefore hope to be able to choose k so as to fulfill the remaining condition on the required line, viz., that the slope is $-5/3$. Now (5.12.3) can be written as

$$(2 + 7k)x - (5 + 2k)y = 61 + 43k$$

so that its slope is $\dfrac{2 + 7k}{5 + 2k}$. Hence we must choose k so that $\dfrac{2 + 7k}{5 + 2k} = -\frac{5}{3}$, i.e.,

$$6 + 21k = -25 - 10k$$
$$31k = -31$$
$$k = -1.$$

For this value of k, (5.12.3) becomes

$$(2x - 5y - 61) - (7x - 2y - 43) = 0$$

i.e., $\qquad 5x + 3y + 18 = 0$, as before.

The advantage of this method is that, without our having to find the point of intersection of the given lines, the condition of containing this point of intersection is met immediately on writing down the combination (5.12.3), which, for two general lines $ax + by + c = 0$ and $px + qy + r = 0$ will be

(5.12.4) $\qquad (ax + by + c) + k(px + qy + r) = 0.$

It then remains only to choose k so that (5.12.4) satisfies whatever other condition is given. This other condition may be that the line should have a given slope, as in our example, or that it contains a given point, or indeed any statement which is sufficient to fix the line. Thus if we are asked for the line containing the point of intersection of the lines given by (5.12.1) and (5.12.2), and also the point $(7, -3)$, then the extra condition is that the coordinates $(7, -3)$ satisfy equation (5.12.3). For this to be so we must have

$$(14 + 15 - 61) + k(49 + 6 - 43) = 0$$

i.e., $\qquad -32 + 12k = 0$

or $\qquad k = \frac{8}{3}.$

With this value of k (5.12.3) becomes $2x - y - 17 = 0$.

Exercise 53

1. l_1, l_2, l_3, are lines all of which contain the point of intersections of the lines $2x - 3y - 1 = 0$ and $5x + y + 3 = 0$. The line l_1 has slope $-\frac{3}{4}$;

l_2 contains the point $(-2, 6)$; l_3 is parallel to the line $12x + 5y = 13$. Find the equations of l_1, l_2, l_3.

2. Find the equations of the lines through the point of intersection of the lines

$$2x - 3y + 1 = 0$$

and
$$5x + 2y - 4 = 0$$

which make an angle A with the line

$$x - 4y = 6$$

where $\tan A = \frac{1}{5}$.

5.13 PERPENDICULAR FORM OF THE EQUATION OF A LINE

Another important result is a formula for the perpendicular distance of a point from a line, but before we can derive this formula we must first consider yet another form of the equation of a line, known as the *perpendicular* form.

Let us take any line in the plane, and let N be the foot of the perpendicular from the origin to the line. We can specify the line uniquely by giving the distance ON (call it p), and the angle (call it α) which ON makes with the x-axis. Using Fig. 5.8 we can readily calculate the intercepts on the x- and y-axes.

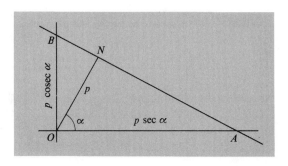

Fig. 5.8

In triangle ONA we have

$$\frac{ON}{OA} = \cos \alpha, \quad \text{i.e., } \frac{OA}{ON} = \sec \alpha$$

and hence $OA = p \sec \alpha$. In triangle ONB we see that angle $BON = \pi/2 - \alpha$, and hence that

$$\frac{ON}{OB} = \cos\left(\frac{\pi}{2} - \alpha\right)$$

from whence we derive

$$OB = p \sec\left(\frac{\pi}{2} - \alpha\right)$$

$$= p \operatorname{cosec} \alpha.$$

Hence from (5.5.3) the equation of the line is

$$\frac{x}{p \sec \alpha} + \frac{y}{p \operatorname{cosec} \alpha} = 1$$

which, on multiplication of both sides by p, can be written

(5.13.1) $\qquad\qquad x \cos \alpha + y \sin \alpha - p = 0.$

This is the perpendicular form of the equation of the line. It is not so very different from the general form $ax + by + c = 0$; all that is special about it is that the coefficients of x and y are respectively the cosine and sine of some angle. Suppose we are given the equation of a line, for instance $5x + 12y - 65 = 0$, how can we convert it into its perpendicular form? Certainly it is not already in perpendicular form, for the coefficients 5 and 12 are not the cosine and sine of any angle. Why not? Because, among other things, the cosine and sine of an angle must satisfy the equation $\cos^2 \alpha + \sin^2 \alpha = 1$, whereas $5^2 + 12^2 \neq 1$. But how can we alter our equation—it is already in a very simple form? There is only one way; we can multiply throughout by a constant (call it λ); for the equation

$$5\lambda x + 12\lambda y - 65\lambda = 0$$

represents the same line as $5x + 12y - 65 = 0$. Can we choose λ so that 5λ and 12λ are the cosine and sine of some angle? If so, then we must have

$$(5\lambda)^2 + (12\lambda)^2 = 1$$

i.e., $\qquad\qquad\qquad 169\,\lambda^2 = 1$

or $\qquad\qquad\qquad\qquad \lambda = \pm \tfrac{1}{13}.$

Thus to do the trick λ can have no values other than $\pm \tfrac{1}{13}$. But will these work? Take $\lambda = \tfrac{1}{13}$ and the equation becomes $\tfrac{5}{13}x + \tfrac{12}{13}y - 5 = 0$. This is indeed in perpendicular form, for $\tfrac{5}{13}$ and $\tfrac{12}{13}$ are the cosine and sine of the angle α defined by the right-angled triangle of Fig. 5.9. If we take $\lambda = -\tfrac{1}{13}$ we get essentially the same equation.

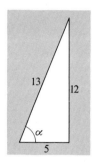

Fig. 5.9

We can now extend this method to the general case. To convert the line $ax + by + c = 0$ to perpendicular form we just divide throughout the equation by $\sqrt{a^2 + b^2}$. This will give

(5.13.2) $$\frac{a}{\sqrt{a^2 + b^2}}x + \frac{b}{\sqrt{a^2 + b^2}}y + \frac{c}{\sqrt{a^2 + b^2}} = 0$$

which is of the form (5.13.1) with $p = -\dfrac{c}{\sqrt{a^2 + b^2}}$ and α given by $\cos \alpha = \dfrac{a}{\sqrt{a^2 + b^2}}$, $\sin \alpha = \dfrac{b}{\sqrt{a^2 + b^2}}$, or (and this is the most convenient way of finding the angle α) $\tan \alpha = \dfrac{b}{a}$.

Exercise 54

1. Write the equations in Exercises 45.1 in perpendicular form.
2. There are two lines that are parallel to the line $3x - 4y + 7 = 0$ and at distance $\frac{7}{2}$ from it. Find their equations.
3. For what values of k is the line

$$5x + 13y - k = 0$$

at distance 4 from the origin?

5.14 DISTANCE OF A POINT FROM A LINE

We now derive the formula for the distance of a point from a line. Let the point be $P \equiv (x_1, y_1)$. The equation of the line can always be put in perpendicular form, and we shall suppose that this has been done, so that the equation is $x \cos \alpha + y \sin \alpha = p$. This is the line NM in Fig. 5.10, and the distance required (call it D) is PM, where PM is perpendicular to the line.

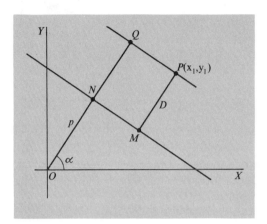

Fig. 5.10

Consider the line which is parallel to the given line and which contains P. If this line meets ON produced in Q (N being defined as for Fig. 5.8) then $QN = PM = D$. The line containing P has the same angle α associated with it as the given line; only its perpendicular distance from the origin is different, being $OQ = p + D$, instead of p. Hence its equation is

(5.14.1) $$x \cos \alpha + y \sin \alpha - (p + D) = 0.$$

The distance D is still to be found, but we know that the point P is on this line, and hence that (x_1, y_1) satisfies (5.14.1). Thus

$$x_1 \cos \alpha + y_1 \sin \alpha - (p + D) = 0$$

or

(5.14.2) $$D = x_1 \cos \alpha + y_1 \sin \alpha - p.$$

In other words, to find the distance of a point from a line we simply substitute its coordinates into the left-hand side of the equation of the line in perpendicular form. If the line is given in the general form $ax + by + c = 0$, we must first use (5.13.2) and then (5.14.2). This gives

$$D = \frac{a}{\sqrt{a^2 + b^2}}x_1 + \frac{b}{\sqrt{a^2 + b^2}}y_1 + \frac{c}{\sqrt{a^2 + b^2}}$$

or

(5.14.3) $$D = \frac{ax_1 + by_1 + c}{\sqrt{a^2 + b^2}}$$

which is the desired formula.

Exercise 55

1. Find the distances of the points $P \equiv (2, 5)$, $Q \equiv (-1, -1)$, $R \equiv (2, -3)$ from each of the lines given by

    (a) $3x + 4y - 7 = 0$
    (b) $5x - 12y + 3 = 0$
    (c) $2x - 3y - 1 = 0$

2. Show that the point $(14, 28)$ is equidistant from the three sides of the triangle whose vertices are the points $(50, 100)$, $(50, -44)$ and $(-46, 28)$
3. Find the length of the perpendicular from
    (a) the origin
    (b) the point $(1, 2)$ to the line $4x - 3y = 20$.
4. Find the length of the perpendicular drawn from the point $(-1, 2)$ to the line $3x - 4y = 7$.

5.15 CHANGE OF COORDINATE SYSTEM

The basic idea from which the material in this chapter has been derived is that by assigning pairs of coordinates to every point of a plane it becomes possible to handle geometrical objects (points, lines, etc.) and to prove geometrical theorems by purely algebraic methods, as opposed to the geometrical methods of Euclid. The coordinates that we assign to a particular point will naturally depend on our choice of coordinate system; points in a plane have no unique coordinates of themselves, but only in relation to the particular choice of origin and axes that we make. It is therefore of importance to ask the question "What is the connection between the coordinates of a point in one coordinate system and the coordinates of the same point in another system?"

In Fig. 5.11 we see two sets of coordinate axes in the same figure. Relative to the axes OX and OY the point P has coordinates (x, y) where $x = ON$ and $y = OM$; relative to the axes $O'X'$ and $O'Y'$ it has coordinates (x', y') where $x' = ON'$ and $y' = OM'$. The question we are asking is thus "What is the relationship between the numbers x, y and the numbers x', y'?" We shall split this question into two parts. We first consider what happens if we merely choose a different origin, without altering the direction of the axes, and then examine the effect of keeping the same origin but altering the directions of the axes.

If we change the origin only then we need to specify where the new origin is to be. This we can do by giving the coordinates of the new origin. Let us say that the new origin is to be the point O' with coordinates (α, β). The new

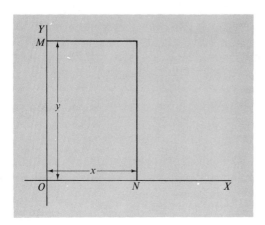

Fig. 5.11

x- and y-axes are lines through O' parallel to the old x- and y-axes. Figure 5.12 shows both sets of axes in the same diagram. The old coordinates of P are (x, y) where

$$x = ON \quad \text{and} \quad y = OM;$$

the new coordinates of P are (x', y') where

$$x' = O'N' \quad \text{and} \quad y' = O'M'.$$

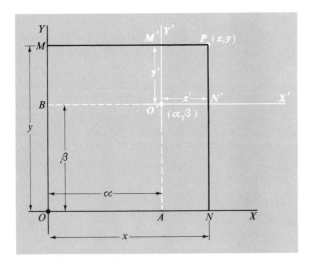

Fig. 5.12

From Fig. 5.12 we see that

$$x = ON = OA + AN = \alpha + x'$$

since $\alpha = OA$ and $AN = O'N' = x'$, while

$$y = OM = OB + BM = \beta + y'$$

since $\beta = OB$ and $BM = O'M' = y'$.

Hence the required relationship between the old and the new coordinates is

(5.15.1)
$$\begin{cases} x = x' + \alpha \\ y = y' + \beta. \end{cases}$$

Suppose we are given the equation

(5.15.2) $$lx + my + n = 0$$

of a line, relative to a certain coordinate system with origin O, and then change the origin to O' with coordinates (α, β). The line itself has not been changed, but because of the change of axes every point of the line will have different coordinates from the ones it had before. Therefore it is unlikely (though it can happen) that the new coordinates will be connected by equation (5.15.2). Instead the connection between the new coordinates (x', y') will, in general, be given by a different equation. What will this equation be?

This question is easily answered. If in (5.15.2) we replace x by $x' + \alpha$ and y by $y' + \beta$ then we obtain the equation connecting x' and y', viz.,

$$l(x' + \alpha) + m(y' + \beta) + n = 0$$
or $$lx' + my' + (l\alpha + m\beta + n) = 0.$$

Thus if the equation of the line is

$$3x - 2y + 7 = 0,$$

and we change the origin to the point $(2, 5)$, then

$$x = x' + 2$$
$$y = y' + 5$$

and the new equation of the line is

$$3(x' + 2) - 2(y' + 5) + 7 = 0$$
or $$3x' - 2y' + 3 = 0.$$

Note that the coefficients of x and y are the same as before. This is reasonable since the new axes are parallel to the old ones, and hence the slope of the line is the same as before. Only the constant term is altered by the change of origin.

Let us now look at the effect of choosing different axes with the same origin. This situation is shown in Fig. 5.13. The new axes, like the old, are orthogonal, i.e., at right angles to each other, and hence the angles XOX' and YOY' are equal. Call the angle θ. We can think of the axes OX' and OY' as having been obtained by rotating the axes OX and OY through an angle θ in the positive (counterclockwise) direction.

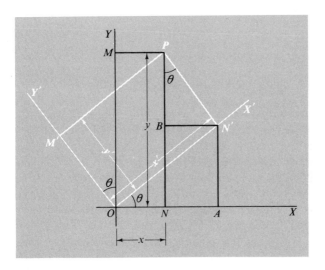

Fig. 5.13

The new coordinates are $x' = ON'$ and $y' = OM'$. Two extra points have been constructed in Fig. 5.13, namely A and B, the feet of the perpendiculars from N' to OX and PN. With their aid we can express x and y in terms of x' and y'. For

$$x = ON = OA - NA$$
$$= OA - BN'.$$

Now $OA/ON = \cos\theta$, so that $OA = ON' \cos\theta = x' \cos\theta$. In the same way we have $BN' = PN' \sin\theta$, since the angle $N'PB$ is also θ. But $PN' = M'O = y'$, so that $BN' = y' \sin\theta$. Hence

$$x = x' \cos \theta - y' \sin \theta.$$

Further,
$$\begin{aligned}
y = OM &= NB + BP \\
&= AN' + BP \\
&= ON' \sin \theta + PN' \cos \theta \\
&= x' \sin \theta + y' \cos \theta,
\end{aligned}$$

by the same sort of argument as before.

We now have the required result. It is

(5.15.3)
$$\begin{cases} x = x' \cos \theta - y' \sin \theta \\ y = x' \sin \theta + y' \cos \theta. \end{cases}$$

As before, we can find the new equation of the line with equation $lx + my + n = 0$ by substituting for x and y in this equation from (5.15.3). We get

$$l(x' \cos \theta - y' \sin \theta) + m(x' \sin \theta + y' \cos \theta) + n = 0$$

or

$$(l \cos \theta + m \sin \theta)x' + (m \cos \theta - l \sin \theta)y' + n = 0.$$

We finally note that we can now cope with any change of coordinate system. For we can change from one system to another by first moving the origin to its new position, and then rotating the axes into their new position (or we could do it the other way round).

Exercise 56

1. Equation (5.15.1) gives the old coordinates in terms of the new. It is very easy to turn these equations round the other way and obtain the equations

$$x' = x - \alpha$$
$$y' = y - \beta$$

for the new coordinates in terms of the old.

Not quite so easy is the task of turning equations (5.15.3) round the other way.

By solving equations (5.15.3) as simultaneous equations in x' and y', using the result $\cos^2 \theta + \sin^2 \theta = 1$, prove that

$$\begin{cases} x' = x \cos \theta + y \sin \theta \\ y' = -x \sin \theta + y \cos \theta. \end{cases}$$

2. If we change the origin without altering the directions of the axes, then in general the equation of a given line will become something different. Under what conditions can the equation of a line be the same before and after a change of origin?

3. Under what conditions, if at all, can a line have the same equation before and after a rotation of the axes?

4. The diagrams of Figs. 5.12 and 5.13 have been drawn for α, β positive and for θ an acute angle. The reader should satisfy himself, by drawing diagrams if necessary, that the two results proved in this section remain true, irrespective of the signs of α and β, and no matter what the angle θ is.

5. If the origin is moved to the point $(-1, 3)$ and the axes are rotated through $30°$ in a counterclockwise direction, find the equations (in the new coordinates) of the lines whose equations in the old coordinates are

(a) $x + y = 5$
(b) $3x - 2y + 7 = 0$
(c) $y = 2x - 15$

5.16 HALF-PLANES

One further observation is necessary before we leave the somewhat cramped plane for the more roomy expanses of three-dimensional space, which we explore in the next chapter. The result of applying (5.14.3) to find the distance of a point from a line may turn out to be either positive or negative. Now there is no significance in the fact that the distance is, say, positive. The distance of $(1, 1)$ from the line $3x - 4y + 5 = 0$ is $\frac{3 - 4 + 5}{5} = \frac{4}{5}$, which is positive; but if we had taken the equation in the form $-3x + 4y - 5 = 0$ then (5.14.3) would give $\frac{-3 + 4 - 5}{5} = \frac{-4}{5}$ — a negative result, even though the point and line are the same as before. However, if we stick to one form of the equation of the line, then reference to Fig. 5.10 and equation (5.14.1) will show that for points on one side of the line D will be positive, and for points on the other side it will be negative. But which side is which will depend on the way in which the equation of the line has been written.

Thus a line divides the points of the plane which are not on it into two sets, one consisting of those points for which D is positive, the other consisting of those points for which D is negative, provided we take the equation of the line in the same form all the time. These two sets are called *half-planes*.

Since we are here concerned only with signs and not the magnitudes, the factor $1/\sqrt{a^2 + b^2}$ in (5.14.3) is not important, and we can therefore say the that the equation $ax + by + c = 0$ represents a line which divides the plane into two half-planes such that for points of one half-plane (the positive half-plane) $ax + by + c > 0$, while for the other (the negative half-plane) $ax + by + c < 0$. The boundary between the two half-planes is the line itself, for which $ax + by + c = 0$. As we have already seen, there is no absolute significance in which half-plane is the positive one, and which is the negative. What *is* significant is the similarity or dissimilarity of the values of the linear function $ax + by + c$. If $ax_1 + by_1 + c$ and $ax_2 + by_2 + c$ are of the same sign, then the points (x_1, y_1) and (x_2, y_2) are on the same side of the line; if they are of opposite sign, then these points are on opposite sides of the line.

Exercise 57 (Chapter Review)

1. Find the 6 distances between the points $A \equiv (3, 2)$, $B \equiv (-1, 4)$, $C \equiv (7, 0)$, $D \equiv (-5, 1)$ taken in pairs.
2. Find the coordinates of each of the two points that make with $A \equiv (-1, 3)$ and $B \equiv (3, 6)$ an equilateral triangle.
3. Find the equation satisfied by the coordinates of points P such that $AP = 2BP$ where $A \equiv (4, -1)$ and $B \equiv (-3, -2)$.
4. Find the point of intersection of
   (a) the line whose intercepts on the axes are 3 and 5
   (b) the line whose intercepts on the axes are 4 and 1.
5. Find the equations of the following lines in the form $lx + my + n = 0$.
   (a) the line through $(5, -5)$ with slope $\frac{2}{5}$
   (b) the line joining $(10, -4)$ to $(3, 7)$
   (c) a line through $(3, 5)$ whose distance from the origin is $\sqrt{2}$. (Two answers)
   (d) the line whose parametric equations are

$$x = 1 + 5t$$
$$y = -2 - 3t$$

6. Find the two lines that are parallel to the line $2x + 3y - 11 = 0$ and are at distance $\frac{5}{3}$ from the point $(-1, 0)$.
7. By equating the corresponding coordinates and solving the two equations in s and t, find the point of intersection of the lines whose parametric equations are

(a)
$$x = 34 + 5s$$
$$y = -2 + 7s$$

(b)
$$x = -5 - 17t$$
$$y = 7 + 8t$$

and the corresponding values of s and t.

8. Three of the following lines are concurrent (i.e., contain a common point). Which are they?
  (a) $x - 2y = 9$
  (b) $x - 3y = 13$
  (c) $5x - 6y = 29$
  (d) $4x + 3y = 8$

9. If $A \equiv (1, 8)$, $B \equiv (9, 4)$, $C \equiv (-3, -1)$ and $D \equiv (6, 11)$ find the ratios in which the point of intersection of AB and CD divides these two line segments.

10. A, B, C are the three points $(6, 4)$, $(-1, 5)$ and $(-1, -3)$. L is the midpoint of BC, M the midpoint of CA, and N the midpoint of AB. Find the coordinates of L, M and N.

If p is the line through L, perpendicular to BC,

q is the line through M perpendicular to CA,

and r is the line through N perpendicular to AB,

prove that p, q and r meet in a point, and find the coordinates of this point.

11. Find the equations of the two lines through the point of intersection of the lines $x - y - 6 = 0$ and $3x - 5y - 16 = 0$ whose distance from the origin is 5.

12. Find what the equation of the line $7x - 5y + 6 = 0$ becomes if
  (a) the origin is changed to the point $(2, 3)$
  (b) the origin is changed to the point $(-1, -4)$
  (c) the axes are rotated counterclockwise through an angle of $45°$
  (d) the axes are rotated clockwise through $30°$

13. In the following examples state whether the two points are on the same or different sides of the given line:
  (a) $(1, -1)$, $(3, -5)$, $2x + 3y + 12 = 0$
  (b) $(0, 0)$, $(3, 5)$, $x - 2y + 7 = 0$
  (c) $(2, -1)$, $(6, 4)$, $2x + 5y - 10 = 0$
  (d) $(1, 3)$, $(4, 1)$, $x + y - 2 = 0$

14. Does the origin lie inside or outside the triangle whose vertices are $(1, 5)$, $(3, -1)$ and $(-3, 2)$?

15. Prove that, if the point (x, y) is equidistant from the axis OY and the point $(4, 0)$, then

$$y^2 = 8(x - 2).$$

16. The coordinates of two given points are $A(-2, 0)$, $B(2, 0)$. Prove that, if $P(x, y)$ is a point such that $PA = 3PB$, then

$$x^2 + y^2 - 5x + 4 = 0.$$

17. A is the point $(1, 3)$ and BC is the line whose equation is

$$5x + 12y = 7.$$

Find the equations of the lines through A
  (a) parallel to BC;
  (b) perpendicular to BC.

18. Calculate which pair of the three points $A(3, 1)$, $B(2, 5)$, $C(-2, -1)$ are on the same side of the straight line $2x - 5y + 7 = 0$.

19. Find the lengths of the sides of the triangle whose angular points are $(5, 1)$, $(-3, 7)$, $(8, 5)$ and show that one of the angles is a right angle.

Suggestions for Further Reading

What the reader will mainly need to do in order to strengthen his grasp of the subject matter of this chapter is to work a large number of examples. Plenty of these can be found in most textbooks on plane coordinate geometry; the following book in the "Outline" series is quite adequate for this purpose.

Rich, B., *Plane Geometry*. New York: Schaum Publishing Company, Ltd.

6 Coordinate Geometry
in Three Dimensions

6.1 INTRODUCTION

The inclusion of this quotation at the head of the chapter may seem like a calculated attempt on the part of the author to discourage the reader before he has even started. Needless to say, this was not the intention; but the difficulty to which this quotation refers is one which the reader will almost certainly discover for himself, so that it will not be out of place to make a comment or two about it before getting down to the main business of the chapter.

It is indeed true that students find the geometry of three-dimensional space to be much more difficult than that of two-dimensional space (the plane). Naturally one would expect it to be somewhat more difficult—on the simple basis of counting dimensions one might expect it to be one half as difficult again!—but if that were all, there would be little to complain of. The fact of the matter is that in three-dimensional space there is, so to speak, so much more room to move around in that a very much greater diversity of geometrical figures, and of relationships between them, becomes possible. Added to this there is the difficulty, mentioned in the quotation, that three-dimensional problems cannot be properly represented on a plane sheet of paper as can problems in plane geometry. Diagrams for figures in space can be drawn, of course, by the use of perspective, but these are not always satisfactory. Solid models are very useful, but cannot very conveniently be supplied in (or even with) a textbook, though the reader who has any facility at that sort of thing may profitably try constructing models to illustrate any ideas considered in this chapter that he may have trouble with.* Even models have their limitations, however, for it is rarely possible to view the whole of a model at once; there will usually be some parts of it hidden by others, whichever way you look at it.

With coordinate geometry the trouble is somewhat mitigated by the fact

* There are several construction kits on the market that are very suitable for this purpose, but even the crudest models (of the toothpick and chewing-gum kind) can be very helpful.

that one can frequently solve problems by concentrating on the algebra and giving little, if any, thought to the geometry which lies behind it. One might well say that the purpose of coordinate geometry is to enable one to do just this, and it is well worthwhile to cultivate the ability. A piece of advice to the reader at this stage, as a help towards acquiring this ability, is that he should note carefully the resemblances between some of the results which we shall prove in this chapter and certain results of the last chapter. Very often the resemblance is close, the result for three-dimensions merely having an extra variable in it; sometimes, because of necessary changes in notation, the resemblance is less obvious. In any case these resemblances will be brought to the attention of the reader, who should note them and compare (or contrast) the proofs with the corresponding proofs in Chapter 5. Doing this will also be of help in making the transition to the geometry of n dimensions which we introduce in the next chapter.

6.2 THE COORDINATE SYSTEM

To set up a coordinate system in three dimensions we need to take three lines (the coordinate axes) through some convenient point O (the origin). We take these axes to be mutually perpendicular, and call them the x-, y- and z-axes. At once we encounter a complication which hardly arose in two dimensions. Suppose we have fixed our x- and y-axes as in Fig. 6.1, where OX and OY denote the positive halves of these axes. The z-axis will then be a line through O perpendicular to the plane of the paper; but which half of this line are we to take as the positive half, and which the negative half? It really does not matter very much—we can choose either—but there is some advantage in making a consistent choice, and it is usual to choose as the positive half of the z-axis the half which, in Fig. 6.1, lies *above* the plane of the paper. The coordinate system which results is called a *right-handed* system of

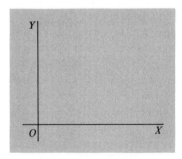

Fig. 6.1

axes;* if the other choice is made we call the resulting system a left-handed system. A set of right-handed coordinate axes will therefore look as in Fig. 6.2 in which the negative portions of the axes are shown dotted.

The x- and y-axes define a plane (the plane of the paper in Fig. 6.1) which is called the xy-plane. Similarly the yz- and zx- planes are defined. These three planes are called the coordinate planes. It is often convenient to think of the xy-plane (or one of the other two coordinate planes) as being "horizontal" and the z-axis as being "vertical." These terms, of course, have no real geometrical significance, but they can be a help to one's intuition when it is necessary to picture what is going on, whether in the mind's eye, or in a diagram or model.

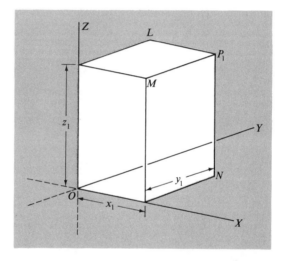

Fig. 6.2

Let us take a point P_1 anywhere in space, and consider the perpendiculars from it to the three coordinate planes. Let the feet of the perpendiculars to the yz-, zx- and xy- planes be L, M and N respectively (as in Fig. 6.2). The lengths P_1L, P_1M and P_1N are called the coordinates of the point P_1. P_1L, being perpendicular to OYZ, is parallel to the x-axis, and likewise P_1M is parallel to OZX and P_1N to OXY. For this reason the lengths P_1L, P_1M, P_1N are called, respectively, the x-, y- and z-coordinates of the point P_1, and if these lengths are denoted by x_1, y_1 and z_1 we say that P_1 is the point whose

* So called because it is possible to arrange the thumb, first finger and second finger of the right hand so that they point in the directions of the x-, y- and z-axes, respectively.

coordinates are (x_1, y_1, z_1). This is often abbreviated to 'P_1 is the point (x_1, y_1, z_1)', or more briefly still, to '$P_1 \equiv (x_1, y_1, z_1)$'.

So far this is a straightforward extension to three dimensions of the definition of coordinates in two dimensions that was given in Chapter 5. We now derive some standard results.

6.3 DISTANCE BETWEEN TWO POINTS

Let $P_1 \equiv (x_1, y_1, z_1)$ and $P_2 \equiv (x_2, y_2, z_2)$ be any two points. There will be three planes that contain the point P_1 and that are parallel to the coordinate planes. They are the planes P_1MN, P_1NL and P_1LM in Fig. 6.3.

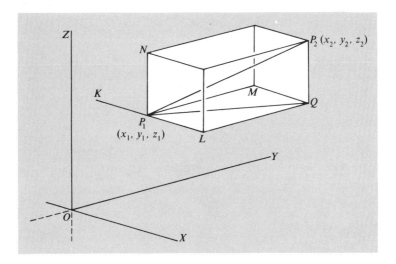

Fig. 6.3

Similarly there are three planes containing the point P_2 and parallel to the coordinate planes. These six planes will define the rectangular box shown in Fig. 6.3, of which the rectangle P_1LQM is one face. The diagonal P_1Q of this face is perpendicular to the side P_2Q of the box, and hence the triangle P_1P_2Q is a right-angled triangle. Applying the theorem of Pythagoras we have

$$P_1P_2{}^2 = P_1Q^2 + QP_2{}^2.$$

But the triangle P_1LQ is also a right-angled triangle, so that $P_1Q^2 = P_1L^2 + QP_2{}^2$. Hence

$$P_1P_2{}^2 = P_1L^2 + LQ^2 + QP_2{}^2$$

(6.3.1)
$$= P_1L^2 + P_1M^2 + P_1N^2$$

since $LQ = P_1M$ and $QP = P_1N$.

Now the face LQP_2 is parallel to the yz-plane, so that L is at the same perpendicular distance x_2 from this plane as P_2 is, i.e., $KL = x_2$. But the perpendicular distance of P_1 from this plane is x_1, i.e., $KP_1 = x_1$. It follows that

$$P_1L = KL - KP_1 = x_2 - x_1.$$

In the same way we can show that $P_1M = y_2 - y_1$ and $P_1N = z_2 - z_1$. Substituting these values into (6.3.1) we obtain the desired result,

(6.3.2) $$P_1P_2{}^2 = (x_2 - x_1)^2 + (y_2 - y_1)^2 + (z_2 - z_1)^2.$$

Apart from the extra term required for the extra variable z, this result is the same as (5.2.1).

Exercise 58

1. Write down the coordinates of the six vertices of the rectangular box of Fig. 6.3.

2. Find the distances between the following pairs of points.

(a) $(-3, 1, 7)$ and $(17, 13, 16)$
(b) $(5, -1, 3)$ and $(-20, -49, -33)$
(c) $(2, 7, 6)$ and $(14, 11, 9)$
(d) $(-1, 0, 5)$ and $(-3, -2, 4)$
(e) $(6, 1, -9)$ and $(26, 13, 0)$
(f) $(2, 1, 3)$ and $(0, -2, -2)$
(g) $(0, 0, 0)$ and $(25, 48, 36)$
(h) $(1, 1, -1)$ and $(-11, -3, -4)$
(i) $(-5, 3, 0)$ and $(0, 5, 3)$
(j) $(-2, -2, -2)$ and $(0, 1, 3)$
(k) $(7, 1, 5)$ and $(1, 0, 6)$
(l) $(-3, 2, 5)$ and $(5, 9, 1)$

6.4 DIRECTION COSINES

In plane geometry we can fix the direction of a line by giving its slope, that is to say by specifying (slightly indirectly) the angle that it makes with the x-axis. In three dimensions the picture is a little more complicated, but no different in principle. Consider first a line through the origin. This line will make certain angles with the three coordinate axes. We may denote by α, β and γ the angles that it makes with the x-, y- and z- axes respectively. Let P be any point of this line, and let L, M, N be the feet of the perpendiculars from P to the x-, y- and z-axes respectively. Finally, let $OP = r$ and we have Fig. 6.4.* Now P and L are equidistant from the yz-plane (since FL, being perpendicular to OX is parallel to this plane). Hence the x-coordinate of P is the distance OL, and from right-angled triangle OPL we see that $OL = r \cos \alpha$. Similarly the y- and z-coordinates of P are $r \cos \beta$ and $r \cos \gamma$, so that $P \equiv (r \cos \alpha,\ r \cos \beta,\ r \cos \gamma)$.

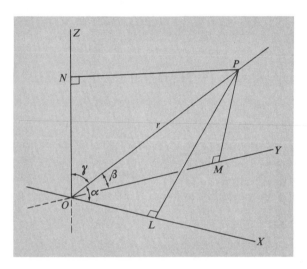

Fig. 6.4

Can the angles α, β and γ take any three values? No; there must be some connection between them. If, for example, $\alpha = \beta = 45°$ then the line must lie in the xy-plane, and hence γ must be $90°$, and cannot be any other angle. The connection between these angles is obtained at once if we use (6.3.1) to to find the distance OP. We find that

$$OP^2 = r^2 \cos^2 \alpha + r^2 \cos^2 \beta + r^2 \cos^2 \gamma.$$

But $OP = r$; hence

(6.4.1) $$\cos^2 \alpha + \cos^2 \beta + \cos^2 \gamma = 1,$$

which is the required relationship between the angles.

    * A little square or parallelogram "□" at the intersection of two lines in a diagram will denote that the lines are perpendicular. This is a useful device with drawings in perspective, in which right-angles do not always look like right-angles.

In specifying the direction of a line it is more usual to work with the cosines of the angles α, β and γ rather than with the angles themselves. We write $l = \cos \alpha$, $m = \cos \beta$, $n = \cos \gamma$ and call l, m, n the *direction cosines* of the line. By virtue of (6.4.1) we have a necessary relation between them, viz.,

(6.4.2) $$l^2 + m^2 + n^2 = 1.$$

If a line through the origin is given then the direction cosines l, m, n are determined and will satisfy (6.4.2). Conversely, given three numbers l, m, n which satisfy (6.4.2), these numbers will be the direction cosines of some line through the origin; for the line joining O to the point (l, m, n) will have just these direction cosines. Thus the direction cosines l, m, n of a line through the origin completely determine the line. In fact they do a little more than this. The points of such a line are all of the form (rl, rm, rn) as we have seen, and the value of r may be positive or negative according to which side of the origin the points are on. Thus the direction cosines not only determine the direction of the line, but also assign to it a *sense*, that is to say that enable us to make a distinction between the two halves into which the line is divided by the origin, calling one *positive* (when $r > 0$) and the other *negative*. Thus the numbers $\frac{2}{3}$, $\frac{2}{3}$ and $\frac{1}{3}$ are the direction cosines of some line (they satisfy (6.4.2)); the numbers $\frac{-2}{3}$, $\frac{-2}{3}$ and $\frac{-1}{3}$ are the direction cosines of what is essentially the same line except that the sense is the reverse of that determined by the first set of direction cosines—what was previously the positive half of the line is now the negative half, and vice versa. It should be remarked, however, that the sense of a line is not usually very important.

It is customary to abbreviate *direction cosines* to "d.c's," and we shall do this from now on when convenient.*

The everyday meaning of "direction" is such that parallel lines have the same direction. We therefore define the d.c's of a line which does not contain the origin to be the same as the d.c's (defined as above) of the parallel line which *does* contain the origin. In this way we can allot a set of d.c's to every line in space, and parallel lines will have the same d.c's (except, possibly, that their signs may be reversed). It follows that the d.c's of a line are not enough to determine the line; we need more information, something that will enable us to pick out the required line from the many parallel lines having the given d.c's. This information is forthcoming if we are told the coordinates of a point on the line.

* If we carry out the same procedure in two dimensions as we have just done in three, we find that the "direction cosines" of a line through the origin in two dimensions are $\cos \psi$ and $\sin \psi$, in the notation of the previous chapter, and these numbers satisfy $\cos^2 \psi + \sin^2 \psi = 1$, which is the analogue of (6.4.1). However, since the single number m, $= \tan \psi$, fixes the line just as well (except for the sense, which is a minor detail) it is not usual to work with these direction cosines separately.

Exercise 59

1.   Which of the following sets of three numbers are the direction cosines of some line?

(a) $-\frac{1}{3}, \ -\frac{2}{3}, \ \frac{2}{3}$

(b) $\frac{1}{\sqrt{3}}, \ -\frac{1}{\sqrt{3}}, \ \frac{1}{\sqrt{3}}$

(c) $\frac{1}{2}, \ \frac{1}{4}, \ \frac{1}{4}$

(d) $-\frac{3}{13}, \ \frac{4}{13}, \ -\frac{12}{13}$

(e) $\frac{2}{3}, \ -\frac{2}{3}, \ -\frac{1}{3}$

(f) $\frac{2}{3}, \ -\frac{1}{3}, \ \frac{2}{3}$

(g) $\frac{3}{5}, \ \frac{4}{5}, \ -\frac{2}{5}$

2.   Using tables of the cosine function, determine the angle that a line makes with the z-axis if the angles that it makes with the x- and y-axes are, respectively

(a)   20° and 80°

(b)   45° and 65°

(c)   60° and 60°

(d)   42° and 73°

6.5   THE LINE HAVING GIVEN D.C'S AND CONTAINING A GIVEN POINT

Let the d.c's of a line be l, m, n, and let $A \equiv (x_1, y_1, z_1)$ be a point on it. Take a general point $P \equiv (x, y, z)$ on the line, and through the point A take lines AU, AV, AW which are parallel to OX, OY and OZ respectively. If OP' is the line through O parallel to AP, then the angle PAU will be the same as the angle $P'OX$. Thus $PAU = \alpha$, and, similarly, $PAV = \beta$ and $PAW = \gamma$. If we now draw the perpendiculars PL, PM and PN from P to AU, AV and AW, respectively, then the heavily printed portion of Fig. 6.5 is a replica of Fig. 6.4. Consequently, if $AP = r$ then $AL = r \cos \alpha$. But AL is the distance of L, and hence of P, from the plane VAW, and this plane, in turn, is at a distance of x_1 from the yz-plane. Hence the distance of P from the yz-plane is $x_1 + r \cos \alpha$, and this is the x-coordinate of P. Thus we obtain

$$x = x_1 + r \cos \alpha = x_1 + rl.$$

Similar results hold for the y- and z-coordinates, and we find that the point P is given by the equations

$$x = x_1 + rl$$

(6.5.1)
$$y = y_1 + rm$$

$$z = z_1 + rn.$$

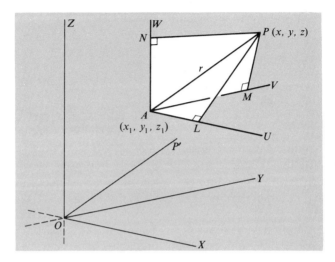

Fig. 6.5

Since these equations express the coordinates of P in terms of a parameter r they are, in effect, parametric equations of the line. They are the analogues of equations (5.7.1) of the last chapter.

If we eliminate r between the first two equations in (6.5.1) we obtain

(6.5.2)
$$\frac{x - x_1}{l} = \frac{y - y_1}{m}$$

while the elimination of r between the second two gives

(6.5.3)
$$\frac{y - y_1}{m} = \frac{z - z_1}{n}.$$

Equations (6.5.2) and (6.5.3) can be written together as

(6.5.4)
$$\frac{x - x_1}{l} = \frac{y - y_1}{m} = \frac{z - z_1}{n}.$$

There is a third equation which can be obtained from (6.5.1) viz., $\frac{x - x_1}{l} =$ $\frac{z - z_1}{n}$, but it gives us no further information, for it is automatically satisfied if (6.5.2) and (6.5.3) are satisfied. Thus (6.5.4) represents essentially *two* equations —the equations of the line.

The analogue of (6.5.4) in two dimensions is (5.4.1), which, as we saw, can be written in the form

$$\frac{x - x_1}{\cos \psi} = \frac{y - y_1}{\sin \psi}$$

which emphasizes the resemblance to (6.5.4).

Exercise 60

1. Write down the equations of the following lines
   (a)   the line through $(5, -3, 7)$ with d.c's $\frac{2}{3}, \frac{1}{3}, -\frac{2}{3}$
   (b)   the line through $(0, 1, -3)$ with d.c's $\frac{1}{\sqrt{2}}, \frac{1}{\sqrt{2}}, 0$
   (c)   the line through $(2, 1, 6)$ with d.c's $-\frac{3}{13}, \frac{4}{13}, \frac{12}{13}$

2. Find the d.c's of the lines joining the pairs of points given in Exercise 58.2.

3. Find the equations of the lines joining the pairs of points given in Exercises 58.2. (Use the results of problem 2 above.)

6.6 LINE JOINING TWO POINTS

Another way in which we may fix the position of a line is to specify two points on it. If the points are $P \equiv (x_1, y_1, z_1)$ and $P_2 \equiv (x_2, y_2, z_2)$, then we may find the equation of the line joining them by first finding its direction cosines and then substituting them in (6.5.4). Let us denote the d.c's by l, m, n, as before. The equations of the line are then given by (6.5.4) and since P_2 lies on the line its coordinates will satisfy these equations. Hence

$$\frac{x_2 - x_1}{l} = \frac{y_2 - y_1}{m} = \frac{z_2 - z_1}{n} = r,$$

where $r = P_1 P_2$. But we already know the distance $P_1 P_2$. It is $(x_2 - x_1)^2 + (y_2 - y_1)^2 + (z_2 - z_1)^2$. Hence the direction cosines of the line are

(6.6.1)

$$\frac{x_2 - x_1}{\sqrt{(x_2 - x_1)^2 + (y_2 - y_1)^2 + (z_2 - z_1)^2}},$$

$$\frac{y_2 - y_1}{\sqrt{(x_2 - x_1)^2 + (y_2 - y_1)^2 + (z_2 - z_1)^2}},$$

and

$$\frac{z_2 - z_1}{\sqrt{(x_2 - x_1)^2 + (y_2 - y_1)^2 + (z_2 - z_1)^2}}.$$

We now know the direction cosines of the line $P_1 P_2$ and a point on it (either P_1 or P_2 will do—we shall take P_1), so we can write down its equations in the form (6.5.4). This would result in the cumbersome expressions of (6.6.1) occurring in the denominators, and this is something that we can avoid. Suppose that λ, μ, and ν are three numbers that are proportional to the d.c's l, m, n of a line, so that $\lambda/l = \mu/m = \nu/n$. Then the equations

(6.6.2) $$\frac{x - x_1}{\lambda} = \frac{y - y_1}{\mu} = \frac{z - z_1}{\nu}$$

will represent the same line as (6.5.4), since they can be obtained from (6.5.4) by multiplying throughout by some constant, and any values x, y, z which satisfy (6.5.4) will automatically satisfy (6.6.2) and vice versa. Hence it is not necessary that the denominators in (6.5.4) should be the d.c's of the line; we can use instead any set of numbers proportional to them. For our present problem this means that we need not use the d.c.'s given in (6.6.1) to write down the required equations of the line $P_1 P_2$, but can use instead the numbers

$$x_2 - x_1, \quad y_2 - y_1 \quad \text{and} \quad z_2 - z_1$$

which are proportional to them. Doing this we obtain

(6.6.3) $$\frac{x - x_1}{x_2 - x_1} = \frac{y - y_1}{y_2 - y_1} = \frac{z - z_1}{z_2 - z_1}$$

as the equations of the line $P_1 P_2$.

Exercise 61

1. There are $\binom{5}{2} = 10$ lines joining five points in pairs. Find the equations of these 10 lines when the 5 points are

$$A \equiv (2, 1, -3)$$
$$B \equiv (7, 0, 5)$$
$$C \equiv (-3, 1, 2)$$
$$D \equiv (5, -4, 3)$$
$$E \equiv (2, 2, -2)$$

2. Which of the following 5 points lie on the line joining $(12, 4, -8)$ to $(-6, 1, 10)$?

    (a) $(18, 11, -7)$

    (b) $(14, 5, 6)$

    (c) $(6, 3, -2)$

    (d) $(18, 1 - 17)$

    (e) $(6, -10, 5)$

6.7 PROJECTION OF A LINE SEGMENT ON ANOTHER LINE

In two dimensions two lines which are not parallel will meet in a point. This is not so in three dimensions, and we refer to two nonparallel lines which do not meet as *skew lines*. Let p and q be two skew lines, and P_1 and P_2 any two points of p, defining a line segment P_1P_2. Let Q_1 and Q_2 be the feet of the perpendiculars to q from P_1 and P_2 respectively. Then the length Q_1Q_2 is called the *projection* of the segment P_1P_2 on the line q. This construction is shown in Fig. 6.6.

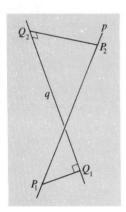

Fig. 6.6

If we replace the line q by some other line q', parallel to q, then the projection of P_1P_2 on q' will be the same as that on q. This is perhaps not immediately obvious, but it is easy to see if we imagine that we have made a model of Fig. 6.6 and then tipped it up so that the line q becomes vertical (Fig. 6.7). Q_1 will then lie in the same horizontal plane as P_2. These planes are shown in Fig. 6.7. Clearly if we take a line q', parallel to q, this line will also be vertical and the new projection (the length of $Q_1'Q_2'$, where Q_1', Q_2' are the feet of the perpendiculars from P_1, P_2 to q') will be same as before, since in either case it is simply the vertical distance between the two planes, the positions of which do not depend on whether we take q or q'. In the same way

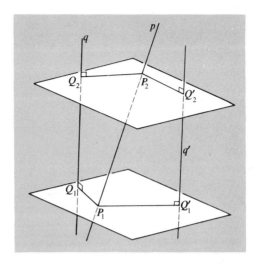

Fig. 6.7

we may show that we may replace the segment P_1P_2 by any segment of equal length on any line parallel to p. Hence we would expect the projection of P_1P_2 on q to depend only on the length P_1P_2 and the directions of the lines p and q. We shall see shortly that this is so.

Let us replace q by the parallel line q'' through P_1, thus obtaining Fig. 6.8. The foot of the perpendicular from P_1 to q'' is, of course, P_1 itself. If Q_2'' is the foot of the perpendicular from P_2 then from Fig. 6.8 the projection is

(6.7.1) $$P_1Q_2'' = P_1P_2 \cos \theta,$$

where θ is the angle between the lines p and q''. It is convenient to call θ

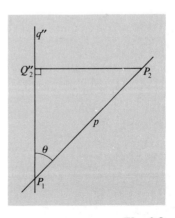

Fig. 6.8

the angle between the original lines p and q also. In other words we may talk about the angle between two lines which do not intersect as being the angle (measured in the usual way) between any two lines, parallel to the given lines, which *do* intersect.

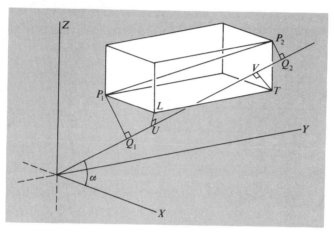

Fig. 6.9

We are now equipped to derive a formula for the projection on a line q whose d.c's are l, m and n of the segment $P_1 P_2$ joining the points $P_1 \equiv (x_1, y_1, z_1)$ and $P_2 \equiv (x_2, y_2, z_2)$. Figure 6.9 shows the rectangular box with sides parallel to the coordinate planes and with P_1 and P_2 as diagonally opposite vertices (compare Fig. 6.3). From the vertices P_1, L, T and P_2 of this box we draw perpendiculars to the given line q, the feet of these perpendiculars being Q_1, U, V and Q_2, respectively. Now $P_1 L = x_2 - x_1$ and the angle between $P_1 L$ and the line q is the same as the angle between OX and q, since $P_1 L$ is parallel to OX. Hence this angle is α, where $\cos \alpha = l$. The projection $Q_1 U$ of $P_1 L$ on q is therefore $(x_2 - x_1) \cos \alpha$, or $l(x_2 - x_1)$. Similarly the projection UV of LT on q is $m(y_2 - y_1)$, and the projection VQ_2 of TP_2 on q is $n(z_2 - z_1)$. Hence the projection $Q_1 Q_2$ of $P_1 P_2$ on q is given by

(6.7.2)
$$Q_1 Q_2 = Q_1 U + UV + VQ_2$$
$$= l(x_2 - x_1) + m(y_2 - y_1) + n(z_2 - z_1).$$

Exercise 62

1. Find the projections of the segments whose endpoints are the pairs of points in Exercise 58.2 on the lines with the following cosines

(a) $\frac{2}{3}$, $\frac{2}{3}$, $-\frac{1}{3}$

(b) $\frac{2}{3\sqrt{6}}$, $\frac{1}{3\sqrt{6}}$, $\frac{7}{3\sqrt{6}}$

(c) $-\frac{12}{13}$, $\frac{4}{13}$, $-\frac{3}{13}$

2. If $A \equiv (2, 1, 3)$, $B \equiv (5, 5, 15)$, $C \equiv (-1, 2, 6)$ and $D \equiv (11, 6, 9)$ find the length of the projection of the segment AB on the line CD. Show that it is the same as the projection of CD on the line AB.

6.8 ANGLE BETWEEN TWO LINES

We have defined the angle between two skew lines as being the angle between any two intersecting lines parallel to the given lines. This angle can therefore depend only on the d.c's of the given lines. Suppose these d.c's are l, m, n and l', m', n', and let us consider the lines q and p through the origin having these d.c's, respectively. The points $(0, 0, 0)$ and $(l', m', n',)$ are points on p, and if we apply (6.7.1) we find that the projection on q of the line segment joining these two points is

$$l(l' - 0) + m(m' - 0) + n(n' - 0) = ll' + mm' + nn'.$$

On the other hand we also know that the projection is $D \cos \theta$, where D is the length of the segment and θ is the angle between p and q. Here $D = 1$, so we finally obtain

(6.8.1) $$\cos \theta = ll' + mm' + nn'$$

which enables us to calculate the angle between two lines whose d.c's are known. It should be noticed that, as we saw in the last chapter, two lines which meet make *two* angles at the point of intersection. The formula just derived will give one of these angles; the other can be deduced immediately since the two angles are supplementary.

An important special case of this result arises when the lines are perpendicular. The angle θ is then a right-angle and $\cos \theta = 0$. Consequently the condition for the two lines to be perpendicular is that their d.c's should satisfy the equation.

(6.8.2) $$ll' + mm' + nn' = 0.$$

6.9 DIRECTION RATIOS OF A LINE

It is often convenient to specify a line not by its three direction cosines, but by three numbers that are proportional to them. We have already seen one example of how three such numbers can be used instead of direction cosines in the derivation of (6.6.2). Such a set of three numbers gives almost as much information as the d.c's themselves, for from them we can almost reconstruct the d.c's (the force of the word "almost" will appear in a moment). Suppose λ, μ and ν are any three numbers; then the numbers

$$l = \frac{\lambda}{\sqrt{\lambda^2 + \mu^2 + \nu^2}}; \quad m = \frac{\mu}{\sqrt{\lambda^2 + \mu^2 + \nu^2}}; \quad n = \frac{\nu}{\sqrt{\lambda^2 + \mu^2 + \nu^2}}$$

will be direction cosines of some line since they satisfy the equation $l^2 + m^2 + n^2 = 1$. Hence the numbers λ, μ and ν will determine a line.

Now the square root $\sqrt{\lambda^2 + \mu^2 + \nu^2}$ may be taken as either positive or negative, so that, strictly speaking, from the numbers λ, μ and ν we may obtain *two* sets of d.c's. These will represent the same line, but will give opposite senses to it. Hence three numbers will determine a line, whose d.c's are proportional to those numbers, but will not allot a sense to it (this is the reason for the "almost" in the last paragraph).

What this amounts to is that what really determine the line are the ratios between its d.c's rather than the d.c's themselves. These ratios are called the direction ratios of the line, and the phrase is often abbreviated to d.r's. Thus the ratios $1 : 2 : 2$ will determine a line, viz., the line whose d.c's are $\frac{1}{3}$, $\frac{2}{3}$ and $\frac{2}{3}$, since $3^2 = 1^2 + 2^2 + 2^2$. These same ratios can be written in many ways, for example as $5 : 10 : 10$, or as $\frac{1}{71} : \frac{2}{71} : \frac{2}{71}$ or $-4 : -8 : -8$ and so on, but all give the same d.c's (disregarding the ambiguity of sign). In writing d.r's we shall write the three numbers with colons in between, to emphasize that the ratios between the numbers rather than the numbers themselves are the important data; that is, we shall say, for example, "the line whose direction ratios are $3 : 1 : 2$" rather than "the line whose direction ratios are 3, 1 and 2." This latter phraseology is common in textbooks but strictly speaking is incorrect.

In many formulae (though by no means all) it does not matter whether we use d.c's or d.r.'s. Thus, instead of (6.5.1) the parametric equations of a line may be written

$$x = x_1 + \lambda t$$
(6.9.1)
$$y = y_1 + \mu t$$
$$z = z_1 + \nu t$$

where $\lambda : \mu : \nu$ are the d.r.'s of the line, (x_1, y_1, z_1) is a point on it, and t is the parameter. Note that the parameter t is not the distance of (x, y, z) from (x_1, y_1, z_1) as was the parameter r in (6.5.1), when we used the d.c's of the line, but is proportional to this distance.* Again, the equations of the same line will be

(6.9.2)
$$\frac{x - x_1}{\lambda} = \frac{y - y_1}{\mu} = \frac{z - z_1}{\nu}$$

* Compare Chapter 5, Exercise 47.3.

which is of the same form as (6.5.4), in which d.c's were used. The d.r's of the line containing (x_1, y_1, z_1) and (x_2, y_2, z_2) are $x_2 - x_1 : y_2 - y_1 : z_2 - z_1$ (see (6.6.1), so that, from (6.9.2) we can derive (6.6.2) immediately. Another important result is that in the condition (6.8.2), for two lines to be perpendicular, d.r's can be used instead of d.c's, i.e., this condition may be written

$$(6.9.3) \qquad \lambda\lambda' + \mu\mu' + \nu\nu' = 0.$$

However, in the formula (6.7.2) for the angle between two lines d.c's *must* be used; if the d.r's of the two lines are given these must first be converted to d.c's.

Exercise 63

1. What are the d.c's corresponding to the following sets of d.r's.
   (a) $5 : 1 : -3$
   (b) $2 : -5 : 4$
   (c) $-1 : -2 : -2$
   (d) $3 : -4 : 12$

2. Find the d.r's of the lines AB and CD where $A \equiv (2, 1, 0)$, $B \equiv (14, 10, -21)$, $C \equiv (3, -1, 6)$ and $D \equiv (7, 2, -1)$ and hence show that they are parallel.

3. Show that the lines in problem 2 above are perpendicular to the lines that make equal angles with the coordinate axes.

4. Find the angles between the lines, taken in pairs, whose direction cosines are

   (a) $-\frac{1}{\sqrt{21}}, -\frac{2}{\sqrt{21}}, \frac{4}{\sqrt{21}}$
   (b) $\frac{3}{\sqrt{14}}, -\frac{1}{\sqrt{14}}, \frac{2}{\sqrt{14}}$
   (c) $\frac{1}{\sqrt{3}}, \frac{1}{\sqrt{3}}, -\frac{1}{\sqrt{3}}$
   (d) $\frac{4}{5}, 0, -\frac{3}{5}$

There are six possible pairs.

5. What equation is satisfied by the d.c's l, m, n of a line if it makes an angle of 60° with the line whose d.c's are $\frac{2}{3}, -\frac{1}{3}$ and $\frac{2}{3}$?

6. Find the angles of the triangle whose vertices are the points $(2, 1, -3)$, $(5, -9, 7)$ and $(6, -3, -2)$.

6.10 POINT DIVIDING A LINE SEGMENT IN A GIVEN RATIO

Let PQ be any line segment, where $P \equiv (x_1, y_1, z_1)$ and $Q \equiv (x_2, y_2, z_2)$, and let $R \equiv (x, y, z)$ be the point which divides PQ in the ratio $\lambda : \mu$. Let

PM, RW and *QN* be the perpendiculars from *P, R* and *Q* to the *xy*-plane, *V* the foot of the perpendicular from *R* to *QN*, and *U* the foot of the perpendicular from *P* to *RW*. (Compare Fig. 5.4.) The triangles *PRU* and *RQV* are similar, and consequently $PR/RQ = UR/VQ$. But $UR = WR - WU = z - z_1$, and $VQ = NQ - NV = NQ - WR = z_2 - z$. Therefore, since

$$\frac{PR}{RQ} = \frac{\lambda}{\mu}$$

we have

$$\frac{z - z_1}{z_2 - z} = \frac{\lambda}{\mu}$$

from which it follows that

$$z = \frac{\mu z_1 + \lambda z_2}{\lambda + \mu}.$$

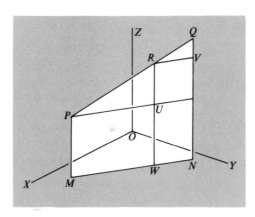

Fig. 6.10

In the same way the *x*- and *y*-coordinates of *R* may be found, and *R* turns out to be the point

(6.10.1) $\left(\dfrac{\mu x_1 + \lambda x_2}{\lambda + \mu}, \dfrac{\mu y_1 + \lambda y_2}{\lambda + \mu}, \dfrac{\mu z_1 + \lambda z_2}{\lambda + \mu} \right).$

The remarks that we made about the corresponding formula in two dimensions (5.9.1) apply equally well here. In particular we can write (6.10.1) as

$(\beta_2 x_1 + \beta_1 x_2, \beta_2 y_1 + \beta_1 y_2, \beta_2 z_1 + \beta_1 z_2)$ where $\beta_1 + \beta_2 = 1,$

or as

(6.10.2) $(t x_1 + (1 - t) x_2, t y_1 + (1 - t) y_2, t z_1 + (1 - t) z_2).$

Finally, this last result can be written in the form of parametric equations for the line joining P_1 and P_2, viz.

$$\begin{aligned} x &= tx_1 + (1 - t)x_2 \\ y &= ty_1 + (1 - t)y_2 \\ z &= tz_1 + (1 - t)z_2 \end{aligned}$$

(6.10.3)

—the three-dimensional analogue of (5.9.5).

As we did at the end of the corresponding section of Chapter 5 (page 187) we can usefully remark here that a point $R \equiv (x, y, z)$ given by (6.10.3) with $0 \le t \le 1$ or (what amounts to the same thing) given by (6.10.1) with $\lambda + \mu = 1, \lambda, \mu \ge 0$, is often called a convex combination of the points $P \equiv (x_1, y_1, z_1)$ and $Q \equiv (x_2, y_2, z_2)$.

Exercise 64

1. Write down the coordinates of the points which divide the following pairs of points:

(a) $(3, 2, -4)$ and $(7. 6, 1)$
(b) $(-3, -2, 6)$ and $(7, 3, -9)$
(c) $(2, 0, 8)$ and $(-5, 14, 1)$

in the ratios $1 : 4$ and $3 : 4$.

2. Write down, and simplify the parametric equations of the line joining the points $(3, -1, \frac{1}{2})$ and $(2, -\frac{1}{4}, 4)$.

6.11 EQUATION OF A PLANE

In three dimensions a plane is determined by one equation; in other words, a plane is a set of points which have the property that their coordinates satisfy a certain equation. We shall see in a moment that this equation is a linear one. Thus in *one* sense (that of being specified by a single linear equation) planes play the same role in three-dimensional geometry as lines do in plane geometry.

Let π be any plane; let N be the foot of the perpendicular from the origin O to π, and let $ON = p$. Let the d.c.'s of the perpendicular ON be l, m, and n, and let P be any point (x, y, z) of the plane π (Fig. 6.11). Now what is it that distinguishes points belonging to π from points that do not belong to π? It is that the line PN is perpendicular to ON. (Imagine Fig. 6.11 tilted so that the line ON is vertical and π horizontal and this statement becomes obvious.) But saying that PN is perpendicular to ON is equivalent to saying that the

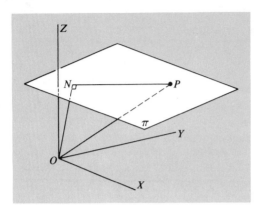

Fig. 6.11

projection of OP on to the line ON is the length ON itself, or p. From this the equation of the plane follows, for the projection of OP on the line ON is, by (6.8.1), simply $lx + my + nz$, and this must therefore be equal to p, if the point P lies on the plane. Hence the required equation is

$$lx + my + nz = p$$

or

(6.11.1) $$lx + my + nz - p = 0.$$

A line perpendicular to a plane (such as the line ON above) is called a *normal* to the plane. All normals to a plane are parallel, and hence have the same direction cosines, and parallel planes will have the same normals. We may therefore describe (6.11.1) as the equation of the plane whose normals have direction cosines l, m, n, and whose (perpendicular) distance from the origin is p. Note that if we omit the term in z, we get an equation in x and y which, with a slight change in notation, is the perpendicular form of the equation of a line in two dimensions [compare equation (5.13.1)].

We see that the equation of a plane is a linear equation. The converse is also true; any linear equation, say

(6.11.2) $$ax + by + cz + d = 0$$

will be the equation of a plane. For although the coefficients of x, y and z in this equation will not, in general, be the direction cosines of a line, as are those in (6.11.1), we can write the equation as

(6.11.3)
$$\frac{a}{\sqrt{a^2 + b^2 + c^2}} x + \frac{b}{\sqrt{a^2 + b^2 + c^2}} y +$$
$$\frac{c}{\sqrt{a^2 + b^2 + c^2}} z = \frac{-d}{\sqrt{a^2 + b^2 + c^2}}$$

in which the coefficients *are* direction cosines, since the sum of their squares is 1. Hence (6.11.3) is the equation of a plane. But since (6.11.2) and (6.11.3) are essentially the same equation—they determine the same set of points— it follows that (6.11.2) is also the equation of a plane. It is the plane whose normals have direction ratios $a : b : c$, and whose distance from the origin is $-d/\sqrt{a^2 + b^2 + c^2}$.

Exercise 65

1.   Find the equation of the plane through the point (3, 0, 4) normal to the line joining the points (2, 0, -5) and (7, -2, 4).
2.   Find the equations of the planes given by the following data:
   (a)   through (-3, 6, 5), normal to the direction with d.r's $2 : -1 : 10$
   (b)   through (-1, -1, 4), normal to the direction with d.r's $21 : 13 : -8$
   (c)   normal to the direction with d.r's $-3 : -4 : 12$ and at distance 5 from the origin. (Two answers.)

6.12   LINE OF INTERSECTION OF TWO PLANES

We have already seen that the line in three dimensions is represented by *two* equations, e.g., (6.5.4). If we consider these two equations separately, i.e., consider, say,

$$\frac{x - x_1}{l} = \frac{y - y_1}{m}$$

and

$$\frac{y - y_1}{m} = \frac{z - z_1}{n}$$

then each of these equations, being linear, will represent a plane. Thus any line can be displayed as the intersection of two planes—it consists of those points which belong to *both* planes. It is not difficult to see that the intersection of two planes which are not parallel is a line. Consider, by way of example, the equations

(6.12.1)
$$3x - 5y + 2z = 1$$
$$2x - y - 8z = 3$$

and let us solve these two equations for two of the variables, say x and y. We write the equations as

$$3x - 5y = 1 - 2z$$
$$2x - y = 3 + 8z,$$

and solve in the usual way to obtain $x = 2 + 6z$ and $y = 1 + 4z$. These two results can be written together as

(6.12.2)
$$\frac{x - 2}{6} = \frac{y - 1}{4} = \frac{z}{1}$$

which is of the form (6.5.4); and any values of x, y, z which satisfy equations (6.12.2) will also satisfy (6.12.1) and vice versa.

This, of course, is only a particular example; but if the reader cares to carry out the same procedure with two general equations, say $ax + by + cz + d = 0$ and $px + qy + rz + s = 0$, he will find that a similar result holds. (See example 2 below, however, for a difficulty which may arise in exceptional cases.)

This method, of solving two equations for two of the variables, enables us to write down the standard equations of any line which is given as the intersection of two planes; but there is a somewhat neater way of doing this. Now that we know that the intersection of, say, $ax + by + cz + d = 0$ and $px + qy + rz + s = 0$ is a line, we shall first find the d.r's of this line, and then find the coordinates of a point on it. This will be sufficient to enable us to write down the standard equations.

Any normal to the first plane has d.r's $a : b : c$, and will be perpendicular to the line of intersection, as it is to any line which lies in the plane. Hence if the d.r's of the line of intersection are $\lambda : \mu : \nu$ we must have

(6.12.3) $a\lambda + b\mu + c\nu = 0.$

Similarly, considering the other plane, we have also

(6.12.4) $p\lambda + q\mu + r\nu = 0.$

Equations (6.12.3) and (6.12.4) look like two equations in *three* unknowns, λ, μ and ν, but this is not so, since we are interested only in the *ratios* $\lambda : \mu : \nu$. If we write the equations as

(6.12.5)
$$a\frac{\lambda}{\nu} + b\frac{\mu}{\nu} = -c$$
$$p\frac{\lambda}{\nu} + q\frac{\mu}{\nu} = -r.$$

we can treat them as two equations in *two* unknowns λ/ν and μ/ν. Solving them we find that

$$\frac{\lambda}{v} = \frac{br - cq}{aq - bp}; \quad \frac{\mu}{v} = \frac{cp - ar}{aq - bp}$$

from which

(6.12.6) $\lambda : \mu : v = br - cq : cp - ar : aq - bp.$

These are the required d.r.'s. It may happen that one expression on the right of (6.12.6) vanishes, or even two of them; but they cannot all vanish. For if $br - cq = cp - ar = aq - bp = 0$, then $a/p = b/q = c/r$ or $a : b : c = p : q : r$, so that the two planes have the same normal, i.e., they are parallel, and this is the case that we have expressly excluded.

Note that the ratios $\lambda : \mu : v$ given by (6.12.6) are those of a line perpendicular to the two directions with d.r.'s $a : b : c$ and $p : q : r$, that is to say, they are the d.r.'s of the common perpendicular of these two directions. This is an important result, quite apart from its application in finding the line of intersection of two planes.

There remains only to find the coordinates of a point of the line. Any point will do, so that we are looking for *any* set of values of x, y, and z which satisfies the equations of both planes. In the absence of any particular reason for doing otherwise, a good procedure is to put $z = 0$ in the equations of the planes, and then solve for x and y. Thus if the equations are $3x - 5y + 2z + 2 = 0$ and $x + 7y - 3z - 8 = 0$, we put $z = 0$ obtaining $3x - 5y + 2 = 0$ and $x + 7y - 8 = 0$, which have the solution $x = 1$, $y = 1$. We have then found a point of the line, viz., $(1, 1, 0)$, since these coordinates satisfy the original equations. In exceptional cases this method may break down. If it does, all that is necessary is to start again, this time puttng $x = 0$ and solving for y and z. If this too breaks down, then putting $y = 0$ instead will certainly work. However, we shall see below how to make sure that the method will work first time.

Having found the d.r.'s of the line of intersection of two planes, and the coordinates of a point on it, we can write down its standard equations immediately. Here are two examples to illustrate the whole procedure.

EXAMPLE 1:

Find the line of intersection of the planes

$$x + 2y + 3z - 3 = 0$$

and

$$5x + 2y - 5z + 5 = 0$$

in the form

$$\frac{x - x_1}{\lambda} = \frac{y - y_1}{\mu} = \frac{z - z_1}{\nu}.$$

From (6.12.6) we find that

$$\lambda : \mu : \nu = -10 - 6 : 15 + 5 : 2 - 10$$
$$= -16 : 20 : -8$$
$$= 4 : -5 : -2.$$

If we put $z = 0$ we have

$$x + 2y = 3 \quad \text{and} \quad 5x + 2y = -5.$$

Subtracting the first equation from the second we have $4x = -8$ or $x = -2$. It follows that $y = \frac{5}{2}$, and hence the point $(-2, \frac{5}{2}, 0)$ lies on both planes and hence on their line of intersection. The required equations can therefore be written as

$$\frac{x + 2}{4} = \frac{y - \frac{5}{2}}{-5} = \frac{z}{-2}.$$

It is not necessary, in finding a point of the line, to follow the method given above. Any point of the line will do, and it may happen that such a point may be "seen" by inspection. Thus in this example we might have noticed that the point $(0, 0, 1)$ lies on both planes, and taken this point instead of $(-2, \frac{5}{2}, 0)$. The equations would then appear in the form

$$\frac{x}{4} = \frac{y}{-5} = \frac{z - 1}{2}.$$

EXAMPLE 2.

Find the equations, in the standard form, of the line of intersection of the planes

$$2x + y + 7z = 8$$

and

$$4x + 2y - 3z = -1.$$

Here we have $\lambda : \mu : \nu = -3 - 14 : 28 + 6 : 4 - 4$
$$= -1 : 2 : 0.$$

If we put $z = 0$ we get $2x + y = 8$ and $4x + 2y = -1$, and these equations are inconsistent; for if $2x + y = 8$ then $4x + 2y$ must be 16, and not -1. Hence the method breaks down. Let us put $x = 0$ instead. We get $y + 7z = 8$ and $2y - 3z = -1$, from which we obtain $y = 1$.

The point $(0, 1, 1)$ therefore lies on the line, and the required equations of the line are

$$\frac{x}{-1} = \frac{y-1}{2} = \frac{z-1}{0}.$$

Note: The zero in the third denominator may look a little odd, but all that it means is that the numerator must also be zero. Hence $z = 1$ for every point of the line. (Little wonder then that putting $z = 0$ gave us a contradiction!) From this we also see that if v turns out to be zero, then we should *not* try putting $z = 0$. Since a similar remark applies to the other variables, we need never make more than one attempt at finding a point of the line, provided that we calculate the d.r's of the line first.

Exercise 66

1. Find the equations, in standard form, of the lines given by the following pairs of planes.

(a) $x - 38y + 5z - 247 = 0$
     $5x - 22y + z - 131 = 0$
(b) $36x - 182y + 135z - 227 = 0$
     $18x - 30z + 49 = 0$
(c) $5x - 15y + 81z - 101 = 0$
     $9x - 27y + 45z - 109 = 0$
(d) $63x + 25y + 90z - 63 = 0$
     $18x + 9y + 35z - 18 = 0$
(e) $3x + 7z + 33 = 0$
     $2x - z - 12 = 0$

2. Show that the planes $3x - 3y - 17z = 0$, $13x - 4y - 23z = 0$ and $29x - 11y - 63z = 0$ pass through one line.

6.13   ANGLE BETWEEN TWO PLANES

The angle between two planes π_1 and π_2 is defined as follows: we take any point A of the line of intersection of the two planes, and consider the line p_1 which lies in the plane π_1, contains the point A, and is perpendicular to the line of intersection; and the similar line p_2 lying in the plane π_2. (See Fig. 6.12.) The angle between π_1 and π_2 is then defined as the angle θ between p_1 and p_2. Clearly it will not matter which point of the line of intersection we take as the point A.

Finding the magnitude of this angle for two planes whose equations are given is straightforward. We simply note that the angle between two planes

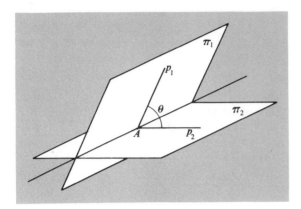

Fig. 6.12

is the same as the angle between their normals. This is clear if we imagine that we are looking at Fig. 6.12 along the line of intersection. The lines p_1 and p_2 and the normals (n_1 and n_2) through A will then appear as in Fig. 6.13.

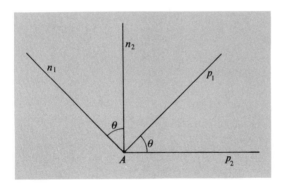

Fig. 6.13

Since the d.r's of the normal to a plane are the coefficients of x, y and z in its equation, and since we already know how to find the angle between two lines whose d.r's are known [see (6.8.1)], the method of finding the angle between two planes is now apparent, and one example will suffice to illustrate it.

EXAMPLE:
Find the angle between the planes

$$x + y - z + 3 = 0$$

and
$$7x + y - 5z - 6 = 0.$$

The d.r's of the normals to these two planes are $1 : 1 : -1$ and $7 : 1 : -5$ respectively. Their d.c's are therefore $\frac{1}{\sqrt{3}}, \frac{-1}{\sqrt{3}}, \frac{-1}{\sqrt{3}}$ and $\frac{7}{5\sqrt{3}}, \frac{1}{5\sqrt{3}}, \frac{-5}{\sqrt{3}}$: consequently, by (6.8.1), we have

$$\cos \theta = \tfrac{7}{15} + \tfrac{1}{15} + \tfrac{1}{3} = \tfrac{13}{15}.$$

Note: The answer to a problem such as this can usually be left in the form $\cos \theta = \tfrac{13}{15}$. The actual value of the angle, in degrees or radians, can always be found from tables of the cosine function. In this example, $\theta = 29°56'$.

Exercise 67

Find the angles between the pairs of planes given in Exercise 66.1.

6.14 PERPENDICULAR DISTANCE OF A POINT FROM A PLANE

Let P be the point (x_1, y_1, z_1) and π the plane $lx + my + nz - p = 0$, where l, m, n are the d.c's of the normal to this plane. We then know that p is the distance from the origin to the plane. [See (6.11).] We now ask for the perpendicular distance of P from this plane. This distance can be found by the method that was used in the last chapter to find the distance of a point from a line in two dimensions, i.e., we consider a plane parallel to π through the point P. Since it has the same normal as π its equation will be of the form

$$lx + my + nz - p' = 0$$

where p' is the distance from the origin of this new plane. The constant p' is easily calculated, for since the point (x_1, y_1, z_1) lies on the new plane we have $lx_1 + my_1 + nz_1 - p' = 0$ or $p' = lx_1 + my_1 + nz_1$. Furthermore, from Fig. 6.14 we see that the distance D between the planes is $p' - p$. This also is the distance of P from the plane π, so that our final result is given by

(6.14.1) $$D = lx_1 + my_1 + nz_1 - p.$$

Thus, if the equation of the plane is given in "perpendicular" form, that is with d.c's as coefficients of x, y and z, then the distance of a point from it is found by merely substituting the coordinates of the points in the left-hand side of the equation of the plane—a procedure which is the exact analogue of that for the corresponding problem in two dimensions. [See (5.14.2).] If the equation of the plane is not given in this form, then it must first be put

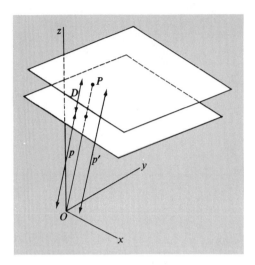

Fig. 6.14

into this form by dividing right through by the square root of the sum of the squares of the coefficients. Hence the perpendicular distance of the point (x_1, y_1, z_1) from the plane $ax + by + cz + d = 0$ is

(6.14.2)
$$D = \frac{ax_1 + by_1 + cz_1 + d}{\sqrt{a^2 + b^2 + c^2}}$$

The observations that we made in section 5.16 about the corresponding formula for two dimensions apply equally well here. D may turn out to be positive or negative, and the plane $ax + by + cz + d = 0$ will divide the whole of the rest of space into two parts, or half-spaces; one consisting of all points for which $ax + by + cz + d$ is positive, and the other consisting of all points for which this expression is negative. The boundary between these two half-spaces is, of course, the plane itself.

Exercise 68

1. Find the distances of the points $(2, 5, 5)$, $(1, 0, -2)$ and $(3, 4, -1)$ from the following planes

    (a) $x + y + z + 6 = 0$

    (b) $3x + 4y - 12z + 1 = 0$

    (c) $2x - y + 2z + 5 = 0$

2. Find the equations of the two planes on which lie all points whose distances from the planes

$$x + 2y - 2z = 4$$

and $$5x - 3y + 4z = 7\sqrt{2}$$

are equal.

3. For the reasons noted in the preceding section the plane $3x - 5y + z + 2 = 0$ divides the space of three dimensions into three disjoint sets—the set of points constituting the plane itself, and two half-spaces. If p denotes the plane, and A and B denote the two half-spaces, where A is the one which contains the origin, determine to which of these sets the following points belong.

(a) $(2, 1, 6)$ (e) $(\frac{1}{4}, \frac{7}{2}, \frac{2}{3})$

(b) $(1, 1, 2)$ (f) $(3, 8, 2)$

(c) $(7, 0, -8)$ (g) $(17, 10, -2)$

(d) $(\frac{2}{5}, \frac{1}{5}, 4)$ (h) $(1, 3, -5)$

Which of these points are on the *origin* side of the plane $3x + 3y - z = 6$?

6.15 INTERSECTION OF A LINE AND A PLANE

If a line is given by an equation of the form

$$\frac{x - x_1}{\lambda} = \frac{y - y_1}{\mu} = \frac{z - z_1}{\nu}$$

then, as we have already seen, any point of the line is of the form

(6.15.1) $$(x_1 + \lambda t, y_1 + \mu t, z_1 + \nu t).$$

If we wish to find which point of the line (if any) lies on a given plane, say $ax + by + cz + d = 0$, then we have only to substitute the above coordinates into the equation of the plane. This gives us an equation

$$a(x_1 + \lambda t) + b(y_1 + \mu t) + c(z_1 + \nu t) + d = 0$$

which we may simplify to

(6.15.2) $$(ax_1 + by_1 + cz_1 + d) + (a\lambda + b\mu + c\nu)t = 0.$$

In general, (6.15.2) gives us the value of the parameter t which corresponds to the point in question; substituting this value in (6.15.1) we obtain the coordinates of this point. Moreover, since (6.15.2) is a simple equation, we also see that there is, in general, only one such point, that is, a line and a plane intersect in a unique point. There is, however, an exceptional case, which arises when $a\lambda + b\mu + cv = 0$. Equation (6.15.2) does not then determine a value of t. It is not difficult to see what happens in this case; if $a\lambda + b\mu + cv = 0$ then the line and the normal to the plane are perpendicular, which is another way of saying that the line is parallel to the plane. Under these circumstances we would not expect the line and plane to intersect, unless the line lies *in* the plane.

Hence a line and a plane that are not parallel to each other will intersect in a single point, and the coordinates of this point can be found by the above method.

Exercise 69

1. Find the equation of the plane through the points $(2, -1, 0)$, $(3, -4, 5)$ parallel to the line $2x = 3y = 4z$.

2. Find the point of intersection of the line

$$\frac{x - 3}{2} = \frac{y + 5}{3} = \frac{z - 1}{-5}$$

and the plane $3x - y - z + 1 = 0$.

3. Find the point in which the line through $(2, 1, 5)$ and $(-1, 1, 3)$ meets the plane which contains the point $(2, 0, 4)$ and is normal to the line joining $(-2, 1, 7)$ and $(6, 15, 1)$.

6.16 INTERSECTION OF THREE PLANES

Suppose we are given three planes. How many points would we expect these three planes to have in common? The answer to this question is simple in the general case, but there are several exceptional cases which have to be considered before a complete answer to the question can be given. Let us consider first a specific example; let us look for points common to the three planes

(6.16.1)
$$3x + 9y - z + 26 = 0$$
$$x - y + 7z - 40 = 0$$
$$2x + y \qquad - 1 = 0$$

Now the coordinates of a point common to the three planes will satisfy each of the equations (6.16.1). We are therefore seeking solutions to this set of three equations, and this we can do by the usual method of eliminating variables (other means will be given later in the book). Adding the second and third of the equations we obtain $3x + 7z - 41 = 0$. Adding 9 times the second equation to the first equation we obtain $12x + 62z - 334 = 0$. The variable y has now been eliminated and we have simultaneous equations in x and z, which we can solve. We find that $x = 2$ and $z = 5$, which means that $y = -3$. Hence there is a unique solution to the equations, giving a unique point of intersection, viz., the point $(2, -3, 5)$.

It is no trouble to show that, *in general*, three planes will intersect in a unique point, as these have done. For the points common to the first two planes will form a line, and this line will intersect the third plane in a unique point. To answer fully the question of the ways in which three planes may intersect we must examine closely the various ways in which this simple argument can break down.

If we are given three planes, it may happen that the planes are all parallel to each other. There are then no common points.

If the planes are not all parallel, then it will be possible to pick out two of the three which are not parallel to each other. These two planes will intersect in a line. This line may (a) meet the other plane in a point, (b) be parallel to the third plane, but not lying in it, or (c) lie in the third plane. If (a), then the point is common to all three planes—this is the general case. If (b), then the planes will form a prism, that is they will all be parallel to some fixed direction. If (c), then the planes all contain a certain line. Finally there are two versions of possibility (b) according as there is, or is not, among the three planes, a pair of parallel planes.

Hence the ways in which three planes may intersect may be summarized as follows:

1. They may intersect in a unique point. Any two of the planes will intersect in a line, and the three lines thus formed will contain the point of intersection.

2. They may form a prism. There is then no common point, and the lines of intersection of the planes taken in pairs will be parallel to each other. In general there will be three such lines, but if two of the planes are parallel, there will be only two lines of intersection, since the parallel planes will not give rise to such a line.

3. The planes may have a line in common. Every point of this line is then common to all three planes.

4. The planes may all be parallel. No two planes will then intersect and, *a fortiori*, there is no point common to all three.

The five possible modes of intersection [two for case (b)] are illustrated in Fig. 6.15.

We have assumed, in what has just gone, that the three planes are *distinct*.

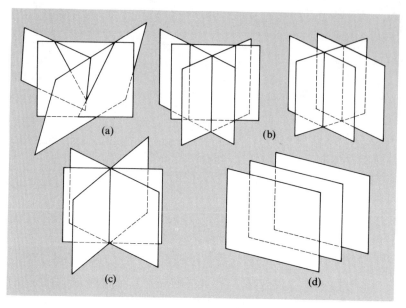

Fig. 6.15

If we take three equations at random it might happen that two of them, or even all three, might represent the same plane. If we take account of this possibility in considering the ways in which three planes can intersect, then we shall have to include some further modes of intersection arising from the coincidence of two or more planes. Geometrically, these further possibilities are of no importance; we shall meet them again, however, in a later chapter when we shall be dealing with a similar problem, but from an algebraic point of view.

Exercise 70

Examine the nature of the intersection of the following sets of planes.

(a) $2x - 5y + z = 3$
$x + y + 4z = 5$
$x + 3y + 6z = 1$

(b) $x + y + z = 6$
$2x + 3y + 4z = 20$
$x - y + z = 2$

(c) $2x + 3y + 4z = 6$
$3x + 4y + 5z = 20$
$x + 2y + 3z = 2$

(d) $3x - y - z = 5$
$2x + 4y + z + 10 = 0$
$6x - 2y + 2z + 9 = 0$

(e) $3x + 4y + 6z = 5$
$6x + 5y + 9z = 10$
$3x + 3y + 5z = 5$

Exercise 71 (Chapter Review)

1. Find the distance of the origin from the plane $6x - 7y - 2z - 14 = 0$.

2. Show that the equations $by + cz + d = 0$, $cz + ax + d = 0$, $ax + by + d = 0$ represent planes parallel to OX, OY, OZ respectively. Find the equations of the planes through the points $(2, 3, 1)$, $(4, -5, 3)$ parallel to the coordinate axes.

3. Find the equation of the plane through $(1, 2, 3)$ parallel to $3x + 4y - 5z = 0$.

4. Prove that the equation of the plane through (α, β, γ) parallel to $ax + by + cz = 0$ is $ax + by + cz = a\alpha + b\beta + c\gamma$.

5. If the axes are rectangular and P is the point $(2, 3, -1)$, find the equation of the plane through P at right angles to OP.

6. Find the distances of the points $(2, 3, -5)$, $(3, 4, 7)$ from the plane $x + 2y - 2z = 9$. Are the points on the same side of the plane?

7. Find where the line $\dfrac{x - 1}{2} = \dfrac{y - 2}{-3} = \dfrac{z - 3}{4}$ meets the plane $2x + 4y - z + 1 = 0$.

8. Find the distance from the point $(3, 4, 5)$ to the point where the line $\dfrac{x - 3}{1} = \dfrac{y - 4}{2} = \dfrac{z - 5}{2}$ meets the plane $x + y + z = 2$.

9. Find the point where the line joining $(2, 1, 3)$, $(4, -2, 5)$ cuts the plane $2x + y - z = 3$.

10. Prove that the equations of the line of intersection of the planes $4x + 4y - 5z = 12$, $8x + 12y - 13z = 32$ can be written $\dfrac{x - 1}{2} = \dfrac{y - 2}{3} = \dfrac{z}{4}$.

11. Find the equations of the line through the point $(1, 2, 3)$ parallel to the line $x - y + 2z = 5$, $3x + y + z = 6$.

12. Prove that the line $\dfrac{x - 3}{2} = \dfrac{y - 4}{3} = \dfrac{z - 5}{4}$ is parallel to the plane $4x + 4y - 5z = 0$.

13. Prove that the planes $2x - 3y - 7z = 0$, $6x - 14y - 13z = 0$, $8x - 31y - 33z = 0$ pass through one line.

14. Find the equation of the plane through $(2, -3, 1)$ normal to the line joining $(3, 4, -1)$, $(2, -1, 5)$.

Suggestions for Further Reading

As with Chapter 5, what is mostly required for the proper understanding of this subject is the working of large numbers of examples. These can be found in

Kindle, J.H., *Plane and Solid Analytical Geometry*. New York: Schaum Publishing Company, 1950.

A more advanced treatment of the subject can be found in

McCrea, H.W., *Analytical Geometry of Three Dimensions*. 2nd ed. Edinburgh: Oliver and Boyd Ltd., 1947.

7 Geometry of

n Dimensions and

Convex Sets

7.1 SPACE OF *n* DIMENSIONS

Let us imagine that we are going on a shopping expedition. We have a certain amount of money to spend and a list of commodities that we wish to buy. How much of each commodity we buy depends upon us. Suppose we buy paint at $1 per quart, cake at $2 per pound, and metal tubing at 50¢ a foot. If we buy x quarts of paint, y pounds of cake and z feet of tubing, our total expenditure will be

$$x + 2y + \tfrac{1}{2}z \text{ dollars.}$$

If we have $10 to spend and do not wish to take any change home, then x, y and z must be chosen so that

$$x + 2y + \tfrac{1}{2}z = 10.$$

More generally, if three commodities cost a, b and c dollars per unit (quart, foot, or whatever it is) and we buy x, y and z units of these commodities respectively, then for our total expenditure to be k dollars we must have

$$(7.1.1) \qquad\qquad ax + by + cz = k.$$

All this is very elementary and, as yet, of no significance.

We have seen in the last chapter how we can represent triples of numbers such as (x, y, z) by points in space. This geometrical representation is often very helpful when we have to deal with such triples, enabling us to "see" clearly results that would otherwise be perceived only with difficulty. As a case in point, the equation (7.1.1), which places a restriction on the amounts of the various commodities that we can buy, is readily interpreted—the point (x, y, z) must lie on the plane whose equation is (7.1.1). The triples of numbers representing all possible purchases of the three commodities do not account for the whole of this plane, however, for in any reasonable interpretation of the amounts bought (x, y, z) we would suppose these quantities to be nonnegative. Thus the points which represent possible shopping lists or budgets will lie on that portion of the plane (7.1.1) that is cut off by the coordinate planes, i.e., by the points of the triangle ABC in Fig. 7.1.

If we did not stipulate that the total expenditure was to be exactly k, but only that it was not to exceed k, then the points representing possible budgets would be given by

(7.1.2)
$$ax + by + cz \leq k$$
$$x \leq 0, y \leq 0, z \leq 0.$$

It is easily verified that the set of all such points is the set of all points of the tetrahedron $OABC$ in Fig. 7.1.

This ability to use geometrical language in describing things that are not in themselves geometrical is very handy, and will be used many times in later chapters, especially Chapters 16 and 17. But in making use of this device we would be greatly handicapped if we allowed ourselves to be restricted to points in space of three dimensions only. To hark back to the domestic example with which this chapter started, it would be a severe limitation if we could not enjoy the advantages of the geometrical terminology in tackling problems with more than three commodities.

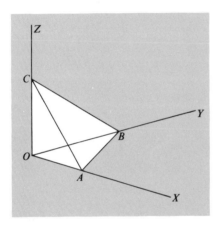

Fig. 7.1

Suppose then that there are four commodities, with prices a, b, c and d dollars per unit. If we buy x, y, z and t units of these respectively, and our total expenditure must be k, then clearly

(7.1.3) $ax + by + cz + dt = k.$

Can we interpret (7.1.3) in geometrical language as we did (7.1.1.)? Since our world is three-dimensional it might seem that this is impossible, or at any rate, that a method of representing sets of four numbers geometrically would be very difficult to picture. Nevertheless the geometrical way of thinking about equations is so convenient that it would be foolish not to try to retain its advantages despite the difficulties. Further, it is not all that easy sometimes to picture what is going on in *three* dimensions (see the quotation from Fred Hoyle at the head of Chapter 6) and we frequently have to rely on algebra to help us out. If we are handling more than three variables at a time we are all the more likely to need to fall back on the algebraic approach when our geometrical imagination fails us, but we can keep to the geometrical nomenclature just the same.

The set of all triples (x, y, z), each interpreted as Cartesian coordinates, makes up a space of three dimensions. The set of possible (x, y, z) in our first shopping example is a subset of this space, called the *positive octant*—it is, so to speak an "eighth" of the whole space. In economic applications this would be called the *commodity space* of the problem. The points of a commodity space, the triples of amounts of the three commodities in this example, are often called *commodity bundles*. If there are four commodities then the commodity space will be a subset of a space of four dimensions. Thus with very little trouble, and in a completely everyday setting, we have introduced that mysterious object, the darling of the early science-fiction writers, a fourth dimension! Once we have decided not to let our geometrical imagination be restricted by any limitation to three dimensions, we may as well go all the way, and instead of considering separately spaces of four, five or six dimensions, etc., consider space of *n* dimensions, where *n* is any positive integer.

7.2 SOME DEFINITIONS

Definition 7.1

A point in *n*-dimensional space is an ordered set of *n* real numbers (its *coordinates*).

Thus in two and three dimensions points are of the form (x, y) and (x, y, z), which agrees with our previous notation. In *n* dimensions we shall write a general point as

$$(x_1, x_2, \ldots, x_n).$$

It is unfortunate that here we have had to turn our previous notation inside-out. Up to now we have used different *letters* for the different coordinates, and used subscripts to distinguish different points. We have written (x_1, y_1, z_1) and (x_2, y_2, z_2), for example, to denote two points of three-dimensional space. We now propose to use subscripts for different coordinates, but it would be difficult to do otherwise, since the number of coordinates is not specified. Since subscripts are earmarked for this purpose, we may as well use different letters for different points, and this is what we shall do.

To be consistent, and to avoid this sudden about-face, we ought logically to have written (x_1, x_2) instead of (x, y), and (x_1, x_2, x_3) instead of (x, y, z), in the two previous chapters. But had we done so we would have been in conflict with the standard notation for problems specifically in two or three dimensions, and would probably have produced more confusion than clarification.

To save us the bother of having to write $(x_1, x_2, \ldots, x_n)$ or $(y_1, y_2, \ldots, y_n)$ every time we wish to refer to a point it is usual to employ an abbreviated notation and write just **x** in place of $(x_1, x_2, \ldots, x_n)$, **y** in place of $(y_1, y_2, \ldots, y_n)$ and so on. This is quite sufficient unless we particularly want to draw attention to the individual coordinates, in which case we can write them out in full. In printing, bold-face type serves to distinguish symbols used in this way; in written work it is usual to underline the symbol.

An ordered set of real numbers of this kind, provided it fulfills certain conditions which we shall set out shortly, is called a *vector*. The bold-face type given above (i.e., **x**) is almost universally used to denote vectors and other kindred objects.

Definition 7.2

The set of points whose coordinates satisfy a linear equation of the form

$$(7.2.1) \qquad a_1 x_1 + a_2 x_2 + \ldots + a_n x_n = k$$

is called a *hyperplane*.

Thus a hyperplane in two dimensions is a line, and in three dimensions is a plane. A hyperplane is therefore a generalization to n dimensions of these familiar geometrical objects. The coefficients $a_1, a_2, \ldots, a_n$ in (7.2.1) form an ordered set of n real numbers, and it is natural to treat this set also as being a vector, and write

$$\mathbf{a} = (a_1, a_2, \ldots, a_n).$$

There would be little point in doing this unless we gained some advantage from it, but it so happens that we do.

We now define the *scalar product*, or simply the *product* of two vectors. It is obtained by multiplying corresponding coordinates in the two vectors, and adding the products. Thus we have

Definition 7.3

If

$$\mathbf{x} = (x_1, x_2, \ldots, x_n)$$

and

$$\mathbf{y} = (y_1, y_2, \ldots, y_n)$$

then the scalar product, written $\mathbf{x} \cdot \mathbf{y}$ *is defined by*

(7.7.2)
$$\mathbf{x} \cdot \mathbf{y} = x_1 y_1 + x_2 y_2 + \ldots + x_n y_n$$
$$= \sum_{i=1}^{n} x_i y_i.$$

Note that the product of two vectors is not itself a vector but a real number. The term *scalar* is often used for a real number in contradistinction to a vector; hence the name of the product. With this notation we can write (7.2.1) in the shorter form

(7.2.3)
$$\mathbf{a} \cdot \mathbf{x} = k.$$

One advantage of this notation is that it contains no reference to *n*, the number of dimensions. Hence (7.2.3) represents the equation of a hyperplane in space of whatever number of dimensions we happen to be currently considering.

Definition 7.4

Addition and subtraction of vectors. If $\mathbf{x} = (x_1, x_2, \ldots, x_n)$ *and* $\mathbf{y} = (y_1, y_2, \ldots, y_n)$ *are any two vectors, we write*

$$\mathbf{x} + \mathbf{y} \quad \text{for} \quad (x_1 + y_1, x_2 + y_2, \ldots, x_n + y_n)$$

and

$$\mathbf{x} - \mathbf{y} \quad \text{for} \quad (x_1 - y_1, x_2 - y_2, \ldots, x_n - y_n).$$

Thus addition and subtraction of vectors is effected by adding or subtracting corresponding coordinates.

Exercise 72

1. If $\mathbf{x} = (1, -3, 4, 7)$
$\mathbf{y} = (2, 9, -4, 8)$
and $\mathbf{z} = (3, 3, 3, -3)$

find the following vectors:

$$\mathbf{x} + \mathbf{y} + \mathbf{z}, \quad \mathbf{x} - \mathbf{y} + \mathbf{z}, \quad \mathbf{x} - \mathbf{z},$$

and the following scalars:

$$\mathbf{x} \cdot \mathbf{z} \quad \mathbf{y} \cdot \mathbf{z}, \quad \mathbf{x} \cdot (\mathbf{y} - \mathbf{z}).$$

2. With the same vectors as in problem 1 above, show that

$$\mathbf{x} \cdot (\mathbf{z} + \mathbf{y}) = \mathbf{x} \cdot \mathbf{z} + \mathbf{x} \cdot \mathbf{y}$$

and that

$$\mathbf{y} \cdot (\mathbf{x} - \mathbf{y} + \mathbf{z}) = \mathbf{y} \cdot \mathbf{x} - \mathbf{y} \cdot \mathbf{y} + \mathbf{y} \cdot \mathbf{z}.$$

3. If $\mathbf{x}$, $\mathbf{y}$ and $\mathbf{z}$ are any three vectors in a space of n dimensions prove that

$$\mathbf{x} \cdot (\mathbf{y} + \mathbf{z}) = \mathbf{x} \cdot \mathbf{y} + \mathbf{x} \cdot \mathbf{z}.$$

7.3 DISTANCE BETWEEN TWO POINTS: DIVISION OF LINE SEGMENTS

The distance D between two points $\mathbf{x} = (x_1, x_2, \ldots, x_n)$ and $\mathbf{y} = (y_1, y_2, \ldots, y_n)$ is defined by

$$D^2 = (x_1 - y_1)^2 + (x_2 - y_2)^2 + \ldots + (x_n - y_n)^2$$

(7.3.1)
$$= \sum_{i=1}^{n} (x_i - y_i)^2.$$

When $n = 2$ this is the same as (5.2.1); when $n = 3$ it is the same as (6.3.2).

Equation (7.3.1) can be written in terms of the vector notation defined above as follows:

$$D^2 = \sum_{i=1}^{n} (x_i - y_i)^2$$

$$= \sum_{i=1}^{n} (x_i^2 - 2x_i y_i + y_i^2)$$

$$= \sum_{i=1}^{n} x_i^2 - 2 \sum_{i=1}^{n} x_i y_i + \sum_{i=1}^{n} y_i^2.$$

Now $\sum_{i=1}^{n} x_i^2 = \mathbf{x} \cdot \mathbf{x}$, by the definition of a scalar product, and we shall agree to write this as $\mathbf{x}^2$. Similarly $\sum_{i=1}^{n} y_i^2 = \mathbf{y} \cdot \mathbf{y}$ can be written as $\mathbf{y}^2$. Then (7.3.1) becomes

(7.3.2)
$$D^2 = \mathbf{x}^2 - 2\mathbf{x} \cdot \mathbf{y} + \mathbf{y}^2.$$

We see from the definition of subtraction for vectors that we can write

(7.3.1) $D^2 = (\mathbf{x} - \mathbf{y})^2.$

From this and (7.3.2) we obtain the vector analogue of a well-known result of elementary algebra, namely

$$(\mathbf{x} - \mathbf{y})^2 = \mathbf{x}^2 - 2\mathbf{x}\cdot\mathbf{y} + \mathbf{y}^2.$$

If the point **y** is the origin, given by the *zero vector*

$$\mathbf{0} = (0, 0, \ldots, 0)$$

then (7.3.1) gives the distance D of x from the origin by

$$D^2 = \mathbf{x}^2$$
(7.3.2) $= x_1{}^2 + x_2{}^2 + \ldots\ldots + x_n{}^2.$

This number $\sum_{i=1}^{n} x_i^2$ is called the *norm* of the vector **x**.

The square root of the norm of a vector **x** is often denoted by $|\mathbf{x}|$. Hence the distance D between two vectors **x** and **y** can be written as $|\mathbf{x} - \mathbf{y}|$, by virtue of (7.3.1).

If $\mathbf{x} = (x_1, x_2, \ldots, x_n)$ is any vector, and λ is a scalar, we define $\lambda\,\mathbf{x}$ to mean the vector $(\lambda\,x_1, \lambda\,x_2, \ldots, \lambda\,x_n)$, and in this way define the product of a scalar and a vector.

In our discussion of geometry of two and three dimensions we tended to use capital letters to denote points of the space; the letter P was a particular favorite. Since the vectors that we are now considering can be interpreted as points in a space of n dimensions we could, if we wished, represent them by capital letters also; or we could continue to use heavy type. We shall in fact do both of these things, and use either a vector symbol like **x** or a capital letter like P to denote a particular point or its vector of coordinates. This is very convenient, for it enables us, for example, to talk about "the point $\lambda P + \mu Q$" instead of having to say "the point whose coordinate vector is

$$\mathbf{z} = \lambda\mathbf{x} + \mu\mathbf{y}$$

where **x** and **y** are the coordinate vectors of P and Q". Using this nomenclature we now state an important theorem.

THEOREM 7.1

IF $\lambda + \mu = 1$ ($\lambda, \mu \geq 0$), THEN THE POINT R WHICH DIVIDES THE SEGMENT PQ IN THE RATIO $\lambda : \mu$ IS GIVEN BY

(7.3.3) $R = \mu P + \lambda Q.$

Thus if
$$P = \mathbf{x} = (x_1, x_2, \ldots, x_n)$$
$$Q = \mathbf{y} = (y_1, y_2, \ldots, y_n)$$
and
$$R = \mathbf{z} = (z_1, z_2, \ldots, z_n)$$
then
$$\mathbf{z} = \mu\mathbf{x} + \lambda\mathbf{y}$$
i.e.,
$$z_i = \mu x_i + \lambda y_i \qquad (i = 1, 2, \ldots, n).$$

PROOF:

This theorem is the analogue in n dimensions of equation (5.9.1) for two dimensions, and (6.10.1) in three. It can be proved most easily by means of a construction similar to those used for the results for $n = 2, 3$. But for those who are wary of placing their trust in diagrams when dealing with more than three dimensions the following, less elegant, but purely algebraic proof is offered. We make full use of the vector notation. From (7.3.1) we have

$$PQ^2 = (\mathbf{x} - \mathbf{y})^2$$

and

$$PR^2 = [\mathbf{x} - (\mu\mathbf{x} + \lambda\mathbf{y})]^2$$
$$= \lambda^2(\mathbf{x} - \mathbf{y})^2, \qquad \text{since } 1 - \mu = \lambda,$$
$$= \lambda^2 PQ^2,$$

while
$$RQ^2 = [\mathbf{y} - (\mu\mathbf{x} + \lambda\mathbf{y})]^2$$
$$= \mu^2 (\mathbf{x} - \mathbf{y})^2, \qquad \text{since } 1 - \lambda = \mu,$$
$$= \mu^2 PQ^2.$$

Hence
$$PR = \lambda PQ \quad \text{and} \quad RQ = \mu PQ$$

whence it follows that $PR : RQ = \lambda : \mu$. We still have to show that R lies on the line segment PQ, but this is easy. For

$$PR + RQ = \lambda PQ + \mu PQ$$
$$= PQ$$

which can happen only if R lies on PQ. (See Exercise 73.3.)

COROLLARY 1. If λ and μ are *any* two numbers, then the point R is given by

$$\frac{\mu}{\lambda + \mu}P + \frac{\lambda}{\lambda + \mu}Q.$$

For $\dfrac{\lambda}{\lambda + \mu}$ and $\dfrac{\mu}{\lambda + \mu}$ are in the same ratio as λ and μ, and their sum is 1.

COROLLARY 2.   If t is a real number between 0 and 1 then any point on the line segment PQ is given by

(7.3.4) $$tP + (1 - t)Q.$$

For t and $1 - t$ can be taken as λ and μ, with sum 1, and the given point is that which divides PQ in the ratio $1 - t : t$.

Exercise 73

(The first three problems below prove, in easy stages, an important result concerning distances in space of any number of dimensions.)

1. By considering the graph of

$$f(t) = at^2 + bt + c$$

where $a > 0$, or otherwise, prove that $f(t)$ is nonnegative for all values of t if, and only if, $b^2 \leq 4ac$.

Show that the function of t

$$F(t) = \sum_{i=1}^{n} (x_i - y_i t)^2$$

is nonnegative for all values of t, irrespective of the values of $x_1, x_2, \ldots, x_n, y_1, y_2, \ldots, y_n$. Deduce that

$$\left(\sum_{i=1}^{n} x_i y_i \right)^2 \leq \left(\sum_{i=1}^{n} x_i^2 \right)\left(\sum_{i=1}^{n} y_i^2 \right).$$

Show that equality holds if, and only if, y_i/x_i is the same for all i; that is, the vector $(y_1, y_2, \ldots, y_n)$ is a scalar multiple of the vector $(x_1, x_2, \ldots, x_n)$.

2.   Let O, P, Q be any three points in space of n dimensions. Choose axes so that O is the origin and let

$$P \equiv (x_1, x_2, \ldots, x_n)$$
$$Q \equiv (y_1, y_2, \ldots, y_n).$$

Prove that the statement $OP + OQ \geq PQ$ is equivalent to

$$\sqrt{\sum_{i=1}^{n} x_i^2} + \sqrt{\sum_{i=1}^{n} y_i^2} \geq \sqrt{\sum_{i=1}^{n} (x_i - y_i)^2}$$

and hence, by squaring, to

$$\sum_{i=1}^{n}(x_i - y_i)^2 \le \sum_{i=1}^{n} x_i^2 + \sum_{i=1}^{n} y_i^2 + 2\left(\sqrt{\sum_{i=1}^{n} x_i^2}\right)\left(\sqrt{\sum_{i=1}^{n} y_i^2}\right).$$

3. Combine the results of problems 1 and 2 above to prove that

$$PQ \le OP + OQ.$$

(In two-and three-dimensional space this is the well-known triangle inequality, that two sides of a triangle are together greater than the third.)

Show that equality holds if, and only if, the triangle OPQ becomes degenerate; that is, O lies on PQ. (Compare the final part of the proof of Theorem 7.1.)

4. Find the coordinates of the points which divide the line segment joining

$$(2, 1, -3, 7) \quad \text{and} \quad (1, 0, 5, -3)$$

in the ratios

   (a) $1 : 5$

   (b) $2 : 3$

   (c) $4 : 1$.

5. Find parametric equations for the line joining the points

$$P \equiv (7, 3, 2, 1, 6) \quad \text{and} \quad Q \equiv (2, 1, 4, 4, 2).$$

If these two points represent *commodity vectors* so that their coordinates are the amounts of five commodities bought by a purchaser, and if the current prices are such that the total expenditure in each case is the same, what can be said about the commodity vectors represented by other points of PQ?

Show that this situation is realized if the prices per unit of the five commodities are given by the *price vector* $(3, 2, 4, 5, 1)$.

6. If P, Q, R are three points in n-dimensional space, and λ, μ, ν are numbers such that $\lambda + \mu + \nu = 1$, prove that the point $\lambda P + \mu Q + \nu R$ is coplanar with (i.e., lies on the same plane as) P, Q and R.

Hint: Write $\lambda P + \mu Q + \nu R$ as $(1 - \nu)\left[\dfrac{\lambda}{1 - \nu}P + \dfrac{\mu}{1 - \nu}Q\right] + \nu R.$

Note: If a, b, c are any three numbers, and $\lambda + \mu + \nu = 1$, then the number $\lambda a + \mu b + \nu c$ can be called a *weighted average* of a, b and c. If $\lambda = \mu = \nu$ this is the ordinary familiar average of the three numbers. In the same way we can conveniently refer to the point $\lambda P + \mu Q + \nu R$

as a *weighted average* of the points P, Q and R. We can similarly have a weighted average of any number of numbers, or of points. Another term for this weighted average of points is *linear combination* of points.

7.4 LINES AND HYPERPLANES

In deriving (7.3.3) and (7.3.4) we assumed that λ and μ were nonnegative and that R was between P and Q. Suppose this is not so, and that R is on the line PQ but external to the segment PQ as in Fig. 7.2. Then, as drawn, Q is between P and R, and we have, from (7.3.4)

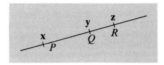

$$\mathbf{y} = u\mathbf{x} + (1 - u)\mathbf{z}$$

where $0 \leq u \leq 1$. This gives us

$$\frac{1}{1-u}\mathbf{y} = \frac{u}{1-u}\mathbf{x} + \mathbf{z}$$

or

(7.4.1) $$z = \frac{-u}{1-u}\mathbf{x} + \frac{1}{1-u}\mathbf{y}.$$

If we write $t = \dfrac{-u}{1-u}$, then (7.4.1) becomes

(7.4.2) $$z = t\mathbf{x} + (1 - t)\mathbf{y}$$

which is the same as (7.3.3) except that t is now negative. If we take R on the other side of PQ we find that we again get (7.4.2) but that this time we have $t > 1$. (Compare section 5.9.)

Hence it appears that if we do not restrict the value of t, then

$$tP + (1 - t)Q$$

gives *all* points of the line joining P and Q. A general point on PQ is therefore $(z_1, z_2, \ldots, z_n)$ where

$$z_i = tx_i + (1 - t)y_i \qquad (i = 1, 2, \ldots, n)$$
$$= y_i + t(x_i - y_i)$$

i.e.,

(7.4.3) $\qquad\qquad z_i = y_i + \lambda_i t \qquad\qquad (i = 1, 2, \ldots, n).$

This set of n equations is the analogue in n dimensions of the parametric equations (5.9.5) for $n = 2$ and (6.5.1) for $n = 3$.

If we eliminate the parameter t from equations (7.4.3) we find that the expressions

$$\frac{z_i - y_i}{\lambda_i} \qquad (i = 1, 2, \ldots, n)$$

are all equal. To bring out more clearly the analogy with the corresponding result in three dimensions, let us replace y_i by α_i (emphasizing that in this context **y** is a constant vector) and z_i by x_i (we can do this since x_i no longer appears in (7.4.3) with its original significance). We then see that

(7.4.4) $\qquad \dfrac{x_1 - \alpha_1}{\lambda_1} = \dfrac{x_2 - \alpha_2}{\lambda_2} = \cdots = \dfrac{x_n - \alpha_n}{\lambda_n}$

These are the equations of a line in n-dimensional space. If we make $n = 3$ we obtain the equations (6.9.2). The constants $\lambda_1, \lambda_2, \ldots, \lambda_n$ are the analogues of direction ratios in three dimensions.

If we put $x_i = \alpha_i + \lambda_i t$ in the equation of a hyperplane, say

$$a_1 x_1 + a_2 x_2 + \ldots + a_n x_n = k$$

we obtain, in general, an equation for t, viz.,

$$\sum_{i=1}^{n} a_i(\alpha_i + \lambda_i t) = k$$

or

$$\left(\sum_{i=1}^{n} a_i \lambda_i\right) t = k - \sum_{i=1}^{n} a_i \alpha_i.$$

In vector notation we have that the result of substituting

$$\mathbf{x} = \boldsymbol{\alpha} + t\boldsymbol{\lambda} \quad \text{in} \quad \mathbf{a} \cdot \mathbf{x} = k \quad \text{is}$$
$$\mathbf{a} \cdot (\boldsymbol{\alpha} + t\boldsymbol{\lambda}) = k$$

or

(7.4.6) $$(\mathbf{a} \cdot \boldsymbol{\lambda})t = k - \mathbf{a} \cdot \boldsymbol{\alpha}.$$

In general (7.4.5) gives a unique value for t, namely

$$t = \frac{k - \Sigma\, a_i\alpha_i}{\Sigma\, a_i\lambda_i} = \frac{k - \mathbf{a} \cdot \boldsymbol{\alpha}}{\mathbf{a} \cdot \boldsymbol{\lambda}}$$

which in turn gives a point lying on both the line $\mathbf{x} = \boldsymbol{\alpha} + t\boldsymbol{\lambda}$ and the hyperplane $\mathbf{a} \cdot \mathbf{x} = k$. Thus we have the result that a line and a hyperplane meet, in general, in one point. However, if $\mathbf{a} \cdot \boldsymbol{\lambda} = 0$ then two things can happen. Either $\mathbf{a} \cdot \boldsymbol{\alpha} = k$, in which case any value of t satisfies (7.4.6); or else $\mathbf{a} \cdot \boldsymbol{\alpha} \neq k$ in which case *no* value of t satisfies (7.4.6). In the first instance the line lies in the hyperplane—every one of its points belongs to the hyperplane; in the second, the line is parallel to the hyperplane and has no points in common with it. This generalizes the results of section 6.15.

Exercise 74

1. Find the points in which the line joining

$$(2, 1, -3, 7, 4) \quad \text{and} \quad (4, 3, 1, -2, 5)$$

meets the coordinate hyperplanes, i.e., the hyperplanes

$$x_1 = 0, \quad x_2 = 0, \quad \ldots, \quad x_5 = 0.$$

2. Prove that

$$\mathbf{a} \cdot (\lambda \mathbf{x} + \mu \mathbf{y}) = \lambda \mathbf{a} \cdot \mathbf{x} + \mu \mathbf{a} \cdot \mathbf{y}.$$

Interpret this when $\mathbf{x}$ and $\mathbf{y}$ are commodity vectors and $\mathbf{a}$ is a price vector (compare Exercise 73.5).

3. Find the point in which the line joining $(3, 0, 0, -3, 1, -4)$ and $(6, -3, -3, -6, -5, -1)$ meets the hyperplane

$$3x_1 + 2x_2 - 5x_3 + 6x_4 + x_5 - x_6 = 2.$$

4. If
$$P \equiv (2, 6, \tfrac{2}{3}, 10)$$
$$Q \equiv (2, 4, 2, 8)$$
$$R \equiv (6, 0, 4, 1)$$
$$S \equiv (3, \tfrac{3}{4}, 4, 4)$$

show that P, Q, R and S are coplanar. Find the point of intersection of the lines PQ and RS.

7.5 VECTOR SPACES

It was stated earlier that there were certain rules that ordered sets of real numbers were required to obey before we would call them vectors. We now have all the definitions needed to set out these rules, so that little remains to do but enumerate them.

A set of mathematical objects, called *vectors*, and denoted by **x**, **y**, etc., is said to form a *vector space V* if

(a) Addition is defined
(b) Multiplication by a scalar is defined

and these operations satisfy the following requirements:

1. For addition:
    (a) if $\mathbf{x} \in V$ and $\mathbf{y} \in V$ then $\mathbf{x} + \mathbf{y} \in V$
    (b) $\mathbf{x} + (\mathbf{y} + \mathbf{z}) = (\mathbf{x} + \mathbf{y}) + \mathbf{z}$ (all $\mathbf{x}, \mathbf{y}, \mathbf{z} \in V$)
    (c) $\mathbf{x} + \mathbf{y} = \mathbf{y} + \mathbf{x}$ (all $\mathbf{x}, \mathbf{y}, \mathbf{z} \in V$)
    (d) there is a zero vector **0** with the property that
         $\mathbf{0} + \mathbf{x} = \mathbf{x} + \mathbf{0} = \mathbf{x}$ for any $\mathbf{x} \in V$
    (e) any vector **x** has a negative, i.e., a vector $-\mathbf{x}$ with the property that $\mathbf{x} + (-\mathbf{x}) = \mathbf{0}$.
2. For multiplication by a scalar:
    (a) if $\mathbf{x} \in V$ then $\lambda\mathbf{x} \in V$
    (b) $0 \cdot \mathbf{x} = 0, \quad 1 \cdot \mathbf{x} = \mathbf{x}$ for all $\mathbf{x} \in V$
    (c) $\lambda(\mu\mathbf{x}) = (\lambda\mu)\mathbf{x}$, for all $\mathbf{x} \in V$
    (d) $\lambda(\mathbf{x} + \mathbf{y}) = \lambda\mathbf{x} + \lambda\mathbf{y}$, for all **x**, **y** in V
    (e) $(\lambda + \mu)\mathbf{x} = \lambda\mathbf{x} + \mu\mathbf{x}$, for all **x**, **y** in V.

The reader should have no difficulty in verifying that if we define a vector as an ordered *n*-tuple of numbers, and define addition and multiplication by a scalar in the manner given earlier in this chapter, then all the above properties hold.

One remark is in order. We defined addition of two vectors in section 7.2, but we did not define the addition of three or more vectors. Rule 1 (*b*) shows that given three vectors **x**, **y**, **z** we can add them two at a time, in any order, and the result is the same. In this way we can extend the definition of vector addition in an obvious way to cover any number of addends—a result which we tacitly assumed in section 7.3.

Note that the scalar product does not appear in these rules. Thus a set of vectors is not prevented from forming a vector space if no scalar product has been defined over it. The scalar product is something extra defined on a set which is already a vector space. Note also that there is nothing in the rules just given which says that the vectors must be ordered n-tuples of real numbers. This is not necessary. *Any* set of objects, no matter what, can be a vector space provided the relevant operations (addition, etc.) are defined and the rules are satisfied. The vectors defined earlier in this chapter (as ordered n-tuples) make up one of the most common examples of a vector space, but by no means the only one.

7.6 INTRODUCTION TO CONVEX SETS

We shall now consider, in a fairly detailed but informal manner, a broad class of geometrical objects known as *convex sets*. These sets, which are subsets of points of an n-dimensional space, are of major importance in economic theory, and are crucial to the two related disciplines of linear programming and the theory of games of strategy which we shall consider in Chapters 16 and 17. A full treatment of convex sets would require more mathematical apparatus than has been given so far in this book, but even at this stage it is quite possible for the student to obtain an understanding of the basic ideas of the theory. Fortunately, too, most of the essential complications are already present in three dimensions (though not in two) so that it is possible to obtain a reasonable grasp of the theory by considering only examples in three dimensions (or sometimes even in two, when these are not misleading). What we shall do, then, is to present the concepts, definitions and theorems of the theory largely as if we were considering only the possibilities $n = 2$ and 3, but interpolate at intervals a discussion of what happens for higher values of n. Usually there is very little to say—the same results hold, irrespective of the value of n—but occasionally a few comments will be necessary.

7.7 INTERSECTION OF HALF-SPACES

Towards the end of Chapter 6, we observed that any linear equation

(7.7.1) $$ax + by + cz + d = 0$$

where a, b and c are not all zero, partitions the (x, y, z)-space into three disjoint sets. If we denote the linear function on the left-hand side of (7.7.1) by l, then we can describe these subsets as follows:

(7.7.2)   (a) D^-, which is the set of points $(x, y, z,)$ such that

$$l(x, y, z) < 0$$

(b) D^0,   which is the set of points (x, y, z) such that

$$l(x, y, z) = 0$$

(c) D^+,   which is the set of points (x, y, z) such that

$$l(x, y, z) > 0.$$

Using the notation in Chapter 2, we may write

$$D^- = \{(x, y, z) \,|\, l(x, y, z) < 0\}$$

and similarly for D^0 and for D^+.

We now introduce two new terms which will be used in the sequel. By the *open negative half-space* associated with a given linear function l, we shall mean the set D^-; and by the *closed negative half-space* associated with the same function, we shall mean the set $F^- = D^- \cup D^0$. *Open positive half-space* and *closed positive half-space* are defined similarly by D^+ and $F^+ = D^+ \cup D^0$, respectively. To put it another way, we can say that an open negative half-space is the set of points (x, y, z) satisfying the *strict linear inequality* $l(x, y, z) < 0$, while the associated closed negative half-space satisfies the linear inequality $l(x, y, z) \leq 0$; and similarly for positive half-spaces.

Now suppose that we have two linear inequalities,* given by

(7.7.3) $a_1 x + b_1 y + c_1 z + d_1 \leq 0$

(7.7.4) $a_2 x + b_2 y + c_2 z + d_2 \geq 0.$

Let us write l_1 for the linear function on the left-hand side of (7.2.3) and l_2 for the corresponding function of (7.7.4). The two inequalities define two closed half-spaces, which we may denote by F_1^- and F_2^+. The intersection D of these two half-spaces has an important algebraic significance. By definition, it is the set of points (x, y, z) satisfying *both* of the linear inequalities (7.7.3) and (7.7.4), and will therefore constitute the *solution* of these simultaneous *linear inequalities*, in exactly the same way as the intersection of the sets $D_1{}^0$ and $D_2{}^0$ is the solution set of the pair of simultaneous *linear equations*

$$l_1(x, y, z) = 0 = l_2(x, y, z).$$

* We have seen earlier (Chapter 5) that it is essentially immaterial which of the half-spaces is nonnegative and which nonpositive. Therefore we can always reverse the "direction" of any inequality by multiplying both sides by -1. It follows that there would be no loss of generality in considering only intersections of negative half-spaces (or positive half-spaces). This has its advantages, but at this stage the more direct method is preferable.

In order to get some idea of what these intersections are like, let us look at Fig. 7.3, which depicts systems of two, three and four linear inequalities in two dimensions. We denote each closed half-space by vertical shading of the side of the line which bounds it; and mark each intersection D by horizontal shading.

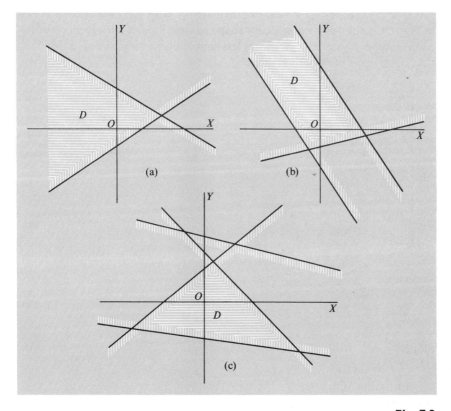

Fig. 7.3

One can easily grasp certain intuitive ideas about the "shape" of D on the basis of these two-dimensional figures, but one's intuition can rapidly become foxed when we move on to three-dimensional space. Notice that the rather special sets

$$X = \{(x, y, z) \,|\, x \geq 0\}$$
$$Y = \{(x, y, z) \,|\, y \geq 0\}$$
$$Z = \{(x, y, z) \,|\, z \geq 0\}$$

are each defined by a linear inequality, since we may consider the set X to be the set of points defined by the inequality.

$$1x + 0y + 0z + 0 \geq 0$$

and similarly for the sets Y and Z. Their intersection is simply the nonnegative octant of three-dimensional space. Let us couple with this set of three inequalities, two of the more usual type, and look at the intersection D of all five sets (which must necessarily be a subset of the nonnegative octant).

The intersection D of the two half-spaces with the nonnegative octant in Fig. 7.4 is not a particularly easy shape to visualize, and if we had considered the intersection of five (or more) less simple linear inequalities, the difficulties would have been even greater.

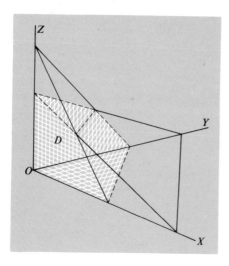

Fig. 7.4

In n-dimensional space the analogue of a plane is a hyperplane with equation

$$a_1x_1 + a_2x_2 + \ldots + a_nx_n + k = 0.$$

Denoting the point $(x_1, x_2, \ldots, x_n)$ by the vector $\mathbf{x}$, and denoting the linear function

$$a_1x_1 + a_2x_2 + \ldots + a_nx_n + k$$

by $l(\mathbf{x})$ we can define half-spaces analogous to (7.7.2) as follows

(7.7.5)

(a) $D^- = \{\mathbf{x} \mid l(\mathbf{x}) < 0\}$

(b) $D^0 = \{\mathbf{x} \mid l(\mathbf{x}) = 0\}$

(c) $D^+ = \{\mathbf{x} \mid l(\mathbf{x}) > 0\}$

D^- and D^+ are the open negative half-space and open positive half-space respectively; D^0 is, of course, the hyperplane itself. The closed half-spaces are defined, as before, by

$$F^- = D^- \cup D^0 \quad \text{and} \quad F^+ = D^0 \cup D^+$$

and correspond to the inequalities $l(\mathbf{x}) \leq 0$ and $l(\mathbf{x}) \geq 0$.

Of special importance among half-spaces are those of the form $x_i \geq 0$ $(i = 1, 2, \ldots, n)$, i.e., which specify that a particular coordinate is not negative. The intersection of all these half-spaces is the set of points having all nonnegative coordinates. It is called the positive *orthant* and corresponds to the positive quadrant in two dimensions and the positive octant in three dimensions.

It goes without saying that it is extremely difficult to "picture" the intersection of half-spaces in n dimensions, and we shall make no attempt to do so.

Exercise 75

1. A region K of the positive octant in space of three dimensions is defined by the inequalities

$$x + y + z \leq 6$$
$$8x + 6y + 3z \leq 24$$
$$12x + 3y + 2z \leq 24.$$

Determine which of the following points lie within the region K

(a) $(1, 1, 1)$ (c) $(\frac{3}{2}, 1, 3)$ (e) $(\frac{1}{2}, 3, \frac{1}{2})$

(b) $(\frac{3}{2}, 1, 2)$ (d) $(2, 1, 3)$ (f) $(\frac{6}{5}, 0, \frac{24}{5})$.

2. In two-dimensional space a region K_1 is defined by the inequalities

$$3x + y \geq 20$$
$$2x + 7y \leq 83$$
$$4x - 5y \leq -5;$$

while a region K_2 is defined by

$$9x + y \geq 77$$
$$3x - 7y \leq -11$$
$$3x + 4y \leq 77.$$

By drawing a diagram, or otherwise, determine whether K_1 and K_2 have any points in common.

3. A region R in two-dimensional space is defined by the inequalities

$$x + y - 2 \geq 0, \quad 2x - 3y + 15 \geq 0,$$
$$4x + 3y - 33 \leq 0, \quad 3x - y - 15 \leq 0,$$
$$x \geq 0, \qquad\qquad y \geq 0.$$

Sketch the region R and determine what portion of the line $2x - 3y + 9 = 0$ lies within R.

7.8 CONVEX SETS

The observation made at the end of section 7.7 implies that, unless possessed of quite phenomenal geometric intuition, we must turn to an algebraic investigation of an intersection like the intersection D of the previous section if we are to arrive at any theorem concerning its properties. One such theorem, which assigns to any such nonempty intersection a simple but remarkable property, we may now demonstrate.

Suppose that we have a finite number of linear inequalities (numbered arbitrarily), and consider any two points, say P_1 and P_2, of the intersection D of the associated half-spaces. If some of the inequalities are *strict*, then the corresponding half-spaces will be open; otherwise they will be closed. We assume that D is not empty, though we allow the possibility that it consists of just one point (i.e., P_1 and P_2 need not be different).

Let P_t be *any* point on the straight line segment joining P_1 and P_2, so that P_t can be expressed by the equation

$$P_t = tP_1 + (1 - t)P_2$$

for some value of t between 0 and 1.

If we take the linear function l_k associated with the k-th inequality, its value at P_t will be given by

$$a_k x_t + b_k y_t + c_k z_t + d_k$$
$$= a_k (tx_1 + (1 - t)x_2) + b_k(ty_1 + (1 - t)y_2) + c_k(tz_1 + (1 - t)z_2) + d_k$$
$$= t(a_k x_1 + b_k y_1 + c_k z_1 + d_k) + (1 - t)(a_k x_2 + b_k y_2 + c_k z_2 + d_k).$$

Hence

$$(7.8.1) \qquad l_k(x_t, y_t, z_t) = tl_k(x_1, y_1, z_1) + (1 - t)l_k(x_2, y_2, z_2).$$

It follows from (7.8.1) that whatever common sign the linear function l_k has at P_1 and at P_2, it will have that sign also at P_t, since the value of l_k at P_t is a simple weighted average of its values at the other two points. Since l_k was arbitrary, the same conclusion must hold for each linear inequality of the set. Put verbally our result is that if two points each satisfy a system of linear inequalities then any point which lies on the line segment joining them also satisfies that system.

Let us again consider what shape the intersection D can have. The most obvious example of a set having the above property is simply the set of points constituting a line segment joining two arbitrary points; but other examples spring readily to mind. In two-dimensional space, the sets of points bounded by a triangle, any parallelogram, or any regular polygon have this property; so does the disk bounded by any circle, the area bounded by an oval, or by an ellipse. In three-dimensional space, a pyramid, a cube and a sphere all have the property. If we remove part of the interior of any one of the areas or volumes mentioned, it ceases to have the property. But we need not resort to such a trick as that. None of the areas illustrated in Fig. 7.5 has the property, as can easily be seen by joining P_1 and P_2 in each illustration by a straight line.

The reader may construct for himself examples in three-dimensional space which fail to have the property.

It remains to name the property that we have described. Any set of points which possesses it will be said to be *convex*. More formally, we adopt

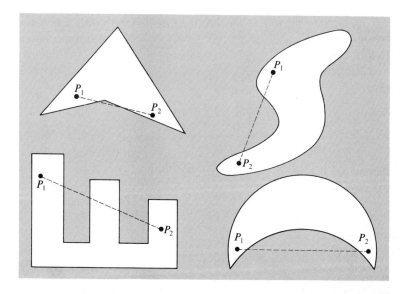

Fig. 7.5

Definition 7.5

A set D of points in n-dimensional space is convex *if, given any two points P_1 and P_2 in D, it is true that any point P_t, expressible in the form $P_t = tP_1 + (1 - t)P_2$ (where $0 \le t \le 1$), also belongs to D.* That is, the line segment P_1P_2 lies entirely in D.*

The definition of a convex set can take many different forms, of which the one we have given is perhaps the simplest. But for some purposes it is more useful to adopt the following definition, which superficially looks somewhat different.

Definition 7.6

A set D of points in n-dimensional space is said to be convex if, given any finite number k of points of D, say $P_1, P_2, \ldots, P_k$, then each point P_t that is expressible in the form

$$P_t = t_1P_1 + t_2P_2 + \ldots \ldots + t_kP_k$$

(where each $t_i \ge 0$ and $t_1 + t_2 + \ldots + t_k = 1$) also belongs to D.

It turns out, however, that Definitions 7.5 and 7.6 are equivalent statements. That 7.6 implies 7.5 follows obviously on putting $k = 2$, but the proof that 7.5 implies 7.6 is not so obvious. We prove it by means of mathematical induction as explained in Chapter 4.

Suppose that we have a set D which is convex according to definition 7.5. We want to show that Definition 7.6 holds for any integral value of $k \ge 2$. We first observe that the result is true if $k = 2$. This followed directly from Definition 7.5. Now let r be any value of k for which the theorem is true.

Take $(r + 1)$ points $P_1, P_2, \ldots, P_{r+1}$ of D and consider the point P_t given by

(7.8.2) $P_t = t_1P_1 + t_2P_2 + \ldots + t_rP_r + t_{r+1}P_{r+1}$

where, as usual, each t_i is nonnegative, and their sum is 1. Does P_t belong to D? If $t_{r+1} = 1$, then $P_t = P_{r+1}$, and P_t certainly belongs to D. If t_{r+1} is less than 1 it follows from (7.8.2) that $t_1 + t_2 + \ldots + t_r$ is positive, equal to t say, so that we may write

(7.8.3) $P_t = t\left(\dfrac{t_1}{t}P_1 + \dfrac{t_2}{t}P_2 + \cdots + \dfrac{t_r}{t}P_r\right) + t_{r+1}P_{r+1}$

---

* Note that in addition to the examples already mentioned, a set consisting of one point is also convex, since the definition does not require P_1 and P_2 to be different. The empty set may or may not be convex, depending on the convention that we choose to adopt. We will suppose it to be convex, although for more advanced work it is usually more convenient to assume the opposite.

Now each t_i/t is nonnegative, and their sum is unity. Hence the point P_u, say, represented by the expression in brackets in (7.8.3), is a weighted average of r points of D and therefore, by our provisional assumption, belongs to D. So we can write [since $t_{r+1} = (1 - t)$]

$$(7.8.4) \qquad P_t = tP_u + (1 - t)P_{r+1}$$

where P_u and P_{r+1} both belong to D. Since D is convex according to Definition 7.5, equation (7.8.4) implies that P_t also belongs to D.

We have therefore proved that *if* a weighted average of any r points of D belongs to D, then a weighted average of any $(r + 1)$ points belongs also. That is, if the theorem is true for $k = r$ it is true for $k = r+1$. This completes the second part of the argument, and hence, by mathematical induction, the theorem is true for all integers $k \geq 2$. We have therefore shown that if D is convex according to Definition 7.5, it is convex according to Definition 7.6. This result justifies the use of the term *convex combination* for weighted averages of the form in (7.8.2). (Compare the use of this phrase on pages 187 and 225.)

Exercise 76

1. If $A \equiv (2, 1)$, $B \equiv (4, 5)$ and $C \equiv (7, 3)$ are three points in two-dimensional space, determine which of the following points can be written as a convex combination (or weighted average) of A, B and C, i.e., can be expressed in the form $\lambda A + \mu B + \nu C$, where λ, μ and ν are nonnegative, and $\lambda + \mu + \nu = 1$.

(a) $(1, 4)$ (d) $(3, 2)$

(b) $(7, 6)$ (e) $(6, 2)$

(c) $(5, 4)$.

2. For those points in problem 1 which are convex combinations of A, B and C determine the values of λ, μ and ν. Draw all five points, and the points A, B and C, on one diagram and interpret geometrically the results of problem 1.

3. Show that the point $P \equiv (4, 3, 6)$ is a convex combination of the points

$$A \equiv (2, 5, 2), \quad B \equiv (7, 3, 9), \quad C \equiv (4, 6, 3) \quad \text{and} \quad D \equiv (3, 1, 7).$$

Interpret this statement geometrically in terms of the position of P relative to the other four points.

4. **(a)** *A*, *B*, *C* are three points in a plane. What is the smallest set of points (i.e., the one of least area) which contains *A*, *B* and *C*?

(b) What is the smallest convex set which contains four given points *A*, *B*, *C* and *D* in three-dimensional space?

7.9 METHODS OF GENERATING CONVEX SETS

The reader will probably have ,noticed that we have now implicitly broadened the concepts from which we started. It is fairly clear that the intersections of a finite number of half-spaces will be convex sets, but they will be rather special types of convex sets, since they will have linear *boundaries*, representing the linear functions defining the half-spaces. But we have mentioned examples of sets—disks, spheres and so on—which clearly have the convexity property, but which equally clearly do not have linear boundaries.

One might then wonder where these 'curved' convex sets come from. Are there other ways of generating convex sets, besides taking the intersection of half-spaces? One such way—which includes the former as a special case—is to take the intersection of any number (finite or infinite) of convex sets. Suppose that we have such a collection $\{D_i\}$, and consider their intersection $D_1 \cap D_2 \cap D_3 \cap \ldots$ If this intersection is empty, then we have agreed to say that the set is convex. If it is not empty, then take any two of its points, say P_1 and P_2 (which need not be different). By construction P_1 and P_2 belong to each set, and so any P_t which is a convex combination of them will also be in each set, and hence in the intersection, which is thereby shown to be convex.

But this still leaves us with the original convex sets. Bearing in mind the example of a circle, which can be looked on as being bounded by its tangents, we might conjecture that *any* convex set can be regarded as the intersection of an "infinite" number of half-spaces. To this inexactly put question can be given only a rough answer, to the effect that this conjecture is broadly true. However, to prove this would involve problems of topology which are beyond the scope of this book. Therefore we will confine our attention to those convex sets defined by the intersection of only a *finite* number of half-spaces.

Before pursuing our investigation of these in greater detail, let us glance at an essentially *different* way of generating convex sets. Suppose that we have a set *S* consisting of a finite number of points scattered about the plane, as in Fig. 7.6., and ask: What is the shortest length of string that would enclose all the points of the set? The answer would of course be found by measuring off the length of the broken line in the figure.

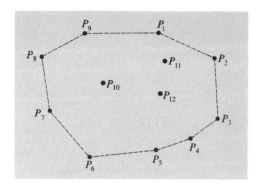

Fig. 7.6

Observe that the polygon $P_1 P_2 P_3 \ldots P_9$ is the boundary of a convex set, a set which is usually described as the *convex hull* (or *convex cover*) of the set of points S from which we started. A little reflection enables us to define the convex hull alternatively as the *smallest* convex set which contains S. The points $P_1, P_2, \ldots, P_9$ are called the *extreme points* of this convex hull, and in a sense "generate" it, since the removal of any one, say P_8, would alter the hull. The adjective *extreme* is apposite for these points, since they alone are not convex combinations of two distinct points in the set.

We can also imagine generating "curved" convex sets by this method. An obvious example is to consider regular polygons, starting with three, four, five,. . . sides, and letting the number of sides—and hence the number of extreme points—increase without limit. Elementary geometry tells us (but does not prove) that the resulting sequence of geometrical figures will tend to the limiting shape of a circle. But here again we shall retreat from the consideration of an unlimited number of operations and stick to sets obtained from a *finite* number of extreme points.

These two different ways of generating convex sets, by intersecting half-spaces and by taking the convex hull of a set, are intimately related. This fact, which is of great importance for the mathematics of classical economic theory, will be proved in the next section, at least for a special but important class of convex sets.

Exercise 77

1. A convex set in two dimensions is determined by the points

$$(2, 6), \quad (4, 1), \quad (8, 4), \quad (5, 4), \quad (4, 6),$$
$$(7, 1), \quad (2, 3), \quad (4, 3) \quad \text{and} \quad (5, 8).$$

Plot these points and draw the convex set (the convex hull) that they

determine. Which of the points are extreme points of this convex hull?
2. R_1 is the convex set in two dimensions determined by the points

$$(7, 5), \quad (11, 9), \quad (10, 14), \quad (8, 5) \quad \text{and} \quad (13, 2);$$

R_2 is the convex set determined by the points

$$(1, 15), \quad (15, 8), \quad (10, 10), \quad (\tfrac{54}{5}, \tfrac{26}{5}) \quad \text{and} \quad (9, 2).$$

Plot these points and draw the convex sets R_1 and R_2. Draw also the intersection $R_1 \cap R_2$, show that it is a convex set, and find its extreme points.

7.10 CONVEX POLYTOPES

From now on we restrict the discussion to convex sets which are the result of intersecting a *finite* number of *closed* half-spaces. These sets are called *convex polyhedral sets*, *convex polyhedra* or *convex polytopes*. From the definition, it follows that all the points in any such set will satisfy a finite collection of (simultaneous) linear inequalities. Conversely, the set of points constituting the "solution" of a finite collection of linear inequalities will be a convex polytope.

Let us ask a question that only has meaning for spaces of not more than three dimensions: what do convex polytopes look like? Any convex polygon is a convex polytope and so are many crystals—especially those cut by jewellers. Each of the shaded sets of Fig. 7.3 is a convex polytope, and so is the three-dimensional shaded set of Fig. 7.4. The simplest such set in a plane is a triangle (Fig. 7.7*a*), and in three-dimensional space is a tetrahedron (Fig. 7.7*b*).

Notice that in the plane the boundary of any convex polygon can be looked on as the union of (1) the set of all extreme points or vertices of the

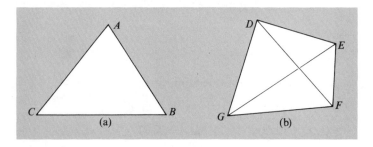

Fig. 7.7

polygon, such as *A*, *B* and *C* in Fig. 7.7*a*; and (2) the set of straight lines joining the vertices, such as *AB*, *BC*, and *AC* in Fig. 7.7*a*, which we shall call *edges*. In three-dimensional space, the boundary of a convex polyhedron is the union of (1) the set of vertices, (2) the set of edges, and (3) the set of *faces*, or plane sections of the boundary, such as *DEF*, *DFG*, *DEG*, and *EFG* in Fig. 7.7*b*. Thus in two-dimensional space a boundary consists of vertices and edges, and in three-dimensional space it consists of vertices, edges and faces. Observe that for any given convex polyhedron the sets (1), (2) and (3) are not disjoint—in the plane the vertices lie in certain edges while in space, in addition, the edges lie in certain faces—but in order to prove the theorems we require, it is very convenient to distinguish these three classes of mathematical objects.

So far we have been referring rather blithely to vertices of convex sets, without giving them a careful definition. This we must now do, with the help of another concept that has been implicit in our treatment. Since a convex polytope in three dimensions is the intersection of a finite number of closed half-spaces, its boundary will consist of faces, which are parts of the planes that define these half-spaces. Where three of these faces intersect, we shall have a vertex, and where two of them intersect we will get an edge. Let us call these planes defining the relevant half-spaces, the *bounding planes* of the convex polyhedral set. In *n* dimensions we shall have, correspondingly, *bounding hyperplanes*.

In the plane the boundary of a convex set will be a polygon consisting of parts of the lines which define the half-planes which define the set. These will be the edges. Where two edges meet we have a vertex of the polygon.*

We may now generalize these observations and define a vertex of a convex polytope *D* as follows:

Definition 7.7

A point P is vertex of a convex polytope D in space of n dimensions if (1) *P belongs to D, and* (2) *P is the common point of intersection of n bounding hyperplanes of D.*

The reader may object that two linear inequalities in space of two dimensions do not define a convex polytope, since there is only one vertex; such a polytope would look like a triangle extending indefinitely far (e.g., as in Fig. 7.3*a*). but a figure like this *is* a convex polytope. What distinguishes the polytopes in Figs. 7.3*a* and 7.3*b* from that in 7.3*c* is that the former are *unbounded*, while the latter is bounded. We may define a *bounded set in n dimensions* ($n \leq 3$)—with enough rigor for our present purposes—as a set which can

---

* From now on we shall use *vertex* in preference to the longer, though more descriptive, term *extreme point*.

be totally enclosed by an n-dimensional sphere of finite radius. More strictly, there is a point P and a distance a such that every point of the set is at distance a or less from P. Thus for D in Fig. 7.3c, we can find a finite two-dimensional sphere (a circle) in which D can be entirely enclosed, while for the corresponding set in each of the other two diagrams, we cannot. In the same way, when $n > 3$, we say that a set of points is bounded if it is a subset of the interior of some n-dimensional sphere, where, by the interior of an n-dimensional sphere we mean the set of all points satisfying an inequality of the form

$$x_1{}^2 + x_2{}^2 + \ldots + x_n{}^2 < a^2.$$

(The set of points satisfying

$$x_1{}^2 + x_2{}^2 + \ldots + x_n{}^2 = a^2$$

is the boundary of the sphere).

We can now proceed to state and prove our fundamental theorem on convex polytopes, a theorem which closely relates the two methods of "building" a convex set.

THEOREM 7.2

LET D BE A BOUNDED CONVEX POLYTOPE, AND LET P_1, P_2, ..., P_k BE ITS VERTICES. THEN

(a) EVERY POINT P THAT IS A CONVEX COMBINATION OF THE VERTICES OF D, BELONGS TO D; AND

(b) EVERY POINT IN D CAN BE WRITTEN AS A CONVEX COMBINATION OF THE VERTICES OF D.

Now we have already proved part (a) of this theorem, since by definition each vertex belongs to D, and it was shown above that Definition 7.6 of convexity is equivalent to Definition 7.5. Since D is by hypothesis convex, part (a) follows. Notice that this proof is valid irrespective of the dimensionality of the space in which we are operating.

Part (b) of the theorem is much more surprising, and correspondingly more difficult to prove. It asserts that we can *always* represent a point of D as a convex combination of a *fixed* number of points of D, viz., of the vertices $P_1, P_2, \ldots, P_k$.

Let us first prove this theorem for $n = 2$, which turns out to be very simple.

To begin with, go back to our "simplest" convex polytopes which are triangles in two-dimensional space, and tetrahedra in three-dimensional space. What are the simplest such sets in space of one dimension? Obviously, segments of a line. But any point on a segment of a line can be represented as a convex combination of the two extreme (end) points of the segment. Does this mean that any point in a triangle can be similarly represented as a convex combination of its three extreme points? The answer is yes, and the proof is simple.

Let the vertices of a triangle be A, B and C as in Fig. 7.8. Any point, such as D, on AC can be represented in the form $tA + (1 - t)C$, for some

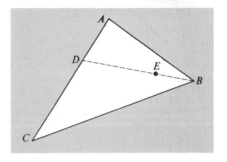

Fig. 7.8

value of t between 0 and 1. But clearly *any* point of the triangle can be represented as a convex combination of B with *some* point of the line AC. Thus a typical point E can be written as $uD + (1 - u)B$, for some value of u between (and including) 0 and 1. Then

$$(7.10.1) \qquad uD + (1 - u)B = utA + u(1 - t)C + (1 - u)B.$$

Each of the coefficients of the right-hand side of (7.10.1) is nonnegative and their sum is $ut + u(1 - t) + (1 - u) = ut + u - ut + 1 - u = 1$. Therefore the right-hand side is a convex combination of the vertices A, B and C, which proves our result.

For the record, let us note that what we have called the "simplest" convex polytopes are usually referred to as *simplexes*. These are more formally defined by the following:

Definition 7.8

An n-simplex in n-dimensional space is a bounded convex polyhedral set that has exactly $(n + 1)$ vertices, and which is not contained in a space of less than n dimensions.

Thus a 1-simplex is a line segment, a 2-simplex is a triangle and a 3-simplex is a tetrahedron; but four coplanar points in 3-space would not form a 3-simplex, since the vertices lie in a plane, i.e., in a two-dimensional subset. By convention, a 0-simplex is a point.

We can now prove part (*b*) of our theorem for $n = 2$. Suppose that we have any convex polygon (as in Fig. 7.9) with vertices $P_1, P_2, \ldots, P_k$. Then we join P_1 to each of the other vertices, as in Fig. 7.9, not counting the two to which it is already joined. In this way we divide the polygon into a number $(k - 2)$ of triangles, and every point on or inside the polygon lies either on or in one of these triangles.

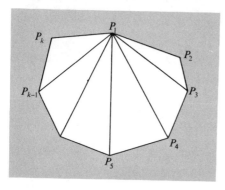

Fig. 7.9

Now we have previously shown that any point in a 2-simplex can be expressed as a convex combination of its vertices. Hence the fact that we have demonstrated that any point of the polygon lies in or on a triangle whose vertices are vertices of the polygon, implies that any such point can be expressed as a convex combination of the vertices of the polygon; in fact, at most three of these vertices will have positive weights t_i.

We can now proceed to the proof of part (*b*) for three dimensions, a proof which utilizes most of our previous work.

Consider any point P of D. Either P is a boundary point or it is not; we treat each case in turn.

1. *P is a boundary point of D.*   Then, by our previous results, P is a vertex of, or on an edge of, or in the interior of, a convex polygon (a face). In any event it may be represented as a convex combination of the extreme points of the convex polygon, and hence of D.

2. *P is not a boundary point of D.*   Draw any line through P. Since D is bounded, this line must strike the boundary of D in at least two points; for if not, D would be unbounded in at least one direction. For the same reason there must be at least one such point on each side of P. Suppose that the line strikes the boundary at *more* than two points. Then let Q_1 and Q_2 be the intersection points nearest to P on each side, and let R be a next nearest point of intersection. Without loss of generality we can take R to be on the same side of P as Q_2 is, as in Fig. 7.10.

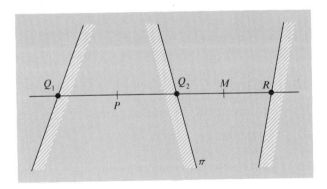

Fig. 7.10

Now every point of Q_1Q_2 lies in the convex set D, by definition of convexity. At Q_2 the line through P intersects at least one of the bounding planes (π in Fig. 7.8), and hence the points a little further beyond Q_2, for example M the midpoint of Q_2R, will be on the "wrong" side of the plane π, and hence will not belong to D. But if Q_2 and R belong to D so does M, since D is convex. Thus we arrive at a contradiction and (by a *reductio ad absurdum* argument) we deduce that a line through P meets the boundary of a bounded convex set in exactly two points.

We have used here a result from Chapter 6, namely that, in general, a line meets a plane in one point. The two possible exceptions do not give any difficulty. We have precluded the possibility of the line being parallel to the plane, since the line does intersect the boundary of D, while if the line lies entirely in one of the bounding planes then P is a point on the boundary, and this possibility has already been considered separately.

Let the line through P meet the boundary in points Q and R. Since they are on the boundary, Q and R will be vertices, or lie on edges, or be in faces. In any event, each may be expressed as convex combinations of at most three extreme points of D; thus

(7.10.1) $Q = t_1A_1 + t_2A_2 + t_3A_3 \qquad 0 \leq t_i \leq 1, t_1 + t_2 + t_3 = 1$

and

(7.10.2) $R = u_1B_1 + u_2B_2 + u_3B_3 \qquad 0 \leq u_j \leq 1, u_1 + u_2 + u_3 = 1$

the As and Bs standing for vertices of D, not necessarily all distinct. If two t_is are zero, then Q is a vertex; if one t_i is zero, Q is on an edge; and if no t_i is zero, Q is in the interior of a face; and similarly for R.

Since P lies on the line joining Q and R, we may write

$$P = vQ + (1 - v)R \qquad (0 \le v \le 1)$$

or

(7.10.3) $\qquad P = vt_1A_1 + vt_2A_2 + vt_3A_3 + (1 - v)u_1B_1$
$$+ (1 - v)u_2B_2 + (1 - v)u_3B_3.$$

Now each coefficient is nonnegative, and

$$v(t_1 + t_2 + t_3) + (1 - v)(u_1 + u_2 + u_3) = v + (1 - v) = 1.$$

Hence P is a convex combination of extreme points of the polyhedron.

Proof for Convex Polyhedra in *n* Dimensions

It is now a relatively simple matter to extend the method of proof just given to prove the general result for convex polytopes in any number of dimensions.

THEOREM 7.3

If D is a bounded convex polytope in space of n dimensions then every point of D can be written as a convex combination of the vertices of D.

PROOF:

The proof is by mathematical induction.

We already know that the theorem is true in spaces of two or three dimensions. Now let p be any integer such that the theorem is true in spaces of p dimensions or less. Consider a convex polytope D in $p + 1$ dimensions, and any point P of D.

Take any line through P. Unless the line meets the boundary of D in at least one point on each side of P the fact that D is bounded is contradicted. To show that the line meets the boundary in exactly two points we reason as before. For we know that in general a line meets a hyperplane in a unique point, so that the previous argument, and Fig. 7.8, can be used in exactly the same way. Let Q and R be the points of intersection.

Q will in general lie in one of the faces of D, but it may be in an "edge" of smaller dimension. In any case it lies within a convex set in a space of dimension p or less. Hence by our induction argument it is a convex combination of vertices of D. Let us write

$$Q \equiv \sum_i t_iA_i \qquad \left(0 \le t_i \le 1, \quad \sum_i t_i = 1\right)$$

where A_i are vertices of D. Similarly we write

$$R \equiv \sum_i u_i B_i \quad \left(0 \leq u_i \leq 1, \ \sum_i u_i = 1 \right).$$

Thus we have $P = vQ + (1 - v)R$, which exhibits P as a convex combination of vertices of D, exactly as in (7.10.3). Hence if the theorem is true for $n = p$ it is true for $n = p + 1$.

Hence by mathematical induction the theorem is true for all integers $n \geq 2$.

Exercise 78 (Chapter Review)

1. Find the point in which the line joining the points $(1, 3, 0, 4, 5)$ and $(-1, 18, 3, 11, 19)$ meets the hyperplane

$$17x_1 + 2x_2 - 5x_3 + 3x_4 - x_5 = 18.$$

2. If $P = (x_1, x_2, \ldots, x_n)$, $Q = (y_1, y_2, \ldots, y_n)$ and O is the origin, then OP and OQ are "perpendicular" if $x_1 y_1 + x_2 y_2 + \ldots + x_n y_n = 0$, i.e., if the scalar product $\mathbf{x} \cdot \mathbf{y}$ is zero. If

$$P \equiv (2, 0, -4, -2)$$
$$Q \equiv (3, -5, -9, -2)$$
$$R \equiv (0, 4, 1, -1)$$

find the coordinates of a point S such that OS is perpendicular to each of OP, OQ, OR.

3. Prove the analogue of result (5.5.3) namely that the equation of the hyperplane which makes intercepts $a_1, a_2, \ldots, a_n$ on the coordinate axes is

$$\frac{x_1}{a_1} + \frac{x_2}{a_2} + \cdots + \frac{x_n}{a_n} = 1.$$

4. What is the intersection of the hyperplanes

$$x_1 \qquad + 3x_3 \qquad = 4$$
$$3x_1 - 2x_2 + 9x_3 + 14x_4 = 10$$
$$x_2 \qquad - 7x_4 = 2$$
$$5x_1 + 3x_2 + 15x_3 - 21x_4 = 26?$$

5. Find for what values of k the distance between the two points $(9, -7, 2k, -28)$ and $(-6, 13, 4, k)$ is 65. For what value of k is the distance a minimum?

6. Draw the convex hull determined by the points (2, 5), (7, 4), (2, 2), (9, 6), (4, 4) and (5, 1) in the plane, and determine which of the following points belong to it

$$(5, 5), (6, 3), (2, 1), (7, 3), (3, 2).$$

7. Draw the convex set in the plane that is determined by the inequalities:

$$3x + y \geq 3$$
$$4x - y \leq 20$$
$$3x + 4y \leq 34$$
$$3x + 2y \leq 30$$
$$x - y \geq -5$$
$$x \geq 0$$
$$y \geq 0.$$

What are the extreme points (vertices) of this set?

8. Draw roughly, in perspective, (or make a model of) the convex hull of the five points

$$(2, 0, 4), \quad (3, 4, 0), \quad (0, 2, 6), \quad (0, 5, 4), \quad (3, 4, 3)$$

in three-dimensional space. The plane $2x + y - z = 0$ intersects this convex hull in a (two-dimensional) convex set. Find the extreme points of this convex set.

9. D is a bounded convex set in space of three dimensions determined by a set of inequalities

$$a_i x + b_i y + c_i z \leq d_i \quad (i = 1, 2, \ldots, k).$$

Prove that an extreme point of D is the point of intersection of three of the k planes

$$a_i x + b_i y + c_i z = d_i \quad (i = 1, 2, \ldots, k)$$

but that the converse is not necessarily true.

Hence find the extreme points of the convex set D determined by the inequalities

(a) $4x + 8y + 5z \leq 20$
(b) $6x + 3y + 2z \leq 18$
(c) $x \geq 0$
(d) $y \geq 0$
(e) $z \geq 0$

by finding all points of intersection of the corresponding planes taken 3 at a time, and rejecting those that do not satisfy all the inequalities.

10. Let

$$l(P) = l(x_1, x_2, \ldots, x_n) = a_1x_1 + a_2x_2 + \ldots + a_nx_n$$

be any linear function defined for every point P of space of n dimensions. Prove that if P and Q are any two points of this space then

$$l(\lambda P + \mu Q) = \lambda l(P) + \mu l(Q).$$

Using this result, or otherwise, prove that if D is a bounded convex polytope in the space and L is the maximum value of $l(P)$ for $P \in D$, then there is an extreme point of D at which l has this greatest value. Under what conditions can this greatest value be attained at more than one point?

Hint: From the finite number of extreme points choose one for which $l(P)$ is greatest. Then show that for no other point can $l(P)$ be greater than this.

Suggestions for Further Reading

There are many good books dealing more or less specifically with the theory of convex sets, which has become an important branch of modern mathematics, but they all go far beyond the requirements of this chapter. The three "Finite Mathematics" books mentioned on page 27 are among those which, along with other topics, deal at a more elementary level with convex sets. These three books should provide enough collateral reading for the reader's needs.

8 Foundations of Analysis,

Sequences, Limits,

Infinite Series and

Continuity

A mathematician and a friend were walking in the country one morning when they saw a flock of sheep in a nearby field. "Look at those sheep," said the mathematician's friend, "They have just been shorn."
"So they have," replied the mathematician, "At least on this *side!"*

8.1 INTRODUCTION—MATHEMATICAL RIGOR

Whenever three or four mathematicians gather together and start to talk about their subject (as they inevitably do), it is seldom very long before the word *rigorous* crops up in the conversation. By a *rigorous treatment* of a subject the mathematician means a treatment in which every step depends logically on what has gone before, in which no hidden assumptions are allowed to creep in at any stage of the logical argument, and in which nothing, however "obvious" it may be, is taken for granted. It is, in short, the attitude taken by the mathematican in the anecdote at the head of this chapter when he refused to commit himself to make any pronouncement concerning those parts of the sheep which he was not able to see.

It goes without saying that mathematical rigor (the rigorous attitude just described) is something which pervades, or should pervade, the whole of mathematics, and it usually does, though mathematicians, being no more infallible than other mortals, sometimes fall short of its very exacting requirements. It is true that in almost any textbook of mathematics you will find phrases like "it is obvious that", or "it is clear that"—indeed we have already met them in this book; but these phrases are to be interpreted as meaning that the proof of the result in question is so simple or straightforward that it is unnecessary for the writer to use up valuable space in writing it out, and that it may be supplied by the reader; it does not mean (or should not) that the result lacks a proof but is going to be assumed despite its absence.

Although mathematical rigor is needed in all branches of mathematics, its presence is more keenly felt in some branches than in others, and perhaps in none so much as in analysis, the branch of mathematics with which this chapter is concerned. However, since this is not a book for mathematicians, we shall permit ourselves the luxury of not insisting on a strict adherence to the requirements of rigor. For example, quite early in this chapter we shall assume without proof a result whose proof would necessitate too great a digression from the subject matter we wish to cover. As far as it lies within the ability of the author to make them so, these lapses from a strictly rigorous approach will be "honest" ones; if a result is given without proof, either the proof will be given later in the book or else references will be given whereby the assiduous reader may look up the proof for himself and assure himself that the author is not trying to pull the wool over his eyes.

Let us start by considering sets of numbers. Suppose we have a set of real numbers. Can we always say that this set has a largest member? If we think in terms of finite sets, it may seem that the answer must be "yes," but a simple example will show that this cannot be the answer for all sets. The set of all positive integers is one example of a set of numbers which does not have any member which is larger than all the others. This is clearly the consequence of the fact that this set is *unbounded* (a term which we shall make precise in a moment), but there are many sets, not thus disqualified, for which the answer is still "no." Take, for example, the set of real numbers between 0 and 1, but not including these two numbers, i.e., the set $\{x \mid 0 < x < 1\}$. If proof is needed that this set has no largest member, we need merely note that the claims of any number x to be the largest member of the set can be demolished by producing the number $\dfrac{1 + x}{2}$ which belongs to the set (since it is < 1), and is greater than x. It is therefore by no means true that a set of numbers has to have a largest member; it may have, but it may not. A *finite* set of numbers, however, will always have a largest member, since we can (in theory at any rate) examine each number in turn, keeping track of which number is currently the largest, until we have examined every number in the set.

In exactly the same way we see that although a finite set of numbers will necessarily have a least number, this may or may not be true of a set in general.

Now although the set $S = \{x \mid 0 < x < 1\}$ does not have a largest member, there is a number, the number 1, which we would be tempted to call the largest member were it not for the annoying fact that this number is not a member of the set. The number 1 clearly has a relationship to the set S which no other number has. Let us look a little more closely at this relationship.

8.2 UPPER AND LOWER BOUNDS

The first property which the number 1 has is that it is at least as large as any member of the set S. Other numbers have this property, for example, 2, 5, 100, etc., but of all the numbers that have this property the number 1 is the least. What is special about the number 1 in relation to the set S is therefore that it is the least of all the many numbers which are at least as large as every member of S. We now extend this approach to apply to any set.

If S is any set of real numbers, we define an *upper bound* of S to be any number which is greater than or equal to every member of S. Thus, 2, 5, 100, etc., are upper bounds of the set $\{x \mid 0 < x < 1\}$. Some sets do not have any upper bounds—the set of positive integers being one example. Sets having no upper bounds are said to be *unbounded above*, while those that do have upper bounds are said to be *bounded above*. It may happen that an upper bound of a set is also a member of the set. Thus 3 is both an upper bound and a member of the set $\{x \,|\, 2 \leqslant x \leqslant 3\}$. We shall see shortly that not more than one upper bound can belong to the set.

It should be clear that if a set is bounded above, i.e., if it has any upper bounds at all, then it will have a lot of them. A set S which is bounded above will therefore determine a set U consisting of its upper bounds. We now ask the question; "Does the set U of upper bounds have a least member?" Since we have already noted that a set of numbers may or may not have a least member, the answer might seem to be "Not necessarily", but we must not overlook the fact that the set U is not just *any* set but is determined in a rather special way, and might therefore have some special properties, as indeed it has. The set U of upper bounds of a given set S will, in fact, always have a least member, and this least member is called the *least upper bound* of the set S.

Now this is a result which clearly needs to be proved, but we shall not supply the proof here. This is the point (mentioned at the beginning of the chapter) where we must depart from a rigorous presentation of the subject in order not to spend a lot of time forging the mathematical tools required for the proof of this result. It will be of value, however, just to indicate wherein the difficulty lies.

The trouble is that, up to now, we have assumed that we know what is meant by a real number, and in particular by an irrational number. This is a colossal assumption. In section 3.2 we glibly defined an irrational number as being a nonterminating, nonrecurring decimal, but this definition will hardly do; it would certainly not satisfy our sheep-watching mathematician, who would be sure to ask exactly what we meant by such a statement. "Even supposing," he might say, "that *you* know what is meant by a nonterminating, nonrecurring decimal, and you have such a number in mind; how do you propose to convey to me what number you are thinking of? To do so you would have to recite an infinite string of digits. And even if you surmount *that* difficulty, how are you going to add two such numbers? You would have to perform an infinite succession of two-digit additions, and life is much too short." No, the concept of nonterminating nonrecurring decimals, useful though it may be for giving us an intuitive idea of what an irrational number is, raises far more problems than it solves, and will not serve to give us a rigorous definition which will avoid the above, and many other, difficulties. A completely new definition of irrational numbers is required, and when this is supplied (as it will be in Chapter 12) it is found that the statement made above, that the set U has a least member, is quite easy to prove. In the meantime we must assume that result without proof.

To save face let us call it an axiom.

Axiom.

Any nonempty set of real numbers which is bounded above has a least upper bound.

A number which is less than or equal to every member of a set S is called a *lower bound* of S. If a set has no lower bounds (the set of negative integers is an example) it is said to be *unbounded below*; a set which has lower bounds is said to be *bounded below*. A set which has a lower bound will have a lot of them, so that a set S which is bounded below will determine a set L consisting of all its lower bounds. This set L will have a largest member, and this largest member is called the *greatest lower bound* of the set S.

That this greatest bound always exists, i.e., that for a set S bounded below the set L of lower bounds always has a largest member, is a cognate result to the axiom which we stated above. Fortunately it is not necessary to take this also on trust; we can prove it from the axiom itself.

Let S' be the set obtained from S by changing the sign of every member of S. That is to say, we define S' by

$$x \in S' \text{ if, and only if, } -x \in S.$$

Now S is bounded below, so that we can find a number A which is less than or equal to every member of S. The number $-A$ will then be greater than or equal to every member of S', so that S' is bounded above. Hence by our axiom it will have a least upper bound—call it M. The number $-M$ is then the greatest lower bound of S. For it is clearly a lower bound of S, and if A is any *other* lower bound of S then $M > A$, since the alternative, $M < A$, would mean that the upper bound $-A$ of S' would be less than the least upper bound $-M$ of S', which is not possible. S therefore has a greatest lower bound.

Hence every set which is bounded below has a greatest lower bound. A set which is bounded both above and below is said to be *bounded*. A bounded set has both a greatest lower bound and a least upper bound.

Exercise 79

1. Prove that if a set has an upper bound which is also a member of the set, then this upper bound is the least upper bound, and that the set does have a greatest member.

2. If M is the least upper bound of a set S, prove that if ϵ is any *positive* number* (however small it may be) then $M - \epsilon$ is less than some member of S, i.e., there is a member x of S such that $M - \epsilon < x$.

3. If m is the greatest lower bound of a set S, and ϵ any positive number, then there is a number y such that $y \, \epsilon \, S$ and $m + \epsilon > y$.

---

* The Greek letter ϵ (epsilon) is traditionally used to denote a small positive number.

4. Say whether or not the following sets of numbers are bounded above or below, and find, where applicable, their greatest lower bounds and least upper bounds.

(a)   the set of numbers of the form $\dfrac{n}{n+1}$ where n is a positive integer.

(b)   The set of numbers of the form $1/m + 1/n$, where m and n are positive integers.

(c)   As for (b) but with the function $1/m - 1/n$.

8.3 FUNCTIONS OF A POSITIVE INTEGRAL VARIABLE—SEQUENCES

In this next section we shall study functions whose arguments take only positive integral values. A function $f(n)$ of this kind will define a *sequence* of numbers, in the following way. To each integer n there is a corresponding value $f(n)$ of the function. We can imagine these values being displayed one after the other thus:

$$f(1), f(2), f(3) \ldots, f(n), \ldots$$

the ordering of these numbers being given by the magnitude of the arguments, so that the value $f(n)$ occurs in the nth place. Thus the function $\dfrac{n+1}{n}$ determines the sequence

$$2, 2\tfrac{1}{2}, 3\tfrac{1}{3}, 4\tfrac{1}{4}, \ldots$$

Many sequences have a property which we describe by saying that the terms become large (or that the function is large) for large values of n. We first make precise what this means. One requirement is that there are terms of the sequence "as large as we like"; that is, that if we think in advance of any number K, however large this number may be, we can find terms of the sequence which are larger than K. This is not all, however, for this says no more than that the set of values of the function (terms of the sequence) form a set that is unbounded above. What is required in addition is that *all* terms from some term onwards are greater than K. The sequence

$$0, 1, 0, 2, 0, 3, 0, 4, 0, \ldots$$

is one whose terms are unbounded above, but it does not satisfy the additional requirement, so that we would not describe its terms as being large for large n. On the other hand the terms of the sequence

$$1, 3, 5, 7, 9, 11, \ldots$$

are large for large n. These remarks will indicate what is required in the definition, which we now give.

Definition 8.1

A function $f(n)$ is said to be "large for large values of n" if, given any number K (however large it may be) there is an integer N such that $f(n) > K$ for all $n > N$, i.e., such that

(8.3.1) $$(n > N) \Rightarrow (f(n) > K).$$

Note. The phrase "however large it may be" is not essential for the definition, but it serves to indicate that the larger K is the larger the value of N will be that has to be chosen to make (8.3.1) true.

This property may be written in another way. We often say that "$f(n)$ tends to infinity as n tends to infinity," or, in symbols, $f(n) \to \infty$ as $n \to \infty$. This phrase must be regarded as being defined *in its entirety* by the definition we have just given; it should not be presumed to ascribe any meaning to the term *infinity* or the symbol ∞.

EXAMPLES:

1. If $f(n) = n^2$ then $f(n)$ is large for large n. For given any number K we can always find a suitable number having the required property; we can, for example, take N to be any integer which is greater than $\sqrt{K}$. For then, if $n > N$ we have

$$f(n) = n^2 > N^2 > K$$

and the requirements of the definition are met.

2. The function $\sqrt{n}$. This is also large for large n. Given K we take N to be any integer greater than K^2. Then $f(n) = \sqrt{n} > \sqrt{N} > K$, if $n > N$.

3. The function $\dfrac{n+1}{n}$. This is not large for large n. For $\dfrac{n+1}{n} < \dfrac{n+n}{n} = 2$, whatever the value of n, so that no value of the function is greater than 2. If therefore, we take $K = 2$ the requirements of the definition cannot be satisfied.

4. The function $f(n) = \frac{1}{2}n \cos^2(n\pi/2)$ is not large for large n. Although there are values of the function which exceed any given number K, nevertheless, for every odd value of n the function has the value zero, so that it cannot be the case that all values from some N onwards exceed a $K > 0$. This function gives rise to the sequence $0, 1, 0, 2, 0, \ldots$ which we met just now.

Let us now see what may happen to a function (besides being large) as its argument becomes large. If we look at the function $\dfrac{n+1}{n}$ we see that as its argument becomes larger and larger, so the values of the function get closer and closer to the value 1. We do, in fact, say that the function *tends to* the number 1 as n tends to infinity (or as n becomes large), but we must, of course, make quite precise what we mean by saying this. The general idea of the function tending to 1 is that however close we may wish the function to be to 1 it will in fact get at least as close as that for all values of n from some value onwards. Thus if we wish the function to be within 1/1000 of the number 1, i.e., to lie between 999/1000 and 1001/1000, then we can find an integer N such that this is the case for all $n > N$. For the function that we are considering we can take $N = 1000$ and this will do the trick. If we wish the function to be even closer to 1, say within 1/1000000 of it, then this will be true for all $n > 1000000$. Similarly, however small the amount by which we restrict the difference between the function and the number 1, we always find that from some value of n onwards the values of the function satisfy the restriction. This is roughly what we mean by saying that the function $\dfrac{n+1}{n}$ tends to 1 as n tends to infinity. One can think of this as a sort of game between two players. The first player thinks of a number greater than 1; the second player then has to think of an integer N such that *all* the numbers $\dfrac{n+1}{n}$ for $n = N + 1$, $N + 2$, etc., lie between 1 and the number thought of by the first player. Can the second player always do this, no matter by how little the first player's number differs from 1? If he can, and in this example he can, then the function tends to 1.

Note: It is not enough that there should be values of the function which obey the restriction; the restriction must be obeyed for all values of n from some value onwards. [Compare the definition of $f(n) \to \infty$.]

It is now no great step to the definition of a function $f(n)$ tending to a number l. The number to which a function tends (if it does tend to some number) is called the limit of the function. In the above example all the values of the function were greater than the limit, but this need not be so. If we had taken the function $\dfrac{n-1}{n}$ (which also tends to 1) then all the values of the function would be less than the limit. In general some values of the function will be greater and some less than the limit. This does not matter, but it means that in talking about the function being within such and such an amount of the limit we must consider the positive difference between the function and the limit, in other words the quantity $|f(n) - l|$. The small number which expresses the required closeness of the function to the limit (1/1000 and 1/1000000 in the paragraph above were examples) is usually denoted by ϵ. We are now ready for our definition.

Definition 8.2

The function $f(n)$ of an integer variable n is said to tend to the limit l as n tends to infinity if, given any positive* number ϵ (however small it may be) there is a number N such that

$$(8.4.1) \qquad\qquad |f(n) - l| < \epsilon \text{ for all } n > N.$$

The number N will, in general, depend on the number ϵ; the smaller ϵ is the larger N will have to be.

Instead of the lengthy phrase "$f(n)$ tends to l as n tends to infinity" we may write $f(n) \to l$ as $n \to \infty$. Another way of expressing the same thing is to say that the limit of $f(n)$ as $n \to \infty$ is l, and this is written as

$$(8.4.2) \qquad\qquad \lim_{n \to \infty} f(n) = l.$$

The reader will observe that there are many resemblances between definitions 8.1 and 8.2. By analogy with (8.4.2) we often use the notation $\lim_{n \to \infty} f(n) = \infty$, to denote that the function $f(n)$ is large for large values of n.

Examples:

1. The function $f(n) = x^n$.
There are four different possibilities here according as x is > 1, $= 1$, between -1 and 1, or less than or equal to -1. We consider each case separately.

Case 1. $x > 1$. Let us write $x = 1 + y$, where y is > 0. Then

$$x^n = (1 + y)^n$$
$$= 1 + ny + \tfrac{1}{2}n(n - 1)y^2 + \ldots, \text{ by the binomial theorem.}$$

In this binomial expansion all the terms are positive. Hence $x^n \geq 1 + ny$. Now given any number K, let N be any integer $> (K - 1)/y$. Then if $n > N$, we have

---

* The term *positive* will always be taken to imply *nonzero*.

$$x^n > 1 + ny$$
$$> 1 + Ny$$
$$> K.$$

Hence if $x > 1$ then $x^n \to \infty$ as $n \to \infty$, by definition 8.1.

Case 2. $x = 1$. Here $f(n) = 1$ for all values of n, so that whatever the value of ϵ we have $|f(n) - 1| = 0 < \epsilon$, for *all* values of n. Definition 8.2 is therefore satisfied and $f(n) \to 1$ as $n \to \infty$.

Case 3. $-1 < x < 1$. If $0 < |x| < 1$ then $1/|x| > 1$. Therefore, by the result for case 1, $\left(\dfrac{1}{|x|}\right)^n \to \infty$ as $n \to \infty$. Given any positive number ϵ we can choose an integer N such that if $n > N$ then

$$\left(\frac{1}{|x|}\right)^n > \frac{1}{\epsilon}, \text{ by definition 8.1.}$$

But if $\left(\dfrac{1}{|x|}\right)^n = 1/|x|^n > 1/\epsilon$, then, taking reciprocals of both sides, we find that $|x|^n < \epsilon$. Hence given any number ϵ we can find an integer N such that $|x^n - 0| = |x|^n < \epsilon$ for all $n > N$. Therefore, by definition 8.2, we have $\lim\limits_{n \to \infty} x^n = 0$, when $-1 < x < 1$. Note that because $|x^n - 0| = |x|^n = |x|^n$, the same proof applies whether x is positive or negative.

Case 4. $x \leq -1$. Here the successive values of $f(n)$ alternate in sign. Hence, if K is any positive number, $f(n)$ will certainly be $< K$ for all odd values of n (if not for others). Hence it is not possible for all values of the function from some value of n onwards to be $> K$. Hence x^n cannot tend to infinity. For a similar reason x^n cannot tend to $-\infty$.*

Clearly the difference between two successive values of x^n is ≥ 2. Hence it is not possible for all values of x^n from some value of n onwards to be within say $\frac{1}{4}$ of a limit l, let alone any smaller number, for then any two such values of x^n would differ from each other by at most $\frac{1}{2}$, which is not so. Hence x^n does not tend to a limit of any kind.

2.   In the foregoing example it was not very difficult to guess what the behavior of the function was going to be in the four different cases. This did not, of course, relieve us of the responsibility of proving the results, even in the trivial case of $x = 1$. Our next example is the function $n^{1/n}$, i.e., the nth root of n. Here it is not at all obvious what the behavior will be. The nth root of a *fixed* number (> 1) decreases as n

---

* We say that $f(n) \to -\infty$ as $n \to \infty$ if the function $-f(n) \to \infty$ as $n \to \infty$.

increases, but here the number whose root is being taken is increasing at the same time. Two opposite tendencies are at work, so to speak, and we must see which will prevail.

Suppose we are given any positive number ϵ. Then we can certainly find an integer N such that $(N - 1)\epsilon^2/2 > 1$, for we have only to choose N to be any integer greater than $2/\epsilon^2 + 1$. Now by the binomial theorem

$$(1 + \epsilon)^n = 1 + n\epsilon + \frac{1}{2}n(n - 1)\epsilon^2 + \cdots$$

and if $n > N$ we have

$$(1 + \epsilon)^n > 1 + n\epsilon + n. \left[\frac{1}{2}(N - 1)\epsilon^2\right] + \text{other positive terms.}$$
$$> 1 + n\epsilon + n$$
$$> n.$$

Hence for $n > N$, $(1 + \epsilon)^n > n$, i.e., $1 + \epsilon > n^{1/n}$ or $n^{1/n} - 1 < \epsilon$. It follows from definition 8.2 that $\lim n^{1/n} = 1$.*

3. The result of Example 2 may be used to find the behavior of the function $x^{1/n}$ when x is a constant > 1. Suppose we are given some positive number ϵ. Then we know from the previous example that we can find an integer N such that $n^{1/n} < 1 + \epsilon$. Now let M be any integer which is larger than both N and x. We can then assert that for $n > M$, $1 < x^{1/n} < n^{1/n} < 1 + \epsilon$, from which it follows that $x^{1/n} \to 1$ as $n \to \infty$.

Exercise 80

1.   By putting $x = 1/z$ show that if $0 < x < 1$ then, also, $x^{1/n} \to 1$ as $n \to \infty$.
2.   We have already seen in Example 1, Case 4 that a function may not tend to a limit of any kind, either finite (definition 8.2) or infinite (definition 8.1). For each of the following functions determine whether it tends to a limit (finite or infinite) and if so find the limit.

(a) $\dfrac{n^2}{1 + n^2}$ (c) $\dfrac{\sin \frac{1}{2}n\pi}{n}$ (e) $\cos \dfrac{n\pi}{2}$

(b) $\dfrac{1}{n(2 + \cos \frac{1}{2}n\pi)}$ (d) $n\left(1 - \dfrac{2}{n}\right)^5$ (f) $(-1)^n n^2$

---

* The vertical lines, indicating the absolute value, which occur in definition 8.2 can be omitted here since $n^{1/n}$ is necessarily greater than 1.

8.5 THEOREMS ON LIMITS OF FUNCTIONS OF AN INTEGER VARIABLE

Frequently a function whose behavior as $n \to \infty$ we wish to determine is some sort of combination (sum, difference, etc.) of functions whose behavior we already know. We now prove a few theorems which apply in this sort of situation. In these proofs we shall make frequent use of the fact that if A and B are any two numbers then

(8.5.1) $$|A + B| \leq |A| + |B|.$$

This is easily seen to be true, since if A and B are of the same sign then $|A + B| = |A| + |B|$, whereas if they are of opposite sign, say $A > 0$ and $B < 0$, then the left-hand side of (8.5.1) is the difference between two positive numbers (A and $-B$) while the right-hand side is their sum. The corresponding result for the product of two numbers is even simpler. It is

(8.5.2) $$|AB| = |A| |B|$$

which is easily verified.

THEOREM 8.1

THE LIMIT OF THE SUM OF TWO FUNCTIONS IS THE SUM OF THEIR LIMITS; I.E., IF $f(n) \to a$ AND $g(n) \to b$ AS $n \to \infty$, THEN $f(n) + g(n) \to a + b$ AS $n \to \infty$.

PROOF:

Let ϵ be any positive number. Then there is an integer N_1 such that $|f(n) - a| < \frac{1}{2}\epsilon$ for $n > N_1$. For since $f(n) \to a$ we can make $|f(n) - a|$ less than *any* positive number by suitable choice of N_1. The reason why we choose to make it less than $\frac{1}{2}\epsilon$ rather than some other number will appear in a moment. In the same way we can find an integer N_2 such that $|g(n) - b| < \frac{1}{2}\epsilon$ for $n > N_2$. Denote by N whichever of the two integers N_1, N_2 is the larger. We may write this as "Let $N = \max(N_1, N_2)$." Then if $n > N$ we have both

$$|f(n) - a| < \frac{1}{2}\epsilon \quad \text{and} \quad |g(n) - b| < \frac{1}{2}\epsilon$$

and hence

$$|(f(n) + g(n)) - (a + b)| = |(f(n) - a) - (g(n) - b)|$$
$$\leqslant |f(n) - a| + |g(n) - b| \qquad \text{by (8.5.1)}$$
$$< \frac{1}{2}\epsilon + \frac{1}{2}\epsilon$$
$$= \epsilon$$

which is what is required to show that $f(n) + g(n) \to a + b$.

Note that if we had used ϵ in place of $\frac{1}{2}\epsilon$ throughout (as one would tend to do) we would have ended up by proving that $|(f(n) + g(n)) - (a + b)|$ was less than 2ϵ. This would have done just as well, since if ϵ represents an arbitrary positive number, 2ϵ will equally well be an arbitrary positive number. These remarks apply to any constant multiple of ϵ. If, in trying to prove that a certain expression can be made as small as we like, we succeed in proving that the expression can be made less than $K\epsilon$, where K is some positive constant and ϵ is any positive number, then this is good enough, for certainly $K\epsilon$ can be made as small as we like by choosing ϵ sufficiently small. We can, if we like, then replace ϵ by ϵ/K throughout the proof, thereby ensuring that the expression is shown to be less than ϵ, and conforming to the letter of definition 8.2, or of some similar definition. It is not really necessary to do this, any more than it is really necessary to put a frame round a painting, but it is usually done, at least in work intended for publication, as it adds a certain elegance to the proof.

If the reader bears this in mind he will have no difficulty in proving the next theorem, which will be left to him as an exercise.

THEOREM 8.2

IF $f(n) \to a$ AS $n \to \infty$, AND k IS ANY CONSTANT, THEN $k f(n) \to ka$ AS $n \to \infty$. (Remember that k may be negative.)

The next theorem is a consequence of Theorems 8.1 and 8.2.

THEOREM 8.3

IF $f(n) \to a$ AND $g(n) \to b$ AS $n \to \infty$, THEN $f(n) - g(n) \to a - b$.

PROOF:

By Theorem 8.2, with $k = -1$, $-g(n) \to -b$ as $n \to \infty$. Hence by Theorem 8.1 the sum of the two functions $f(n)$ and $-g(n)$, i.e., $f(n) - g(n)$ will tend to $a + (-b) = a - b$.

Before proving the next theorem, which concerns products of functions, we need to prove a subsidiary result. A result which is proved as a preliminary to proving a more important result is often called a *lemma*.

LEMMA.

A FUNCTION WHICH TENDS TO A FINITE LIMIT IS BOUNDED; THAT IS TO SAY, THE
SET OF VALUES OF THE FUNCTION FOR $n = 1, 2, 3, \ldots$ IS A BOUNDED SET.

PROOF:

Let $\lim\limits_{n \to \infty} f(n) = a$. Then given any positive number ϵ there is an integer N
such that if $n > N$ then $a - \epsilon < f(n) < a + \epsilon$. Let us give some definite
value to ϵ, say 1. Then for $n >$ some N all values of $f(n)$ lie between
$a - 1$ and $a + 1$. Consider now the *finite* set $\{f(1), f(2), f(3), \ldots, f(N),$
$a - 1, a + 1\}$. Since this set is finite it has a greatest member and a least
member. Call these M and m, respectively. We can then assert that
$m < f(n) < M$ for *all* values of n. For if $n \leq N$, then $f(n)$ belongs to the
above set whose members all lie between m and M; whereas if $n > N$
then $f(n)$ lies between $a - 1$ and $a + 1$ which, in turn, lie between m and
M. This proves that the set of values of $f(n)$ is bounded.

We can now prove the next theorem.

THEOREM 8.4

IF $f(n) \to a$ AND $g(n) \to b$ AS $n \to \infty$, THEN THE PRODUCT $f(n)\, g(n) \to ab$ AS
$n \to \infty$.

PROOF:

By the lemma the function $f(n)$ is bounded. We can therefore find a
number K such that all values of $f(n)$ lie between $-K$ and K, i.e., such
that $|f(n)| < K$, for all n.

   To prove the theorem we must show that $|f(n)\, g(n) - ab|$ can be
made as small as we like. Now

$$
\begin{aligned}
|f(n)g(n) - ab| &= |f(n)g(n) - bf(n) + bf(n) - ab| \\
&\leq |f(n)g(n) - bf(n)| + |bf(n) - ab| &&\text{by (8.5.1),} \\
&= |f(n)||g(n) - b| + |b||f(b) - a| &&\text{by (8.5.2).}
\end{aligned}
$$

Thus, since $|f(n)| < K$, we have

(8.5.3) $|f(n)g(n) - ab| < K|g(n) - b| + |b||f(n) - a|$.

Now $|g(n) - b|$ can be made as small as we like; therefore given any
positive number ϵ we can find an integer N_1 such that

(8.5.4) $$|g(n) - b| < \frac{\epsilon}{2K}$$

for all $n > N_1$. Similarly there is an integer N_2 such that

(8.5.5) $$|f(n) - a| < \frac{\epsilon}{2|b|}.$$

Consequently if $n > \max(N_1, N_2)$ then, from (8.5.3) we have

$$|f(n)g(n) - ab| < K.\frac{\epsilon}{2K} + |b|\frac{\epsilon}{2|b|}$$
$$= \frac{1}{2}\epsilon + \frac{1}{2}\epsilon$$
$$= \epsilon.$$

This is what is required for the proof of the theorem.

THEOREM 8.5

IF $f(n) \to a$, AND $g(n) \to b(\neq 0)$ AS $n \to \infty$, AND IF, FURTHER, $g(n) \neq 0$ FOR ANY VALUE OF n, THEN THE FUNCTION $f(n)/g(n)$ TENDS TO A LIMIT AS $n \to \infty$, AND THIS LIMIT IS a/b.

PROOF:

Since $g(n) \to b$ we can make $|g(n) - b|$ as small as we like. In particular we can find an integer N_1 such that $|g(n) - b| < \frac{1}{2}|b|$ for $n > N_1$. This means that $g(n)$ lies between $b/2$ and $3b/2$*, and that, consequently, $|g(n)| > \frac{1}{2}|b|$, for $n > N_1$.

We must now show that the expression $\left|\dfrac{f(n)}{g(n)} - \dfrac{a}{b}\right|$ can be made as small as we please. We write

$$\left|\frac{f(n)}{g(n)} - \frac{a}{b}\right| = \left|\frac{bf(n) - ag(n)}{bg(n)}\right|$$
$$= \frac{|bf(n) - ag(n)|}{|b||g(n)|}$$
$$< \frac{2}{|b|^2}|bf(n) - ag(n)| \text{ for } n > N_1, \text{ since } |g(n)| > \frac{1}{2}|b|.$$

We now turn our attention to $|bf(n) - ag(n)|$. We have

$$|bf(n) - ag(n)| = |b\{f(n) - a\} - a\{g(n) - b\}|$$
$$\leq |b||f(n) - a| + |a||g(n) - b|, \text{ by (8.5.1)}.$$

Both $|f(n) - a|$ and $|g(n) - b|$ can be made as small as we like for sufficiently large values of n. We can therefore find an integer N_2 such

---

* If b is negative, $\frac{1}{2}b$ will be the larger number.

that $|bf(n) - ag(n)| < \frac{1}{2}b^2\epsilon$ for $n > N_2$ whatever the value of $\epsilon(> 0)$ may be.

If $n > \max(N_1, N_2)$ we then have

$$\left| \frac{f(n)}{g(n)} - \frac{a}{b} \right| < \frac{2}{b^2} |bf(n) - ag(n)|$$

$$< \frac{2}{b^2} \cdot \frac{1}{2} b^2 \epsilon = \epsilon.$$

which proves the theorem.

Exercise 81

1. The reader may like to try his hand at proving the following theorems.

THEOREM 8.6

IF $f(n) \to \infty$ AND $g(n)$ TENDS EITHER TO INFINITY OR TO A FINITE POSITIVE LIMIT, THEN $f(n) g(n) \to \infty$.

THEOREM 8.7

IF $f(n) \to \infty$ AND $g(n) \to b(> 0)$, AND IF $g(n) \neq 0$ FOR ANY VALUE OF n, THEN $f(n)/g(n) \to \infty$.

THEOREM 8.8

IF $f(n) \to a$ AND $g(n) \to \infty$, AND IF $g(n) \neq 0$ FOR ANY VALUE OF n, THEN $f(n)/g(n) \to 0$.

THEOREM 8.9

IF $f(n) \to a \, (> 0)$ AND $g(n) \to 0$, AND IF $g(n) > 0$ FOR ANY VALUE OF n, THEN $f(n)/g(n) \to \infty$.

2. The foregoing theorems enable us to determine the limits of a great number of functions. The following are a few examples.

(a) $\dfrac{1 - 2/n}{3 + 5/n^2}$

(b) $\left(\dfrac{n-1}{n}\right)^2 n^{1/n}$

(c) $\dfrac{n^3 + 3n^2 + 7}{6n^3 + 1}$

(*Hint:* divide top and bottom by n^3 before applying the theorems)

(d) $\dfrac{4n^4 - 3n + 15}{7n^4 - 6n^3 - 5n + 4}$

3. Give examples to show that no deduction can be made about the behavior of $f(n)/g(n)$ from the statement that $f(n)$ and $g(n)$ tend to infinity.

4. Prove that if $P(n)$ and $Q(n)$ are polynomials of degrees r and s respectively, whose leading coefficients (i.e., those of the highest powers of x) are positive, then, as $n \to \infty$

$$\frac{P(n)}{Q(n)} \to \infty \qquad \text{if } r > s$$

$$\to 0 \qquad \text{if } r < s$$

$$\text{and} \to l \qquad \text{if } r = s, \text{ where } l \text{ is a real number.}$$

8.6 MONOTONIC SEQUENCES

A function f of an integral variable, or a sequence, is said to be *monotonic* if either $f(n + 1) \geq f(n)$ for all n (in which case it is *monotonic increasing*) or $f(n + 1) \leq f(n)$ for all n (in which case it is *monotonic decreasing*). These two phrases are often abbreviated to *increasing* and *decreasing*, respectively. If the signs $\geq$ and $\leq$ are replaced by strict inequalities ($>$ and $<$) the functions or sequences thus defined are called *strictly monotonic* (*strictly increasing* and *strictly decreasing*, respectively).

The following theorem on monotonic functions is extremely important.

THEOREM 8.10

IF $f(n)$ IS AN INCREASING FUNCTION, THEN, AS $n \to \infty$, EITHER $f(n) \to \infty$ OR $f(n)$ TENDS TO A FINITE LIMIT.

PROOF:

There are two possibilities: either the function $f(n)$ is bounded, or it is unbounded. (These two statements are contradictory). We first see what happens if $f(n)$ is unbounded.

If $f(n)$ is unbounded then, given any number K, there is a value N of n such that $f(N) > K$ (since otherwise K would be an upper bound). But $f(n)$ is increasing, so that if $n > N$ we have $f(n) \geq f(N) > K$. This means that $f(n) \to \infty$ as $n \to \infty$.

Let us now suppose that $f(n)$ is bounded, and let M denote the least upper bound of the set of values of $f(n)$. If ϵ is any positive number then $M - \epsilon$ is not an upper bound of $f(n)$, and there will be some integer N such that $f(N) > M - \epsilon$. (Compare Exercise 79.2.) Now since $f(n)$ is increasing, if $n > N$ we have

$$f(n) \geq f(N) > M - \epsilon$$

whereas, since M is an upper bound we have

$$f(n) \leq M < M + \epsilon.$$

Hence for $n > N$ the values $f(n)$ satisfy $|f(n) - M| < \epsilon$. Thus the function $f(n)$ tends to the limit M.

This completes the proof of Theorem 8.10. An immediate consequence is

THEOREM 8.11

IF $f(n)$ IS A DECREASING FUNCTION THEN, AS $n \to \infty$, EITHER $f(n) \to -\infty$ OR $f(n)$ TENDS TO A FINITE LIMIT.

PROOF:

Apply Theorem 8.10 to the function $-f(n)$, which is an increasing function.

A very important application of Theorem 8.10 is the following.

THEOREM 8.12

THE FUNCTION $\left(1 + \dfrac{1}{n}\right)^n$ TENDS TO A FINITE LIMIT AS $n \to \infty$.

PROOF:

Write $f(n) = \left(1 + \dfrac{1}{n}\right)^n$. Then, by the binomial theorem

$$f(n) = 1 + n \cdot \frac{1}{n} + \frac{1}{2}n(n-1)\frac{1}{n^2} + \frac{1}{6}n(n-1)(n-2)\frac{1}{n^3} + \cdots$$

$$= 1 + 1 + \frac{1}{2} \cdot 1 \cdot \left(1 - \frac{1}{n}\right) + \frac{1}{6} \cdot 1 \cdot \left(1 - \frac{1}{n}\right) \cdot \left(1 - \frac{2}{n}\right) + \cdots$$

$$(n + 1 \text{ terms}).$$

Similarly

$$f(n+1) = 1 + 1 + \frac{1}{2} \cdot 1 \cdot \left(1 - \frac{1}{n+1}\right) +$$

$$\frac{1}{6} \cdot 1 \cdot \left(1 - \frac{1}{n+1}\right)\left(1 - \frac{2}{n+1}\right) \cdots \quad (n + 2 \text{ terms}).$$

Let us compare these expressions for $f(n)$ and $f(n + 1)$. The expression for $f(n + 1)$ has an extra term at the end, which is positive. Comparing the other terms in $f(n + 1)$ with the corresponding terms in $f(n)$ we see that where the expression for $f(n)$ has $1 - \dfrac{1}{n}$, $1 - \dfrac{2}{n}$, $1 - \dfrac{3}{n}$ etc. that for $f(n + 1)$ has the larger numbers $1 - \dfrac{1}{n + 1}$, $1 - \dfrac{2}{n + 1}$, $1 - \dfrac{3}{n + 1}$, etc. Thus the expression for $f(n + 1)$ has terms which are greater than the corresponding terms in the expression for $f(n)$, and an extra positive term at the end for good measure. Hence $f(n + 1) > f(n)$, i.e., $f(n)$ is an increasing function. By Theorem 8.10, therefore, either $f(n) \to \infty$, or $f(n) \to$ a finite limit. We shall eliminate the first possibility by showing that $f(n)$ is bounded. Now

$$f(n) = 1 + 1 + \frac{1}{2!}\left(1 - \frac{1}{n}\right) + \frac{1}{3!}\left(1 - \frac{1}{n}\right)\left(1 - \frac{2}{n}\right) + \cdots$$

(8.6.1) $$< 1 + 1 + \frac{1}{2!} + \frac{1}{3!} + \cdots + \frac{1}{n!} \quad \text{(for } n \geq 2\text{)}$$

since the numbers $1 - \dfrac{1}{n}$, $1 - \dfrac{2}{n}$, ... are all less than 1, so that the right-hand side of (8.6.1) has been increased. We proceed to make it even larger by replacing 3! by 2^2, 4! by 2^3, and generally $r!$ by 2^{r-1}, thus decreasing the denominators of all except the first three terms. We thus see that

$$f(n) < 1 + 1 + \frac{1}{2} + \left(\frac{1}{2}\right)^2 + \left(\frac{1}{2}\right)^3 + \cdots + \left(\frac{1}{2}\right)^{n-1}.$$

This expression on the right is a geometrical progression (with an extra term at the beginning), and its sum* is $3 - (\frac{1}{2})^n$. Hence for all values of n we have $f(n) \leq 3 - (\frac{1}{2})^n < 3$. Hence $f(n)$ is bounded above, and thus tends to a finite limit by Theorem 8.10. This proves Theorem 8.12.

From (8.6.1), in which all the terms on the right-hand side are positive, it follows that $\left(1 + \dfrac{1}{n}\right)^n > 2$ for all n. Consequently the limit to which this function tends is a number between 2 and 3. This number, which is of fundamental importance in mathematics, is denoted by the letter e. The value of e correct to 5 decimal places is 2.71828.

Exercise 82

1. Prove that $x^n/n^k \to \infty$ $(x > 1, k \geq 0)$.
2. Prove that $n^k x^n \to \infty$ $(x > 1, k > 0)$. Thus if $x > 1$ then $x^n n^k \to \infty$ for all values of k, both positive and negative.

* See Exercise 31.2.

3. If $-1 < x < 1$, show that $x^n n^k \to 0$ for all values of k (positive and negative).

4. For what values of x does the function $f(n) = x^n/n^n$ tend (a) to a finite limit, (b) to infinity?

5. Let x_1 and A be any two positive real numbers. We define a sequence by

$$x_2 = \frac{1}{2}\left(x_1 + \frac{A}{x_1}\right)$$

$$x_3 = \frac{1}{2}\left(x_2 + \frac{A}{x_2}\right)$$

and in general

$$x_{n+1} = \frac{1}{2}\left(x_n + \frac{A}{x_n}\right).$$

Prove that x_n tends to a finite limit, and show that this limit is $\sqrt{A}$.

6. A sequence is defined as in the previous exercise but using the equation $x_{n+1} = (A + x_n)^{1/2}$. Prove that x_n tends to a finite limit as $n \to \infty$. What is this limit?

7. Let a_1 and g_1 be any two positive real numbers. We define two sequences $\{a_n\}$ and $\{g_n\}$ as follows:

$$a_2 = \frac{1}{2}(a_1 + g_1); \qquad g_2 = a_1 g_1$$

$$a_3 = \frac{1}{2}(a_2 + g_2); \qquad g_3 = a_2 g_2$$

and in general $a_{n+1} = \frac{1}{2}(a_n + g_n)$; $g_{n+1} = a_n g_n$. Show that each of these sequences tends to a finite limit. Then show that they tend to the *same* limit.

8. (This one is tricky.)

Find $\lim\limits_{m \to \infty}\left\{\lim\limits_{n \to \infty}[\cos(m!\,\pi x)]^{2n}\right\}$, where x is a real number.

Hint: The value of this double limit will depend on whether x is a rational or an irrational number. Consider the two cases separately.

8.7 SERIES—CONVERGENCE AND DIVERGENCE

Before we carry on with the next topic it will be worthwhile to think back to the time when we made our first acquaintance with mathematics—when we first learned how to add. We learned that $2 + 3$ was 5 and that $3 + 2$ was also 5, and it probably seemed obvious that these two different sums should give the same number. The general property of numbers of which this is an instance is called the *commutative law of addition* and can be written

$$a + b = b + a \text{ for all numbers } a \text{ and } b.$$

The pure mathematician will protest that this law is not so obvious as it seems; but it would take us too deeply into the fundamentals of mathematics to give a proof of it.

Having acquired a certain amount of practice at adding two numbers we were then taught to add three numbers. This was very easy; we added two of them (this we knew how to do) and then added the remaining one. Nothing could be simpler, but our sheep-watching mathematician is not satisfied. "How do you know that the answer you will get will not depend on which of the three numbers you choose to add first?" he asks. "If you add a and b first, and then add in the third number c, you will get a result which we may denote by $(a + b) + c$; whereas if you add b and c first the result may be denoted by $a + (b + c)$. Are you sure these two results are the same?" The statement which answers this objection by saying that the two results *are* the same is known as the *associative law of addition*. It states that $a + (b + c) = (a + b) + c$ for all numbers a, b and c. Again, it is not possible here to give a proof of this law.

Armed with the commutative and associative laws of addition we are able to "add up" any finite set of numbers, that is to say to associate with any finite set of numbers a unique number called the sum of the set. We do this by adding any two numbers in the set, then adding a third, and so on until all numbers of the set have been added. Moreover the two laws show that the answer we get does not depend on the order in which we take the numbers when we add them.

This is all very elementary and familiar. It begins to get more interesting when we ask to what extent it is possible to generalize this idea of the "sum" of a set of numbers to include sets that are not finite. In other words, if we are given an infinite set of numbers, can we associate with that set of numbers a single number which we may call the sum of the set (i.e., a number which will have at least some, if not all, of the properties that the sum of a finite set of numbers has)? The previous definition of "sum" will not apply any longer, since it would entail performing an infinite number of additions. This does not mean that there might not be some subterfuge whereby this difficulty could be met, but certainly the answer to the query is not immediately apparent. In fact, the answer is that, in general, it is not possible to associate a "sum" with an infinite set of numbers, but that there are sets for which it is possible. To a large extent, what we shall now be concerned with is the study of these special kinds of sets.

We have already seen that a function $f(n)$ of an integer variable defines a sequence $f(1), f(2), f(3), \ldots, f(n), \ldots$. We can write this sequence a little more compactly as $u_1, u_2, u_3, \ldots, u_n, \ldots$, where $u_n = f(n)$. Here we have a set of numbers, but a sequence is more than just a set of numbers, for with these numbers there is associated an ordering, given by the natural ordering of the subscripts. This ordering is important, and we must not forget that it is there. Let us now see how far we can get by proceeding as if we had a finite set only to deal with; that is, we shall start by adding the numbers up in the usual way—first adding u_1 and u_2, then adding in u_3, then u_4, and so on. By means of this step by step process we obtain at each stage what is called a *partial sum*. We write

$$s_1 = u_1$$
$$s_2 = u_1 + u_2$$
$$s_3 = u_1 + u_2 + u_3$$
$$\cdot \ \cdot \ \cdot \ \cdot \ \cdot$$
$$s_n = u_1 + u_2 + \cdots + u_n$$
$$\cdot \ \cdot \ \cdot \ \cdot \ \cdot$$

where s_1, s_2, etc., are the partial sums.

Now whatever the value of n we can find the corresponding partial sum s_n by means of a finite number of additions. We have therefore defined a perfectly good function s_n of n, since its value for any given value of n can be found. We now ask whether this function will tend to a limit as $n \to \infty$. If it does, then we call the limit to which it tends the "sum" of the infinite sequence $u_1, u_2, \ldots$. When we are interested mainly in the sum of a sequence (if it has one) rather than in the sequence itself, it is customary to call the sequence a *series* and to indicate our intentions by using plus signs instead of commas to separate the terms. Thus we write a series as

$$u_1 + u_2 + u_3 \cdots + u_n + \cdots$$

This indicates that the addition process just described is going to be applied to these terms. In short, we use the word *sequence* if we are interested in the behavior of the terms themselves; *series* if we are interested in the behavior of their partial sums.

As a simple example of this process of *summing* an infinite series we may take the series

$$1 + \frac{1}{2} + \frac{1}{4} + \frac{1}{8} + \cdots + \left(\frac{1}{2}\right)^{n-1} + \cdots$$

The nth partial sum is

$$s_n = 1 + \frac{1}{2} + \frac{1}{4} + \cdots + \left(\frac{1}{2}\right)^{n-1}$$
$$= 2 - \left(\frac{1}{2}\right)^{n-1}$$

by the formula for the sum of a geometrical progression. Now as $n \to \infty$, $\left(\frac{1}{2}\right)^{n-1} \to 0$, as we have already seen. Therefore $s_n \to 2$, and the sum of the series is therefore 2.

This result is readily generalized. An important series, of which we shall make much use, is the geometric series (an infinite geometric progression)—the series $a + ar + ar^2 + ar^3 + \cdots + ar^{n-1} + \cdots$, of which our first example is a special case. We assume that $r \neq 1$. Hence the nth partial sum is

$$s_n = a + ar + ar^2 + \cdots + ar^{n-1}$$
$$= \frac{a(1 - r^n)}{1 - r} \quad \text{(sum of a geometrical progression)*}$$
$$= \frac{a}{1 - r} - \left(\frac{a}{1 - r}\right)r^n \quad \text{(provided } r \neq 1\text{).}$$

Now $\frac{a}{1 - r}$ is some constant, and if $|r| < 1$, then $r^n \to 0$ as $n \to \infty$. Hence for $|r| < 1$, $s_n \to a/(1 - r)$ as $n \to \infty$. On the other hand if $r \leq -1$ or $r > 1$ then r^n does not tend to a finite limit as $n \to \infty$, and hence neither does s_n. In the remaining case, $r = 1$, the nth partial sum is na, so that $s_n \to \infty$ as $n \to \infty$.

If the partial sums of a series tend to a finite limit as $n \to \infty$, we say that the series is *convergent;* if they do not, we say the series is *divergent.* Thus the geometric series given above is convergent when $|r| < 1$, but is divergent otherwise. Clearly, only a convergent series can have a sum in the sense in which we have defined it.

As another example of a convergent series we may take the series

$$\frac{1}{1.2} + \frac{1}{2.3} + \frac{1}{3.4} + \cdots + \frac{1}{n(n + 1)} + \cdots$$

Here the nth term of the series is $\frac{1}{n(n + 1)} = \frac{1}{n} - \frac{1}{n + 1}$. Consequently the nth partial sum is

$$s_n = \left(1 - \frac{1}{2}\right) + \left(\frac{1}{2} - \frac{1}{3}\right) + \left(\frac{1}{3} - \frac{1}{4}\right) + \cdots + \left(\frac{1}{n} - \frac{1}{n + 1}\right)$$
$$= 1 - \frac{1}{n + 1}.$$

As $n \to \infty$ so $s_n \to 1$. Hence this series is convergent, and its sum is 1.

* See equation (4.7.3) and Exercise 31.2.

In these two examples it was possible to find an explicit expression for the general partial sum s_n. This is not usually possible, and when it is not, other methods have to be used to find out whether the series converges or diverges. One common reason why a series diverges is that $s_n \to \infty$ as $n \to \infty$. It is often possible to show that this is so even though we cannot find the value of s_n explicitly. Take, for example, the series

$$1 + \frac{1}{2^k} + \frac{1}{3^k} + \frac{1}{4^k} + \cdots$$

We shall show that if $k \leq 1$ this series is divergent.

The nth partial sum is

$$s_n = 1 + \frac{1}{2^k} + \frac{1}{3^k} + \frac{1}{4^k} + \cdots + \frac{1}{n^k}.$$

We now bracket these terms together as follows

$$1 + \left(\frac{1}{2^k}\right) + \left(\frac{1}{3^k} + \frac{1}{4^k}\right) + \left(\frac{1}{5^k} + \frac{1}{6^k} + \frac{1}{7^k} + \frac{1}{8^k}\right) + \cdots$$

where the successive pairs of brackets enclose 1, 2, 4, 8, terms, and so on. In general there will be some terms left over at the end of the sum which will not fill up a bracket of the required length. If so, we simply leave these terms out, thereby obtaining an expression which is less than the original partial sum. Hence we may write

$$(8.7.1) \quad s_n \geq 1 + \left(\frac{1}{2^k}\right) + \left(\frac{1}{3^k} + \frac{1}{4^k}\right) + \left(\frac{1}{5^k} + \frac{1}{6^k} + \frac{1}{7^k} + \frac{1}{8^k}\right) + \cdots$$
$$+ (\cdots),$$

where the last pair of brackets encloses 2^m terms, and m tends to ∞ as $n \to \infty$.*

Now in each bracketed expression there is a certain number of terms, and if we replace each of these terms by the smallest term occurring within these brackets the right-hand side of (8.7.1) will be decreased still further. This gives us

$$s_n \geq 1 + \left(\frac{1}{2^k}\right) + \left(\frac{1}{4^k} + \frac{1}{4^k}\right) + \left(\frac{1}{8^k} + \frac{1}{8^k} + \frac{1}{8^k} + \frac{1}{8^k}\right) + \cdots$$
$$(8.7.2) \quad = 1 + \frac{1}{2^k} + \frac{2}{4^k} + \frac{4}{8^k} + \cdots + \frac{2^m}{2^{(m+1)k}}.$$

Except for the first term, the right-hand side of (8.7.2) is a geometric progression with common ratio $(1/2)^{k-1}$, which is ≥ 1 since $k < 1$. Hence the

---

* The reader may care to verify that m will be the largest integer such that $m(m + 1)/2 \leq n$.

sum of this progression can be made as large as we like by taking m sufficiently large, and it follows that s_n can be made as large as we like by taking n sufficiently large. Consequently $s_n \to \infty$ as $n \to \infty$, and the series is divergent.

In a rather similar way we may show that if $k > 1$ the series is convergent. We bracket the terms of the nth partial sum as follows:

$$s_n = 1 + \left(\frac{1}{2^k} + \frac{1}{3^k} \right) + \left(\frac{1}{4^k} + \frac{1}{5^k} + \frac{1}{6^k} + \frac{1}{7^k} \right) + \cdots$$

if necessary filling out the last bracketed expression to the required length by *adding* the next few terms of the series. This will make the right-hand side larger, and we write

$$s_n \le 1 + \left(\frac{1}{2^k} + \frac{1}{3^k} \right) + \left(\frac{1}{4^k} + \frac{1}{5^k} + \frac{1}{6^k} + \frac{1}{7^k} \right) + \cdots + (\cdots).$$

In each bracketed expression replace each term by the largest term within the brackets; this makes the right-hand side even larger, and we have

$$s_n < 1 + \left(\frac{1}{2^k} + \frac{1}{2^k} \right) + \left(\frac{1}{4^k} + \frac{1}{4^k} + \frac{1}{4^k} + \frac{1}{4^k} \right) + \cdots$$

(8.7.3) $$= 1 + \frac{1}{2^{k-1}} + \frac{1}{4^{k-1}} + \cdots + \frac{1}{2^{m(k-1)}}.$$

The right-hand side is now a geometric progression in which the common ratio is $(\frac{1}{2})^{k-1}$, which is < 1 if $k > 1$. From the formula for the sum of a geometric progression it follows that the right-hand side of (8.7.3) is always less than $2^{k-1}/(2^{k-1} - 1)$, whatever the value of m. Consequently the sequence $\{s_n\}$ is bounded above by this number. Since this sequence is clearly monotonic increasing it will tend to a limit, by Theorem 8.10. This proves that the series is convergent.

Exercise 83

1. Let s_n denote the nth partial sum of the series

$$\sum_{r=0}^{\infty} rx^r = x + 2x^2 + 3x^3 + 4x^4 + \cdots$$

By considering the expression $(1 - x)s_n$, or otherwise, show that

$$s_n = \frac{x}{(1 - x)^2} \{ 1 - (n + 1)x^n + nx^{n+1} \}.$$

Hence show that the infinite series converges if $|x| < 1$ and diverges otherwise. Find the value of the sum of the series when it converges.

We met this series in Chapter 4 (section 16) with $x = \frac{1}{2}$. Verify the value of 2 that we found there (by another method) for the sum of this infinite series.

2. A number written in the decimal system can be regarded as a series, either finite or infinite. Thus a number less than 1 will be of the form

$$\cdot a_1 a_2 a_3 a_4 \cdots$$

where each a_i is a digit 0, 1, 2, ..., 9. This stands for the sum of the series

$$\frac{a_1}{10} + \frac{a_2}{10^2} + \frac{a_3}{10^3} + \frac{a_4}{10^4} + \cdots$$

Prove (a) that if the decimal expression terminates, i.e., all the a_i from some decimal place onwards are zeros, then the sum of the series is a rational number. Find this number.

(b) that the recurring decimal

$$\cdot \dot{a}_1 a_2 a_3 \cdots \dot{a}_n$$

which stands for

$$\cdot a_1 a_2 a_3 \cdots a_n a_1 a_2 a_3 \cdots a_n a_1 a_2 a_3 \cdots a_n a_1 \cdots$$

is the sum of the geometric series

$$\frac{a_1 a_2 a_3 \cdots a_n}{10^n} + \frac{a_1 a_3 a_3 \cdots a_n}{10^{2n}} + \frac{a_1 a_2 a_3 \cdots a_n}{10^{3n}} + \cdots$$

Hence find the rational number that this recurring decimal represents.

3. Prove that the series

$$\frac{1}{1.2} + \frac{1}{3.4} + \frac{1}{5.6} + \cdots + \frac{1}{(2n+1)(2n+2)} + \cdots$$

is convergent, and that its sum lies between 0 and 1.

8.8 COMPARISON TESTS

The last two examples indicate that it is often possible to prove that a series is convergent or divergent by comparing its terms (or combinations of them) with the terms of another series whose convergency or divergency is already known. We shall express this idea of comparison between series in the form of two general theorems, but before doing so we need to extend the symbolism

$$\sum_{r=1}^{n} f(r)$$

meaning "the sum of $f(r)$ from $r = 1$ to n," that is,

$$f(1) + f(2) + f(3) + \cdots + f(n).$$

The same notation is taken over for infinite series, and we write, for example

$$\sum_{r=1}^{\infty} u_r = u_1 + u_2 + u_3 + \cdots$$

the ∞ symbol merely indicating that the expression stands for an infinite series. Very often it is not necessary to indicate the values assumed by the integer r (or whatever other letter is used) and if nothing is written above or below the sigma "$\sum$" then the range of values of r is taken to be evident from the context. Thus in the present context, for example, $\sum u_r$ will always mean $\sum_{r=1}^{\infty} u_r$, unless otherwise stated.

THEOREM 8.13

IF $\sum u_r$ IS CONVERGENT AND $0 \leq v_r \leq u_r$ FOR ALL VALUES OF r, THEN $\sum v_r$ IS CONVERGENT.

Note: By virtue of the condition $0 \leq v_r \leq u_r$, this theorem applies to series of non-negative terms only.

PROOF:

Let $s_n = \sum_{r=1}^{n} u_r$, and $t_n = \sum_{r=1}^{n} v_r$, be the partial sums of the two series. Then by the condition of the theorem, $t_n \leq s_n$ for all values of n. Since $\sum u_r$ is convergent s_n tends to some limit K, say, as $n \to \infty$. Further, since s_n is an increasing function of n (the terms of the series are positive) $s_n \leq K$ for all n. But t_n is also an increasing function, and is bounded above, since $t_n \leq s_n \leq K$. Hence t_n tends to a finite limit, by Theorem 8.10, and thus $\sum v_r$ is convergent.

The condition "$v_r \leq u_r$ for all r" can be described by saying that the series $\sum v_r$ is "term for term less than" the series $\sum u_r$. Thus Theorem 8.13 states that a series of non-negative terms which is term for term less than a convergent series is convergent.

THEOREM 8.14

IF $\sum u_r$ IS DIVERGENT AND $v_r \geq u_r \geq 0$ FOR ALL VALUES OF r, THEN $\sum v_r$ IS DIVERGENT.

PROOF:

If $\sum v_r$ were convergent then $\sum u_r$ would be term for term less than this convergent series and hence would be convergent, by Theorem 8.13. But $\sum u_r$ is given to be divergent. Thus $\sum v_r$ cannot be convergent, and is therefore divergent.

We may paraphrase Theorem 8.14 by saying that a series which is term for term greater than a divergent series of non-negative terms is divergent.
The usefulness of these two theorems can be extended by means of a simple observation, viz., that the addition of a finite number of terms to a series, or the removal of a finite number of terms from it does not affect its convergency or divergency. This is not difficult to see. Suppose, for example, that we have a convergent series, and that we remove the first 10 terms. Let the sum of these 10 terms be k. Then the partial sums of the new series will be those of the old series, each diminished by k. If the original sequence of partial sums tended to a limit l, then the new sequence of partial sums will tend to $l - k$; if the original partial sums did not tend to a finite limit then neither will the new partial sums. Thus the convergency or divergency of the series is not affected by the removal of the 10 terms, or indeed of any finite number of terms. In the same way it is not affected if a finite number of terms is added to the series.
It appears therefore that the requirement $0 \leq v_r \leq u_r$ for *all* r in Theorem 8.13 is unnecessarily stringent, since it will not matter if there is a finite number of exceptions to this condition. Suppose this is so. Then we can find a number N sufficiently large that all the exceptions to the condition occur for values of $r < N$. (This is possible if, and only if, the number of exceptions is finite.) Remove the first N terms of $\sum u_r$ and $\sum v_r$. The two truncated series will now satisfy the requirements of Theorem 8.13 to the letter. But any conclusions concerning the convergency or otherwise of $\sum_{r=N+1}^{\infty} v_r$ will apply equally well to the whole series $\sum v_r$. Thus Theorem 8.13 (and, similarly, Theorem 8.14) can still be applied when there is a finite number of exceptions to the condition contained in it.
When a statement (like $0 \leq v_r \leq u_r$) depending on an integer r holds for all except a finite number of values of r, we say that it holds for "almost all values of r" or just "for almost all r." Thus we can enlarge the scope of Theorems 8.13 and 8.14 by restating them as follows:

THEOREM 8.13A

IF $\sum u_r$ IS CONVERGENT, AND $0 \leq v_r \leq u_r$ FOR ALMOST ALL r, THEN $\sum v_r$ IS CONVERGENT.

THEOREM 8.14A

IF $\sum u_r$ IS DIVERGENT, AND $v_r \geq u_r \geq 0$ FOR ALMOST ALL r, THEN $\sum v_r$ IS DIVERGENT.

EXAMPLE: The exponential series.

This is the series

$$(8.8.1) \qquad \sum_{r=0}^{\infty} \frac{x^r}{r!} = 1 + \frac{x}{1!} + \frac{x^2}{2!} + \cdots + \frac{x^r}{r!} + \cdots$$

We prove first

THEOREM 8.15

THE EXPONENTIAL SERIES IS CONVERGENT FOR ALL POSITIVE VALUES OF x.

PROOF:

Consider a particular value of x, and let N be any integer $> x$. Then if $r > N$, say $r = N + s$ where $s \geq 1$, we have

$$\frac{x^r}{r!} = \frac{x^{N+s}}{(N+s)!}$$

$$= \frac{x^N}{N!} \cdot \frac{x^s}{(N+1)(N+2)\cdots(N+s)}$$

$$< \frac{x^N}{N!} \cdot \frac{x^s}{\underbrace{N \cdot N \cdots N}_{s \text{ factors}}}$$

$$= \frac{x^N}{N!} \cdot \frac{x^s}{N^s}$$

Thus if we leave out the first $N + 1$ terms of the exponential series we obtain a series which is term for term less than the geometric series $\frac{x^N}{N!} \sum_{s=1}^{\infty} \left(\frac{x}{N}\right)^s$, which converges, since $x/N < 1$. This truncated series is therefore convergent by Theorem 8.13, and hence the original series is convergent.

Tests for convergence such as those of Theorems 8.13A and 8.14A are called *comparison tests*, since they depend on making a comparison between the given series and a series whose behavior is already known. Any suitable series may be used with which to compare the series whose behavior is required, and in a great number of examples the geometric series is suitable for this purpose, as it was for the exponential series. When this is the case it is possible to whittle down the testing procedure to a bare minimum. The test that results is known as *d'Alembert's test*.

THEOREM 8.16

(D'ALEMBERT'S TEST FOR CONVERGENCY.) IF $\sum u_r$ IS A SERIES OF POSITIVE TERMS, AND FOR SOME $k < 1$ THE CONDITION

(8.8.2) $$\frac{u_{r+1}}{u_r} \leq k$$

IS SATISFIED FOR ALMOST ALL r, THEN THE SERIES $\sum u_r$ IS CONVERGENT.

PROOF:

If (8.8.2) holds for almost all r then it will hold for all r greater than some integer N. Then

$$u_{N+1} \leq k u_N$$
$$u_{N+2} \leq k u_{N+1} \leq k^2 u_N$$
$$u_{N+3} \leq k u_{N+2} \leq k^3 u_N$$
$$\text{etc.}$$

and in general $u_{N+s} \leq k^s u_N$. Hence the series

(8.8.3) $$u_N + u_{N+1} + u_{N+2} \cdots + u_{N+s} \cdots$$

is term for term less than the series

(8.8.4) $$u_N + k u_N + k^2 u_N \cdots + k^s u_N \cdots$$

which is a convergent geometric series (since $k < 1$). Hence (8.8.3) is convergent by Theorem 8.13. Since (8.8.3) is simply the original series with a finite number of terms removed, the original series must be convergent also.

EXAMPLE 1:

Let us see how this test applies to the exponential series. We have $u_r = x^r/r!$. Hence

$$\frac{u_{r+1}}{u_r} = \frac{x^{r+1}}{(r+1)!} \cdot \frac{r!}{x^r} = \frac{x}{r+1}.$$

Now $x/(r+1)$ will be less than $\frac{1}{2}$, say, for almost all values of r; for the values of r which make it greater than $\frac{1}{2}$ are those for which $r < 2x - 1$, and for each value of x there is only a finite number of values of r for which this inequality holds. Hence, for almost all values of r, $u_{r+1}/u_r < \frac{1}{2}$. Thus the requirements of d'Alembert's test are met, and the series is therefore convergent.

Note that in this example we could have taken *any* positive number less than 1 to be the number k. There is not usually such a wide range of possibilities for this number, as the next example will show.

EXAMPLE 2:

The series $\sum \dfrac{r(r+1)}{2^r}$. Here we have

$$\frac{u_{r+1}}{u_r} = \frac{(r+1)(r+2)}{2^{r+1}} \cdot \frac{2^r}{r(r+1)}$$
$$= \frac{r+2}{r} \cdot \frac{1}{2} = \frac{1}{2} + \frac{1}{r}.$$

It is clear that if $r > 4$ then $u_{r+1}/u_r < 3/4$. The requirements of the test can therefore be met, and the series is convergent. Here we had less choice for the number k since no number less than $1/2$ would serve.

It is very important to note that the number k of the test has to be *strictly* less than 1; we cannot choose it to be equal to 1 and still expect the test to work. The reason for this is that if all we can prove is that u_{r+1}/u_r is < 1 then the only geometric series that we can compare our series with is one with common ratio 1 (or more) and this series is no good, since it does not converge. That the test sometimes fails if k is taken as 1 is shown by the example of the series $1 + 1/2 + 1/3 + 1/4 + \cdots$. For this series $u_r = 1/r$, and thus $u_{r+1}/u_r = r/(r+1)$. This is certainly always less than 1, but the series, as we have already seen, is divergent.

This should not be too surprising; for although $r/(r+1)$ is always less than 1, nevertheless, whatever number k, less than 1, we may choose, $r/(r+1)$ will exceed k for sufficiently large values of r. Hence Theorem 8.16 gives us no reason to expect the series to converge.

From Theorem 8.14A we can deduce a test for divergency.

THEOREM 8.17

(D'ALEMBERT'S TEST FOR DIVERGENCY).

IF $\sum u_r$ IS A SERIES OF POSITIVE TERMS, AND $\dfrac{u_{r+1}}{u_r} \geq 1$ FOR ALMOST ALL r,

THEN $\sum u_r$ IS DIVERGENT.

PROOF:

The proof is similar to that of Theorem 8.16, except that we have

$$u_{N+1} \geq u_N$$
$$u_{N+2} \geq u_{N+1} \geq u_N$$
$$u_{N+3} \geq u_{N+2} \geq u_N \quad \text{etc.}$$

and the truncated series is term for term greater than the series $u_N + u_N + u_N \ldots$ which is clearly divergent. From this the theorem follows.

Note that we do not have any trouble with the number k in this test since the geometric series with common ratio 1 *is* suitable as a comparison series:

EXAMPLE:

The series $\sum \dfrac{2^r}{r(r+1)}$. Here

$$\frac{u_{r+1}}{u_r} = \frac{2r}{r+2} = 2 - \frac{4}{r+2}.$$

This is certainly ≥ 1 for all $r \geq 2$, so that the condition of the theorem is satisfied, and the series is therefore divergent.

Exercise 84

1. The conditions for the two d'Alembert tests are only sufficient conditions, not necessary ones. That is to say a series may converge without satisfying the condition $u_{r+1}/u_r \leq k < 1$. Show that the series

$$\frac{1}{2^2} + \frac{1}{2} + \frac{1}{2^4} + \frac{1}{2^3} + \frac{1}{2^6} + \frac{1}{2^5} + \cdots$$

which is obtained from a geometric series (with common ratio $\frac{1}{2}$) by interchanging successive pairs of terms, does not satisfy the condition of theorem, but is convergent.

2. If a series $\sum u_r$ is convergent then $u_r \to 0$ as $r \to \infty$. (Why?) This is a necessary condition for convergency, but not a sufficient one. (The series $\sum 1/r$ satisfies this condition, but is not convergent). The fact that the terms of a convergent series tend to 0 as $r \to \infty$ may not always be apparent from an inspection of the first few terms of the series. Thus the first few terms of the exponential series with $x = 600$ are

$$1 + 600 + 180{,}000 + 36{,}000{,}000 + 5{,}400{,}000{,}000 + \cdots$$

Satisfy yourself that the terms of this series do, in fact, tend to 0 (despite the apparently unpromising start), and explain why it is that they do so.

3. Prove that the series

$$\frac{1}{A+1^k} + \frac{1}{A+2^k} + \frac{1}{A+3^k} + \cdots + \frac{1}{A+n^k} + \cdots \qquad (A>0)$$

is convergent if $k > 1$, and divergent if $k \le 1$.

8.9 SERIES OF TERMS OF MIXED SIGNS

So far we have considered only series whose terms were all non-negative. Series of terms of mixed sign are usually more difficult to handle, but there is one special kind of series whose convergence we can prove very easily.

THEOREM 8.18

IF u_r IS A DECREASING FUNCTION OF r, AND $u_r \to 0$ AS $r \to \infty$, THEN THE SERIES

$$\sum (-1)^{r-1} u_r = u_1 - u_2 + u_3 - u_4 + \cdots$$

IS CONVERGENT.

PROOF:

We first see what the partial sums s_n are when n is an odd integer. They are

$$s_1 = u_1;$$
$$s_3 = u_1 - u_2 + u_3$$
$$= s_1 - (u_2 - u_3)$$
which is $\le s_1$, since $u_2 - u_3 \ge 0;$
$$s_5 = u_1 - u_2 + u_3 - u_4 + u_5$$
$$= s_3 - (u_4 - u_5)$$
$$\le s_3 \quad \text{since } u_4 - u_5 \ge 0;$$

and so on.

Carrying on in this way we see that

$$s_1 \ge s_3 \ge s_5 \ge s_7 \ge \cdots$$

i.e., that the sequence $\{s_{2k+1}\}$ is a decreasing sequence.

Similarly, if we consider the partial sums for n even, we have

$$s_2 = u_1 - u_2;$$
$$s_4 = u_1 - u_2 + u_3 - u_4$$
$$= s_2 + (u_3 - u_4)$$
$$\geq s_2 \quad \text{since } u_3 - u_4 \geq 0;$$
$$s_6 = s_4 + (u_5 - u_6)$$
$$\geq s_4, \quad \text{and so on.}$$

Consequently the sequence $\{s_{2k}\}$ is an increasing sequence. Moreover,

(8.9.1) $$s_{2k+1} = s_{2k} + u_{2k+1} \geq s_{2k} \geq s_2$$

so that the decreasing sequence $\{s_{2k+1}\}$ is bounded below (by s_2). It there-fore follows that this sequence of odd partial sums tends to a finite limit, by Theorem 8.11. Similarly $s_{2k} = s_{2k-1} - u_{2k} \leq s_{2k-1} \leq s_1$, so that the increasing sequence $\{s_{2k}\}$ is bounded above, by s_1, and therefore tends to a finite limit, by Theorem 8.10. Denote $\lim\limits_{k \to \infty} s_{2k+1}$ and $\lim\limits_{k \to \infty} s_{2k}$ by K and L, respectively.

Now although these two sequences of partial sums tend to finite limits the sequence of *all* the partial sums will not tend to a limit if K and L are different numbers, but will do so if they are equal. To complete the proof, therefore, we must show that $K = L$. This is easy. The left-hand side of (8.9.1) is s_{2k+1}, which tends to K as $k \to \infty$. The right-hand side is the sum of two terms, one of which (s_{2k}) tends to L while the other (u_{2k+1}) tends to 0, by the conditions of the theorem. Thus the right-hand side tends to L, and it follows that $K = L$. This proves the theorem.

EXAMPLE 1:

The series $1 - 1/2^k + 1/3^k - 1/4^k + \dots$ is convergent if $k > 0$. The proof of this is left to the reader. Note in particular that the series $1 - 1/2 + 1/3 - 1/4 + 1/5 - \dots$ is convergent.

EXAMPLE 2:

The exponential series is convergent for negative values of the argu-ment. If $x < 0$ then the exponential series $\sum x^r/r!$ can be written as $1 - y/1! + y^2/2! - y^3/3! + y^4/4! \dots$ where $y = -x$ is a positive number. By the methods used in the proof of Theorem 8.15 it follows that the sequence $\sum y^r/r!$ decreases from some value of r onwards (viz., for $r > y$) and tends to 0. Thus the requirements of Theorem 8.18 are met if we remove a finite number of terms from the series (namely, those for which

$r < y$). Since the removal of these terms does not affect the convergency of the series, it follows that the given series is convergent.

This result together with the result of Theorem 8.15 shows that the exponential series is convergent for *all* real values of the argument x which occurs in it. Thus given a value of x we can always sum the series. This series therefore defines a function, viz., the function whose value corresponding to a given value of x is the sum of the exponential series with that particular value of x. This function is called the exponential function, and can be denoted by exp (x).

Exercise 85

1. Prove that for a series $\sum (-1)^r u_r$ satisfying the conditions of Theorem 8.18 the difference between the sum of the series and the partial sum $s_n = \sum_{r=1}^{n} (-1)^r u_r$ is numerically less than u_{n+1}.

Use this result to find the sum of the series

$$1 - \frac{1}{2^2} + \frac{1}{2^2 \cdot 4^2} - \frac{1}{2^2 \cdot 4^2 \cdot 6^2} + \cdots$$

correct to 4 decimal places.

Hint: Work to 6 places and round off the final answer.

2. Calculate the values of exp (x) correct to four decimal places when $x = -2, -1, 1$ and 2. If we denote exp (1) by the letter e, show that, to four decimal places,

$$\exp(2) = e^2, \quad \exp(-1) = e^{-1}, \quad \exp(-2) = e^{-2}.$$

What would you be tempted to conjecture from these observations?

8.10 ABSOLUTE CONVERGENCE

Theorem 8.18 shows that the following two series are convergent.

(8.10.1) $$1 - \frac{1}{2^2} + \frac{1}{3^2} - \frac{1}{4^2} + \frac{1}{5^2} - \cdots$$

(8.10.2) $$1 - \frac{1}{2} + \frac{1}{3} - \frac{1}{4} + \frac{1}{5} - \cdots$$

If in these series, we make all the terms positive, then the first series becomes $1 + 1/2^2 + 1/3^2 + 1/4^2 + \ldots$ which is convergent, while the second becomes $1 + 1/2 + 1/3 + 1/4 + \ldots$ which is divergent. A series which is convergent when all its terms are made positive is said to be *absolutely convergent*. A more formal definition is

Definition 8.3

A series $\sum u_r$ is said to be absolutely convergent if the series $\sum |u_r|$ is convergent.

A series that is convergent but not absolutely convergent is said to be semi-convergent or conditionally convergent.

Notice that the above definition does not say anything about whether the series $\sum u_r$ itself is convergent or divergent. The reason for this is that it is not necessary, as the following theorem shows.

THEOREM 8.19

AN ABSOLUTELY CONVERGENT SERIES IS CONVERGENT.

Note: There is a tendency for students meeting this theorem for the first time to think that it is "obvious," but to think this is to be misled by the terminology employed. The nomenclature suggests that a series that is absolutely convergent is like a convergent series "only more so"! But there is nothing in the definition alone which says anything like this. $\sum u_r$ and $\sum |u_r|$ are two *different* series, and there is no immediately obvious reason why the first should not diverge while the second converges. That this cannot happen is something that must be proved. Had the theorem not been true, the terminology "absolutely convergent," with its suggestion of "convergent and more so," would probably never have come into use.

PROOF:

Consider the series x_r where

$$x_r = u_r \quad \text{if} \quad u_r \geq 0,$$
$$x_r = 0 \quad \text{if} \quad u_r \leq 0,$$

i.e., the series obtained from the series $\sum u_r$ by replacing every negative term by zero. This series of non-negative terms is term for term less than the series $\sum |u_r|$ which is given to be convergent. Hence $\sum x_r$ is convergent.

Similarly, the series $\sum y_r$ where

$$y_r = 0 \quad \text{if} \quad u_r \geq 0$$
$$y_r = -u_r \quad \text{of} \quad u_r < 0$$

is also a series of non-negative terms, and is term for term less than $\sum |u_r|$. Hence it is convergent. Now $u_r = x_r - y_r$, so that

$$\sum_{r=1}^{n} u_r = \sum_{r=1}^{n} x_r - \sum_{r=1}^{n} y_r.$$

But the two partial sums on the right-hand side of this equation each tend to a finite limit; therefore their difference tends to a finite limit (Theorem 8.3); hence the partial sum on the left-hand side tends to a finite limit. It follows that $\sum u_r$ is convergent.

REARRANGEMENTS OF SERIES

Let us take the terms of the series

$$1 - \tfrac{1}{2} + \tfrac{1}{3} - \tfrac{1}{4} + \tfrac{1}{5} - \tfrac{1}{6} + \cdots$$

and write them down in a different order, as follows:

$$(1 - \tfrac{1}{2}) - \tfrac{1}{4} + (\tfrac{1}{3} - \tfrac{1}{6}) - \tfrac{1}{8} + (\tfrac{1}{5} - \tfrac{1}{10}) + \tfrac{1}{12} + (\tfrac{1}{7} - \tfrac{1}{14}) + \cdots$$

We must first check that this new series contains the same terms as the original one—that we have not left any out, or put in new terms that were not there before. The terms which occur first in each of the bracketed expressions are those with odd denominator. Those occurring second in the bracketed expressions are those whose denominator is divisible by 2 but not by 4. The unbracketed terms are those whose denominators are multiples of 4. Thus all the terms of the original series are accounted for, and all terms of the second series occur somewhere in the first series. Thus we have what we may call a rearrangement of the original series.

If we simplify each bracketed expression, we obtain

$$\tfrac{1}{2} - \tfrac{1}{4} + \tfrac{1}{6} - \tfrac{1}{8} + \tfrac{1}{10} - \tfrac{1}{12} + \tfrac{1}{14} - \cdots$$

from which it is seen that the sum of the rearranged series is half that of the original series!

Now this may appear to the reader to be paradoxical, or even impossible. These two series have exactly the same terms, yet the sum of one is twice the sum of the other. How can this be? If this thought has occurred to the reader he should recall (or refer back to) the definition of the sum of an infinite series. This definition, by means of partial sums, depends fundamentally on the order in which the terms of the series occur. If we alter the order of the terms then we will get a different sequence of partial sums, and there is no reason to expect that the two sequences should have the same limit. If the rearrangement affects only a finite portion of the series, so that terms beyond, say, the Nth term are left in their original places, then the sum of the series *will* be the same as before, since all the partial sums after the Nth will be unchanged. But when the rearrangement makes itself felt, so to speak, throughout the series (as is the case in our example) it should not be a matter of surprise to find that the sum of the series has been affected. Indeed we should rather express surprise that there are some series whose sum is *not* affected by any rearrangement, however drastic. This is a property of series that are absolutely convergent, as we shall now prove.

THEOREM 8.20

IF $\sum u_r$ IS ABSOLUTELY CONVERGENT, AND ITS SUM IS l THEN ANY REARRANGEMENT OF $\sum u_r$ IS CONVERGENT TO THE SAME SUM.

PROOF:

We first prove this theorem for series of non-negative terms. Let $\sum u_r$ be the series and let $\sum u'_r$ be the rearrangement of it, where u'_r denotes the new rth term. Let s_m be the mth partial sum for the series $\sum u_r$, and s'_n the nth partial sum for $\sum u'_r$. Let l be the sum of $\sum u_r$, and l' that of $\sum u'_r$.

Consider the partial sum $s'_n = u'_1 + u'_2 + u'_3 + \ldots + u'_n$. Each of these terms occurs somewhere in the series $\sum u_r$, i.e., each is a u_r for some value of r. Let m be the largest of the values of r for which $u_r = u'_s$ with $s \leq n$. The partial sum $s_m = u_1 + u_2 + \ldots u_m$ is then the sum of a set of numbers that includes all the terms $u'_1, u'_2, \ldots u'_n$ as well as others (in general). Hence $s_m \geq s'_n$. Now s_m is an increasing function of m (since all the terms are non-negative) and tends to l. Hence $s'_n \leq s_m \leq l$. Thus s'_n is an increasing function which is bounded above, and therefore it tends to a finite limit. $\sum u'_r$ is therefore convergent.

The limit, l', to which $s_n{}'$ tends cannot be $> l$. For if we had $l' - l = a > 0$, then by the definition of "$s'_n \to l'$" we could find an integer N such that $l' - \frac{1}{2}a < s'_n < l' + \frac{1}{2}a$ for $n > N$, and this gives the contradiction $s'_n > l' - \frac{1}{2}a > l$. Hence $l' \leq l$.

If we now repeat the above argument, but interchange the roles of the series $\sum u_r$ and $\sum u'_r$, we arrive at the result $l \leq l'$. For these results to be compatible we must have $l' = l$, which proves the rest of the theorem for series of non-negative terms.

We now turn to series whose terms are not all non-negative. In the proof of Theorem 8.19 we split the series $\sum u_r$ into two series $\sum x_r$ and $\sum y_r$, each of which was a series of non-negative terms. If the sums of these two series are denoted by x and y respectively then the sum of $\sum u_r$

is $x - y$. If we rearrange the series $\sum u_r$ then the series $\sum x'_r$ and $\sum y'_r$ obtained from the rearranged series $\sum u'_r$ in the same way will be merely rearrangements of $\sum x_r$ and $\sum y_r$. By the first part of the proof the sums of these series will be the same as those of $\sum x_r$ and $\sum y_r$. Hence

$$\sum u'_r = \sum x'_r - \sum y'_r = \sum x_r - \sum y_r = \sum u_r$$

which proves the theorem.

If $\sum u_r$ is not an absolutely convergent series the above proof breaks down because the two series $\sum x_r$ and $\sum y_r$ are not then convergent. Thus it is not so surprising that the sum of a conditionally convergent series depends on the order of the terms. The extent of this dependence is indicated by the following theorem.

THEOREM 8.21

IF $\sum u_r$ IS A CONDITIONALLY CONVERGENT SERIES, AND l IS ANY REAL NUMBER, THEN THERE IS A REARRANGEMENT OF $\sum u_r$ WHOSE SUM IS l. IN OTHER WORDS A CONDITIONALLY CONVERGENT SERIES CAN BE REARRANGED TO GIVE ANY SUM WHATEVER.

PROOF:

We first make a few observations.

1. The positive terms of a conditionally convergent series taken by themselves, form an infinite series. For if there were only a finite number of positive terms, and if u_N were the last, then the series $u_N + u_{N+1} + u_{N+2} + \ldots$ would be a convergent series of negative terms. It would therefore be absolutely convergent, and this would mean that the original series was absolutely convergent.

Similarly the series consisting of the negative terms alone is an infinite series.

2. These two series, of positive and negative terms, are divergent. For if both were convergent then the original series would be absolutely convergent, which it is not. On the other hand, if one were divergent and one convergent then the original series would be divergent. For example, if it were the positive series which were diverged, then by taking enough terms of the original series we could arrange to include sufficient positive terms for their sum to be as large as we like; whereas the contribution of the negative terms is limited by the sum of the (convergent) infinite series of negative terms. Hence the partial sums of the original series could be made as large as we like, which contradicts the fact that this series was given as convergent.

These two series will be the same as the series $\sum x_r$ and $\sum (-y_r)$ that we had previously, except that we shall omit all the zero terms which occur. We now continue with the proof.

For the sake of definiteness let us suppose that l is positive; the proof of the theorem is trivially modified if it is not. We rearrange the given series as follows. We start by writing down the positive terms of the series in order from the beginning, and continue until the sum of those terms that we have written down first exceeds the number l. Since the series of positive terms alone is a divergent series the number l is sure to be exceeded sooner or later. When this happens we stop writing down positive terms and start writing down the negative terms, again taking them in the order in which they occur in the given series, and continue until the sum of all the terms written down becomes less than l. This must happen eventually since the series of negative terms is divergent. We then go back to writing down positive terms in order until l is again exceeded; then write negative terms until the sum is less than l again, and so on.

We must first show that the series that we get in this way is a rearrangement of the original series. This is easy, for we have not written down any terms that were not in the original series, and every term of the original series will be written down sooner or later since the process described above does not come to an end.

We must now show that the rearranged series will tend to l. Since the original series is convergent $u_r \to 0$ as $r \to \infty$. Given any number ϵ, there is an integer N such that $u_n < \epsilon$ if $n < N$. Let us suppose that when l is exceeded for the Nth time m terms have been written down altogether. If S_n denotes the sum of the first n terms written down, i.e., the nth partial sum for the rearranged series, then if $n > m$, $S_n - l < \epsilon$. For the last term to be added in must have been a u_r with $r \geq N$, and hence $< \epsilon$; and this means that if $S_n - l > \epsilon$, then $S_{n-1} - l > 0$, i.e., the previous partial sum already exceeded l. But this is a contradiction since negative terms are taken *as soon as* the partial sum exceeds l. In the same way we get a contradiction if we assume that $l - S_n > \epsilon$ for $n > m$. Thus for $n > m$ we have $|S_n - l| < \epsilon$. This proves that the series converges to the limit l.

Exercise 86

1.   Prove that a conditionally convergent series can be rearranged so that it diverges to $+\infty$, i.e., so that its partial sum $s_n \to \infty$. (Take care that all the negative terms of the original series are included in the rearrangement.)

2. Following the method used in the previous section for proving that a conditionally convergent series can be rearranged to have any sum, find the first 20 terms in a rearrangement of the series

$$1 - \tfrac{1}{2} + \tfrac{1}{3} - \tfrac{1}{4} + \tfrac{1}{5} - \cdots$$

which sums to the value 0.

8.11 LIMITS OF A FUNCTION OF A REAL VARIABLE

In the early part of this chapter we considered what was meant by the *limit* of a function of an integral variable n, as n tended to infinity. We now extend the idea of a limit to functions whose arguments are not integers but real numbers. The first two definitions that we give are very straightforward.

Definition 8.4

A function $f(x)$ of a real variable x is said to tend to infinity as x tends to infinity (written $f(x) \to \infty$ as $x \to \infty$, or $\lim_{x \to \infty} f(x) = \infty$) if, for any real number K, however large, there is a number X such that $f(x) > K$ for all $x > X$.

Definition 8.5

A function $f(x)$ of a real variable is said to tend to a limit l as $x \to \infty$ (written $f(x) \to l$ as $x \to \infty$, or $\lim_{x \to \infty} f(x) = l$) if, given any positive number ϵ, however small, there is a number X such that $|f(x) - l| < \epsilon$ for all $x > X$.

These definitions parallel exactly definitions 8.1 and 8.2, and we need say no more about them.

With functions of an integral variable, all limits were taken "as $n \to \infty$", and a moment's reflection will show that there can be no other kind of limit with such functions. But when we come to functions of a real variable we can talk about a function $f(x)$ tending to a limit as x tends, not to infinity, but to some finite number x_0, say. Thus we say that the function x^2 tends to 4 as x tends to 2; that $\sin x$ tends to 1 as x tends to $\pi/2$, etc.—to mention two very elementary examples. The general idea of what is meant by this is that as x gets closer and closer to the value x_0, so $f(x)$ gets closer and closer to some number a; and that we can make $f(x)$ come as close to a as we like by taking x to be sufficiently close to x_0.

Now the idea of "closeness to x_0" can be rigorously conveyed by saying that x does not differ from x_0 by more than some positive number, which is customarily denoted by δ. Such a statement would usually be written as

$|x - x_0| < \delta$, but since we are talking about what happens as x *tends* to x_0 it will be in our interests not to assume anything about what happens when x is actually *equal* to x_0. It is for this reason that, in the definition that we shall shortly give, the idea of "closeness to x_0" is expressed by an inequality of the form $0 < |x - x_0| < \delta$, in which the possibility $x = x_0$ is excluded. This may seem a minor point, but in fact it is of the utmost importance.

The closeness of the function $f(x)$ to the limit a is expressed by specifying that $f(x)$ must not differ from a by more than some positive number ϵ, and, as before, this is written $|f(x) - a| < \epsilon$. These preliminary remarks should indicate the "reasonableness" of the following definition.

Definition 8.6

A function $f(x)$ of a real variable is said to tend to a limit a as x tends to x_0 ($f(x) \to a$ as $x \to x_0$) if, given any positive number ϵ, a positive number δ can be found such that if $0 < |x - x_0| < \delta$, then $|f(x) - a| < \epsilon$.

At this stage a diagram may possibly be of some help. Let us consider the graph of $y = f(x)$, shown in Fig. 8.1. The limit a is shown and also the value

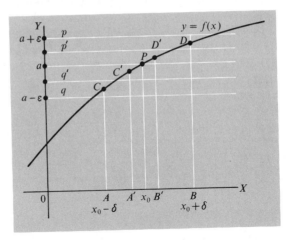

Fig. 8.1

x_0 of x. If we want to show that $f(x) \to a$ as $x \to x_0$ then, according to the definition, we must show that we can make $f(x)$ differ from a by less than any given number ϵ (however small it may be) by taking x to be sufficiently near to x_0. Now if $f(x)$ is within ϵ of a, i.e., $a - \epsilon < f(x) < a + \epsilon$, then the corresponding point of the graph will lie between the lines p and q on the diagram. The question is then that of finding, if possible, an interval on the x-axis (AB in the diagram) of the form $x_0 - \delta < x < x_0 + \delta$, such that the portion of the graph corresponding to values of x in this interval (CD in the diagram) will lie wholly between the lines p and q. All this with the possible exception of the point (P) which corresponds to $x = x_0$ which, by our definition, does not come into the story. Clearly for the function shown in the figure this is possible. If we take a smaller value of ϵ the lines p and q are closer together, and in general a smaller interval $(A'B')$ will have to be taken in order that the corresponding portion of the graph $(C'D')$ will lie between the new lines p' and q'. If this requirement can be met no matter how small ϵ is, i.e., no matter how close p and q are above and below the line $y = a$, then our definition enables us to assert that $f(x) \to a$ as $x \to x_0$.

The statement that $f(x) \to a$ as $x \to x_0$ can also be written as $\lim_{x \to x_0} f(x) = a$.

Let us further illustrate the definition by a specific example. We shall prove that $x^2 \to 1$ as $x \to 1$. This, of course, is "obvious," but it will be as well to see how the definition works with an obvious example before applying it to other less obvious ones. We have to show that if ϵ is any positive number we can make the function x^2 lie between $1 - \epsilon$ and $1 + \epsilon$ for all values of x in a range of values either side of $x = 1$, i.e., in intervals $1 - \delta < x < 1$ and $1 < x < 1 + \delta$, where δ is some suitable chosen positive number. This is clearly quite easy to do. If $\epsilon < 3$ then taking $\delta = \frac{1}{3}\epsilon$ will do the trick; for if

$$1 - \frac{1}{3}\epsilon < x < 1 + \frac{1}{3}\epsilon$$

then

$$\left(1 - \frac{1}{3}\epsilon\right)^2 < x^2 < \left(1 + \frac{1}{3}\epsilon\right)^2.$$

But $(1 - \frac{1}{3}\epsilon)^2 = 1 - \frac{2}{3}\epsilon + \frac{1}{9}\epsilon^2 > 1 - \epsilon$, and $(1 + \frac{1}{3}\epsilon)^2 = 1 + \frac{2}{3}\epsilon + \frac{1}{9}\epsilon^2 < 1 + \epsilon$, since $\frac{1}{9}\epsilon^2 < \frac{1}{3}\epsilon$. Hence we have

$$1 - \epsilon < x^2 < 1 + \epsilon \quad \text{or} \quad |x^2 - 1| < \epsilon,$$

which is what we wanted. We do not have to worry about what happens if $\epsilon \geq 3$, for the value of δ which serves when ϵ is taken to be say 2 (and there is such a value—for example, $\delta = \frac{2}{3}$) will certainly serve for any larger value of ϵ. Thus the requirements of the definition are met whatever the value of ϵ, and we conclude that $x^2 \to 1$ as $x \to 1$.

It will be seen that, in this example, the limit to which the function *tends* as $x \to 1$ is in fact the *value* of the function *at* $x = 1$. It is very often the case that $\lim_{x \to x_0} f(x)$ turns out to be just $f(x_0)$, but this is by no means always the case, and we now give two examples illustrating this point.

In the first example we ensure that the statement $\lim\limits_{x \to x_0} f(x) \neq f(x_0)$ is true by deliberately making $f(x_0)$ different from $\lim\limits_{x \to x_0} f(x)$. Take the function defined by

$$f(x) = x \quad \text{if } x \neq 0,$$
$$f(x) = 1 \quad \text{if } x = 0.$$

Here $\lim\limits_{x \to 0} f(x) = 0$; for given any positive number ϵ we can find a positive number δ such that $|f(x) - 0| < \epsilon$ for $0 < |x - 0| < \delta$, i.e., $|f(x)| < \epsilon$ for $0 < |x| < \delta$. It is sufficient to take $\delta = \epsilon$. Note how the definition of the limit excludes the value $x = 0$ at which the function is, so to speak, anomalous. Hence $\lim\limits_{x \to 0} f(x)$ and $f(0)$ are not the same, since the first is 0 and the second 1. A glance at the graph of this function (Fig. 8.2) will show that this exception to what one might expect to happen corresponds to a "break" in the graph. More of this later.

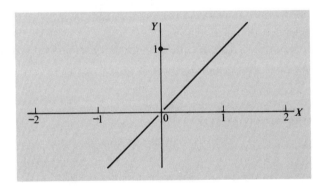

Fig. 8.2

Our second example will be the function

(8.11.1)
$$f(x) = \frac{x^2 - 1}{x - 1}.$$

For most values of x the expression on the right gives a unique value for the function $f(x)$, but if $x = 1$ the expression reduces to $0/0$ which has no meaning. This expression therefore defines a function of x at all values of x except the value 1, but at this value the function is undefined. It is true that when $x \neq 1$, $f(x) = \dfrac{x^2 - 1}{x - 1} = x + 1$ so that one might think that the function *ought* to have the value 2 when $x = 1$. But whatever we think it ought to do, the fact remains that (8.11.1) which defined the function does *not* ascribe the value 2 to the function when $x = 1$ simply because it does not ascribe *any* value to the function for that value of x. There is nothing to stop us from *extending* the function to the value $x = 1$ by means of an extra definition pertaining to this value alone, but we are under no compulsion to do this, and even if we do we are not bound to make $f(1) = 2$—we could make it any other number just as well. But all this is rather beside the point, for if we did so extend the function we would end up with what was strictly speaking a *different* function (since, for example, the domain of the functional relation would be different, including, as it would, the value 1 of x which did not belong to the domain of the original functional relation). For this reason the functions $\dfrac{x^2 - 1}{x - 1}$ and $x + 1$ are *different* functions, though admittedly they are not *very* different!

Despite the fact that $\dfrac{x^2 - 1}{x - 1}$ is not defined at $x = 1$ there is no difficulty about finding its limit as x tends to 1. For the definition of this limit does not involve in any way the value of the function at $x = 1$ and therefore the behavior of the function as x tends to 1 will be the same as that of the function $x + 1$. After these preliminary remarks it is very easy to show that $\lim\limits_{x \to 1} \dfrac{x^2 - 1}{x - 1} = 2$ and we leave it to the reader to complete the proof of this.

ONE-SIDED LIMITS

Occasionally we meet a function that does not tend to a limit a as x tends to x_0, and so does not satisfy the requirement of definition 8.6, but which will satisfy a rather similar requirement, viz.,

$$|f(x) - a| < \epsilon \text{ for } 0 < x - x_0 < \delta.$$

That is to say that the function satisfies the requirement for tending to a limit provided we restrict ourselves to values of x that are greater than x_0 instead of considering values "on both sides" of it. Such a function is said to tend to a as x tends to x_0 *from above*. This state of affairs is usually written, rather whimsically, as "$f(x) \to a$ as $x \to x_0 + 0$" or as "$\lim\limits_{x \to x_0 + 0} f(x) = a$." In the same sort of way we can define "$f(x)$ tends to a as x tends to x_0 from below," or $f(x) \to a$ as $x \to x_0 - 0$ or $\lim\limits_{x \to x_0 - 0} f(x) = a$, by the requirement that, given any positive ϵ, there is a number δ such that $|f(x) - a| < \epsilon$ for $0 < x_0 - x < \delta$. In practice one usually writes "$x \to x_0 +$" in place of "$x \to x_0 + 0$" and, similarly "$x \to x_0 -$" for "$x \to x_0 - 0$."

EXAMPLE:

Take the function given by

$$f(x) = 1 \quad \text{if } x > 0$$
$$f(0) = 0$$
$$f(x) = -1 \quad \text{if } x < 0$$

whose graph is shown, diagrammatically, in Fig. 8.3. It is readily seen that $\lim_{x \to 0} f(x)$ does not exist, i.e., that $f(x)$ does not tend to a limit as

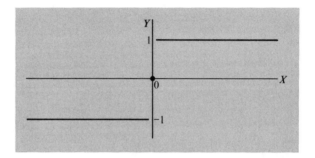

Fig. 8.3

$x \to 0$. On the other hand the "one-sided" limits that we defined above *do* exist. We have, in fact,

$$\lim_{x \to 0+} f(x) = 1$$

and

$$\lim_{x \to 0-} f(x) = -1.$$

Note that neither of these one-sided limits is equal to the value of the function *at* $x = 0$.

Directly from the definition of *limit* and *one-sided limit* we get the following theorem.

THEOREM 8.21

IF $\lim_{x \to x_0+} f(x)$ AND $\lim_{x \to x_0-} f(x)$ BOTH EXIST, AND ARE EQUAL, THEN THE ORDINARY LIMIT $\lim_{x \to x_0} f(x)$ ALSO EXISTS, AND HAS THE SAME VALUE AS THE ONE-SIDED LIMITS.

Amid all this talk of *limits*, *one-sided limits* and so on, we should not overlook the fact that a function may not tend to *any* sort of limit as x tends to some given value, i.e., $\lim\limits_{x \to x_0} f(x)$ may not *exist*. Take, for example, the function $y = \sin 1/x$, part of the graph of which is shown in Fig. 8.4. This function is not defined at $x = 0$, and does not tend to any limit as $x \to 0$. It is not difficult to see intuitively why this is so; in any interval of δ either side of $x = 0$ there will be "undulations" of the graph (since these undulations "crowd together" as x becomes smaller and smaller) so that in any such interval there will be values of x for which y is 1, and values for which it is -1. This precludes any possibility of our being able to meet the requirements for the existence of a limit. Let us make this rigorous.

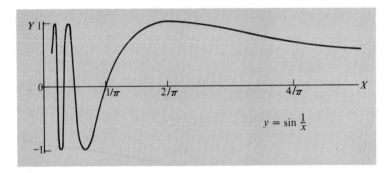

Fig. 8.4

The set of values of x given by $0 < |x| < \delta$ will include the values $x = \delta/2$ and $x = \delta/4$, and the set of corresponding values of y will include all the numbers of the form $\sin \theta$, for $2/\delta < \theta < 4/\delta$. If $\delta < 1/\pi$ then this range of values for θ is of length more than 2π, and hence will include at least one value of θ for which $\sin \theta = 1$, and at least one value for which $\sin \theta = -1$. It is therefore impossible for all values of the function to be within ϵ of any fixed number if ϵ is anything less than 1. Hence we cannot meet the requirements of definition 8.6, and this function does not tend to any limit at all as $x \to 0$.

We now prove some theorems on limits of functions of a real variable that are precise analogues of Theorems 8.1 through 8.5. The first three are so straightforward that we shall not bother to give the proofs.

THEOREM 8.22

IF $\lim\limits_{x \to x_0} f(x) = a$ AND $\lim\limits_{x \to x_0} g(x) = b$, THEN $\lim\limits_{x \to x_0} [f(x) + g(x)]$ EXISTS AND IS $a + b$.

THEOREM 8.23

IF $\lim\limits_{x \to x_0} f(x) = a$ AND k IS A CONSTANT, THEN $\lim\limits_{x \to x_0} kf(x)$ EXISTS, AND IS ka.

THEOREM 8.24

IF $\lim_{x \to x_0} f(x) = a$ AND $\lim_{x \to x_0} g(x) = b$, THEN $\lim_{x \to x_0} [f(x) - g(x)]$ EXISTS AND IS $a - b$.

THEOREM 8.25

IF $\lim_{x \to x_0} f(x) = a$ AND $\lim_{x \to x_0} g(x) = b$, THEN $\lim_{x \to x_0} f(x)g(x)$ EXISTS AND IS ab.

PROOF:

Since $f(x) \to a$ we shall certainly have

$$a - 1 < f(x) < a + 1 \text{ for } 0 < |x - x_0| < \text{some number } \delta_1.$$

Therefore there is a number K such that $|f(x)| < K$ for $0 < |x - x_0| < \delta_1$. Now

$$(8.11.2) \qquad |f(x)g(x) - ab| = |f(x)\{g(x) - b\} + b\{f(x) - a\}|$$
$$\leq |f(x)||g(x) - b| + |b||f(x) - a|$$
$$< K|g(x) - b| + |b||f(x) - a|.$$

Given any positive number ϵ we can find a number δ_2 such that $|f(x) - a| < (1/2|b|)\epsilon$ for $0 < |x - x_0| < \delta_2$ and a number δ_3 such that $|g(x) - b| < \epsilon/2K$ for $0 < |x - x_0| < \delta_3$. From (8.11.2) it then follows that

$$|f(x)g(x) - ab| < K. \epsilon/2K + |b|. \epsilon/2|b|$$
$$= \epsilon \text{ for } 0 < |x - x_0| < \min(\delta_1, \delta_2, \delta_3)$$

from which the theorem follows.

THEOREM 8.26

IF $\lim_{x \to x_0} f(x) = a$, $\lim_{x \to x_0} g(x) = b \neq 0$, THEN $\lim_{x \to x_0} f(x)/g(x)$ EXISTS AND IS a/b.

PROOF:

Without any loss of generality we may assume that $b > 0$. (If it is not, then take instead the functions $-f(x)$ and $-g(x)$).

Since $g(x) \to b$ we can find a number δ_1 such that

$$|g(x) - b| < \tfrac{1}{2}b \quad \text{for} \quad 0 < |x - x_0| < \delta_1.$$

Hence if $0 < |x - x_0| < \delta_1$ we have $g(x) > \tfrac{1}{2}b$.

Now $\qquad \left| \dfrac{f(x)}{g(x)} - \dfrac{a}{b} \right| = \left| \dfrac{bf(x) - ag(x)}{bg(x)} \right|$

$$= \frac{|\,b(f(x) - a) - a(g(x) - b)\,|}{|\,b\,|\,g(x)}$$

$$< \frac{b\,|\,f(x) - a\,| + |\,a\,|\,|\,g(x) - b\,|}{\frac{1}{2}b^2}$$

(8.11.3) $$= \frac{2}{b}\,|\,f(x) - a\,| + \frac{2\,|\,a\,|}{b^2}\,|\,g(x) - b\,|.$$

Given any positive number ϵ we can find a number δ_2 such that $|\,f(x) - a\,| < \frac{1}{2}b\epsilon$ for $0 < |\,x - x_0\,| < \delta_2$, and a number δ_3 such that $|\,g(x) - b\,| < (b^2/4\,|\,a\,|)\epsilon$ for $0 < |\,x - x_0\,| < \delta_3$. Equation (8.11.3) then gives

$$\left|\frac{f(x)}{g(x)} - \frac{a}{b}\right| < \frac{2}{b}\cdot\frac{b}{4}\epsilon + \frac{2\,|\,a\,|}{b^2}\cdot\frac{b^2}{4\,|\,a\,|}\epsilon$$

$$= \frac{1}{2}\epsilon + \frac{1}{2}\epsilon$$

$$= \epsilon, \text{ for } 0 < |\,x - x_0\,| < \min\,(\delta_1, \delta_2, \delta_3),$$

which proves the theorem.

8.12 CONTINUITY

The example that we gave above, of a function $f(x)$ for which $\lim\limits_{x \to x_0} f(x)$ existed but was not $f(x_0)$, was a somewhat artificial one—we deliberately defined it in such a way that the limit would turn out to be different from the value $f(x_0)$. Examples of this sort of behavior *can* arise naturally, though they are not frequently met. The statement $\lim\limits_{x \to x_0} f(x) = f(x_0)$ therefore expresses something about the function which is characteristic of most of the functions that commonly occur in practice. To this "something" we give the name *continuity*, a term which is made precise by the following definitions.

Definition 8.7

A function is said to be "continuous at $x = x_0$" if $\lim\limits_{x \to x_0} f(x) = f(x_0)$.

Definition 8.8

A function which is continuous at every value x_0 belonging to a set of values X is said to be a continuous function, or just to be continuous, over (or in) the set X. A function that is continuous at all values of its argument is said to be a continuous function.

Thus if X is a set of values of x, and if a function $f(x)$ is continuous at every value x_0 in X, i.e., if $x_0 \in X \Rightarrow [f(x)$ is continuous at $x = x_0]$, then we say that the function is continuous over (or in) the set X. The commonest examples of sets X that arise in practice are *intervals*. The set of values of x satisfying $a \le x \le b$, where a and b are real numbers, is called a *closed interval*. It is often denoted by $[a, b]$. The set of values $\{x \mid a < x < b\}$ is called an *open interval* and denoted by (a, b). Thus a closed interval contains the end values a and b, while an open interval does not.

EXAMPLES:

The function $f(x) = x^2$ is continuous, since it can be shown that for any value x_0 the function is continuous at $x = x_0$. (The reader should verify this.)
   The function given by

$$f(x) = 1, \text{ if } x \ne 0$$
$$f(0) = 0$$

is continuous over the set of nonzero values of x. It is not continuous at $x = 0$ since $\lim_{x \to 0} f(x) = 1$, while $f(0) = 0$.

Exercise 87

1. Show that the following functions are continuous.
    (a) x^n (where n is a positive integer)
    (b) $|x|$
    (c) $\dfrac{1}{1 + x^2}$
2. Find for what values of x the following functions are not continuous.
    (a) $\dfrac{1}{3x}$
    (b) $x - [x]$
    (c) $\dfrac{1}{x^2 - 1}$
3. Show that if the functions $f(x)$ and $g(x)$ are continuous at $x = x_0$ then the functions $f(x) + g(x)$ and $f(x)g(x)$ are also continuous at $x = x_0$; and that $f(x)/g(x)$ is continuous at $x = x_0$ provided that $g(x_0) \ne 0$. Hence show that any polynomial function is continuous; and that any *rational function* (i.e., a function of the form $P(x)/Q(x)$ where $P(x)$ and $Q(x)$ are polynomials) is continuous except for those values (if any) of the argument for which the denominator vanishes.
    Hint: Use Theorems 8.22, 8.25 and 8.26.

4. Evaluate the following limits:

(a) $\lim\limits_{x \to 9} \dfrac{3 - \sqrt{x}}{9 - x}$

(c) $\lim\limits_{x \to a} \dfrac{x^4 - a^4}{x - a}$

(b) $\lim\limits_{x \to 2} \left\{ \dfrac{x^5 - 32}{x - 2} + \dfrac{x^3 - 8}{x - 2} \right\}$

(d) $\lim\limits_{\theta \to \pi/2} \dfrac{1 - \sin \theta}{1 - \tan \theta/2}$

Hint: Use the result $\sin \theta = \dfrac{2t}{1 + t^2}$, where $t = \tan \dfrac{\theta}{2}$.

5. Evaluate the following one-sided limits.

(a) $\lim\limits_{x \to 1 + 0} \dfrac{1 + |x - 1|}{x}$

(d) $\lim\limits_{x \to 3 - 0} \dfrac{x - [x]}{x^2}$

(b) $\lim\limits_{x \to 2 + 0} \{x - 2\}^3$

(e) $\lim\limits_{x \to 1 - 0} \dfrac{1}{1 + 2^{1/(1-x)}}$

(c) $\lim\limits_{x \to 1 + 0} \dfrac{1}{1 + 2^{1/(1-x)}}$

where $[x]$ denotes the integral part of n (as defined in Chapter 3, page 66).

8.13 SOME THEOREMS ON CONTINUITY

The definition of continuity given above (Definition 8.6) contained the statement $\lim\limits_{x \to x_0} f(x) = f(x_0)$. If we replace this statement by its definition (Definition 8.5) we obtain an alternative statement of the definition of continuity at $x = x_0$.

Definition 8.9

A function $f(x)$ is continuous at the value $x = x_0$ if, given any positive number ϵ, there is a number δ such that $|f(x) - f(x_0)| < \epsilon$ for $|x - x_0| < \delta$.

It can be seen from the above examples (especially if the reader draws the graphs of the functions concerned) that the idea of a continuous function corresponds very closely to the idea of an unbroken curve. In these examples the values of x at which the functions are not continuous correspond to obvious "breaks" in the graph of the function. This parallel between continuity of the function and the absence of breaks in the graph of the function enables us to form an intuitive picture of what is meant by continuity, but it must be remarked that the parallel is not exact. For instance, if we ask whether it is possible for a function to be continuous at one value only, the "unbroken curve" idea would prompt us to give the answer "No." Yet it is quite possible for this to happen. Consider the function given by

$$f(x) = x \quad \text{if } x \text{ is a rational number;}$$
$$f(x) = 0 \quad \text{if } x \text{ is an irrational number.}$$

Given any positive number ϵ we can find a number δ such that $|f(x) - f(0)|$ $= |f(x)| < \delta$. Indeed, we have only to take $\epsilon = \delta$ to ensure that this is so. Therefore, from definition 8.9 it follows that this function is continuous at $x = 0$. The reader should verify that this function is not continuous anywhere else.

We now prove four important results concerning continuous functions.

THEOREM 8.28

A FUNCTION WHICH IS CONTINUOUS IN A CLOSED INTERVAL IS BOUNDED IN THAT INTERVAL.

PROOF:

Let the interval in question be $a \leq x \leq b$. Consider the set S of values $\xi (\leq b)$ of x which have the property that the function is bounded in the interval $a \leq x \leq \xi$. This set is not the empty set since it certainly contains the value a, even if no other value of x. Denote by x_0 the least upper bound of the set S, and let $f(x_0) = y_0$. We shall assume for the moment that $x_0 \neq a$ and $x_0 \neq b$. Then since the function is continuous at $x = x_0$, given any positive number ϵ (for the sake of definiteness we can take $\epsilon = 1$) there is a number δ such that for $|x - x_0| < \delta$ we have $|f(x) - y_0| < \epsilon$. In other words, with our choice of ϵ we have $y_0 - 1 < f(x) < y_0 + 1$ for $x_0 - \delta < x < x_0 + \delta$. Furthermore, since $a < x_0 < b$ we can choose δ so that $a < x_0 - \delta < x < x_0 + \delta < b$. Now $x_0 - \delta$ belongs to S (since otherwise the least upper bound of S would be $< x_0 - \delta$, which is not the case), so that $f(x)$ is bounded in the interval $a \leq x \leq x_0 - \delta$. The function is also bounded in the interval $x_0 - \delta < x \leq x_0 + \delta/2$ (since it lies between $y_0 - 1$ and $y_0 + 1$ for these values). Hence the function is bounded in the interval $a \leq x \leq x_0 + \delta/2$. In the special case $x_0 = a$, continuity at a implies that $f(x)$ is bounded for $a \leq x \leq a + \delta/2$—the same result. But this means that $x_0 + \delta/2$ belongs to S, which contradicts the statement that x_0 is the least upper bound of the set S. This contradiction disappears if $x_0 = b$, since all numbers in S are $\leq b$. Thus only if the theorem is true ($x_0 = b$) do we avoid a contradiction.

In theorem 8.28 it is necessary that we specify the interval over which the function is continuous to be a *closed* interval; we cannot make do with an open one. For example, the function $1/x$ is continuous in the open interval $0 < x < 1$ but is not bounded in the interval.

THEOREM 8.29

A FUNCTION WHICH IS CONTINUOUS IN A CLOSED INTERVAL ATTAINS ITS UPPER AND LOWER BOUNDS IN THE INTERVAL, i.e., IF M AND m ARE THE LEAST UPPER

BOUND AND GREATEST LOWER BOUND OF THE VALUES OF THE FUNCTION $f(x)$
IN THE INTERVAL $a \leq x \leq b$, THEN THERE IS A VALUE OF x, SAY x_1, IN THIS INTER-
VAL FOR WHICH $f(x_1) = M$, AND ANOTHER, SAY x_2, FOR WHICH $f(x_2) = m$.
THERE MAY, OF COURSE, BE MORE THAN ONE VALUE OF x FOR WHICH THE
FUNCTION ATTAINS THE BOUND.

PROOF:

We shall prove the result for the least upper bound M; the result for the
greatest lower bound m is precisely similar.

Suppose that there is no value x_1 of x for which $f(x_1) = M$. Consider

the function $\dfrac{1}{M - f(x)}$. This function is continuous in the interval

$a \leq x \leq b$ because of the continuity of $f(x)$ (compare Exercise 87.3
above). Hence by Theorem 8.28 it is bounded in this interval. But
this is not possible; for since we can make $M - f(x)$ as small as we like
[otherwise a smaller number than M would be an upper bound of $f(x)$]

we can make $\dfrac{1}{M - f(x)}$ as large as we like.

This contradiction results from the continuity of $\dfrac{1}{M - f(x)}$ which can

be asserted only if the denominator of this function does not vanish.
Thus we arrive at a contradiction if we assume that for no value of x
is $f(x) = M$. There must therefore be at least one value of x in the interval for which this is so, and the theorem is true.

THEOREM 8.30

IF A FUNCTION $f(x)$ IS CONTINUOUS AT $x = x_0$ AND $f(x_0) > 0$, THEN $f(x)$ IS
POSITIVE FOR ALL VALUES OF x IN SOME "NEIGHBORHOOD" OF x_0, i.e., IN AN
OPEN INTERVAL $x_0 - \delta < x < x_0 + \delta$ FOR SOME POSITIVE NUMBER δ.

PROOF:

In Definition 8.8 take ϵ to be the positive number $f(x_0)$. Then, since $f(x)$
is continuous at $x = x_0$ there is a number δ such that

$$|f(x) - f(x_0)| < f(x_0) \text{ for } |x - x_0| < \delta$$

that is, if $x_0 - \delta < x < x_0 + \delta$, then

$$f(x_0) - f(x_0) < f(x) < f(x_0) + f(x_0).$$

The left-hand half of this inequality simply states that $f(x) > 0$, which is
the result of the theorem.

THEOREM 8.31

A CONTINUOUS FUNCTION TAKES, BETWEEN ANY PAIR OF ITS VALUES ALL THE
INTERMEDIATE VALUES, i.e., IF $f(a) < k < f(b)$ THEN THERE IS A VALUE c OF x
BETWEEN a AND b SUCH THAT $f(c) = k$.

The graphical interpretation of this theorem (see Fig. 8.5) is that if the
graph of the function is a continuous (unbroken) curve through the points
$[a, f(a)]$ and $[b, f(b)]$, then somewhere between these two points the curve will
cut the line $y = k$, where $f(a) < k < f(b)$. Looked at graphically this result
is obvious, but we shall exercise our usual caution and give a rigorous proof.
For simplicity we shall assume that $a < b$.*

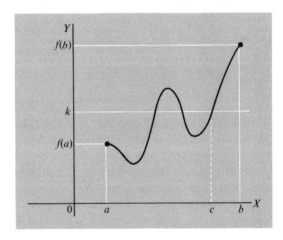

Fig. 8.5

PROOF:

Consider the set S of all values of x between a and b such that $f(x) < k$.
Let c be the least upper bound of S. There are then three possibilities to
consider, viz., (1) $f(c) < k$, (2) $f(c) > k$ and (3) $f(c) = k$. We consider
each case in turn.

Case 1. $f(c) < k$. In this case the function $k - f(x)$ is positive at
$x = c$ and therefore, by Theorem 8.30, it will be positive in some
neighborhood of $x = c$. But this would mean that $k - f(x)$ would be
positive, (i.e., that $f(x)$ would be less than k) for values of x greater than c.
This contradicts the definition of c as the least upper bound of the set S.
Hence the assumption $f(c) < k$ leads to a contradiction.

* The proof needs only slight modification if $b < a$.

Case 2. $f(c) > k$.   Here the function $f(x) - k$ is positive at $x = c$, and hence is positive in a neighborhood of $x = c$. But this means that there is a number ξ less than c such that $f(x) - k$ is positive for all values of x in the interval $\xi < x < c$, which therefore contains no elements of S. This again contradicts the definition of c as the least upper bound of S.

Case 3.   Since these three cases are the only possibilities, and since the other two have been shown to give rise to contradictions, the remaining possibility, $f(x) = c$, must be true. This proves the theorem.

Note that if the graph of $f(x)$ crosses the line $y = k$ more than once, the number c which we defined in the proof of the theorem will, in fact, be the largest value of x for which this happens. This is of no consequence, since all that the theorem states, and all that we have to prove, is that the graph crosses the line somewhere.

Exercise 88

1.   For given values of a and b we need to know that the function $f(x)$ is continuous in the closed interval $[a, b]$ in order to be able to prove Theorem 8.31. Construct an example showing that the result of the theorem need not be true if $f(x)$ is known to be continuous only in the open interval (a, b).
2.   Construct an example to show that the converse of Theorem 8.31 is not true. That is, find a function which is not continuous in an interval $[a, b]$ but which assumes, in this interval all the values intermediate between $f(a)$ and $f(b)$.

8.14   THE HAM SANDWICH THEOREM

This has been a long chapter, and one in which many new and difficult concepts have been introduced. It will have been hard going for those who have not met these concepts before (and possibly for some who have!), and the reader may feel the need for a little light relief. For this reason, and with no other excuse, we conclude the chapter with an amusing application of Theorem 8.31. This application has nothing to do with economics but it will perhaps serve a useful purpose in demonstrating how an "obvious" result such as Theorem 8.31 may have fairly immediate consequences that are far from being obvious. Since this example is included mainly as a diversion, we shall not be too fussy about being strictly rigorous.

We consider two areas in a plane, each being bounded by a smooth and, of course, continuous closed curve, as shown in Fig. 8.6. The areas must be

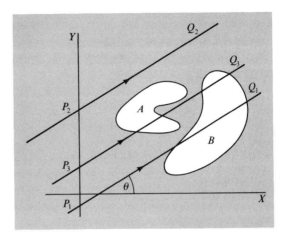

Fig. 8.6

in one piece, but can otherwise be of any shape. We propose to show that there is a line in the plane which bisects *both* these areas, i.e., which divides each of them into two parts of equal area. We would hardly say that this was an obvious result, yet it is a fairly direct consequence of Theorem 8.31.

We call the areas A and B, and take any convenient pair of perpendicular lines in the plane, as x- and y- axes. Let us consider lines given by parametric equations of the form $x = t \cos \theta$, $y = c + t \sin \theta$, where t is the parameter and c and θ are, for the present, constants. By taking the equations of the lines in this form we not only determine the lines but also ascribe to each line a sense, which we can denote in the figure by arrows, as shown. Now if c has a large value, the corresponding line will lie entirely "above" the area A (as $P_2 Q_2$ in Fig. 8.6); if c has a much smaller value (possibly negative) then the line will lie entirely "below" the area A (as $P_1 Q_1$ in Fig. 8.6); whereas for some intermediate values of c the line will intersect A (as the line $P_3 Q_3$, for example). With each value of c we now associate a number, obtained as follows: we look along the line in the direction of the arrow, and see how much more of the area A there is to the right of us than to the left, i.e., we calculate $R_A - L_A$ where R_A is the area of that portion (possibly all) of A which lies to the right of the line, and L_A is the area of that portion which lies to the left of it. This number is a function of c (we are keeping θ fixed, so that these lines are all parallel). Let us denote this function by $\phi(c)$. It is not difficult to satisfy ourselves that $\phi(c)$ is a continuous function of c.

Now if the line is like $P_1 Q_1$, the whole of the area A lies to the left of it, and consequently $\phi(c_1) < 0$; while for a line like $P_2 Q_2$ we have $\phi(c_2) > 0$. From Theorem 8.31 it then follows that there is a value of c between c_1 and c_2 for which $\phi(c)$ takes the intermediate value 0. This means that there is a line making with the x-axis an angle of θ and bisecting the area A. We can say a little more than this—we can assert that there is only one such line. For $\phi(c)$ and $\phi(c')$ cannot be equal if $c \neq c'$, for these two values differ by the area of the portion of A enclosed between them.* Thus there is one and only one value of c for which $\phi(c) = 0$, and thus only one line for a given value of θ which bisects A.

We now turn our attention to the area B. For each value of θ we have a line bisecting A and making an angle θ with the x-axis. We look along this line in the direction of the arrow, and see by how much the portion of B to the right exceeds in area the portion of B to the left, i.e., we calculate $R_B - L_B$ (with an obvious notation). The number that we arrive at will be a function of θ, which we can denote by $F(\theta)$. Now because of the uniqueness of the bisector of A in a given direction the bisecting lines corresponding to $\theta = 0$ and $\theta = \pi$ will be the same line. Only the sense ascribed to the line will be different. Consequently $F(\pi) = -F(0)$, for the portion of B that was on the right for $\theta = 0$ is on the left for $\theta = \pi$. We can imagine that the line starts out parallel to the x-axis, and gradually increases its inclination to this axis, always adjusting its position so that it continues to bisect the area A. When it has rotated through a half-turn it will be back in its original position (because of the uniqueness of the bisector) but its sense will have been reversed. Figure 8.7 will help to make this clear.

We can easily satisfy ourselves that $F(\theta)$ is a continuous function of θ. Since $F(\pi) = -F(0)$ then one of the numbers $F(\pi)$, $F(0)$ is positive and the other negative.† Whichever way round it is, Theorem 8.31 enables us to deduce that for some value of θ between 0 and π the function $F(\theta)$ assumes the

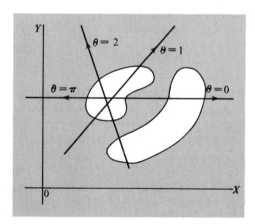

Fig. 8.7

* Since A is in one piece this enclosed area cannot be zero unless both lines are outside, and on the same side of the area A. But in this case neither line can be a candidate for the bisector of A.

† Unless both are zero, in which case there is nothing more to prove.

intermediate value 0. The bisector of A which corresponds to this value of θ will therefore also bisect the area B. Thus the existence of a line bisecting both areas has been demonstrated.

It is of interest to remark that although we have shown that such a line exists, the method by which we did so does not give us any method of actually constructing this line in a given example. The above result is a pure *existence theorem*, asserting nothing more than that something exists having such and such properties. Existence theorems are of quite common occurrence in higher mathematics for the very good reason that it is advisable (and often obligatory for a rigorous treatment) to make sure that something exists before you start to talk about it.

The result that we have just proved has an analogue in three dimensions, viz., that given any *three* chunks of space there exists a plane which bisects all three of these volumes. This is a well-known theorem, usually referred to as the *Ham Sandwich Theorem*. The idea behind the name is that given a somewhat surrealist sandwich consisting of two pieces of bread of arbitrary shape, and a piece of ham (also of arbitrary shape) situated anywhere in space, it is possible with a single straight cut of a knife to bisect all three pieces. We shall not give a proof of this result, but we invite the reader to try his hand at it, along the lines of the two-dimensional analogue proved above.

Exercise 89 (Chapter Review)

1. Find the limits of the following functions as $n \to \infty$ (n an integer).

    (a) $\dfrac{n^3 - 3n + 7}{3(n + 1)^3}$

    (b) $\dfrac{\sin n\pi}{n^{5/2}}$

    (c) $\dfrac{n^n}{n!}$ *Hint:* Consider the ratio of two successive values of the function.

    (d) $\dfrac{n^2 + [n]}{[n]^2}$

    (e) $\dfrac{\sqrt{4n^2 + 6n - 7}}{3n + 8}$

2. A sequence $\{u_n\}$ is defined by

    $$u_{n+1} = \sqrt{au_n + b}$$

    where a and b are positive constants, and u_1 is arbitrary. What can be said about $\lim\limits_{n \to \infty} u_n$?

3. If $f(n) < g(n) < h(n)$ for all values of n, prove that $\lim\limits_{n \to \infty} f(n) \le \lim\limits_{n \to \infty} g(n) \le \lim\limits_{n \to \infty} h(n)$. Give an example to show that in this result "$\le$" cannot be replaced by "$<$."

    Deduce that if $\lim\limits_{n \to \infty} f(n) = \lim\limits_{n \to \infty} h(n) = A$, then $\lim\limits_{n \to \infty} g(n) = A$.

4. The Fibonacci sequence is defined as follows.

$$u_0 = u_1 = 1$$
$$u_{n+2} = u_{n+1} + u_n \quad (n \geq 0).$$

Prove that $\lim\limits_{n \to \infty} \dfrac{u_{n+1}}{u_n} = \dfrac{1}{2}(\sqrt{5} + 1)$.

5. Prove that if the sequence $\{a_n\}$ tends to a finite limit A, and if

$$b_n = \frac{1}{n}(a_1 + a_2 + \cdots + a_n)$$

then $\lim\limits_{n \to \infty} b_n = A$. By reference to the sequence $\{(-1)^n\}$, or otherwise, show that $\lim\limits_{n \to \infty} b_n$ may exist even though $\lim\limits_{n \to \infty} a_n$ does not.

6. Determine whether the following series are convergent or divergent.

(a) $\sum \dfrac{59^n}{(2n)!}$

(c) $\sum \dfrac{59^{n^2}}{(n!)^{73}}$

(b) $\sum \dfrac{59^{n^2}}{(n!)^n}$

(d) $\sum \dfrac{n!}{7^{5n}}$

7. Determine for what values of x the following series are convergent.

(a) $\sum \dfrac{n!}{n^n} x^n$

(c) $\sum \dfrac{n^n}{n!} x^n$

(b) $\sum \dfrac{x^n}{n^3}$

(d) $\sum \dfrac{x^{n^2}}{n!}$

8. Why is the function $f(x) = \dfrac{x^3 - 2x^2 - 4x + 3}{x^2 - x - 6}$ not continuous at $x = 3$?
Define a function $g(x)$ such that
(a) $g(x) = f(x)$ if $x \neq 3$,
(b) $g(x)$ is continuous at $x = 3$.

9. At what points are the following functions not continuous?

(a) $\dfrac{x^2 - 11x + 6}{x^2 - 3x - 10}$

(c) $\dfrac{\sin x}{(1 + \tan x)^2}$

(b) $(x - [x])^2$

(d) $\dfrac{1 + x}{3 - 4 \sin^2 x}$

In (b) and (d) assume that $\sin x$ is continuous, and that $\tan x$ is continuous except when x is an odd multiple of $\pi/2$. The proof of these statements requires a result to be found in the next chapter (see page 342).

10. Prove that any polynomial $P(x)$ is a continuous function. Hence show that if a and b are any two numbers such that

$$P(a) < 0 < P(b)$$

then, between a and b, there is in general, an odd number of values ξ of x such that $P(\xi) = 0$.

By determining the sign of the polynomial

$$P(x) = 15x^4 - 166x^3 + 585x^2 - 722x + 240$$

where $x = 0, 1, 2, 3, 4, 5$, and 6, or by drawing a graph, or both, determine *roughly* (i.e., to within 1/4 or so) the locations of the roots of the equation $P(x) = 0$.

11. If $f(x)$ and $g(x)$ are two continuous functions such that
  (a) $f(0) = 23, \quad g(0) = 5,$
  (b) $\lim\limits_{x \to \infty} f(x) = 10, \quad \lim\limits_{x \to \infty} g(x) = 15,$
prove that there is some positive value ξ of x such that $f(\xi) = g(\xi)$.

12. C is a closed curve in the plane. P is a point inside C such that any line through P meets C in exactly two points. Prove that there is a chord of C which is bisected at P.

Suggestions for Further Reading

The term "Analysis" is used in mathematics in a sense which includes not only the subject matter of this chapter but also that of the next two chapters (and quite a bit more besides). Hence, books listed in the "Suggestions for Further Reading" for Chapters 9 and 10 (Calculus) will also be useful in the study of this chapter. Below are some suggestions for books that are not primarily concerned with calculus as an end in itself.

One old book (the first edition was published in 1908) which is still extremely good (it is written by a master of the art, and has been revised several times) is

Hardy, G.H., *A Course in Pure Mathematics*. 10th ed. New York: Cambridge University Press, 1960.

Also recommended are:

Apostol, T., *Mathematical Analysis*. Reading, Massachusetts: Addison-Wesley Co., Inc., 1957.

Burkhill, J.C., *A First Course in Mathematical Analysis*. New York: Cambridge University Press, 1962.

Phillips, E.G., *A Course of Analysis*. New York: Cambridge University Press, 1948.

Ribenboim, P., *Functions, Limits and Continuity*. New York: John Wiley & Sons, Inc., 1964.

These all go well beyond the reader's present needs, but there is no compulsion to read them from cover to cover, and it is quite easy to pick out what is relevant to the present course. The same comment applies to the following standard work on infinite series:

Bromwich, T. J. I'A., *Infinite Series*. New York: The Macmillan Company, 1955.

Rather more manageable is the following pocket-book on the same subject:

Hyslop, J.M., *Infinite Series*. Edinburgh: Oliver and Boyd Ltd., 1960.

A delightful, and eminently readable introduction to higher mathematics—a book the reader should not fail to look at—that covers a great deal of the material of this chapter among many other topics, is

Courant, R., and Robbins, H., *What is Mathematics?* New York: Oxford University Press, 1941.

9 Differential

Calculus

The differential calculus has its origins in the attempt to solve a straightforward, though not altogether simple, problem; namely to construct the tangent to a given curve at a given point. Let us see exactly what this involves. To draw a tangent to a circle is easy—we have merely to draw a line through the required point perpendicular to the radius at that point, as shown in Fig. 9.1, where the tangent at P is the line through P which is perpendicular to OP. When we come to consider what is meant by the tangent to a more complicated curve at a given point, however, it turns out that a number of difficulties arise.

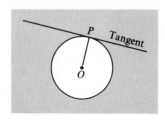

Fig. 9.1

Let P be a point on a curve C, which is, shall we say, the graph of a continuous function $f(x)$. (See Fig. 9.2.) If Q is some other point on the curve, there is no difficulty about drawing the chord PQ, and the slope of the chord is easily found by the methods of Chapter 5. Knowing the slope of PQ we can write down the equation of PQ, or anything else that we wish to say about it.

If we take several points Q_1, Q_2, ... getting closer and closer to P (as in Fig. 9.2) then the successive lines PQ_1, PQ_2, ... will approximate more

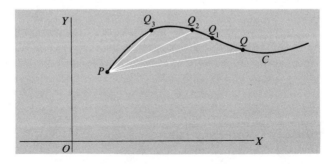

Fig. 9.2

and more to our intuitive idea of what we mean by the "tangent" at P to the curve. It thus appears that we must expect the notion of the "tangent" to a curve at a point to be linked with the idea of *tending to a limit*—the sort of thing that we were largely concerned with in the last chapter. We therefore look to see in what way precisely these ideas are connected.

Let P be a point on the graph of a continuous function $f(x)$, and let x_0 be its x-coordinate. Let Q be the point whose x-coordinate is $x_0 + h$, where h may be positive or negative. (In Fig. 9.3, h is taken to be positive.)

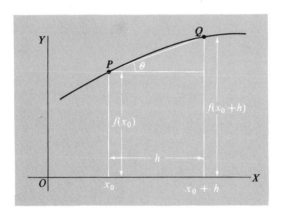

Fig. 9.3

The slope of the chord PQ is then

$$\tan \theta = \frac{f(x_0 + h) - f(x_0)}{h}$$

as is easily seen. We are therefore interested in what happens as Q gets closer and closer to P, i.e., as h tends to 0. Now if the slope $\dfrac{f(x_0 + h) - f(x_0)}{h}$ tends to a limit as $h \to 0$ we shall call the limit the *slope of the tangent at P*, and then define the tangent at P to be the line through P which has this particular slope.

In fact we shall not be greatly concerned with tangents to curves in this book but rather with the property which the above geometrical example has served to introduce—the property that the limit

$$\lim_{h \to 0} \frac{f(x_0 + h) - f(x_0)}{h}$$

exists. This property is an extremely important one, for which special terms are used. We now define these terms.

Definition 9.1

If $\dfrac{f(x_0 + h) - f(x_0)}{h}$ tends to a finite limit as $h \to 0$ this limit is called the derivative of $f(x)$ at $x = x_0$, and the function $f(x)$ is said to be differentiable at $x = x_0$.

Definition 9.2

If $f(x)$ is differentiable at every value of x, then $f(x)$ is said to be a differentiable function, or simply, to be differentiable.

Definition 9.3

If $f(x)$ is differentiable we can form from it a new function whose value at $x = x_0$ is defined as the derivative of $f(x)$ at $x = x_0$. [Since the function $f(x)$ is differentiable, the new function is defined at every value of x.] This new function is called the derived function of $f(x)$, or simply the derivative of $f(x)$. It is denoted by $f'(x)$.

Distinguish carefully between the *derivative* of a function (Definition 9.3), and the derivative of a function *at a point* (Definition 9.1); also between *differentiable* and *differentiable at a point*. The following examples should help to make these distinctions clear.

EXAMPLES:
Consider the function $f(x) = x^2$.

We have
$$\frac{f(x_0 + h) - f(x_0)}{h} = \frac{(x_0 + h)^2 - x_0^2}{h}$$
$$= \frac{2x_0 h + h^2}{h}$$
$$= 2x_0 + h.$$

Now this clearly tends to $2x_0$ as $h \to 0$, so that, for example, the derivative of x^2 at $x = 1$ is 2; the derivative of the function at $x = -3$ is -6, and so on. In general the derivative of this function at $x = x_0$ is the number $2x_0$. The function therefore possesses a derivative, i.e., is differentiable, for every value of x. It is thus a differentiable function, and its derived function, or derivative, is the function $2x$. In this case, therefore, we can write $f'(x) = 2x$.

The function $f(x) = 5$, i.e., the function which takes the value 5 for every value of x, is differentiable. In fact its derivative at every point is zero, since $f(x_0 + h)$ and $f(x_0)$ are equal. Thus the derived function

$f'(x) = 0$ in this example. The same is true if any other constant is used in place of the constant 5.

The function $f(x) = |x|$ is not differentiable at the point $x = 0$. For if $h > 0$ then

$$\frac{f(h) - f(0)}{h} = \frac{|h| - 0}{h} = \frac{h}{h} = 1$$

whereas if $h < 0$ then

$$\frac{|h| - 0}{h} = \frac{-h}{h} = -1.$$

Thus the (two-sided) limit cannot exist, and the function is thus not differentiable at the point $x = 0$. It is easily verified that it is differentiable everywhere else.

THEOREM 9.1

IF A FUNCTION IS DIFFERENTIABLE AT $x = x_0$ THEN IT IS CONTINUOUS AT $x = x_0$.

PROOF:

Let the function $f(x)$ be differentiable at $x = x_0$, and let K be its derivative at $x = x_0$. Then $\dfrac{f(x_0 + h) - f(x_0)}{h} \longrightarrow K$ as $h \to 0$. If K is positive then, given any positive number ϵ we have

$$K - \epsilon < \frac{f(x_0 + h) - f(x_0)}{h} < K + \epsilon$$

for h sufficiently small. In particular, with $\epsilon = K$ we have

$$0 < \frac{f(x_0 + h) - f(x_0)}{h} < 2K$$

from which it follows that

(9.1.1) $|f(x_0 + h) - f(x_0)| < 2 |K| \cdot |h|.$

Inequality (9.1.1) is true also when $K < 0$, though some of the signs in the previous stages need to be changed when demonstrating this. If $K = 0$, then $|f(x_0 + h) - f(x_0)| < |h| \epsilon$. Thus whatever the value of K, we can make $|f(x_0 + h) - f(x_0)|$ as small as we like by taking h sufficiently small. This means that the function $f(x)$ is continuous at $x = x_0$, since $\lim\limits_{h \to 0} f(x_0 + h) = f(x_0)$.

The converse of this theorem is not true; a function which is continuous at a point may well fail to be differentiable there. This is seen to be so from the example of the function $|x|$, which we have already considered. This function is continuous at $x = 0$, but, as we have seen, it is not differentiable there.

Exercise 90

1. Show that
   (a) the function $f(x) = x$ is differentiable, and its derivative is 1.
   (b) the function $f(x) = x^3$ is differentiable and its derivative is $3x^2$.
2. Prove that the function $f(x) = x - [x]$ (where $[x]$ denotes the integral part of x, as on page 66) is differentiable except at the points where x is an integer. (*Hint:* Draw a graph of this function first.)
3. Determine whether the function $f(x) = x^2 \sin(1/x)$ is differentiable or not at $x = 0$, when $f(0)$ is defined as 0.

9.2 DERIVATIVES OF STANDARD FUNCTIONS

The functions that we shall now consider will all be differentiable. The symbol x_0 can therefore stand for *any* value of x, and for convenience we shall drop the subscript zero and write simply x.

THE FUNCTION x^n, WHERE n IS AN INTEGER

Here we have

$$\frac{f(x + h) - f(x)}{h} = \frac{(x + h)^n - x^n}{h}$$

$$= \frac{(x + h)^n - x^n}{(x + h) - x}.$$

From the formula for the sum of a geometric progression we readily verify that the sum

$$a^{n-1} + a^{n-2}b + a^{n-3}b^2 + \dots + a^2b^{n-3} + ab^{n-2} + b^{n-1}$$

where a and b are any two numbers is $\dfrac{a^n - b^n}{a - b}$.* If we put $a = x + h$ and $b = x$ we have

$$\frac{(x + h)^n - x^n}{(x + h) - x} = (x + h)^{n-1} + (x + h)^{n-2}x + \dots + (x + h)x^{n-2} + x^{n-1}.$$

* See Chapter 4, Exercise 31.4.

Now on the right-hand side of this equation we have n terms, each of which tends to x^{n-1} as $h \to 0$. Thus as $h \to 0$ the limit of the right-hand side is nx^{n-1}. These statements are consequences of Theorems 8.22 and 8.25. Thus $\lim_{h \to 0} \dfrac{(x + h)^n - x^n}{h} = nx^{n-1}$, so that the derivative of x^n is nx^{n-1}, i.e., if $f(x) = x^n$ then $f'(x) = nx^{n-1}$.

TRIGONOMETRIC FUNCTIONS

If $f(x) = \sin x$ then

$$f'(x) = \lim_{h \to 0} \frac{\sin (x + h) - \sin x}{h}.$$

Now there is a well-known result to the effect that

$$\sin A - \sin B = 2 \cos \frac{A + B}{2} \sin \frac{A - B}{2}$$

where A and B are any numbers.* Writing $A = x + h$ and $B = x$ we have

$$\sin (x + h) - \sin x = 2 \cos (x + \tfrac{1}{2}h) \sin \tfrac{1}{2}h \cdot$$

Let us put $\theta = \tfrac{1}{2}h$ so that $\theta \to 0$ as $h \to 0$. We may then write

(9.2.1) $$f'(x) = \lim_{\theta \to 0} \cos (x + \theta) \frac{\sin \theta}{\theta}.$$

Here we have two functions: $\cos (x + \theta)$, whose limit as $\theta \to 0$ is $\cos x$, and $\dfrac{\sin \theta}{\theta}$ whose limit as $\theta \to 0$ remains to be found.

To find this limit we refer to Fig. 9.4 which shows an angle of θ radians

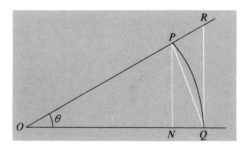

Fig. 9.4

* See Chapter 3, Exercise 23.3.

subtended by the arc PQ of a circle of radius r. N is the foot of the perpendicular from P to OQ. The area of the segment OPQ is $r^2\theta/2$ since it is $\theta/2\pi$ of the area of the whole circle (since the circumference of the whole circle subtends an angle of 2π, and the area of the whole circle is πr^2). The area of the triangle OPQ is $\frac{1}{2} OQ \cdot PN = \frac{1}{2} r \cdot r \sin \theta = \frac{1}{2} r^2 \sin \theta$. The area of the triangle OQR, where QR is perpendicular to OQ, is $\frac{1}{2} OQ \cdot QR = \frac{1}{2} r \cdot r \tan \theta$ (since $QR/OQ = \tan \theta$). From the way that these areas include, or are included in, each other we see that

$$\tfrac{1}{2}r^2 \sin \theta < \tfrac{1}{2}r^2\theta < \tfrac{1}{2}r^2 \tan \theta.$$

Dividing by $\frac{1}{2} r^2 \sin \theta$ we obtain

(9.2.2) $$1 < \frac{\theta}{\sin \theta} < \frac{\tan \theta}{\sin \theta} = \frac{1}{\cos \theta}.$$

Now as $\theta \to 0$ so $\cos \theta \to 1$ and therefore $1/\cos \theta \to 1$. Consequently, $\theta/\sin \theta$, which is sandwiched between 1 and $1/\cos \theta$, must also tend to 1. It follows that its reciprocal, $\sin \theta/\theta$, also tends to 1 as $\theta \to 0$. This was the limit that we wanted, and we can carry on from equation (9.2.1). This now reduces to

$$f'(x) = \cos x \cdot 1 = \cos x$$

by the result for the limit of a product of two functions (Theorem 8.25).
   Thus the derivative of $\sin x$ is $\cos x$.
   In much the same way we can find the derivative of $\cos x$. The main steps are as follows:

$$\lim_{h \to 0} \frac{\cos (x + h) - \cos x}{h} = \lim_{h \to 0} \frac{2 \sin (x + \frac{1}{2}h) \sin (-\frac{1}{2}h)}{h}$$

$$= \lim_{\theta \to 0} - \sin (x + \theta) \frac{\sin \theta}{\theta}$$

$$= -\sin x.$$

Thus the derivative of $\cos x$ is $- \sin x$.

Exercise 91

1. Prove that if $f(x) = \sin ax$ then $f'(x) = a \cos ax$.
2. Prove that if $f(x) = \cos (ax + b)$ then $f'(x) = -a \sin (ax + b)$.

3. By multiplying numerator and denominator of the expression

$$\frac{\sqrt{x+h} - \sqrt{x}}{h}$$

by $\sqrt{x+h} + \sqrt{x}$, prove that if $f(x) = \sqrt{x}$, then $f'(x) = \dfrac{1}{2\sqrt{x}}$.

9.3 RULES FOR DIFFERENTIATION

The process of finding the derivative of a function is known as *differentiation;* we "differentiate" the function, and the result of this operation is to obtain the derived function, or derivative. In order to be able to differentiate a large variety of functions we now prove certain rules whereby the differentiation of complicated functions can be reduced to the differentiation of simpler functions. The first two rules are easy to prove, and we leave them as exercises for the reader.

RULE 1.

Differentiation of the sum of two functions.

If $f(x)$ and $g(x)$ are differentiable then so is the function $f(x) + g(x)$, and its derivative is $f'(x) + g'(x)$. In other words the derivative of a sum is the sum of the derivatives. This rule applies equally well to the sum of three or more functions.

RULE 2:

Differentiation of a constant multiple.

If $f(x)$ is differentiable, then so is $k f(x)$, where k is a constant, and its derivative is $k f'(x)$.

These two rules enable us to differentiate any polynomial. The derivative of each term in the polynomial

$$a_0 x^n + a_1 x^{n-1} + \ldots + a_{n-1} x + a_n$$

can be found by rule 2 (since we know the derivative of x^n), and the derivative of the whole polynomial is then their sum, by rule 1.

Exercise 92

1. Find the derivatives of the following functions:
   (a) $3x^2 + 2x - 7$

(b) $5x^3 + 2x^2 - 9x - 4$

(c) $\frac{3}{4}x^5 - \frac{7}{2}x^3 - 5x$

(d) $10x^9 - 17x^5 + 9x^3 - 8x$

(e) $(x^2 + 3x + 2)(x^2 - x + 1)$ (expand before differentiating)

(f) $(x^3 - 2x + 5)(x^2 - \frac{1}{2}x + 1)$

(g) $2 \sin x + 9 \cos x$

(h) $2x^2 + 7 \sin x - \frac{1}{2}\cos x$

(i) $3(x^3 + 5x - 1) + 12 \cos x$

(j) $7x^2 - \sin x + 3 \cos x$

9.4 ALTERNATIVE NOTATIONS

It will help us in our statements and proofs of the remaining rules if we first look at an alternative notation for derivatives. This notation arose from the historical development of the subject, and the simplest way of introducing it is to recapitulate briefly this development. We take another look at the problem of finding the slope of a tangent to a curve.

The slope of the chord PQ in Fig. 9.3 was $\dfrac{f(x+h) - f(x)}{h}$. The denominator h is the difference of the x cordinates of P and Q, and this can be thought of as an *increment* in the value of x. It became customary to denote this increment in x by the symbol δx.* Note that we say "symbol" and not "symbols," for δx is to be regarded as defined in its entirety as a single symbol; it is not the product of δ and x. The denominator $f(x + h) - f(x)$ is the difference of the y-coordinates of P and Q, and this we can think of as the change in y consequent on the change δx in x. We can call it the increase or the *increment* in y (on the understanding that it may be negative, and hence a decrease rather than an increase) consequent on the increment δx in x, and denote it by the symbol δy. Using this notation we write the expression $\dfrac{f(x + h) - f(x)}{h}$ as $\dfrac{\delta y}{\delta x}$. The derivative of $y = f(x)$ will then be defined by $\lim\limits_{\delta x \to 0} \dfrac{\delta y}{\delta x}$. All that we have done so far in this chapter can be done just as well with this alternative notation, and, in many textbooks the derivative of x^2, for example, is found in some such manner as the following.

The function is

(9.4.1) $$y = x^2.$$

* Frequently a capital delta "Δ" is used, so that the increment is written as Δx.

If x increases by δx the corresponding increase in y is δy, where

(9.4.2)
$$y + \delta y = (x + \delta x)^2$$
$$= x^2 + 2x\delta x + (\delta x)^2.$$

Subtracting (9.4.1) from (9.4.2) we have

$$\delta y = 2x\delta x + (\delta x)^2.$$

Dividing both sides by δx (which is assumed to be nonzero) we obtain

(9.4.3)
$$\frac{\delta y}{\delta x} = 2x + \delta x.$$

The required derivative is then $\lim\limits_{\delta x \to 0} (2x + \delta x)$, which is clearly $2x$—a result that we have already discovered.

There is nothing new in all this; we have merely found the derivative by what is essentially the same process, but using a different notation. This notation has certain advantages, as we shall see, but one thing so far is lacking; we need a symbol for the derivative itself, that is to say, a symbol to correspond with the $f'(x)$ of our previous notation, which will serve for the derivative of an unspecified function. Historically what was required was something to denote that the limit of $\frac{\delta y}{\delta x}$ as $\delta x \to 0$ had been taken. The device that was used was to alter δ to d, so that what had been $\frac{\delta y}{\delta x}$ became $\frac{dy}{dx}$ after the limit had been taken. This is a convenient device, but it must be noticed that because of the intervention of the limiting process we cannot assume that $\frac{dy}{dx}$ is the quotient of two quantities dy and dx, in the same way that $\frac{\delta y}{\delta x}$ is the quotient of the two quantities δy and δx. For the moment then we shall regard $\frac{dy}{dx}$ as being a single composite symbol which is defined in its entirety by $\frac{dy}{dx} = \lim\limits_{\delta x \to 0} \frac{\delta y}{\delta x}$. Thus if $y = f(x)$ then $\frac{dy}{dx}$ means $f'(x)$. We note in passing that $\frac{dy}{dx}$ is read as "$d\,y$ by $d\,x$."

This notation has its advantages. For one thing it contains the symbol (y) which serves to represent the function. If we like, we can replace this symbol by the expression for a particular function. Thus if we wish to express the derivative of $\sin x$ we can say that it is $\frac{dy}{dx}$ where $y = \sin x$. Alternatively we may condense this to $\dfrac{d(\sin x)}{dx}$, meaning the same thing. Similarly we may denote the derivative of $15x^4$ by $\dfrac{d(15x^4)}{dx}$. This is not possible with the $f'(x)$ notation; certainly not with a function like $15x^4$. One could, presumably, write $\sin'x$ for the derivative of $\sin x$ but this is never, in fact, done.

If the function in question is given by some complicated expression which would not go well in the "numerator" of $\dfrac{dy}{dx}$ it can be placed on the line following the rest of the $\dfrac{dy}{dx}$ symbol. Thus the derivative of $\dfrac{x^2+1}{x^2-1}\sin x$ can be written as

$$\frac{d}{dx}\left\{\frac{x^2+1}{x^2-1}\sin x\right\},$$

meaning the same as "$\dfrac{dy}{dx}$, where $y = \dfrac{x^2+1}{x^2-1}\sin x$."

This is all a matter of notation; no new ideas have been introduced in the last few paragraphs.

In many old textbooks (and regrettably in a few that are not so old) will be found a treatment of the above method of finding derivatives which is not only nonrigorous but also misleading. In these books the reader, having arrived at equation (9.4.3) is told to put $\delta x = 0$. This yields the result $2x$, which is, in fact, the right answer. But it is not possible to assert that this is the right answer, since there is now no way of relating the result to what has gone before. What went before depended essentially on the datum that δx was not zero, and without this assumption we could not have divided by δx and obtained (9.4.3). We cannot suddenly do an about-face and put $\delta x = 0$; if we do we are no longer talking about the same problem.

In the early days, soon after its invention, the calculus came under heavy fire from many critics because of various inadequacies in the treatment then given to it, and the philosopher Bishop Berkeley among others attacked it for its dependence on the fallacious argument mentioned in the last paragraph. These controversies were settled once the notion of a limit was put on a firm footing, but it is instructive to see why it is that the fallacious argument should give the right answer.

The original expression for $\dfrac{\delta y}{\delta x}$ was $\dfrac{(x+\delta y)^2-x^2}{\delta x}$, and in this expression it is not possible to put $\delta x = 0$, for the numerator and denominator will then both vanish. In fact, this expression, which is really a function of *two* variables, x and δx, is not defined for $\delta x = 0$. Nevertheless we can still talk about the limit of this expression, for fixed x, as δx tends to zero, since the definition of the limit does not require that we know the value of the expression *at* $\delta x = 0$, or even that the expression be defined for this value of δx. (Compare the example in Chapter 8, section 11.)

When we divide by δx to obtain $2x + \delta x$, we thereby obtain a function which agrees with the previous function for all nonzero values of δx (we are treating x as fixed throughout). Consequently its limit as $\delta x \to 0$ will be the same as the limit of $\dfrac{(x + \delta x)^2 - x^2}{\delta x}$ as $\delta x \to 0$. However, $2x + \delta x$ is strictly speaking a different function since, unlike $\dfrac{(x + \delta x)^2 - x^2}{\delta x}$, it *is* defined at $\delta x = 0$. What is more, it is continous at $\delta x = 0$ so that its limit as $\delta x \to 0$ is the same as its value at $\delta x = 0$. Thus we see that the fallacious process of putting $\delta x = 0$ gives us the right answer simply because the function $2x + \delta x$ is continuous at $\delta x = 0$, so that the limit as $\delta x \to 0$ and the value at $\delta x = 0$ coincide.

This serves to emphasize how important it was that the definition of $\lim\limits_{x \to x_0} f(x)$ (Definition 8.5) should not require that $f(x)$ be defined at $x = x_0$.

What we have said about the special case $f(x) = x^2$ applies equally well to other functions. The process of dividing top and bottom by δx (so as to get a function that is defined at $\delta x = 0$) and then putting $\delta x = 0$, though strictly speaking fallacious, can be used as a sort of short cut to obtaining the derivative. There is little harm in this provided the reader knows what he is doing.

Exercise 93

1. Use the δx notation to find the derivative of x^n.
2. Use the δx notation to obtain the derivative of $\sin 2x$.

9.5 FURTHER RULES FOR DIFFERENTIATION

Now that we have introduced this "d" notation, we shall use it to obtain some further rules for differentiation, rather more simply than would be possible using the other notation.

RULE 3:

Differentiation of a product.

Suppose that the function to be differentiated is the product of two functions, $u(x)$ and $v(x)$ (or simply u and v for brevity) whose derivatives are known. We consider the changes in the functions y, u and v when x

changes from a given value of x to $x + \delta x$, and we denote these changes by δy, δu, δv respectively. Thus $\delta u = u(x + \delta x) - u(x)$, and δv and δy are defined similarly. We then have

$$y + \delta y = (u + \delta u)(v + \delta v)$$
$$= uv + v\delta u + u\delta v + \delta u\delta v.$$

Therefore, since $y = uv$, we have

$$\delta y = v\delta u + u\delta v + \delta u\delta v.$$

Dividing by δx we get

(9.5.1) $$\frac{\delta y}{\delta x} = v\frac{\delta u}{\delta x} + u\frac{\delta v}{\delta x} + \delta u\frac{\delta v}{\delta x}.$$

Thus $\frac{\delta y}{\delta x}$, whose limit as $\delta x \to 0$ we wish to find, is the sum of three terms, whose limits can be found. For the limit of $v\frac{\delta u}{\delta x}$ is $v\frac{du}{dx}$; the limit of $u\frac{\delta v}{\delta x}$ is $u\frac{dv}{dx}$; and the limit of $\delta u\frac{\delta v}{\delta x}$ is 0, since $\delta u \to 0$ while $\frac{\delta v}{\delta x}$ tends to a finite limit. Hence we obtain $\frac{dy}{dx} = \lim\limits_{\delta x \to 0} \frac{\delta y}{\delta x}$, which is given by

(9.5.2) $$\frac{dy}{dx} = v\frac{du}{dx} + u\frac{dv}{dx}.$$

In deriving (9.5.2) from (9.5.1) we have assumed that the functions u and v are differentiable. Thus (9.5.2) is the rule (Rule 3) for the derivative of a product $y = uv$, where u and v are differentiable functions.

RULE 3:

$$\frac{d(uv)}{dx} = \frac{du}{dx}v + u\frac{dv}{dx}.$$

It is often convenient to remember this rule in verbal form as "derivative of the first times the second, plus first times derivative of the second."

EXAMPLE 1:

Put $$u = x^2 + 1 \quad \text{and} \quad v = x^3 + 5x^2 - 3.$$

Then $$\frac{d}{dx}(uv) = 2x(x^3 + 5x^2 - 3) + (x^2 + 1)(3x^2 + 10x)$$

$$= 5x^4 + 20x^3 + 3x^2 + 4x.$$

We could, of course, have differentiated this function without using Rule 3, by first getting rid of the brackets. If we multiply $x^2 + 1$ by $x^3 + 5x^2 - 3$ we get $x^5 + 5x^4 + x^3 + 2x^2 - 3$, of which the derivative is $5x^4 + 20x^3 + 3x^2 + 4x$, as before.

EXAMPLE 2:

Applying Rule 3 we find that the derivative of $x^2 \sin x$ is $2x \sin x + x^2 \cos x$. Here the use of Rule 3 is almost essential.

Exercise 94

1. Find the derivatives of the following functions:
   (a) $(x^3 + 7x - 5) \cos x$
   (b) $5 \cos x + 2x \sin x$
   (c) $(3x - 5) \sin x + (x^3 - 2) \cos x$
   (d) $\sin^2 x$

2. Prove Rule 3 using the $f'(x) = \lim\limits_{h \to 0} \dfrac{f(x + h) - f(x)}{h}$ definition for the derivative. (This should emphasize the advantages of using the "d" notation.)

3. Prove that if u, v, w are differentiable functions of x then

$$\frac{d}{dx}(u\,v\,w) = \frac{du}{dx}v\,w + u\frac{dv}{dx}w + u\,v\frac{dw}{dx}.$$

Generalize this result, and supply a proof for the general case.

4. Using the result given in Problem 3 above, differentiate the following functions:
   (a) $x^2 \sin x \cos x$
   (b) $(x^2 + 1)(3x^3 - 5x^2 + 7x - 2) \sin x$
   (c) $(9x^{17} - 25x^4 + 137) \sin^2 x$
   (d) $\sin^n x$.

9.6 DERIVATIVE OF THE QUOTIENT OF TWO FUNCTIONS

We now consider a function of the form $y = \dfrac{u(x)}{v(x)}$, where u and v are differentiable functions. As before we denote by δy, δu, δv the increments in y, u and v respectively consequent on a change δx in x. Then

$$y + \delta y = \frac{u + \delta u}{v + \delta v}.$$

Hence, since $y = u/v$, we have

$$\delta y = \frac{u + \delta u}{v + \delta v} - \frac{u}{v}$$

$$= \frac{v(u + \delta u) - u(v + \delta v)}{v(v + \delta v)}$$

$$= \frac{v\delta u - u\delta v}{v(v + \delta v)}.$$

Thus $\dfrac{\delta v}{\delta x} = \dfrac{v\dfrac{\delta u}{\delta x} - u\dfrac{\delta v}{\delta x}}{v(v + \delta v)}.$

Now this is the quotient of two functions whose limits as $\delta x \to 0$ can be easily found. The limit of the numerator is $v\dfrac{du}{dx} - u\dfrac{dv}{dx}$ while the limit of the denominator is just v^2. Hence

(9.6.1) $$\lim_{\delta x \to 0} \frac{\delta y}{\delta x} = \frac{v\dfrac{du}{dx} - u\dfrac{dv}{dx}}{v^2}$$

by Theorem 8.26.

There is a slight complication here in that we have to leave out of account those values of x, if any, at which the function v takes the value 0. For these values of x the expression on the right-hand side of (9.6.1) is not defined; but then neither is the function u/v itself, so that these values of x do not come into the picture at all. With this proviso we may state Rule 4 for the derivative of a quotient as follows:

RULE 4:

$$\frac{d}{dx}\left(\frac{u}{v}\right) = \frac{v\dfrac{du}{dx} - u\dfrac{dv}{dx}}{v^2}.$$

There is little difficulty in remembering this result. We have a v^2 in the denominator, and the numerator can be remembered as "Bottom times the derivative of the top, minus top times the derivative of the bottom."

EXAMPLE 1:

From Rule 4 we see that the derivative of the function $\dfrac{x^2 + 1}{x^3 + 2x - 7}$ is

$$\frac{(x^3 + 2x - 7)2x - (x^2 + 1)(3x^2 + 2)}{(x^3 + 2x - 7)^2}$$

$$= \frac{-x^4 - x^2 - 14x - 2}{(x^3 + 2x - 7)^2}.$$

EXAMPLE 2:

Rule 4 enables us to find the derivative of x^k when k is a negative integer. If we write $k = -n$, where n is a positive integer, then the derivative of $x^k = \dfrac{1}{x^n}$ will be

$$\frac{x^n \cdot 0 - 1 \cdot n x^{n-1}}{x^{2n}} = \frac{-n x^{n-1}}{x^{2n}}$$

$$= -n\, x^{-n-1}$$

$$= k\, x^{k-1}.$$

Thus the result previously obtained for the derivative of x^k for k a positive integer holds equally well when k is a negative integer. We note in particular that the derivative of $1/x$ is $-1/x^2$.

EXAMPLE 3:

We now find the derivative of $\tan x$. If $y = \tan x = \dfrac{\sin x}{\cos x}$, then, by Rule 4, dy/dx is given by

$$\frac{\cos x \cos x - \sin x(-\sin x)}{\cos^2 x}$$

$$= \frac{\cos^2 x + \sin^2 x}{\cos^2 x}$$

$$= \frac{1}{\cos^2 x} \qquad (\text{since } \cos^2 x + \sin^2 x = 1)$$

$$= \sec^2 x.$$

Thus the derivative of $\tan x$ is $\sec^2 x$.

Exercise 95

1. Show that the derivative of $\sec x$ is $\sec x \tan x$.
2. Find the derivatives of $\operatorname{cosec} x$ and $\cot x$.
3. Find the derivatives of the following functions:

   (a) $\dfrac{\sin x}{x^3 + 1}$

   (b) $\dfrac{1 + \cos x}{1 - \cos x}$

(c) $\dfrac{x^9 - 3x^6 + 4x^2}{(x + 1)(3x^3 - 5x + 7)}$

(d) $\dfrac{(x^2 + 1) \cos x}{x^3 + \sin x}$

9.7 FUNCTION OF A FUNCTION

Suppose that we have two functional relationships given by $y = f(u)$ and $u = g(x)$. Then given any value of x there corresponds a unique value $g(x)$ of u. To this value of u there corresponds a unique value $f(u)$ of y. Thus a given value of x determines a unique value of y, so that we have a functional relationship between x and y; y is a function of x. We may write $y = f[g(x)]$. A function defined in this way can be called a *function of a function*. As examples we may cite the functions

$$\sin (x + 5) \qquad (y = \sin u, \quad u = x + 5)$$
$$\cos^2 x \qquad (y = u^2, \quad u = \cos x)$$
$$5 \tan x + \tan^3 x \quad (y = 5u + u^3, \quad u = \tan x).$$

If we know the derivatives of the functions f and g then we can find the derivative of $f[g(x)]$ by means of Rule 5, which we shall quote first and prove afterwards.

RULE 5:
Derivative of a function of a function.
    If f and g are differentiable functions, and $y = f[g(x)]$, then

$$\frac{dy}{dx} = \frac{dy}{du} \cdot \frac{du}{dx}.$$

A straightforward "proof" of this result goes as follows: Let δu be the change in u consequent on a change δx in x; and let δy be the change in y consequent on the change δu in u. Then

(9.7.1) $\qquad \dfrac{\delta y}{\delta x} = \dfrac{\delta y}{\delta u} \cdot \dfrac{\delta u}{\delta x} \qquad$ for all values of δx.

Now the right-hand side of (9.7.1) is the product of two expressions; one is $\dfrac{\delta y}{\delta u}$ whose limit as $\delta x \to 0$ is $\dfrac{dy}{du}$ (since if $\delta x \to 0$ then $\delta u \to 0$), and the other is $\dfrac{\delta u}{\delta x}$ which tends to $\dfrac{du}{dx}$ as $\delta x \longrightarrow 0$. By Theorem 8.25 the limit of the product is the product of the limits, and we have

$$\frac{dy}{dx} = \lim_{\delta u \to 0} \frac{\delta y}{\delta u} \cdot \lim_{\delta x \to 0} \frac{\delta u}{\delta x}$$

$$= \frac{dy}{du} \cdot \frac{du}{dx}$$

—the required result.

(The reader may like to see if he can discover for himself the fallacy in this proof before reading further.)

The proof just given is, unfortunately, not quite good enough. The snag is that the change δu in u consequent on the change δx in x may be zero for some values of δx, and if this is so then the quotient $\dfrac{\delta y}{\delta u}$ is not defined. We must therefore tinker with the proof a little in order that it may hold water.

We first note that there is hardly any difficulty if the value of $\dfrac{du}{dx}$ is nonzero at the value of x that we are considering. Suppose that at a typical value of x, say $x = a$, we have $g'(a) = K$, where K is nonzero. Then there will be a number α such that

$$(9.7.2) \qquad \left| \frac{\delta u}{\delta x} - K \right| < \tfrac{1}{3}K \qquad \text{for } 0 < |\delta x| < \alpha.$$

This follows from the definition of $\lim\limits_{\delta x \to 0} \dfrac{\delta u}{\delta x} = \dfrac{du}{dx} [= g'(a)]$, i.e., Definition 9.5, with $K/3$ in place of ϵ. From (9.7.2) it follows that for $0 < |\delta x| < \alpha$ the quotient $\dfrac{\delta u}{\delta x}$ is not zero, and therefore that δu cannot be zero. Thus the simple argument that we gave above will serve when $g'(a) \neq 0$. What this amounts to is that by considering sufficiently small values of δx we can "get inside" all the awkward values of δx for which the corresponding δu is zero.

If $g'(a) = 0$ then there remains the possibility that, however small a range $0 < |\delta x| < \alpha$ we take, we cannot exclude all the values of δx for which $\delta u = 0$. Fortunately there is another way out of the difficulty. The derivative of $f[g(x)]$ that we are concerned with is the limit as $\delta x \longrightarrow 0$ of

$$\frac{\delta y}{\delta x} = \frac{f(u + \delta u) - f(u)}{\delta x},$$

and for those values of δx (if any) which make $\delta u = 0$ this expression is certainly zero, since $f(u + \delta u) = f(u)$. (Note that δx is never zero.) For those values of δx for which δu is not zero it is legitimate to write

(9.7.3) $$\frac{\delta y}{\delta x} = \frac{\delta y}{\delta u} \cdot \frac{\delta u}{\delta x}.$$

Then since $\lim\limits_{\delta u \to 0} \dfrac{\delta y}{\delta u}$ exists, the first factor on the right-hand side of (9.7.3) is bounded in absolute value for sufficiently small δu; and the second factor $\dfrac{\delta u}{\delta x}$ can be made as small as we like by taking δx sufficiently small (since $\lim\limits_{\delta x \to 0} \dfrac{\delta u}{\delta x} = \dfrac{du}{dx} = 0$).

Hence $\dfrac{\delta y}{\delta x}$ can be made as small as we like for sufficiently small values of δx which do not make δu vanish. But if δu *does* vanish then $\dfrac{\delta y}{\delta x}$ is not only small, it is actually zero. Thus the awkward values of δx do not, in fact, present any difficulty. We deduce that $\dfrac{\delta y}{\delta x} \to 0$ as $\delta x \to 0$, so that when $g'(a) = 0$ we still have

$$\frac{dy}{dx} = \frac{dy}{du} \cdot \frac{du}{dx}$$

as before.

This completes the proof of Rule 5. The rule is easily remembered since, in the form in which we have given it, it "looks as if" the dus in $\dfrac{dy}{du}$ and $\dfrac{du}{dx}$ cancel each other.

Examples. Let us find the derivatives of the functions that we gave earlier.

(1) $\sin(x + 5)$.   Here $y = \sin u$, so that $\dfrac{dy}{du} = \cos u$

and $u = x + 5$, so that $\dfrac{du}{dx} = 1$.

Thus $\dfrac{dy}{dx} = \cos u \cdot 1$, or $\dfrac{d}{dx} \sin(x + 5) = \cos(x + 5)$, since $u = x + 5$.

(2) $\cos^2 x$.   Here $y = u^2$, so that $\dfrac{dy}{du} = 2u$;

and $u = \cos x$, so that $\dfrac{du}{dx} = -\sin x$.

Thus $\dfrac{dy}{dx} = -2u \sin x$ or $\dfrac{d}{dx} \cos^2 x = -2 \cos x \sin x$.

(3) $5 \tan x + \tan^3 x$.   Here $y = 5u + u^3$, and $u = \tan x$.

Hence $\dfrac{dy}{du} = 5 + 3u^2$, and $\dfrac{du}{dx} = \sec^2 x$.

Thus $\dfrac{d}{dx}(5 \tan x + \tan^3 x) = (5 + 3u^2) \sec^2 x$

$$= (5 + 3 \tan^2 x) \sec^2 x.$$

Once some practice has been gained in differentiating functions of functions it is no longer necessary to introduce the intermediate variable u; or rather, this introduction of u can be done entirely mentally, and need not be written down. An example will show how this can be done. Suppose we wish to differentiate the function $y = \tan(x^2 + 1)$; we reason to ourselves as we go, something like this: "This function is the tangent of something. The derivative of "tan" is "sec²," so we must have sec² of this something." And we write down $\sec^2(x^2 + 1)$, since the "something" (the "u" of the previous method) is $x^2 + 1$. We continue, "Now we must differentiate the something." The derivative of $x^2 + 1$ is $2x$. We then write down $2x$ to multiply the expression previously recorded, and we have the answer.

The important thing about this procedure is that in the first stage the function $u = g(x)$, which may possibly be some complicated expression in x, is treated as a single variable, and the first differentiation is performed accordingly.* For the next stage we go "inside the brackets" [which we can imagine as enclosing the function $g(x)$, even if none are actually there], and differentiate again.

It may happen that the function $g(x)$ inside the brackets is itself a function of a function. If so we merely carry the process a stage further. For example; to differentiate the function $\sin[\tan(5x^2 + 2x)]$ we first differentiate it as sine of something, and therefore write down the cosine of that something, i.e., we write down $\cos[\tan(5x^2 + 2x)]$. We then go inside the brackets to differentiate $\tan(5x^2 + 2x)$, and to do this we differentiate it first as the tangent of something, writing down therefore the "sec²" of the something, i.e., $\sec^2(5x^2 + 2x)$. Finally we go inside the brackets again to differentiate $5x^2 + 2x$. The product of the three results, viz.,

$$\cos[\tan(5x^2 + 2x)] \sec^2(5x^2 + 2x) (10x + 2),$$

is then the required derivative.

Exercise 96

1. If y is a function of u, which is a function of v, which is a function of w, which is a function of x, show that

---

* The suggestion to call the expression "something" was intended to emphasize this treatment of the expression as if it were a single symbol.

$$\frac{dy}{dx} = \frac{dy}{du} \cdot \frac{du}{dv} \cdot \frac{dv}{dw} \cdot \frac{dw}{dx}$$

2. Without writing down any intermediate steps, find the derivatives of the following functions:

(a) $(2x^2 - 5)^3$. (Check your answer by expanding this expression and differentiating the resulting polynomial.)

(b) $\tan(x^5 + 4x^3 - x - 7)$

(c) $\sin(1 + \cos x)$

(d) $(\sin x + \cos x)^{-2}$

(e) $(\cos x + x \sin x)^{-4}$

(f) $\sin^2(2x + 5)$

(g) $\sin(\sin 2x)$

(h) $\tan(x^2 + \sin^2 x)$

9.8 INVERSE FUNCTIONS

At the beginning of Chapter 3 we saw how the functional relation $y = f(x)$ was another way of writing a statement of the form $x \, R \, y$. Now if $x \, R \, y$ is a true statement then $y \, R' \, x$ is a true statement (see page 49); but this converse relation R' may not be a functional relation, and a simple example will show why. Suppose $x \, R \, y$ is the functional relation $y = x^2$; then we have, for example, $2 \, R \, 4$, since the relation is satisfied by $x = 2$ and $y = 4$. We also have $(-2) \, R \, 4$, since $x = -2$ and $y = 4$ also satisfy the relation. Looking at the converse relation we see that we have both $4 \, R' \, 2$ and $4 \, R' \, (-2)$, which shows that R' is not a functional relation, since the essential thing about such relations was that for any element in the domain there should be at most one element in the range having the given relation to it. Here 4 has the relation R' to both 2 and -2.

Thus the converse of a functional relation is not necessarily a functional relation. If it *is*, we say that the function defined by it is the *inverse* of the original function. For example, if $y = 2x + 3$ then $x = \frac{1}{2}(y - 3)$. Denoting the first of these functions by $x \, R \, y$ we will denote the second by $y \, R' \, x$, and in the example both R and R' are functional relations. We may therefore say that the functions $2t + 3$ and $\frac{1}{2}(t - 3)$ are inverse functions.

(Notice that in order to avoid confusion we have written "t" for the independent variable in each case. This is one of the places where a notation for functions which does not require the inclusion of the independent variable, would be useful. See the discussion on page 64).

Now this definition of an inverse function is too restrictive. If we used it exclusively, then very few everyday functions would have inverses, not even (as we have seen) the simple function x^2. In order to make the concept of an inverse function more useful, we must soften the definition of "function" a little. The definition that we gave in Chapter 3 was, strictly speaking, the definition of a *single-valued function*, but since, up to now, we have not had occasion to consider any other kind of function we have not insisted on the use of this longer, but more accurate, terminology. We must now consider what are called *many-valued functions*.

A many-valued function is defined in much the same way as a single-valued function, except we now allow the possibility that there may be more than one value y of the function corresponding to a given value of x. This would bring us back to the original general definition of a relation except that many-valued functions are required to satisfy a condition, rather difficult to put into words, which we may vaguely describe by saying that it must be possible to sort out, in some way, the various values corresponding to a given value of x. This requirement, which we shall not attempt to describe precisely, will be illustrated by the examples which follow shortly below.

There are many advantages to be gained from extending the concept of a function in the manner just mentioned, and these make it worthwhile to do so. Nevertheless, many-valued functions are often difficult to handle, and for that reason mathematicians will frequently try to restore some measure of "single-valuedness" to these functions, by stipulating, in one way or another, that, of the several values of y corresponding to a given x, they are going to choose a particular one and forget about the others. It is in the possibility of doing this that the distinction between many-valued functions and relations in general really resides. A few examples should make this point clear.

We would be tempted to say that the functions t^2 and $\sqrt{t}$ were inverse functions, in the sense that if $x = y^2$ then $y = \sqrt{x}$. But $\sqrt{x}$ is not a function according to our original definition for the reasons just given—there are *two* values of $\sqrt{x}$ corresponding to $x = 4$, for example, 2 and -2. In $\sqrt{x}$ we have an example of a many-valued function (in this case a *two*-valued function). We quite often emphasize the fact that $\sqrt{x}$ is a two-valued function by writing it as $\pm \sqrt{x}$. If we now make a convention that we will always take the *positive* square root of x, we arrive at a function (which we could denote by $+\sqrt{x}$ though the $+$ is usually omitted) satisfying our original definition of the word function. This is to say that by restricting ourselves to one only of the two possible values of $\sqrt{x}$, we have arrived at a single-valued function, $+\sqrt{x}$. In the same way, if we decided to choose always the negative root of x we would define a single-valued function, denoted by $-\sqrt{x}$. Either of the functions $+\sqrt{t}$ and $-\sqrt{t}$ is an inverse of the function t^2 for values of t for which they are defined. They are not, at present, defined for negative values of the argument t.

As a second example let us take the function *sine*. Consider the functional relationship $x = \sin y$. It does not, as it stands, define y as a function of x, since there are many values of y corresponding to any given value between -1 and $+1$ of x. In fact, if y_0 is any value of y satisfying $x_0 = \sin y_0$, where x_0 is any given value of x, then all the numbers of the form $y_0 + 2k\pi$, where k is an integer, will also correspond to this value of x. Thus $x = \sin y$ defines y as a multivalued function of x for which there is an infinite number of values of y corresponding to a value of x, and not just two as in the last example. The graph of this multivalued function will look like Fig. 9.5.

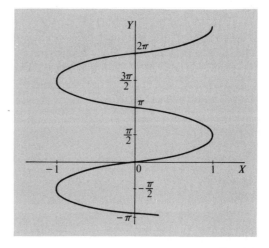

Fig. 9.5

Looking at this figure, however, we see that of all the many values of y corresponding to a given value of x there is one and only one which lies in the range $-\frac{1}{2}\pi \leq y \leq \frac{1}{2}\pi$. If we therefore make a convention that we will always take the value of y which lies within this range, then we shall have obtained a single-valued function defined over $-1 \leq x \leq 1$. This single-valued function is denoted by $\sin^{-1} x$. Thus $\sin^{-1} x$ is the number between $-\frac{1}{2}\pi$ and $\frac{1}{2}\pi$ whose sine is x. The notation $\sin^{-1} x$ may seem a little strange to the reader who has not met it before; there is some reasonableness in the objection that since $\sin^2 x$ means the square of $\sin x$, $\sin^{-1} x$ ought to mean the reciprocal of $\sin x$. It is perhaps for this reason that some books use the notation $\arcsin x$ where we have used $\sin^{-1} x$. On the other hand, the reciprocal of $\sin x$ already has a symbol, viz., $\operatorname{cosec} x$, and this fact means that we can, if we wish, use $\sin^{-1} x$ to denote something else, without causing any ambiguity.* Since the notation "$\arcsin x$" seems to be going out of fashion, we shall use $\sin^{-1} x$ for the inverse of the sine function, and a similar notation for the

---

* Note that expressions like $\sin^{-2} x$, $\sin^{-1/2} x$, etc., which *would* be ambiguous, are never used.

inverses of the other trigonometrical functions, and of other functions too. The graph of the single-valued function $\sin^{-1}x$ is given in Fig. 9.6.

The inverse of the cosine function will be a many-valued function. We can convert it into a single-valued function however by stipulating that the value of y satisfying $\cos y = x$ for a given value of x is to be that value which lies between 0 and π. (Note that if we took y between $-\frac{1}{2}\pi$ and $\frac{1}{2}\pi$ as above, we would still not have arrived at a single-valued function.) The single-valued function thus obtained will be denoted by $\cos^{-1}x$. Its graph is given in Fig. 9.7.

Perhaps the most important of these inverse trigonometrical functions is the inverse of the function *tangent*. It is denoted by $y = \tan^{-1}x$, and the value of y satisfying $x = \tan y$ is chosen to lie between $-\frac{1}{2}\pi$ and $\frac{1}{2}\pi$ as with the inverse of sine. The graph of $\tan^{-1}x$ is given in Fig. 9.8. Note that this function is defined for *all* values of x, and not just in the range $-1 \leq x \leq 1$, as were $\sin^{-1}x$ and $\cos^{-1}x$.

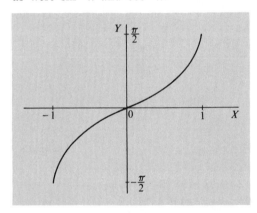

Fig. 9.6

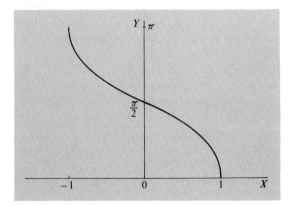

Fig. 9.7

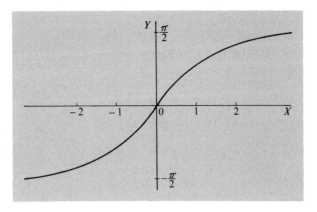

Fig. 9.8

The inverse functions $\cot^{-1}$, $\sec^{-1}$ and $\operatorname{cosec}^{-1}$ are defined in a similar way, but are less used.

We shall now derive the rule for finding the derivative of an inverse function when we know the derivative of the function of which it is the inverse. We shall assume that the inverse function has been rendered single-valued by a convention such as those just given.

Suppose that $\phi(t)$ is the inverse of the function $f(t)$. This means that if $y = \phi(x)$ then $x = f(y)$, which means, in turn, that $x = f[\phi(x)]$. Thus the function-of-a-function $f[\phi(x)]$ simply reduces to the function x, and its derivative must therefore be just 1. Using the rule for the derivative of a function of a function, we see that

$$f'[\phi(x)] \cdot \phi'(x) = 1, \quad \text{or} \quad f'(y) \cdot \phi'(x) = 1.$$

writing this last equation in the "d"-notation we have

$$\frac{dx}{dy} \cdot \frac{dy}{dx} = 1.$$

Hence the derivatives of the inverse functions turn out to be reciprocals of each other. Note however that dx/dy will be a function of y, while dy/dx will be a function of x; but y and x will be connected by the relation $y = \phi(x)$ [or $x = f(y)$]. Thus we have

RULE 6:

Derivative of an inverse function.

If $y = \phi(x)$ defines the function ϕ as the inverse of the function f, where $x = f(y)$, then

$$\phi'(x) = \frac{dy}{dx} = 1 \Big/ \frac{dx}{dy} = \frac{1}{f'(y)}.$$

Substituting $\phi(x)$ for y then gives $\phi'(x)$ as a function of x.

EXAMPLE 1:

If $x = y^q$, where q is an integer, then the derivative of x with respect to y is known. It is $\frac{dx}{dy} = qy^{q-1}$.

The inverse function is given by $y = x^{1/q}$, —the qth root of x. This function is two-valued if q is an even integer, in which case we make it single-valued by taking the positive root only. We then have

$$\frac{dy}{dx} = 1 \Big/ \frac{dx}{dy} = \frac{1}{qy^{q-1}} = \frac{1}{q} y^{-q+1}.$$

Since $y = x^{1/q}$ we may write

(9.8.1)
$$\frac{dy}{dx} = \frac{1}{q}(x^{1/q})^{-q+1}$$
$$= \frac{1}{q} x^{\frac{1}{q}-1}.$$

This result enables us to find the derivative of any rational power of x. For if $y = x^{p/q}$, where p and q are integers, then y can be written as a function of a function, viz., $y = (x^{1/q})^p$, and Rule 5 then gives

$$\frac{dy}{dx} = p(x^{1/q})^{p-1} \cdot \frac{d}{dx}(x^{1/q})$$
$$= p(x^{1/q})^{p-1} \cdot \frac{1}{q} x^{\frac{1}{q}-1} \quad \text{(by 9.8.1)}$$
$$= \frac{p}{q} x^{\frac{p-1}{q}+\frac{1}{q}-1}$$
$$= \frac{p}{q} x^{\frac{p}{q}-1}.$$

Since, by previous results, the above reasoning holds just as well for p negative as for p positive, we have the following general result: if k is a rational number, positive or negative, then the derivative of x^k is kx^{k-1}.

EXAMPLE 2:

Inverse trigonometric functions.

If $x = \sin y$, then $\frac{dx}{dy} = \cos y$. Hence $\frac{dy}{dx} = \frac{1}{\cos y}.$

But $\cos y = \sqrt{1 - \sin^2 y} = \sqrt{1 - x^2}$. Hence if $y = \sin^{-1} x$ we have $\dfrac{dy}{dx} = \dfrac{1}{\sqrt{1 - x^2}}$, or $(1 - x^2)^{-\frac{1}{2}}$.

Similarly if $x = \cos y$, then $\dfrac{dx}{dy} = -\sin y = -\sqrt{1 - x^2}$. Hence the derivative of $\cos^{-1} x$ is $-(1 - x^2)^{-\frac{1}{2}}$.

If $x = \tan y$, then $\dfrac{dx}{dy} = \sec^2 y = 1 + \tan^2 y = 1 + x^2$. From this it follows that the derivative of $\tan^{-1} x$ is $\dfrac{1}{1 + x^2}$.

Exercise 97

1. From the results just given it follows that the derivative of $\sin^{-1} x + \cos^{-1} x$ is zero. Why would you expect the derivative of this function to vanish?

2. Find the derivatives of $\sec^{-1} x$, $\operatorname{cosec}^{-1} x$ and $\cot^{-1} x$.

3. If the function $\phi(x)$ is defined by $y = \phi(x)$ where $x = y \sin y$, show that the derivative of $\phi(x)$ can be written in the following forms:

$$\frac{1}{y\cos y + \sin y}, \qquad \frac{1}{y\left(1 - \dfrac{x^2}{y^2}\right)^{1/2} + \dfrac{x}{y}}, \qquad \frac{y}{y\sqrt{y^2 - x^2} + x}.$$

Note: It is not possible, here, to get rid of y in the answer, thus expressing the derivative in terms of standard functions only. We could, of course, replace y by $\phi(x)$. It is the exception, rather than the rule, that the derivative of an inverse function can be written in a form which does not involve the inverse function itself.

9.9 HIGHER DERIVATIVES

We have seen how, starting from a function $f(x)$ we can associate with it, under certain conditions, another function—its derivative $f'(x)$. Not all functions have derivatives, as we have seen, but most of those that one meets in practice do have derivatives, and the rules that we have given so far enable us to find what these derivatives are for a very wide class of functions.

It is frequently the case that the function $f'(x)$ itself has a derivative—the derivative of the derivative of the original $f(x)$. This is called the second derivative of $f(x)$ and is denoted by $f''(x)$. The derivative of $f''(x)$, if it has one, is called the third derivative of $f(x)$ and is denoted by $f'''(x)$ The 4th, 5th, ..., nth derivatives are similarly defined. To avoid too many dashes the 4th, 5th, etc., derivatives are usually denoted by $f^{iv}(x)$, $f^v(x)$, etc., using Roman numerals, while a general nth derivative is denoted by $f^{(n)}(x)$.

These "higher" derivatives can also be written using an extension of the "d" notation. If $y = f(x)$ then $f'(x)$ is written as $\dfrac{d}{dx} f(x)$ or $\dfrac{dy}{dx}$. The second derivative is $\dfrac{d}{dx}\left[\dfrac{d}{dx} f(x)\right]$, and this is contracted to $\dfrac{d^2}{dx^2} f(x)$ or $\dfrac{d^2 y}{dx^2}$. (Read as "dee-2-y-by-dee-x-squared"). The third derivative is written as $\dfrac{d^3}{dx^3} f(x)$, or $\dfrac{d^3 y}{dx^3}$, and so on. The nth derivative will be denoted by $\dfrac{d^n y}{dx^n}$.

Once the technique of finding the derivative of a function has been acquired, the finding of higher derivatives presents no new difficulties. It is simply a question of performing the operation of taking the derivative several times instead of just once.

EXAMPLE 1:

Find the successive derivatives of the function $\tan^{-1} x$. We know that the first derivative of this inverse function is $\dfrac{1}{1 + x^2}$ or $(1 + x^2)^{-1}$. Differentiating this as a function of a function we obtain the second derivative of $\tan^{-1} x$, viz.,

$$\frac{d^2}{dx^2} \tan^{-1} x = -2x \cdot (1 + x^2)^{-2}.$$

To find the third derivative we must differentiate again. We have a produce of two functions, the second of which is a function of a function. Using Rules 3 and 5 we find that

$$\frac{d^3}{dx^3} \tan^{-1} x = -2 \cdot (1 + x^2)^{-2} + 2x \cdot 2(1 + x^2)^{-3} \cdot 2x$$

$$= (6x^2 - 2)(1 + x^2)^{-3}.$$

In the same way the fourth and higher derivatives can be found.

EXAMPLE 2:

Find the successive derivatives of the function

$$7x^5 + 23x^4 - 17x^3 + 2x^2 - 118x + 237.$$

Denoting the function by y we have

$$\frac{dy}{dx} = 35x^4 + 92x^3 - 51x^2 + 4x - 118,$$

$$\frac{d^2y}{dx^2} = 140x^3 + 276x^2 - 102x + 4,$$

$$\frac{d^3y}{dx^3} = 420^2 + 552x - 102,$$

$$\frac{d^4y}{dx^4} = 840x + 552,$$

$$\frac{d^5y}{dx^5} = 840.$$

The sixth, and all higher derivatives will be zero. In the last section (page 363), we proved that

$$\frac{dy}{dx} \cdot \frac{dx}{dy} = 1.$$

The student should beware of assuming that a similar result holds for higher derivatives. Thus it is *not* in general true that

$$\frac{d^2y}{dx^2} \cdot \frac{d^2x}{dy^2} = 1.$$

For the relationship between $\frac{d^2y}{dx^2}$ and $\frac{d^2x}{dy^2}$ see Exercise 98.3.

Exercise 98

1. Prove that the $(n + 1)$th derivative of a polynomial of degree n is zero.

2. Prove by mathematical induction that the nth derivative of $\tan^{-1} x$ (for $n \geq 1$) is of the form

$$T_n(x) \cdot (1 + x^2)^{-n},$$

where $T(x)$ is a polynomial of degree $n - 1$. Find also the relation between $T_{n+1}(x)$ and $T_n(x)$.

3. Using the "function of a function" rule in the form

$$\frac{dF}{dy} = \frac{dF}{dx}\frac{dx}{dy}$$

show that

$$\frac{d^2x}{dy^2} = -\left(\frac{dy}{dx}\right)^{-3}\frac{d^2y}{dx^2}.$$

We have already seen that the derivative of a product of two functions u and v is given by $\frac{d}{dx}(u\,v) = \frac{du}{dx}v + u\frac{dv}{dx}$. The second derivative of this product will be

$$\frac{d^2}{dx^2}(u\,v) = \frac{d}{dx}\left(\frac{du}{dx}v\right) + \frac{d}{dx}\left(u\frac{dv}{dx}\right)$$

(9.10.1)
$$= \frac{d^2u}{dx^2}v + \frac{du}{dx}\frac{dv}{dx} + \frac{du}{dx}\frac{dv}{dx} + u\frac{d^2v}{dx^2}$$

$$= \frac{d^2u}{dx^2}v + 2\frac{du}{dx}\frac{dv}{dx} + u\frac{d^2v}{dx^2}.$$

If we work out the third derivative of the product we find that

(9.10.2) $$\frac{d^3}{dx^3}(u\,v) = \frac{d^3u}{dx^3}v + 3\frac{d^2u}{dx^2}\frac{dv}{dx} + 3\frac{du}{dx}\frac{d^2v}{dx^2} + u\frac{d^3v}{dx^3}.$$

Now the coefficients 1, 2, 1 and 1, 3, 3, 1 on the right-hand sides of (9.10.1) and (9.10.2) look familiar; they are the binomial coefficients $\binom{2}{0}, \binom{2}{1}, \binom{2}{2},$ and $\binom{3}{0}, \binom{3}{1}, \binom{3}{2}, \binom{3}{3}$. This suggests that in the expression for the nth derivative of the product uv the binomial coefficients $\binom{n}{0}, \binom{n}{1}$, etc. would occur. We shall prove this and thereby prove

THEOREM 9.2

LEIBNIZ'S THEOREM.

$$\frac{d^n}{dx^n}(u\,v) = \frac{d^nu}{dx^n}v + \binom{n}{1}\frac{d^{n-1}u}{dx^{n-1}}\frac{dv}{dx} + \binom{n}{2}\frac{d^{n-2}u}{dx^{n-2}}\frac{d^2v}{dx^2} + \cdots$$

$$+ \binom{n}{r}\frac{d^{n-r}u}{dx^{n-r}}\frac{d^rv}{dx^r} + \cdots + u\frac{d^nv}{dx^n}.$$

PROOF:

To make the proof a little less unwieldy we shall write the single symbol "D" in place of the collection of symbols "$\frac{d}{dx}$." Thus we shall write Du, Dv for $\frac{du}{dx}$ and $\frac{dv}{dx}$; we shall also write D^2u, D^3u, ..., D^ru, etc., for $\frac{d^2u}{dx^2}$,

$\dfrac{d^3u}{dx^3}, \ldots \dfrac{d^r u}{dx^r}$, etc., and similarly with the derivatives of v. This is a convenient notation that we shall make much more use of at a later stage.

The proof is by mathematical induction.

We first observe that, by Rule 3, the theorem is true when $n = 1$. Now consider any value m of n for which the theorem is true, i.e., such that

(9.10.3) $$D^m(u\,v) = \sum_{r=0}^{m} \binom{m}{r} D^{m-r}u\,D^r v.$$

[The right-hand side of (9.10.3) is a convenient way of writing the right-hand side of the statement of the theorem (with $n = m$) provided we interpret $D^0 u$ and $D^0 v$ as standing for the functions u and v themselves.]

Differentiating both sides of (9.10.3) we find that

$$D^{m+1}(u\,v) = \sum_{r=0}^{m} \binom{m}{r}\{D^{m-r+1}u\,D^r v + D^{m-r}u\,D^{r+1}v\} \quad \text{(by Rule 3)}$$

$$= \binom{m}{0} D^{m+1}u\,D^0 v \qquad + \binom{m}{0} D^m u\,Dv$$

$$+ \binom{m}{1} D^m u\,Dv \qquad + \binom{m}{1} D^{m-1}u\,D^2 v$$

$$+ \binom{m}{2} D^{m-1}u\,D^2 v \qquad + \binom{m}{2} D^{m-2}u\,D^3 v$$

$$\cdots \qquad \cdots \qquad \cdots \qquad \cdots$$

$$+ \binom{m}{m-1} D^2 u\,D^{m-1}v \qquad + \binom{m}{m-1} Du\,D^m v$$

$$+ \binom{m}{m} Du\,D^m v \qquad + \binom{m}{m} D^0 u\,D^{m+1}v.$$

If we now take the terms together as indicated by the dotted lines, we can write this right-hand side as

(9.10.4) $\quad D^{m+1}u\,D^0 v + \displaystyle\sum_{r=1}^{m} \left\{\binom{m}{r} + \binom{m}{r-1}\right\} D^{m+1-r}u\,D^r v + D^0 u\,D^{m+1}v.$

But we have already proved (in Chapter 4) that $\binom{m}{r} + \binom{m}{r-1} = \binom{m+1}{r}$. Consequently we can simplify the summation in (9.10.4), and after including the two end terms in the same summation, we obtain

$$D^{m+1}(u\,v) = \sum_{r=0}^{m+1} \binom{m+1}{r} D^{m+1-r}u\,D^r v.$$

This last result states that the theorem is true if $n = m + 1$. Hence we have shown that if m is a value of n for which the theorem is true, then $m + 1$ is also a value for which it is true. Hence by mathematical induction, the truth of the theorem is established for all positive integral values of n.

The reader will observe that this proof is exactly similar to the proof of the Binomial Theorem given in Chapter 4, section 9.

Exercise 99

1. Find the nth derivatives of the following functions by means of Leibniz's theorem:

    (**a**) $x^3 \sin x$

    (**b**) $(x^4 - 16x + 3)\cos x$

    (**c**) $\sin^2 x$

2. Show that

$$\frac{d^n}{dx^n}\{x^m f(x)\} = \sum_{r=0}^{n} \binom{n}{r} \frac{m!\, x^{m-r}}{(m-r)!} f^{(n-r)}(x).$$

3. Prove that

$$\frac{d^n}{dx^n}(1 - 4x)^{-1/2} = \frac{(2n)!}{n!}(1 - 4x)^{-n-1/2}$$

and hence show that

$$\frac{d^n}{dx^n}\left\{\frac{y}{\sqrt{1 - 4x}}\right\} = \sum_{r=0}^{n} \binom{n}{r} \frac{(2r)!\,\dfrac{d^{n-r}y}{dx^{n-r}}}{r!(1 - 4x)^{n+1/2}}$$

9.11 AN IMPORTANT DERIVATIVE

In Chapter 8 we introduced, almost in passing, a function which we denoted by $\exp(x)$, and which was defined by the infinite series

(9.11.1) $\exp(x) = 1 + x + \dfrac{x^2}{2!} + \dfrac{x^3}{3!} + \cdots + \dfrac{x^r}{r!} + \cdots$

Since the series is always convergent, the function $\exp(x)$ is defined for all values of x. This function is one of the most important functions in the whole of mathematics. The reasons for its importance will unfold gradually as we go on, but one of them will become apparent when we seek to find its derivative, as we shall now do.

By the definition of the function we have

(9.11.2) $\dfrac{\exp(x+h) - \exp(x)}{h} = \dfrac{1}{h} \sum\limits_{r=1}^{\infty} \dfrac{(x+h)^r - x^r}{r!}$

the summation being from $r = 1$, since for $r = 0$ the term is zero. The general term in the right-hand side of (9.11.2) is

$$\dfrac{(x+h)^r - x^r}{h \cdot r!} = \dfrac{1}{h \cdot r!}\left\{ rhx^{r-1} + \binom{r}{2}h^2 x^{r-2} \right.$$
$$\left. + \binom{r}{3}h^3 x^{r-3} + \cdots + rh^{r-1}x + h^r \right\}$$
$$= \dfrac{x^{r-1}}{(r-1)!} + \dfrac{h}{r!}\left\{ \binom{r}{2}x^{r-2} + \binom{r}{3}hx^{r-3} + \cdots + rh^{r-3} + h^{r-2} \right\}$$

Therefore, since $\sum\limits_{r=1}^{\infty} \dfrac{x^{r-1}}{(r-1)!} = \exp(x)$, we have

$$\dfrac{\exp(x+h) - \exp(x)}{h} - \exp(x) = \sum\limits_{r=1}^{\infty} \dfrac{h}{r!}\left\{ \binom{r}{2}x^{r-2} \right.$$

(9.11.3)
$$\left. + \binom{r}{3}hx^{r-3} + \cdots + rh^{r-3}x + h^{r-2} \right\}.$$

Now if $|h| < 1$ then the part of (9.11.3) which is enclosed in braces is less than

$$\binom{r}{2}|x|^{r-2} + \binom{r}{3}|x|^{r-3} + \cdots + r|x| + 1$$

which, in turn is less than

$$(1 + |x|)^r.$$

Hence the whole of the right-hand side of (9.11.3) is numerically less than $\sum\limits_{r=1}^{\infty} \dfrac{h}{r!}(1 + |x|)^r$, which is less (by $|h|$) than $|h|\exp(1 + |x|)$.

Hence for any given value of x, we can assert that

(9.11.4) $$\left| \frac{\exp(x+h) - \exp(x)}{h} - \exp(x) \right| < M|h|$$

where $M = \exp(1 + |x|)$, and is a constant, and $|h| < 1$. It follows that the left-hand side of (9.11.4) tends to 0 as $h \to 0$, i.e., that

$$\lim_{h \to 0} \frac{\exp(x+h) - \exp(x)}{h} = \exp(x).$$

We have therefore succeeded in finding the derivative of $\exp(x)$; it is $\exp(x)$—the function is its own derivative.

The reader may have noticed that if we take the series (9.11.1) and differentiate each of its terms, the resulting series will be identical to the original. He may be tempted to ask whether this would not serve as a short proof that the derivative of $\exp(x)$ is $\exp(x)$. The answer is "no"; at least, not without justifying the assumptions that lie hidden in this argument. If a function of x is defined by a series whose terms depend on x, say $f(x) = \sum_{r=1}^{\infty} u_r(x)$, it does not follow that $f'(x) = \sum_{r=1}^{\infty} u_r'(x)$. For one thing, there is no guarantee that the series $\sum_{r=1}^{\infty} u_r'(x)$ will even be convergent. For a wide class of series [to which (9.11.1) happens to belong] this term by term differentiation *is* justifiable, but it would take us a good bit out of our way to prove the theorem which justifies it; and since this is the only occasion on which we might have used this theorem, it is simpler to find the derivative of $\exp(x)$ by the *ad hoc* argument given above.

Exercise 100

1. Show that if a is a constant

$$\frac{d}{dx} \exp(ax) = a \exp(x).$$

Use Leibniz's theorem to show that

$$\frac{d^n}{dx^n} \{\exp(ax) \cdot f(x)\} = \exp(ax) \sum_{r=0}^{n} \binom{n}{r} a^{n-r} f^{(r)}(x).$$

2. Show that the nth derivative of $\exp(\tfrac{1}{2}x^2)$ is of the form

$$H_n(x) \exp(\tfrac{1}{2}x^2)$$

where $H_n(x)$ is a polynomial. Find the relation between $H_{n+1}(x)$ and $H_n(x)$.

3. Prove by mathematical induction that

$$D^n(x^{n-1}e^{1/x}) = (-1)^n \frac{e^{1/x}}{x^{n+1}}.$$

Hint: Use Leibniz's theorem to prove that

$$D^{n+1}(xy) = (n+1)D^n y + x D^{n+1} y,$$

then apply this result.

9.12 ROLLE'S THEOREM, THE MEAN VALUE THEOREM AND TAYLOR'S THEOREM

Let us consider the graph of a function, and two points P and Q on it at which the values of the function are the same. Suppose, for example, that $f(a) = f(b)$, as in Fig. 9.9. Then it is clear intuitively that if the graph between

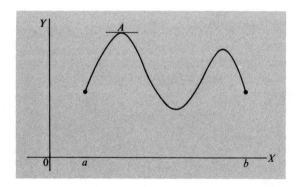

Fig. 9.9

P and Q is reasonably smooth there will be a point (A) at which the tangent to the graph is parallel to the x-axis, i.e., at which $f'(x) = 0$. This result, expressed more rigorously, is our next theorem.

THEOREM 9.3. ROLLE'S THEOREM

IF $f(x)$ IS A FUNCTION WHICH IS CONTINUOUS IN THE CLOSED INTERVAL $a \leq x \leq b$, AND DIFFERENTIABLE IN THE OPEN INTERVAL $a < x < b$, AND IF $f(a) = f(b)$, THEN THERE IS A NUMBER ξ BETWEEN a AND b, FOR WHICH $f'(\xi) = 0$.

PROOF:

Since $f(x)$ is continuous in the closed interval $[a, b]$, it is bounded in that interval (Theorem 8.28). Let M and m be its least upper bound and greatest lower bound over the interval.

If $M = m$ then the function must be constant over the interval, and the theorem is trivial. We may therefore suppose that $M > m$, and therefore that one of these numbers (at least) is different from $f(a)$. Let us suppose that $M > f(a)$.

Now $f(x)$ attains its least upper bound M for some value of x in $[a, b]$ (Theorem 8.29). Let ξ be a value of x for which $f(\xi) = M$. Then since $f(x) \leq f(\xi)$ for all values of x, it follows that $\dfrac{f(x) - f(\xi)}{x - \xi}$ is negative for $x > \xi$, and positive for $x < \xi$. Hence

$$\lim_{x \to \xi + 0} \frac{f(x) - f(\xi)}{x - \xi} \leq 0, \quad \text{and} \quad \lim_{x \to \xi - 0} \frac{f(x) - f(\xi)}{x - \xi} \geq 0.$$

But since $f(x)$ is differentiable at $x = \xi$, these one-sided limits must be equal, and this can happen only if each is zero. Hence $f'(\xi) = \lim_{x \to \xi} \dfrac{f(x) - f(\xi)}{x - \xi} = 0$, which is the theorem.

If $M = f(a)$ but $m \neq f(a)$, we consider a value ξ for which $f(\xi) = m$. The argument is almost the same as before.

Another important theorem, rather similar to Rolle's Theorem and readily deducible from it, is the following.

THEOREM 9.4.   THE MEAN VALUE THEOREM

IF $f(x)$ IS CONTINUOUS IN THE CLOSED INTERVAL $a \leq x \leq b$, AND DIFFERENTIABLE IN THE OPEN INTERVAL $a < x < b$, THEN THERE IS A VALUE ξ BETWEEN a AND b FOR WHICH $f'(\xi) = \dfrac{f(b) - f(a)}{b - a}$. [FIGURE 9.10 ILLUSTRATES THE GEOMETRICAL INTERPRETATION OF THIS THEOREM, i.e., THAT AT SOME POINT BETWEEN THE POINTS P AND Q ON THE GRAPH OF $f(x)$ THERE IS A POINT (A) AT WHICH THE TANGENT IS PARALLEL TO THE CHORD PQ].

PROOF:

Let us denote by $\phi(x)$ the function $f(x) - f(a) - k(x - a)$, where k is a constant. Whatever the value of k may be we have $\phi(a) = 0$, and we shall choose k so that we have also $\phi(b) = 0$. To do this we must make $k = \dfrac{f(b) - f(a)}{b - a}$, as is easily verified. We now apply Rolle's theorem to the function $\phi(x)$ [since the conditions on continuity and differentiability are satisfied by $\phi(x)$], and we deduce that there is a value such that

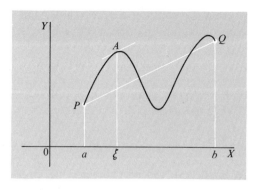

Fig. 9.10

$\phi'(\xi) = 0$. But from the definition of $\phi(x)$ we have $\phi'(x) = f'(x) - k$. Hence

$$f'(\xi) = k = \frac{f(b) - f(a)}{b - a}.$$

An important, though seemingly trivial, consequence of the mean value theorem is that a differentiable function whose derivative is zero (at all points) must be a constant. We already know that the derivative of a constant is zero; what we now have to prove is the converse of this result.

Suppose $f'(x) = 0$ for all values of x, but that $f(x)$ is not the same for all values of x. Then we can find two different values of x, say a and b, such that $f(a) \neq f(b)$. Since $f(x)$ is given to be differentiable everywhere the conditions for the mean value theorem are satisfied, and we deduce the existence of a value $x = \xi$ such that $f'(\xi) = \dfrac{f(b) - f(a)}{b - a}$, which means that $f'(\xi)$ is nonzero. This contradicts the datum that $f'(x)$ was zero for all x. Hence $f(x)$ must be a constant.

We shall make immediate use of this result to derive a further property of the exponential function. Consider the function

$$\exp(x) \cdot \exp(a - x).$$

We can find its derivative by Rule 2; it is

$$\exp(x) \cdot -\exp(a - x) + \exp(x) \cdot \exp(a - x) = 0$$

for all values of x. Thus the derivative vanishes everywhere and we deduce that

(9.12.1) $\exp(x) \cdot \exp(a - x) = K$

where K is a constant. So far so good; but what is this constant? We can find it by taking any particular value of x, say $x = 0$. This gives us

$$\exp(0) \cdot \exp(a) = K$$

or $K = \exp(a)$, since $\exp(0) = 1$. If we now write y for $a - x$, equation (9.12.1) becomes

(9.12.2) $$\exp(x) \cdot \exp(y) = \exp(x + y)$$

an important, indeed characteristic, property of the exponential function.

We shall have more to say about the exponential function shortly, but for the moment we must return to the line of development that started with Rolle's theorem and the Mean Value theorem.

By applying to a rather more complicated function the sort of treatment that we gave to the function $\phi(x)$ in the proof of the mean value theorem we can prove the following very important theorem.

THEOREM 9.5   TAYLOR'S THEOREM

LET $f(x)$ BE A FUNCTION WHICH IS DIFFERENTIABLE $n + 1$ TIMES IN THE OPEN INTERVAL $a < x < b$, AND WHOSE DERIVATIVES, UP TO THE nTH, ARE CONTINUOUS IN THE CLOSED INTERVAL $a \leq x \leq b$. THEN THERE IS A NUMBER ξ BETWEEN a AND b SUCH THAT

(9.12.3)
$$f(b) = f(a) + \frac{b-a}{1!}f'(a) + \frac{(b-a)^2}{2!}f''(a) + \cdots$$
$$+ \frac{(b-a)^n}{n!}f^{(n)}(a) + \frac{(b-a)^p(b-\xi)^{n-p+1}}{n!\,p}f^{(n+1)}(\xi)$$

WHERE p IS ANY GIVEN POSITIVE NUMBER.

PROOF:

Consider the expression

(9.12.4)
$$f(x) + \frac{b-x}{1!}f'(x) + \frac{(b-x)^2}{2!}f''(x) + \cdots$$
$$+ \frac{(b-x)^n}{n!}f^{(n)}(x) + k(b-x)^p.$$

When $x = b$ the expression (9.12.4) takes the value $f(b)$, whatever the value of k. We can choose k so that the value of (9.12.4) is also $f(b)$ when $x = a$, i.e., so that

$$f(a) + \frac{b-a}{1!}f'(a) + \frac{(b-a)^2}{2!}f''(a) + \cdots$$

(9.12.5)

$$+ \frac{(b-a)^n}{n!}f^{(n)}(a) + k(b-a)^p = f(b)$$

since the equation just written is a simple equation for k, and the coefficient of k is nonzero. Let us put this value of k in (9.12.4) and call the resulting expression $F(x)$.

From the conditions of the theorem it is seen that $F(x)$ is continuous in $[a, b]$ and differentiable in (a, b). Hence, applying Rolle's theorem to it, we deduce that there is a number ξ in the interval such that $F'(\xi) = 0$. But from the definition of $F(x)$ we have

$$F'(\xi) = f'(\xi) + \frac{b-\xi}{1!}f''(\xi) - f'(\xi)$$

$$+ \frac{(b-\xi)^2}{2!}f'''(\xi) - \frac{b-\xi}{1!}f''(\xi)$$

(9.12.6)

$$+ \cdots$$

$$+ \frac{(b-\xi)^n}{n!}f^{(n+1)}(\xi) - \frac{(b-\xi)^{n-1}}{(n-1)!}f^{(n)}(\xi)$$

$$- kp(b-\xi)^{p-1}$$

which reduces to

$$\frac{(b-\xi)^n}{n!}f^{(n+1)}(\xi) - k\,p(b-\xi)^{p-1}$$

since most of the terms in (9.12.6) cancel. Since this last expression is zero, we have

(9.12.7)
$$k = \frac{(b-\xi)^{n-p+1}}{n!\,p}f^{(n+1)}(\xi)$$

If we now substitute in (9.12.5) the value of k given by (9.12.7) we obtain (9.12.3)—the result to be proved.

If we denote the last term in (9.12.3) by R_{n+1}, and put $b - a = h$, we can write Taylor's theorem in a more compact form, as follows:

(9.12.8)
$$f(a + h) = f(a) + \frac{h}{1!}f'(a) + \frac{h^2}{2!}f''(a) + \cdots$$

$$+ \frac{h^n}{n!}f^{(n)}(a) + R_{n+1}.$$

Exercise 101

1. If $f(x)$ is differentiable in $[a, b]$ and $f'(x) > 0$ for $a < x < b$ prove that $f(x)$ is an increasing function in (a, b), i.e., that if $a < x_1 < x_2 < b$ then $f(x_1) < f(x_2)$.

2. Give examples to show that the condition that $f(x)$ be continuous in the closed interval $[a, b]$ in Rolles theorem and the Mean Value theorem cannot be omitted.

3. Give examples to show that the condition that $f(x)$ be differentiable in the open interval (a, b) cannot be omitted from Rolle's theorem or the Mean Value theorem.

4. What does Taylor's theorem become when

  (a) $f(x) = \exp(x)$, $a = 0$
  (b) $f(x) = x^k$, $a = 1$?

9.13 APPLICATIONS

If we can show, for a given function $f(x)$, that the *remainder term* R_{n+1} tends to 0 as $n \longrightarrow \infty$, then it will follow that the series

$$f(a) + \frac{h}{1!}f'(a) + \frac{h^2}{2!}f''(a) + \cdots = \sum_{r=0}^{\infty} \frac{h^r}{r!}f^{(r)}(a)$$

converges, and that its sum is $f(a + h)$. We shall have obtained an expansion for $f(a + h)$ (regarded as a function of h, a being constant) as a series of terms in ascending powers of h—a *power series* in h. Such power series expansions are of great importance and we shall now find the expansions of some common functions.

EXAMPLE 1:

The function $\sin x$.

Let us put $a = 0$ in (9.12.8), and take p (which is at our disposal) to be $n + 1$. The remainder term is then

$$R_{n+1} = \frac{h^{n+1}}{(n+1)!}f^{(n+1)}(\xi).$$

(This is known as Lagrange's form of the remainder.)

Now $f^{(n+1)}(x)$ is one of the functions $\sin x$, $\cos x$, $-\sin x$ or $-\cos x$. We can certainly assert, therefore, that $|f^{(n+1)}(\xi)| \leq 1$, whatever value ξ may turn out to have. Hence

$$|R_{n+1}| \leq \frac{|h|^{n+1}}{(n+1)!}.$$

It is an easy matter to show that this tends to zero as $n \longrightarrow \infty$, so we can assert that the infinite series

(9.13.1) $$f(0) + \frac{h}{1!}f'(0) + \frac{h^2}{2!}f''(0) + \cdots$$

sums to $f(h) = \sin h$. The values of $f^{(r)}(0)$ are readily found: $f(0) = \sin 0 = 0$; $f'(0) = \cos 0 = 1$; $f''(0) = -\sin 0 = 0$; $f'''(0) = -\cos 0 = -1$; $f^{\text{iv}}(0) = \sin 0 = 0$, and so on, repeating the four values 0, 1, 0, -1. Substituting these values in (9.13.1), and (for convenience) replacing h by x, we have the result

(9.13.2) $$\sin x = x - \frac{x^3}{3!} + \frac{x^5}{5!} - \frac{x^7}{7!} + \cdots = \sum_{r=0}^{\infty} (-1)^r \frac{x^{2r+1}}{(2r+1)!}$$

With an almost identical argument we can prove that

(9.13.3) $$\cos x = 1 - \frac{x^2}{2!} + \frac{x^4}{4!} - \frac{x^6}{6!} + \cdots = \sum_{r=0}^{\infty} (-1)^r \frac{x^{2r}}{(2r)!}$$

(The reader should prove this for himself.)

EXAMPLE 2:

The binomial series.

We now consider the function x^k, where k is a rational number. The continuity and differentiability conditions of Taylor's theorem are then satisfied for any interval which does not contain zero. We take $a = 1$ and $p = 1$. The remainder is then

$$R_{n+1} = \frac{(b-a)(b-\xi)^n}{n!} f^{(n+1)}(\xi).$$

(This is known as Cauchy's form of the remainder.) In the present example we have

$$f(x) = x^k; \quad f(1) = 1$$
$$f'(x) = kx^{k-1}; \quad f'(1) = k$$
$$f''(x) = k(k-1)x^{k-2}; \quad f''(1) = k(k-1)$$
$$f^{(n)}(x) = k(k-1)(k-2)\ldots(k-n+1)x^{k-n}$$
$$f^{(n)}(1) = k(k-1)(k-2)\ldots(k-n+1).$$

Moreover

$$R_{n+1} = \frac{h.(1+h-\xi)^n}{n!} k(k-1)(k-2)\ldots(k-n)\xi^{k-n-1}$$

$$= \frac{k(k-1)(k-2)\ldots(k-n)}{n!} \left(\frac{1+h-\xi}{\xi h}\right)^n h^{n+1}\xi^{k-1}$$

and we want to show that R_{n+1} tends to zero.

If h is positive then ξ (which lies between 1 and $1 + h$) is > 1. Hence

$$\xi < 1 + h < \xi(1 + h)$$

or

$$0 < 1 + h - \xi < \xi h.$$

Hence

(9.13.4) $$0 < \frac{1 + h - \xi}{\xi h} < 1.$$

If h is negative, but > -1, then $0 < 1 + h < \xi < 1$. Hence $\xi(1 + h) < 1 + h$ or $\xi h < 1 + h - \xi$. Dividing by ξh (and reversing the inequality sign, since ξh is negative) we obtain

(9.13.5) $$\frac{1 + h - \xi}{\xi h} < 1.$$

We can also write $\xi(1 - h) > \xi > 1 + h$

whence $-\xi h > 1 + h - \xi$

or

(9.13.6) $$-1 < \frac{1 + h - \xi}{\xi h}.$$

From (9.13.4), (9.13.5) and (9.13.6) we deduce that if $-1 < h < 1$,

then $\left| \dfrac{1 + h - \xi}{\xi h} \right| < 1$. Clearly the term $\left(\dfrac{1 + h - \xi}{\xi h} \right)^n$ in R_{n+1} is not

going to give any trouble. The factor ξ^{k-1} in R_{n+1} lies between 1 and $(1 + h)^{k-1}$ and hence is positive and bounded if $h > -1$. It will be sufficient therefore to prove that the remaining factor in R_{n+1}, viz.,

$$A_{n+1} = \frac{k(k - 1)(k - 2) \ldots (k - n)}{n!} h^{n+1}, \text{ tends to zero as } n \to \infty.$$

To do this we compare $|A_{n+1}|$ with $|A_n|$. We have

$$\left| \frac{A_{n+1}}{A_n} \right| = \left| \frac{k(k - 1)(k - 2) \ldots (k - n)h^{n+1}}{k(k - 1)(k - 2) \ldots (k - n + 1)h^n} \cdot \frac{(n - 1)!}{n!} \right|$$

$$= \left| \frac{k - n - 1}{n + 1} \right| |h|.$$

Now for large values of n, in fact for $n \geq k - 1$,

$$\left| \frac{k-n-1}{n+1} \right| = \frac{n+1-k}{n+1} < 1.$$

Therefore, if N is an integer greater than $k - 1$ we have

$$A_{N+1} < A_N |h|$$
$$A_{N+2} < A_{N+1}|h| < A_N |h|^2 \text{ etc.}$$

and, in general, $A_{N+r} < A_N |h|^r$. From this it follows that $A_n \to 0$ as $n \to \infty$, provided that $|h| < 1$.

This completes the proof that the remainder $R_{n+1} \to 0$ as $n \to \infty$. We can now assert that the series

$$1 + \frac{h}{1!}k + \frac{h^2}{2!}k(k-1) + \frac{h^3}{3!}k(k-1)(k-2) + \cdots$$

converges if $|h| < 1$, and that its sum is $(1+h)^k$. It is convenient to extend the meaning of the symbol $\binom{k}{r}$ to the case where k is not an integer, using it as a short form of $\frac{1}{r!}k(k-1)(k-2)\ldots(k-r+1)$, with the convention that $\binom{k}{0} = 1$. Writing x in place of h we arrive at the important result

(9.13.7)
$$(1+x)^k = 1 + \binom{k}{1}x + \binom{k}{2}x^2 + \binom{k}{3}x^3 + \cdots$$
$$= \sum_{r=0}^{\infty} \binom{k}{r}x^r,$$

provided $|x| < 1$. The series on the right of (9.13.7) is known as the binomial series.

If k is an integer, then $\binom{k}{r} = 0$ if $r > k$. Equation (9.13.7) then reduces to the binomial theorem almost as given in Chapter 3. Note however that the first term in the "binomial" is a "1" and must be a "1" if (9.13.7) is to be applied for nonintegral values of k.

Exercise 102

1. Use (9.13.7) to find the value of $(1.05)^{1/2}$ correct to 4 places of decimals. (Be sure to show that the sum of *all* the omitted terms of the series is not going to affect the fourth place.)

2. Find similarly the value of $(1.03)^{3/4}$ to 6 places of decimals.

3. Prove that $(1 - x)^{-k} = \sum_{r=0}^{\infty} \binom{k + r - 1}{r} x^r$, provided $|x| < 1$.

4. Use the Taylor series for $\sin x$ to calculate $\sin (0.1)$ correct to 4 places of decimals.

Calculate also $\cos (0.1)$ by means of the Taylor series for $\cos x$, and verify that $\cos^2 (0.1) + \sin^2 (0.1) = 1$.

9.14 MAXIMA AND MINIMA

So far in this chapter we have dealt with the mere mechanics of finding the derivatives of functions and with theorems having no very obvious useful application, either in economics or elsewhere. We now come to some results which are readily seen to have a practical significance. We shall see how to use our previous results to find the *maximum* and *minimum values* of a function.

We must first make quite sure that we know exactly what it is that we are trying to find, since the terms *maxima* and *minima* will be used in a sense somewhat different from what one might expect. When we say, for example, that a function has a maximum at the point (or value) $x = x_0$, we shall *not* necessarily mean that the value $f(x_0)$ of the function is larger than any other value that the function may assume. We shall mean that $f(x_0)$ is the largest value of the function *in the immediate neighborhood of the value* x_0. The difference in meaning between these two possible interpretations of the word *maximum* can be seen from Fig. 9.11. The value of the function at $x = x_0$ (the point P of the graph) is greater than its values immediately to the left or right of $x = x_0$. We therefore say that the function has a maximum at $x = x_0$. This statement is not falsified by the existence of other values of

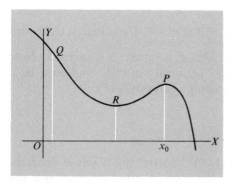

Fig. 9.11

x at which the function assumes values greater than its value at $x = x_0$, such as is represented by the point Q in Fig. 9.11. Thus when we talk about a maximum we shall mean a *local maximum*, as opposed to what we might call an *overall* maximum.* In the same way *minimum* will mean *local minimum* as, for example, at the point R in Fig. 9.11. These remarks are made precise by the following definitions.

Definition 9.4

A function $f(x)$ is said to have a maximum at $x = a$ if there is a positive number δ such that $f(a + h) < f(a)$ for all h satisfying $0 < |h| < \delta$. In other words $f(a)$ is larger than any other value of the function over the interval $(a - \delta, a + \delta)$.

Definition 9.5

A function $f(x)$ is said to have a minimum at $x = a$ if there is a positive number δ such that $f(a + h) > f(a)$ for all h satisfying $0 < |h| < \delta$.

EXAMPLES:
The functions x^2, $|x|$, x^4, $\sin^2 x$ all have a minimum at $x = 0$.
The functions $-x^2$, $\cos x$, $-|x|$, all have a maximum at $x = 0$.
The function defined by

$$f(x) = 0 \quad \text{if} \quad x \neq 0$$
$$f(0) = 1$$

has a maximum at $x = 0$.

It will be seen from some of these examples (especially the last) that it is not necessary for a function to be differentiable, or even continuous, for it to be able to have a maximum or a minimum at a given point. However, when a function *is* differentiable (possibly many times) there is a routine method for determining what values of its argument give maximum and minimum values of the function. This is accomplished by the following theorem.

THEOREM 9.6

IF A FUNCTION $f(x)$ SATISFIES THE FOLLOWING CONDITIONS:
1. $f(x)$ IS $(n + 1)$ TIMES DIFFERENTIABLE AT $x = a$, WHERE n IS A POSITIVE INTEGER,
2. $f^{(n+1)}(x)$ IS CONTINUOUS AT $x = a$,

---

* Or, to the use the current O.K. word, a *global* maximum.

3. $f'(a) = f''(a) = f'''(a) = \ldots = f^{(n)}(a) = 0$,   AND
4. $f^{(n+1)}(a) \neq 0$;

THEN IF n IS AN EVEN INTEGER, $f(x)$ HAS NEITHER A MAXIMUM NOR A MINIMUM AT $x = a$, WHILE IF n IS AN ODD INTEGER $f(x)$ HAS A MAXIMUM AT $x = a$ IF $f^{(n+1)}(a) < 0$, AND A MINIMUM IF $f^{(n+1)}(a) > 0$.

According to this theorem the question of what happens at $x = a$ depends on the *lowest* derivative of $f(x)$ which does not vanish at $x = a$, and, if this derivative is an even one, on whether it is positive or negative.

PROOF:

By Taylor's theorem we can write

(9.14.1)
$$f(a + h) = f(a) + \frac{h}{1!}f'(a) + \frac{h^2}{2!}f''(a) + \cdots$$
$$+ \frac{h^n}{n!}f^{(n)}(a) + \frac{h^{n+1}}{(n+1)!}f^{(n+1)}(\xi)$$

which reduces to ˙

(9.14.2)
$$f(a + h) - f(a) = \frac{h^{n+1}}{(n+1)!}f^{(n+1)}(\xi)$$

by virtue of condition (3) of the theorem.

Let us suppose that $f^{(n+1)}(a)$ is positive. Then by Theorem 8.30, $f^{(n+1)}(x)$ is positive in a neighborhood, say $(a - \delta, a + \delta)$ of $x = a$. Hence, if we take $|h| < \delta$, we can be sure that $f^{(n+1)}(\xi)$ is also positive, whatever ξ may be. (ξ may be different for different values of h but is always less than h in magnitude.) It follows from (9.14.2) that $f(a + h) - f(a)$ has the same sign as h^{n+1}. Thus it is always positive (for $|h| < \delta$) if n is an *odd* integer, which means that $f(x)$ has a minimum value at $x = a$; but if n is *even* then $f(a + h) - f(a)$ is positive if h is positive and negative if h is negative. Under these circumstances $f(x)$ can have neither a maximum nor a minimum at $x = a$.

If $f^{(n+1)}(a)$ is negative the argument goes in much the same way, and we deduce that the function has a maximum if n is an odd integer, but that if n is an even integer the function has neither a maximum nor a minimum. This completes the proof of the theorem.

We can now see how to find the maximum and minimum values of a given function. Provided the function is differentiable, we must certainly have $f'(x) = 0$ at any value at which the function has a maximum or minimum. Our first step is therefore to find all values of x for which $f'(x) = 0$, i.e., we solve the equation $f'(x) = 0$. This will give us a number of candidates for points at which $f(x)$ is a maximum or minimum. (These are called *stationary points* and the corresponding values of the function are called its *stationary* values. All maxima and minima will be included among them).

The statements in the last paragraph are certainly true if the function $f(x)$ is defined for all values of x. If it is defined only over a more restricted domain, say over a closed interval, then we must be more circumspect in considering the possibility of, say, a maximum at one of the boundary points of the domain (an endpoint of the interval). According to the strict interpretation of Definition 9.4 there cannot be a maximum at such a point; for it is implicit in that definition that the function is defined on either side of the alleged maximum. Yet if we relax the definition, and require $f(a + h) < f(a)$ to apply only when $f(a + h)$ is defined, we may get more than we want. For example, the function defined by $f(x) = x$ in the interval $[0, 1]$ and undefined elsewhere would then have a maximum at $x = 1$ (and a minimum at $x = 0$). This may or may not be what we want; it is a matter of interpretation. We are on safe ground if we assert our results on maxima and minima only for interior points of the domain of the function; that is, points lying in an open interval over which the function is defined.

We then take each of these values in turn and see whether the function satisfies the conditions of Theorem 9.6. In the general case the second derivative will not vanish at these points, and we then deduce that the function has a maximum if $f''(a)$ is negative, and a minimum if $f''(a)$ is positive. If $f''(a) = 0$ we have to go on differentiating until we find a derivative that does not vanish at $x = a$, but it is only exceptionally that one needs to go beyond the second derivative.

Note. In many elementary textbooks on calculus the condition $f'(x) = 0$ for a maximum or minimum point is derived from the geometrical argument that at a maximum or minimum the slope of the graph of the function will be zero. It is further shown that as x increases through a maximum the slope decreases, so that $f''(x) < 0$; while for a minimum it increases, so that $f''(x) > 0$. So far so good. But often the possibility of $f''(x)$ and possibly higher derivatives being zero is either ignored or postponed. The condition $f'(x) = 0$ is *necessary* for a maximum or minimum, but is not *sufficient:* to obtain a sufficient condition we must allow for the consideration of what happens to the higher derivatives. For a full treatment, such as we have tried to give here, the use of Taylor's theorem is essential. This explains why the topic of maxima and minima—perhaps one of the commonest applications of elementary calculus—has come rather late in the chapter.

EXAMPLE 1:

To find the maximum and minimum values of the function

$$f(x) = x^3 + 3x^2 - 9x - 5.$$

Here we have $f'(x) = 3x^2 + 6x - 9$. Thus $f'(x)$ will be zero if

$$x^2 + 2x - 3 = 0$$

i.e., if $x = -3$ or $x = 1$. These then are the only two values of x at which a maximum or a minimum *might* occur. We take them in turn. If $x = -3$ then $f''(x) = 6x + 6 = -12$. Since $f''(x)$ is therefore negative we have a maximum at $x = -3$. If $x = 1$ then $f''(x) = +12$, and we therefore have a minimum at this value. The corresponding values of the function are 22 and -10 respectively.

These values are shown in Fig. 9.12 which gives the graph of the

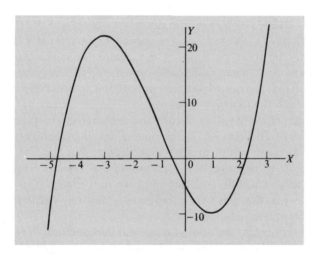

Fig. 9.12

function from $x = -5$ to $x = 3$. This figure also illustrates the geometrical argument that at a maximum or minimum the slope of the tangent is zero, ie., the tangent is "horizontal" (parallel to the x-axis). This is a necessary condition for either a maximum or minimum, but Theorem 9.6 and, in particular, the next example show that it is not sufficient.

EXAMPLE 2:

The function $f(x) = x^3$. Here $f'(x) = 3x^2$, which is zero only when $x = 0$. The point $(0,0)$ of the graph of this function is therefore the only stationary value. It is not, however, a maximum or a minimum since $f''(x) = 6x$ is also zero at $x = 0$, and the next derivative $f'''(x) = 6$ and therefore does not vanish. Thus the appropriate value of n in Theorem 9.6 is $n = 2$—an even number, and the deduction from the theorem is that the point in question is neither a maximum nor a minimum.

The graph of this function is given in Fig. 9.13. Note that the curve is indeed "horizontal" when $x = 0$, i.e., the slope is zero. The reason why there is no maximum or minimum is that the curve crosses over the tangent (the x-axis) at this point instead of remaining on one side of the tangent, as in the previous example. A point at which a curve crosses over the tangent at the point is called a *point of inflexion.** The function x^3 is the simplest function having such a point.

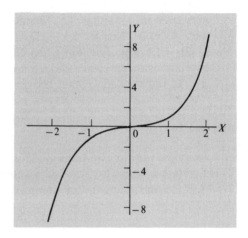

Fig. 9.13

EXAMPLE 3:

The function $f(x) = (x - 1)^6$. For this function we have $f'(x) = 6(x- 1)^5$, which is zero only for $x = 1$. Hence this is the only value of x at which a maximum or minimum might occur. We can readily verify that $f'(1) = f''(1) = f'''(1) = f^{iv}(1) = f^v(1) = 0$, but that $f^{vi}(1) = 720$. Hence we have $n = 5$ in Theorem 9.6, and the fact that $f^{vi}(1)$ is positive shows that this function has a minimum at $x = 1$. The minimum value is, of course, 0.

EXAMPLE 4:

The functions in the above examples were polynomials, and there was no difficulty about the existence of the derivatives or their continuity. Let us consider one example where we have to be careful about conditions (1) and (2) of Theorem 9.6. We consider the function defined by

---

* The only points of inflexion that may cause trouble when finding maxima and minima are those where the tangent is parallel to the x-axis. The rather loose definition of "inflexion" (curve crossing the tangent) just given applies even when this is not the case.

$$f(x) = (x - 1)^3 - 12x + 30 \quad \text{if } x < 3;$$
$$f(x) = (5 - x)^3 + 12x - 42 \quad \text{if } x > 3;$$

and

$$f(3) = 2.$$

This function is clearly differentiable any number of times except possibly at the value $x = 3$. The function $f(x)$ is differentiable at $x = 3$, and $f'(3) = 0$, so that the point $x = 3$, $y = 2$ is therefore a candidate for a maximum or minimum. The function $f'(x)$ is also differentiable at $x = 3$, and $f''(x)$ is given by $6(x - 1)$ if $x < 3$, by $6(5 - x)$ if $x > 3$, while $f''(3) = 12$. The function $f''(x)$ is not differentiable at $x = 3$, but it is continuous there. Hence the requirements of Theorem 9.6 are met with $n = 1$, and we deduce that this function has a minimum at $x = 3$.

Exercise 103

1. Verify the various assertions made in the treatment of Example 4 above. This function has another stationary value. Find it and determine what it is.
2. Find the maximum and minimum values of

$$y = (x - 1)^6 (x^2 - 10x + 17)$$

3. Show that the function $\dfrac{\exp(ax)}{1 + x^2}$ has a maximum and a minimum if $|a| < 1$, but no stationary value if $|a| \geq 1$.
4. A conical cup is formed from a circular disk of radius a from which a sector has been removed. Show that the greatest volume the cup can have is $\dfrac{2\pi a^3}{9\sqrt{3}}$.

9.15 MORE ABOUT THE EXPONENTIAL FUNCTION

We now derive yet a further property of the exponential function, and in order to simplify our expressions we shall denote the number $\exp(1)$, that is, the sum of the series $1 + \dfrac{1}{1!} + \dfrac{1}{2!} + \dfrac{1}{3!} + \ldots$, by the letter e. [The reader should now object that we have already used this letter on page 294 for $\lim\limits_{n \to \infty} \left(1 + \dfrac{1}{n}\right)^n$. We shall therefore anticipate a later result by saying that the two numbers will turn out to be the same.]

Since $\exp(1) = e$, we have, by (9.12.2),
$$\exp(2) = \exp(1)\cdot\exp(1) = e^2,$$
$$\exp(3) = \exp(2)\cdot\exp(1) = e^3,$$

whence it follows that $\exp(n) = e^n$, where n is any positive integer. (The proof of this is a simple exercise in mathematical induction.)

Suppose q is a positive integer. Then by repeated applications of (9.12.2) we have

$$\underbrace{\exp\left(\frac{1}{q}\right)\cdot\exp\left(\frac{1}{q}\right)\cdot\exp\left(\frac{1}{q}\right)\cdots\exp\left(\frac{1}{q}\right)}_{q \text{ factors}} = \exp\left(\frac{1}{q}+\frac{1}{q}+\frac{1}{q}+\cdots+\frac{1}{q}\right)$$
$$= \exp(1).$$

Hence $\exp\left(\frac{1}{q}\right)$ is a number whose qth power is $\exp(1)$, or e. From this we deduce not, strictly speaking, that $\exp\left(\frac{1}{q}\right) = e^{1/q}$, but that $\exp\left(\frac{1}{q}\right)$ is *one* of the values of $e^{1/q}$. For instance, if q is an even integer there will be two numbers whose qth power is e; one positive and one negative. Clearly $\exp\left(\frac{1}{q}\right)$ must be positive, from its definition as a series. So long as we are dealing with real numbers we can write $\exp\left(\frac{1}{q}\right) = e^{1/q}$ on the understanding that $e^{1/q}$ stands for the positive qth root of e.*

Now let p be another positive integer. Then we have

$$\underbrace{\exp\left(\frac{1}{q}\right)\cdot\exp\left(\frac{1}{q}\right)\cdot\exp\left(\frac{1}{q}\right)\cdots\exp\left(\frac{1}{q}\right)}_{p \text{ factors}} = \exp\left(\frac{1}{q}+\frac{1}{q}+\frac{1}{q}\cdots+\frac{1}{q}\right)$$
$$= \exp\left(\frac{p}{q}\right)$$

using (9.12.2) again. Hence $(e^{1/q})^p = \exp\left(\frac{p}{q}\right)$, or

(9.15.1) $$\exp\left(\frac{p}{q}\right) = e^{p/q}.$$

It rather looks as though $\exp(x)$ is turning out to be e^x, at least for rational values of x, but we have not yet considered negative values of x. Applying (9.12.2) again we have

---

* This caution in dealing with fractional powers is even more necessary when we come to consider complex numbers. See Chapter 13.

$$\exp\left(\frac{p}{q}\right) \cdot \exp\left(-\frac{p}{q}\right) = \exp\left(\frac{p}{q} - \frac{p}{q}\right)$$
$$= \exp(0)$$
$$= 1.$$

Thus $\exp\left(-\dfrac{p}{q}\right) = \dfrac{1}{\exp\left(\dfrac{p}{q}\right)} = \dfrac{1}{e^{p/q}} = e^{-p/q}$. Hence it is indeed true that

(9.15.2) $$\exp(x) = e^x$$

for all rational values (positive and negative) of x.

EXAMPLES:

To underline what we have just done, let us work out a few values of $\exp(x)$. The value of $\exp(1)$, e itself, is given by the sum of the series $1 + 1 + \dfrac{1}{2!} + \dfrac{1}{3!} + \ldots$ The first ten terms of this series are, to 6 decimal places

1.000000
1.000000
.500000
.166667
.041667
.008333
.001389
.000198
.000025
.000003

and their sum is   2.718282

It now looks as though the remaining terms are going to be small enough not to affect our answer to the degree of accuracy to which we are working, but we must make certain of this. For although the individual terms are small from this point on, it might yet happen that taken all together they may add up to something appreciable. (After all, there is an infinite number of them!)* The terms that we have omitted are the terms of the series

$$\frac{1}{10!} + \frac{1}{11!} + \frac{1}{12!} + \frac{1}{13!} + \cdots = \frac{1}{10!}\left(1 + \frac{1}{11} + \frac{1}{11\cdot12} + \frac{1}{11\cdot12\cdot13} + \cdots\right)$$

---

* If the reader adds up the first few terms of the series $1 + \frac{1}{2} + \frac{1}{3} + \frac{1}{4} + \ldots$ he will reach a stage where it "looks as though" the remaining terms are negligible. Nevertheless the series is *divergent*!

which is term for term less than the series

$$\frac{1}{10!}\left(1 + \frac{1}{11} + \frac{1}{11^2} + \frac{1}{11^3} + \cdots\right)$$

which is a convergent geometric series. Its sum is

$$\frac{1}{10!} \cdot \frac{1}{1 - \frac{1}{11}} = \frac{1}{10!} \cdot \frac{11}{10}$$

This number is certainly less than .0000004 and so will not affect the sixth place of decimals.

We would be wrong to assume, however, that our answer 2.718282 was correct to six decimal places. For we have rounded off the various summands to this number of places, and the errors introduced by this rounding off will certainly affect the sixth place, if not the fifth. In this example, with only seven roundings-off, the fifth place will not be affected. We can therefore assert that, to five places, $e = 2.71828\ldots$

Let us now calculate $e^{1/2} = 1 + \frac{1}{2} + \frac{1}{2!}\left(\frac{1}{2}\right)^2 + \frac{1}{3!}\left(\frac{1}{2}\right)^3 + \cdots$

The first eight terms are

$$
\begin{aligned}
&1.000000 \\
&.500000 \\
&.125000 \\
&.020833 \\
&.002604 \\
&.000260 \\
&.000022 \\
&\underline{.000002}
\end{aligned}
$$

and their sum is 1.648721

As before, we can verify that the sum of all the omitted terms is less than .0000005, and cannot therefore affect the sixth place of decimals. The sixth place will be affected by rounding-off errors, but not the fifth. We therefore assert that $e^{1/2} = 1.64872\ldots$

If the theory that preceded these computations is correct, then 1.648721 ought to be the square root of 2.71828, to five-figure accuracy. This is the case, as the reader can verify if he cares to slog out the square of 1.64872, or (better) to look it up in tables of squares.

We have seen that when x is a rational number, $\exp(x) = e^x$, i.e., the number e raised to the rational power x. This can be found by multiplications and the taking of roots—and we have been assuming that we know what is meant by these operations as applied to real numbers. Suppose we take x

to be an irrational number, let us say $x = \sqrt{2}$. Then, under the same assumptions, we can evaluate $\exp(\sqrt{2})$ by means of the series for $\exp(x)$ to any required degree of accuracy. Will this number $\exp(\sqrt{2})$ turn out to be the same as $e^{\sqrt{2}}$?

Before we waste time in fruitless speculation, let us pick on the weak point in this question. The point is that we cannot tell whether $\exp(\sqrt{2})$ is the same as $e^{\sqrt{2}}$ because *we do not know what $e^{\sqrt{2}}$ is*. As has been mentioned already, it is assumed that the reader has some sort of idea of what is meant by an irrational number, and by the usual arithmetical operations applied to those numbers. This is in order to facilitate the development of the subject up to the present point. But we are going to draw the line at assuming that we have any such intuitive notion of what is meant by raising an irrational number to an irrational power. Not that there would be any great difficulty in cooking up some such idea; indeed there is a fairly obvious one ready to hand, namely that the number a^b, where a and b are real (and hence possibly irrational) numbers, is the real number which is approximated more and more closely by numbers of the form A^B where A and B are rational numbers approximating to a and b. Unfortunately this idea is not very easy to make rigorous, and we therefore abandon it in favor of a more profitable line of development.

The present situation then is that the function $\exp(x)$ is defined for *all* real values of x, while e^x is defined only for rational values of x; but for rational values of x the two functions are the same. We now complete the definition of e^x by defining it for irrational x to be the same as $\exp(x)$. The two functions are now the same for all real values of x.

The question we asked just now then becomes meaningless—$e^{\sqrt{2}}$ is the same as $\exp(\sqrt{2})$ *by definition*, and our results concerning the function $\exp(x)$ simply show that $\exp(x)$ behaves in the sort of way that you would expect a function of the form e^x to behave; for example, the result $\exp(x + y) = \exp(x) \cdot \exp(y)$ becomes $e^{x+y} = e^x \cdot e^y$—a result that should not surprise us, since it is merely a slightly more general form of the law of indices that was given on page 85.

To conclude this section let us summarize our results on the exponential function, adding a few easily-proved corollaries for good measure. The function is defined by

$$e^x = \exp(x) = \sum_{r=0}^{\infty} \frac{x^r}{r!}$$
$$= 1 + x + \frac{x^2}{2!} + \frac{x^3}{3!} + \frac{x^4}{4!} + \cdots$$

Two important properties of the function are

$$\exp(x + y) = \exp(x) \cdot \exp(y) \text{ or } e^{x+y} = e^x \cdot e^y$$

and $\dfrac{d}{dx}(e^x) = e^x$, — the function is its own derivative.

The corollaries are:

1. $e^x = 1 + x + \dfrac{x^2}{2} + \cdots > 1 + x \quad$ for $x > 0$.

Hence $e^x \to \infty$ as $x \to \infty$.

2. $e^{-x} \to 0$ as $x \to \infty$ (or $e^x \to 0$ as $x \to -\infty$). This follows at once from (1).

3. e^x is always positive. For if $x \geq 0$ then $e^x \geq 1$, and if $x < 0$ then e^x is the reciprocal of e^{-x} which is positive.

4. e^x is an increasing function. For if $x_1 > x_2$, then

$$\frac{e^{x_1}}{e^{x_2}} = \frac{\exp(x_1)}{\exp(x_2)} = \exp(x_1 - x_2)$$
$$> 1 \quad (\text{since } x_1 - x_2 > 0).$$

Hence $e^{x_1} > e^{x_2}$, i.e., e^x is an increasing function.
The graph of the function e^x is shown in Fig. 9.14.

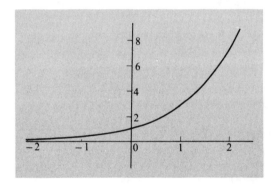

Fig. 9.14

Exercise 104

1. From the exponential series calculate the values, to four places of decimals, of $\exp\left(\tfrac{1}{3}\right)$, $\exp\left(-\tfrac{1}{2}\right)$ and $\exp\left(-\tfrac{1}{3}\right)$.

Note: When x is negative, the omitted terms of the series form a series of terms decreasing in magnitude and alternating in sign. Since the sum of such a series is numerically less than the first term, we need only ensure that the first omitted term is small enough that it will not affect the answer to the degree of accuracy required. Thus this part of the calculation is easier than when x is positive.

Check your answers by multiplying $\exp\left(-\tfrac{1}{2}\right)$ and $\exp\left(\tfrac{1}{2}\right)$, and $\exp\left(-\tfrac{1}{3}\right)$ and $\exp\left(\tfrac{1}{3}\right)$.

2. Expand the following functions as power series in x as far as the term in x^6.

(a) $e^{-(1/2)x^2}$

(b) $\frac{1}{2}(e^x + e^{-x})$

(c) e^{x-x^2}

(d) $\frac{1}{2}(e^x - e^{-x})$

9.16 THE LOGARITHM FUNCTION

We now look at the inverse of the exponential function. Given a positive number x we can always find a real number y such that $x = e^y$. This is clear (at least intuitively) from Fig. 9.14. More strictly, since $e^y \to 0$ as $y \to -\infty$ and $e^y \to \infty$ as $y \to \infty$, we can certainly find numbers y_1 and y_2 such that $e^{y_1} < x < e^{y_2}$. Then since e^y is a differentiable (and hence continuous) function of y, it will assume the intermediate value x for some value of y between y_1 and y_2 (by Theorem 8.31). Moreover, from the fact that e^y is an increasing function of y, it follows readily that there is only one such value of y. Hence the equation $x = e^y$ defines y as a single-valued function of x. We call this function the logarithm function; we say that y is the logarithm of x, and we write $y = \log x$.

The important properties of this function follow almost at once.

1. We have $\log 1 = 0$, since $e^0 = 1$.

2. Let $y_1 = \log x_1$ and $y_2 = \log x_2$. Then $x_1 = e^{y_1}$ and $x_2 = e^{y_2}$, from which it follows that $x_1 x_2 = e^{y_1} \cdot e^{y_2} = e^{y_1 + y_2}$.

This may be rewritten as $y_1 + y_2 = \log (x_1 x_2)$, or

$$(9.16.1) \qquad \log (x_1 x_2) = \log x_1 + \log x_2$$

i.e., the logarithm of the product of two numbers is the sum of their logarithms. This result extends to the product of any number of factors. We have

$$(9.16.2) \qquad \log (x_1 x_2 x_3 \ldots x_n) = \log x_1 + \log x_2 + \log x_3 + \ldots$$
$$+ \log x_n$$

3. In much the same way as in (2) we can show that

$$(9.16.3) \qquad \log \left(\frac{x_1}{x_2} \right) = \log x_1 - \log x_2$$

and, in particular, that

(9.16.4) $$\log\left(\frac{1}{x}\right) = -\log x.$$

4. If $y = \log x$, then $x = e^y$, and hence

$$\frac{dx}{dy} = e^y = x.$$

Therefore $\frac{dy}{dx} = \frac{1}{x}$. Thus the derivative of the function $\log x$ is the function x^{-1}, or $\frac{1}{x}$.

5. Directly from the definition of $\log x$ we observe that

$$a = e^{\log a}.$$

If we raise both sides of this equation to the power b, where b is a rational number we obtain

$$a^b = (e^{\log a})^b = e^{b \log a}.$$

This suggests the way out of the difficulty about raising a real number to the power of another real number which we mentioned on page 392. If a and b are real numbers, a being positive, we *define* a^b to be the number $e^{b \log a} = \exp(b \log a)$. This definition coincides with the usual definition of a^b where b is a rational number, but is applicable even when b is irrational. (See the exercise below.)

Exercise 105

1. If p and q are integers, and a is a positive real number, show that $\log a^{p/q} = \frac{p}{q} \log a$.

2. Using the definition given above of a^b show that the following laws of indices hold:

$$a^b \cdot a^c = a^{b+c}; \quad (a^b)^c = a^{bc}; \quad a^b/a^c = a^{b-c}$$

3. Now that the derivative of the function $\log x$ has been found we can differentiate any function built up from *elementary functions,* viz., rational powers, the trigonometric functions and their inverses, and the exponential and logarithmic functions. Below is a summary of the derivatives of these functions. The reader should satisfy himself that he can prove all these results.

y	dy/dx
x^k	kx^{k-1} (k rational)
$\sin x$	$\cos x$
$\cos x$	$-\sin x$
$\tan x$	$\sec^2 x$
$\cot x$	$-\mathrm{cosec}^2 x$
$\sec x$	$\sec x \tan x$
$\mathrm{cosec}\, x$	$-\mathrm{cosec}\, x \cot x$
e^{kx}	ke^{kx}
$\sin^{-1}(x/a)$	$1/\sqrt{a^2 - x^2}$
$\cos^{-1}(x/a)$	$-1/\sqrt{a^2 - x^2}$
$\tan^{-1}(x/a)$	$a/(a^2 + x^2)$
$\log x$	$1/x$

9.17 THE LOGARITHMIC SERIES

Given a real number x we can calculate the value of e^x to any degree of accuracy by means of the series (9.11.1). The definition of $\log x$ as the inverse of the function e^x does not provide us with an easy method of finding the value of $\log x$ for a given x. We therefore look for a series from which logarithms could be calculated if one should wish to do so. (It is probably not necessary to add that, in practice, one finds these values by consulting tables of the functions, and not by personal calculation. Thus the series we shall derive is of the theoretical rather than practical interest, but important nonetheless.)

If $f(x) = \log x$ we have

$$f'(x) \ \ = x^{-1}$$
$$f''(x) \ = -x^{-2}$$
$$f'''(x) = 2x^{-3}$$

and, in general, $f^{(n)}(x) = (-1)^{n+1}(n-1)!\, x^{-n}$. From this we have $f(1) = 0$ and $f^{(n)}(1) = (-1)^{n+1}(n-1)!$

From Taylor's theorem with Cauchy's form of the remainder we have

$$\log(1 + h) = h - \frac{h^2}{2} + \frac{h^3}{3} - \cdots + (-1)^{n+1}\frac{h^n}{n} + R_n$$

where

$$R_n = \frac{h(1 + h - \xi)^n}{n!} f^{(n+1)}(\xi).$$

Now since $1 < \xi < 1 + h$ we may write $\xi = 1 + \theta h$, where $0 < \theta < 1$. Then

$$R_n = \frac{h(h - \theta h)^n}{n!}(-1)^n \cdot \frac{n!}{(1 + \theta h)^{n+1}}$$

$$= \frac{h^{n+1}(1 - \theta)^n(-1)^n}{(1 + \theta h)^{n+1}}$$

Therefore
$$|R_n| = \frac{|h|^{n+1}}{|1 + \theta h|}\left|\frac{1 - \theta}{1 + \theta h}\right|^n.$$

The reader should verify that, if $-1 < h < 1$, then $\left|\dfrac{1 - \theta}{1 + \theta h}\right| < 1$, and $1 + \theta h > 1 - |h|$. It follows then that

$$|R_n| < \frac{1}{1 - |h|}|h|^{n+1}$$

and that therefore $R_n \longrightarrow 0$ as $n \longrightarrow \infty$. This proves that the sum of the infinite series

$$h - \frac{h^2}{2} + \frac{h^3}{3} - \cdots + (-1)^{n+1}\frac{h^n}{n} + \cdots$$

is $\log (1 + h)$, or, using x in place of h, that

(9.17.1) $$\log (1 + x) = x - \frac{x^2}{2} + \frac{x^3}{3} - \frac{x^4}{4} + \cdots (-1)^{n+1}\frac{x^n}{n} + \cdots$$

provided $|x| < 1$.

If $h = 1$ then $1 + \theta h > 1$, and it is still true that $R_n \longrightarrow 0$. Hence (9.17.1) also holds for $x = 1$, though it does not hold for $x = -1$, since the series is then divergent. This series, with $x = 1$, viz.

$$1 - \tfrac{1}{2} + \tfrac{1}{3} - \tfrac{1}{4} + \cdots$$

was the one that we made great use of in Chapter 8. We now see that its sum is $\log 2$.

From (9.17.1) we can, in theory, calculate the logarithms of numbers. From (9.17.1) directly we can only calculate the logarithms of positive numbers less than (or equal to) 2, since we need $-1 < x \leq 1$, but by various subterfuges we could calculate the logarithms of larger numbers. Thus we might write $\log 3 = \log \tfrac{9}{5} + \log \tfrac{5}{3}$, and then calculate the two terms on the right-hand side by means of (9.17.1), thus obtaining $\log 3$. But in point of fact (9.17.1) is not very useful for computation, although it is the starting point for the derivation of other series which are (see the examples below).

Term by term differentiation of the right-hand side of (9.17.1) yields the series $1 - x + x^2 - x^3 + \cdots = (1 + x)^{-1}$, the derivative of $\log(1 + x)$. This fact may enhance our confidence in the truth of (9.17.1) but, in the absence of some general result on differentiation of series, which we shall not consider, it cannot be used to prove it. (Compare the similar remarks on page 372.)

EXAMPLES:

The series (9.17.1) does not converge very rapidly unless x is quite small. Thus to use it to calculate $\log 2$ would be impracticable, since a thousand or so terms would have to be taken in order to get a result correct to only three decimal places! A little manipulation will give us a series that is more suitable for computation. Since

$$\log(1 + x) = x - \frac{x^2}{2} + \frac{x^3}{3} - \frac{x^4}{4} + \cdots$$

we have $$\log(1 - x) = -x - \frac{x^2}{2} - \frac{x^3}{3} - \frac{x^4}{4} - \cdots$$

provided that $|x| < 1$. Subtracting the second of these from the first we have

(9.17.2)
$$\log\left(\frac{1 + x}{1 - x}\right) = \log(1 + x) - \log(1 - x)$$
$$= 2\left\{x + \frac{x^3}{3} + \frac{x^5}{5} + \cdots\right\}$$

since the even powers of x cancel. To find the logarithm of a number N we first find a number x such that $\dfrac{1 + x}{1 - x} = N$ and then use (9.17.2) with this value of x. It is easily verified that x is given by $x = \dfrac{N - 1}{N + 1}$.

Thus if $N = 2$, we have $x = \frac{1}{3}$. For this value of x the series (9.17.2) converges fairly rapidly and (as the reader should check for himself) the first four terms give a result which is accurate to four decimal places.

Exercise 106

1. Carry out the computation of $\log 2$ just mentioned, and show that the sum of the terms omitted is not large enough to affect the fourth decimal place. (Remember to take six places throughout the computation, to avoid round-off errors.)

2. Calculate log 1.5 to four places using (9.17.2) with $x = \frac{1}{5}$.

3. Calculate log 3 to four places using (9.17.2).

4. See how well your results obtained above satisfy the equation

$$\log 2 + \log 1.5 = \log 3.$$

9.18 LOGARITHMS TO ANY BASE

Now that we have defined exponential expressions of the form a^b we can talk about logarithms in a more general sense. If $x = a^y$, where a is a positive real number, then we write $y = \log_a x$, and say that y is the logarithm of x *to the base a.* Thus it appears that the logarithms that we defined in the last section were logarithms to the base e, and we could have written them in the form $\log_e x$. Now in practical work, where logarithms are used to simplify arithmetic, the logarithms used are those to the base 10. This is purely for convenience since, with this base, numbers which differ only in the placing of their decimal points (such as 7.186 and 718.6) will have the same decimal parts to their logarithms; only the integral parts will differ and these can be supplied by inspection. However, for theoretical work, such as we are concerned with, logarithms to the base e (also known as natural logarithms or Naperian logarithms) are more important, and it has become conventional to assume that where no base is indicated the base e is implied. (Compare the convention for trigonometric functions, where radian measure is assumed unless otherwise stated.) Thus we can continue to write $\log x$ instead of $\log_e x$ if we wish.

As with logarithms to the base e, we prove that

$$\log_a(x_1 x_2) = \log_a x_1 + \log_a x_2$$

$$\log_a\left(\frac{x_1}{x_2}\right) = \log_a x_1 - \log_a x_2$$

$$\log_a x_1^{x_2} = x_2 \log_a x_1$$

We also have the following results: If $y = \log_a x$, then

$$x = a^y = e^{y \log_e a} \text{ from which } \log_e x = y \log_e a.$$

Hence

$$(9.18.1) \qquad\qquad \log_e x = \log_a x \cdot \log_e a$$

which gives the method for changing from natural logarithms to logarithms to another base a, and vice versa.

Exercise 107

1. Prove the slightly more general result, viz.,

$$\log_b x = \log_a x \cdot \log_b a$$

and deduce that the numbers $\log_a b$ and $\log_b a$ are reciprocals.

2. Expand the following functions as power series in x as far as the term in x^6.

(a) $\log(1 - x^3)$

(b) $\log(1 - x + x^2)$

(c) $\sqrt{x} \log \left(\dfrac{1 + \sqrt{x}}{1 - \sqrt{x}} \right)$

(d) $\log \left\{ e^{x^2/2} \sqrt{1 - 6x} \right\}$

9.19 THE NUMBER e AGAIN

We must now tidy up one loose end. We have not yet shown that the number $\exp(1)$, which we denoted by e, is the same number as $\lim\limits_{n \to \infty} \left(1 + \dfrac{1}{n}\right)^n$, which we introduced in Chapter 8, and also denoted by e. We shall now remedy this omission.

If $f(n) = \left(1 + \dfrac{1}{n}\right)^n$, then $\log f(n) = n \, \log\left(1 + \dfrac{1}{n}\right)$, which, for $n > 1$, is

$$n\left\{\frac{1}{n} - \frac{1}{2}\left(\frac{1}{n}\right)^2 + \frac{1}{3}\left(\frac{1}{n}\right)^3 - \cdots\right\} \quad \text{by (9.17.1)}$$

$$= 1 - \frac{1}{2n} + \frac{1}{3n^2} - \frac{1}{4n^3} + \cdots.$$

Now it is fairly obvious that the sum of this series will tend to 1 as $n \to \infty$; but it would not be rigorous to assume that we can let n tend to infinity in each term and then sum the resulting series. The reason is that here we have *two* limiting processes—one giving the sum of the series for a given value of n (this will be a limit as $N \to \infty$, where N is the number of terms of the series that are taken to give the partial sums), and the other being the limit of the sums as $n \to \infty$. To assume that we can let $n \to \infty$ in each term is tantamount to assuming that it does not matter in which order these two limiting operations are performed. Unfortunately, it often *does* matter, as the example in Exercise 108 below shows. Consequently a *proof* that our function $\log f(n)$ tends to 1 is very necessary. Fortunately, it is easily supplied.

We have $\log f(n) = 1 - \left(\dfrac{1}{2n} - \dfrac{1}{3n^2}\right) - \left(\dfrac{1}{4n^3} - \dfrac{1}{5n^4}\right) - \cdots$ and therefore
$f(n) < 1$, since each of the bracketed expressions is positive. On the other hand

$$\log f(n) = 1 - \frac{1}{2n} + \left(\frac{1}{3n^2} - \frac{1}{4n^3}\right) + \left(\frac{1}{5n^4} - \frac{1}{6n^5}\right) + \cdots > 1 - \frac{1}{2n}$$

since, again, the bracketed terms are all positive.

Hence $1 - \dfrac{1}{2n} < \log f(n) < 1$, from which it follows at once that $\log f(n)$
$\to 1$ as $n \to \infty$. From the continuity of the function $\exp(x)$ it now follows
that since $\log f(n)$ can be made as near to 1 as we like by taking n sufficiently
large, $f(n) = \exp[\log f(n)]$ can be made as near to $\exp(1)$ as we like by taking
n sufficiently large. (The strict proof, using epsilons and deltas is not difficult
to construct.) Hence $\lim_{n \to \infty} f(n) = \exp(1)$, which is what we wanted to show.

Exercise 108

1. Show that the sum of N terms of the series

$$\frac{n}{n(n+1)} + \frac{n}{(n+1)(n+2)} + \frac{n}{(n+2)(n+3)} + \cdots$$

in which the Nth term is $\dfrac{n}{(n+N-1)(n+N)}$, is $\dfrac{N}{n+N}$, and deduce that,
if $\phi(n)$ denotes the sum of the corresponding infinite series then $\phi(n) = 1$
for all values of n.

Note that if we let $n \to \infty$ in each term of the above series, each term
tends to 0, and we get a series whose sum is zero. On the other hand $\lim_{n \to \infty} \phi(n)$
$= 1$, clearly. This example should show the need for caution in stating that
$\lim_{n \to \infty} \log f(n) = 1$ in the proof given above.

It is easy to pinpoint the reason for the behavior of this series. It is that
$\lim_{n \to \infty}\left\{\lim_{N \to \infty} \dfrac{N}{n+N}\right\} = 1$, while $\lim_{N \to \infty}\left\{\lim_{n \to \infty} \dfrac{N}{n+N}\right\} = 0$. The reader should verify
these two results, which show that the order in which limiting operations
are performed is often, though not always, important.

2. By writing $\log\left(1 + \dfrac{1}{n}\right)$ as $-\log \dfrac{n}{n+1}$, or otherwise, prove that
$\log\left(1 + \dfrac{1}{n}\right)^n = 1 - \displaystyle\sum_{r=1}^{\infty} \dfrac{1}{r(r+1)(n+1)^r}$. Deduce that, when n is large

$$\left(1 + \frac{1}{n}\right)^n = e\left(1 - \frac{1}{2n} + \frac{11}{24n^2} - \cdots\right).$$

9.20 PARTIAL DIFFERENTIATION

At the beginning of this chapter we gave the definition

$$f'(x_0) = \lim_{h \to 0} \frac{f(x_0 + h) - f(x_0)}{h}$$

for the derivative at $x = x_0$ of the function $f(x)$ of a single variable. It is natural to ask whether it is possible to define the derivative of a function of two or more variables. It is not difficult to see that this is not possible. The quotient

$$\frac{f(x_0 + h) - f(x_0)}{h}$$

is the difference in the value of the function at two points (values of x) divided by the distance between those points. If we have a function $z = f(x, y)$ of two variables, and M and N are two points in the xy-plane, and MP and NQ the corresponding ordinates (values of z) then we are led by analogy to consider the ratio

(9.20.1) $$\frac{QN - PM}{\text{distance } MN}$$

(see Fig. 9.15).

Let us consider the plane parallel to the z-axis through M and N (and hence also through P and Q). This plane will intersect the surface $z = f(x, y)$ in some

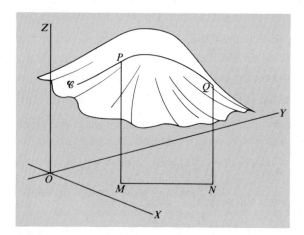

Fig. 9.15

curve ($\mathscr{C}$ in Fig. 9.15). The limit of the ratio (9.20.1), as the distance MN tends to zero is the slope of this curve at P. Now clearly this slope cannot in general be a unique number—it will depend on which plane through MP is taken. If M is the point (x_0, y_0) and N the point $(x_0 + h, y_0 + k)$, for example then (9.20.1) becomes

(9.20.2)
$$\frac{f(x_0 + h, y_0 + k) - f(x_0, y_0)}{\sqrt{h^2 + k^2}}$$

which depends on h and k and the ratio between them. If for example we make $k = 2h$ always, we shall get a certain slope—(9.20.2) becomes a function of h alone and we can find the limit as $h \to 0$—but if the ratio of h to k has some other value the limit will be different again. All this is common knowledge to anyone who has stood on a hillside. If you face the summit and go forwards the slope of the path you take is large; if you turn at right angles to this path then the slope is usually small. For directions in between, the slope has some intermediate value.

Hence there is no one number that we can associate with $f(x, y)$ at a given point (x_0, y_0), analogous to the unique derivative of function of one variable. Instead the situation is rather more complicated. If we agreed not to change the value of y, there would be no difficulty; in fact if y remains fixed at y_0, then $f(x, y_0)$ is a function of one variable x only, and hence may have a derivative in the previously defined sense. In the same way, if we keep x fixed at x_0, then $f(x, y)$ becomes a function $f(x_0, y)$ of y alone and we can again consider its derivative with respect to y (if it has one). These two derivatives (if they exist) are called *partial derivatives* at the point (x_0, y_0) and are denoted by $f_x(x_0, y_0)$ and $f_y(x_0, y_0)$ respectively. The subscript variable indicates the variable with respect to which the differentiation is being performed, and all other variables are assumed to be left constant. This notation corresponds to the notation $f'(x_0)$ used earlier.

We can write partial derivatives also in the counterpart of the "d" notation already introduced. A different sort of "d" is used to indicate that the derivatives are partial, and we write $\dfrac{\partial z}{\partial x}$ and $\dfrac{\partial z}{\partial y}$ for the partial derivatives. This notation has the disadvantage that it does not indicate the point at which the partial derivatives are being evaluated; that is, it does not mention the point (x_0, y_0) as does the previous notation. As a remedy for this, one often writes $\dfrac{\partial z}{\partial x_0}$ and $\dfrac{\partial z}{\partial y_0}$ instead of $f_x(x_0, y_0)$ and $f_y(x_0, y_0)$. This notation is not entirely satisfactory, but has its uses. Thus we have that at the point (x_0, y_0)

(9.20.3)
$$\frac{\partial z}{\partial x} = f_x(x_0, y_0) = \lim_{h \to 0} \frac{f(x_0 + h, y_0) - f(x_0, y_0)}{h}$$

and

$$(9.20.4) \qquad \frac{\partial z}{\partial y} = f_y(x_0, y_0) = \lim_{h \to 0} \frac{f(x_0, y_0 + k) - f(x_0, y_0)}{k}$$

These are known as the partial derivatives of z with respect to x and y respectively.

Thus to compute a partial derivative with respect to x we simply treat y (or any other variable) as if it were a constant. If $z = x^2 + y^2$, then $\frac{\partial z}{\partial x} = 2x$, since the y^2 (being regarded as a constant) disappears when the differentiation is performed.

If
$$z = x^3y + 7x^2y^2 + 4xy^3 + 10y^4$$

then
$$\frac{\partial z}{\partial x} = 3x^2y + 14xy^2 + 4y^3$$

and
$$\frac{\partial z}{\partial y} = x^3 + 14xy^2 + 12xy^2 + 40y^3.$$

If
$$z = \log\left(\frac{y}{x}\right) = \log y - \log x$$

$$\frac{\partial z}{\partial x} = -\frac{1}{x}; \qquad \frac{\partial z}{\partial y} = \frac{1}{y}$$

and so on.

If we differentiate partially with respect to x the function $\frac{\partial z}{\partial x}$ we obtain a second partial derivative, which can be written as $\frac{\partial}{\partial x}\left(\frac{\partial z}{\partial x}\right)$ or, more usually, as $\frac{\partial^2 z}{\partial x^2}$. We can also differentiate $\frac{\partial z}{\partial x}$ with respect to y, in which case we would obtain $\frac{\partial}{\partial y}\left(\frac{\partial z}{\partial x}\right)$ which can be written as $\frac{\partial^2 z}{\partial y \partial x}$. These higher partial derivatives can also be written as f_{xx} and f_{xy} respectively, using the original notation, with $z = f(x, y)$. In the same way we can define $\frac{\partial^2 z}{\partial y^2}$ and $\frac{\partial^2 z}{\partial x \partial y}$, or f_{yy} and f_{yx}. Thus if $z = x^2 \log y + e^x y$ then

$$\frac{\partial z}{\partial x} = 2x \log y + e^x y; \qquad \frac{\partial z}{\partial y} = \frac{x^2}{y} + e^x.$$

$$\frac{\partial^2 z}{\partial x^2} = 2 \log y + e^x y; \qquad \frac{\delta^2 z}{\partial y \partial x} = \frac{2x}{y} + e^x.$$

$$\frac{\partial^2 z}{\partial x \partial y} = \frac{2x}{y} + e^x; \qquad \frac{\partial^2 z}{\partial y^2} = -\frac{x^2}{y^2}.$$

In a similar way still higher derivatives are defined. Thus

$$f_{xxy} = \frac{\partial}{\partial y}\left\{\frac{\partial}{\partial x}\left(\frac{\partial z}{\partial x}\right)\right\} \quad \text{or} \quad \frac{\partial^3 z}{\partial y \partial x^2}$$

$$f_{xyxy} = \frac{\partial}{\partial y}\left(\frac{\partial}{\partial x}\left[\frac{\partial}{\partial y}\left(\frac{\partial z}{\partial x}\right)\right]\right) \quad \text{or} \quad \frac{\partial^4 z}{\partial y \partial x \partial y \partial x}$$

and so on. Derivatives that require differentiation with respect to both variables are known as *mixed* derivatives.

Exercise 109

1. If $z = \sin(x + y) + 2e^x \cos y$ find the partial derivatives

   (a) $\dfrac{\partial z}{\partial x}$ (d) $\dfrac{\partial z}{\partial y}$ (g) $\dfrac{\partial^2 z}{\partial x^2}$

   (b) $\dfrac{\partial^2 z}{\partial x \partial y}$ (e) $\dfrac{\partial^2 z}{\partial y \partial x}$ (h) $\dfrac{\partial^3 z}{\partial x^2 \partial y}$

   (c) $\dfrac{\partial^3 z}{\partial x \partial y^2}$ (f) $\dfrac{\partial^4 z}{\partial x^2 \partial y^2}$.

2. Prove that if $z = x^m y^n$ then

$$\frac{\partial^2 z}{\partial x \partial y} = \frac{\partial^2 z}{\partial y \partial x}$$

and deduce that the same is true if z is any polynomial in x and y, i.e., any sum of terms of the form $A\,x^m y^n$, where A is a constant.

3. Prove that $f_{xy}(x, y) = f_{yx}(x, y)$ when $f(x, y)$ is
   (a) $x^2 \log y - e^{3x+y}\cos(xy)$
   (b) $\tan(x^2 + 3xy + 7y^2)$
   (c) $e^{x^2/y^2}(x + y)^2$.

4. (The preceding two problems indicate that for many functions $f_{xy} = f_{yx}$, i.e., that the order in which the differentiations are performed in a mixed partial derivative is not important. This problem shows that the matter is not as simple as that.)

$$\text{If } f(x, y) = \begin{cases} xy\left(\dfrac{x^2 + y^2}{x^2 - y^2}\right) & \text{if } x \text{ and } y \text{ are not both zero} \\ 0 & \text{if } x = y = 0 \end{cases}$$

find the functions $f_x(x, y)$ and $f_y(x, y)$. Show that at the point $(0, 0)$ the two mixed derivatives f_{xy} and f_{yx} are not the same, i.e., that $f_{xy}(0, 0) \neq f_{yx}(0, 0)$.

9.21 GEOMETRICAL INTERPRETATION

It is fairly easy to interpret partial derivatives geometrically. If we keep y fixed at y_0, then we are considering only those points which lie on the plane whose equation is $y = y_0$. Thus $\dfrac{\partial z}{\partial x}$ is the slope of the curve in which this plane meets the surface $z = f(x, y)$—the curve PQ_1 in Fig. 9.16. Similarly $\dfrac{\partial z}{\partial y}$ is the slope of the curve in which the surface is cut by the plane $x = x_0$ i.e., the curve PQ_2 in Fig. 9.16.

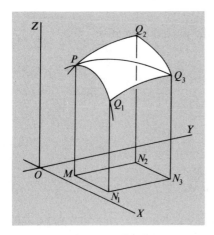

Fig. 9.16

If we think of the z-axis as being vertical, the y-axis as pointing northwards, the x-axis as pointing eastwards, and the surface as being a hill, then $\dfrac{\partial z}{\partial x}$ is the slope of the path we travel if we move eastwards from where we are standing, while $\dfrac{\partial z}{\partial y}$ is the slope of the path we travel on if we go northwards from where we are. The next question is "Suppose we travel in some other direction, what will the slope be?"

To find this we first observe that if we travel a short distance in some other direction, say from M to N_3 in the xy-plane, we can think of this movement as a combination of an increase in the x-coordinate of $h = MN_1$ and an increase in the y-coordinate of $k = MN_2$. The change in the value of the function is

$$(9.21.1) \qquad\qquad f(x_0 + h, y_0 + k) - f(x_0, y_0).$$

Now this can be written as

(9.21.2) $f(x_0 + h, y_0 + k) - f(x_0, y_0 + k) + f(x_0, y_0 + k) - f(x_0, y_0)$

since the two middle terms cancel.

Since we are interested in the slope of a certain "path" or curve on the surface, we must consider the ratio of the change in height [given by (9.21.2)] to the horizontal distances travelled, namely h and k. For this reason we write (9.21.2) as

(9.21.3)
$$\frac{f(x_0 + h, y_0 + k) - f(x_0, y_0 + k)}{h} h$$
$$+ \frac{f(x_0, y_0 + k) - f(x_0, y_0)}{k} k.$$

So far we have done nothing but manipulate the expression (9.21.1). We now observe that (9.21.3) is of the form $Ah + Bk$ and that A and B are quotients rather like those used to define the partial derivatives $\frac{\partial z}{\partial x}$ and $\frac{\partial z}{\partial y}$. We have in fact

$$\lim_{h \to 0} \frac{f(x_0 + h, y_0 + k) - f(x_0, y_0 + k)}{h} = \frac{\partial z}{\partial x}$$

at the point $(x_0, y_0 + k)$ and

$$\lim_{k \to 0} \frac{f(x_0, y_0 + k) - f(x_0, y_0)}{k} = \frac{\partial z}{\partial y}$$

at the point (x_0, y_0) provided these limits exist.

If both h and k tend to zero then, under suitable conditions of continuity of $\frac{\partial z}{\partial x}$, these two limits become $\frac{\partial z}{\partial x}$ and $\frac{\partial z}{\partial y}$ at the point (x_0, y_0). There are some fine points which we should consider if we want a rigorous treatment of the subject, but we shall not attempt to be rigorous at this point. If we write δz for the change in z, given by (9.21.3), and δx, δy in place of h, k, then it looks as though there is a relation between δz, δx, δy, and the partial derivatives. We can write (9.21.3) as

$$\delta z = A\delta x + B\delta y,$$

but we cannot write

$$\delta z = \frac{\partial z}{\partial x}\delta x + \frac{\partial z}{\partial y}\delta y$$

if δz, δx, δy are actual nonzero changes in z, x and y, for the coefficients A and B are only equal to $\dfrac{\partial z}{\partial x}$ and $\dfrac{\partial z}{\partial y}$ when we proceed to the limit. Clearly we must have a way of dealing with the quantities δx, δy and δz as these tend to zero. Let us first see what is, in fact, done, and then why it is done and what it means. What is done is to write

(9.21.4)
$$dz = \frac{\partial z}{\partial x} dx + \frac{\partial z}{\partial y} dy.$$

Now in this equation we must interpret carefully the symbols dz, dx and dy. They cannot represent zero changes in z, x and y, for then equation (9.21.4) though true, would be unhelpful. What then do they mean? Let us take a simple example, namely,

$$z = x^2 y^3$$

and (x_0, y_0) as the point $(2, 1)$. Then

$$
\begin{aligned}
\delta z &= f(2 + \delta x, 1 + \delta y) - f(2, 1) \\
&= (2 + \delta x)^2 (1 + \delta y)^3 - 2^2 1^3 \\
&= [4 + 4\delta x + (\delta x)^2][1 + 3\,\delta y + 3(\delta y)^2 + (\delta y)^3] - 4 \\
&= 4\delta x + 12\delta y + \text{terms involving products or powers of } \delta x \text{ and } \delta y.
\end{aligned}
$$

Note also that at the point $(2, 1)$, $\dfrac{\partial z}{\partial x} = 2xy^3 = 4$ and $\dfrac{\partial z}{\partial y} = 3x^2y^2 = 12$. Hence if δx and δy are small we have

$$\delta z = \left(\frac{\partial z}{\partial x}\right)\delta x + \left(\frac{\partial z}{\partial y}\right)\delta y \text{ approximately}$$

since the discrepancy between the two sides is given by terms in products and powers of δx and δy, which will be small compared to δx and δy.

Now there are different ways in which we can let δx and δy tend to zero. We might, for example, decide to make $\delta y = 2\delta x$. If so, then the point $N_3 = (2 + \delta x, 1 + \delta y)$ will tend to the point $M = (2, 1)$ along the line $y - 1 = 2(x - 2)$. If this is the case then

$$\delta z = 4\,\delta x + 24\,\delta x + \text{the other terms}$$

or

$$\frac{\delta z}{\delta x} = 28 + \text{terms in } \delta x, \text{ etc.}$$

and the ratio $\dfrac{\delta z}{\delta x} \to 28$ as δx and $\delta y \to 0$. Thus if we fix the ratio $\delta x : \delta y$ as $1 : 2$, then the ratio $\delta z : \delta x$ is 28, in the limit. Thus, in the limit we have $\delta x : \delta y : \delta z = 1 : 2 : 28$. If, on the other hand, we fix the ratio of $\delta x : \delta y$ at $-3 : 5$, shall we say, then it is easily verified that in the limit we have

$$\delta x : \delta y : \delta z = -3 : 5 : 48.$$

Thus as the variable point $(2 + \delta x, 1 + \delta y)$ tends to the fixed point $(2, 1)$ in some specific way so the ratios $\delta x : \delta y : \delta z$ tend to certain limiting ratios. These ratios are not fixed once and for all, but depend on how the point $(2, 1)$ is approached.

The interpretation of (9.21.4) is now as follows. The *differentials* (as they are called) dx, dy, dz are defined to be numbers (not necessarily small) whose ratios are the limiting ratios of the three numbers δx, δy and δz as the point $(x_0 + \delta x, y_0 + \delta y)$ tends to (x_0, y_0) in some way. Thus since, in our example, if we let M_3 tend to M in one way (along a line of slope 2) the ratios $\delta x : \delta y : \delta z$ become $1 : 2 : 28$, it is permissible to take

$$dx = 1$$
$$dy = 2$$
$$dz = 28.$$

One could also take $dx = 2000$, $dy = 4000$, $dz = 56000$, and so on. Another set of possible values is $dx = -3$, $dy = 5$, $dz = 48$ as we have seen. The result (9.21.4) then says that (in this particular example)

$$dz = 4\,dx + 12\,dy$$

and it is easily verified that the given sets of values do satisfy this equation.

This use of differentials, defined in this rather odd and apparently arbitrary manner, is useful, because it enables us to obtain several results from equation (9.21.4). For example, if we fix the ratio of δx to δy, thus fixing a line along which M_3 tends to M, then δz becomes a function of δx. If we divide (9.21.4) by dx (which we can do, since it is a nonzero quantity) we obtain

$$(9.21.5) \qquad \frac{dz}{dx} = \frac{\partial z}{\partial x} + \frac{\partial z}{\partial y}\frac{dy}{dx}.$$

Now $\dfrac{dz}{dx}$ is the quotient of two numbers that have the same ratio as $\delta z : \delta x$ has in the limit. Hence, by definition, $\dfrac{dz}{dx} = \lim\limits_{\delta x \to 0} \dfrac{\delta z}{\delta x}$, in other words

$\frac{dz}{dx}$ (as the quotient of two differentials) is the same as the derivative $\frac{dz}{dx}$ (as originally defined). Hence the notation is at least consistent. In the same way the quotient $\frac{dy}{dx}$ is the same as the derivative $\frac{dy}{dx}$, being the ratio of δy to δx (which we supposed to have been fixed) or, more generally, it is the slope of the line along which M_3 tends to M.

The interpretation of (9.21.4) is thus as follows. If z is a function of x and y, say $f(x, y)$, and y is a function of x, say $g(x)$, then z is a function of x. We have in fact

$$z = f(x, y) = f[x, g(x)].$$

Equation (9.21.5) then gives us an expression for the derivative $\frac{dz}{dx}$.

EXAMPLE 1:

If $z = x^2 + y^2$ and $y = \cos x$, find $\frac{dz}{dx}$, at a general point (x_0, y_0).

We have

$$\frac{\partial z}{\partial x} = 2x_0 \quad \text{at} \quad (x_0, y_0)$$

and

$$\frac{\partial z}{\partial y} = 2y_0 \quad \text{at} \quad (x_0, y_0).$$

Also

$$\frac{dy}{dx} = -\sin x_0 \quad \text{at} \quad (x_0, y_0).$$

Hence, at this point

$$\frac{dz}{dx} = 2x_0 - 2y_0 \sin x_0.$$

EXAMPLE 2:

If $z = 3x^2y + 2y^3$ and $y = x^2 + 6$ find $\frac{dz}{dx}$ at a general point.

We have
$$\frac{\partial z}{\partial x} = 6xy, \quad \frac{\partial z}{\partial y} = 3x^2 + 6y^2$$

and
$$\frac{dy}{dx} = 2x.$$

Hence
$$\frac{dz}{dx} = 6xy + (3x^2 + 6y^2) \cdot 2x.$$

EXAMPLE 3:

If $z = \sin(x + y)$ and y is an unspecified function of x, obtain an expression for $\frac{dz}{dx}$.

Here we have

$$\frac{dz}{dx} = \frac{\partial z}{\partial x} + \frac{\partial z}{\partial y}\frac{dy}{dx}$$

$$= \cos(x+y) + \cos(x+y)\frac{dy}{dx}.$$

Since the function that y is of x is not specified the answer must be left like this. Note that we could obtain these results in other ways. For example, in Example (2) we could substitute $y = x^2 + 6$ in $3x^2y + 2y^3$ and differentiate z directly as a function of x. Or we could use the rule for differentiating a function (z) of a function (y). Thus in Example (3) we could differentiate $\sin(x+y)$ as the sine of "something," obtaining the cosine of that something, and then multiply by the derivative of that something, $x + y$, which is

$$1 + \frac{dy}{dx}.$$

This gives us $\cos(x+y)\left(1 + \frac{dy}{dx}\right)$ which is the same as before. Equation (9.21.5) shows, however, that when we do this, the coefficient of $\frac{dy}{dx}$ in the result is in fact $\frac{\partial z}{\partial y}$, while the terms not involving $\frac{dy}{dx}$ make up $\frac{\partial z}{\partial x}$.

If instead of y being given as a function of x we have x as a function of y, then we would naturally divide (9.21.4) by dy to obtain

$$\frac{dz}{dy} = \frac{\partial z}{\partial x}\frac{dx}{dy} + \frac{\partial z}{\partial y}$$

for the derivative $\frac{dz}{dy}$.

EXAMPLE 4:

If $z = \frac{x^2}{y}$ and $x = y^2 + 7$ then

$$\frac{dz}{dy} = \frac{\partial z}{\partial x}\frac{dx}{dy} + \frac{\partial z}{\partial y}$$

$$= \frac{2x}{y} \cdot 2y - \frac{x^2}{y^2}$$

$$= 4x - \frac{x^2}{y^2}$$

$$= 4(y^2 + 7) - \frac{(y^2 + 7)^2}{y^2}.$$

Yet another result that can be obtained from (9.21.4) is this: if z is a function of x and y, and x and y are functions of another variable t, then

(9.21.6)
$$\frac{dz}{dt} = \frac{\partial z}{\partial x}\frac{dx}{dt} + \frac{\partial z}{\partial y}\frac{dy}{dt}.$$

This follows from (9.21.4) by dividing both sides by the differential dt (defined in a similar way).

EXAMPLE 5:

If $z = x^3 + y^3$ and $x = 4t^3 + t$, $y = 3t^2 + 1$ find $\frac{dz}{dt}$.

We could, of course substitute for x and y obtaining

$$z = (4t^3 + t)^3 + (3t^2 + 1)^3,$$

and differentiate in a straightforward manner. The result will be the same as that obtained by using (9.21.6), which is

$$\frac{dz}{dx} = 3x^2(12t^2 + 1) + 3y^2(6t)$$

$$= 3(4t^3 + t)^2(12t^2 + 1) + 3(3t^2 + 1)^2 \cdot 6t.$$

Again we see that the formula is equivalent to the use of the rule for the derivative of a function of a function.

Finally we note that this notion of differentials can be applied to functions of other than two variables. For functions of one variable it is more or less trivial. By analogy we would define dx and dy to be any two numbers whose ratio is the ratio $\delta x : \delta y$ as these two numbers tend to zero. Hence dx and dy are any two numbers which satisfy

$$dy = f'(x)\,dx$$

at the particular point under consideration. Dividing by dx we have $\frac{dy}{dx} = f'(x)$ which shows that we can, after all, treat $\frac{dy}{dx}$ as a quotient, although we forebore from doing this earlier, and defined $\frac{dy}{dx}$ in its entirety as a single symbol. Since the ratio $dx : dy$ is fixed once x is fixed the concept of a differential does not help very much in the discussion of derivatives of functions of one variable only.

If we have a function u of several variables, say

$$u = u(x, y, z \ldots)$$

then we can prove a result analogous to (9.21.4). It is

(9.21.7) $$du = \frac{\partial u}{\partial x}dx + \frac{\partial u}{\partial y}dy + \frac{\partial u}{\partial z}dz + \cdots$$

with du, dx, dy, dz, ... defined in an analogous manner. As before we can divide by, say, dx and obtain

$$\frac{du}{dx} = \frac{\partial u}{\partial x} + \frac{\partial u}{\partial y}\frac{dy}{dx} + \frac{\partial u}{\partial z}\frac{dz}{dx} + \cdots$$

Or we can divide by another differential dt, where t is a variable of which x, y, z, ... are functions, and write

$$\frac{du}{dt} = \frac{\partial u}{\partial x}\frac{dx}{dt} + \frac{\partial u}{\partial y}\frac{dy}{dt} + \frac{\partial u}{\partial z}\frac{dz}{dt} + \cdots$$

exactly as before.

Before leaving this section we must say some more about a result that is strongly suggested by Exercise 109.2, 3, 4, namely that for many functions $f(x, y)$ of two variables it is true that

(9.21.8) $$f_{xy}(x_0, y_0) = f_{yx}(x_0, y_0)$$

for any point (x_0, y_0). Exercise 109.4, however, shows that it is not always true. Now it would take us too much out of our way to examine rigorously the circumstances under which (9.21.8) is true, but we can note two results, references for the proofs of which are given in the notes at the end of this chapter. They are

THEOREM A

IF f_x AND f_y EXIST IN A NEIGHBORHOOD OF (x_0, y_0) AND ARE DIFFERENTIA-BLE AT (x_0, y_0); THEN $f_{xy} = f_{yx}$ THERE.

THEOREM B

IF f_x, f_y, f_{yx} EXIST IN A NEIGHBORHOOD OF THE POINT (x_0, y_0) AND f_{yx} IS CONTINUOUS AT (x_0, y_0); THEN f_{xy} EXISTS AT (x_0, y_0) AND $f_{xy} = f_{yx}$ THERE.

The reader will see from these two, slightly different, conditions for the equality of f_{xy} and f_{yx}, that provided no difficulties arise over the continuity or differentiability of $f(x, y)$ and its derivatives we can safely assert that $f_{xy} = f_{yx}$. We shall therefore assume, unless the contrary is stated, that any function we are dealing with satisfies these requirements, and that we can interchange the order of the partial differentiations as we wish.

Exercise 110

1. If $z = (x^2 + y^2)e^{y/x}$ obtain

$$\frac{\partial z}{\partial x}, \quad \frac{\partial z}{\partial y}, \quad \frac{\partial^2 z}{\partial x^2}, \quad \frac{\partial^2 z}{\partial x \partial y}, \quad \text{and} \quad \frac{\partial^2 z}{\partial y^2}.$$

2. Obtain the same five partial derivatives as in problem 1 for the following functions

   (a) $\dfrac{x^2 + y^2}{1 - x - y}$ (c) $\exp(\log x \log y)$

   (a) $\tan(x^2 + y^2)$ (d) $(1 - x^2 y)^{-\frac{1}{2}}$

3. Show that for all the functions in problems 1 and 2

$$\frac{\partial^2 z}{\partial x \partial y} = \frac{\partial^2 z}{\partial y \partial x}.$$

4. The generalized Cobb-Douglas demand function is of the form

$$y = A x_1^{\alpha_1} x_2^{\alpha_2} \ldots x_n^{\alpha_n}$$

where $\alpha_1, \alpha_2, \ldots, \alpha_n$ and A are constants. Show that

$$\frac{\partial y}{\partial x_i} = \frac{\alpha_i}{x_i} y$$

and that

$$\frac{\partial^2 y}{\partial x_i \partial x_j} = \frac{\alpha_i \alpha_j}{x_i x_j} y \quad (i \neq j)$$

$$\frac{\partial^2 y}{\partial x_i^2} = \frac{\alpha_i (\alpha_i - 1)}{x_i^2} y$$

5. In a study for incentives for building, the function

$$E = \frac{(Rp - T)^{0.86} W}{C^{0.86}}$$

was obtained. Find the partial derivatives

$$\frac{\partial E}{\partial R}, \quad \frac{\partial E}{\partial p}, \quad \frac{\partial^2 E}{\partial T^2}, \quad \frac{\partial^2 E}{\partial p \partial C}, \quad \text{and} \quad \frac{\partial^3 E}{\partial p \partial W \partial C}$$

6. If $z = x e^{y/x} + y^2/x$, prove that

$$x \frac{\partial z}{\partial x} + y \frac{\partial z}{\partial y} = z$$

and that

$$x^2\frac{\partial^2 z}{\partial x^2} + 2xy\frac{\partial^2 z}{\partial x \partial y} + y^2\frac{\partial^2 z}{\partial y^2} = 0.$$

7. If u and v are functions of x and y, and are related by an equation of the form

$$v = f(u)$$

where f is a function of a single variable prove that

$$\frac{\partial u}{\partial x}\frac{\partial v}{\partial y} = \frac{\partial u}{\partial y}\frac{\partial v}{\partial x}.$$

Hint: Differentiate $v = f(u)$ partially with respect to x, and with respect to y, and eliminate $f'(u)$ between the two results.

8. If $x = r \cos \theta$, $y = r \sin \theta$ and $u = r^\alpha \cos \theta$ calculate

$$\frac{\partial u}{\partial x}, \quad \frac{\partial u}{\partial y}, \quad \frac{\partial^2 u}{\partial x^2}, \quad \frac{\partial^2 u}{\partial y^2}.$$

If $\frac{\partial^2 u}{\partial x^2} + \frac{\partial^2 u}{\partial y^2} = 0$ find the possible values of α.

9. If z is a function of x and y, and x and y are functions of u and v, show that

$$\frac{\partial z}{\partial u} = \frac{\partial z}{\partial x}\frac{\partial x}{\partial u} + \frac{\partial z}{\partial y}\frac{\partial y}{\partial u}$$

and

$$\frac{\partial z}{\partial v} = \frac{\partial z}{\partial x}\frac{\partial x}{\partial v} + \frac{\partial z}{\partial y}\frac{\partial y}{\partial v}.$$

If $x = \frac{1}{2}(u^2 - v^2)$ and $y = uv$, show that

(a) $u\dfrac{\partial z}{\partial v} - v\dfrac{\partial z}{\partial u} = 2\left(x\dfrac{\partial z}{\partial y} - y\dfrac{\partial z}{\partial x}\right)$

(b) $\dfrac{\partial^2 z}{\partial u^2} + \dfrac{\partial^2 z}{\partial v^2} = (u^2 + v^2)\left(\dfrac{\partial^2 z}{\partial x^2} + \dfrac{\partial^2 z}{\partial y^2}\right)$

10. If $w = f(y - z, z - x, x - y)$ prove that

$$\frac{\partial w}{\partial x} + \frac{\partial w}{\partial y} + \frac{\partial w}{\partial z} = 0.$$

We can illustrate some uses of the notation for partial derivatives by reference to the theory of consumer preference. It is a matter of common observation that a consumer, presented with two commodity vectors (as defined in section 7.1), will usually prefer one to the other. Thus given the alternatives of receiving either

(1) 3 Cadillacs, 2 oil wells and 5 dancing girls,

or (2) 4 Cadillacs, 3 oil wells and 7 dancing girls,

a consumer (if that is the right word) would probably prefer (2) to (1). There remains a possibility that he might prefer neither of two given vectors; he might well be unable to express a preference, one way or the other, between (for example)

(1) 4 boxes of chocolates and 1 cream cake,

and (2) 2 boxes of chocolates and 2 cream cakes.

This basic concept is concisely expressed in the notation of relations developed in Chapter 2. Let us write xRy to mean that the commodity vector y is not preferred to x by the consumer in question—this thus includes the possibility that x and y are indifferent, i.e., the consumer is unable to express a preference. Then it is easily verified that this relation R is reflexive. If we assume (and we shall) that the consumer is consistent, in the sense that if he does not prefer y to x and does not prefer z to y then he will not prefer z to x, then we see that R is also transitive; for we have

$$xRy \ \& \ yRz \longrightarrow xRz.$$

Moreover R is connected in the set of commodity vectors being considered; either xRy or yRx (possibly both) since either the consumer displays a preference or he does not. Thus the set of commodity vectors is completely weakly ordered by the relation R.

As in Chapter 2 page 58, we can define another relation E by

$$xEy \equiv xRy \ \& \ yRx,$$

which is the relation of indifference, neither x nor y being preferred to the other. E is an equivalence relation and partitions the set of commodity vectors into equivalence classes, each of which is a set of vectors which, to the consumer, are of equal value or utility, and among which he expresses no preferences. We can also define a relation P by

$$xPy \equiv xRy \ \& \ \sim yRx$$

meaning that x is strictly preferred to y (indifference being excluded). P is then a complete ordering of the equivalence classes under the relation E, which we can conveniently call indifference classes of commodity vectors. This result was proved in Chapter 2 page 59.

Now the relations R, E and P just defined are similar to the relations "$\geq$", "$=$" and "$>$" of ordinary arithmetic, and it is natural to ask whether it is possible to associate a number with each commodity vector in such a way that two vectors are related by R, E or P whenever the associated numbers are related by $\geq$, $=$ or $>$. In short, can we define single-valued function $u(x_1, x_2, \ldots, x_n)$ of the commodity vector $x = (x_1, x_2, \ldots, x_n)$ in such a way that

(9.22.1) $$(x_1\, x_2, \ldots, x_n)\, R\, (y_1, y_2, \ldots, y_n)$$

if, and only if,

$$u(x_1, x_2, \ldots, x_n) \geq u(y_1, y_2, \ldots, y_n)$$

or, more briefly, that

(9.22.2) $$x R y \text{ if, and only if, } u(x) \geq u(y)?$$

It is easily verified that if (9.22.2) holds then also

$$x E y \text{ if, and only if, } u(x) = u(y)$$

and

$$x P y \text{ if, and only if, } u(x) > u(y).$$

The function $u(x)$ is usually called a *utility function*.

Now leaving aside the question of whether a utility function can be defined having the property (9.22.1), let us see what other properties such a function might be expected to have. We first note that if we increase the amount of one commodity, keeping the other amounts fixed, we would expect the new commodity vector to be preferred to the old. Thus if $h > 0$, we would expect

$$(x_1, x_2, \ldots, x_i + h, \ldots, x_n)\, P\, (x_1, x_2, \ldots, x_i, \ldots, x_n)$$

and hence

(9.22.3) $$u(x_1, x_2, \ldots, x_i + h, \ldots, x_n) \geq u(x_1, x_2, \ldots, x_i, \ldots, x_n).$$

This may not always be so. Thus, of the first two commodity vectors mentioned in this section, a consumer *might* prefer (1) to (2) if, for example, the garaging of the fourth Cadillac raised insurmountable difficulties. Insofar, however, as the consumer in question does not become satiated with any commodity, (9.22.3) is a reasonable assertion. By considering what happens as $h \to 0$ we see that (9.22.3) implies that

$$(9.22.4) \qquad \frac{\partial u}{\partial x_i} > 0 \qquad \text{for } i = 1, 2, \ldots, n$$

assuming of course that u is a differentiable function of x_i. The quantity $\frac{\partial u}{\partial x_i}$ is the rate at which utility increases with increase of the ith commodity. One would expect that this rate would decrease with x_i. To revert again to the first example of this section, an increase in the number of dancing girls from 2 to 3 might well make a considerable difference to the value of $u(x)$, whereas the acquisition of yet one more dancing girl by someone who already had 117 might well pass almost unnoticed! Thus we expect $\frac{\partial u}{\partial x_i}$ to be a *decreasing* function of x_i, and hence expect that

$$(9.22.5) \qquad \frac{\partial^2 u}{\partial x_i^2} < 0 \qquad \text{for } i = 1, 2, \ldots, n.$$

Note that our introduction of derivatives assumes that $x_1, x_2, \ldots, x_n$ are real numbers rather than integers. This means that our example is, to this extent, inappropriate; Cadillacs, oil wells and dancing girls tend to come in integral amounts. But then quantities that appear in economic problems, including money itself, are integral variables anyway. The assumption that the amounts of commodities are real numbers is an idealization of the situation. It is made to simplify the mathematics, but is basically unrealistic, as anyone who has attempted to buy $\sqrt{2}$ tons of pig iron will be able to confirm.

There is not much one can say *a priori* about the mixed derivatives $\frac{\partial^2 u}{\partial x_i \partial x_j} (i \neq j)$; one would hesitate to predict how the increase in utility from the acquisition of another Cadillac would be affected by one's holding in oil wells.

In the early days of consumer preference theory economists tended to postulate the existence of some sort of "absolute" utility. This is equivalent to postulating a unique utility function. Now unique functions of commodity vectors can be defined, an obvious one being the cash value of the vector at current market prices. If p_i is the price of the ith commodity then this cash value is

$$C = p_1 x_1 + p_2 x_2 + \ldots + p_n x_n$$

and is a unique function of the commodity vector $(x_1, x_2, \ldots, x_n)$ for given prices p_i. Attempts to define a similar unique function corresponding to the more nebulous concept of "utility" failed, as was inevitable. For the data are mere preferences, not quantitative measurements, and any function which represents these preferences [in the sense of (9.22.2)] is as good a utility function as any other such function.

Thus, for example, if $u(x)$ satisfies the condition (9.22.2) then so does $u(x) + A$, where A is any constant. More generally, if $f(t)$ is any increasing function of t, that is, a function such that

$$(9.22.6) \qquad t_1 > t_2 \rightarrow f(t_1) > f(t_2)$$

and if $u(x)$ is a utility function, then so also is

$$(9.22.7) \qquad v(x) = f[u(x)].$$

For if xPy then $u(x) > u(y)$ and hence $v(x) > v(y)$ by (9.22.6) and (9.22.7). If u and v are related as in (9.22.7), what can we say about them without knowing anything more about the function f than that it is an increasing function? Let us differentiate both sides of (9.22.7) partially with respect to x_i. We have

$$\frac{\partial v}{\partial x_i} = f'(u)\frac{\partial u}{\partial x_i}$$

or

$$\frac{\partial v}{\partial x_i} \bigg/ \frac{\partial u}{\partial x_i} = f'(u).$$

Hence we can assert that the ratios $\dfrac{\partial v}{\partial x_i} \bigg/ \dfrac{\partial u}{\partial x_i}$ $(i = 1, 2, \ldots, n)$ are all equal. We can also assert that this common ratio is positive, since the fact that f is an increasing function means that $f'(u) > 0$.

If we take a particular commodity vector ξ and consider the set of all vectors that are indifferent to ξ we obtain one of the indifference classes, defined by

$$(9.22.8) \qquad u(x_1, x_2, \ldots, x_n) = k$$

where $k = u(\xi_1, \xi_2, \ldots, \xi_n)$ is the utility of ξ. The geometrical interpretation of (9.22.8) is a *hypersurface* in the commodity space; for two commodities it is a curve (an indifference curve or locus) in the plane, while for three commodities it is a surface in three dimensions. Thus we have a whole family of hypersurfaces, corresponding to the various possible values of k. These hypersurfaces

(1) fill up, or cover, the whole commodity space (since any point of the space gives a value of u, and hence of k)

(2) are mutually disjoint (since u is single-valued).

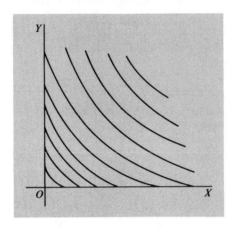

Fig. 9.17

Figure 9.17 shows a selection from a family of indifference curves that might be obtained in this way, and which cover the commodity space—in this case the positive quadrant. Note that a change from $u(x)$ to $v(x) = f[u(x)]$ will not alter the indifference hypersurfaces, it merely alters the numbers that are associated with them. Thus

$$u(x_1, x_2, \ldots, x_n) = k$$

is the same as

$$v(x_1, x_2, \ldots, x_n) = f(k),$$

and the hypersurface associated with the number k, when u was the utility function, becomes associated with the number $f(k)$ instead.

Starting with $k = u(x_1, x_2, \ldots, x_n)$ we obtain

$$\delta k = \frac{\partial u}{\partial x_1}\delta x_1 + \frac{\partial u}{\partial x_2}\delta x_2 + \cdots + \frac{\partial u}{\partial x_n}\delta x_n$$

approximately. [See the derivation of (9.21.4).] If the δx_i represent changes from the point $(x_1, x_2, \ldots, x_n)$ to a neighboring point *on the same indifference hypersurface*, then k is unchanged and $\delta k = 0$. Thus we have

(9.22.9) $\dfrac{\partial u}{\partial x_1}\delta x_1 + \dfrac{\partial u}{\partial x_2}\delta x_2 + \cdots + \dfrac{\partial u}{\partial x_n}\delta x_n = 0$ approximately, which relates the rates of change $\dfrac{\partial u}{\partial x_i}$ to the changes δx_i consequent on a small displacement within an indifference class.

There are occasions when the contribution of one commodity to the utility of the commodity vector is not affected by the amounts of the other commodities. This will be so, for example, if the utility function is of the form

$$(9.22.10) \qquad u(x_1, x_2, \ldots, x_n) = q_1(x_1) + q_2(x_2) + \ldots + q_n(x_n)$$

where each q_i is a function of a single variable. There was a tendency among some economists in the past to assume that all utility functions were essentially of this form, though it is not difficult to show that this supposition is untenable. Consider the utility (under any definition) of the commodity vector

(20 left shoes, 0 right shoes);

it will certainly be small—hardly better than no shoes at all. The same is true of the commodity vector

(0 left shoes, 20 right shoes).

Yet the utility of the vector

(20 left shoes, 20 right shoes)

will be quite high, especially if the shoes match! This sort of behavior is inconsistent with the supposition of a utility function of the form (9.22.10); the two commodities are not *independent*—the presence of one enhances the utility of the other. Naturally, commodities often *are* independent in this sense but we cannot assume that this will always be the case.

Let us consider two commodities only, and see if we can characterize those utility functions for which the commodities are independent. If the utility function is

$$(9.22.11) \qquad u(x, y) = q(x) + r(y)$$

this is easy. We have

$$\frac{\partial u}{\partial y} = r'(y)$$

[since $q(x)$ is a constant when we differentiate with respect to y], and hence

$$(9.22.12) \qquad \frac{\partial^2 u}{\partial x \partial y} = 0$$

since $r(y)$ does not contain x. Conversely since the derivative of $\dfrac{\partial u}{\partial y}$ with respect to x is zero, $\dfrac{\partial u}{\partial y}$ must be a function of y alone. Hence if u contains any terms in x these must vanish on differentiating with respect to y. This leads us back to (9.22.11).

It does not follow however that if the utilities of x and y are independent then the utility function is necessarily like (9.22.11). It could, as we have seen, be of the more general form

$$(9.22.13) \qquad v(x, y) = f\{q(x) + r(y)\}$$

where f is an increasing function. Can we characterize functions of this form? We can, though it is more difficult.

We rewrite (9.22.13) as

$$(9.22.14) \qquad q(x) + r(y) = \phi\{v(x, y)\}$$

where ϕ is the function inverse to f. Then from the form of the left-hand side of (9.22.14) we know that

$$\frac{\partial^2}{\partial x \partial y}[\phi\{v(x, y)\}] = 0$$

as in (9.22.12). Now

$$\frac{\partial}{\partial x}[\phi\{v(x, y)\}] = \phi'(v)v_x$$

by the rule for differentiating a function of a function. Differentiating with respect to y and equating the result to zero we obtain

$$\phi''(v)v_y v_x + \phi'(v)v_{xy} = 0,$$

which can be rearranged as

$$-\frac{\phi''(v)}{\phi'(v)} = \frac{v_{xy}}{v_x v_y}.$$

Thus it appears that $\dfrac{v_{xy}}{v_x v_y}$ is a function of v, and we can write for simplicity

$$(9.22.15) \qquad \frac{v_{xy}}{v_x v_y} = \psi(v).$$

If we differentiate (9.22.15) with respect to x, we obtain

$$\frac{\partial}{\partial x}\left(\frac{v_{xy}}{v_x v_y}\right) = \psi'(v)v_x$$

or

(9.22.16)
$$\frac{1}{v_x} \frac{\partial}{\partial x} \left(\frac{v_{xy}}{v_x v_y} \right) = \psi'(v).$$

If we now repeat this procedure, but with x and y interchanged, we obtain

(9.22·17)
$$\frac{1}{v_y} \frac{\partial}{\partial y} \left(\frac{v_{xy}}{v_x v_y} \right) = \psi'(v).$$

Since the right-hand sides of (9.22.16) and (9.22.17) are equal, so are the left-hand sides, and we obtain the following result

(9.22.18)
$$\frac{1}{v_x} \frac{\partial}{\partial x} \left(\frac{v_{xy}}{v_x v_y} \right) = \frac{1}{v_y} \frac{\partial}{\partial y} \left(\frac{v_{xy}}{v_x v_y} \right)$$

which does not contain the function ϕ (or, for that matter, q or r). Hence to test whether x and y contribute independently to a given utility function $v(x, y)$ we can calculate the appropriate partial derivatives of v and see if they satisfy (9.22.18). If they do not, then it follows that $v(x, y)$ cannot be of the form (9.22.13). If they do, then although this may encourage us to believe that x and y are independent we cannot at this stage assert that they are, since all we have proved is that *if* the commodities are independent then (9.22.18) holds. We shall not pursue here the question of whether the converse is also true.

Demand Functions.   Another example of how partial derivatives can arise in economic theory is provided by the study of demand functions. Let $f(x_1, x_2, x_3, \ldots)$ be a function giving the demand for a commodity in terms of certain measurable quantities $x_1, x_2, \ldots$ relating to that commodity, (such as price, standard, time, etc.). It is of interest to the economist to know how the demand is affected by a change in one only of the variables x_i, that is, what change δf in the demand results from a change δx_i in the variable x_i (the other variables remaining constant). The ratio of these changes is $\frac{\delta f}{\delta x_i}$ which in the limit becomes the partial derivative $\frac{\partial f}{\partial x_i}$. This is also apparent from the derivation of equation (9.21.4); for if the changes δx_i are small we have

$$\delta f = \frac{\partial f}{\partial x_i} \delta x_i + \frac{\partial f}{\partial x_2} \delta x_2 + \cdots$$

approximately, and this reduces to

$$\delta f = \frac{\partial f}{\partial x_i} \delta x_i$$

if all variables except x_i remain fixed. ($\delta x_j = 0$ if $j \neq i$).

More often than not the economist is not so much interested in the actual changes $(\delta f, \delta x_i)$ as in the ratio of the change to the present amount, that is, in the ratios $\delta f/f$, $\delta x_i/x_i$. These are known as *relative changes* or, when multiplied by 100, as *percentage changes*. One great advantage of using relative or percentage changes, rather than the actual changes, is that these quantities are ratios and therefore do not depend on the units in which the measurements were made. The ratio of the relative change in f to the relative change in x_i (other variables remaining fixed) is

$$\frac{\delta f}{f} \bigg/ \frac{\delta x_i}{x_i} = \frac{x_i}{f} \frac{\delta f}{\delta x_i}$$

which becomes $\dfrac{x_i}{f} \dfrac{\partial f}{\partial x_i}$ in the limit. This ratio is known as the x_i *elasticity* of f.

As an example, consider the function

(9.22.19) $$f = 1.058 x_1^{0.136} x_2^{-0.727} x_3^{0.914} x_4^{0.816}$$

obtained by R. Stone for the consumption of beer in the United Kingdom for 1920-1938. Here f is the quantity consumed, x_1 the aggregate real income, x_2 the retail price of beer, x_3 the average retail price of other commodities, and x_4 is a measure of the strength of the beer. It is easily verified that

$$\frac{x_2}{f} \frac{\partial f}{\partial x_2} = -0.727$$

so that the *price elasticity* of the demand for beer is -0.727. This is a constant in this example, but in general it would be a function of the x_i's. Thus an increase of 1 percent in the price of beer would cause a *decrease* of 0.727 percent in the demand, if the other variables do not change.

These elasticities are usually more easily found by the often-used device of taking the logarithm of the demand function before differentiating. Thus we can write (9.22.19) as

$$\log f = \log 1.058 + 0.136 \log x_1 - 0.727 \log x_2 + 0.94 \log x_3$$
$$+ 0.816 \log x_4.$$

On differentiating both sides of this equation with respect to x_1 we obtain

$$\frac{1}{f} \frac{\partial f}{\partial x_1} = 0.136 \frac{1}{x_1}$$

whence we easily see that the income elasticity of the demand is

$$\frac{x_1}{f} \frac{\partial f}{\partial x_1} = 0.136,$$

again a constant. Thus a 1 percent rise in income will cause an increase of 0.136 percent in the demand for beer.

Exercise 111

1. Show that the function

$$u(x, y) = Ax^\alpha y^\beta$$

satisfies the condition (9.22.18) for a utility function for which the variables x and y are independent. Can you see how to write this function in the form

$$f\{q(x) + r(y)\}?$$

2. Show that the function

$$u(x, y) = (x + 1)(y + 1)$$

satisfies the requirements $u_x > 0$, $u_y > 0$ and that $u_{xx} = u_{yy} = 0$, for all points of the commodity space $x > 0$, $y > 0$. On one and the same diagram draw the indifference loci for $u(x, y) = 1, 3, 5$.

3. If $u(x, y) = f\{q(x) + r(y)\}$, show that for points satisfying $u(x, y) = k$ where k is a constant, we have

$$f'\{q(x) + r(y)\}q'(x) + f'\{q(x) + r(y)\}r'(y)\frac{dy}{dx} = 0,$$

and hence that

$$\frac{dy}{dx} = -\frac{q'(x)}{r'(y)}.$$

Hence show that if the indifference loci for a utility function of this type are to have the usual properties (as seen in Fig. 9.17) that $\frac{dy}{dx}$ is negative and $\frac{d^2y}{dx^2}$ is positive, then

    (a) $r'(y)$ and $q'(x)$ have the same sign,

    (b) $\dfrac{q''(x)[r'(y)]^2 - [q'(x)]^2 r''(y)}{r'(y)^3}$ is negative

4. Show that for a general utility function $u(x, y)$, conditions (a) and (b), of problem 3 become

    (a) u_x and u_y have the same sign,

    (b) $\dfrac{u_y^2 u_{xx} - 2u_x u_y u_{xy} + u_x^2 u_{yy}}{u_y^3}$ is negative.

5. A demand function for wheat in the U.S.A. (1921-34) was obtained by Schultz in the form

$$\log f = 1.0802 - 0.2143 \log p - 0.00358t - 0.00163t^2$$

where p is the price and t is time. What is the price elasticity of wheat? What is the interpretation of this figure?

9.23 TAYLOR'S THEOREM FOR A FUNCTION OF TWO OR MORE VARIABLES

Taylor's Theorem, which we derived in section 9.12, gave the value of a function at $x = x_0 + h$ in terms of h and the values of the function and its derivatives at $x = x_0$; that is to say it gave $f(x_0 + h)$ as a power series in h in which the coefficients depended on the function and its derivatives at x_0, provided that the remainder term tended to 0 as $n \rightarrow \infty$. It is natural to ask whether there is a corresponding theorem for a function of, for instance, two variables; that is, a theorem that would yield an expression for the value $f(x_0 + h, y_0 + k)$ of a function f at a point $(x_0 + h, y_0 + k)$ near to a given point (x_0, y_0). There *is* such a theorem, and we shall now consider it. Let it be noted however that in order not to get bogged down in tiresome detail we shall not consider the question of what conditions will suffice for the remainder term to tend to zero; indeed we shall not even introduce a remainder term, but assume that the function that we are dealing with is one for which the remainder term can be shown to vanish. Furthermore, for the limited though important use that we shall make of the theorem, it will be enough to consider only a very few terms in the expansion.

Our aim is to obtain an expression for $f(x_0 + h, y_0 + k)$. If we treat this as a function of x alone (keeping y fixed at the value $y_0 + k$) then we can use Taylor's Theorem, as given in section 9.12, and we obtain

$$(9.23.1) \quad \begin{aligned} f(x_0 + h, y_0 + k) = {} & f(x_0, y_0 + k) + h f_x(x_0, y_0 + k) \\ & + \tfrac{1}{2} h^2 f_{xx}(x_0, y_0 + k) + \ldots \end{aligned}$$

Naturally, the derivatives $f_x, f_{xx}, \ldots$ are partial derivatives. We now apply the same theorem in turn to the functions

$$f(x_0, y_0 + k), \quad f_x(x_0, y_0 + k), \quad f_{xx}(x_0, y_0 + k), \text{ etc.,}$$

as functions of y alone, x being fixed at $x = x_0$. We have

$$\begin{aligned} f(x_0, y_0 + k) &= f(x_0, y_0) + k f_y(x_0, y_0) + \tfrac{1}{2} k^2 f_{yy}(x_0, y_0) + \cdots \\ (9.23.2) \quad f_x(x_0, y_0 + k) &= f_x(x_0, y_0) + k f_{xy}(x_0, y_0) + \ldots \\ f_{xx}(x_0, y_0 + k) &= f_{xx}(x_0, y_0) + k f_{xxy}(x_0, y_0) + \ldots \end{aligned}$$

and so on. Substituting these expressions into (9.23.1) we obtain

(9.23.3)
$$\begin{aligned}
f(x_0 + h, y_0 + k) = &f(x_0, y_0) + hf_x(x_0, y_0) + kf_y(x_0, y_0) \\
&+ \tfrac{1}{2}h^2 f_{xx}(x_0, y_0) + hk f_{xy}(x_0, y_0) \\
&+ \tfrac{1}{2}k^2 f_{yy}(x_0, y_0) + \text{terms in higher powers} \\
&\text{and products of } h \text{ and } k,
\end{aligned}$$

provided we make the assumption that $f_{xy} = f_{yx}$.

If we take it as read that the function and its partial derivatives are evaluated at the point (x_0, y_0) we can write (9.23.3) in the form

$$\begin{aligned}
f(x_0 + h, y_0 + k) = &f + h\frac{\partial f}{\partial x} + k\frac{\partial f}{\partial y} \\
&+ \frac{1}{2}\left\{ h^2\frac{\partial^2 f}{\partial x^2} + 2hk\frac{\partial^2 f}{\partial x \partial y} + k^2\frac{\partial^2 f}{\partial y^2} \right\} \\
&+ \cdots
\end{aligned}$$

If we have a function of three variables, then it can be shown in the same way that

(9.23.4)
$$\begin{aligned}
f(x_0 + h, y_0 + k, z_0 + l) = &f + h\frac{\partial f}{\partial x} + k\frac{\partial f}{\partial y} + l\frac{\partial f}{\partial z} \\
&+ \frac{1}{2}\left\{ h^2\frac{\partial^2 f}{\partial x^2} + k^2\frac{\partial^2 f}{\partial y^2} + l^2\frac{\partial^2 f}{\partial z^2} \right. \\
&\left. + 2kl\frac{\partial^2 f}{\partial y \partial z} + 2lh\frac{\partial^2 f}{\partial z \partial x} + 2hk\frac{\partial^2 f}{\partial x \partial y} \right\} \\
&+ \text{terms in higher powers of } h, k, l,
\end{aligned}$$

where it is understood that f and its derivatives are evaluated at (x_0, y_0, z_0). The corresponding expressions for functions of yet more variables are similar.

If in (9.23.3) the quantities h and k are sufficiently small, then the terms in h^2, hk and k^2 as well as those in higher powers of h and k will be small in comparision to those in h and k. Hence we can say that

$$f(x_0 + h, y_0 + k) - f(x_0, y_0) \text{ is approximately } h\frac{\partial f}{\partial x} + k\frac{\partial f}{\partial y}.$$

This is simply equation (9.21.4) so we have succeeded in generalizing that result.

One further form of this result, known as Maclaurin's form of Taylor's Theorem, is easily deduced. If the point (x_0, y_0) is the origin, i.e., $x_0 = y_0 = 0$, then we have

$$f(h, k) = f(0, 0) + hf_x(0, 0) + kf_y(0, 0)$$
$$+ \tfrac{1}{2}(h^2 f_{xx}(0, 0) + 2hk f_{xy}(0, 0) + k^2 f_{yy}(0, 0))$$
$$+ \text{ higher terms.}$$

Now h and k are any numbers, and we can just as well write them as x and y. If we do this, and take it as understood that the partial derivatives are evaluated at the origin then we have

(9.23.5)
$$f(x, y) = f + \frac{\partial f}{\partial x}x + \frac{\partial f}{\partial y}y$$
$$+ \frac{1}{2}\left(\frac{\partial^2 f}{\partial x^2}x^2 + 2\frac{\partial^2 f}{\partial x \partial y}xy + \frac{\partial^2 f}{\partial y^2}y^2\right)$$
$$+ \text{ terms in higher powers of } x \text{ and } y,$$

which gives the value of the function in terms of its value and those of its derivatives at $x = y = 0$, at least in so far as the appropriate conditions for convergence of the (double) series on the right-hand side of (9.23.5) are met.

EXAMPLE 1:

Expand $\dfrac{1 + x + xy}{1 + x + y}$ as a double power series in x and y as far as the terms of the second degree.

We have

$$f = \frac{1 + x + xy}{1 + x + y} = 1 \quad \text{at} \quad (0, 0)$$

$$f_x = \frac{(1 + x + y)(1 + y) - (1 + x + xy)}{(1 + x + y)^2}$$

$$= \frac{2y + y^2}{(1 + x + y)^2} = 0 \quad \text{at} \quad (0, 0)$$

$$f_y = \frac{(1 + x + y)x - (1 + x + xy)}{(1 + x + y)^2}$$

$$= \frac{x^2 - 1}{(1 + x + y)^2} = -1 \quad \text{at} \quad (0, 0)$$

$$f_{xx} = \frac{-2(2y + y^2)}{(1 + x + y)^3} = 0 \quad \text{at} \quad (0, 0)$$

$$f_{xy} = \frac{(1 + x + y)^2(2 + 2y) - 2(2y + y^2)(1 + x + y)}{(1 + x + y)^4}$$

which reduces to

$$\frac{2(1 + x)^2(1 + y) + 2xy^2}{(1 + x + y)^4} = 2 \quad \text{at} \quad (0, 0)$$

$$f_{yy} = \frac{-2(x - 1)}{(1 + x + y)^3} = 2 \quad \text{at} \quad (0, 0).$$

Hence, by (9.23.5) we have

(9.23.6) $$f = 1 - y + 2xy + y^2 + \cdots$$

Now this may seem a very laborious method of obtaining a simple result like (9.23.6); certainly there are easier ways of reaching it. We could for example use the expansion (9.13.7) for $\dfrac{1}{1 + t}$ to write the function as

$$f = (1 + x + xy)\{1 - (x + y) + (x + y)^2 - \cdots\}$$

whence we obtain

$$f = 1 + x + xy - (1 + x)(x + y) + (x + y)^2 + \cdots$$

on dropping terms of higher degree than the second. This then reduces to the same result.

EXAMPLE 2:

Expand $(1 - xy)^{\frac{1}{2}}e^{3x+y}$ in the same way as the previous example.

Here we have

$$f_x = -\frac{1}{2}y(1 - xy)^{-\frac{1}{2}}e^{3x+y} + 3(1 - xy)^{\frac{1}{2}}e^{3x+y}$$

$$= 3 \quad \text{when} \quad x = y = 0.$$

$$f_y = -\frac{1}{2}x(1 - xy)^{-\frac{1}{2}}e^{3x+y} + (1 - xy)^{\frac{1}{2}}e^{3x+y}$$

$$= 1 \quad \text{when} \quad x = y = 0.$$

The second derivatives of f are clearly going to be pretty complicated functions, but since we shall not need to find any higher derivatives we can save ourselves trouble by ignoring any terms that we can see are going to vanish when we put $x = y = 0$. These are, for example, the terms containing a factor x or y, and any terms whose derivation requires the differentiation of $(1 - xy)$ to some power (since this introduces a factor

x or y). In this way we find without much difficulty that

$$f_{xx} = 3(1 - xy)^{\frac{1}{2}} \cdot 3e^{3x+y} + \text{terms that vanish}$$
$$= 9 \quad \text{when} \quad x = y = 0,$$
$$f_{xy} = -\frac{1}{2}(1 - xy)^{-\frac{1}{2}}e^{3x+y} + 3(1 - xy)^{\frac{1}{2}}e^{3x+y}$$
$$+ \text{vanishing terms}$$
$$= \frac{5}{2} \quad \text{when} \quad x = y = 0,$$

and

$$f_{yy} = (1 - xy)^{\frac{1}{2}}e^{3x+y} + \text{vanishing terms}$$
$$= 1 \quad \text{when} \quad x = y = 0.$$

Applying (9.23.3) we now obtain

$$f = 1 + 3x + y + \frac{9}{2}x^2 + \frac{5}{2}xy + \frac{1}{2}y^2 + \cdots$$

Again we could obtain this result more easily from the expansions of the functions $(1 - t)^{\frac{1}{2}}$ and e^t. We write

$$f = \left(1 - \frac{1}{2}xy + \cdots\right)\left\{1 + (3x + y) + \frac{1}{2}(3x + y)^2 + \cdots\right\}$$

which reduces easily to the same result.

In point of fact it is more often true than not that the expansion of a function of two variables which is defined in terms of standard functions is more easily obtained by methods other than the use of Taylor's Theorem for two variables. Nevertheless the theorem has its uses. For example, if we have calculated the appropriate derivatives then we can easily give the expansion of the function about *any* given point (x_0, y_0) simply by making the appropriate substitutions. Where the theorem really pays off, however, is in those applications where we do not know the point about which the function is to be expanded; that is, where we are looking for a point for which the expansion has certain properties. How this can arise in practice will appear in the next section.

9.24 MAXIMA AND MINIMA OF FUNCTIONS OF MORE THAN ONE VARIABLE

We have already seen in section 9.14 how maxima and minima of a function of one variable can be found. We now ask how to find maxima and minima of functions of two or more variables. A function $z = f(x, y)$ has a maximum at the point (x_0, y_0) if $f(x, y)$ has a larger value at (x_0, y_0) than at at any other point in some neighborhood of (x_0, y_0); more formally, if there is a neighborhood, say a rectangle, defined by

$$x_0 - \epsilon_1 < x < x_0 + \epsilon_1$$
$$y_0 - \epsilon_2 < y < y_0 + \epsilon_2,$$

where ϵ_1 and ϵ_2 are positive numbers, for all points of which, other than (x_0, y_0) itself, we have $f(x, y) < f(x_0, y_0)$.

Another way of wording this is to say that $f(x_0, y_0) - f(x_0 + h, y_0 + k)$ must be positive for all small values of h and k, i.e., $f(x_0 + h, y_0 + k) - f(x_0, y_0)$ must be negative.

Now we see from (9.22.3) that

$$
\begin{aligned}
f(x_0 + h, y_0 + k) - f(x_0, y_0) &= hf_x(x_0, y_0) \\
&\quad + kf_y(x_0, y_0) \\
&\quad + \text{higher powers of } h \text{ and } k,
\end{aligned}
$$

(9.24.1)

and for sufficiently small values of h and k the sign of the right-hand side of (9.24.1) is that of

$$(9.24.2) \qquad hf_x(x_0, y_0) + kf_y(x_0, y_0),$$

provided that (9.24.2) is not always zero. But unless this happens there is no hope whatever of the left-hand side of (9.24.1) being always negative (which is what we want for a maximum). For if (9.24.2) is negative at $(x_0 + h, y_0 + k)$ it will be positive at $(x_0 - h, y_0 - k)$ and the condition cannot be satisfied. In order that the requirements for a maximum should be satisfied (9.24.2) must be zero for all values of h and k and this can happen only if $f_x(x_0, y_0) = f_y(x_0, y_0) = 0$. Thus we have:

THEOREM 9.7

A NECESSARY CONDITION FOR $f(x, y)$ TO HAVE A MAXIMUM AT (x_0, y_0) IS THAT THE PARTIAL DERIVATIVES OF THE FIRST ORDER $\left(\text{i.e., } f_x \text{ AND } f_y, \text{ OR}\right.$ $\dfrac{\partial z}{\partial x}$ AND $\left.\dfrac{\partial z}{\partial y}\right)$ BOTH VANISH AT (x_0, y_0). [As before (see page 385) a certain amount of caution is necessary if $f(x, y)$ is not defined for all x and y. If this is the case we assert Theorem 9.7 and similar results only for points (x_0, y_0) in the interior of the domain of the function f.]

Note that this is a *necessary* condition only; it is necessarily true if there is a maximum at (x_0, y_0), but its truth alone does not guarantee that there is a maximum. It is not *sufficient* for the existence of a maximum.

A minimum of $f(x, y)$ at (x_0, y_0) is defined exactly as for a maximum, except that the function is smaller at (x_0, y_0) than at any neighboring point. It is easily verified that the vanishing of the first-order partial derivatives is also a necessary condition for the existence of a minimum.

If $f_x = f_y = 0$ at (x_0, y_0) then, at that point, we have

$$(9.24.3) \qquad f(x_0 + h, y_0 + k) - f(x_0, y_0) = \frac{1}{2}h^2 f_{xx} + hk f_{xy} + \frac{1}{2}k^2 f_{yy}$$
$$+ \text{ higher powers of } h \text{ and } k,$$

from (9.22.3), and for h and k small the sign of the right-hand side is determined by that of

$$(9.24.4) \qquad \frac{1}{2}h^2 f_{xx} + hk f_{xy} + \frac{1}{2}k^2 f_{yy}.$$

For a maximum this must be always negative. At once we see that f_{xx} must be negative; for otherwise (9.23.4) would be positive for $k = 0$ and h as near zero as we like. In the same way f_{yy} must be negative. But this is not enough; we must show that (9.24.4) is not positive for *any* combination of values of h and k. If we write (9.24.4) as

$$(9.24.5) \qquad \frac{1}{2}k^2\left\{f_{xx}\left(\frac{h}{k}\right)^2 + 2f_{xy}\left(\frac{h}{k}\right) + f_{yy}\right\}$$

we see that we must prove that the quadratic expression

$$(9.24.6) \qquad f_{xx}\lambda^2 + 2f_{xy}\lambda + f_{yy}$$

is not positive for any value of $\lambda \, (= h/k)$. For this it is sufficient to prove that (9.24.6) is never zero, and the condition for this is that

$$(9.24.7) \qquad f_{xx}f_{yy} > f_{xy}^2.$$

(Compare Exercise 73.1.)

For a minimum we also need (9.24.7), since we still need (9.24.4) to remain with the same sign; but now f_{xx} and f_{yy} will both be positive.

A point (x_0, y_0) at which $f_x = f_y = 0$ is called a *stationary point* for the function $f(x, y)$, and $f(x_0, y_0)$ is the corresponding *stationary value*. This will correspond to a point at which the surface $z = f(x, y)$ is "level", i.e., parallel to the xy-plane. This may or may not be a maximum or a minimum.

To sum up, we have the following results:

For $f(x, y)$ to have a maximum or a minimum at a point (x_0, y_0) we must have

1. $f_x(x_0, y_0) = f_y(x_0, y_0) = 0$. In other words (x_0, y_0) must be a stationary point. Further if

2. $f_{xx}f_{yy} > f_{xy}^2$ at (x_0, y_0) then we have either a maximum or a minimum at that point. Which it is depends on the sign of f_{xx}—negative for a maximum, positive for a minimum. Naturally f_{xx} and f_{yy} have the same sign since $f_{xx}f_{yy} > f_{xy}^2 \geq 0$.

Condition (2) can also be written as $\left(\dfrac{\partial^2 z}{\partial x^2}\right)\left(\dfrac{\partial^2 z}{\partial y^2}\right) > \left(\dfrac{\partial^2 z}{\partial x \partial y}\right)^2.$

If condition (2) does not hold, then (9.24.5) will be positive for some points near (x_0, y_0) and negative for others. Neither a maximum nor a minimum is then possible. If f_{xx}, f_{yy} and f_{xy} are all zero then the reasoning given above does not apply and the behavior of the function is still in doubt. We shall not examine the conditions under which a function has a maximum or minimum when all second-order partial derivatives vanish [but see example (3) below].

The following examples should illustrate the results obtained in this section.

EXAMPLE 1:

Find the maxima and minima of the function

$$f(x, y) = 3x^4 + 4x^3 - 36x^2 + y^3 - 12y + 9.$$

We have

$$f_x = 12x^3 + 12x^2 - 72x$$
$$f_y = 3y^2 - 12.$$

The only possible candidates for maximum and minimum points are those which satisfy $f_x = 0$ and $f_y = 0$. Hence we have

$$x(x - 2)(x + 3) = 0$$

and

$$y^2 - 4 = 0,$$

and we have to examine the six points

$$(0, 2), (0, -2), (2, 2), (2, -2), (-3, 2) \text{ and } (-3, -2).$$

Now

$$f_{xx} = 36x^2 + 24x - 72$$
$$f_{yy} = 6y$$
$$f_{xy} = 0.$$

Hence

$$f_{xx}f_{yy} - f_{xy}^2 = 72y(3x^2 + 2x - 6)$$

and this must be positive for a maximum or minimum point. This rules out the points $(0, 2)$, $(2, -2)$ and $(-3, -2)$. Looking at the sign of f_{xx} (or better, f_{yy}) we see that

$(0, -2)$ is a maximum point,

$(2, 2)$ is a minimum point,

and

$(-3, 2)$ is a minimum point.

EXAMPLE 2:

Show that the function

$$f(x, y) = x^2 + 2xy - 3y^2 - 2x + 2y + 5$$

has a single stationary value, and determine its nature.

We must have

$$f_x = 2x + 2y - 2 = 0.$$
$$f_y = 2x - 6y + 2 = 0.$$

These equations have the unique solution $(2, -1)$, which is the only stationary point. To determine its nature we calculate

$$f_{xx}f_{yy} - f_{xy}^2 = 2. -6 - 2^2$$

which is negative. Hence the stationary point is neither a maximum nor a minimum.

If we consider the surface

$$z = x^2 + 2xy - 3y^2 - 2x + 2y + 5$$

and think of the z-axis as being vertical, then the point $x = 2$, $y = -1$ (for which $z = 4$) is a point at which the surface is parallel to the xy-plane. In geographical terms it is a point at which the surface is "level," but is not a maximum (the top of a hill) or a minimum (bottom of a depression). It corresponds, in fact, to a *col, pass* or *saddle*, at which, for example, the land to the east and west rises, while that to the north and south falls away. Figure 9.18 gives an idea of what the surface looks like in the neighborhood of a saddle point.

We see from Fig. 9.18 that if we leave a saddle point we may go up

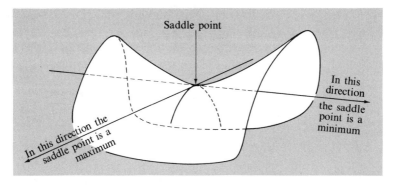

Saddle point

In this direction the saddle point is a minimum

In this direction the saddle point is a maximum

Fig. 9.18

or down according to the direction that we take, and that the saddle point will be a maximum for some directions through it, and a minimum for others.

EXAMPLE 3:

Investigate the stationary values of the function

$$z = x^6 + y^6.$$

Here we have

$$z_x = 6x^5$$
$$z_y = 6y^5$$

so that the only stationary value is $x = y = 0$. But at the point $(0, 0)$ all of z_{xx}, z_{xy}, z_{yy} vanish. Hence the criterion (9.24.7) is not applicable.

However, since z cannot be negative, and is zero only when $x = y = 0$ it is clear that $(0, 0)$ is a minimum. This shows that a function may have maxima and minima even when the criterion (9.24.7) breaks down. If this happens then, in the absence of a more detailed theory of the subject, one simply has to use *ad hoc* methods to sort out what happens in the neighborhood of the stationary point, which is what we have done here.

Exercise 112

1. Show that the stationary points for the function

$$x^2y - 4x^2 - y^2$$

are $(0, 0)$, $(2\sqrt{2}, 4)$ and $(-2\sqrt{2}, 4)$, and prove that two of these are saddle points. What is the other one?

2. Find the stationary points for the following functions, and determine their nature.

    (a) $x^4 - 4x^3 + 4x^2 - 3y^2 + 6y$

    (b) $x^2 y + 2x^2 - 2xy + 3y^2 - 4x + 7y$

    (c) $\log(x^2 + y^2) - x - 2y$.

3. For a function $f(x, y, z)$ of 3 variables a necessary condition for a maximum or minimum at a point is that

$$f_x = f_y = f_z = 0$$

at the point. Find all such points for the functions

    (a) $\sqrt{2}\log(x^2 + y^2 + z^2) - 3(x^3 + y^3 + z^3)$

    (b) $\dfrac{1}{xyz}\exp(x + y + z)$.

4. Prove the theorem contained in the first sentence of problem 3.

9.25 AN APPLICATION

A frequent occupation of the mathematical economist is that of constructing a *mathematical model* of an economic structure. By this is meant the formulation of a set of equations, inequalities, or other mathematical conditions, in which the variables are related in a manner analogous to the way in which the economic variables (price, population, demand, etc.) which they represent are related to each other. We shall meet several models of economic situations in the chapters that follow, and these will serve to clarify the (purposely) inadequate definition of "model" just given. For the moment we will consider what is perhaps the simplest of all possible models.

An economist may observe that two quantities such as, for example, supply and price of a certain commodity, which he is able to measure with a fair degree of accuracy, seem to be proportional to each other, or perhaps (to be rather more general) that they seem to be related in a linear manner. If this were exactly true, and if x and y denote the two variables, then we would have a relation of the form

(9.25.1) $y = mx + d$

between them. Now the economist is not sanguine enough to expect that the quantities he measures will conform *exactly* to equation (9.25.1); for one thing there will probably be some inaccuracies in the measurement of these quantities, and even so, exceptional circumstances might cause y to have a value somewhat different from the value given by (9.25.1). Equation

(9.25.1) is a mathematical model of the system being studied, a system consisting of two quantities and the relation between them. Like any model it portrays some but not all of the characteristics of the situation of which it is a model. Whether it is a good model or a bad one depends on how well it approximates to the situation it represents. Thus if we took pairs of values of x and y, measured in the economic field, plotted them, and obtained a graph looking like Fig. 9.19 we would conclude that although there is clearly some functional relation connecting x and y, at least approximately, this relation is not a linear one, and that the linear model (9.25.1) is not a good one. The points show a tendency to lie on a curve, but not on a straight line. On the other hand if the graph turned out like Fig. 9.20 then we would deem the linear model (9.25.1) to be a reasonably good one, provided that the values of m and d were chosen so that (9.25.1) was the equation of a line such as that shown in the Fig. 9.20, near which the plotted points obviously tend to lie.

In building a mathematical model of an economic structure one almost always introduces parameters whose values are not yet known. Thus in

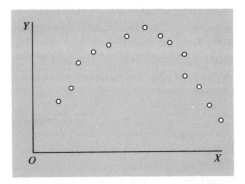

Fig. 9.19

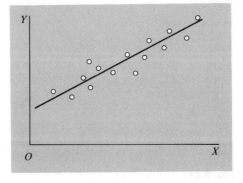

Fig. 9.20

assuming a linear relation between our two variables we automatically
introduced the two parameters m and d, whose values are at our disposal.
The question of deciding whether the model is a good one or not is thus a
question of whether the parameters can be chosen so that the model portrays
the economic system reasonably accurately. In our case, can we choose m and
d so that the line $y = mx + d$ is a "good fit" to the points plotted in Fig.
9.20? We do not expect to be able to choose them so that the points lie exactly
on the line—our model is a simple one which inevitably ignores many factors
which will cause the value of y to differ from its value according to (9.25.1);
but we will be satisfied if the points are all fairly near the line. We shall
now see how to find the "best" values of m and d.

Let there be n observed points (x_i, y_i) $(i = 1, 2, \ldots, n)$. Then corresponding to the value of x_i of x we would expect the corresponding value of y to
be $mx_i + d$. Since it is, in fact, y_i there is a discrepancy, or *deviation*, of

$$mx_i + d - y_i$$

between the value predicted by the model and the observed value, and this
deviation will not, in general, be zero. Thus we have n deviations which, in
some sense or other, we want to make as small as we can by a suitable choice
of m and d. These deviations are the lengths of the short thick lines in Fig.
9.21. Now in choosing m and d to make a particular deviation small we
could well make others large, and it might seem that we should aim to make
the *sum* of the n deviations as small as possible, i.e., that we should try
to minimize

$$\sum_{i=1}^{n} (mx_i + d - y_i).$$

But this would be pointless; for we might have many large positive deviations
cancelling many large negative deviations to give a small sum of deviations.
Somehow we must prevent positive and negative deviations from cancelling

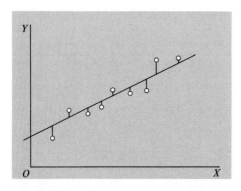

Fig. 9.21

to give a misleadingly small total. One way would be to take the absolute value of each deviation, i.e., to choose m and d to minimize

$$\sum_{i=1}^{n} |mx_i + d - y_i|$$

but this, though sound in theory, is not easy to put into practice. An easier method is to square each deviation (thus obtaining a positive quantity) and minimize the sum

$$S = \sum_{i=1}^{n} (mx + d - y_i)^2.$$

The line thus obtained is said to be the line of *best fit* in the *least squares sense*.

Now S is a function of two variables m and d only, (the x_i and y_i are constants since they are the observed data). We therefore apply the results of the last section to find the minimum value of S. We have

$$\frac{\partial S}{\partial m} = \sum_{i=1}^{n} 2(mx_1 + d - y_i)x_i$$

and

$$\frac{\partial S}{\partial d} = \sum_{i=1}^{n} 2(mx_i + d - y_i),$$

and for a stationary point these must both vanish. Hence we obtain

(9.25.2) $$m \sum_{i=1}^{n} x_i^2 + d \sum_{i=1}^{n} x_i - \sum_{i=1}^{n} x_i y_i = 0$$

and

(9.25.3) $$m \sum_{i=1}^{n} x_i + d \sum_{i=1}^{n} 1 - \sum_{i=1}^{n} y_i = 0.$$

If we write

$$\bar{x} = \frac{1}{n} \sum_{i=1}^{n} x_i \quad \text{and} \quad \bar{y} = \frac{1}{n} \sum_{i=1}^{n} y_i$$

so that $\bar{x}$ and $\bar{y}$ are the means, or averages, of the xs and the ys, we can write (9.25.2) and (9.25.3) as

(9.25.4) $$m.\frac{1}{n} \sum x_i^2 + d\bar{x} = \frac{1}{n} \sum x_i y_i$$

(9.25.5) $$m\bar{x} + d = \bar{y}.$$

Equation (9.25.5) shows that the line of best fit contains the point $(\bar{x}, \bar{y})$. If we multiply (9.25.5) by $\bar{x}$ and subtract from (9.25.4) we obtain

$$(9.25.6) \qquad m\left(\frac{1}{n}\sum x_i^2 - \bar{x}^2\right) = \frac{1}{n}\sum x_i y_i - \bar{x}\bar{y}$$

which gives us the value of m for the line of best fit. We can express this rather more elegantly by, in effect, shifting the origin of our graph to the point $(\bar{x}, \bar{y})$ in the manner of section 5.15, and considering the quantities $x_i - \bar{x}$ and $y_i - \bar{y}$. We have

$$\frac{1}{n}\sum (x_i - \bar{x})^2 = \frac{1}{n}\sum (x_i^2 - 2x_i\bar{x} + \bar{x}^2)$$

$$= \frac{1}{n}\sum x_i^2 - 2\bar{x}\cdot\frac{1}{n}\sum x_i + \frac{1}{n}\sum \bar{x}^2$$

$$= \frac{1}{n}\sum x_i^2 - 2\bar{x}\cdot\bar{x} + \bar{x}^2$$

$$= \frac{1}{n}\sum x_i^2 - \bar{x}^2.$$

Similarly

$$\frac{1}{n}\sum (y_i - \bar{y})^2 = \frac{1}{n}\sum y_i^2 - \bar{y}^2,$$

and

$$\frac{1}{n}\sum (x_i - \bar{x})(y_i - \bar{y}) = \frac{1}{n}\sum x_i y_i - \bar{x}\cdot\frac{1}{n}\sum y_i - \bar{y}\cdot\frac{1}{n}\sum x_i + \bar{x}\bar{y}$$

$$= \frac{1}{n}\sum x_i y_i - \bar{x}\bar{y}.$$

Hence (9.25.6) can be written as

$$(9.25.7) \qquad m\cdot\frac{1}{n}\sum (x_i - \bar{x})^2 = \frac{1}{n}\sum (x_i - \bar{x})(y_i - \bar{y})$$

or as

$$(9.25.8) \qquad m = \frac{\sum (x_i - \bar{x})(y_i - \bar{y})}{\sum (x_i - \bar{x})^2}.$$

Thus from the observed data we can find the line which best fits them, in the least squares sense that we have been considering. It is the line through $(\bar{x}, \bar{y})$ whose slope is given by (9.25.8).

Note that (9.25.8) is not symmetrical in x and y. If we were to regard x as a function of y, instead of the other way round, and look for the line

$$x = m'y + d'$$

of best fit, we would find that it contained the point $(\bar{x}, \bar{y})$ as before (this *is* symmetrical in x and y), but we would have

(9.25.9) $$m' = \frac{\sum (x_i - \bar{x})(y_i - \bar{y})}{\sum (y_i - \bar{y})^2}$$

which does not give the same line (this would require $m' = 1/m$). We thus have *two* best fit lines for what appears to be the same set of data.

The solution to this mystery is that in deriving (9.25.8) we considered the deviations of the ys from the predicted values. This is tantamount to assuming that the reason why the plotted points do not lie exactly on the line is that their y-coordinates are in error, rather than the x-coordinates. The derivation of (9.25.9) made the opposite assumption, so that it is not surprising that a different line was obtained. The two lines given by (9.25.8) and (9.25.9) are called the *regression lines* of y upon x and of x upon y, respectively. It frequently happens that one variable (x or y) can be measured to some accuracy, while the measuring of the other is uncertain, or subject to random errors; according to which it is, so one regression line or the other is used. If neither variable can be accurately measured a more complicated technique may be required. (See problem 113.6 below.)

In the course of the above treatment we have introduced certain often-used quantities whose names the reader may as well know. The quantities

$$\frac{1}{n} \sum (x_i - \bar{x})^2 \quad \text{and} \quad \frac{1}{n} \sum (y_i - \bar{y})^2$$

are known as the *variances* of the xs and the ys respectively. The square roots of these quantities are known as the *root mean square deviations* of the xs and the ys, and are written σ_x and σ_y. They are therefore defined by

$$\sigma_x^2 = \frac{1}{n} \sum (x_i - \bar{x})^2; \quad \sigma_y^2 = \frac{1}{n} \sum (y_i - \bar{y})^2.$$

One further technical term is worth introducing at this stage. We define a quantity r, called the *correlation coefficient* by

(9.25.10) $$r = \frac{\frac{1}{n} \sum (x_i - \bar{x})(y_i - \bar{y})}{\sigma_x \sigma_y}.$$

Using (9.25.10) we can write (9.25.8) as

$$m = r \cdot \frac{\sigma_y}{\sigma_x}$$

and the equation of the regression line of y on x becomes

$$\frac{y - \bar{y}}{\sigma_y} = r\frac{x - \bar{x}}{\sigma_x}.$$

(The reader should verify these two results.)

The correlation coefficient r provides a measure of how closely the plotted points lie on a straight line, i.e., of how suitable is the linear model that we have constructed. If the points all lie *exactly* on a line, then $r = +1$ if the slope of the line is positive, and is -1 if the slope is negative. For less perfect "correlation" between x and y the coefficient r will take some intermediate value. (See Exercise 113.2 below.)

Exercise 113

1. Refer to Exercise 73 where it is proved that

$$(\sum \xi_i^2)(\sum \eta_i^2) \geq (\sum \xi_i\eta_i)^2.$$

By putting $\xi_i = x_i - \bar{x}$ and $\eta_i = y_i - \bar{y}$ prove that $0 \leq |r| \leq 1$.

2. If $y_i = mx_i + d$ for all i, and $m > 0$ show that $r = 1$; and that if $m < 0$ then $r = -1$.

3. Find the best straight line to fit the following pairs of values of x and y

x	1	2	3	4	5	6	7	8	9	10
y	0.8	1.4	2.1	2.6	3.5	4.3	4.7	5.6	6.2	6.8

Calculate the correlation coefficient for these data.

4. We are given sets of values (x_i, y_i, z_i) $(i = 1, 2, \ldots, n)$ of three variables and we postulate that these variables are related approximately by a linear relationship of the form

$$z = ax + by + c$$

which is the equation of a plane. We follow the method of this section to find the plane which best fits the data in a least squares sense by choosing a, b, c, so as to minimize the sum of the squares of the deviations, viz.,

$$\sum_{i=1}^{n} (ax_i + by_i + c - z_i)^2.$$

Show that the problem of finding the minimum point reduces to that of solving three linear equations for a, b, and c. [Do not attempt to show that the (unique) stationary value that you obtain is in fact a minimum.]

5. It is postulated that two variables x and y are related approximately by a quadratic relationship of the form

$$y = ax^2 + bx + c.$$

Show that the problem of finding the "best" values of a, b, c (in the least squares sense) to fit a given set of points (x_i, y_i) ($i = 1, 2, \ldots, n$) reduces to that of solving three linear equations in a, b, c

Hint: Minimize $\sum_{i=1}^{n} (ax_i^2 + bx_i + c - y_i)^2.$

6. In fitting a regression line we use the regression line of y on x if x can be measured accurately but y is subject to error. Thus we take as the deviations (whose sum of squares we are minimizing) the "vertical" distances of the plotted points from the line (see Fig. 9.21). If x is subject to error but not y, we take the "horizontal" distances and get the regression line of x on y. Neither procedure is justified if *both* x and y are subject to error. One possible procedure in that case would be to take the perpendicular distances from the line. (See Fig. 9.22.)

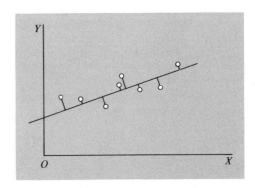

Fig. 9.22

Find the values of m and d which minimize the sum of squares of the perpendicular distances of (x_i, y_i) from the line $y = mx + d$.

Hints: Show that the expression to be minimized is

$$S = \sum \frac{(mx_i + d - y_i)^2}{m^2 + 1}.$$

Next show that $\dfrac{\partial S}{\partial d} = 0$ gives the same condition as before namely that $(\bar{x}, \bar{y})$ lies on the line. Now put $\xi_i = x_i - \bar{x}$ and $\eta_i = y_i - \bar{y}$, so that S becomes

$$\frac{1}{m^2 + 1} \sum (m\xi_i - \eta_i)^2,$$

and show that $\dfrac{\partial S}{\partial m} = 0$ becomes

$$m\sigma_x^2 + (m^2 - 1) \cdot \frac{1}{n} \sum \xi_i \eta_i - m\sigma_y^2 = 0,$$

and hence that

$$\frac{m^2 - 1}{m} = \frac{\sigma_y^2 - \sigma_x^2}{r\sigma_x \sigma_y}.$$

Prove that this procedure gives *two* values for m, corresponding to perpendicular lines. One of these lines is the one we want. What is the significance of the other?

9.26 RESTRICTED MAXIMA AND MINIMA

We often need to find the maximum or minimum values of a function of, let us say, three variables when these variables are not permitted to take all possible sets of values, but are restricted by some relation which must hold between them. To be more specific, we may wish to find the (local) maxima and minima of a function

(9.26.1) $$F = \phi(x, y, z)$$

when x, y and z are not independent but are restricted to satisfy the equation

(9.26.2) $$\psi(x, y, z) = 0.$$

There is a convenient technique for handling problems of this type, and we shall now examine it.

We first observe that we can tackle the problem by sheer brute force as follows. We start by regarding equation (9.26.2) as defining z as a function of x and y. Clearly this is sometimes possible (for example, if

$$\psi(x, y, z) \equiv x^2 + y^2 - xz = 0$$

then $z = (x^2 + y^2)/x$ but it is by no means obvious that it is always so. There is, however, a theorem—the *implicit function theorem*—which says that under certain fairly general conditions a relationship of the form $\psi(x, y, z)$ *does* define z as a function of x and y over some suitable domain. We shall not prove this theorem, or even consider what the conditions for its validity are, but assume that ψ is such that the theorem applies. (The reader can consult the suggested reading at the end of the chapter for further information about this theorem.) If we now substitute this function for z in (9.26.1) then F becomes a function of two independent variables x and y. [It was for this resaon that we wrote both F and ϕ in (9.26.1); ϕ is a function of three variables but F is now a function of only two, x and y.] Treating F as a function of a function we find that

$$(9.26.3) \qquad F_x = \phi_x + \phi_z z_x$$

and

$$(9.26.4) \qquad F_y = \phi_y + \phi_z z_y.$$

Both (9.26.3) and (9.26.4) must vanish if we are to have a maximum or a minimum.

Now although $\psi(x, y, z) = 0$ theoretically expresses z as a function of x and y it would in general be difficult to obtain this expression in practice, and find from it the functions z_x and z_y that are called for in (9.26.3) and (9.26.4). Fortunately we can obtain them indirectly. Since $\psi(x, y, z) = 0$ determines z, given x and y, the function $\psi(x, y, z)$ is a constant (namely zero) for the values of x, y and z that we are considering. Hence its derivatives vanish and we have

$$(9.26.5) \qquad \begin{aligned} \psi_x + \psi_z z_x &= 0 \\ \psi_y + \psi_z z_y &= 0 \end{aligned}$$

from which we can obtain z_x and z_y, viz., $z_x = -\dfrac{\psi_x}{\psi_z}$ and $z_y = -\dfrac{\psi_y}{\psi_z}$, provided $\psi_z \neq 0$.

For a stationary value we must have $F_x = F_y = 0$, and if we substitute for z_x and z_y from (9.26.5) we obtain

$$\phi_x + \phi_z\left(-\frac{\psi_x}{\psi_z}\right) = 0$$

$$\phi_y + \phi_z\left(-\frac{\psi_y}{\psi_z}\right) = 0$$

whence

(9.26.6)
$$\frac{\phi_x}{\psi_x} = \frac{\phi_y}{\psi_y} = \frac{\phi_z}{\psi_z}.$$

Note that although we introduced asymmetry into the problem by choosing z arbitrarily as the variable to express in terms of the other two, the final result is symmetrical. This is not surprising, since there was nothing special about z—we could have chosen x or y instead, provided $\psi_x \neq 0$ or $\psi_y \neq 0$ as the case may be.

We can obtain this result in another way which preserves the symmetry throughout. Let us consider the function

$$G = \phi + \lambda\psi$$

where λ is an undetermined constant multiplier. Then

$$G_x = \phi_x + \lambda\psi_x, \quad G_y = \phi_y + \lambda\psi_y \quad \text{and} \quad G_z = \phi_z + \lambda\psi_z.$$

If we make these three derivatives zero we have

$$\phi_x + \lambda\psi_x = \phi_y + \lambda\psi_y = \phi_z + \lambda\psi_z = 0$$

whence

$$\frac{\phi_x}{\psi_x} = \frac{\phi_y}{\psi_y} = \frac{\phi_z}{\psi_z}$$

as before.

Thus in order to find the restricted maxima and minima of the function $\phi(x, y, z)$ we merely go through the motions of finding the stationary values of the function $\phi + \lambda\psi$; that is, we equate its partial derivatives to zero. We must then eliminate λ, and from what remains determine the possible candidates for maximum and minimum points. This method is known as the method of *Lagrange multipliers*. Some examples will indicate how it works in practice.

EXAMPLE 1:

Find the minimum value of $x^2 + y^2 + z^2$ subject to the restriction $x + y + z = 1$.

Here we have

$$\phi(x, y, z) = x^2 + y^2 + z^2$$

and

$$\psi(x, y, z) = x + y + z - 1.$$

We form the function

$$G = x^2 + y^2 + z^2 + \lambda(x + y + z - 1)$$

and equate to zero its partial derivatives. We have

$$G_x = 2x + \lambda = 0$$
$$G_y = 2y + \lambda = 0$$
$$G_z = 2z + \lambda = 0$$

whence $x = y = z = \frac{1}{2}\lambda$. Hence, x, y and z are equal. But since $x + y + z = 1$ we must have $x = y = z = \frac{1}{3}$. Hence the only candidate for a stationary point is $(\frac{1}{3}, \frac{1}{3}, \frac{1}{3})$, and this is easily seen to give a minimum. The geometrical interpretation of this problem is that it asks for the point on the plane $x + y + z = 1$ for which the square of the distance to the origin is a minimum. Clearly this point is the foot of the perpendicular from the origin to the plane, and the methods of Chapter 6 will confirm the answer we have just obtained.

EXAMPLE 2:
What is the maximum value of xy^2z^3 if $x + y + z = 12$?
    Here we have $G = xy^2z^3 + \lambda(x + y + z - 12)$, which gives

$$y^2z^3 + \lambda = 0$$
$$2xyz^3 + \lambda = 0$$
$$3xy^2z^2 + \lambda = 0$$

on differentiating and equating to zero. From these equations we obtain $\frac{2x}{y} = 1$ and $\frac{3y}{2z} = 1$, or $x : y : z = 1 : 2 : 3$. Since $x + y + z = 12$ we must have $x = 2$, $y = 4$, $z = 6$. Hence the only stationary value is $(2, 4, 6)$. It is not difficult to show that it gives a maximum.

EXAMPLE 3:
Find the maxima and minima of $x^2 + y^2 - 3x + 15y$ subject to the restriction $(x + y)^2 = 4(x - y)$.
    Here $\phi = x^2 + y^2 - 3x + 15y$ and $\psi = (x + y)^2 - 4(x - y)$. We consider the function

$$G = x^2 + y^2 - 3x + 15y + \lambda\{(x + y)^2 - 4(x - y)\}$$

which gives

$$G_x = 2x - 3 + \lambda\{2(x + y) - 4\} = 0$$
$$G_y = 2y + 15 + \lambda\{2(x + y) + 4\} = 0.$$

Eliminating λ we have

(9.26.7) $$\frac{2x - 3}{2x + 2y - 4} = -\lambda = \frac{2y + 15}{2x + 2y + 4}$$

or

$$(2x - 3)(2x + 2y + 4) = (2y + 15)(2x + 2y - 4)$$

which reduces to

(9.26.8) $$x^2 - y^2 - 7x - 7y + 12 = 0.$$

We must remember that we are considering only points satisfying

(9.26.9) $$(x + y)^2 = 4(x - y)$$

so we must find the solutions to these last two equations. For general equations of the second degree, like these, this would not be an easy task, but in this particular case there is an easy method.

We write $2\xi = x + y$ and $2\eta = x - y$, so that $x = \xi + \eta$ and $y = \xi - \eta$. Equations (9.26.8) and (9.26.9) then become

$$2\xi\eta - 7\xi + 6 = 0 \quad \text{and} \quad \xi^2 = 2\eta.$$

It follows that

(9.26.10) $$\xi^3 - 7\xi + 6 = 0.$$

Now $\xi = 1$ is an obvious solution of (9.26.10), and since

$$\xi^3 - 7\xi + 6 = (\xi - 1)(\xi^2 + \xi - 6)$$

the other solutions are the solutions of $\xi^2 + \xi - 6 = 0$ namely -3 and 2. For these three values of ξ we obtain the following pairs (x, y):

$$\left(\frac{3}{2}, \frac{1}{2}\right) \quad ; \quad \left(\frac{3}{2}, -\frac{15}{2}\right) \quad ; \quad (4, 0).$$

This example is rather too complicated to enable us to see easily by *ad hoc* methods whether these points are maxima or minima or not. We therefore approach the problem a little more systematically.

As with unrestricted maxima and minima we consider the expression

(9.26.11) $$\frac{1}{2}(G_{xx}h^2 + 2G_{xy}hk + G_{yy}k^2)$$

where h and k are small changes in x and y. This gives approximately the change in the function G as we go to a neighboring point $(x + h, y + k)$. But h and k are not now independent. For since

$$d\psi = \psi_x dx + \psi_y dy$$

and $\psi = 0$ for all x and y, we have, in the limit

(9.26.12) $$\psi_x h + \psi_y k = 0.$$

Now $\psi_x = 2(x + y) - 4$ and $\psi_y = 2(x + y) + 4$. Also

$$G_{xx} = 2 + 2\lambda = G_{yy}$$

and

$$G_{xy} = 2\lambda.$$

Consequently for the point $(\frac{3}{2}, \frac{1}{2})$, when $\lambda = -2$ (from 9.26.7)* we have $\psi_x = 0, \psi_y = 4, G_{xx} = G_{yy} = -2$ and $G_{xy} = -4$. Thus $k = 0$ (from (9.26.12), and (9.26.11) reduces to

$$\frac{1}{2} \cdot -2 \cdot h^2 = -h^2$$

which is always negative. Thus any change from the point $(\frac{3}{2}, \frac{1}{2})$ consistent with the restriction $(x + y)^2 = 4(x - y)$ produces a decrease in the function. Hence this point is a maximum.

At the point $(\frac{3}{2}, -\frac{15}{2})$, $\lambda = 0$ and (9.26.12) becomes

$$-16h - 8k = 0$$

so that $k = -2h$. The increase in the function is, from (9.26.11),

$$\frac{1}{2}(2h^2 + 0 + 2k^2)$$

and this is clearly positive, whatever the relation between h and k. Hence the point $(\frac{3}{2}, -\frac{15}{2})$ is a minimum.

At the point $(4, 0)$, $\lambda = -\frac{5}{4}$ and (9.26.12) becomes

$$4h + 12k = 0$$

---

* Note that the left-hand expression in (9.26.7) becomes indeterminate; both numerator and denominator vanish. The value of λ is found without trouble from the right-hand side however.

so that $k = -\frac{1}{3}h$. The increase (9.26.11) becomes

$$\frac{1}{2}\left(-\frac{1}{2}h^2 - \frac{5}{2}hk - \frac{1}{2}k^2\right)$$

which reduces to $\frac{5}{36}h^2$ when we put $k = -\frac{1}{3}h$. Since this is necessarily positive the point (4, 0) gives a minimum.

We may remark in passing (though we shall not make use of the fact) that if we want to maximize or minimize a function $\phi(x_1, x_2, \ldots, x_n)$ subject to *several* restrictions, $\psi_1 = 0$, $\psi_2 = 0$, $\ldots$, $\psi_k = 0$, we can do so in a precisely similar way, All that is necessary is to introduce the appropriate number of Lagrange multipliers. Thus we consider the function

$$G = \phi + \lambda_1\psi_1 + \lambda_2\psi_2 + \cdots + \lambda_k\psi_k,$$

differentiate it with respect to each of the x_is, eliminate the multipliers from the equations thus obtained, and determine the values of the x_is. Note that we obtain n equations from the differentiations which, together with the k restrictions, give the right number of equations for determining the x_is and the multipliers.

In our discussion of consumer preference in section 9.22 we took no account of the regrettable fact that bundles of consumer goods are seldom handed to us on a plate, but have to be bought. Thus given the choice of buying one or other of two vectors of commodities a consumer may choose the one of lesser utility because he cannot afford the one of greater utility. If he acts rationally, what he will try to do is to acquire the commodity vector which has the highest utility of those that he can afford. In mathematical terms he will try to maximize $u(x_1, x_2, \ldots, x_n)$ subject to the restriction

$$p_1x_1 + p_2x_2 + \cdots + p_nx_n = k,$$

where p_i is the price of the ith commodity and k is his total budget.

This is a problem in restricted maxima and minima. As above, we consider

$$G = u(x_1, x_2, \ldots, x_n) + \lambda \left(\sum_{i=1}^n p_ix_i - k\right)$$

and equate to zero the partial derivatives. We obtain

(9.26.13) $$\frac{\partial u}{\partial x_i} = -\lambda p_i \qquad (i = 1, 2, \ldots, n).$$

Thus for the optimum choice the quantities $\frac{\partial u}{\partial x_i}$ (called *marginal utilities*) are proportional to the prices.

If we change to a different utility function $v = f(u)$ then (9.26.13) becomes

$$\frac{\partial v}{\partial x_i} = f'(u)\frac{\partial u}{\partial x_i} = -\lambda p_i.$$

Thus the same result holds; but the quantity $-\lambda$ (often called the *marginal utility of income*) will have a different value.

We have, in equations (9.26.13) plus the constraint $\sum p_i x_i = k$, a set of $(n + 1)$ equations in the $(n + 1)$ variables $x_1, x_2, \ldots, x, \lambda$. We could try to solve these equations for the demands x_i and the marginal utility of income λ as explicit functions of the prices p_i and income k, but this would usually be difficult, since these equations will not in general be linear. Indeed, there is no guarantee, even then, that they would have a solution at all. Even a set of $(n + 1)$ *linear* equations in $(n + 1)$ unknowns does not necessarily have a solution, as we shall see in Chapter 14. Nevertheless, we might expect that, under favorable circumstances, these $(n + 1)$ equations would at least implicitly define a set of functions—the x_i and λ as functions of the p_i and k—which determine how the consumer's demand for the various commodities depends on the prices of those commodities and the consumer's income. These functions are demand functions, similar to those that we considered earlier in this chapter.

Exercise 114

1. Find the minimum value of $x^2 + y^2 + z^2$ subject to the restriction $xyz = 1$.

2. A closed rectangular box is to be made from 30 square inches of sheet metal. What is the greatest volume that the box can have?

Hint: Show that the problem is that of maximizing the volume xyz subject to $yz + zx + xy = 15$, and solve by using Lagrange multipliers.

Exercise 115 (Chapter Review)

1. Find from first principles, using the δx notation, the derivatives of
   (a) $\cos(\pi x)$
   (b) $\frac{1}{3}x^3 - 8$.

2. Find the derivatives of the following functions
   (a) $\sin 3x \cos 5x$ (c) $(x^2 + 9x - 7)\sin 2x$
   (b) $\dfrac{x^2 + 2\sin x}{1 - \cos 3x}$ (d) $\dfrac{(1 + \cos x)(3 + \sin^2 x)}{1 - \tan x}$

3. Differentiate with respect to x

(a) $\sec(x^{3/4})$

(c) $\sin(\sin(\sin x))$

(b) $\sqrt{x^2 + x \sin x}$

(d) $\dfrac{\sin^2(x - \tan x)}{1 + \sec^2(3 - 2x)}$

4. (a) If y is defined as a function of x by the equation

$$y = \sin(x + y)^2$$

find $\dfrac{dy}{dx}$.

(b) Find the derivative of $\tan^{-1} \dfrac{2\sqrt{x}}{1 - x}$ in its simplest form.

(c) Differentiate the functions

$$\cos^{-1}\{2x\sqrt{1 - x^2}\} \quad \text{and} \quad \tan^{-1}\left(\frac{\cos x - \sin x}{\cos x + \sin x}\right)$$

5. If $y = e^{ax} \cos bx$ show that

$$\frac{d^n y}{dx^n} = r^n e^{ax} \cos(bx + n\theta)$$

where $r^2 = a^2 + b^2$ and $\tan \theta = \dfrac{b}{a}$.

Hint: Use mathematical induction.

6. Obtain expressions for the nth derivatives of

$$e^x \sin x \quad \text{and} \quad \frac{\sin x}{x}.$$

7. (a) Use Taylor's theorem to show that

$$\tan\left(\frac{\pi}{4} + x\right) = 1 + 2x + 2x^2 + \frac{8}{3}x^3 + \frac{10}{3}x^4 + \cdots$$

8. Find the maxima and minima (if any) of the following functions.

(a) $(x - 1)^2(x + 2)$

(c) $2x^3 - 3x^2 - 36x + 10$

(b) $4x^3 - 18x^2 + 27x - 7$

(d) $x^5 - 15x^3 + 13$

9. Discuss the maxima and minima of the function

$$\frac{\sin^2 x}{\sin\left(x + \frac{\pi}{3}\right)\sin\left(x + \frac{2\pi}{3}\right)}$$

10. If $y = x^m \log x$, find $\dfrac{d^n y}{dx^n}$, where $n < m$, and show that it contains x^{m-n} as a factor.

11. Find the derivative of the function

$$f(x) = \sin x \tan x - 2 \log \sec x$$

and hence show that $f(x)$ increases as x increases between 0 and π.
   Show that the function

$$2 \sin x \tan x - 5 \log \sec x$$

has a minimum value for $0 < x < \dfrac{\pi}{2}$.

12. Find the partial derivatives of the first and second order of the following functions

(a) $\dfrac{x^2 + y^2}{x + y}$

(c) $(x + y) \cos \dfrac{x}{y}$

(b) $(x - y) \sin (3x + 2y)$

13. If $u = 2xy$, $v = \dfrac{1}{y}$ and F is a function of u and v, find

$$F_x, F_y, F_{xx}, F_{xy}, F_{yy},$$

in terms of x, y and the partial derivatives of F with respect to u and v.
   Express the equation

$$F_{xx} + 8xy^2 F_x + 8y(1 - y^2)F_y + 16x^2y^2 F = 0$$

in a form which involves only F, u and v.

14. If $x = r \cos \theta$ and $y = r \sin \theta$, find the following derivatives

(a) $\dfrac{\partial r}{\partial x}, \dfrac{\partial r}{\partial y}, \dfrac{\partial \theta}{\partial x}, \dfrac{\partial \theta}{\partial y}.$

(b) $\dfrac{\partial x}{\partial \theta}, \dfrac{\partial y}{\partial \theta}, \dfrac{\partial x}{\partial r}, \dfrac{\partial y}{\partial r}.$

   Note that, for example, $\dfrac{\partial x}{\partial r}$ and $\dfrac{\partial r}{\partial x}$ are not reciprocals, since in $\dfrac{\partial x}{\partial r}$ the "other" variable—the one that is taken to remain constant—is θ, whereas for $\dfrac{\partial r}{\partial x}$ it is y.

15. In Exercise 111.4 we saw that for the indifference loci of the utility function $u(x, y)$ to have the general shape of Fig. 9.17 the following conditions are necessary.

(a) u_x and u_y have the same sign,

(b) $\dfrac{u_y^2 u_{xx} - 2u_x u_y u_{zy} + u_x^2 u_{yy}}{u_y^3} < 0.$

Show that if $u(x, y)$ has these two properties then so does the utility function

$$v(x, y) = \phi\{u(x, y)\},$$

whatever the function ϕ may be.

16. (a)   Find the maxima and minima of the function

$$x^3 + y^3 - 3x - 12y + 20.$$

(b)   Show that the function

$$(x + y)e^{x^2 + y^2}$$

has no stationary values.

17.   Show that the function $\cos(x^2 + y^2)$ is stationary at an infinite number of points (x, y), viz., at all points for which

$$x^2 + y^2 = k\pi \qquad (k = 0, \pm 1, \pm 2, \ldots).$$

(These points form concentric circles in the xy-plane.)

Which, if any, of these points are maximum or minimum points?

18.   Find the maximum and minimum values of xy subject to the restriction $x^2 + xy + y^2 = 1$.

Suggestions for Further Reading

In addition to the books recommended for Chapter 8 the reader can profitably refer to the following:

Apostol, T.M., *Calculus.* 2 vols. New York: Blaisdell Publishing Company, 1962.

Flett, T. M., *Mathematical Analysis.* New York: McGraw Hill Book Company, 1965.

Kaplan, W., *Advanced Calculus.* Reading, Massachusetts: Addison-Wesley, Inc., 1952.

Lamb, H., *An Elementary Course of Infinitesimal Calculus.* New York: Cambridge University Press, 1961.

The following book in the "Outline" series is also recommended:

Ayres, F., *Differential and Integral Calculus.* New York: Schaum Publishing Company, 1950.

10   Integration

10.1   AREAS

We saw in the last chapter how the idea of the derivative of a function arose from a fairly simple geometrical concept, namely that of the tangent to a curve at a point. In much the same way our next topic, *integration*, also arose from the need to solve a geometrical problem of common occurrence, namely the determination of areas. The formulae for the areas of squares, rectangles, triangles and other rectilinear figures have been known for millenia. The Greeks, and in particular Archimedes, discovered formulae for the areas of many curvilinear figures, but mainly by *ad hoc* methods that did not admit of a great deal of generalization. It was not until the invention of the integral calculus in the 17th century that the determination of areas of figures became largely a matter of routine.

Let us consider the following problem: Fig. 10.1 shows a portion of the graph of some function $f(x)$; P and Q are two points of the graph, and PM and QN are the ordinates through P and Q corresponding to $x = a$ and $x = b$. It is required to find the area which is shaded in the figure, i.e., the area "underneath" the arc PQ of the curve. It is clear that if we can solve this problem then we can find the areas of a wide range of geometrical figures. The fact that the shaded area has three rectilinear sides and only one curvilinear

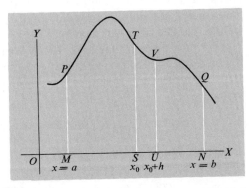

FIG. 10.1

457

side does not matter since, given a fairly general curvilinear figure, such as that of Fig. 10.2, we can cut it up into a number of pieces each of which presents us with a problem of the kind that we are considering. This figure also shows how, for a wide variety of areas at least, this cutting up can be done in a way that justifies us in assuming that the function $f(x)$ mentioned above is single-valued.

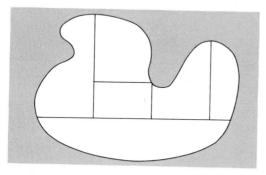

Fig. 10.2

Whether this can be done for *any* area is quite another matter. It would be as well to make it clear now that in talking about integration we shall not be able to make even a near-rigorous approach to the subject, let alone give a completely rigorous treatment. To do so would lead us too far away from the general trend and purpose of the book. A little further on we shall see many ways in which our treatment fails to be rigorous.

We start by considering the area between the ordinate at $x = a$ and the ordinate for a general value of x (denoted in Fig. 10.1 by ST). For different values of x this ordinate will assume different positions, and the area under the curve, lying between it and the ordinate MP, will have different values. This area is thus a function of x. Let us denote this function by $F(x)$. Our problem will certainly be solved if, given $f(x)$, we can find the function $F(x)$, for the required area will then be $F(b)$.

We now compare the area $MPTS$ "between" the ordinates $x = a$ and $x = x_0$ (with an obvious interpretation of the word "between") with the area $MPUV$ between $x = a$ and $x = x_0 + h$, where h is some number. Then we have

$$F(x_0) = \text{Area } MPTS$$
$$F(x_0 + h) = \text{Area } MPUV.$$

Therefore

$$F(x_0 + h) - F(x) = \text{Area } MPUV - \text{Area } MPTS$$
$$= \text{area of the strip } STVU.$$

This strip $STVU$ is an area which is bounded in part by the curvilinear segment, TV, so we cannot yet find its area. However, if M and m are the upper and lower bounds of $f(x)$ in the range $x_0 \leq x \leq x_0 + h$, we can certainly say that

$$mh \leq \text{area } STVU \leq Mh$$

or

$$m \leq \frac{\text{area } STVU}{h} \leq M.$$

Now as $h \to 0$ we would expect that M and m would both $\to f(x_0)$. If we assume that this is so, we see that

(10.1.1.)
$$\lim_{h \to 0} \frac{\text{area } STVU}{h} = f(x_0)$$

and hence

(10.1.2)
$$\lim_{h \to 0} \frac{F(x_0 + h) - F(x_0)}{h} = f(x_0).$$

We now see that the left-hand side of (10.1.2) is the derivative of $F(x)$ at $x = x_0$. The outcome of all this is therefore that the function $F(x)$ has this property—its derivative at $x = x_0$ is $f(x_0)$, i.e., $F'(x_0) = f(x_0)$. Since x_0 was any point in the range $a \leq x \leq b$ we see that

(10.1.3)
$$F'(x) = f(x)$$

for all values of x in this range.

Thus the problem of finding areas leads to the following problem: given a function $f(x)$ defined over a certain range of values of x, does there exist (and can we find) a function $F(x)$ whose derivative over that range is $f(x)$?

Now we must remark that in what has just gone we have made some very powerful assumptions. We have, for example, assumed that M and m will tend to $f(x_0)$ as $h \to 0$. This is clearly a rash assumption, as many examples will show. The function $\sin 1/x$, which we have already met, is one which fluctuates wildly in the neighborhood of $x = 0$, and for values in the range $0 < x < h$ the values of M and m are 1 and -1 respectively. Thus no matter how we defined this function at $x = 0$ (it is not defined by the formula $\sin 1/x$), M and m would not both tend to $f(0)$.

Again, take as the function $f(x)$ the function which takes the value 1 for all values of x. It is clear that the corresponding function $F(x)$ is just the function x, if we take $a = 0$. Now consider the function $g(x)$ defined by

$$g(x) = 1 \quad \text{if } x \neq 0$$
$$g(0) = 0,$$

which differs from the previous $f(x)$ only at a single point. Intuitively it is clear that this trivial difference cannot affect the area under the curve between two ordinates, whatever you may choose to mean by "area," so that the function $F(x)$ must be the same for $f(x)$ as for $g(x)$. But if this is so then we cannot have both $F'(x) = f(x)$ and $F'(x) = g(x)$, for these two statements imply that $F'(0)$ is both 0 and 1. Something is wrong somewhere; clearly there is much more to the relationship between $f(x)$ and $F(x)$ than meets the eye.

In point of fact we shall not be greatly concerned with the determination of areas, and we are on fairly safe ground if we confine ourselves to the problem which has arisen as a result of our talking around the matter of areas in the previous paragraphs, namely "Given a function $f(x)$, find (if possible) a function $F(x)$ such that $F'(x) = f(x)$ over some given range." The process of going from the function $f(x)$ to the function $F(x)$ will be called *integration*, and $F(x)$ will be called an *integral* of $f(x)$.

There are still a few loose ends to tidy up before we get to the main business of the chapter. First of all, the function $F(x)$ is not unique. For if $F(x)$ is a function whose derivative is $f(x)$, then so is $F(x) + A$, where A is any constant.

Thus $F(x)$ is ambiguous to the extent of an arbitrary constant. If we are finding the area between given ordinates at $x = a$ and $x = b$ this constant will be determined; for $F(x)$ gives the area between the ordinate at $x = a$ and the ordinate at a general value of x, and therefore it must be zero when $x = a$. Thus if $G(x)$ is any function whose derivative is the given function $f(x)$, then

$$F(x) = G(x) + A.$$

But we must have $F(a) = 0$, and therefore $A = -G(a)$. The required area between $x = a$ and $x = b$ is then

$$F(b) = G(b) + A$$
$$= G(b) - G(a).$$

Alternatively we can say that $F(x)$, whichever of the possibilities we choose, represents the area from *some* ordinate, say $x = \xi$, to a general ordinate x. The area between $x = a$ and $x = b$ will then be the difference between the area from $x = \xi$ to $x = b$ and that from $x = \xi$ to $x = a$, as in Fig. 10.3. This gives us $F(b) - F(a)$ as the area immediately.

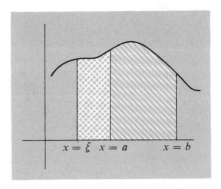

$x = \xi$ $x = a$ $x = b$

Fig. 10.3

Let us summarize what has so far been suggested by the consideration of the problem of finding areas. We are given a function $f(x)$, and we wish to find, if possible, a function $F(x)$ whose derivative is $f(x)$, (at least in some given range $a \leq x \leq b$). We have seen that if there is one such function then there are many others, which differ from it merely by an added constant. We may add here that any such function is called an *indefinite integral* of the function $f(x)$, and is denoted by the rather odd set of symbols

$$\int f(x)\, dx.$$

We also saw that expressions of the form $F(b) - F(a)$ are of importance in at least one application, viz., the finding of areas. This expression is called the *definite integral* of $f(x)$ between the limits a and b. It is denoted by

$$\int_a^b f(x)\, dx,$$

and if a and b are constants, it is itself a constant, not a function. The function $f(x)$, in both the indefinite and definite integral, is called the *integrand; a* and b are called the *limits* of integration. The symbols $\int \ldots dx$ which effectively mean "the integral of" the function which they enclose, always go together, and should be treated as one composite symbol.

The way in which this peculiar notation arose is worth a brief mention. The early pioneers of the integral calculus imagined the figure of which they were finding the area to be divided up into a large number of strips, as in Fig. 10.4. These strips were of the same width, which they denoted by δx, since the intention was to let $\delta x \to 0$. (Compare Chapter 9, page 348). The area of the strip between x and $x + \delta x$ was therefore *approximately* $f(x)\, \delta x$. The total area was regarded as being obtained by a summation process, and was denoted by

$$\sum_a^b f(x)\, \delta x \quad \text{or} \quad \underset{a}{\overset{b}{S}} f(x)\, \delta x,$$

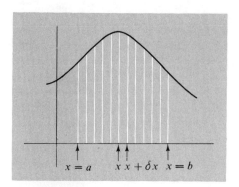

Fig. 10.4

using the $\sum$ notation that we have already met, or the letter S as an alternative to $\sum$. The "δx" naturally became "in the limit" the symbol "dx", and in the course of time the "S" became elongated to form the integral sign $\int$.

The process of finding integrals (either definite or indefinite) of functions is known as *integration*, and by defining it in this way, i.e., as a sort of *inverse* of differentiation, we avoid many of the difficulties that were mentioned earlier. It does mean, however, that a lot of functions will not have integrals, including many, like the function $g(x)$ mentioned on page 460, which, one feels, ought to have integrals. This is the price we must pay for not being rigorous. Luckily, for the purposes of this book, this disadvantage will not be of any consequence.

Exercise 116

1. Divide the interval $[0, 1]$ along the x-axis into n equal parts (each of length $1/n$) and through the points of division draw ordinates to the curve $y = x^2$. This divides the area under this curve into n vertical strips, rather as in Fig. 10.4.

Show that the area of the rth strip lies between $\dfrac{1}{n^3}(r - 1)^2$ and $\dfrac{1}{n^3}r^2$.
Deduce that the area under the curve, between $x = 0$ and $x = 1$, lies between $\dfrac{1}{n^3}\sum\limits_{s=1}^{n-1} s^2$ and $\dfrac{1}{n^3}\sum\limits_{s=1}^{n} s^2$.

2. Using the result

$$\sum_{s=1}^{n} s^2 = \frac{1}{6}n(n + 1)(2n + 1)$$

show that if A is the area under the curve in problem 1, then

$$\frac{1}{6}\left(1 - \frac{1}{n}\right)\left(2 - \frac{1}{n}\right) \leq A \leq \frac{1}{6}\left(1 + \frac{1}{n}\right)\left(2 + \frac{1}{n}\right).$$

By letting $n \longrightarrow \infty$ deduce that $A = \tfrac{1}{3}$.

3. By the same argument show that the area under the curve $y = x^2$ between $x = 0$ and $x = a$ is $\frac{1}{3}a^3$.

Note: This exercise indicates that areas of curvilinear figures can sometimes be found without using any new techniques; but for curves of any complexity such methods, even if feasible, would be laborious.

10.2 INTEGRALS OF SIMPLE FUNCTIONS

Since integration is, so to speak, differentiation in reverse, we can immediately write down the integrals of many functions by reference to the table of derivatives given on page 396. Thus to find the value of

$$\int_3^5 x^2 \, dx$$

we note that, since the derivative of x^3 is $3x^2$, the derivative of $\frac{1}{3}x^3$ is x^2. Hence the indefinite integral, call it $F(x)$, of x^2 is

$$F(x) = \int x^2 \, dx = \frac{1}{3}x^3 + A$$

where A is an arbitrary constant. Hence the definite integral in question is given by

$$\int_3^5 x^2 \, dx = F(5) - F(3)$$
$$= \frac{1}{3} \times 5^3 - \frac{1}{3} \times 3^3$$
$$= \frac{98}{3}.$$

This calculation is usually written in the following form:

(10.2.1) $$\int_3^5 x^2 \, dx = \left[\frac{1}{3}x^3 \right]_3^5$$
$$= \frac{1}{3} \times 5^3 - \frac{1}{3} \times 3^3,$$

where the right-hand side of (10.2.1) is a convenient way of symbolizing the intermediate stage at which the function $F(x)$ has been found, but the limits of integration have not yet been substituted.

Here are a few more examples which the reader should check for himself.

1. $$\int_{-1}^{1} x^4 \, dx = \left[\frac{1}{5} x^5 \right]_{-1}^{1}$$
$$= \frac{1}{5} \cdot 1 - \frac{1}{5}(-1)$$
$$= \frac{2}{5}.$$

2. $$\int_{0}^{3} (x^3 + e^x) \, dx = \left[\frac{1}{4} x^4 + e^x \right]_{0}^{3}$$
$$= \frac{1}{4} \cdot 3^4 + e^3 - 0 - 1$$
$$= \frac{77}{4} + e^3.$$

3. $$\int_{-1}^{2} e^{5x} dx = \left[\frac{1}{5} e^{5x} \right]_{-1}^{2}$$
$$= \frac{1}{5} e^{10} - \frac{1}{5} e^{-5}.$$

4. $$\int_{0}^{\pi/2} \sin x \, dx = [-\cos x]_{0}^{\pi/2}$$
$$= -0 + 1$$
$$= 1.$$

5. $$\int_{0}^{\pi/4} \sec^2 x \, dx = [\tan x]_{0}^{\pi/4}$$
$$= 1 - 0$$
$$= 1.$$

When finding indefinite integrals one should always remember about the arbitrary added constant. To be on the safe side it is best to include this constant explicitly in the indefinite integral. Thus we write

$$\int x^3 \, dx = \frac{1}{4} x^4 + A$$

$$\int \sin x \, dx = - \cos x + A$$

$$\int \cos x \, dx = \sin x + A$$

$$\int (x^4 + 5x^3 - 7x^2 + 3x - 5)\, dx = \frac{1}{5}x^5 + \frac{5}{4}x^4 - \frac{7}{3}x^3 + \frac{3}{2}x^2 - 5x + A$$

and so on.

From the definition of integration as the reverse of differentiation it follows that

$$\int (f(x) + g(x))\, dx = \int f(x)\, dx + \int g(x)\, dx$$

and that

$$\int af(x)\, dx = a \int f(x)\, dx$$

where a is any constant. Thus if we have a list of integrals of a certain number of standard functions, the integral of any function which is a linear combination of these standard functions is readily found. A short list of this kind is the following:

$f(x)$	$\int f(x)\, dx$
x^k	$\dfrac{1}{k+1}x^{k+1} + A$
	(provided $k \neq -1$)
x^{-1}	$\log x + A$
$\sin x$	$-\cos x + A$
$\cos x$	$\sin x + A$
$\sec^2 x$	$\tan x + A$
$e^{\lambda x}$	$\dfrac{1}{\lambda}e^{\lambda x} + A$

We shall add more functions to this list as we go along.

Exercise 117

1. Integrate the following functions
    (a) $\sqrt{x}$
    (b) $\sqrt{3x - 4}$
    (c) $4x^3 + 6x^2 - 5x + 7$
    (d) $2 \cos x + \sin x$
    (e) $5e^x + 7 \sec^2 x - \dfrac{1}{x}$

2. Evaluate the following definite integrals:

(a) $\displaystyle\int_0^\pi \sin x \, dx$ (c) $\displaystyle\int_0^{\pi/3} \sec^2 x \, dx$

(b) $\displaystyle\int_{a^2}^{b^2} \frac{1}{x} \, dx$ (d) $\displaystyle\int_2^3 e^{2x} \, dx$

3. The *cost stream* for a firm is a function $C(t)$ which expresses the rate at which money is being payed out by the firm at any given time t. From the definition of a definite integral deduce that the total payment made by a firm between time $t = t_0$ and $t = t_1$ is

$$\int_{t_0}^{t_1} C(t) \, dt.$$

Show similarly that if $R(t)$ is the *receipts stream* or rate at which money is being received, then the total amount received between $t = t_0$ and $t = t_1$ is

$$\int_{t_0}^{t_1} R(t) \, dt.$$

Deduce that if $P(t)$, the *profit stream*, is defined as $R(t) - C(t)$, then the profit made by the firm between $t = t_0$ and $t = t_1$ is

$$\int_{t_0}^{t_1} P(t) \, dt = \int_{t_0}^{t_1} \{R(t) - C(t)\} \, dt$$

$$= \int_{t_0}^{t_1} R(t) \, dt - \int_{t_0}^{t_1} C(t) \, dt.$$

10.3 INTEGRATION BY SUBSTITUTION

The remainder of this chapter will be concerned with various tricks for finding the integrals of functions whose integrals are not apparent on inspection.

Very often we can evaluate an integral by regarding the variable with respect to which the integral is being taken (and which we shall usually denote by x, as in the examples so far) as being itself a function of another variable, say t. Suppose $F(x)$ is an indefinite integral of $f(x)$, so that

$$\frac{d}{dx} F(x) = f(x),$$

and we regard x as a function of t, say $x = g(t)$; then F will be a function of t, viz., $F[g(t)]$. Differentiating this with respect to t we have

$$\frac{d}{dt} F[g(t)] = F'[g(t)]g'(t)$$

$$= f[g(t)]g'(t).$$

Hence $F[g(t)]$ is the integral of $f[g(t)]g'(t)$. This we may write as

(10.3.1)
$$\int f[g(t)]g'(t)\, dt = F[g(t)]$$

$$= F(x)$$

and we have expressed the required integral $F(x)$ in terms of the integral of another function $f[g(t)]g'(t)$. We have, in fact,

(10.3.2)
$$\int f(x)\, dx = \int f[g(t)]g'(t)\, dt$$

and it may well happen that the integral on the right-hand side of (10.3.2) is easily evaluated, in which case we will have found the required integral. The finding of an integral by this method is known as *integration by substitution* since it is effected by substituting $g(t)$ for x.

The method of getting from the left-hand side of (10.3.2) to the right-hand side can be easily remembered in the following way. If in the integral $\int f(x)\, dx$ we make the substitution $x = g(t)$, then $f(x)$ becomes $f[g(t)]$. At the same time we have

$$\frac{dx}{dt} = g'(t)$$

which we shall write in an alternative form $dx = g'(t)\, dt$. Making these two substitutions, for $f(x)$ and dx, we obtain

$$\int f[g(t)]\, g'(t)\, dt$$

which is the right-hand side of (10.3.2). It is hardly necessary to remark that since $\int \ldots dx$ is really a single, composite symbol it is not logically justifiable to juggle with part of it (viz., "dx") in this way. However, this juggling is not meant to prove anything; it is simply a rule for constructing the new integral, and as such (but only as such) it will be very useful. A few examples should make the method clear.

EXAMPLE 1.

To find $\int (3x + 5)^{10}\, dx$.

We could, of course, expand $(3x + 5)^{10}$ by the binomial theorem into a polynomial, and integrate term by term; but this would be very laborious. Instead, let us set $3x + 5 = t$, that is $x = \frac{1}{3}(t - 5)$. We then have that $(3x + 5)^{10}$ becomes t^{10} and since $dx/dt = \frac{1}{3}$ we have $dx = \frac{1}{3}\,dt$. Substituting in the integral we obtain

$$\int t^{10} \cdot \frac{1}{3}\,dt, \text{ which is } \frac{1}{33}\,t^{11} + A.$$

Hence, reverting to our original variable x, we have

$$\int (3x + 5)^{10}\,dx = \frac{1}{33}\,(3x + 5)^{11} + A.$$

EXAMPLE 2.

To find $\int \dfrac{1}{\sqrt{a^2 - x^2}}\,dx$.

A useful substitution for integrands in which expressions like $\sqrt{a^2 - x^2}$ appear is $x = a \sin \theta$. Making this substitution in the above integral we have

$$\frac{1}{\sqrt{a^2 - x^2}} = \frac{1}{\sqrt{a^2 - a^2 \sin^2 \theta}} = \frac{1}{a \cos \theta},$$

since $1 - \sin^2 \theta = \cos^2 \theta$; we also have $dx = a \cos \theta\,d\theta$. Hence

$$\int \frac{1}{\sqrt{a^2 - x^2}}\,dx = \int \frac{1}{a \cos \theta} \cdot a \cos \theta\,d\theta.$$

$$= \int d\theta$$

$$= \theta + A.$$

Now $\theta = \sin^{-1}(x/a)$, so that, reverting to the original variable x, we have

$$\int \frac{1}{\sqrt{a^2 - x^2}}\,dx = \sin^{-1}\frac{x}{a} + A.$$

This is not a new result, since

$$\frac{d}{dx}\left(\sin^{-1}\frac{x}{a}\right) = \frac{1}{\sqrt{a^2 - x^2}}$$

occurs in the list of derivatives in Chapter 9, page 396.

EXAMPLE 3.

To find $\int \sqrt{1 - x^2} \, dx$.

Using the same substitution as in Example 2, but with $a = 1$, we put $x = \sin \theta$. We then have $\sqrt{1 - x^2} = \cos \theta$ and $dx = \cos \theta \, d\theta$. On substitution, the integral becomes

$$\int \cos \theta \cdot \cos \theta \, d\theta = \int \cos^2 \theta \, d\theta.$$

Now $2 \cos^2 \theta = 1 + \cos 2\theta$ (see Exercise 23.4) and we can therefore write the integral as

$$\frac{1}{2} \int (1 + \cos 2\theta) \, d\theta = \frac{1}{2} \int d\theta + \frac{1}{2} \int \cos 2\theta \, d\theta.$$

The first integral is just $\frac{1}{2}\theta$. In the second we can substitute $\theta = \frac{1}{2}\phi$, $d\theta = \frac{1}{2}d\phi$ and obtain*

$$\frac{1}{4} \int \cos \phi \, d\phi = \frac{1}{4} \sin \phi + A$$

$$= \frac{1}{4} \sin 2\theta + A$$

$$= \frac{1}{2} \sin \theta \cos \theta + A.$$

Hence

$$\int \sqrt{1 - x^2} \, dx = \frac{1}{2} \theta + \frac{1}{2} \sin \theta \cos \theta + A$$

$$= \frac{1}{2} \sin^{-1} x + \frac{1}{2} x \sqrt{1 - x^2} + A.$$

EXAMPLE 4.

To find $\int \dfrac{2x + 3}{x^2 + 3x + 5} \, dx$.

Let us put $t = x^2 + 3x + 5$. This gives us t as a function of x, rather than the other way around, but as it happens it is not necessary to find

---

* Very simple substitutions like this can often be made mentally, without the need for writing them out explicitly.

x in terms of t. We have

(10.3.3) $$\frac{dt}{dx} = 2x + 3$$

whence, as we saw in Chapter 9,

$$\frac{dx}{dt} = \frac{1}{2x + 3}$$

or

$$dx = \frac{1}{2x + 3}\, dt.$$

Note that x still appears on the right-hand side, whereas we are supposed to be substituting for x in terms of t. If we perform what substitutions we can, we find that

$$\int \frac{2x + 3}{x^2 + 3x + 5}\, dx = \int \frac{2x + 3}{t} \cdot \frac{1}{2x + 3}\, dt$$
$$= \int \frac{1}{t}\, dt.$$

The unsubstituted xs have disappeared! This is a consequence of the way that the integrand was chosen. Had any xs remained in the new integral, they would have had to be expressed in terms of t before the method could proceed further.

We could have shortened this substitution by observing from (10.3.3) that

$$(2x + 3)\, dx = dt.$$

Writing the original integral in the equivalent form

$$\int \frac{(2x + 3)\, dx}{x^2 + 3x + 5},$$

we obtain

$$\int \frac{(2x + 3)\, dx}{x^2 + 3x + 5} = \int \frac{dt}{t}$$
$$= \log t + A$$
$$= \log(x^2 + 3x + 5) + A.$$

This last result has an obvious generalization. We know that if $u(x)$ is a differentiable function, then

$$\frac{d}{dx} \log u(x) = \frac{u'(x)}{u(x)}.$$

Hence

$$\int \frac{u'(x)}{u(x)} \, dx = \log u(x) + A.$$

Thus if the integrand is a fraction whose numerator is the derivative of the denominator, the integral can be found immediately.

EXAMPLE 5.

To find $\int \frac{dx}{1 - x^2}$.

The denominator of this integrand factorizes into two factors, $1 + x$ and $1 - x$, and this enables us to express it as the sum of two simpler fractions. We have in fact

$$\frac{1}{1 - x^2} = \frac{1/2}{1 + x} + \frac{1/2}{1 - x}$$

as the reader can readily verify. Thus

$$\int \frac{dx}{1 - x^2} = \frac{1}{2} \int \frac{dx}{1 + x} + \frac{1}{2} \int \frac{dx}{1 - x}$$

$$= \frac{1}{2} \log (1 + x) - \frac{1}{2} \log (1 - x) + A$$

$$= \frac{1}{2} \log \frac{1 + x}{1 - x} + A.$$

If the integral to be found is a definite integral, and a substitution is made, it is important to keep track of what happens to the limits. One method, shown in Example 1 below, would be to find the indefinite integral and then, when this has been expressed in terms of the original variable, to find its values at the two limits and take the difference. This method may require some unnecessary work however. Alternatively, where we make a substitution, the limits can be replaced by the corresponding values of the new variable. If this is done it is not necessary to change back to the original variable in order to evaluate the integral. Example 2 given below should make this clear.

Note that when limits appear in an integral, they are assumed to be values of the variable that appears in the expression of the integral. If they are values of some other variable then it must be made clear what variable they are values of. Example 3 shows how this is done.

1. To find $\displaystyle\int_{3/2}^{9/2} \frac{dx}{\sqrt{6x - x^2}}$.

We can write this as $\displaystyle\int_{3/2}^{9/2} \frac{dx}{\sqrt{9 - (x - 3)^2}}$, and if we now make the substitution $t = x - 3$, we get $\displaystyle\int \frac{dt}{\sqrt{9 - t^2}} = \sin^{-1} \frac{t}{3} + A$. Hence

$$\int \frac{dx}{\sqrt{6x - x^2}} = \sin^{-1}\left(\frac{x - 3}{3}\right) + A.$$

Putting in the limits for x, we have

$$\int_{3/2}^{9/2} \frac{dx}{\sqrt{6x - x^2}} = \left[\sin^{-1}\left(\frac{x - 3}{3}\right)\right]_{3/2}^{9/2}$$

$$= \sin^{-1}\left(\frac{1}{2}\right) - \sin^{-1}\left(-\frac{1}{2}\right)$$

$$= \frac{\pi}{3}.$$

2. To find $\displaystyle\int_1^2 \frac{x\,dx}{\sqrt{16 - x^4}}$.

If we make the substitution $t = x^2$, then $x\,dx = \frac{1}{2}\,dt$. The values of t corresponding to $x = 1$ and $x = 2$ are 1 and 4. Hence

$$\int_1^2 \frac{x\,dx}{\sqrt{16 - x^4}} = \int_1^4 \frac{\frac{1}{2}\,dt}{\sqrt{16 - t^2}}$$

$$= \frac{1}{2}\left[\sin^{-1}\frac{t}{4}\right]_1^4$$

$$= \frac{1}{2}\left[\sin^{-1} 1 - \sin^{-1}\frac{1}{4}\right]$$

$$= \frac{\pi}{4} - \frac{1}{2}\sin^{-1}\frac{1}{4}.$$

Note that, using this method, we do not have to go back to the original variable x.

3. To find $\int_1^e \dfrac{dx}{x\{1 + (\log x)^2\}}$.

We shall make the substitution $t = \log x$, so that $dt = dx/x$, but for the sake of example we shall not find the values of t corresponding to the limits. This being so we must, if we put the limits in at all, indicate that they are values of x, and not of t—the variable in terms of which the integral is currently being expressed. This we can do as follows

$$\int_1^e \frac{dx}{x\{1 + (\log x)^2\}} = \int_{x=1}^{x=e} \frac{dt}{1 + t^2}$$

$$= \left[\tan^{-1} t\right]_{x=1}^{x=e}$$

$$= \left[\tan^{-1} (\log x)\right]_1^e$$

$$= \tan^{-1} 1 - \tan^{-1} 0$$

$$= \frac{\pi}{4}.$$

Here we had to return to the original variable.

Exercise 118

1. Find the indefinite integral $\int \dfrac{x \, dx}{\sqrt{4 - x^2}}$ in two different ways:
   (a) by substituting $x = 2 \sin \theta$
   (b) by substituting $t = x^2$
2. Show that

$$\frac{3x - 5}{(x - 1)(x - 2)} = \frac{2}{x - 1} + \frac{1}{x - 2}$$

and hence find

$$\int \frac{3x - 5}{(x - 1)(x - 2)} \, dx.$$

3. By putting $x = 2 \tan \theta$ find the value of $\int_0^2 \dfrac{dx}{4 + x^2}$.

4. Find $\int \dfrac{dx}{\sqrt{x + a} - \sqrt{x - a}}$.

Hint: multiply top and bottom in the integrand by $\sqrt{x + a} + \sqrt{x - a}$.

5. By means of the substitution $x^2 = a^2 \cos 2\theta$, show that

$$\int_0^a x \sqrt{\frac{a^2 - x^2}{a^2 + x^2}} \, dx = \frac{1}{4} a^2 (\pi - 2).$$

6. The cost stream of a certain firm throughout a year is given by the function

$$C(t) = 1000t(12 - t) \text{ dollars per month}$$

where t is the time in months. (See Exercise 117.3.)

The stream of receipts for the same firm during the same year is given by

$$R(t) = 20(t^3 - 30t^2 + 300t + 1000) \text{ dollars per month.}$$

Sketch these two curves and, by finding the area between them, calculate the total net profits earned during the year.

10.4 INTEGRATION BY PARTS

If U and V are functions of x then

$$(10.4.1) \qquad \frac{d}{dx}(UV) = \frac{dU}{dx}V + U\frac{dV}{dx}$$

as we proved in the last chapter. If we integrate both sides of (10.4.1) we obtain

$$\int \frac{d}{dx}(UV)\, dx = \int V\frac{dU}{dx}\, dx + \int U\frac{dV}{dx}\, dx.$$

Now the integral on the left is just UV, since it is the integral of a derivative. Hence we may integrate both sides and write

$$(10.4.2) \qquad \int U\frac{dV}{dx}\, dx = UV - \int V\frac{dU}{dx}\, dx.$$

Suppose we consider the particular case when $U = \tan^{-1} x$ and $V = x$. Then (10.4.2) becomes

$$\int \tan^{-1} x \cdot 1 \cdot dx = (\tan^{-1} x)x - \int x\frac{1}{1 + x^2}\, dx$$

since $\dfrac{dU}{dx} = \dfrac{d}{dx}(\tan^{-1} x) = \dfrac{1}{1 + x^2}.$ Hence we have

$$\int \tan^{-1} x \, dx = x \tan^{-1} x - \int \frac{x\, dx}{1 + x^2}$$

$$= x \tan^{-1} x - \frac{1}{2} \log (1 + x^2) + A,$$

and we have found a new integral, namely that of $\tan^{-1} x$.

Clearly equation (10.4.2) has potentialities as a method for finding integrals. It can be used to find integrals of products, i.e., integrals of the form $\int f(x)g(x)\,dx$. We take $f(x)$ as U and $\int g(x)\,dx$ as V, so that $g(x)=dV/dx$. Equation (10.4.2) then becomes

$$(10.4.3) \qquad \int f(x)g(x)\,dx = f(x)\int g(x)\,dx - \int \left\{\int g(x)\,dx\right\} f'(x)dx.$$

It will be seen that the usefulness of this method will depend on whether the function $g(x)$ is easily integrated, and if so, whether the second integral on the right-hand side of (10.4.3) can be found. Frequently these integrals can be determined and (10.4.2) then enables new integrals to be found. The method embodied in equation (10.4.2) is known as *integration by parts*.

EXAMPLE 1.

To find $\int x \sin x \, dx$.

Here the integrand is a product, and to obtain our U and V one of the factors must be integrated [the $\int g(x)dx$ of (10.4.3)]; but which one? If we choose the first, and write

$$U = \sin x; \quad \frac{dV}{dx} = x; \quad V = \frac{1}{2}x^2$$

then we have

$$\int U\frac{dV}{dx}\,dx = \int \sin x \cdot x \, dx$$

$$= \sin x \cdot \frac{1}{2}x^2 - \int \frac{1}{2}x^2 \cos x \, dx$$

$$= \frac{1}{2}x^2 \sin x - \frac{1}{2}\int x^2 \cos x \, dx$$

and we are worse off than before, since we have an integral with a trigonometric function and a *higher* power of x. Let us try it round the other way; let us take instead

$$U = x, \quad \frac{dV}{dx} = \sin x, \quad V = -\cos x.$$

Then

$$\int x \sin x \, dx = -x \cos x - \int (-\cos x) \cdot 1 \cdot dx$$

$$= -x \cos x + \int \cos x \, dx$$

$$= -x \cos x + \sin x + A.$$

which solves the problem.

This example illustrates that there are many ways in which we can express an integrand as the product of two factors, and for each such expression two ways of choosing U and V. Consequently it frequently needs a certain amount of ingenuity to apply integration by parts in a way that will yield a useful result, and not give an integral that is worse than the one with which we started. It should be remembered, too, that there is no guarantee that one can evaluate an integral by means of integration by parts (or indeed of any other method); there is no sure-fire method of integrating a product of two functions as there is for differentiating a product.

The evaluation of an integral may require the use of the method of integration by parts several times over, possibly in conjunction with one or more integrations by substitution. The next example illustrates a repeated use of integration by parts.

EXAMPLE 2.

To find $\int x^2 \sin x \, dx$.

In order that the integral on the right-hand side of (10.4.2) should be simpler than the original, we must take

$$U = x^2, \quad \frac{dV}{dx} = \sin x, \quad V = -\cos x.$$

Then from (10.4.2) we have

$$\int x^2 \sin x \, dx = -x^2 \cos x - \int (-\cos x) \, 2x \, dx$$

$$= -x^2 \cos x + 2 \int x \cos x \, dx$$

and the problem is reduced to that of finding $\int x \cos x \, dx$. For this integral we take

$$U = x, \quad \frac{dV}{dx} = \cos x, \quad V = \sin x,$$

and we have

$$\int x \cos x \, dx = x \sin x - \int \sin x \, dx$$

$$= x \sin x - \cos x + A.$$

Combining these two results, we have

$$\int x^2 \sin x \, dx = -x^2 \cos x + 2x \sin x - 2 \cos x + 2A.$$

EXAMPLE 3.

To find $\int \sin^n x \, dx$, where n is an integer.

We shall split up the integrand, and express it as $\sin^{n-1} x \cdot \sin x$. Take $U = \sin^{n-1} x$ and $V = -\cos x$ (so that $dV/dx = \sin x$). Then we have

$$\int \sin^{n-1} x \sin x \, dx = -\sin^{n-1} x \cos x$$

$$+ \int \{(n-1) \sin^{n-2} x \cdot \cos x\} \cos x \, dx$$

(differentiating $U = \sin^{n-1} x$ as a function of a function). Thus

$$\int \sin^n x \, dx = -\sin^{n-1} x \cos x + (n-1) \int \sin^{n-2} x \cos^2 x \, dx.$$

It may seem that we are not much better off, but if we now make the substitution $\cos^2 x = 1 - \sin^2 x$ in the right-hand integral, we obtain

$$\int \sin^n x \, dx = -\sin^{n-1} x \cos x + (n-1) \int \sin^{n-2} x (1 - \sin^2 x) \, dx$$

(10.4.4)
$$= -\sin^{n-1} x \cos x + (n-1) \int \sin^{n-2} x \, dx$$

$$-(n-1) \int \sin^n x \, dx.$$

Notice that the integral we are trying to find has now turned up on the right-hand side. We can rewrite (10.4.4) as

(10.4.5) $$n \int \sin^n x \, dx = -\sin^{n-1} x \cos x + (n-1) \int \sin^{n-2} x \, dx$$

and the unknown integral now occurs only on the left-hand side.

Equation (10.4.5) enables us to express our integral in terms of a similar integral with a lower power of $\sin x$. By successive application of (10.4.5) we can express our integral in terms of the integral of either $\sin x$ or $\sin^0 x = 1$. Thus to find $\int \sin^5 x \, dx$ we observe that

$$5 \int \sin^5 x \, dx = -\sin^4 x \cos x + 4 \int \sin^3 x \, dx$$

and

$$3 \int \sin^3 x \, dx = -\sin^2 x \cos x + 2 \int \sin x \, dx$$

by putting $n = 5$ and $n = 3$ in (10.4.5). Combining these two results we have

$$\sin^5 x \, dx = -\frac{1}{5} \sin^4 x \cos x - \frac{4}{15} \sin^2 x \cos x - \frac{8}{15} \cos x + C$$

where C is an arbitrary constant.

In the same way we can write

$$4 \int \sin^4 x \, dx = -\sin^3 x \cos x + 3 \int \sin^2 x \, dx$$

and

$$2 \int \sin^2 x \, dx = -\sin x \cos x + \int 1 \cdot dx$$
$$= -\sin x \cos x + x + A.$$

Hence $\int \sin^4 x \, dx = -\frac{1}{4} \sin^3 x \cos x - \frac{3}{8} \sin x \cos x + \frac{3}{8} x + C$. Clearly, for any value of n we can find $\int \sin^n x \, dx$ by repeated applications of (10.4.5), reducing the power of $\sin x$ by 2 with each application. An equation like (10.4.5) is called a *reduction formula*.

Integration by parts can be used with definite integrals as well as indefinite integrals. All that is necessary is that the limits be inserted where appropriate, and this includes the term UV. Thus we write

$$\int_a^b U \frac{dV}{dx} dx = \left[UV \right]_a^b - \int_a^b V \frac{dU}{dx} dx$$

Example

$$\int_0^{\pi/4} x \sec^2 x \, dx = \left[x \tan x \right]_0^{\pi/4} - \int_0^{\pi/4} \tan x \, dx$$

(10.4.6)
$$= \frac{\pi}{4} - \int_0^{\pi/4} \tan x \, dx.$$

Here we took $U = x$, $V = \tan x$.

The integral of $\tan x$ is one that we have not yet had, but it is easily found. For we can write $\tan x = \sin x / \cos x$, and observe that in this second expression the numerator is the derivative of the denominator except for a change of sign. This suggests looking at $\log (\cos x)$. Now

$$\frac{d}{dx} \log (\cos x) = \frac{1}{\cos x} (-\sin x) = -\tan x.$$

Hence

$$\int \tan x \, dx = -\log \cos x$$

$$= \log \sec x \qquad \left(\text{since } \frac{1}{\cos x} = \sec x\right).$$

Thus

$$\int_0^{\pi/4} x \sec^2 x \, dx = \frac{\pi}{4} - \Big[\log \sec x\Big]_0^{\pi/4}$$

$$= \frac{\pi}{4} - \left[\log \sec \frac{\pi}{4} - \log \sec 0\right]$$

$$= \frac{\pi}{4} - \log \sqrt{2} \qquad (\text{since } \sec 0 = 1)$$

$$= \frac{\pi}{4} - \frac{1}{2} \log 2.$$

Exercise 119

1. Find the integral

$$\int \sec^{-1} x \, dx$$

Hint: Take $U = \sec^{-1} x$, $V = x$.

2. Show that

$$\int e^{ax} \cos bx \, dx = \frac{1}{a} e^{ax} \cos bx + \frac{b}{a} \int e^{ax} \sin bx \, dx$$

and that

$$\int e^{ax} \sin bx \, dx = \frac{1}{a} e^{ax} \sin bx - \frac{b}{a} \int e^{ax} \cos bx \, dx.$$

Deduce that

$$\int e^{ax} \cos bx \, dx = \frac{e^{ax}}{a^2 + b^2} [a \cos bx + b \sin bx] + C$$

and find the corresponding expression for $\int e^{ax} \sin bx \, dx$.

3. If $I_m = \int x^m e^{ax} \, dx$, where $a \neq 0$, show that

$$I_m = \frac{1}{a} x^m e^{ax} - \frac{m}{a} I_{m-1}.$$

Hence evaluate

$$\int_0^1 x^4 e^{3x}\, dx.$$

4. If $J_m = \int_0^1 x^{4m+1}(1 - x^4)^{1/2}\, dx$, where m is a positive integer, prove that

$$2(m + 1)J_m = (2m - 1)J_{m-1}.$$

Evaluate $\int_0^1 x^{13}(1 - x^4)^{1/2}\, dx$.

5. By putting $x = \tan \theta$ find the integral

$$\int \tan^{-1}\left(\frac{1 - x}{1 + x}\right) dx.$$

10.5 SOME USEFUL TRICKS—HYPERBOLIC FUNCTIONS

The methods of substitution and integration by parts will not suffice to enable us to find the integrals of all functions, not even all the functions built up from the standard elementary functions such as polynomials, sine, cosine, logarithm, and so on. This is not the result of our not having considered enough methods, it is characteristic of the integration process; there are some quite simple functions which just cannot be integrated in terms of elementary functions. The function $\log x/x$ (which is simple enough) is one such; another is the function $e^{-x^2/2}$ which is of fundamental importance in statistical theory and probably familiar to many readers of this book; and there are many others.

This is in marked contrast to the rules of differentiation, which enable us to differentiate any function which is formed from standard functions with known derivatives. Knowing the derivatives of $\log x$ and x we can readily find that of $\log x/x$, but this information is of no use in finding the integral of this function. Fortunately, in this book we shall not have need to consider integrals that cannot be expressed in terms of elementary functions, but the reader should be aware of the possibility.

An interesting question arises "How can one tell if the integration of a function is possible in terms of known functions?" In effect, one cannot; but it is as well to have at hand a battery of methods, techniques and tricks to use on a recalcitrant integral in order that one should not give up too soon the attempt to find it. For this reason the rest of this chapter will be devoted to an account of several such techniques.

HYPERBOLIC FUNCTIONS.

We define two functions as follows:

$\sinh x = \frac{1}{2}(e^x - e^{-x})$ (usually pronounced "shine")

$\cosh x = \frac{1}{2}(e^x + e^{-x})$.   (pronounced as "cosh")

The function names *sinh* and *cosh* suggest a connection with the trigonometric functions, and we shall see shortly what that connection is. The graphs of these two functions have the following form:

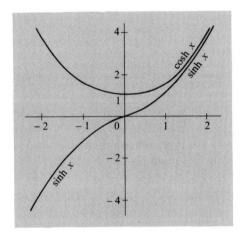

Fig. 10.5

Certain relationships between these functions are easily proved. For example, writing $(\sinh x)^2$ as $\sinh^2 x$, and so on, we have

$$\cosh^2 x = \frac{1}{4}(e^x + e^{-x})^2$$

$$= \frac{1}{4}(e^{2x} + 2 + e^{-2x})$$

and

$$\sinh^2 x = \frac{1}{4}(e^{2x} - 2 + e^{-2x}).$$

Subtracting,

(10.5.1)
$$\cosh^2 x - \sinh^2 x = \frac{1}{4}[2 - (-2)]$$

$$= 1.$$

This is analogous to the formula $\cos^2 x + \sin^2 x = 1$ for the functions sine and cosine. Again we have

$$\sinh 2x = \frac{1}{2}(e^{2x} - e^{-2x})$$

and

$$\cosh x \sinh x = \frac{1}{4}(e^x + e^{-x})(e^x - e^{-x})$$

$$= \frac{1}{4}(e^{2x} - e^{-2x}).$$

Hence

(10.5.2) $\sinh 2x = 2 \cosh x \sinh x.$

Equations (10.5.1) and (10.5.2) (analogous to $\sin 2x = 2\cos x \sin x$) show that there is indeed some formal resemblance between these two new functions and the trigonometric functions $\sin x$ and $\cos x$. Differentiation brings out a further resemblance. We have

$$\frac{d}{dx}(\sinh x) = \frac{1}{2}\frac{d}{dx}(e^x - e^{-x})$$

$$= \frac{1}{2}(e^x + e^{-x})$$

$$= \cosh x$$

and

$$\frac{d}{dx}(\cosh x) = \frac{1}{2}(e^x - e^{-x})$$

$$= \sinh x.$$

Thus each is the derivative of the other. The functions $\sin x$ and $\cos x$ are also derivatives of each other but with a change of sign for the derivative of $\cos x$.

The functions $\sinh x$ and $\cosh x$ (together with some others that we shall introduce later) are known as hyperbolic functions. Many of their properties parallel very closely the properties of the trigonometric functions, being either exactly the same, or the same except for changes of sign. To appreciate this similarity between the two sets of functions, and to gain some practice in manipulating the hyperbolic functions the reader should prove the results of the following exercise, and compare them with the corresponding trigonometrical results.

Exercise 120

Prove the following

(a) $\sinh(x \pm y) = \sinh x \cosh y \pm \cosh x \sinh y$.

(b) $\cosh(x \pm y) = \cosh x \cosh y \pm \sinh x \sinh y$.

(c) $\cosh 2x = \cosh^2 x + \sinh^2 x$
$$= 1 + 2\sinh^2 x$$
$$= 2\cosh^2 x - 1.$$

(d) $2\sinh x \cosh y = \sinh(x + y) + \sinh(x - y)$.

(e) $\cosh x + \cosh y = 2\cosh \dfrac{x + y}{2} \sinh \dfrac{x - y}{2}$.

10.6 SUBSTITUTIONS USING HYPERBOLIC FUNCTIONS

Just as the substitution $x = a \sin \theta$ turned out to be useful in the integration of functions containing $\sqrt{a^2 - x^2}$, so the substitution $x = a \sinh t$ will help in the integration of functions containing $\sqrt{a^2 + x^2}$. For example, with this substitution

$$\int \frac{dx}{\sqrt{a^2 + x^2}} = \int \frac{a \cosh t \, dt}{\sqrt{a^2 + a^2 \sinh^2 t}}$$

$$= \int \frac{a \cosh t \, dt}{a \cosh t}$$

$$= \int dt = t + A.$$

Hence

$$\int \frac{dx}{\sqrt{a^2 + x^2}} = \sinh^{-1} \frac{x}{a} + A.$$

This is analogous to the result $\displaystyle\int \frac{dx}{\sqrt{a^2 - x^2}} = \sin^{-1} \frac{x}{a} + A$ that we had before. In this case, however, we can go further, for the inverse function $\sinh^{-1} \dfrac{x}{a}$ $\left(\text{unlike } \sin^{-1} \dfrac{x}{a}\right)$ can be expressed in terms of more familiar functions. If

$$t = \sinh^{-1} \frac{x}{a}$$

then

$$\frac{2x}{a} = 2 \sinh t = e^t - e^{-t}.$$

This can be written as

$$e^{2t} - \frac{2x}{a} e^t - 1 = 0.$$

This equation is a quadratic equation in e^t, and if we solve it, using the standard formula for the solution of a quadratic equation, we obtain

(10.6.1)
$$e^t = \frac{x}{a} \pm \sqrt{\frac{x^2}{a^2} + 1}$$
$$= \frac{1}{a} [x \pm \sqrt{x^2 + a^2}].$$

We may assume that a is positive, in which case we must take the "+" sign; for otherwise the right-hand side of (10.6.1) would be negative, while the left-hand side must necessarily be positive. Hence

$$e^t = \frac{1}{a}(x + \sqrt{x^2 + a^2})$$

or

$$t = \log (x + \sqrt{x^2 + a^2}) - \log a.$$

Hence

(10.6.2) $$\sinh^{-1} \frac{x}{a} = \log(x + \sqrt{x^2 + a^2}) - \log a$$

and our integral simplifies to

$$\int \frac{dx}{\sqrt{a^2 + x^2}} = \log (x + \sqrt{x^2 + a^2}) + C$$

where $C = A - \log a$ is an arbitrary constant.

The result to remember here is

(10.6.3) $$\sinh^{-1} x = \log (x + \sqrt{x^2 + 1})$$

from which (10.6.2) can be obtained by replacing x by x/a.

If in the integral $\int \dfrac{dx}{\sqrt{x^2 - a^2}}$ we put $x = a \cosh t$ we obtain the integral $\int dt$ again, by virtue of the fact that $a^2 \cosh^2 t - a^2 = a^2 \sinh^2 t$. Hence

$$\int \frac{dx}{\sqrt{x^2 - a^2}} = \cosh^{-1} \frac{x}{a} + A.$$

Now if $t = \cosh^{-1} x/a$ then following our previous reasoning we have

$$e^{2t} - \frac{2x}{a} e^t + 1 = 0$$

from which we deduce that

(10.6.4)
$$\begin{aligned} e^t &= \frac{x}{a} \pm \sqrt{\frac{x^2}{a^2} - 1} \\ &= \frac{1}{a} [x \pm \sqrt{x^2 - a^2}]. \end{aligned}$$

Here there is no objection to either of the signs. We notice, however, that

$$(x + \sqrt{x^2 - a^2})(x - \sqrt{x^2 - a^2}) = x^2 - (x^2 - a^2) = a^2$$

so that $(x + \sqrt{x^2 - a^2})/a$ and $(x - \sqrt{x^2 - a^2})/a$ are reciprocals of each other. Their logarithms will therefore be equal and opposite, and we may write

(10.6.5) $\cosh^{-1} \dfrac{x}{a} = t = \pm \{\log (x + \sqrt{x^2 + a^2}) - \log a\}.$

The result to remember here is

(10.6.6) $\cosh^{-1} x = \pm \log (x + \sqrt{x^2 - 1}).$

The function $\cosh^{-1} x$ is two-valued, but this should occasion no surprise, since a glance at the graph of $y = \cosh x$ will show that there are two equal and opposite values of x which give the same value of the function.

OTHER HYPERBOLIC FUNCTIONS.

By analogy with the trigonometric functions, we define

$$\tanh x = \frac{\sinh x}{\cosh x}$$

$$\text{sech } x = \frac{1}{\cosh x}$$

$$\text{coth } x = \frac{1}{\tanh x}$$

$$\text{cosech } x = \frac{1}{\sinh x}. \quad *$$

The reader should verify the following results

$$\frac{d}{dx}(\tanh x) = \text{sech}^2 x$$

$$\frac{d}{dx}(\text{sech } x) = -\text{sech } x \tanh x$$

$$\tanh^2 x + \text{sech}^2 x = 1$$

$$\text{coth}^2 x - \text{cosech}^2 x = 1.$$

Exercise 121

1. Find the derivatives of the inverse functions
   (a) $\tanh^{-1} x$
   (b) $\coth^{-1} x$
   (c) $\text{sech}^{-1} x$
   (d) $\text{cosech}^{-1} x$
2. Show that by writing

$$ax^2 + 2bx + c = a\left(x + \frac{b}{a}\right)^2 + \left(c - \frac{b^2}{a}\right)$$

and making a linear substitution, we can reduce the integral

$$\int \frac{dx}{\sqrt{ax^2 + 2bx + c}}$$

to one of the following types
   (a) $\displaystyle\int \frac{dt}{\sqrt{t^2 + p^2}}$
   (b) $\displaystyle\int \frac{dt}{\sqrt{t^2 - p^2}}$
   (c) $\displaystyle\int \frac{dt}{\sqrt{p^2 - t^2}}$

* These function names are pronounced as written except for *tanh*, which is pronounced "than," with the "th" sounding as in "thank."

where p is a constant. (We assume that $ax^2 + 2bx + c$ is not negative for *all* values of x.)

3. Evaluate the three integrals in problem 2.

4. By writing

$$\alpha x + \beta = \frac{\alpha}{a}(ax + b) + \beta - \frac{b}{a}\alpha$$

show how to evaluate an integral of the form

$$\int \frac{\alpha x + \beta}{\sqrt{ax^2 + 2bx + c}}\, dx.$$

Evaluate

(a) $\displaystyle \int \frac{2x + 1}{\sqrt{x^2 + 6x + 5}}\, dx$

(b) $\displaystyle \int \frac{3x - 7}{\sqrt{2x^2 + 2x + 5}}\, dx.$

5. By considering the derivative of $\dfrac{x}{(a^2 + x^2)^{n-1}}$, prove that

$$\int \frac{dx}{(a^2 + x^2)^n} = \frac{x}{2(n-1)a^2(a^2 + x^2)^{n-1}} + \frac{2n-3}{(2n-2)a^2}\int \frac{dx}{(a^2 + x^2)^{n-1}}.$$

Deduce that

$$(2n - 2)\int \operatorname{sech}^{2n-1} x\, dx = \operatorname{sech}^{2n-2} x \sinh x$$
$$+ (2n - 3)\int \operatorname{sech}^{2n-3} x\, dx.$$

Find the integral of $\operatorname{sech}^5 x$.

10.7 INTEGRALS WITH TRIGONOMETRIC FUNCTIONS.

A substitution that frequently works when the integral contains only trigonometric functions is $t = \tan \frac{1}{2}x$. This is a consequence of the following two identities

$$\sin x = \frac{2\tan \frac{1}{2}x}{1 + \tan^2 \frac{1}{2}x}$$

$$= \frac{2t}{1 + t^2}$$

and

$$\cos x = \frac{1 - \tan^2 \frac{1}{2}x}{1 + \tan^2 \frac{1}{2}x}$$

$$= \frac{1 - t^2}{1 + t^2}$$

which the reader should be able to verify. *Hint:* Use $\sin 2\theta = 2 \sin \theta \cos \theta$, $\cos 2\theta = \cos^2 \theta - \sin^2 \theta$ and $\cos^2 \theta + \sin^2 \theta = 1$.

Since $x = 2 \tan^{-1} t$ we also write

$$dx = \frac{2}{1 + t^2} dt.$$

An example will illustrate the use of this substitution.

EXAMPLE:

To find

$$\int \frac{dx}{3\sin x + 4\cos x}.$$

Putting $t = \tan \frac{1}{2}x, \sin x = \frac{2t}{1 + t^2}, \cos x = \frac{1 - t^2}{1 + t^2}$ and $dx = \frac{2}{1 + t^2} dt,$

we have

$$\int \frac{dx}{3\sin x + 4\cos x} = \int \frac{1}{\dfrac{6t}{1 + t^2} + \dfrac{4(1 - t^2)}{1 + t^2}} \cdot \frac{2\,dt}{1 + t^2}$$

$$= \int \frac{2dt}{4 + 6t - 4t^2}$$

$$= \int \frac{dt}{(1 + 2t)(2 - t)}$$

$$= \int \frac{1}{5}\left(\frac{2}{1 + 2t} + \frac{1}{2 - t}\right) dt$$

$$= \frac{1}{5} \log (1 + 2t) - \frac{1}{5} \log (2 - t) + A$$

$$= \frac{1}{5} \log \frac{1 + 2t}{2 - t} + A$$

(10.7.1) $$= \frac{1}{5} \log \frac{1 + 2 \tan \frac{1}{2}x}{2 - \tan \frac{1}{2}x}.$$

This integral can be evaluated by another method. If we let α denote the angle of a $3 - 4 - 5$ triangle, as shown in Fig. 10.6, then

$$3\sin x + 4\cos x = 5\left(\frac{3}{5}\sin x + \frac{4}{5}\cos x\right)$$

$$= 5(\sin\alpha\sin x + \cos\alpha\cos x)$$

$$= 5\cos(x - \alpha).$$

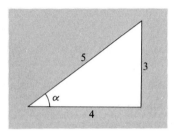

Fig. 10.6

Hence, since $\int \sec x\,dx = \log(\sec x + \tan x)$, we have

(10.7.2) $$\int \frac{dx}{3\sin x + 4\cos x} = \frac{1}{5}\log[\sec(x - \alpha) + \tan(x - \alpha)] + C.$$

It follows that the right-hand sides of (10.7.1) and (10.7.2) must agree, i.e., that

$$\frac{1 + 2\tan\frac{1}{2}x}{2 - \tan\frac{1}{2}x} \quad\text{and}\quad \sec(x - \alpha) + \tan(x - \alpha)$$

are the same function, or at most differ only by a constant. A direct verification of this equivalence is possible but tedious. Anyone who enjoys manipulating trigonometrical expressions can try this as an exercise.

The second method used above is one that is frequently employed in other connections also. It stems from the fact that an expression of the form

$$a\cos x + b\sin x$$

can be written as a multiple of a single trigonometric function. For

$$a\cos x + b\sin x = \sqrt{a^2 + b^2}\left\{\frac{a}{\sqrt{a^2 + b^2}}\cos x + \frac{b}{\sqrt{a^2 + b^2}}\sin x\right\}.$$

Now $\dfrac{a}{\sqrt{a^2 + b^2}}$ and $\dfrac{b}{\sqrt{a^2 + b^2}}$ are respectively the cosine and sine of an angle, viz, the angle in the triangle of Fig. 10.7. Denoting the angle by α we have

$$a \cos x + b \sin x = \sqrt{a^2 + b^2}\, \{\cos \alpha \cos x + \sin \alpha \sin x\}$$
$$= \sqrt{a^2 + b^2}\, \cos (x - \alpha).$$

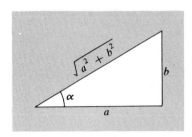

Fig. 10.7

Exercise 122

1. Find the following integrals by means of the substitution $t = \tan \frac{1}{2}x$.

(a) $\displaystyle\int \frac{dx}{2 + \sin x}$

(b) $\displaystyle\int \frac{dx}{1 + \sin x + \cos x}$

(c) $\displaystyle\int_{\pi/3}^{\pi/2} \operatorname{cosec} x\, dx$

(d) $\displaystyle\int_{0}^{\pi/2} \frac{dx}{(1 + \cos x)^2}$

(e) $\displaystyle\int_{0}^{\pi/2} \frac{dx}{3 + 5 \cos x}$

(f) $\displaystyle\int \frac{dx}{\sin x - \cos x}.$

2. By writing the expressions of the form

$$a \cos x + b \sin x$$

as a single sine or cosine (as in the last section) evaluate the integrals

(a) $\displaystyle\int \frac{dx}{\sin x + \cos x}$

(b) $\displaystyle\int \frac{5 \cos x - 12 \sin x}{\sqrt{5 \sin x + 12 \cos x}}\, dx$

(c) $\displaystyle\int (3 \sin x - 5 \cos x)^2\, dx$

3. By making the indicated trigonometric substitutions evaluate the following integrals

(a) $\displaystyle\int x \tan^{-1}(x + 1)\, dx \qquad (x = \tan \theta - 1)$

(b) $\displaystyle\int \left(\frac{a}{a - x}\right)^{1/2} dx \qquad (x = a \sin^2 \theta)$

(c) $\displaystyle\int_a^b \frac{dx}{(b-x)(x-a)}$ $\qquad (x = a \cos^2 \theta + b \sin^2 \theta)$

(d) $\displaystyle\int_{1/2}^1 \sin^{-1}(\sqrt{x})\, dx$ $\qquad (x = \sin^2 \theta)$

10.8 A SUMMARY OF DERIVATIVES AND INTEGRALS

It will be convenient for future reference to collect together the derivatives and integrals that we have obtained in this and the preceding chapter. We exhibit these in the form of the table below. This table therefore includes and supplements the table of derivatives given on page 396 and the short table of integrals given on page 465.

Exercise 123 (Chapter Review)

1. In Chapter 13 we shall prove the following result:

$$\sum_{r=1}^{n} \sin r\theta = \sin \frac{n+1}{2}\theta \sin \frac{n\theta}{2} \Big/ \sin \frac{\theta}{2}.$$

Taking this result on trust, use the method of Exercise 116 to find the area under the curve $y = \sin x$ from $x = 0$ to $x = a$, without using integration.

2. Integrate the following functions

(a) $7x^6 - 5x^4 + 3x^2 - 15$

(b) $\dfrac{x^2}{2x-1}$

(c) $x^3 \sqrt{1-x}$

(d) $25x^2 + 5 \cos x - 9e^{-2x}$

(e) $\cosh^2 x$

(f) $\dfrac{1}{\sqrt{x-1} + \sqrt{x+1}}$

3. Evaluate the definite integrals

(a) $\displaystyle\int_1^2 (4x^3 - 2x^2 + 7)\, dx$

(b) $\displaystyle\int_0^{\pi/4} (\cos x + \tan x)\, dx$

(c) $\displaystyle\int_0^{1/2} \frac{(x+1)\, dx}{3 - 2x - x^2}$

(d) $\displaystyle\int_{-1}^3 (x^2 - 1)^2 \, dx$

(e) $\displaystyle\int_0^a \frac{dx}{\sqrt{a^2 + x^2}}$

(f) $\displaystyle\int_0^X \frac{dx}{1 + x^2}$

4. Prove that as $X \to \infty$, so $\displaystyle\int_0^X \frac{dx}{1 + x^2} \to \frac{1}{2}\pi$.

It is customary to write $\displaystyle\int_0^\infty \frac{dx}{1 + x^2}$ for $\displaystyle\lim_{X \to \infty} \int_0^X \frac{dx}{1 + x^2}$, so that we can write $\displaystyle\int_0^\infty \frac{dx}{1 + x^2} = \frac{1}{2}\pi$. An integral such as this is called an *improper integral* of the first kind, a general example being $\displaystyle\int_a^\infty f(x)\, dx$, defined as $\displaystyle\lim_{X \to \infty} \int_a^X f(x)\, dx$.

Function	Derivative	Integral
$x^k \quad (k \neq -1)$	kx^{k-1}	$\dfrac{x^{k+1}}{k+1} + A$
x^{-1}	$-x^{-2}$	$\log x + A$
e^{kx}	ke^{kx}	$\dfrac{1}{k}e^{kx} + A$
$\sin x$	$\cos x$	$-\cos x + A$
$\cos x$	$-\sin x$	$\sin x + A$
$\tan x$	$\sec^2 x$	$\log \sec x + A$
$\cot x$	$-\operatorname{cosec}^2 x$	$\log \sin x + A$
$\operatorname{cosec} x$	$-\operatorname{cosec} x \cot x$	$\log(\operatorname{cosec} x + \cot x) + A$
$\sec x$	$\sec x \tan x$	$\log(\sec x + \tan x) + A$
$\sin^{-1} x$	$\dfrac{1}{\sqrt{1-x^2}}$	$x \sin^{-1} x + \sqrt{1-x^2} + A$
$\cos^{-1} x$	$-\dfrac{1}{\sqrt{1-x^2}}$	$x \cos^{-1} x - \sqrt{1-x^2} + A$
$\tan^{-1} x$	$\dfrac{1}{1+x^2}$	$x \tan^{-1} x - \frac{1}{2}\log(1+x^2) + A$
$\cot^{-1} x$	$-\dfrac{1}{1+x^2}$	$x \cot^{-1} x + \frac{1}{2}\log(1+x^2) + A$
$\operatorname{cosec}^{-1} x$	$-\dfrac{1}{x\sqrt{x^2-1}}$	$x \operatorname{cosec}^{-1} x + \log(x + \sqrt{x^2-1}) + A$
$\sec^{-1} x$	$\dfrac{1}{x\sqrt{x^2-1}}$	$x \sec^{-1} x - \log(x + \sqrt{x^2-1}) + A$
$\cosh x$	$\sinh x$	$\sinh x + A$
$\sinh x$	$\cosh x$	$\cosh x + A$
$\tanh x$	$\operatorname{sech}^2 x$	$\log \cosh x + A$
$\coth x$	$\operatorname{cosech}^2 x$	$\log \sinh x + A$
$\operatorname{sech} x$	$\operatorname{sech} x \tanh x$	$\log(\operatorname{sech} x + \tanh x) + A$
$\operatorname{cosech} x$	$\operatorname{cosech} x \coth x$	$\log(\operatorname{cosech} x + \coth x) + A$
$\cosh^{-1}x = \pm\log(x + \sqrt{x^2-1})$	$\dfrac{1}{\sqrt{1+x^2}}$	$x \cosh^{-1} x - \sqrt{1+x^2} + A$
$\sinh^{-1} x = \log(x + \sqrt{x^2+1})$	$\dfrac{1}{\sqrt{x^2-1}}$	$x \sinh^{-1} x - x\sqrt{x^2-1} + A$
$\tanh^{-1} x = \frac{1}{2}\log\left(\dfrac{1+x}{1-x}\right)$	$\dfrac{1}{1-x^2}$	$x \tanh^{-1} x + \frac{1}{2}\log(1-x^2) + A$
$\coth^{-1} x = \frac{1}{2}\log\left(\dfrac{x+1}{x-1}\right)$	$\dfrac{1}{1-x^2}$	$x \coth^{-1} x + \frac{1}{2}\log(x^2-1) + A$
$\operatorname{sech}^{-1} x$	$\dfrac{1}{x\sqrt{1-x^2}}$	$x \operatorname{sech}^{-1} x - \sin^{-1} x + A$
$\operatorname{cosech}^{-1} x$	$\dfrac{1}{x\sqrt{1+x^2}}$	$x \operatorname{sech}^{-1} x - \sin^{-1} x + A$
$\operatorname{cosech}^{-1} x$	$\dfrac{1}{x\sqrt{1+x^2}}$	$x \operatorname{cosech}^{-1} x - \cosh^{-1} x + A$
$\log x$	x^{-1}	$x \log x - x + A$

The reader should check through this list and verify any results that have not appeared before, as well as any that he is not sure of. All the integrals can be found by the methods given in this chapter.

5. Evaluate the following improper integrals.

(a) $\displaystyle\int_1^\infty e^{-x}\,dx$ (b) $\displaystyle\int_0^\infty xe^{-x^2/2}\,dx$.

6. Use the method of integration by parts to find the integral of the following functions.

(a) $\log x$ (d) $x^2 e^{-3x}$ (g) $x\cos 4x$

(b) $\sec^{-1} x$ (e) $x^2\tan^{-1} x$ (h) $x^2\log x$

(c) $\sinh^{-1} x$ (f) $x\cos^2 x$

7. Show that $\displaystyle\int_0^a \frac{dx}{\sqrt{p-x}} = 2\{\sqrt{p} - \sqrt{p-a}\}$ where p is a positive

constant, and $a < p$. Deduce that $\displaystyle\lim_{a\to p-0}\int_0^a \frac{dx}{\sqrt{p-x}} = 2\sqrt{p}$.

This limit can be written as $\displaystyle\int_0^p \frac{dx}{\sqrt{p-x}}$, the limiting process being implied by the fact that the integrand tends to ∞ as x tends to the upper limit from below. In general, if $f(x) \longrightarrow \infty$ as $x \longrightarrow b$ from below, then the definite integral $\displaystyle\int_a^b f(x)\,dx$ is interpreted to mean $\displaystyle\lim_{X\to b-0}\int_a^X f(x)\,dx$. Such an integral is called an *improper integral of the second kind*. (There is, of course, no guarantee that this limit will exist in general.)

8. Evaluate the following improper integrals

(a) $\displaystyle\int_0^6 \frac{dx}{\sqrt{36-x^2}}$

(b) $\displaystyle\int_0^{16} x^{-1/4}\,dx$

(c) $\displaystyle\int_{3/2}^2 \frac{dt}{\sqrt{(2-t)(t-1)}}$

Note: In 8(*b*) it is the lower limit at which the integrand becomes infinite, and this integral will be interpreted in an analogous way as

$$\lim_{X\to+0}\int_X^{16} x^{-1/4}\,dx.$$

9. By means of the substitution $x = (1-y)/(1+y)$, or otherwise, show that if $I(p,q,r) = \int_0^1 x^p(1-x)^q(1+x)^r\,dx$ $(p \geq 0, q \geq 0)$, then $I(p,q,r) = 2^{q+r+1}I(q, p, -p-q-r-2)$. Hence evaluate

(a) $\displaystyle\int_0^1 \frac{x^2(1-x)^3}{(1+x)^9}\,dx$

(b) $\displaystyle\int_0^{\pi/2} \frac{\cos^3\theta\sin^3\theta\,d\theta}{(1+\cos^2\theta)^3}$

(c) $\displaystyle\int_0^1 \frac{x(1-x)}{(1+x)^3}\,dx$

10. If $I_{m,n} = \int \cos^m x \sin nx \, dx$ prove that

$$(m + n)I_{m,n} = -\cos^m x \cos nx + mI_{m-1,n-1}.$$

Hence show that $\int_0^{\pi/4} \cos^2 x \sin 4x \, dx = \dfrac{5}{12}$.

11. Evaluate the following integrals

(a) $\displaystyle\int_0^{\log 2} \frac{dx}{\sinh x + 5 \cosh x}$

(c) $\displaystyle\int_2^3 \frac{dx}{x + \sqrt{x^2 - 1}}$

(b) $\displaystyle\int \frac{dx}{x(1 + x^2)^{3/2}}$

(d) $\displaystyle\int_0^1 \frac{dx}{\sqrt{2x^2 + 3}}$

12. Prove that if m and n are unequal positive integers then

$$\int_0^\pi \sin mx \sin nx \, dx = \int_0^\pi \cos mx \cos nx \, dx = 0;$$

while if $m = n$ we have

$$\int_0^\pi \sin^2 nx \, dx = \int_0^\pi \cos^2 mx \, dx = \frac{1}{2}\pi \qquad (m > 0).$$

13. If $Y > X > 0$, and $p > k$, show that

$$I = \int_X^Y e^{-t} t^{k-1} \, dt = \int_X^Y e^{-t} t^p \frac{dt}{t^{p-k+1}}$$

and deduce, from the fact that $e^{-t} t^p \longrightarrow 0$ as $t \longrightarrow \infty$ that

$$I < M \int_X^Y \frac{dt}{t^{p-k+1}}$$

where M is a constant. Hence show that the improper integral $\int_X^\infty e^{-t} t^{k-1} \, dt$ exists and tends to zero as $X \longrightarrow \infty$. It follows then that the improper integral $\Gamma(k) = \int_0^\infty e^{-t} t^{k-1} \, dt$ exists. The function of k thus defined is called the *Gamma function*.

By integration by parts show that

$$\Gamma(k + 1) = k \, \Gamma(k)$$

and deduce that if n is a positive integer, $\Gamma(n + 1) = n!$

14. The integral $B(p, q) = \int_0^1 t^{p-1}(1 - t)^{q-1} \, dt$ is known as the *Beta function*. It is improper if $p < 1$ or $q < 1$, but it can be shown that the integral exists if $p > 0, q > 0$.

Show that

(a) $p \, B(p, q) = (q - 1) \, B(p + 1, q - 1)$

(b) $B(p + 1, q) = B(p, q) - B(p, q + 1)$

(c) $B(p, q + 1) = \dfrac{q}{p + q} B(p, q).$

Suggestions for Further Reading

The suggested reading for this chapter consists mainly of the same books as those recommended for Chapter 9, since the tendency is for textbooks to cover both differential and integral calculus. The reader may like to note the following book which deals specifically with integration:

Gillespie, R.P., *Integration*. Edinburgh: Oliver and Boyd Ltd., 1959.

The reader who wishes to examine a more rigorous approach to the material of this chapter can look up the subject of "Riemann Integration" in some of the books already mentioned, or in

Sokolnikoff, I.S., *Advanced Calculus*. New York: McGraw-Hill Book Co., 1939, Chapter IV.

11 Differential Equations

A mathematician and a sailor were once ship-wrecked on an island on which grew two coconut trees, each bearing a single cocount.

On the first morning the sailor volunteered to get breakfast. With great skill he climbed to the top of one of the trees, neatly detached the coconut, and brought it down. The mathematician looked on admiringly, and noted carefully how the sailor had solved this problem.

The next morning it was the mathematician's turn to provide breakfast. He was about to climb the same tree that the sailor had climbed, when he realized that there was no coconut at the top. Yesterday's method would not work! Accordingly he struggled to the top of the other tree, and brought down the coconut. He then climbed the first tree, and with a piece of string from his pocket, tied the coconut firmly to its top. Climbing down, he announced triumphantly, "See! I have reduced the problem to the preceding one!"

11.1 DEFINITION AND CLASSIFICATION

It frequently happens that the mathematical formulation of a problem in economics, physics or many other branches of science, results in an equation in which appears, together with some of its derivatives, an unknown function (which is to be found). Such an equation is called a *differential equation*.

By a solution of a differential equation is meant any function which satisfies the equation, i.e., which, when substituted in place of the unknown function, converts the differential equation into an identity.

Thus, to take a simple example, the equation $f'(x) + x + f(x) = 0$ is a differential equation, since it contains the unknown function $f(x)$ and its first derivative $f'(x)$. We could also write it as

$$(11.1.1) \qquad \frac{dy}{dx} + x + y = 0$$

meaning the same thing. If we replace y by $1 - x$ in (11.1.1), the left-hand side becomes

$$-1 + x + (1 - x)$$

which is identically zero. Thus $1 - x$ is a solution of (11.1.1). It is not the only solution, however, for if we replace y by $1 - x + Ae^{-x}$, where A is any constant, then the left-hand side of (11.1.1) becomes

$$-1 - Ae^{-x} + x + 1 - x + Ae^{-x}$$

which is again identically zero. Thus any function of the form $1 - x + Ae^{-x}$ is a solution of (11.1.1).

We shall first give a broad classification of differential equations, then we shall see how they can arise in practice, and finally we shall consider various methods of finding solutions of differential equations.

If the unknown function is a function of two or more variables, then the derivatives which occur will necessarily be partial derivatives. The differential equation is then known as a *partial differential equation*. Thus the equation $(x^2 + 3) \, \partial F/\partial x + \partial F/\partial y = xy$ is a partial differential equation. We shall not have occasion to consider partial differential equations in this book, and therefore will not say any more about them.

If the function is of a single variable the derivatives that occur will be ordinary derivatives, and the equation is then called an *ordinary differential equation*. Thus (11.1.1) is an ordinary differential equation, and so are the following:

(11.1.2)
$$\left(\frac{dy}{dx}\right)^3 + 3xy\frac{dy}{dx} + x^2 + y^2 = 0$$

(11.1.3)
$$\frac{d^3y}{dx^3} + \frac{d^2y}{dx^2} + (x^2 - 5)\frac{dy}{dx} + (x^3 + x)\,y = 0$$

(11.1.4)
$$x^2y\left(\frac{d^2y}{dx^2}\right)^2 + (x^2 + 7y^2)\frac{d^2y}{dx^2} + 3\frac{dy}{dx} = 7x + 1.$$

The next classification of differential equations is by the order of the highest derivative which occurs in the equation. This order is called the order of the equation, so that a differential equation is of order n if the derivative d^ny/dx^n occurs in it, but $d^{n+1}y/dx^{n+1}$ does not. Equation (11.1.1) is therefore of order 1, and so is (11.1.2). Equation (11.1.3) is of order 3, and (11.1.4) is of order 2.

Finally we classify these equations according to their degree. The degree of a differential equation is the highest power to which the derivative of highest order (the one that determines the order of the equation) occurs. Equations (11.1.1) and (11.1.3) are therefore of first degree; equation (11.1.2) is of degree 3, and (11.1.4) is of degree 2.

We may express this classification in a different, but equivalent way, as follows. An equation of order n will be of the form

$$F\left(x, y, \frac{dy}{dx}, \frac{d^2y}{dx^2}, \dots, \frac{d^ny}{dx^n}\right) = 0$$

where F is a polynomial in d^ny/dn^n, with coefficients that are functions of the other arguments of F.* The degree of this polynomial is the degree of the equation.

* Differential equations like

$$\frac{d^2y}{dx^2} + \sin\left(\frac{d^2y}{dx^2}\right) = 1 - x + \frac{dy}{dx}$$

for example, that are not expressible in this form are so rarely encountered that we can ignore them here.

A differential equation of order n and of the first degree will be of the form

$$\frac{d^n y}{dx^n} = G\left(x, y, \frac{dy}{dx}, \ldots, \frac{d^{n-1} y}{dx^{n-1}}\right)$$

where G is some function of the indicated variables.

In particular, an equation of the first order and first degree will be of the form

(11.1.5) $$\frac{dy}{dx} = \phi(x, y).$$

Exercise 124

1. Give the order and degree of the following differential equations

   (a) $x\dfrac{d^2 y}{dx^2} + 5(x + y)\dfrac{dy}{dx} + 3xy = x^2$

   (b) $\left(\dfrac{dy}{dx}\right)^3 + 2x\left(\dfrac{dy}{dx}\right)^2 - y\dfrac{dy}{dx} + x^2 = y^2$

   (c) $\dfrac{d^3 y}{dx^3} - y = \dfrac{dy}{dx}$

   (d) $\dfrac{dy}{dx} + 2x^2 - y\sin x = 0$

   (e) $\left(\dfrac{d^2 y}{dx^2}\right)^4 - 6\left(\dfrac{dy}{dx}\right)^5 + xy = 20$

2. (a) Show that $y = \log x$ and $y = x^3$ are solutions of the differential equation

   $$x^2(3\log x - 1)\frac{d^2 y}{dx^2} - x(6\log x - 1)\frac{dy}{dx} + 9y = 0.$$

   (b) Show that $y = e^{5x}$ and $y = \sin x$ are solutions of the differential equation

   $$(5\sin x - \cos x)\frac{d^2 y}{dx^2} - 26\sin x\frac{dy}{dx} + 5(\sin x + 5\cos x)y = 0.$$

11.2 HOW DIFFERENTIAL EQUATIONS ARISE

Now although it might be an interesting mental pastime to write down equations containing x, y, dy/dx, etc., and then try to find functions of x which would satisfy the resulting differential equations, there would be little point in our discussing ways and means of doing this if there were no practical advantages to be obtained from this sort of occupation. Let us see then how differential equations can arise in practice.

Suppose we have a functional relationship which contains in it some constant which is not given any specific value. For example we may take the relationship $y = Ae^{2x}$. This equation really represents not *one* functional relationship, but a whole collection of them, of which $y = 7.2e^{2x}$, $y = 100e^{2x}$, $y = 15e^{2x}$, are different examples. In short, there is one functional relationship for every value of A. We can call such a collection a *family* of functions, and if we think geometrically (and it is often helpful to do so), we can refer to the collection of graphs of these functions as a family of curves.

For each of these functional relationships A denotes a constant, but its value will differ according to which function of the family we are talking about. This idea of a "variable constant" is something which we have met before, in Chapter 5 (page 181) where we called it a *parameter*.

If we differentiate both sides of the equations $y = Ae^{2x}$ we obtain

$$\frac{dy}{dx} = 2Ae^{2x},$$

and since we now have two equations, we can eliminate A between them. This gives us the differential equation

(11.2.1) $$\frac{dy}{dx} = 2y,$$

of the first order and first degree.

What is the significance of equation (11.2.1)? Since it does not contain the parameter A, it must be a statement which can be asserted of *any* function of the family that we are considering. Geometrically (11.2.1) represents a property that is common to all the curves of the family, and it is easy to see what this is. It is that the slope (dy/dx) of the curve at any point is twice the ordinate (y) of that point. Figure 11.1 shows some of the members of this family of curves.

Equation (11.2.1) is equivalent to the geometrical statement that for any point P, the slope at P is twice the ordinate PN of P.

To take another example; the family of curves

(11.2.2) $$y = Ax^2 + x + A^2$$

can be represented by a differential equation. Differentiating with respect to x we have

(11.2.3) $$\frac{dy}{dx} = 2Ax + 1.$$

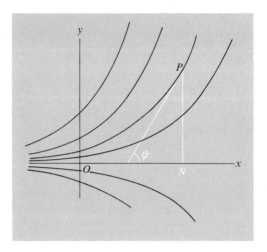

Fig. 11.1

For convenience we shall write y' in place of dy/dx, and (11.2.3) can then be written

$$A = \frac{y' - 1}{2x}.$$

Substituting in (11.2.2) we obtain

$$y = \frac{y' - 1}{2x}x^2 + x + \left(\frac{y' - 1}{2x}\right)^2$$

i.e.,

$$4x^2y = 2x^3(y' - 1) + 4x^3 + (y' - 1)^2$$

which reduces to

$$y'^2 + 2(x^3 - 1)y' - 4x^2y + 2x^3 + 1 = 0.$$

We again have a first order differential equation, this time of the second degree.

These examples suggest that when we attempt the reverse process, that of going from the differential equation to the functions that satisfy it, we would expect that, for a first order equation, there would be a parameter in the solution; in other words that there is a whole family of solutions to a first order equation. This is indeed true in general, though we shall not give the proof of this result. In this context, the parameter that appears when a first order differential equation is solved is usually called an *arbitrary constant*. It is similar to the constant that appears when integration is performed.

If our functional relationship contains *two* parameters, we shall need three equations in order to eliminate them, and these can be obtained only by differentiating twice. Thus if

$$y = Ae^x + Bx$$

then

$$y' = Ae^x + B$$

and

$$y'' = Ae^x$$

Thus

$$B = y' - y''$$

and therefore

$$y = y'' + (y' - y'')x$$

which can be written as

$$(11.2.4) \qquad (x - 1)y'' - xy' + y = 0.$$

Because of the two differentiations that have been performed, (11.2.4) is a differential equation of the second order.

Another example (an important one) is that of the functional relationship

$$y = A \cos qx + B \sin qx$$

where q is some constant (not a parameter).

We derive

$$y' = -qA \sin qx + qB \cos qx$$
$$y'' = -q^2 A \cos qx - q^2 B \sin qx.$$

Hence

$$(11.2.5) \qquad y'' = -q^2 y.$$

We shall have occasion to use (11.2.5) later on.

It would seem then that the solution of a differential equation of the second order will contain two arbitrary constants. This is indeed so, and a solution containing two arbitrary constants is called the *general solution* of the differential equation. Moreover, for the kind of differential equations with which we shall be concerned, any particular solution is obtainable from the general solution by giving specific values to these constants. A similar remark applies to equations of general order n; the general solution has n arbitrary constants.

The above examples are of use in that they indicate something about the form of the solutions of a differential equation, but they do not shed much light on the question of how differential equations might arise in the first place.

The usual way in which differential equations arise is a fairly straightforward one. We often have occasion to consider quantities that vary with time, such as the profits of a firm, or the cost-of-living index. These quantities are therefore functions of time. Often we are interested in the rate at which these quantities are changing, and this rate of change will be expressed by the derivative of the function with respect to time. In setting up a mathematical model of an economic or physical process, we may observe that there are relationships between these rates of change and certain quantities which enter into the problem, and when these relationships are expressed in mathematical terms, they will be in the form of differential equations. Some specific examples of this will be given in the next section.

Exercise 125

1. Find the differential equations that result when the parameters A, B, etc., are eliminated from the following families of functions

  (a) $y = (x + A)^3$

  (b) $y = A \log x + \dfrac{1}{A}$

  (c) $y = Ae^{3x} + Be^{2x} + \sin x$

  (d) $y = A \log x + Be^x + A^2$

  (e) $y = Ae^{-2x} + B \cos 5x + C \sin 5x$

  (f) $y = A \sec x + B \tan x$

2. What geometrical property of the family of its solutions does the differential equation

$$(y + 1)y' = x - 1$$

express? Show that the general solution of this equation is

$$(y + 1)^2 = (x - 1)^2 + A$$

and sketch the graphs of the solutions for which $A = -1$ and 1.

11.3 EXPONENTIAL GROWTH AND DECAY

As an example of how differential equations may arise in economics let us consider the rate at which a firm expands, measured, shall we say, by the rate of increase of its total output. This will normally depend on a great many factors. Some of these, such as the initiative shown by the managing director, will be very intangible, while others will be more objective and measurable, such as the price of output, the prices of inputs, the number of firms in the same industry, the profit level, and so on. One of the things that will certainly have some effect on the rate of growth of a firm will be the size of the firm itself. If we make the rather unrealistic assumption that this is the *only* factor that influences the rate of growth of the firm, then the rate of growth will be some function of the size. In mathematical terms, this would be expressed as

$$(11.3.1) \qquad \frac{dy}{dt} = \phi(y)$$

where y is some suitable quantitative measure of the size of the firm, t is time, and ϕ is some function whose precise nature will depend on the circumstances of the problem. One would expect, however, that within certain limits a large firm would tend to grow faster than a small one, and one would therefore expect that $\phi(y)$ would be an increasing function of y over some range of values of y. On the other hand, it could well be that if y became too large, then ϕ might decrease, and possibly become negative; if the firm became so big as to be unwieldy, its growth might be arrested, and it might even start to decrease in size. Be that as it may (11.3.1) is a differential equation, and its solution (y as a function of t) will describe how the size of the firm will vary with time.

If we make the further assumption that $\phi(y)$ is directly proportional to y, i.e., that $\phi(y) = ky$ where k is a positive constant, then (11.3.1) becomes

$$(11.3.2) \qquad \frac{dy}{dt} = ky.$$

It is readily verified that this equation is satisfied by $y = Ae^{kt}$; and since this result contains the arbitrary constant A, $y = Ae^{kt}$ is the general solution. This solution gives us a picture of the growth of the firm; it grows larger at an ever-increasing rate as time goes on, increasing indefinitely. (See Fig. 11.2.)

Clearly the assumption $\phi(y) = ky$ cannot hold all the time in any practical case, but it may well be a close approximation to the truth within certain limits. The law of growth exemplified by equation (11.3.2) is one hav-

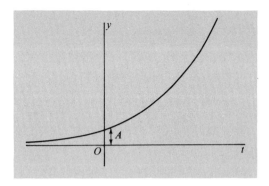

Fig. 11.2

ing widespread application, and the manner of growth which it describes is known as *exponential growth*. The graph (Fig. 11.2) is the *exponential curve*. With such a curve the constant A, given by the length marked A in Fig. 11.2, is the size of the firm (or whatever it is) at time $t = 0$.

If the rate of change of a quantity y *decreases* in proportion to the value of y, we have a different law, that of *exponential decay*. This assumption is a close approximation to the truth for the cooling of a cup of coffee—the temperature decreases at a rate proportional to its excess over room temperature; it is even better for the decay of radioactive substances—the rate at which decay occurs is proportional to the amount of radioactive substance still left. In mathematical terms we have

(11.3.3) $$\frac{dy}{dt} = -\lambda y$$

where λ is a positive constant, and the solution is $y = Ae^{-\lambda t}$. The graph of this function is given in Fig. 11.3. This is an exponential decay curve.

Associated with this function there is an interval of time called the

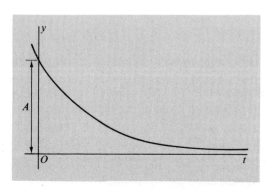

Fig. 11.3

half-life; it is the time required for the quantity y to decrease to half its value. If H denotes this time, then we must have

$$\frac{1}{2}A = Ae^{-\lambda H}$$

that is

$$-\lambda H = \log\left(\frac{1}{2}\right)$$

or

$$\lambda H = \log 2.$$

Thus $H = \log 2/\lambda$. Alternatively we can write $\lambda = \log 2/H$, which gives us an interpretation of the constant λ in terms of the time H, which has rather more appeal to the intuition.

One would not expect that economic processes would conform to laws as simple as these with the exactitude displayed by the above two examples from physics, but the laws of exponential growth and decay nevertheless provide a good approximation to a quantitative description of many economic processes.

If the reader has now grasped the idea of what a differential equation is, what is meant by *solution* and *general solution* of such an equation, and has at least an inkling of how differential equations might arise in economic problems, it is time to look at some of the methods whereby solutions to differential equations can be found. This is largely a routine matter, a question of learning certain "tricks of the trade" which yield the desired result. It will be seen that all these tricks are aimed at reducing the problem in hand to a previous problem that has already been solved (though with rather more point to them than the antics of the mathematician in the chapter heading!).

Exercise 126

1. In one economic model of income and investment the assumptions are made that consumption (C) is proportional to income (Y) and that induced investment (I) is proportional to the rate of change in income. Show that these assumptions can be expressed mathematically by the equations

$$C = cY; \qquad I = v\frac{dY}{dt}$$

where c and v are constants.

Show further that these equations, taken in conjunction with the equation $Y = C + I + A$, where A, a constant, is autonomous investment, yield the differential equation

$$v\frac{dY}{dt} - sY + A = 0$$

on elimination of C and I, where $s = 1 - c$. We can note here that this equation exhibits the equality of investment and savings; $v(dY/dt) + A$ being investment, and sY being savings.

2. In another model the assumptions are that

$$\frac{dI}{dt} = -k(I - J)$$

$$J = v\frac{dY}{dt}$$

$$Z = C + I + A$$

and

$$\frac{dY}{dt} = -\lambda(Y - Z)$$

where J is the potential rate of investment and Z the total demand.

By eliminating Z, J and I from these equations and $C = cY$, obtain the differential equation

$$\frac{d^2 Y}{dt^2} + (s\lambda + k - kv\lambda)\frac{dY}{dt} + ks\lambda Y = k\lambda A.$$

3. A law of growth of the form

$$y = \frac{k}{1 + ae^{-\lambda t}}$$

is sometimes proposed as a more realistic model of growth of organisms, populations, etc., than the straight exponential law of growth. Show that it has the advantage of not tending to infinity as $t \longrightarrow \infty$, but tends instead to a finite value.

Show that if k and λ are taken to be constants, so that a is the only parameter, then functions of this kind are characterized by the differential equation

$$\frac{dy}{dt} = \frac{\lambda}{k}y(y - k).$$

11.4 METHODS FOR THE SOLUTION OF
DIFFERENTIAL EQUATIONS

For our purposes it will be necessary to consider only a few types of differential equations of the first order, and only one type (though a very important one) of equation of the second order. We shall deal with those of the first order first, and shall denote by Roman numerals the various types of equation. This numbering will be purely for our own convenience; it does not follow any standard convention.

TYPE I. DIRECT INTEGRATION.

Differential equations of the form $dy/dx = \phi(x)$, can be solved by direct integration, and the general solution is

$$y = \int \phi(x)\, dx + A.$$

This general solution contains an arbitrary constant, as we would expect. This type of differential equation needs no further explanation, since the integration of standard functions and many others was dealt with in the last chapter.

TYPE II. SEPARATION OF VARIABLES.

If a functional relationship between y and x is given in the form

$$f(y) = g(x)$$

then on differentiation we have

$$(11.4.1) \qquad f'(y) \cdot \frac{dy}{dx} = g'(x)$$

the left-hand side being differentiated as a function of a function. Using the differential notation, as in Chapter 9, we can write (11.4.1.) as

$$f'(y)\, dy = g'(x)\, dx.$$

Reversing this argument we see that if we can get a differential equation into the form

$$(11.4.2) \qquad P(y)\, dy = Q(x)\, dx$$

then we can find the relationship between y and x by integrating both sides, since if $P(y) = f'(y)$ and $Q(x) = g'(x)$, then the relation $f(y) = g(x) +$ constant will yield the correct differential equation (11.4.2). Thus we may write

$$(11.4.3) \qquad\qquad \int P(y)\,dy = \int Q(x)\,dx.$$

In this way we have reduced the problem to a double application of Type I. In (11.4.3) all the xs are on one side, and all the ys are on the other; hence this method is known as that of *separation of variables*.

Example 1.

The differential equation

$$\frac{dy}{dx} = \frac{x^2}{y}$$

can be written in the form

$$y\,dy = x^2\,dx.$$

On integrating both sides, we obtain

$$(11.4.4) \qquad\qquad \tfrac{1}{2}y^2 = \tfrac{1}{3}x^3 + A$$

where A is an arbitrary constant. Note that it would be pointless to include an arbitrary constant on *both* sides of (11.4.4), for

$$\tfrac{1}{2}y^2 + B = \tfrac{1}{3}x^3 + C$$

would be the same as

$$\tfrac{1}{2}y^2 = \tfrac{1}{3}x^3 + (C - B),$$

and if C and B are arbitrary then so is $C - B$. Hence $C - B$ might just might just as well be denoted by a single symbol, as in (11.4.4).

Example 2.

The equation $\dfrac{dy}{dx} = \dfrac{y}{x}$.

The variables can be separated as follows:

$$\frac{dy}{y} = \frac{dx}{x}.$$

Integrating both sides we have

$$(11.4.5) \qquad\qquad \log y = \log x + C.$$

This is the solution of the equation, but is not in its most convenient form. If we take antilogarithms of both sides we get

$$y = x \cdot e^c.$$

Now e^c is just some constant, and we can replace it by a single letter, say A. The solution of the equation then reads

(11.4.6) $y = Ax$

—the equation of a straight line through the origin.

In work of this kind it is often possible to anticipate the taking of antilogarithms at a later stage. Thus in solving the above equation we could write the constant in (11.4.5) as $\log A$, rather than C, and from

$$\log y = \log x + \log A$$

proceed directly to $y = Ax$.

There are one or two points of interest in this example. The equality $A = e^c$ would seem to suggest that A must be a *positive* constant whereas it is readily verified that (11.4.6) satisfies the differential equation even if A is negative. One way of resolving this difficulty is to observe that the derivative of $\log(-x)$ is also $1/x$. Indeed, if it is known that x is negative this is the form in which we must write $\int \frac{dx}{x}$. Further, if we wish to be quite general and give an expression for $\int \frac{dx}{x}$ which will be meaningful for both positive *and* negative values of x we must write

$$\int \frac{dx}{x} = \log|x|$$

a form which automatically ensures that the number that we are taking the logarithm of is positive. Thus we should really write the solution of the equation in the form

(11.4.7) $|y| = A|x|$

where we *can* now assume that A is positive (it cannot, in fact, be otherwise).

Any solution of the form (11.4.6) will be consistent with (11.4.7) but the converse is not true. For example the functional relation

(11.4.8) $y = |x|$

is consistent with (11.4.7) but not of the form (11.4.6). The solution (11.4.8) is not differentiable at the origin however (it was given in Chapter 9 as an example of a function not differentiable at a certain point), and if we add a condition, which would often be required in an example arising in a practical application, that the solution is to be everywhere differentiable, then (11.4.7) is equivalent to (11.4.6). This peculiar behavior arises from the fact that at the point $x = y = 0$ the expression for dy/dx, viz., y/x, becomes indeterminate. Such a point is known as a *singularity* of the family of solutions, and from Fig. 11.4, which depicts this family, it will be seen that the usual property, that at each point of the plane the slope of the integral curve through that point is determined, breaks down at this particular point.

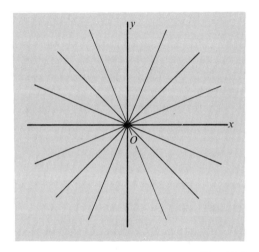

Fig. 11.4

The other way out of the difficulty just mentioned is by the use of complex numbers. These will be discussed in Chapter 13, so we shall not pursue this topic further at present.

Exercise 127

1. Solve the following differential equations

(a) $x^2 dx = \dfrac{1}{y} dy$

(b) $\cot x \cdot dy + y\, dx = 0$

(c) $\dfrac{dy}{dx} = \dfrac{2x - 5}{3y + 7}$

(d) $(1 - \log y) \dfrac{dy}{dx} = 2 \tan x \sec^2 x$

2. Show that the general solution of the differential equation

$$\frac{dy}{dx} = \left(\frac{1-y^2}{1-x^2}\right)^{1/2}$$

can be written as

$$\sin^{-1} y = \sin^{-1} x + C$$

or as

$$y = x\sqrt{1-A^2} + A\sqrt{1-x^2}.$$

3. Show that the equation

$$\frac{dy}{dx} = \frac{y(x+y)}{x^2}$$

is not of Type II, but that if we make the substitution $y = xv$ (with $dy/dx = x\,dv/dx + v$) the resulting equation is of Type II. Hence solve the original equation.

4. By the same method as in problem 3, solve the equations

(a) $x\dfrac{dy}{dx} - y = \sqrt{x^2 + y^2}$

(b) $(x + 2y)dy = (2x + y)\,dx$

(c) $\dfrac{dy}{dx} = \dfrac{3x - 5y}{5x + 2y}$

(d) $2\left(\dfrac{dy}{dx} - \dfrac{y}{x}\right) \sin\left(\dfrac{y}{x}\right) = x^4.$

5. In Chapter 8, section 22, we saw that the condition for a utility function to be of the form $f\{q(x) + r(y)\}$ was that $v_{xy}/v_x v_y$ should be a function of v. Show that when $v = Ax^\alpha y^\beta$ (as in Exercise 111.1) this function is v^{-1}. Writing $z = \phi'(v)$, where ϕ is the function inverse to f, show that

$$\frac{z'(v)}{z(v)} = -\frac{1}{v},$$

an equation of Type II. Solve this equation and hence find the function $\phi(v)$.

6. As in problem 5 show that the function $v = x\sqrt{1-y^2} + y\sqrt{1-x^2}$ is expressible in the form $f\{q(x) + r(y)\}$, and express it in this form.

11.5 INTEGRATING FACTORS

TYPE III. LINEAR DIFFERENTIAL EQUATIONS

A linear differential equation of the first order is one of the following form

(11.5.1) $$K(x)\frac{dy}{dx} + L(x)\,y = M(x)$$

where K, L and M are (as the notation implies) functions of x alone. Linear equations are so called because they are linear in dy/dx and y, which appear only to the first degree. Of course, this will not in general be true of x.

If we assume that the domain of values of x for which we are interested in the solution of the differential equation does not include any value of x for which $K(x) = 0$, then we can divide both sides of (11.5.1) by $K(x)$ and obtain

(11.5.2)
$$\frac{dy}{dx} + Py = Q$$

P and Q being functions of x alone, and this we shall take as the standard form of this type of equation.

A clue to the method for solving linear equation comes from the consideration of the derivative of an expression of the form $\mu(x)y$, where $\mu(x)$ is some function of x. We observe that

(11.5.3)
$$\frac{d}{dx}[\mu(x)y] = \mu(x)\frac{dy}{dx} + \mu'(x)y.$$

Now the right-hand side of (11.5.3) looks something like the left-hand side of (11.5.1). If, in fact, it happened that $L(x)$ was the derivative of $K(x)$, then these two expressions would be equivalent, and we would be able to write

$$K(x)\frac{dy}{dx} + L(x)y = \frac{d}{dx}[K(x)y].$$

This would bring us very close to the solution, but it would, of course, be only in exceptional circumstances that $L(x)$ would be the derivative of $K(x)$. However, we do have a certain amount of freedom in how we write the equation; we could multiply the standard form (11.5.2) all through by a function of x, say $\mu(x)$, and we would have essentially the same equation. Might it not be possible that, by choosing this factor $\mu(x)$ in a suitable fashion, the coefficient of y would indeed turn out to be the derivative of the coefficient of dy/dx?

Let us see. If we multiply (11.5.2) through by an as yet unknown function $\mu(x)$, we get

(11.5.4)
$$\mu\frac{dy}{dx} + (\mu P)y = \mu Q.$$

We want to know if the coefficient of y (viz., μP) can be the derivative of the coefficient of dy/dx (viz., μ). If so, then

(11.5.5)
$$\mu P = \frac{d\mu}{dx}$$

in other words we must have

$$\frac{d\mu}{\mu} = P\,dx.$$

This is a differential equation of Type II, and its solution is

$$\log \mu = \int P \, dx$$

or

$$\mu = e^{\int P \, dx}.$$

Since P is known, μ can be found by a simple integration. With μ chosen in this way we know that (11.5.4) can be written

$$\frac{d}{dx}[\mu y] = \mu Q.$$

This is essentially a differential equation of Type I (with μy as the dependent variable) and can be solved by one further integration.

$$\mu y = \int \mu Q \, dx.$$

The method for these equations is therefore to find first $\int P \, dx$. The function $\mu(x)$ (known as the *integrating factor*) is then $e^{\int P \, dx}$. Multiplying the standard form of the equation by this integrating factor gives a left-hand side which is the derivative of $\mu(x)y$. The solution of the equation then follows by an integration. Note that in finding the integrating factor the constant of integration when $\int P \, dx$ is found can be ignored. For, if included, it will simply supply a multiplied constant, which will make no essential difference to the equation or the method by which it is solved.

EXAMPLE 1:

$$\frac{dy}{dx} + \frac{y}{x} = x^2.$$

Here $P = 1/x$, and we have

$$\log \mu = \int \frac{dx}{x} = \log x,$$

and

$$\mu = x.$$

Multiplying the given equation by μ, i.e., by x, we have

$$x \frac{dy}{dx} + y = x^3.$$

If we have not made any mistakes, the left-hand side should now be the derivative of μy. As a check on our working let us see if this is so. We have

$$\frac{d}{dx}[xy] = x\frac{dy}{dx} + y$$

which is correct. Our equation is therefore

$$\frac{d}{dx}[xy] = x^3.$$

Integrating both sides, we get

$$xy = \int x^3\, dx + A$$
$$= \frac{1}{4}x^4 + A.$$

Thus the final solution is

$$y = \frac{1}{4}x^3 + \frac{A}{x}.$$

EXAMPLE 2:

$$y' - y \tan x = \sec^3 x.$$

Here we have $P = -\tan x$, and $\int P\, dx = -\int \tan x\, dx = \log \cos x$. Consequently the integrating factor is $\cos x$. Multiplying, we have

$$y' \cos x - y \sin x = \sec^2 x$$

or

$$\frac{d}{dx}[y \cos x] = \sec^2 x.$$

Integration of both sides gives

$$y \cos x = \int \sec^2 x\, dx = \tan x + A$$

which is the solution. It could also.be written as

$$y = \tan x \sec x + A \sec x.$$

Example 3:

$$y' - xy = 1.$$

In this deceptively simple-looking example $P = -x$, and $\mu = e^{-\frac{1}{2}x^2}$, since $\int P\, dx = -\frac{1}{2}x^2$. Hence we write

$$e^{-\frac{1}{2}x^2}y' - xe^{-\frac{1}{2}x^2}y = e^{-\frac{1}{2}x^2}$$

or

$$\frac{d}{dx}[e^{-\frac{1}{2}x^2}y] = e^{-\frac{1}{2}x^2}.$$

Integration yields

$$e^{-\frac{1}{2}x^2}y = \int e^{-\frac{1}{2}x^2}dx + A$$

but the integral on the right-hand side cannot be evaluated further. We can therefore do no more than write:

$$y = e^{\frac{1}{2}x^2}\int e^{-\frac{1}{2}x^2}\,dx + Ae^{\frac{1}{2}x^2}$$

as the solution.

Note: There is a polite convention among mathematicians that a differential equation shall be deemed to have been solved if the solution is expressed in terms of integrals, even though the integrals may not be expressible in terms of standard functions, and even though the form of the solution may not be of much help to anyone wanting to make use of it. In this sense the above equation has been "solved." Notice how a simple-looking equation like this one can turn out to be one for which a solution in terms of standard functions is not possible.

Exercise 128

1. Show that the following first order equations are linear, find the appropriate integrating factor, and hence obtain the general solution

(a) $\dfrac{dy}{dx}\cos x + y \sin x = x \sin 2x + x^2$

(b) $\sin x \dfrac{dy}{dx} - y = 1 - \cos 2x$

(c) $3x\dfrac{dy}{dx} - 2y = x^2 - x$

(d) $(1 + x^2)\dfrac{dy}{dx} + 4xy = \tan^{-1} x$

(e) $\sin x\dfrac{dy}{dx} + y \cos x = 2x(y \sin x - x^2)$

(f) $(1 + x)^2\dfrac{dy}{dx} + (1 - x^2)y = x$

(g) $x\dfrac{dy}{dx} - y(x + 2) = x^3 - x^4$

Note: Remember to start by converting the equation to the standard Type III form.

2. The equation

$$\dfrac{dy}{dx} + 2y = 2xy^{3/2}$$

is not a linear equation. Show that the substitution $y = v^{-2}$ converts it into a linear equation, and hence find the general solution.

3. Show that a substitution of the form $y = v^k$, where k is suitably chosen, will convert the differential equation

$$\dfrac{dy}{dx} + y \cot x = y^2 \sin^2 x$$

into a Type III equation. Find the value of k, and solve the equation.

4. As in problem 3, solve the equations

(a) $\dfrac{dy}{dx} + xy = x^3y^2$

(b) $x\dfrac{dy}{dx} = 2y + x^{n+1}y - x^n y^{3/2}$

5. The tangent at a general point P of a curve meets the x-axis at the point Q. If the ordinate of P is always k times the length OQ (where O is the origin) show that

$$\dfrac{dy}{dx} = \dfrac{ky}{kx - y}$$

for points on the curve. Hence show that the equation of the curve is of the form

$$Ay = e^{-kx/y}.$$

11.6 EQUATIONS OF THE SECOND ORDER

We shall consider only one type of equation of second order, namely the *linear equation*. This is an equation of the form

$$K(x)\frac{d^2y}{dx^2} + L(x)\frac{dy}{dx} + M(x)y = N(x)$$

where K, L, M, N are functions of x alone. Clearly, by dividing through by $K(x)$, we can write this as

(11.6.1) $$\frac{d^2y}{dx^2} + P\frac{dy}{dx} + Qy = R$$

where P, Q, R are functions of x alone. The name *linear*, as before, comes from the fact that the equation is *linear* in y and its derivatives, that is, that y, y' and y'' occur only to the first power.

Even this equation, specialized though it is, is too general for our purposes, and we shall shortly make the further restriction that P and Q are in fact constants. However, the preliminary results that we shall need are true of the general linear equation (11.6.1), so we shall not place any such restriction on P and Q until we have to.

Let us first see what these preliminary results are.

THEOREM 11.1

IF $u(x)$ AND $v(x)$ ARE TWO DIFFERENT SOLUTIONS OF THE EQUATION

(11.6.2) $$y'' + Py' + Qy = 0$$

THEN $Au(x) + Bv(x)$, WHERE A AND B ARE ARBITRARY CONSTANTS, IS THE GENERAL SOLUTION OF THIS EQUATION.

Note that we are talking here about an equation whose right-hand side is zero. Such an equation is known as a *reduced linear equation* of order 2.

PROOF:

If $y = Au + Bv$ then

$$y'' + Py' + Qy = Au'' + Bv'' + P(Au' + Bv') + Q(Au + Bv)$$
$$= A(u'' + Pu' + Qu) + B(v'' + Pv' + Qv) = 0.$$

For $u'' + Pu' + Qu = 0$ and $v'' + Pv' + Qv = 0$ since u and v are solutions. Hence $y = Au + Bv$ is a solution, and since it contains two arbitrary constants it is the general solution.

Theorem 11.1 requires that *two* solutions be known. If it should happen that we know only one solution then we can make use of the next theorem to find a second solution.

THEOREM 11.2

IF $u(x)$ IS ONE SOLUTION OF THE REDUCED EQUATION (11.6.2), ANOTHER, DIFFERENT, SOLUTION CAN BE FOUND BY MEANS OF THE SUBSTITUTION $y = u(x)z$.

PROOF:

If we put $y = u(x)z$ then

$$y' = u(x)z' + u'(x)z$$

and

$$y'' = u(x)z'' + 2u'(x)z' + u''(x)z.$$

Substituting in the differential equation, we have

$$(uz'' + 2u'z' + u''z) + P(uz' + u'z) + Quz = 0,$$

or

$$uz'' + (2u' + Pu)z' + (u'' + Pu' + Qu)z = 0.$$

But since $u'' + Pu' + Q = 0$, we have

$$uz'' + (2u' + Pu)z' = 0$$

which is a differential equation in z'. For the sake of convenience let us put $z' = p$, then we can write this equation as

$$u\frac{dp}{dx} + (2u' + Pu)p = 0$$

which is a Type II equation; the variables can be separated as

$$\frac{dp}{p} = -\frac{2u' + Pu}{u} dx$$

and the solution for p can be found. A further integration then gives z, and hence the other solution $y = uz$ of the original equation.

THEOREM 11.3

IF $\psi(x)$ IS ANY SOLUTION OF THE LINEAR EQUATION (11.6.1) AND $u(x)$ AND $v(x)$ ARE TWO DIFFERENT SOLUTIONS OF THE CORRESPONDING REDUCED EQUATION (11.6.2), THEN THE GENERAL SOLUTION OF (11.6.1) IS $y = Au(x) + Bv(x) + \psi(x)$, WHERE A AND B ARE ARBITRARY CONSTANTS.

PROOF:

If $y = Au + Bv + \psi$ then

$$y'' + Py' + Qy = A(u'' + Pu' + Qu) + B(v'' + Pv' + Qv)$$
$$+ (\psi'' + P\psi' + Q)$$
$$= R.$$

For the first two terms vanish as before, and the third term is R, since it satisfies (11.6.1). Thus $Au(x) + Bv(x) + \psi(x)$ is a solution, and, having two arbitrary constants, it is the general solution.

By saying that $u(x)$ and $v(x)$ are *different* solutions of the reduced equation we merely mean that one is not a constant multiple of the other. If, for example, $y = x^2$ were a solution of a reduced linear equation, then $y = 25x^2$ would also be a solution (because of the absence of a term on the right-hand side, the 25 would appear in each term, and the equation would be satisfied). However the solutions x^2 and $25x^2$ would not be considered as *different* solutions. Clearly if $u(x)$ is a constant multiple of $v(x)$ then $Au(x) + Bv(x)$ is simply an arbitrary multiple of $v(x)$; we would therefore have only one arbitrary constant, and the solution would not be the general one.

The solution $\psi(x)$ of the linear equation is any *one* solution, and it will not contain any arbitrary constants. It is called a *particular solution* or (more usually) a *particular integral* of the equation (11.6.1). The function $Au(x) + Bv(x)$, the general solution of the reduced equation (11.6.2), which together with the particular integral makes up the solution of (11.6.1), is called the *complementary function*.

The observant reader will have noticed that Theorem 11.3 contains Theorem 11.1 as a special case (with $R = 0$) so that we could have omitted the first of these theorems. There is no harm in taking Theorem 11.3 in two stages, however, and this procedure will serve to underline the fact that in what follows we shall first consider in some detail the problem of finding complementary functions, and then, quite separately, apply ourselves to the problem of finding particular integrals. As already mentioned, we shall do this only for equations in which P and Q are constants. Theorem 11.2 will be used to deal with an awkward borderline case that will arise.

A point which frequently causes difficulty can conveniently be dealt with here. Theorem 11.3 states that if we have found *any* particular integral of an equation (11.6.1), then by adding this to the complementary function—the general solution of the reduced equation (11.6.2)—we get the general solution of the original equation (11.6.1). This seems at first sight to imply that there are many "general solutions" of the original equation, namely, one for every particular integral that we may happen to find.

This is not so. It is a consequence of Theorem 11.3 (and can easily be proved directly) that the difference between any two particular integrals is a solution of the reduced equation. Thus if $\psi_1(x)$ and $\psi_2(x)$ are any two particular integrals, then $\psi_2(x) = \psi_1(x) + Pu(x) + Qv(x)$, where P and Q are constants. If we form the general solution of (11.6.1) from these two particular integrals and the complementary function, we get, in the first case,

$$(11.6.3) \qquad\qquad Au(x) + Bv(x) + \psi_1(x),$$

and in the second

$$(11.6.4) \qquad Au(x) + Bv(x) + \psi_1(x) + Pu(x) + Qv(x)$$
$$= (A + P)u(x) + (B + Q)v(x) + \psi_1(x).$$

Since A, B, P and Q are arbitrary, (11.6.3) and (11.6.4) are just different ways of saying the same thing. Hence we end up with the same general solution no matter what particular integral we happen to have found.

Exercise 129

1. Show that $\sin x$ and $\tan x$ are solutions of the differential equation

$$\sin x \cdot y'' - \sin x(1 + 2\sec^2 x)y' + (\sin x + \cos x + 2\sec x)\, y = 0,$$

and write down the general solution.

2. Show that e^x is a solution of the differential equation

$$y'' - (2 - \tan x)y' + (1 - \tan x)y = 0.$$

Use Theorem 11.2 to show that another solution is $e^x v$ where

$$v'' + v' \tan x = 0.$$

Hence obtain v and write down the general solution of the equation.

3. Prove that $x^{3+\sqrt{5}}$ and $x^{3-\sqrt{5}}$ are solutions of

$$x^2 \frac{d^2 y}{dx^2} - 5x \frac{dy}{dx} + 4y = 0$$

and that $-x$ is a solution of

$$x^2 \frac{d^2y}{dx^2} - 5x\frac{dy}{dx} + 4y = x.$$

Hence find the general solution of this latter equation.

11.7 EQUATIONS WITH CONSTANT COEFFICIENTS

From now on we assume that P and Q are constant, and thus take our equation in the form

(11.7.1) $\qquad ay'' + by' + cy = R(x)$

where a, b and c are constants.

We first consider the reduced equation

(11.7.2) $\qquad ay'' + by' + cy = 0.$

One method of finding solutions to a differential equation is to try out some standard function (preferably with at least one parameter thrown in to make it more general) and see if it can be made to fit. This is a hit-or-miss method but it often works. Let us try it now. There is no point in discussing those functions which do not fit, but high up in any list of standard functions will be the function $e^{\lambda x}$ (where λ is a parameter at our disposal). We try $y = e^{\lambda x}$ as a possible solution of equation (11.7.2). Substituting for y in the left-hand side, we obtain

$$a\lambda^2 e^{\lambda x} + b\lambda e^{\lambda x} + ce^{\lambda x} = e^{\lambda x}(a\lambda^2 + b\lambda + c).$$

We would like this to be zero, and since $e^{\lambda x} \neq 0$ this can happen only if

(11.7.3) $\qquad a\lambda^2 + b\lambda + c = 0.$

Hence if the value of λ is taken to be one of the roots of the quadratic equation (11.7.3) then $e^{\lambda x}$ will be a solution of (11.7.2). This quadratic equation is called the *auxiliary equation* corresponding to the differential equation, and since it will usually have two roots we have high hopes of being able to find the two *different* solutions that we require. There are possible snags, however, and we recognize the following three cases:

1. the roots of the auxiliary equation are real and distinct,
2. the roots of the auxiliary equation coincide,

3. the roots of the auxiliary equation are not real, i.e., the solution of the quadratic equation involves the square root of a negative number.

We shall take each of these cases in turn.

Case 1.   The roots of the auxiliary equation are real and distinct.
Applying the formula for the roots of a quadratic equation to the auxiliary equation (11.7.3) we obtain

$$\lambda = \frac{-b \pm \sqrt{b^2 - 4ac}}{2a}.$$

If $b^2 > 4ac$ then the roots are real and distinct. Let us call these roots λ_1 and λ_2. Then $e^{\lambda_1 x}$ and $e^{\lambda_2 x}$ are solutions of the reduced equation (11.7.2) and are certainly different. From Theorem 11.1 the general solution of (11.7.2) is

$$y = Ae^{\lambda_1 x} + Be^{\lambda_2 x},$$

and that is that.

One example will suffice to illustrate this, the simplest of the three cases. If the differential equation is

$$6y'' + y' - 2y = 0$$

and we put $y = e^{\lambda x}$, we obtain

$$6\lambda^2 e^{\lambda x} + \lambda e^{\lambda x} - 2e^{\lambda x} = 0,$$

and for equality we must have

$$6\lambda^2 + \lambda - 2 = 0.$$

This is the auxiliary equation; its solutions are $\lambda = \frac{1}{2}$ and $\lambda = -\frac{2}{3}$, as the reader can check. Hence the general solution of this differential equation is

$$y = Ae^{x/2} + Be^{-2x/3}$$

where A and B are arbitrary constants.

Case 2.   The roots of the auxiliary equation are coincident.
If this happens then the left-hand side of the auxiliary equation is a perfect square; and if the one and only root is $\lambda = \alpha$, the auxiliary equation can be written in the form

(11.7.4) $$\lambda^2 - 2\alpha\lambda + \alpha^2 = 0$$

and the original equation will be of the form

(11.7.5)
$$\frac{d^2y}{dx^2} - 2\alpha\frac{dy}{dx} + \alpha^2 y = 0.$$

From what has gone before it is clear that $e^{\alpha x}$ will be a solution of (11.7.5), but since α is the only root we do not have at hand two different solutions from which to construct the general solution. This does not mean that there is no other solution, merely that the previous method does not give it to us. We must find it for ourselves, and Theorem 11.2 gives us a means to do this. Theorem 11.2 tells us that, having found the solution $e^{\alpha x}$, we can find another solution by making the substitution $y = e^{\alpha x}z$.

With this substitution we have

$$y' = e^{\alpha x}z' + \alpha e^{\alpha x}z$$

and

$$y'' = e^{\alpha x}z'' + 2\alpha e^{\alpha x}z' + \alpha^2 e^{\alpha x}z.$$

Substituting in (11.7.5) we have

$$e^{\alpha x}z'' + 2\alpha e^{\alpha x}z' + \alpha^2 e^{\alpha x}z - 2\alpha(e^{\alpha x}z' + \alpha e^{\alpha x}z) + \alpha^2 e^{\alpha x}z$$
$$= e^{\alpha x}z'' = 0.$$

Thus $z'' = 0$. This means that $z' = A$ (a constant) and hence $z = Ax + B$, where B is some other constant, by two successive integrations. Hence it would seem that $y = (Ax + B)e^{\alpha x}$ is also a solution and it is readily verified that this is so. Since this solution contains two arbitrary constants it must be the general solution. It is the solution derived (in the manner of Theorem 11.1) from the solution $y = e^{\alpha x}$, which we started with, and the independent solution $y = xe^{\alpha x}$.

EXAMPLE:

$$y'' - 6y' + 9y = 0.$$

Here the auxiliary equation is $\lambda^2 - 6\lambda + 9 = 0$ which has the single (repeated) root $\lambda = 3$. Hence $y = e^{3x}$ is one solution. By what has been done above, a different solution is $y = xe^{3x}$. Hence the general solution is $y = Axe^{3x} + Be^{3x}$ or

$$y = (Ax + B)e^{3x}.$$

Case 3.   The roots of the auxiliary equation are not real.

This case occurs when the expression under the square root in the formula for the roots of the auxiliary equation is negative, that is, when $b^2 - 4ac < 0$. Numbers which involve square roots of negative numbers are called complex numbers, and form the subject matter of Chapter 13. Fortunately we can get by for the purposes of this chapter without requiring any of the results of Chapter 13, though we shall anticipate slightly the notation used in it.

We first notice that we can write

$$\sqrt{b^2 - 4ac} = \sqrt{(-1) \cdot (4ac - b^2)}$$
$$= \sqrt{-1} \cdot \sqrt{4ac - b^2}.$$

At least, this will be so if we assume that expressions like $\sqrt{-1}$ can be manipulated like ordinary numbers. Whether they can or not need not concern us here. At present we are simply groping for a solution to our differential equation, and the method by which we hit upon it will not matter—we shall test by direct substitution whether it really is a solution.

Suppose we write $p = -b/2a$, $q = \sqrt{4ac - b^2}/2a$ and denote $\sqrt{-1}$ by i. Then the roots of the auxiliary equation will be

$$\lambda = p \pm iq.$$

If we just push blindly ahead, and follow the method of case (1) then we can say that the two solutions are

$$e^{(p+iq)x} \quad \text{and} \quad e^{(p-iq)x}$$

or, if we assume the usual laws of indices to hold,

$$e^{px} \cdot e^{iqx} \quad \text{and} \quad e^{px} \cdot e^{-iqx}$$

whatever these expressions may mean. Whatever the interpretation that we give to the awkward $i = \sqrt{-1}$, it certainly looks as though the solutions are multiples of e^{px}. Let us assume that this is so and see where it gets us.

We shall let $y = e^{px}z$, just as we did for case (2). We have

$$y' = e^{px}z' + pe^{px}z$$
$$y'' = e^{px}z'' + 2pe^{px}z' + p^2e^{px}z,$$

and the differential equation (11.7.2) becomes

$$e^{px}(az'' + 2apz' + ap^2z + bz' + bpz + cz) = 0$$

or

$$az'' + (2ap + b)z' + (ap^2 + bp + c)z = 0.$$

Now $$2ap + b = 0$$

and

$$ap^2 + bp + c = a \cdot \frac{b^2}{4a^2} - \frac{b^2}{2a} + c$$

$$= \frac{1}{4a}(4ac - b^2)$$

$$= aq^2$$

by the definition of p and q. Thus our equation reduces to

(11.7.6) $$z'' + q^2z = 0.$$

This equation is one which we have seen before. It was used as an example in section 11.2, and its general solution is known. It is

$$z = A \cos qx + B \sin qx.$$

We have therefore found the solution (and it is the general solution) of the equation for case (3). With p and q defined as above, the solution is

$$y = e^{px}(A \cos qx + B \sin qx).$$

The skeptical reader may convince himself, by direct substitution in the differential equation, that this is indeed the solution.

EXAMPLE:
The equation

$$36y'' + 36y' + 13y = 0.$$

The auxiliary equation is

$$36\lambda^2 + 36\lambda + 13 = 0,$$

and the formula for the solution of a quadratic equation gives

$$\lambda = \frac{-36 \pm \sqrt{36^2 - 4.36.13}}{72}$$

$$= \frac{-36 \pm 6\sqrt{36 - 52}}{72}$$

$$= \frac{-6 \pm \sqrt{-16}}{12}$$

$$= \frac{-6 \pm 4\sqrt{-1}}{12}$$

$$= \frac{-1}{2} \pm \frac{1}{3}i.$$

Using the result just obtained, we see that the general solution is

$$y = Ae^{-x/2}\left(A \cos \frac{x}{3} + B \sin \frac{x}{3}\right).$$

We may note in passing that this seems to indicate that the rather odd expressions

$$e^{iqx} \quad \text{and} \quad e^{-iqx}$$

have some connection with the functions $\cos qx$ and $\sin qx$. For the moment it would be as well to leave it at that. Further elucidation awaits us in Chapter 13.

Exercise 130

1. Obtain the general solutions to the differential equations
   (a) $y'' - 8y' + 15y = 0$
   (b) $2y'' + 3y' - 2y = 0$
   (c) $15y'' - 7y' - 4y = 0$
   (d) $6y'' - 29y' + 35y = 0$
2. Determine to which of the three cases considered in the last section each of the following differential equations belongs, and find the solutions of these equations.
   (a) $y'' - 4y' + 13y = 0$
   (b) $y'' + 18y' + 77y = 0$
   (c) $y'' + 10y' + 25y = 0$
   (d) $y'' - 2y' + 2y = 0$
   (e) $y'' - 2y' + 10y = 0$

11.8 FINDING PARTICULAR INTEGRALS —THE "D" OPERATOR

We are now at the stage where we can find the general solution of any reduced linear equation with constant coefficients. We therefore turn to equations which are not reduced, i.e., for which $R(x)$ is not identically zero. By

Theorem 11.3 the solution of such an equation will follow if we can find any particular integral, and to this end we introduce what, at first sight, seems to be merely a change in the notation for derivatives, but which is really something rather more deep.

Instead of the notations dy/dx or y' for the derivative of y we shall write Dy. The letter "D" is what is known as an *operator;* it "operates" on whatever follows it to produce something else, in this case the derivative. In more mathematical terms it is a functional relationship whose domain is the set of all differentiable functions, and whose range is the set of their derivatives. We could, if we wished, write the result of the operator acting on y as $D(y)$, to preserve the analogy with the functional notation, $f(x)$ for example; but it is more convenient to omit the parentheses.

Thus the differential operator "D" acts on a function to produce its derivative. We have $Dx^2 = 2x$, $D \sin x = \cos x$ and, in general, $Df(x) = f'(x)$. Further, if we write $D \cdot Df(x)$, this will mean "the derivative of the derivative of $f(x)$", in other words $f''(x)$. By analogy with algebra we shall write this as $D^2f(x)$, and interpret "D^2" as the operator which yields the *second* derivative of the function which follows it. The higher "powers" of D are defined similarly and denote the operations of taking higher derivatives.

Using this notation, the typical linear differential equation of the second order, not reduced, with constant coefficients can be written as

(11.8.1) $aD^2y + bDy + cy = R(x)$.

We shall write the left-hand side of this in an alternative form, viz.,

(11.8.2) $aD^2y + bDy + cIy$

where I denotes the *identity* operator, that is, the operator whose action on its operand (the function that follows it and on which it acts) is to leave it exactly as it was.

We now go a stage further and rewrite (11.8.2) as

(11.8.3) $(aD^2 + bD + cI)y$.

There is no question here of whether it is permissible to take the y outside parentheses in this way; it is purely a matter of notation. What we are saying is that by the expression

$$(aD^2 + bD + cI)$$

we shall mean the operator which, acting on y, gives expression (11.8.2). In other words (11.8.2) and (11.8.3) are, *by definition*, different ways of writing the same thing.

To simplify our expressions we shall omit the identity operator I in what follows, and write (11.8.3) as

(11.8.4) $(aD^2 + bD + c)y.$

By rights we ought not to do this, since we thereby introduce possible confusion between the number c, and the operator cI, but in practice it is unlikely that any confusion will arise on this account. We should bear in mind, nevertheless, that the "c" in (11.8.4) is strictly speaking not a number but an operator—the operator which, acting on a function y, yields the function cy.

The notation introduced so far suggests that we are going to handle the operator D and its powers in much the same way as we handle ordinary algebraic quantities. This is true; for it turns out that, within certain limits which we shall prescribe, the operator D does obey the familiar laws of algebra, and hence can legitimately be treated in the same way as an algebraic variable. It is up to us to establish the extent to which this is true, and this is what we shall do first.

The expression $(aD + b)y$ stands for $a(dy/dx) + by$, and this is some function of x. If we act on this function with the operator $(pD + q)$, i.e., if we form the expression

$$(pD + q)z, \quad \text{where } z = (aD + b)y$$

we have

$$(pD + q)z = p\frac{dz}{dx} + qz$$

$$= p\frac{d}{dx}\left\{a\frac{dy}{dx} + by\right\} + q\left\{a\frac{dy}{dx} + b\right\}$$

$$= pa\frac{d^2y}{dx^2} + pb\frac{dy}{dx} + qa\frac{dy}{dx} + qb$$

(11.8.5) $$= (paD^2 + pbD + qaD + qb)y.$$

This is the result of acting with the operator $(pD + q)$ on $(aD + b)y$, and this we would naturally write as

$$(pD + q)\{(aD + b)y\}.$$

Let us now consider

$$(aD + b)\{(pD + q)y\}.$$

Treating this in the same way, we find that it is the same as

$$(apD^2 + bpD + qaD + bq)y$$

i.e., the same as the right-hand side of (11.8.5). Hence

$$(pD + q)\{(aD + b)y\} \quad \text{and} \quad (aD + b)\{(pD + q)y\}$$

are seen to mean the same thing. We can write these as

$$(pD + q)(aD + b)y \quad \text{and} \quad (aD + b)(pD + q)y.$$

The operator $(pD + q)(aD + b)$ will be called the *product* of the operators $pD + q$ and $aD + b$. Thus it is seen that the product of factors of the form $pD + q$ does not depend on the order in which the factors occur. Thus the commutative law of multiplication applies, and

$$(pD + q)(aD + b) = (aD + b)(pD + q)$$

both being equal to the operator

$$apD^2 + (bp + aq)D + bq.$$

In the same sort of way we can prove that the product of two operators that are *polynomials* in D does not depend on the order in which the factors are taken. That is, if $F(D)$ and $G(D)$ are polynomials in D, then

$$F(D)\,G(D) = G(D)\,F(D)$$

that is

(11.8.6) $$F(D)\{G(D)y\} = G(D)\{F(D)y\}.$$

Thus these operators, the expressions in D, can be multiplied just as if D were an algebraic quantity. Note that the restriction to constant coefficients is very important. If we allow functions of x to occur in the operators, then the results here will not usually hold. For example, it is not true that

$$(D + x)\{(D + 3x)y\} = (D + 3x)\{(D + x)y\}.$$

For the left-hand side is

$$(D + x)\left\{\frac{dy}{dx} + 3xy\right\} = \frac{d}{dx}\left\{\frac{dy}{dx} + 3xy\right\} + x\left\{\frac{dy}{dx} + 3xy\right\}$$

$$= \frac{d^2y}{dx^2} + 3x\frac{dy}{dx} + 3y + x\frac{dy}{dx} + 3x^2y$$

$$= (D^2 + 4xD + 3)y + 3x^2y.$$

whereas the right-hand side is

$$(D + 3x)\{(D + x)y\} = \frac{d}{dx}\left\{\frac{dy}{dx} + xy\right\} + 3x\left\{\frac{dy}{dx} + xy\right\}$$

$$= \frac{d^2y}{dx^2} + x\frac{dy}{dx} + y + 3x\frac{dy}{dx} + 3x^2y$$

$$= (D^2 + 4xD + 1)y + 3x^2y;$$

not the same as before. However, provided that the expressions in D are polynomials (or later, power series) in D with constant coefficients we can invert the order of multiplication.

Exercise 131

1. By carrying out the indicated differentiations find

$$(1 + D + D^2 + D^3 + D^4 + D^5 + D^6) \sin 2x.$$

Act on the result with the operator $1 - D$, and show that the result is the same as

$$(1 - D^7) \sin 2x.$$

2. If $F(D)$ stands for the operator $(1 + \frac{2}{3}D)$, and $Q(D)$ stands for the operator

$$\left(1 - \frac{2}{3}D + \frac{4}{9}D^2 - \frac{8}{27}D^3 + \frac{16}{81}D^4\right)$$

verify directly that

$$F(D) \cdot Q(D)x^4 = Q(D) \cdot F(D)x^4$$

$$= \left(1 - \frac{32}{243}D^5\right)x^4$$

$$= x^4.$$

11.9 PARTICULAR INTEGRALS WHEN $R(x)$ IS A POLYNOMIAL

Suppose we have a polynomial $F(D)$ in D with constant coefficients. In general it will be possible to expand the expression $1/F(D)$ formally as a power series in D. By "formally" we mean not only that we are going to treat D as if it were an ordinary algebraic quantity but also that we shall not consider any questions of convergence or divergence. Thus, in this formal way we shall write

$$\frac{1}{1-D} = 1 + D + D^2 + \cdots$$

(11.9.1) $$\frac{1}{4+D^2} = \frac{1}{4} \cdot \frac{1}{1+\frac{1}{4}D^2}$$

$$= \frac{1}{4}\left(1 - \frac{D^2}{4} + \frac{D^4}{16} - \frac{D^6}{64} + \cdots\right)$$

and so on. Let us denote such a power series expansion of $1/F(D)$ by $Q(D)$. The important property of the series Q is then that

$$F(D)\cdot Q(D) = 1,$$

where, as we noted, the right-hand side is not the number 1 but the identity operator.

We are now in a position to see how such a formal expansion of $1/F(D)$ can help us to find particular integrals, and why it is that we can ignore the question of convergence of the power series.

Let us consider the differential equation

(11.9.2) $$\frac{d^2y}{dx^2} + y = x^3, \quad \text{or} \quad (D^2 + 1)y = x^3.$$

The complementary function can be found by the methods given above; it is

$$A\cos x + B\sin x.$$

To find a particular integral we first expand $1/(1 + D^2)$ as a formal power series in D, since $F(D) = D^2 + 1$. We obtain

$$Q(D) = \frac{1}{1+D^2} = 1 - D^2 + D^4 - D^6 + \cdots .$$

We now operate with the operator $Q(D)$ on both sides of (11.9.2) and we have

$$Q(D)\cdot(D^2 + 1)y = Q(D)x^3$$

or

$$y = Q(D)\cdot x^3 \quad \text{since} \quad Q(D)\cdot F(D) = 1.$$

Hence

$$y = (1 - D^2 + D^4 - D^6 + \cdots)x^3$$
$$= x^3 - D^2x^3 + D^4(x^3) \cdots$$
$$= x^3 - 6x.$$

since all derivatives of x^3 after the third are zero. The particular integral that we have found is thus $y = x^3 - 6x$, and the complete solution of the differential equation is

$$y = A \cos x + B \sin x + x^3 - 6x.$$

The reader may object (quite properly) that to operate with an operator containing an unlimited number of terms is, to say the least, stretching our definition of operators rather to the limit, since we defined only polynomials in D. It is easy to justify the above procedure, however, when the function on the right-hand side of the differential equation is a polynomial in x. To see this, let us look again at equation (11.9.2).

Suppose we had taken $Q(D)$ to be

$$1 - D^2 + D^4 - D^6$$

instead of the infinite series. Operating on both sides of (11.9.2) with this $Q(D)$ (to which there is no objection) we then get

$$(1 - D^2 + D^4 - D^6)(1 + D^2)y = (1 - D^2 + D^4 - D^6)x^3$$

which reduces to

$$y - D^8y = x^3 - 6x.$$

Remembering that we are not looking for a general solution but for any particular integral, we see that

$$y = x^3 - 6x$$

will fill the requirements. By taking enough terms of the series into $Q(D)$ we have arranged that the extra term $-D^8y$ on the left-hand side is zero, being a high derivative of $x^3 - 6x$, our proposed particular integral.

In general, if the right-hand side $R(x)$ of the differential equation is a polynomial in x, we can choose $Q(D)$ to be a polynomial containing a certain number of terms of the formal expansion of $1/F(D)$, and we then have, say,

$$Q(D) \cdot F(D) = 1 + \text{terms in high powers of } D.$$

The result of acting with $Q(D)$ on the right-hand side $R(x)$ will be a polynomial of the same degree as $R(x)$. Hence by taking enough terms into $Q(D)$ we can arrange that all the terms in $Q(D) \cdot F(D)$, except the 1, will give zero when operating on $R(x)$. It then follows that $y = Q(D) \cdot R(x)$ is a particular integral.

All this is equivalent to saying that if $R(x)$ is a polynomial we can take $Q(D)$ to be the infinite series (defined by $Q(D) \cdot F(D) = 1$). No troubles over convergence will arise since $Q(D)$ gives only a finite number of nonzero terms when it acts on $R(x)$. Naturally, if $R(x)$ were *not* a polynomial—if it were e^x or $\sin x$ say—then $Q(D) \cdot R(x)$ would not terminate, and we *would* have to concern ourselves with the question of convergence. We shall take pains not to get ourselves in this sort of situation.

Exercise 132

1. Find particular integrals of the following differential equations
   (a) $y'' - 3y' - 4y = x^3$
   (b) $y'' + 4y' + 4y = \frac{1}{2}(x^2 + 3x)$
   (c) $y'' + 16y = x^5 - 5$
   (d) $y'' - 6y' + 34y = 2x^2$

2. Although (as mentioned at the end of Section 9) we shall avoid having to manipulate infinite series when using the operator $Q(D)$, it is worth remarking that we may still obtain a particular integral even if we do have to use the full series for $Q(D)$. Thus for the equation

$$\left(\frac{1}{2}D^2 - 1\right)y = \sin x$$

we would be led to write

$$y = \frac{1}{\frac{1}{2}D^2 - 1} \sin x$$

$$= -\left(1 + \frac{1}{2}D^2 + \frac{1}{4}D^4 + \cdots\right)\sin x.$$

Show that this gives an infinite series which can be summed, and that its sum is $-\frac{2}{3}\sin x$. Verify that this is a particular integral.

On the other hand, if we did the same with the equation

$$(2D^2 - 1)y = \sin x$$

we would end up with

$$-(1 + 2D^2 + 4D^4 + 8D^6 + \cdots)\sin x.$$

Show that the series obtained does not converge. Despite this there is a simple particular integral of this second equation. What is it?

11.10 FURTHER PARTICULAR INTEGRALS

By the methods given above we can find particular integrals of equations of the type

$$a \frac{d^2y}{dx^2} + b \frac{dy}{dx} + cy = R(x)$$

where $R(x)$ is a polynomial, and hence we can solve the equation completely. It is quite a restriction on the right-hand side, however, to limit it to being a polynomial, and we therefore seek means of extending our methods to cope with a more general function $R(x)$. We shall first see what to do if $R(x)$ is of the form $e^{px}\phi(x)$, where $\phi(x)$ is a polynomial.

Let us consider the expression $D^n[e^{px}\phi(x)]$. This is the nth derivative of a product, and by Leibniz's theorem (see section 9.10) we have,

$$(11.10.1) \qquad D^n[e^{px}\phi(x)] = \sum_{r=0}^{n} \binom{n}{r} D^r(e^{px}) D^{n-r}\phi(x).$$

Now $D^r(e^{px}) = p^r e^{px}$, so that the right-hand side of (11.10.1) becomes

$$\sum_{r=0}^{n} \binom{n}{r} p^r e^{px} D^{n-r}\phi(x)$$

$$= e^{px} \sum_{r=0}^{n} \binom{n}{r} p^r D^{n-r}\phi(x)$$

$$= e^{px} \left\{ \sum_{r=0}^{n} \binom{n}{r} p^r D^{n-r} \right\} \phi(x)$$

where we have taken out between braces the operator which acts on $\phi(x)$. This operator can be written as

$$(D + p)^n$$

by the binomial theorem, and we have

$$D^n\{e^{px}\phi(x)\} = e^{px}(D + p)^n\phi(x).$$

Hence if $Q(D)$ is a power of D, we have

$$(11.10.2) \qquad Q(D)\{e^{px}\phi(x)\} = e^{px}Q(D + p)\phi(x)$$

and it is easy to see that the same is true if $Q(D)$ is a polynomial in D; for it will be true of each separate term in the operator. It is perhaps not so easy to see that we will be justified in applying (11.10.2) even when the operator is an infinite series. But let us suspend our skepticism for a few lines and push on.

We can crystallize (11.10.2) into the following rule: "the factor e^{px} can be moved from the operand over to the left of the operator, provided D is replaced by $D + p$."

Applying this to the equation

$$F(D) \cdot y = e^{px}\phi(x)$$

we have

(11.10.3) $\quad\quad \begin{aligned} y &= Q(D)e^{px}\phi(x) \quad (Q \text{ defined as before}) \\ &= e^{px}Q(D + p)\phi(x) \end{aligned}$

as a likely candidate for a particular integral. Now unless we are unlucky (a possibility which we shall consider later) we can express $Q(D + p)$ as a power series in D, and since $\phi(x)$ is a polynomial, the right-hand side of (11.10.3) will have only a finite number of nonzero terms, and hence can be found.

Let us return to the question of whether the use of (11.10.2) is justified when $Q(D)$ is an infinite series. The only relevant question is "Is (11.10.3) really a solution of the equation?" This we can easily check by seeing if it satisfies it. If

$$y = e^{px}Q(D + p)\phi(x)$$

then

$$\begin{aligned} f(D)y &= F(D)\{e^{px}Q(D + p)\phi(x)\} \\ &= e^{px}F(D + p)Q(D + p)\phi(x) \end{aligned}$$

by virtue of our rule [or (11.10.2)] applied to the operator $F(D)$, which is a polynomial. But

$$F(D + p)Q(D + p) = 1$$

from the definition of Q.* Hence

$$F(D)y = e^{px}\phi(x)$$

and the equation is indeed satisfied. Thus this method does lead to a solution.

---

* Since the operand is a polynomial the justification given in the last section for using the infinite series $Q(D)$ applies also here.

The following examples should make the method clear.

EXAMPLE 1:

$$D^2y + 2Dy - 15y = e^{2x}x.$$

Here

$$F(D) = D^2 + 2D - 15 = (D - 3)(D + 5).$$

Thus

$$y = \frac{1}{(D - 3)(D + 5)} \cdot e^{2x}x$$

$$= e^{2x}\frac{1}{(D - 1)(D + 7)} \cdot x,$$

applying the rule and adding 2 to D when we make e^{2x} cross to the other side of the operator. We now must expand $1/(D - 1)(D + 7)$ in powers of D. We can write

$$\frac{1}{(D - 1)(D + 7)} = \frac{A}{D - 1} + \frac{B}{D + 7}$$

where A and B are to be determined. We have, in fact

$$1 = A(D + 7) + B(D - 1)$$

whatever D may be. Hence

$$A + B = 0$$

and

$$7A - B = 1$$

whence

$$A = \frac{1}{8}; \quad B = -\frac{1}{8}.$$

Thus

$$y = \frac{1}{8}e^{2x}\left\{\frac{1}{D - 1} - \frac{1}{D + 7}\right\}x$$

$$= \frac{1}{8}e^{2x}\left\{-(1 + D + D^2 + \cdots) - \frac{1}{7}\left(1 - \frac{D}{7} + \frac{D^2}{49}\cdots\right)\right\}x,$$

since

$$\frac{1}{D+7} = \frac{1}{7(1+D/7)} = \frac{1}{7}\left(1+\frac{D}{7}\right)^{-1}.$$

Hence

$$y = \frac{1}{8}e^{2x}\left\{-x - 1 - \frac{1}{7}\left(x - \frac{1}{7}\right)\right\}$$

which reduces to

$$-\frac{1}{49}e^{2x}(7x+6)$$

which is the required particular integral.

EXAMPLE 2:

$$D^2y + Dy - 3y = e^{-4x}$$

or

$$(D^2 + D - 3)y = e^{-4x}.$$

Here we have

$$y = \frac{1}{(D-1)(D+2)}e^{-4x}$$

$$= e^{-4x}\frac{1}{(D-5)(D-2)}\cdot 1$$

adding -4 to D as we bring the operand over.

Notice that we must provide something for the operator to operate on, and in this case it will be the number 1. Now when a polynomial (or power series) $F(D)$ in D acts on the number 1, any nonzero power of D will automatically yield a zero result. Hence the only term in $F(D)$ which gives a nonzero result when acting on 1 is the constant term, and the constant term of a polynomial or power series, if not obvious, can be found by making the variable zero. Hence we see that

$$F(D)\cdot 1 = F(0).$$

In our particular case, therefore,

$$\frac{1}{(D-5)(D-2)}\cdot 1 = \frac{1}{(-5)(-2)}.$$

Thus we have

$$y = \frac{1}{10}e^{-4x}$$

as the particular integral.

Notice that we have carefully avoided putting anything other than a 1 in the denominator of the above operators. We did not, for example, write

$$\frac{e^{-4x}}{(D-5)(D-2)} \cdot 1$$

Such an expression would be ambiguous, for it would not be clear whether the e^{-4x} was a part of the operand, or whether it was a factor multiplying the result of the operation (as is in fact the case).

Exercise 133

1. Find particular integrals of the following differential equations.
    (a) $y'' - 5y' + 6y = x^2 e^x$
    (b) $y'' - 3y' - 4y = xe^{2x} + x^3$
    Note: If $R(x)$ is the sum of two or more functions we can treat these functions separately, and add the particular integrals.
    (c) $y'' + 8y' + 16 = 5xe^{-2x}$
    (d) $9y'' - 9y' - 4y = (1 + x^2)e^x$
2. Find the general solution of the equation

$$(D-a)(D-b)y = x^2 e^{px}$$

where a, b and p are real numbers and $p \neq a$, $p \neq b$.

11.11 AN EXCEPTIONAL CASE

In the preceding section we have once again reduced a problem [that of coping with a right-hand side of the form $e^{px}\phi(x)$] to a preceding problem (coping with a polynomial). There are occasions when the method breaks down, however, as the following example shows.

EXAMPLE:

$$D^2 y + 2Dy - 15y = e^{3x}x.$$

Here we have

$$y = \frac{1}{(D-3)(D+5)}e^{3x}x$$

and on applying the rule (11.10.2) we obtain

$$y = e^{3x} \frac{1}{D(D+8)} \cdot x.$$

Now $1/D(D+8)$ cannot be expanded in non-negative powers of D because of the presence of the D in the denominator, and our previous method therefore will not apply. Postponing the decision of what to do about this D, we can still deal with the rest of the fraction. For

$$\frac{1}{D+8} \cdot x = \frac{1}{8}\left(1 - \frac{D}{8} + \cdots\right)x$$

$$= \frac{x}{8} - \frac{1}{64}.$$

We thus have

$$y = e^{3x} \frac{1}{D}\left(\frac{x}{8} - \frac{1}{64}\right)$$

and we must now decide what is meant by the operator "$1/D$."

If the normal rules of algebra still apply we would expect to have

$$\frac{1}{D} \cdot D = 1.$$

In other words, the operation "$1/D$" applied to the derivative of a function would give the function itself. This strongly suggests (to say the least) that the operation performed by $1/D$ is that of taking the integral of the operand. We might ask "which integral," for a function has many integrals according to which value of the arbitrary constant is taken. But we are searching for a particular integral of the differential equation—any particular integral—so it will not matter what value we give to the arbitrary constant. Thus we might as well take it to be zero, for simplicity's sake.

To return to our example. If we interpret "$1/D$" as the operator which takes the integral of the operand, then

$$y = e^{3x} \frac{1}{D}\left(\frac{x}{8} - \frac{1}{64}\right)$$

$$= e^{3x}\left(\frac{x^2}{16} - \frac{x}{64}\right)$$

and this is a particular integral, as can readily be verified.

EXAMPLE:

$$D^2y - 4Dy + 4y = e^{2x}(x^2 + 1)$$

Here

$$(D^2 - 4D + 4)y = e^{2x}(x^2 + 1)$$

$$y = \frac{1}{(D - 2)^2}e^{2x}(x^2 + 1)$$

$$= e^{2x}\frac{1}{D^2}(x^2 + 1)$$

by application of (11.10.2). Now if $1/D$ means "integrate," then $1/D^2$ will mean "integrate twice." If we do this (leaving out arbitrary constants), then

$$y = e^{2x}\frac{1}{D}\left(\frac{x^3}{3} + x\right)$$

$$y = e^{2x}\left(\frac{x^4}{12} + \frac{x^2}{2}\right).$$

Hence the particlar integral is $\frac{1}{12}e^{2x}x^2(x^2 + 6)$.

Exercise 134

1. Find particular integrals for the following differential equations.
   (a) $y'' - 3y' + 2 = x^2e^{2x}$
   (b) $y'' - 9y = xe^{3x}$
   (c) $9y'' + 12y' + 4y = (2x^2 + 5)e^{-2x/3}$
   (d) $y'' + 5y' + 4y = xe^{-4x} + 3x^2e^{-x}$
2. By writing $f(x)$ as $e^{\beta x}e^{-\beta x}f(x)$ show that

$$\frac{1}{D - \beta}f(x) = e^{\beta x}\frac{1}{D}\{e^{-\beta x}f(x)\}.$$

By repeating this process once more show that a particular integral of

$$(D - \alpha)(D - \beta)y = f(x)$$

is

$$y = e^{\alpha x}\frac{1}{D}\left[e^{(\beta - \alpha)x}\frac{1}{D}\{e^{-\beta x}f(x)\}\right].$$

This shows that it is possible to give a general expression for any particular integral, though usually this is not a practical way of obtaining particular integrals.

3. Use the general result of problem 2 to find a particular integral when $f(x) = e^{\beta x}$. Verify the result by finding a particular integral by the method of this section.

11.12   THE FUNCTIONS SIN qx AND COS qx

If the right-hand side of a differential equation of the kind we are considering is of the form sin qx or cos qx or a combination of these, the methods so far considered will not work; but since these functions are of common occurrence it would be a pity if we were unable to cope with them too. We shall therefore derive a method of doing this.

We start by observing that

(11.12.1)
$$D^2(\sin qx) = -q^2 \sin qx$$
$$D^2(\cos qx) = -q^2 \sin qx.$$

Hence, when acting on either of these functions, the operator D^2 has the same effect as the *constant* operator $-q^2$. Hence, remembering that our operators are all polynomials or power series in D [even if not explicitly written as such, as when $Q(D) = 1/(1 - D)$, for example] we can always replace D^2 by $-q^2$ when the operator acts on sin qx or cos qx. Let us see, by means of two examples, how this helps us.

EXAMPLE 1:

The equations $D^2y + 4y = \cos 3x$.

Here we have

$$(D^2 + 4)y = \cos 3x$$

whence

$$y = \frac{1}{D^2 + 4} \cos 3x.$$

Replacing D^2 by -3^2, we have

$$y = \frac{1}{-9 + 4} \cos 3x$$
$$= -\frac{1}{5} \cos 3x,$$

which is a particular integral, as can easily be verified.

Obviously the simplicity of this example was due to the fact that only D^2 occurred in the operator. If D occurs also then the calculation is slightly more involved but still not difficult.

EXAMPLE 2:

The equation $D^2y - 5Dy - y = \sin 2x$.

Here we have

$$(D^2 - 5D - 1)y = \sin 2x$$

and hence

$$y = \frac{1}{D^2 - 5D - 1} \sin 2x.$$

We first replace D^2 by $-2^2 = -4$, according to (11.12.1). This gives us

$$y = \frac{1}{-4 - 5D - 1} \sin 2x$$

$$= \frac{1}{5} \cdot \frac{1}{1 + D} \sin 2x.$$

The operator is now a fraction with D appearing in the denominator. Now a D in the numerator would not bother us—a polynomial operator gives us no trouble since there is only a finite number of terms. Nor are we worried by even powers of D in the denominator—they can be replaced. Bearing these two points in mind we can see that a simple little trick will solve our difficulties. We multiply the operator, top and bottom, by $1 - D$. This gives

$$\frac{1 - D}{(1 + D)(1 - D)} = \frac{1 - D}{1 - D^2},$$

and there is now nothing we cannot cope with.

We therefore have

$$y = -\frac{1}{5} \cdot \frac{1 - D}{1 - D^2} \sin 2x$$

$$= -\frac{1}{5} \cdot \frac{1 - D}{1 - (-4)} \sin 2x$$

[applying (11.12.1) again]

$$= -\frac{1}{25}(1 - D) \sin 2x$$

$$= -\frac{1}{25}(\sin 2x - 2 \cos 2x)$$

which is a particular integral.

This trick is quite general. After the first application of (11.12.1) the operator will be of the form

$$\frac{1}{aD + b} \qquad (a, b \text{ constants})$$

which we replace by

$$\frac{aD - b}{(aD + b)(aD - b)} = \frac{aD - b}{a^2 D^2 - b^2}.$$

A second application of (11.12.1) in the denominator then yields a polynomial operator.

There is no guarantee, of course, that the function on the right-hand side of a linear differential equation with constant coefficients will be of one of the types given above. If it is not, then the methods just given will not apply. There are many other methods of finding particular integrals, but we shall not be able to consider them here. Nevertheless, the equations that the reader is now able to solve include a very large proportion of the sorts of equations that occur in mathematical economics. Furthermore the reader should not now find much difficulty in learning more advanced techniques (should the occasion arise) from the extensive literature on the subject.

Exercise 135

1. Solve the following differential equations.
   (a) $y'' - 4y' + 4y = \sin 2x$
   (b) $y'' - 6y' + 9y = \sin 2x + 5xe^{3x}$
   (c) $y'' - 4y' - 5y = \cos x$
   (d) $y'' + 4y = x^2 + 3 \cos x$

2. Find particular integrals for the following differential equations.
   (a) $2\dfrac{d^2y}{dx^2} + \dfrac{dy}{dx} - y = e^x \sin 2x$

   (b) $\dfrac{d^2y}{dx^2} + 7\dfrac{dy}{dx} + 10y = e^{-2x} \cos 3x$

   (c) $\dfrac{d^2y}{dx^2} - 4\dfrac{dy}{dx} + 8y = 5e^{2x} \sin 2x.$

   Hint: consider the effect of differentiating $x \cos 2x$ twice.

   (d) $\dfrac{d^2y}{dx^2} + 25y = 10(\sin 5x + \cos 5x)$

11.13 SIMULTANEOUS DIFFERENTIAL EQUATIONS

If we are given an equation which contains two unknown functions f and g and their derivatives, it may be possible to derive a connection between the two functions. Thus if we know that

$$f'(x) = g'(x) \cdot f(x)$$

i.e., that

$$(11.13.1) \qquad \frac{f'(x)}{f(x)} = g'(x),$$

then by integrating both sides of (11.13.1) we obtain

$$\log f(x) = g(x) + A$$

which gives us a relationship between the functions f and g. Given one of these functions we can find the other. It requires two equations, however, to determine both the functions—a situation which is analogous to the well-known fact in elementary algebra that two (algebraic) equations are required to solve for two unknowns.

Two equations which involve two functions and their derivatives are known as *simultaneous differential equations*. More generally we could have n simultaneous differential equations in n unknown functions and their derivatives. In this section we shall touch very briefly on the subject of sets of simultaneous differential equations, confining ourselves to just one fairly simple, but important, set. For convenience we shall denote the two unknown functions by x and y, and denote by t the independent variable of which they are functions.

The equations that we shall consider are the following:

$$a_1 \frac{dx}{dt} + b_1 x + p_1 \frac{dy}{dt} + q_1 y = f_1(t)$$

$$a_2 \frac{dx}{dt} + b_2 x + p_2 \frac{dy}{dt} + q_2 y = f_2(t)$$

where a_1, b_1, p_1, etc., are constants, and f_1 and f_2 are functions of t. If we use "D" in place of "d/dt" these equations assume the form

$$(11.13.2) \qquad (a_1 D + b_1)x + (p_1 D + q_1)y = f_1$$

$$(11.13.3) \qquad (a_2 D + b_2)x + (p_2 D + q_2)y = f_2.$$

We now have to consider how to deduce, from (11.13.2) and (11.13.3), what the functions x and y are.

The usual technique for solving algebraic equations in two unknowns is to eliminate one of the unknowns by adding multiples of the two equations. Thus if we had to solve

$$(11.13.4) \qquad 2x + y = 5$$

$$(11.13.5) \qquad 9x - 2y = 3$$

we would multiply equation (11.13.4) by 2 and add the result, $4x + 2y = 10$, to (11.13.5) obtaining

$$13x = 13.$$

Thus y has been eliminated, and we have a simple equation for x, from which we see that $x = 1$. In order to find y, we must express y in terms of x from one of the equations. From equation (11.13.4) we have

$$y = 5 - 2x,$$

whence $y = 3$. This process will be quite familiar to the reader.

We can use a similar technique on equations (11.13.2) and (11.13.3), but we shall run into a snag if we are not careful. Suppose we decide to eliminate x. We can do this by operating on both sides of (11.13.2) with the operator $(a_2 D + b_2)$, obtaining

$$(11.13.6) \qquad (a_2 D + b_2)(a_1 D + b_1)x + (a_2 D + b_2)(p_1 D + q_1)y$$
$$= (a_2 D + b_2)f_1$$

and both sides of (11.13.3) with the operator $(a_1 D + b_1)$ obtaining

$$(11.13.7) \qquad (a_1 D + b_1)(a_2 D + b_2)x + (a_1 D + b_1)(p_2 D + q_2)y$$
$$= (a_1 D + b_1)f_2.$$

Now, since

$$(a_2 D + b_2)(a_1 D + b_1)x = (a_1 D + b_1)(a_2 D + b_2)x,$$

we can eliminate x by subtracting (11.13.7) from (11.13.6). The equation that results will be of the form

$$(11.13.8) \qquad F(D) \cdot y = \phi(t)$$

where $F(D)$ is a second-degree operator, the coefficients in which are constants which we do not need to specify in detail.

Equation (11.13.8) is a second order equation of the kind that we have been considering in the last few sections, and we shall assume that we can solve it by the standard methods. This gives us y as a function of t which will, of course, contain the usual two arbitrary constants. Now we must find the function x, and we can do this by going back to one of the original equations, say (11.13.2). Since y is now a known function we can substitute for it in (11.13.2) and obtain an equation of the form

$$(11.13.9) \qquad (a_1 D + b_1)x = \psi(x).$$

This is a linear equation in x, and we shall assume that we can solve it by the methods of section 11.7. But (and here is the snag) in doing so we shall automatically introduce yet another arbitrary constant, making three in all. Now we would not expect to get three arbitrary constants in the solution to a pair of equations like this, and in point of fact only two can occur.

What this means is that the three constants that we have obtained are not all arbitrary; there is a connection between them. What this connection is can be found by substituting the functions that we now have into the remaining equation—the one that was not used to find the second variable. We can then retain two of them, and express the third in terms of these two.

EXAMPLE:
Solve

$$(11.13.10) \qquad Dx + 2y = \sin 2t$$

$$(11.13.11) \qquad 2x - Dy = -\cos 2t.$$

If we operate on (11.13.11) with D we get

$$(11.13.12) \qquad 2Dx - D^2y = 2 \sin 2t.$$

We now eliminate x by subtracting twice (11.13.10) from (11.13.12). We get the differential equation

$$D^2y + 4y = 0$$

whose solution

$$(11.13.13) \qquad y = A \cos 2t + B \sin 2t$$

contains two arbitrary constants, as expected.

If we now substitute for y in the first equation (11.13.10) we obtain

$$Dx = \sin 2t - 2A \cos 2t - 2B \sin 2t$$

from which we obtain

$$x = -\frac{1}{2}\cos 2t - A \sin 2t + B \cos 2t + C,$$

and a third "arbitrary" constant has appeared. If we now substitute for x and y in the second equation (11.13.11)—this is something that we would very probably do anyway, as a check on the accuracy of our solution—we have

$$-\cos 2t - 2A \sin 2t + 2B \cos 2t + 2C$$
$$-(-2A \sin 2t + 2B \cos 2t) = -\cos 2t$$

which shows that C must be zero, and is therefore not arbitrary at all.

This was a particularly simple example. In more complicated examples we would find that the equations were satisfied only if some relation held between A, B and C such as, for example, $3A + 5B - 7C = 12$. We can then eliminate any one of these three constants according to our convenience.

Now although this method works quite well, and gives the correct answers, it is rather a waste of time to obtain our results in terms of three constants and then have to eliminate one of them by a substitution which might well be laborious. If there were no alternative we could learn to live with a method like this, but fortunately it is quite easy to arrange our manipulations of the equations so as to avoid this snag altogether.

The germ of the method can be seen in the simple example that we have just considered. If, after finding y as a function of t, we had substituted in the second equation (11.13.11) instead of the first we would have obtained

$$x = \frac{1}{2} Dy - \frac{1}{2}\cos 2t$$

$$= -A \sin 2t + B \cos 2t - \frac{1}{2}\cos 2t$$

which contains no further "arbitrary" constant, since no integration has been necessary. If, as in this example, one equation gives us x as a function of y, without containing Dx, then this situation holds and the snag does not arise. If both equations contain a term in Dx, then what we must do is to take a combination of the given equations which does not contain such a term.

We see that this can be done for our original pair of equations by multiplying (11.13.2) by a_2, (11.13.3) by a_1, and subtracting one equation from the other. This eliminates the term in Dx, and gives x as a function of y which can be found (when y is known) without the need of integration. We have not eliminated x altogether, but this can be done by substituting for x in one of the equations. We then solve the resulting second order equation for y, and then find x. This way, only two arbitrary constants are introduced. An example should make the method clear.

EXAMPLE:

Solve the equations

$$\frac{dx}{dt} + x - 2\frac{dy}{dt} = \cos t$$

$$\frac{dx}{dt} - \frac{dy}{dt} + 6y = 0.$$

These may be written as

$$(D + 1)x - 2Dy = \cos t$$

and

$$Dx - (D - 6)y = 0.$$

Subtracting the second from the first we obtain

$$x = (D + 6)y + \cos t$$

which gives x in terms of y with no integration. Thus

$$Dx = (D^2 + 6D)y - \sin t,$$

and on substituting in the second equation we obtain

$$(D^2 + 5D + 6)y = \sin t,$$

which is of a form that we know how to solve.

The auxiliary equation is

$$\lambda^2 + 5\lambda + 6 = 0$$

whence $\lambda = -2$ or -3. The complementary function is thus

$$Ae^{-2t} + Be^{-3t}.$$

To find a particular integral we consider

$$\frac{1}{D^2 + 5D + 6} \sin t = \frac{1}{-1 + 5D + 6} \sin t$$

$$= \frac{1}{5} \cdot \frac{1}{D + 1} \sin t$$

$$= \frac{1}{5} \cdot \frac{D - 1}{D^2 - 1} \sin t$$

$$= \frac{1}{5} \cdot \frac{D - 1}{-1 - 1} \sin t$$

$$= -\frac{1}{10}(D - 1) \sin t$$

$$= \frac{1}{10}(\sin t - \cos t).$$

Hence

$$y = Ae^{-2t} + Be^{-3t} + \frac{1}{10}(\sin t - \cos t).$$

From this we obtain

$$x = (D + 6)y + \cos t$$

$$= -2Ae^{-2t} - 3Be^{-3t} + \frac{1}{10}(\cos t + \sin t)$$

$$+ 6Ae^{-2t} + 6Be^{-3t} + \frac{6}{10}(\sin t - \cos t) + \cos t$$

$$= 4Ae^{-2t} + 3Be^{-3t} + \frac{1}{10}(5 \cos t + 7 \sin t).$$

11.14 STABILITY OF SOLUTIONS OF DIFFERENTIAL EQUATIONS

In this chapter we have been mainly concerned with techniques for the solution of differential equations. But when a differential equation has been set up as a model of some economic process it is not enough merely to find its solution; we must also interpret the solution, and see how it relates to its economic background.

Just what is entailed by this will depend on the particular economic situation under consideration, but one thing we shall almost certainly want to know about a solution is whether it tends to equilibrium, and, if so, whether that equilibrium is *stable* (in a sense to be defined shortly). We touched on the borders of this topic in section 11.3 when we remarked that the exponential law of growth could not represent an actual economic situation for all time, since it implies that the dependent variable (the size of the firm, or whatever it may be) continually increases without limit and can become indefinitely large. We shall now put this observation in a more general setting.

Let us consider the differential equation

(11.14.1) $y'' + 5y' + 6y = 24.$

It is easily verified that

(11.14.2) $y = 32e^{-2t} + 4$

is a solution of this equation. Now as $t \to \infty$, so $e^{-2t} \to 0$, and hence $y \to 4$. We say that the solution tends to the equilibrium value 4 as t becomes large. Again, the solution

(11.14.3) $y = 4e^{-3t} \cos 2t + 1$

of the differential equation

(11.14.4) $y'' + 6y' + 13y = 13$

also tends to an equilibrium value, namely 1, as $t \to \infty$. Unlike the previous example however, where the solution was a decreasing function of t, approaching the equilibrium value from above, here the solution oscillates above and below this equilibrium value. We can say that the solution (11.14.3) is *damped oscillatory*, since the oscillations die away, or become damped. In contrast, the solution (11.14.2) of (11.14.1) can be described as *damped nonoscillatory* or *damped monotonic*. These two kinds of behavior are illustrated in Figs. 11.5 and 11.6.

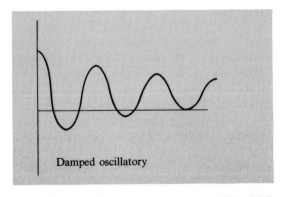

Damped oscillatory

Fig. 11.5

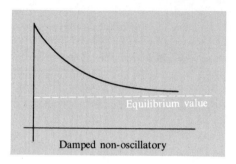

Fig. 11.6

Not all solutions of a differential equation are of this kind. The solution may be *explosive*, meaning that as $t \to \infty$ the dependent variable y can assume arbitrarily large absolute values, that is, it is not bounded. This may happen because it tends to infinity, as with the solution

(11.14.5) $y = 5e^{2t}$

of the differential equation

(11.14.6) $y'' + y' - 6y = 0,$

which clearly behaves in this fashion.

There are other possibilities. Consider the solution

(11.14.7) $y = 7e^{3t} \sin t$

of the differential equation

(11.14.8) $y'' - 6y' + 10y = 0.$

Here the solution oscillates between positive and negative values, but instead of being damped, these oscillations increase in magnitude. Because of the changing sign the function does not tend to infinity as $t \to \infty$, but it is nevertheless described as explosive, and is said to be *explosive oscillatory*. These two kinds of explosive solutions are illustrated in Figs. 11.7 and 11.8.

There is yet another kind of behavior. A solution may fail to tend to any equilibrium value, and yet not be explosive. Such is the case with the solution

(11.14.9) $y = 3 \cos 5t$

of the differential equation

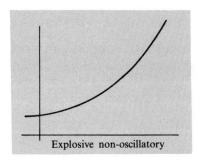

Explosive non-oscillatory

Fig. 11.7

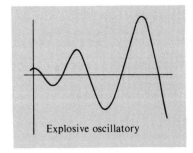

Explosive oscillatory

Fig. 11.8

(11.14.10) $$y'' + 25y = 0.$$

This solution oscillates between the values -3 and $+3$, and hence is bounded; yet it does not tend to a limit. This is illustrated in Fig. 11.9.

Since *any* differentiable function is a solution of *some* differential equation it is clear that we can meet almost any kind of behavior among solutions of differential equations in general. The above examples are fairly typical of solutions of second order linear equations with constant coefficients, and they are all particular solutions of equations of that kind. The choice of *particular*

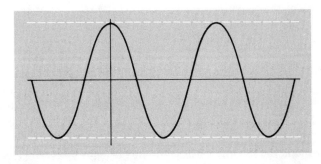

Fig. 11.9

rather than general solutions was deliberate; for when we consider general solutions we have to face a complication —the question of stability.

Suppose that we have come to the conclusion, on theoretical grounds, that the growth of a business firm is determined by the differential equation

(11.14.11) $y'' - 2y' + 15y = 30,$

and have observed that the growth of one particular firm seems to obey the rule

(11.14.12) $y = 25e^{-3t} + 2.$

Since (11.14.12) is a particular solution of (11.14.11) our observation accords with the theory. We might therefore expect the size of the firm to decrease exponentially to the equilibrium value of 2, in accordance with (11.14.12), but if we did we would certainly be disappointed.

The general solution of (11.14.11) is

(11.14.13) $y = Ae^{5t} + Be^{-3t} + 2.$

To apply this result to a particular firm we must supply the values of A and B that are appropriate to that firm, and these can be found only from observations of how the firm is actually growing. If we measure the size of the firm at two different times, then, by substituting for y and t in (11.14.13) we obtain two equations from which A and B can be determined. Thus if y is observed to be 4 when $t = 1$, and is 3 at time $t = 2$, then we have

$$4 = Ae^5 + Be^{-3} + 2$$
$$3 = Ae^{10} + Be^{-6} + 2$$

which can be solved to find A and B. These given values of y at specific times are often called *initial values* (since they frequently relate to the beginning of the range of values of y in which we are interested). Alternatively, the conditions that y has certain values at certain times are known as *boundary conditions*.

Now although A and B can thus be found by measuring y at certain times, these measurements can be determined only to a certain degree of accuracy, probably only to within 0.1 percent or so, at best, and a similar limitation on accuracy will therefore affect the constants A and B. In particular, although the particular solution (11.14.12) is one for which $A = 0$, we could not assert that A is *exactly* zero; the most that we can say is that it appears to be very small. But if A is not *exactly* zero, even though it is very small, the term Ae^{5t} will eventually become large, and will swamp the bounded term Be^{-3t}. Hence the solution will *not* tend to equilibrium.

Even if we suppose that, by some fluke, the solution we are interested in gets under way with values that make A exactly zero, any perturbation of this solution, that is, any slight change in the value of y due to extraneous factors, will cause the solution to continue in a slightly different way, determined by new values of A and B, and the solution will no longer tend to equilibrium.

A solution of this kind is said to be *unstable*. Although in theory it tends to equilibrium [in our example the equilibrium solution is $y = 2$, to which (11.14.12) tends] this equilibrium cannot be realized in practice since any inaccuracies or fluctuations in the values of y from those dictated by this solution, no matter how small they may be, will inevitably prevent equilibrium from being reached.

The term *unstable* has been taken over from a similar concept in applied mathematics—one which provides a helpful analogy in the present discussion. A perfectly symmetrical needle could, in theory, be stood on its point on a hard flat tabletop, and remain balanced. For if it is *absolutely* vertical there is nothing to make it fall in one direction rather than another. This is, theoretically, an equilibrium position; but it is a position of *unstable* equilibrium. For any disturbance, however small, will destroy this equilibrium; once the needle departs, however slightly, from the vertical position it must inevitably topple.

In contrast to the above example let us look at the the general solution of equation (11.14.1). It is

$$(11.14.14) \qquad y = Ae^{-2t} + Be^{-3t} + 4$$

and y tends to the equilibrium value 4 as $t \to \infty$. What is more, this is true no matter what A and B may happen to be. Thus no perturbations, not even large ones, can make a solution of this equation become explosive. Such a solution is said to be *stable*. This is analogous to a weight suspended by a string. Its equilibrium position is directly below the point of suspension, and if it is disturbed from this position the weight will perform damped oscillations about the equilibrium position until it eventually comes to rest there again.

We are now able to assess the stability or instability of solutions of differential equations of the second order with constant coefficients. These consist of a complementary function and a particular integral. Now if the particular integral does not tend to a finite limit as $t \to \infty$ (and this is something that we can check when we have found it) then there is no possibility of equilibrium for the general solution. If the particular integral tends to a finite limit, then the behavior of the general solution depends on that of the complementary function. Now, as we have already seen in section 11.7, the complementary functions can be of the following three types:

$$(11.14.15) \qquad Ae^{\alpha t} + Be^{\beta t}$$

(11.14.16) $\qquad\qquad (At + B)e^{\alpha t}$

(11.14.17) $\qquad\qquad e^{pt}(A \cos qt + B \sin qt),$

and we now consider each of these in turn.

If $\alpha < 0$ and $\beta < 0$ in (11.14.15) then this function tends to 0 whatever A and B may be. Hence this solution is stable. If either α or β is positive (or both) then the solution is unstable, since this is precisely the type of equation that we considered above.

In (11.14.16), if α is positive then the solution is explosive; but if α is negative then the solution tends to equilibrium since $te^{\alpha t} \longrightarrow 0$ for $\alpha < 0$.

In (11.14.17) the factor $A \cos qt + B \sin qt$ is bounded. Hence the solution will tend to zero if $p < 0$ but will assume arbitrarily large positive and negative values if $p > 0$.

Hence, if neither root of the auxiliary equation $a\lambda^2 + b\lambda + c = 0$ is zero, the solutions of the equation $ay'' + by' + cy = 0$ are stable under the following necessary and sufficient conditions:

1. (When $b^2 \geq 4ac$) the roots of the auxiliary equation are both negative;
2. (When $b^2 < 4ac$) $p = -b/a$ is negative.

If one root only of the auxiliary equation is zero, it can be shown that there is a solution which tends to equilibrium and that this solution may be stable. If both roots are zero, no stable equilibrium is possible. If $p = 0$ in (2) above, then there is no equilibrium solution. (See Exercise 136.15.)

Similar remarks on stability apply to solutions of two simultaneous differential equations. If the equations are each first order, then the solutions for the two dependent variables are of the same general form as those given above, and the same criteria for stability apply.

Exercise 136 (Chapter Review)

1. Show that the general solution of the differential equation

$$\frac{dy}{dx} = \frac{2xy}{x^2 + y^2}$$

can be found by means of the substitution $y = vx$, and is $x^2 - y^2 = Cy$.
2. Solve the following equations
    (a) $\dfrac{dy}{dx} = \dfrac{1 + 2y}{3 + 5x}$

(b) $\dfrac{dy}{dx} = \dfrac{x^2(1 - y^2)}{y(x^2 + 1)}$

(c) $\dfrac{dy}{dx} = \dfrac{\sin y}{\sin x}$

(d) $(y^2 - 4)^{1/2} \, dx + (x^2 - 9)^{1/2} \, dy = 0$

3. If $\dfrac{dy}{dx} = \dfrac{P(x, y)}{Q(x, y)}$, and $P(x_0, y_0) = Q(x_0, y_0) = 0$ for some point (x_0, y_0) then the slope is not determined at (x_0, y_0). If we take (x_0, y_0) to be the origin, then, by Taylor's theorem we have

$$P(x, y) = \alpha x + \beta y + \text{terms in higher powers of } x \text{ and } y,$$

$$Q(x, y) = \lambda x + \mu y + \text{terms in higher powers of } x \text{ and } y.$$

We would therefore expect that for small values of x and y the solutions of the equation would approximate to those of the equation

$$\frac{dy}{dx} = \frac{\alpha x + \beta y}{\lambda x + \mu y}.$$

Show that this equation can be reduced to a Type I equation by the substitution $y = vx$, and solve it in the following cases:

   (a) $\alpha = 6, \quad \beta = 4, \quad \lambda = 3, \quad \mu = 2$
   (b) $\alpha = 2, \quad \beta = 5, \quad \lambda = -5, \quad \mu = 7$
   (c) $\alpha = 3, \quad \beta = 2, \quad \lambda = 1, \quad \mu = 2$
   (d) $P(x, y) = \sin(x + y); \quad Q(x, y) = 1 - (1 + x)^2(1 - y)$

4. Solve the following linear equations.

   (a) $y' + y \cot x = \sin x$

   (b) $x\dfrac{dy}{dx} + 2y = (x - 2)e^x$

   (c) $y' + (\cot x + 1)y = \cot x - 1$

   (d) $y' \log x + \dfrac{2}{x}y = 1$

5. Show that the following equations can be converted to linear form by a substitution of the form $y = v^k$, for suitably chosen k, and hence solve them.

   (a) $\dfrac{dy}{dx} + y \cot x = y^2 \sin^2 x$

   (b) $5x\dfrac{dy}{dx} + 6y = 2xy^{-3/2}$

   (c) $y' - y \operatorname{cosec} 2x = y^3$

6. Solve completely the following linear second order equations.

   (a) $y'' - 5y' + 4y = 2x^2 e^{4x} - 5e^x$
   (b) $6y'' + 11y' - 10y = 19(e^{2x/3} - e^{-5x/2})$
   (c) $3y'' - 5y' + 2y = 2x^2 - 3 \sin 2x + 7e^{-x}$

(d) $y'' - 2y' + y = 15e^x \cos x$

(e) $y'' - 9y = 2 \cosh 3x + 5x^2$

7. Find the solution of the differential equation

$$\frac{d^2y}{dx^2} - 2\frac{dy}{dx} + y = 12x^2e^x$$

for which $y = 1$ when $x = 0$, and $y = 0$ when $x = 1$.

8. Show that by writing $y = zx^{-1/2}$, the equation

$$4x^2\frac{d^2y}{dx^2} + 4x\frac{dy}{dx} + (4x^2 - 1)y = 0$$

can be transformed into a linear equation with constant coefficients. Hence solve it, and find the solution for which $y = 0$ when $x = \frac{1}{2}\pi$, and $y = 1$ when $x = \pi$.

9. A possibility that may arise in finding particular integrals, and that is not covered by the methods of Section 11.12. is that of expressions of the form

$$\frac{1}{D^2 + q^2} \sin qx \quad \text{and} \quad \frac{1}{D^2 + q^2} \cos qx.$$

Application of the standard method will give a zero in the denominator.

By considering the expressions $(D^2 + q^2)x \sin qx$ and $(D^2 + q^2)x \cos qx$, show that

$$\frac{1}{D^2 + q^2} \cos qx = \frac{1}{2q}x \sin qx, \text{ and } \frac{1}{D^2 + q^2} \sin qx = -\frac{1}{2q}x \cos qx.$$

10. By means of a substitution of the form $y = zx^n$ solve the equation

$$x^2\frac{d^2y}{dx^2} + 2x(x + z)\frac{dy}{dx} + 2(x + 1)^2y = e^{-x} \cos x.$$

Show that all solutions tend to zero as $x \longrightarrow \infty$.

Hint: Use the result of problem 9.

11. Solve the simultaneous differential equations

$$\frac{dy}{dt} - 2x = \cos 2t; \quad \frac{dx}{dt} + 2y = \sin 2t.$$

12. Solve the following pairs of simultaneous differential equations

(a) $\frac{dx}{dt} + 2x - 5\frac{dy}{dt} - 7y = e^x$

$2\frac{dx}{dt} - 7x + \frac{dy}{dt} + 19y = 2e^x$

(b) $2\dfrac{dx}{dt} + 36x + \dfrac{dy}{dt} + 3y + 10e^{2t} = 0$

$5\dfrac{dx}{dt} + 22x + \dfrac{dy}{dt} + 2y + 7e^{-2t} = 0$

13. Find that solution of the pair of equations

$$(D - 1)x + (2D - 5)y = \sin 2t + \cos 2t$$
$$(D - 2)x + (3D - 6)y = 9e^t$$

for which $x = 0$ and $y = 1$ when $t = 0$.

14. Two grocery stores A and B are in competition with each other. The stocks in hand in the two stores at any given time t are $x(t)$ and $y(t)$, respectively. Each storekeeper spends a certain proportion of his sales on advertising, and it may be assumed that the amount thus spent varies directly with the stocks in hand, the factor of proportionality being the same for both stores.

The sales of each store depend on two things: (1) the degree of confidence of the customers in the ability of the store to supply their wants; and (2) the extent of its advertising compared with the other store. In mathematical terms we can assume that the rate of increase of stock is the sum of two terms, one proportional to the existing stock (with different factors of proportionality for the two stores), and the other proportional to the difference in the amounts spent on advertising between the two stores (the factor of proportionality being the same for both stores).

Show that this situation can be summarized by the equations

$$\frac{dx}{dt} = \alpha x + p(x - y)$$

$$\frac{dy}{dt} = \beta y + p(y - x)$$

where α, β and p are positive constants.

Obtain second-order differential equations for the functions $x(t)$ and $y(t)$ separately.

Hence, or otherwise, show that there is no oscillating solution of these equations. Show further that for each of x and y the only equilibrium solution is the trivial one, $x = y = 0$, and that this solution is unstable.

15. Prove the assertions made at the end of section 11.14, namely that if one root of the auxiliary equation of a linear second order differential equation with constant coefficients is zero, then stable equilibrium is possible, but not if both roots are zero; and that if $b^2 < 4ac$, then the special case $p = -b/a = 0$ is one for which no equilibrium is possible.

Suggestions for Further Reading

Most books on differential and integral calculus also cover the topic of differential equations to some degree or other. Thus the books recommended for Chapters 9 and 10 will also supplement the study of the present chapter.

Recommended among books specifically concerned with differential equations are the following:

Ayres, F., *Theory and Problems of Differential Equations.* New York: Schaum Publishing Company, 1952.

Hochstadt, H., *Differential Equations, A Modern Approach.* New York: Holt, Rinehart and Winston, Inc., 1964. This is a more advanced book, but the earlier chapters will repay study.

Ince, E. L., *Integration of Ordinary Differential Equations.* Edinburgh: Oliver and Boyd Ltd., 1963.

A book which covers the material of this chapter and also that of Chapter 15 is

Chorlton, F., *Ordinary Differential and Difference Equations.* Princeton, N.J.: D. Van Nostrand Company, Inc., 1965.

12 Rational Numbers

and Real Numbers

12.1 FOUNDATIONS—FROM INTEGERS TO RATIONAL NUMBERS

It has been remarked several times already that we do not have a satisfactory definition of a real number. In this chapter we shall remedy this defect,
but first we should perhaps ask whether it is really necessary.

It is true that one can quite easily "make do" without a proper definition of a real number. There are many books which go into advanced mathematics without tackling the problems of what is meant by a number like
$\sqrt{2}$ or π; sometimes the fact that there is a problem at all is not even mentioned. It might seem therefore that the subject-matter of this chapter could
just as well be omitted, and this is possibly true. Yet there is a lot to be said for
getting the problem settled, and although we cannot afford to be completely
rigorous throughout this book, it is as well to have at least the basic parts of
the subject on a firm foundation. What is more, the successive definitions of
rational, real and (in the next chapter) complex numbers give a fine example
of the building up of a mathematical structure by a process of generalization
of one mathematical system to another, larger, system.

Now we must start somewhere, and to avoid digging down too deeply
into the foundations of mathematics, let us assume that we know what we
mean by the integers both positive and negative, and the laws of arithmetic

562

(addition, multiplication, etc.) that apply to them. Although the integers have been familiar to us since our early school days, their definition is not as easy as one might think. For an account see the references in the notes at the end of this chapter. Our first task will be to define the rational numbers, or fractions.

Let us consider the set of all ordered pairs of integers of the form

$$(p, q)$$

where p is an integer which may be positive, negative or zero, and q is a nonzero integer. Eventually we are going to relate this pair to the rational number which we would normally write as p/q, but not just yet; there are several things we must do first. We must, for example, define a notion of equivalence between these integer pairs.

EQUIVALENCE OF INTEGER-PAIRS.

We shall say that the integer-pairs (p, q) and (a, b) are equivalent if, and only if,

$$pb = qa,$$

and we shall denote this equivalence by $(p, q) \sim (a, b)$. Note that to tell whether two pairs are equivalent requires only the operation of multiplication of integers, which we are assuming we know all about.

It is obvious that $(p, q) \sim (p, q)$, and that if $(p, q) \sim (a, b)$ then $(a, b) \sim (p, q)$. Hence the relation "$\sim$" is reflexive and symmetric.

It is also transitive; for if

$$(p, q) \sim (a, b)$$

and

$$(a, b) \sim (l, m)$$

then

$$pb = qa \quad \text{and} \quad am = bl.$$

Thus $pbm = qam = qbl$, from which it follows that $pm = ql$. For $b(pm - ql) = 0$, and if the product of two integers is zero then at least one of them is zero; but we know that $b \neq 0$. Hence

$$(p, q) \sim (l, m).$$

Thus this relation is an equivalence relation and like all equivalence relations it divides the set over which it is defined into equivalence classes. We shall call these equivalence classes the *rational numbers*. Thus the pairs

$$(2,4), \quad (10,20), \quad (260,520), \quad (3,6)$$

all belong to the same equivalence class, since they are all equivalent.

For the time being we shall denote the rational number—i.e., the equivalence class—to which the pair (p, q) belongs by the symbol $[p/q]$. A point which must not be overlooked here is that this designation for a rational number is not unique. For example, the symbols

$$\left[\frac{2}{4}\right], \quad \left[\frac{10}{20}\right], \quad \left[\frac{260}{520}\right], \quad \left[\frac{3}{6}\right]$$

all designate the *same* rational number. They denote in fact the rational number which we would describe in words as "one half." We shall return to this point later.

Having defined our rational numbers as equivalence classes of pairs of integers, we now define certain operations on them. It will be convenient to denote rational numbers by single symbols, such as X, Y, etc.

DEFINITION OF ADDITION.

We define the sum of the rational number $X = [p/q]$ and the rational number $Y = [a/b]$ to be the rational number

$$\left[\frac{pb + qa}{qb}\right]$$

i.e., the equivalence class to which the pair $(pb + qa, qb)$ belongs. We can write the sum as $X + Y$.

What this definition says is that to form the sum of two rational numbers, you take a pair from each, say (p, q) and (a, b), and form the pair $(pb + qa, qb)$. The sum is then the rational number to which this pair belongs. The question naturally arises, "*Which* pairs must we take from the given rational numbers, or does it not matter?" If $X + Y$ is to depend only on X and Y (as would appear reasonable) then it ought not to matter which pairs we choose to represent these rational numbers, the sum should be the same. Let us see if this is so.

Suppose $(p, q) \sim (p', q')$ and $(a, b) \sim (a', b')$, and suppose we take

$$X = \left[\frac{p'}{q'}\right] \quad \text{and} \quad Y = \left[\frac{a'}{b'}\right].$$

Then X and Y are the same as before. Now

$$pq' = qp'$$

(12.1.1) $$ab' = ba'.$$

Therefore since

$$(pb + qa, qb) \sim (pq'bb' + q'qab', q'qbb')$$

on multiplication by $q'b'$ of both integers in the left-hand pair, we can substitute from (12.1.1) to obtain

$$(pb + qa, qb) \sim (p'qbb' + q'qba', qq'bb')$$
$$\sim (p'b' + q'a', q'b').$$

The last equivalence results, essentially, from the taking out of the nonzero factor qb, and can be checked by direct appeal to the definition of equivalence. Hence the pair we obtain from (p', q') and (a', b') is equivalent to the pair we obtained before, and the rational number that it belongs to is therefore the same as before. Hence our definition of the sum of two rational numbers X and Y depends only on X and Y, and not on which pairs we have chosen from them. The definition is thus unambiguous.

DEFINITION OF MULTIPLICATION.

We define the product XY, or $X \cdot Y$ of two rational numbers $X = [p/q]$ and $Y = [a/b]$ to be the rational number $[pa/qb]$.

Again we need to show that this product does not depend on the pairs (p, q) and (a, b) chosen to represent X and Y. If we take two different pairs (p', q') and (a', b') to represent X and Y respectively, we have

$$XY = \left[\frac{p'a'}{q'b'}\right]$$

from the definition. It is clear that

$$(pa, qb) \sim (p'a', q'b')$$

for $pq'ab' = qp'ba'$ since $pq' = qp'$ and $ab' = ba'$. Hence the rational number XY is the same as before.

We can now state a few laws concerning rational numbers.

1. The commutative law of addition. For any two rational numbers X and Y

$$X + Y = Y + X.$$

2. The commutative law of multiplication. For any two rational numbers X and Y

$$XY = YX.$$

3. The associative law of addition. For any rational numbers X, Y, Z

$$X + (Y + Z) = (X + Y) + Z.$$

4. The associative law of multiplication. For any rational numbers $X, Y, Z,$

$$X(YZ) = (XY)Z.$$

These associative laws enable us to extend the definitions of *sum* and *product* to any number of rational numbers. We write the sums and products as

$$X + Y + Z + \cdots$$

and $XYZ \cdots$

since, by the laws, parentheses are unnecessary.

5. The distributive law of multiplication over addition. For any rational numbers X, Y, Z,

$$X(Y + Z) = XY + XZ.$$

These laws all need to be proved, but the proof of most of them will be left as an exercise to the reader. We shall prove one of them, however, by way of an example.

To prove the associative law of addition [number (3) above] let

$$X = \left[\frac{p}{q}\right], \qquad Y = \left[\frac{a}{b}\right], \qquad Z = \left[\frac{l}{m}\right].$$

Then

$$Y + Z = \left[\frac{am + bl}{bm}\right]$$

and

$$X + (Y + Z) = \left[\frac{p}{q}\right] + \left[\frac{am + bl}{bm}\right]$$
$$= \left[\frac{pbm + q(am + bl)}{qbm}\right]$$
$$= \left[\frac{pbm + qam + qbl}{qbm}\right].$$

On the other hand

$$X + Y = \left[\frac{pb + qa}{qb}\right]$$

so that

$$(X + Y) + Z = \left[\frac{pb + qa}{qb}\right] + \left[\frac{l}{m}\right]$$
$$= \left[\frac{(pb + qa)m + qbl}{qbm}\right]$$
$$= \left[\frac{pbm + qam + qbl}{qbm}\right]$$

which is the same as before. The other laws can be proved in a similar way.

Besides these "laws" there are some other observations that can be made concerning rational numbers. Among the various rational numbers there is one which contains the pair (0,1). This we shall call the *zero* rational number, and denote it by the single symbol 0. It is easily verified that for any rational number X

$$X + 0 = X$$

and

$$X \cdot 0 = 0.$$

There is also a rational number, which we shall denote by 1 and call "one" which is the equivalence class to which the pair (1,1) belongs. This rational number has the property that for any rational number X

$$X \cdot 1 = X.$$

As can be seen, these rational numbers 0 and 1 behave very much like the *integers* 0 and 1.

Given a rational number X, there is always a rational number Y such that $X + Y = 0$. For if $X = [p/q]$ then $Y = [-p/q]$ has this property. This rational number Y is called the negative of X and denoted by $-X$.

Given a rational number X, which is not 0, there is a rational number Y such that $XY = 1$. For if $X = [p/q]$, then $Y = [q/p]$ has this property. Note that $[q/p]$ does not denote a rational number if $p = 0$, which is why we had to stipulate that X was not 0. The rational number Y is called the inverse of X and is denoted by X^{-1}.

We have now built up the system of rational numbers on the foundation of what we know about the integers. The rational number $[p/q]$ is what we would normally write as the fraction p/q, and from now on we shall so write it. It is useful however to bear in mind that when we write $\frac{2}{3}$ we really mean $[\frac{2}{3}]$, that is to say, the rational number (equivalence class) of which the pair (2,3) is just one representative, others being $(-4, -6)$, (20, 30) and so on. The failure to make the distinction between a particular member of a rational number—a particular representation of it, if you like—and the rational number itself, is responsible for many misunderstandings. Many a school-teacher has been floored by a query such as "Please teacher, if $\frac{2}{3}$ is the same as $\frac{4}{6}$, why is the denominator of $\frac{2}{3}$ not the same as the denominator of $\frac{4}{6}$?" We can easily give the answer, which is that $\frac{2}{3}$ and $\frac{4}{6}$ are *not* the same, if by "same" we mean that the pairs (2,3) and (4,6) are *identical*. They are not identical, but they *are* equivalent, so that they belong to the same equivalence class. Thus $\frac{2}{3}$ and $\frac{4}{6}$ are different ways of denoting the same equivalence class. In the same way

$$\frac{2}{3}, \quad \frac{10}{15}, \quad \frac{120}{180}, \quad \frac{-6}{-9}$$

are all *different* ways of denoting the *same* rational number.

Exercise 137

Prove the various laws for rational numbers given in the preceding section.

12.2 THE RATIONAL INTEGERS

Of particular importance are the rational numbers of the form $[n/1]$. These are called the rational integers. According to our notation they would normally be written as $n/1$, but in practice they are further simplified and written just as n. Thus $[17/1]$ can be written as $17/1$ or just 17. Yet the rational integer $17/1$ (or 17) is not the same as the integer 17. How could it be? It is an equivalence class of pairs of integers whereas the integer 17 is—just an integer. Nevertheless we can see some connection betweeen these two; we can set up a correspondence between the integers and the rational integers by making the integer n and the rational integer $[n/1]$ correspond to each other. This correspondence is clearly "one-to-one," and moreover is one which preserves the operations of addition and multiplication, in the sense that any operation performed with corresponding elements of the two sets (integers and rational numbers) will yield corresponding results, Thus just as

$$3 + 5 \cdot 7 = 38$$

so

$$\left[\frac{3}{1}\right] + \left[\frac{5}{1}\right] \cdot \left[\frac{7}{1}\right] = \left[\frac{38}{1}\right]$$

and the two results, 38 and $[38/1]$, correspond.

Thus any operations performed on the integers are *paralleled* by similar operations performed on the corresponding rational numbers. It is because of this *isomorphism*, as it is called, between the integers and the rational numbers that we can use the existing terms "addition" and "multiplication" for the operations we defined above, without creating confusion. For the same reason we can, for practical purposes, ignore the distinction between the integers and the rational integers. Looked at in this way the integers are seen as special cases of the rational numbers or, what amounts to the same thing, the system of rational numbers is seen to be a generalization of the system of the integers. We shall denote the set of all rational numbers by R.

The process by which we have arrived at the system of rational numbers is one of frequent occurrence. A completely abstract system is built up by means of arbitrary definitions. Resemblances are observed (or quite possibly engineered) between the system and certain intuitive concepts (in this case our intuitive idea of what a "fraction" is). Certain special elements of the system (the rational integers) are seen to be closely related to the elements of the familiar system. Thus a generalization of the familiar system is born.

Exercise 138

1. The term "one-to-one" was used in the last section. Although this term is more or less self-explanatory we can define it formally as follows:

A correspondence between two sets A and B is said to be one-to-one if

(1) to every element of A corresponds exactly one element of B,

(2) every element of B corresponds to at least one element of A,

(3) no element of B corresponds to more than one element of A.

A correspondence which satisfies (1) but not necessarily (2) or (3) is called a *mapping* of the set A *into* the set B. If we give this correspondence a symbol, say f, we can write it as

$$f: A \to B.$$

Clearly a mapping is simply another name for a function as defined on page 63. In a mapping every element of A has a corresponding element in B (called its *image*) onto* which it maps. Since (2) and (3) are not required to hold, there may be elements of B which are not images of any element of A, and it may happen that two or more elements of A map onto the same element of B.

If (1) and (2) hold, but not necessarily (3) we have a mapping of A *onto* B; every element of B has some element of A mapped onto it.

The following correspondences are from A to B where $A = B =$ the set of integers (positive, negative and zero), and "$a \to b$" means "a maps onto b" or "b corresponds to a." Determine which of these correspondences are one-to-one, which of the rest are mappings, and whether they are *into* or *onto*.

(a) $a \to a^2$

(b) $a \to -a$

(c) $a \to 0$, if a is even
   $a \to 1$, if a is odd

(d) $a \to b$, if b divides a

(e) $a \to 2a$.

2. Prove that if a mapping $f: A \to B$ satisfies condition (3) of problem 1 but not necessarily (2), then it sets up a one-to-one correspondence between A and some subset of B. Example (e) of problem 1 has this property; what is the subset of B in this case?

---

* The word *onto* is a technical term, and its spelling as a single word (contrary to usual grammatical practice) is standard.

3. Let T be the set of all ordered triads of real numbers, and U the set of all *unordered* pairs of real numbers. Let an ordered triad (a, b, c) in T and an unordered pair (α, β) in U correspond if α and β are the roots of the quadratic equation

$$ax^2 + bx + c = 0.$$

Show that this correspondence is not one-to-one. If we introduce an equivalence relation "$\sim$" between the elements of T, defined by

$$(a, b, c) \sim (p, q, r) \text{ if, and only if, } \frac{a}{p} = \frac{b}{q} = \frac{c}{r}$$

show that there is a one-to-one correspondence between the equivalence classes thus defined over the set T and the elements of U.

12.3 DEDEKIND SECTIONS

We shall meet a further generalization in our next task which is to define a *real number*. We should first show that the rational numbers above are not sufficient for our needs, and we shall do this by showing that there is no rational number whose square is the (rational) integer 2. The proof (which goes back to Pythagoras) is by a *reductio ad absurdum* argument.

Let us suppose that there is a rational number whose square is 2; let it be p/q. If p and q have a highest common factor d, then the pair $(p/d, q/d)$ also represents the rational number in question. There is no harm, then, in assuming that p and q have no common factor, for if they do we can replace them with another pair of integers (p/d and q/d) which do not. In the language of arithmetic we may assume that p/q is in its *lowest terms*.

We want to have

$$\left(\frac{p}{q}\right)^2 = 2$$

that is

(12.3.1) $$p^2 = 2q^2.$$

Now the right-hand side of (12.3.1) is an even integer, so that the left-hand side must also be even. Hence p must be even, since the square of an odd number is odd. Let us therefore write $p = 2r$, where r is an integer. Substituting in (12.3.1) we obtain

$$4r^2 = 2q^2$$

or

$$2r^2 \ = \ q^2.$$

This shows that the right-hand side, and hence q, is even. But now we have succeeded in proving that both p and q are even, and hence have a common factor 2. This is impossible since p and q were chosen initially so that they had no common factors.

Hence the assumption that there is a rational number whose square is 2 leads to a contradiction. We deduce that there can be no such rational number; the equation $x^2 = 2$ has no solutions for x a rational number.

Nevertheless, we have a feeling that there ought to be some sort of number whose square is 2, and it is to fill this obvious gap, and many others in the rational number system, that we shall define real numbers. Let us first look more closely at the equation $x^2 = 2$.

Although, as we have seen, there is no rational number x for which $x^2 = 2$, there are rational numbers x for which $x^2 > 2$ (for example, $x = 3$), and others for which $x^2 < 2$ (for example, $x = 1$). We can therefore form two sets of rational numbers, L and U, as follows:

1. L is the set of all rational numbers x such that $x \leqslant 0$ or $x^2 < 2$. This we can abbreviate to

$$L = \{x \, | \, x \in R \ \& \ (x \leqslant 0 \ \lor \ x^2 < 2)\}.$$

2. U is the set of all positive rational numbers whose square is > 2. This can be written

$$U = \{x \, | \, x \in R \ \& \ x > 0 \ \& \ x^2 > 2\}.$$

We may make the following observations concerning these two sets.
1. Every rational number belongs either to L or to U. Briefly $R = L \cup U$.
2. No rational number belongs to both L and U, i.e., $L \cap U = \emptyset$.
3. Neither L nor U is empty.
4. If $x \in L$ and $y \in U$, then $x < y$.
This is clearly true if x is negative or zero, since y is necessarily positive. If x is positive then

$$x^2 < 2 < y^2$$

whence $x < y$.

We have divided the rational numbers into two disjoint sets L and U, having the properties given above. This pair of sets was obtained by considering the equation $x^2 = 2$, but clearly there will be other pairs of sets of rational numbers which have these properties. Let us briefly recapitulate in symbolic form what these properties are.

1. $R = L \cup U$
2. $L \cap U = \emptyset$
3. $L \neq \emptyset \; \& \; U \neq \emptyset$
4. $(x \in L \; \& \; y \in U) \rightarrow (x < y)$

A pair of sets (L, U) of rational numbers having properties (1) through (4) is called a *Dedekind section*. With a slight modification to be described shortly, these Dedekind sections will be taken as the real numbers.

Now the reader may object at this point and say that this is not what *he* understands by a real number. To this objection the only answer, and one which has been given before, is that unless the reader has met this sort of thing already he will almost certainly not have any really satisfactory concept of what a real number is. The intuitive idea of an infinite decimal, or something like that, though good enough for some purposes, will not stand up to rigorous analysis without a lot of scaffolding. (This scaffolding *could* be supplied if we wanted to tackle the problem from this angle, but we are adopting a different method of attack.) What we need is to build up a mathematical system which will have all the properties that we want the real numbers to have; in particular, the set of real numbers when we define it must contain some elements that we can associate with the rational numbers, just as, when we defined the rational numbers, we were able to associate some of them with the integers. The use of Dedekind sections is one way of doing this. There are others.

12.4 DEFINITION OF REAL NUMBERS

The slight modification which was mentioned concerns those sections in which either L has a greatest member or U has a least member. This will happen, for example, if we construct a section corresponding to the real number whose square is 4, in the same way that the section that we first considered corresponded to the real number whose square was 2. In this case we are in doubt as to whether to put the rational number 2 into L or U, since its square is exactly 4. There are thus two sections which appear to have equal right to represent the real number we are looking for; they can be written as

$$L = \{x \mid x \in R \; \& \; x \leq 2\}$$
$$U = \{x \mid x \in R \; \& \; x > 2\}$$

and

$$L' = \{x \mid x \in R \ \& \ x < 2\}$$
$$U' = \{x \mid x \in R \ \& \ x \geq 2\}.$$

These sections are not the same, since L is different from L' and U from U', but we shall regard them as equivalent. This notion of equivalence must be defined precisely before we go any further.

EQUIVALENCE OF TWO SECTIONS.

Two sections (L, U) and (L', U') are equivalent if L and L' differ by at most one element (rational number).

If $L = L'$ then clearly $U = U'$ (since $L \cup U = R$ and $L' \cup U' = R$), and the two sections are in fact identical. If L and L' differ by a single element a, where $a \in L$ but $a \notin L'$, then $a \in U'$. Hence a is greater than any element of L', and hence is the greatest element in L. For the same reason it is the least element of U'.

Thus we see that two equivalent sections are either identical, or differ only in that the greatest element of the lower part of one is the least element of the upper part of the other. This equivalence between sections is easily seen to be an equivalence relation and hence divides the set of all sections into equivalence classes. These equivalence classes will be called the *real numbers*.

By the definition of equivalence a real number may contain one or two sections; there are no other possibilities. Thus there are two kinds of real numbers, to which we can give names.

Definition 12.1

A real number which contains two sections is called a rational real number.

Definition 12.2

A real number which contains only one section is called an irrational real number.

We are now faced with much the same situation as when we set up a correspondence between the integers and the rational integers. We can set up a correspondence between the rational numbers and the rational real numbers, in the following way. A rational real number contains two sections which differ in the placing (into L or U) of just one element a (a rational number). To this rational number a we make correspond the rational real number which consists of the two sections

$$L = \{x \mid x \in R \ \& \ x \leq a\}, \qquad U = \{x \mid x \in R \ \& \ x > a\}$$

and

$$L' = \{x \,|\, x \in R \;\&\; x < a\}, \qquad U' = \{x \,|\, x \in R \;\&\; x \geq a\}.$$

Conversely, for any rational real number there is a rational number a related to it in this way.

As with the relationship between rational integers and integers, it is permissible, for practical purposes, to identify the rational real numbers and the rational numbers, even though strictly speaking they are quite different concepts. This identification will be of use however only if we can extend the definitions of addition and multiplication to cover the real numbers in general, in such a way that when these definitions are applied to the *rational* (real) numbers we get the same results as the former definitions applied to rational numbers. The definition of addition and multiplication of real numbers is therefore our next task. For convenience we shall often use single symbols α, β, etc., to denote real numbers.

Exercise 139

1. Show that the sets L and U defined by

$$U = \{x \,|\, x \in R \;\&\; x > 0 \;\&\; x^2 - x - 1 > 0\}$$
$$L = R - U$$

where R is the set of rational numbers, form a Dedekind section defining an irrational number. What is this irrational number?

2. Let L be a subset of the rational numbers defined by "$x \in L$ if, and only if, there is an integer n such that $x < \dfrac{n}{n+1}$." Define U as $R - L$. Show that (L, U) is a Dedekind section. What real number is this section?

3. If $U = \{x \,|\, x \in R \;\&\; x^4 > 8\}$ and $L = \{x \,|\, x \in R \;\&\; x^4 < 8\}$

show that (L, U) is not a section. What modifications to the definitions of U and L are needed in order that (L, U) should define the irrational number $8^{1/4}$?

12.5 ADDITION OF REAL NUMBERS

Given two real numbers α and β represented by sections (L_1, U_1) and (L_2, U_2), respectively, we define the sum $\alpha + \beta$ to be the real number containing the section (L, U) defined as follows.

The rational number $x \in L$ if there is an $x_1 \in L_1$ and an $x_2 \in L_2$ such that $x = x_1 + x_2$. If this is not so, then $x \in U$.

Thus L consists of all rational numbers that can be expressed as the sum of a rational number occurring in L_1 and one occuring in L_2. U consists of all those not so expressible. Clearly $L \cup U = R, L \cap U = \emptyset$ and neither L nor U is empty. Hence, in order to verify that (L, U) is indeed a section, we need only show that

$$(x \in L \text{ and } y \in U) \to (x < y).$$

This turns out to be easy. There are rational numbers x_1 in L_1 and x_2 in L_2 such that

$$x = x_1 + x_2.$$

Let $y_2 = y - x_1$. Then if $y \leq x$

$$y_2 \leq x - x_1 = x_2.$$

But $x_2 \in L_2$, and therefore the smaller number y_2 must also belong to L_2. Then we can express y as

$$y = x_1 + y_2$$

the sum of an element of L_1 and one of L_2. It follows that $y \in L$. Thus if $y \leq x, y \notin U$, and therefore if $x \in L$ and $y \in U$, we must have $x < y$.

This completes the proof that (L, U) is a section, which defines a real number, the *sum* of α and β. We denote it by $\alpha + \beta$.

The real number containing the section

$$L = \{x \,|\, x < 0\}$$
$$U = \{x \,|\, x \geq 0\}$$

($x \in R$ being understood), is called the *zero* real number. Its other member is obtained by putting the rational number 0 into L instead of U. We shall denote this real number by the same symbol 0. It is easily proved that if α is any real number

$$\alpha + 0 = \alpha.$$

Given a real number $\alpha = (L, U)^*$ we can define another real number (L', U') as follows.

The rational number $x \in L'$ if, and only if, $-x \in U$; otherwise $x \in U'$.

---

* From now on, for convenience, we shall write $\alpha = (L, U)$ where we should properly say "α, one of whose elements is (L, U)." Since at most two sections can make up a real number, and these almost identical, this will cause no confusion.

This defines a real number which we call the negative of α, and denote by $-\alpha$. From our definition of addition of real numbers it is easy to prove that

$$\alpha + (-\alpha) = 0.$$

A real number (L, U) which is not zero, and in which the rational number 0 belongs to L is called a *positive real number*. Those for which $0 \in U$ are the *negative real numbers*.

The reader should satisfy himself of the truth of the following results: If α, β, γ are any three real numbers, then

$$\alpha + \beta = \beta + \alpha$$

and

$$\alpha + (\beta + \gamma) = (\alpha + \beta) + \gamma.$$

In consequence of these results we may write the common expression as $\alpha + \beta + \gamma$. The extension to the sum of any number of real numbers is obvious.

Subtraction. We define the *difference* of two real numbers as follows

$$\alpha - \beta \text{ is } \alpha + (-\beta).$$

12.6 MULTIPLICATION OF POSITIVE REAL NUMBERS

If $\alpha = (L_1, U_1)$ and $\beta = (L_2, U_2)$ are any two *positive* real numbers, we define the product (L, U) as follows:
$y \in U$ if there is a $y_1 \in U_1$ and a $y_2 \in U_2$ such that $y = y_1 y_2$. Otherwise $y \in L$. (y_1, y_2 and y are rational numbers).

Clearly $L \cup U = R$ and $L \cap U = \emptyset$ and neither L nor U is empty. If $x \in L$ and $y \in U$ then $x < y$. For if not, then $x > y$ (since they cannot be equal). Now $y = y_1 y_2$ with $y_1 \in U_1$ and $y_2 \in U_2$. Consider the rational numbers y_1 and xy_2/y; since $xy_2/y > y_2$ it follows that $xy_2/y \in U_2$. Now x can be written as the product of the rational numbers y_1 and xy_2/y (since $y_1 y_2 = y$) and the first of these is in U_1, the second in U_2. From the definition of multiplication it follows that $x \in U$. But $x \in L$, so we have a contradiction. It follows that (L, U) is a section of the rational numbers, and defines a real number, the product of α and β. We write it $\alpha \cdot \beta$ or just $\alpha\beta$.

It has been convenient to restrict the definition of multiplication to that of positive real numbers. Its extension to all real numbers is accomplished by the following statements

$$\alpha(-\beta) = -(\alpha\beta)$$
$$(-\alpha)\beta = -(\alpha\beta)$$
$$(-\alpha)(-\beta) = \alpha\beta$$
$$\alpha\cdot 0 = 0\cdot\alpha = 0.$$

Because of the correspondence between the rational numbers and the rational real numbers we can make the symbols for the rational numbers serve to denote the corresponding real numbers. In particular we can denote by 1 the real number (L, U) where

$$x \in L \text{ if } x \leq 1, \text{ and } \in U \text{ otherwise.}$$

This real number has the property, that for any real number α

$$\alpha\cdot 1 = 1\cdot\alpha = \alpha.$$

The reader should verify this result, and also the following:

$$\alpha\beta = \beta\alpha$$
$$(\alpha\beta)\gamma = \alpha(\beta\gamma)$$
$$\alpha(\beta + \gamma) = \alpha\beta + \alpha\gamma.$$

If $\alpha = (L, U)$ is a positive real number, we can define another real number $\beta = (L', U')$ by

$$x \in L', \text{ and only if, } x \leq 0 \text{ or } \frac{1}{x} \in U; \text{ otherwise } x \in U'.$$

We can easily check that (L', U') *is* a section of the rational numbers, and thus defines a real number, called the inverse or reciprocal of α. We denote it by α^{-1}. It has the property that

$$\alpha\cdot\alpha^{-1} = 1.$$

If α is negative we define its inverse to be $-(-\alpha)^{-1}$. The inverse is not defined if α is the zero real number, this therefore being the only real number with no inverse.

Having defined inverses, the definition of division is easy. We define the quotient β/α to be

$$\beta\cdot\alpha^{-1}.$$

Since

$$\alpha^{-1} = 1 \cdot \alpha^{-1}$$

we can write

$$\alpha^{-1} = \frac{1}{\alpha}$$

and this notation is often more convenient.

12.7 INEQUALITIES

If $\alpha - \beta$ is negative we say α is less than β or $\alpha < \beta$. In a similar way we can define the familiar expressions $\alpha \leq \beta$ and $\alpha \geq \beta$. Notice that if

$$\alpha = (L, U) \text{ and } \beta = (L', U')$$

then $L \subset L'$ if, and only if, $\alpha \leq \beta$; and $U \subset U'$ if, and only if, $\alpha \geq \beta$. Another result that we shall use presently is that if $L \cap U'$ is nonempty and contains more than one element, then $\alpha > \beta$.

Note: It is not enough to say that $L \cap U'$ is nonempty, since we might have $\alpha = \beta$, with (L, U) and (L', U') two different sections for the same real number.

Exercise 140

1. Write down definitions for subsets L and U of the set of rational numbers such that (L, U) is a section, and defines the irrational number $2\sqrt{3} + 3\sqrt{2}$.

2. Prove from first principle that if α, β and γ are real numbers, and $\alpha = \beta + \gamma$, then $\beta = \alpha - \gamma$.

3. Prove similarly that if $\alpha = \beta\gamma$ then $\beta = \alpha/\gamma$.

4. If α is a positive real number > 1 and $\alpha = (L', U')$, show that the subsets of R defined by

$$U = \{x \mid x > 0 \ \& \ x^q \in U'\}$$
$$L = R - U$$

where q is an integer, form a section (L, U) defining a real number which we may write as $\alpha^{1/q}$.

How would you extend this definition of the qth root to positive numbers less than 1?

5. Show that from the definition of multiplication of real numbers, and from the result of problem 4 above we can define a rational power of a real number, viz., α^a where α is a real number and a is rational.

In Chapter 8 we were obliged to take as an axiom the statement that every set of real numbers which is bounded above has a least upper bound. At that stage we were not able to prove this statement, but with our real numbers properly defined we can now do so rigorously.

THEOREM 12.1.

IF S IS A NONEMPTY SET OF REAL NUMBERS, AND S IS BOUNDED ABOVE, THEN S HAS A LEAST UPPER BOUND. THAT IS, THERE IS A REAL NUMBER M SUCH THAT M IS GREATER THAN OR EQUAL TO ANY ELEMENT OF S (AND IS THEREFORE AN UPPER BOUND OF S), AND ANY NUMBER LESS THAN M IS EXCEEDED BY SOME ELEMENT OF S (AND IS THEREFORE NOT AN UPPER BOUND).

PROOF:

Let us denote a typical element of S by

$$\alpha_\sigma = (L_\sigma, U_\sigma).$$

We define a section (L, U) of the rational numbers as follows: A rational number x is in L if there is an α_σ such that $x \in L_\sigma$; otherwise it is in U. In other words L is to be the union of all the L_σ, and it follows that U will be the intersection of all the U_σ.

Clearly we again have $L \cup U = R$ and $L \cap U = \emptyset$. U is nonempty, and this is where the datum that S is bounded above comes in. For there is a real number, say $\mu = (L', U')$, which is larger than any α_σ. It follows then, that since $\mu > \alpha_\sigma$, $U' \subset U_\sigma$. Hence U, the intersection of the U_σ contains the whole of U' and cannot therefore be empty.

In order to show that (L, U) is a section, we must show that if $x \in L$ and $y \in U$ then $x < y$. Since $x \in L$ there must be an α_{σ_1}, such that $x \in L_{\sigma_1}$. Since $y \in U$ we have $y \in U_{\sigma_1}$ (since U is contained in every U_σ). Hence, since $(L_{\sigma_1}, U_{\sigma_1})$ is a section, it follows that $x < y$.

This completes the proof that (L, U) is a section. Let the real number that it defines be denoted by M. We now show that M has the properties that we are looking for.

1. For any α_σ we have $U \subset U_\sigma$. Hence $\alpha_\sigma \leq M$ for all α_σ. Hence M is an upper bound.

2. Let $x = (L',U')$ be a real number less than M. Then $U \subset U'$ and there are at least *two* rational numbers w_1 and w_2 which are in U' but not in U. The larger of these rational numbers must belong to some L_{σ_1} (otherwise it would belong to U) and the smaller will therefore also belong to this L_{σ_1}. Hence the intersection $L_{\sigma_1} \cap U'$ has at least two elements. From what we said above it follows that $\alpha_{\sigma_1} > x$. This proves that any real number less than M is *not* an upper bound, and hence that the number M that has been defined is indeed the *least* upper bound. Thus the least upper bound always exists under the conditions of the theorem, and the theorem is proved.

We started with integers and built up the system of rational numbers. By considering sections of rational numbers we defined real numbers. We may well ask whether the process stops here. Could we not define yet another type of number by considering sections of *real* numbers? Unfortunately there is a theorem (the Dedekind section theorem) which we shall not prove, which states, in effect, that you get nothing essentially new from considering sections of real numbers. There is thus no scope for further generalization along these lines.

Nevertheless we have not yet reached the end of our game of building up number systems; there still remains what is perhaps the most important number system of all to be dealt with. This is the system of complex numbers which we shall consider in the next chapter.

Exercise 141 (Chapter Review)

1. Prove that any terminating decimal, or recurring decimal is a rational number.

2. Prove that the cube of a rational number cannot be 2, i.e., that $\sqrt[3]{2}$ is irrational.

3. If p, q, m and n are integers such that

$$\left(\frac{p}{q}\right)^2 = \frac{m}{n}$$

prove that m and n are both squares of integers. Deduce that the square root of a rational number which is not of the form u^2/v^2 (where u and v are integers) is irrational.

4. Define the subset U of R by

$$x \in U \equiv x^2 > -1;$$

and let $L = R - U$. Is (L, U) a Dedekind section?

5. If α, β, γ are real numbers prove from first principles that

    (a) $0 = -0$

(b) If $\alpha \leq \beta$ and $\beta < \gamma$ then $\alpha < \gamma$

(c) $\alpha - \beta = -(\beta - \alpha)$

(d) $\alpha(\beta + \gamma) = \alpha\beta + \alpha\gamma$

6. Now that the usual arithmetical operations (addition, etc.) have been defined for the real numbers as we have defined them in this chapter, the reader should satisfy himself that all the more extended operations that were introduced in Chapters 8 and 9 (definition of limits, summation of series, definition of derivatives, and so on) now rest on a firm footing; and that the mathematical structure built up in those two chapters is now based only on the assumption that we know what the integers are and have defined the four basic operations of arithmetic.

Suggestions for Further Reading

For further reading on the subject matter of this chapter, and especially the topic of Dedekind sections, the reader can profitably consult:

Hardy, G. H. *A Course in Pure Mathematics.* 10th ed. New York: Cambridge University Press, 1960.

Also recommended is "Appendix I" of

Bromwich, T. J. I'A., *Infinite Series.* New York: The MacMillan Company, 1955.

and the Appendix of

Ribenboim, P., *Functions, Limits and Continuity.* New York: John Wiley & Sons, Inc., 1964.

All three of these books were mentioned in the suggested reading for Chapter 8. Note also the following:

Landau, E., *Foundations of Analysis.* New York: Chelsea Press, 1951.

13 Complex Numbers

13.1 DEFINITIONS

In this chapter we shall define numbers of yet another kind, called *complex numbers*, which are a generalization of the real numbers. Our search for a suitable definition of the real numbers was prompted by the observation that there is no rational number x satisfying the equation $x^2 = 2$. In the same sort of way, our reason for extending the number system still further is prompted by the observation that there is no real number x which satisfies the equation $x^2 = -1$. It is clear that there is no such real number, for the square of a real number (whether positive or negative) is necessarily positive, and cannot therefore be -1.

It will not do simply to postulate the existence of some "number" whose square is -1, any more than it was any use postulating the existence of a number whose square is 2; this might give us the illusion of knowing what we were talking about, but that would be all. In order to settle this problem satisfactorily we must proceed by what may at first seem a rather roundabout method.

We shall define a new sort of mathematical object, called a complex number. The definition is very simple.

Definition 13.1

A complex number is an ordered pair of real numbers.

Thus the pairs $(2, 3)$, $(4, 27/2)$, $(\sqrt{2}, 5)$ and $(1, \pi)$ are typical complex numbers, according to the definition. The force of the word *ordered* is, of course, that the pair (a, b) is not the same as the pair (b, a), if $a \neq b$.

In the two number systems derived in the last chapter we had to consider what was meant by saying that two numbers were equal, and this led to the consideration of certain equivalence relations between the objects that we had defined. We do not meet this complication here; equality for complex numbers is the same as identity. That is to say:

Definition 13.2

Two complex numbers (a,b) and (c,d) are said to be equal if $a = c$ and $b = d$. We then write $(a, b) = (c, d)$.

It will often be convenient to denote a complex number by a single symbol, and for the time being we shall use the capital letters X, Y, Z, etc., for this purpose.

We shall not get very far with this new number system until we have defined certain operations to be performed on them. There are two of these—addition and multiplication—and we now define them.

Definition of addition.

The sum of two complex numbers (a, b) and (c, d) is defined to be the complex number $(a + c, b + d)$. We shall use the ordinary $+$ sign to denote this operation. Thus we have

$$(3, -9) + (2, 7) = (5, -2)$$

$$\left(\frac{5}{2}, \pi\right) + (e^2, 1.7) = \left(\frac{5}{2} + e^2, \pi + 1.7\right)$$

and so on.

The following laws are easily verified:

(13.1.1) $\qquad X + Y = Y + X$ *(Commutative law)*

(13.1.2) $\qquad X + (Y + Z) = (X + Y) + Z$ *(Associative law)*

where X, Y, Z are any three complex numbers. In consequence of the associative law we can write $X + Y + Z$ for either side of (13.1.2) above. This clearly extends to the sum of four or more complex numbers.

Definition of multiplication.

The product of two complex numbers (a, b) and (c, d) is defined to be the complex number

$$(ac - bd, ad + bc).$$

We shall use the symbol $\cdot$, or simple juxtaposition to denote this operation. Thus

$$(3, -9)\cdot(2, 7) = (3 \times 2 + 9 \times 7, 3 \times 7 + 2 \times -9)$$
$$= (69, 3)$$

$$\left(\frac{5}{2}, \pi\right)\cdot(e^2, 1.7) = \left(\frac{5}{2}e^2 - 1.7\pi, \pi e^2 - 4.25\right)$$

and so on.

The *commutative law*

(13.1.3) $\qquad\qquad\qquad XY = YX$

is easily verified to hold for any two complex numbers X and Y. The *associative law* also holds, but this is not so obvious, and we give the proof. Suppose

$$X_i = (a_i, b_i) \qquad (i = 1, 2, 3)$$

are any three complex numbers, then

$$\begin{aligned}
X_1 \cdot (X_2 \cdot X_3) &= (a_1, b_1) \cdot (a_2 a_3 - b_2 b_3, a_2 b_3 + a_3 b_2) \\
&= (a_1 a_2 a_3 - a_1 b_2 b_3 - a_2 b_1 b_3 - a_3 b_1 b_2, \\
&\qquad a_1 a_2 b_3 + a_1 a_3 b_2 + a_2 a_3 b_1 - b_1 b_2 b_3).
\end{aligned}$$

On the other hand

$$\begin{aligned}
(X_1 \cdot X_2) \cdot X_3 &= (a_1 a_2 - b_1 b_2, a_1 b_2 + a_2 b_1) \cdot (a_3, b_3) \\
&= (a_1 a_2 a_3 - a_3 b_1 b_2 - a_1 b_2 b_3 - a_2 b_1 b_3, \\
&\qquad a_1 a_2 b_3 - b_1 b_2 b_3 + a_1 a_3 b_2 + a_2 a_3 b_1).
\end{aligned}$$

Since the two results are the same, we have

$$(13.1.4) \qquad\qquad X_1 \cdot (X_2 \cdot X_3) = (X_1 \cdot X_2) \cdot X_3$$

for any three complex numbers X_1, X_2 and X_3. We write either side of (13.1.4) as $X_1 \cdot X_2 \cdot X_3$.

Another law, the *distributive law*, involves both multiplication and addition. It is that

$$(13.1.5) \qquad\qquad X_1 \cdot (X_2 + X_3) = X_1 \cdot X_2 + X_1 \cdot X_3.$$

Using the same notation as before we have

$$\begin{aligned}
X_1 \cdot (X_2 + X_3) &= (a_1, b_1) \cdot (a_2 + a_3, b_2 + b_3) \\
&= [a_1(a_2 + a_3) - b_1(b_2 + b_3), b_1(a_2 + a_3) + a_1(b_2 + b_3)].
\end{aligned}$$

While

$$\begin{aligned}
X_1 \cdot X_2 + X_1 \cdot X_3 &= (a_1, b_1) \cdot (a_2, b_2) + (a_1, b_1) \cdot (a_3, b_3) \\
&= (a_1 a_2 - b_1 b_2, b_1 a_2 + a_1 b_2) + (a_1 a_3 - b_1 b_3, a_1 b_3 + a_3 b_1) \\
&= (a_1 a_2 - b_1 b_2 + a_1 a_3 - b_1 b_3, b_1 a_2 + a_1 b_2 + a_1 b_3 + b_1 a_3).
\end{aligned}$$

Since the right-hand sides are equal, the distributive law is proved.

It certainly looks as though our complex numbers have much in common with more familiar numbers. This resemblance is strengthened by the following additional properties.

There is a *zero complex number*, namely the pair $(0, 0)$ (which we can denote by **0**). This clearly has the property that

$$X + \mathbf{0} = X$$

for any complex number X. For if $X = (a, b)$ then

$$(a, b) + (0, 0) = (a, b) = X.$$

There is a complex number, denoted by **1**, which has the property enjoyed by the real number 1, namely that

$$\mathbf{1} \cdot X = X$$

for any X. This is the complex number $\mathbf{1} = (1, 0)$. For if $X = (a, b)$, then

$$\mathbf{1} \cdot X = (1, 0) \cdot (a, b)$$
$$= (a, b) = X.$$

Further, any complex number $X = (a, b)$ has a negative, i.e., there is a complex number, which can be denoted by $-X$, with the property that

$$X + (-X) = \mathbf{0}.$$

This is obvious; the complex number $-X$ is just $(-a, -b)$.

Finally, every complex number X, except **0**, has an inverse; that is, a complex number X^{-1} such that

$$X \cdot X^{-1} = \mathbf{1}.$$

This is not as obvious as the existence of the complex number $-X$, but is easily shown. If $X = (a, b)$ where $b \neq 0$, suppose that $X^{-1} = (p, q)$. Then we must have

$$(a, b) \cdot (p, q) = (1, 0)$$

or

$$(ap - bq, bp + aq) = (1, 0).$$

Hence

$$ap - bq = 1$$
$$bp + aq = 0.$$

These are simple equations for p and q, which we can solve in the usual way. Since $p = -aq/b$ from the second equation, we have, from the first equation,

$$-\frac{a^2q}{b} - bq = 1$$

i.e.,

$$q(a^2 + b^2) = -b.$$

Thus we have $q = \dfrac{-b}{a^2 + b^2}$ and $p = \dfrac{a}{a^2 + b^2}$.

The inverse of $X = (a, b)$ is therefore the complex number

$$\left(\frac{a}{a^2 + b^2}, \frac{-b}{a^2 + b^2}\right)$$

which we denote by X^{-1}. When $b = 0$, $X^{-1} = (a^{-1}, 0)$.

We see, therefore, that these complex numbers that we have defined, together with the given rules for addition and multiplication, follow the usual laws of arithmetic. The system that we have set up resembles very closely the system of real numbers with its definition of addition and subtraction.

We are now going to change to a different notation. The notation used so far, though excellent for the purposes of setting up the system in a general manner, is not too suitable for practical purposes. The new notation is more convenient, but it might have been misleading had we used it right from the start.

Exercise 142

1. Verify the laws given in this section relating to the operations of addition, multiplication, etc., for complex numbers.
2. Find the inverses of the following complex numbers.

    (a) $(1, -2)$ (c) $(3, 8)$ (e) $(-5, 21)$

    (b) $(6, 3)$ (d) $(-7, -4)$ (f) $(\pi, 2.71)$

3. If $Z = (x, y)$ and $A = (p, q)$ are complex numbers related by the equation $Z^2 = A$, show that

$$x^2 - y^2 = p \quad \text{and} \quad 2xy = q.$$

Hence show that there are two complex numbers Z whose square is A.

13.2 NOTATION AND TERMINOLOGY

A number of the form $(a, 0)$, in which the second member of the pair is zero, will be called a *real complex number*. It is easily seen that the operations of addition and multiplication performed on real complex numbers will parallel exactly the operations of the same name performed on real numbers. Thus, just as $2.54 + 7.66 = 10.20$, so

$$(2.54, 0) + (7.66, 0) = (10.20, 0);$$

and just as $2.5 \times 3 = 7.5$ so

$$(2.5, 0) \cdot (3, 0) = (7.5, 0).$$

In general

$$(a, 0) + (p, 0) = (a + p, 0)$$

and

$$(a, 0) \cdot (p, 0) = (ap, 0).$$

We can therefore set up a correspondence between real complex numbers and real numbers; we arrange that

$$(a, 0) \text{ corresponds to } a,$$

and this correspondence has the property that the operations of addition and multiplication are preserved. For this reason, although the real number a and the real complex number $(a, 0)$ are conceptually two completely different things, it is permissible to identify these two concepts and to treat them as being essentially the same thing. This is on a par with what we did when we identified the rational integers with the corresponding integers, and the rational real numbers with the rational numbers.

Allowing this identification, we can say that the real numbers are special cases of complex numbers, namely those for which the second number in the pair is zero. We can also drop the word *complex* from the phrase *real complex number*. Looking at this situation in another way, we can say that the complex numbers form a generalization of the real numbers, which they include as special cases.

Since we are now identifying $(a, 0)$ with the real number a, it is unnecessary to have an alternative notation, and we can let "a" stand for either concept according to the context. In particular, the zero complex number $(0, 0)$ can be identified with the real number zero, and written with the same symbol "0".

The complex number $(0,1)$ next engages our attention. It is not a real complex number, and we shall denote it by a special symbol "i". We now observe that any complex number can be expressed in terms of real numbers and the symbol i. For

$$(a, b) = (a, 0) + (0, b)$$
$$= (a, 0) + (0, 1) \cdot (b, 0)$$

by the rules for addition and multiplication. Hence in the new notation which we are developing, we would write

$$(a, b) = a + i \cdot b$$

or simply

$$(a, b) = a + ib, \quad \text{or} \quad a + bi.$$

Let us look at this number i a little more closely. If we square it, we get

$$i \cdot i = (0, 1) \cdot (0, 1)$$
$$= (-1, 0)$$

by the rule for multiplication. Hence i^2 is the real number -1. Thus it turns out that although there is no *real* number whose square is the real number -1, there is a complex number whose square is the real (complex) number -1. We have extended the system of real numbers to a larger system in which the equation $x^2 + 1 = 0$ *does* have a solution.

Using the new notation, we can easily describe the operations of addition and multiplication. All we have to do to find the sum or product of two complex numbers is to treat the symbols that occur in them (e.g., a, b, i, etc.) just as we would if they were real numbers, but whenever i^2 occurs we can replace it by -1.

Thus

$$(a + ib) + (c + id) = a + ib + c + id$$
$$= (a + c) + i(b + d)$$

which correctly expresses the rule for addition; and

$$(a + ib) \cdot (c + id) = ac + i(bc + ad) + i^2 bd$$
$$= (ac - bd) + i(bc + ad)$$

which gives the rule for multiplication.

We can now essay a description of the complex number system in terms of this new notation. A complex number is an expression of the form $a + ib$, where a and b are real numbers. Addition and multiplication are defined exactly as if the symbols involved were real numbers, except that i^2 can be replaced by -1.

EXAMPLES:

$$(2 + 3i) + (7 - 9i) = 9 - 6i;$$

$$\left(\frac{4}{7} + \frac{1}{2}i\right) + \left(\frac{10}{7} + 5i\right) = 2 + \frac{11}{2}i;$$

$$(3 - 4i)\cdot(5 + 2i) = 15 - 20i + 6i - 8i^2$$
$$= 15 - 14i + 8$$
$$= 23 - 14i;$$

$$\left(\frac{7}{2} + 5i\right)\left(4 - \frac{2i}{5}\right) = 14 + 20i - \frac{7}{5}i - 2i^2$$
$$= 16 + \frac{93}{5}i.$$

If z denotes the complex number $x + iy$, then the real number x is called the *real part* of z, while the real number y is called the *imaginary part* of z. A complex number whose imaginary part is zero is called a *real complex number*, or, more loosely (as we have seen), a *real number;* a complex number whose real part is zero is said to be an *imaginary number.* Our first statement about complex numbers, after we had defined them, was to the effect that two complex numbers are equal if, and only if, they are identical—in other words, if, and only if, their real parts are the same, and likewise their imaginary parts. For example, if we know that

$$(x + \alpha) + i(y + \beta) = \alpha^2 + i\beta^2$$

then we can deduce that

$$x + \alpha = \alpha^2$$

and

$$y + \beta = \beta^2.$$

This process is known as *equating real and imaginary parts*, and is one which we shall have great occasion to use. Clearly a single equation between complex numbers is equivalent to two equations between real numbers—one obtained by equating real parts, the other by equating imaginary parts.

Exercise 143

1. Express each of the following equations as a pair of equations in x and y, where $z = x + iy$, and solve equation (a).
   (a) $z^2 + z + 1 = 0$
   (b) $(2 + i)z^2 - iz + (3 + 5i) = 0$

2. If $z = x + iy$ and $w = u + iv$, obtain u and v as functions of x and y when

(a) $w = z^3$

(c) $w = z(2z + 1)(1 - 2z)$

(b) $w = (2 + i + z)^2$

(d) $w = (z + i)^2 + \dfrac{1}{z}$

3. Show that the two criteria given at the end of section 11.14 for the stability of solutions of second order differential equations with constant coefficients can be combined into a single criterion, as follows:

The solutions of the differential equation $ay'' + by' + cy = 0$ are stable if, and only if, the roots of the auxiliary equation $a\lambda^2 + b\lambda + c = 0$ have negative real parts.

13.3 CONJUGATE COMPLEX NUMBERS

If $z = x + iy$, the complex number $x - iy$ is known as the *complex conjugate* (or just the *conjugate*) of z, and is denoted by $\bar{z}$.

If $z_1 = x_1 + iy_1$, $z_2 = x_2 + iy_2$ and $z = z_1 + z_2$ then clearly

$$\bar{z} = \text{the conjugate of } (x_1 + x_2) + i(y_1 + y_2)$$
$$= (x_1 + x_2) - i(y_1 + y_2)$$
$$= (x_1 - iy_1) + (x_2 - iy_2)$$
$$= \bar{z}_1 + \bar{z}_2.$$

Hence the conjugate of the sum of two (and therefore of any number of) complex numbers is the sum of their conjugates.

If $w = z_1 z_2$, we have

$$w = (x_1 x_2 - y_1 y_2) + i(x_1 y_2 + x_2 y_1)$$

and

$$\bar{w} = (x_1 x_2 - y_1 y_2) - i(x_1 y_2 + x_2 y_1)$$
$$= (x_1 - iy_1) \cdot (x_2 - iy_2)$$
$$= \bar{z}_1 \cdot \bar{z}_2.$$

Consequently the conjugate of the product of two complex numbers is the product of their conjugates. This easily extends to products of any number of complex numbers. In particular, the conjugate of z^n is $\bar{z}^n$. It follows that if $P(z)$ is a polynomial in z with complex coefficients, say

$$P(z) = a_n z^n + a_{n-1} z^{n-1} + \cdots + a_1 z + a_0$$

then

$$\overline{P(z)} = \bar{a}_n \bar{z}^n + \bar{a}_{n-1} \bar{z}^{n-1} + \cdots + \bar{a}_1 \bar{z} + \bar{a}_0.$$

In the special case when the coefficients a_i are all real we have $\bar{a}_i = a_i$, and hence

$$\overline{P(z)} = a_n \bar{z}^n + a_{n-1} \bar{z}^{n-1} + \cdots + a_1 z + a_0$$
$$= P(\bar{z}),$$

—a result that we shall use shortly.

THE MODULUS.

If $z = x + iy$ then

$$z\bar{z} = (x + iy)(x - iy)$$
$$= x^2 + y^2.$$

The real number $\sqrt{x^2 + y^2}$ (the positive root being taken) is known as the *modulus* of z. It is denoted by $|z|$. Thus we write

(13.3.1) $$z\bar{z} = |z|^2.$$

Notice that this notation is consistent with that used in Chapter 8 for the modulus of a real number, for if z happens to be a real number then $|z|$ is simply this real number, but taken with a positive sign. An immediate result is that the modulus of the product of two complex numbers is the product of their moduli. For

$$|z_1 z_2|^2 = (z_1 z_2)\overline{(z_1 z_2)}$$
$$= (z_1 z_2)(\bar{z}_1 \bar{z}_2)$$
$$= z_1 \bar{z}_1 \, z_2 \bar{z}_2$$
$$= |z_1|^2 \, |z_2|^2.$$

THEOREM 13.1.

IF $z_1 z_2 = 0$, THEN EITHER $z_1 = 0$ OR $z_2 = 0$.

PROOF:

If $z_1 z_2 = 0$ then

$$|z_1 z_2| = |z_1| \cdot |z_2| = 0.$$

Since $|z_1|$ and $|z_2|$ are real numbers, it follows that either $|z_1|$ or $|z_2|$ is zero.

Hence either $z_1 = 0$ or $z_2 = 0$.

DIVISION OF COMPLEX NUMBERS.

We have seen that every complex number has an inverse. Division of one complex number by another is therefore possible. Since

$$z\bar{z} = |z|^2$$

we have

$$z \cdot \frac{1}{|z|^2}\bar{z} = 1$$

so that the inverse of z can be written

$$z^{-1} = \frac{1}{|z|^2}\bar{z}.$$

More generally, if we denote the quotient z_1/z_2 by w so that we have

$$z_1 = wz_2$$

then

$$z_1\bar{z}_2 = wz_2\bar{z}_2$$
$$= w|z_2|^2.$$

Now $|z_2|^2$ is a real number, which is not zero if $z_2 \neq 0$. Hence

$$w = \frac{1}{|z_2|^2}z_1\bar{z}_2$$

and we have an expression for the quotient. Effectively what we have done is to multiply the numerator and denominator of z_1/z_2 by $\bar{z}_2$ thus obtaining

$$\frac{z_1}{z_2} = \frac{z_1\bar{z}_2}{z_2\bar{z}_2} = \frac{z_1\bar{z}_2}{|z_2|^2}.$$

EXAMPLE:

Express $\dfrac{3 + 5i}{4 + 5i}$ in the form $a + bi$.

We have

$$\frac{3 + 5i}{4 + 5i} = \frac{(3 + 5i)(4 - 5i)}{(4 + 5i)(4 - 5i)}.$$

(This is the multiplication by $\bar{z}_2$ that we performed above.) On simplification this becomes

$$\frac{12 + 20i - 15i + 25}{16 + 25} = \frac{37 + 5i}{41}$$

$$= \frac{37}{41} + \frac{5}{41}i.$$

By successive applications of this method together with the rule for addition and multiplication of complex numbers, any expression in complex numbers can be reduced to the form $a + bi$, where a and b are real numbers.

EXAMPLE:

Simplify $\dfrac{2 + 9i}{3 - 4i} + \dfrac{(3 - i)(2 + i)}{7 - i} + \dfrac{3}{25}(27 + 11i)$.

We have

$$\frac{2 + 9i}{3 - 4i} + \frac{(3 - i)(2 + i)}{7 - i} + \frac{3}{25}(27 + 11i)$$

$$= \frac{(2 + 9i)(3 + 4i)}{(3 - 4i)(3 + 4i)} + \frac{(3 - i)(2 + i)(7 + i)}{(7 - i)(7 + i)} + \frac{3}{25}(27 + 11i)$$

$$= \frac{6 + 27i + 8i - 36}{9 + 16} + \frac{(6 - 2i + 3i + 1)(7 + i)}{49 + 1} + \frac{3}{25}(27 + 11i)$$

$$= \frac{-30 + 35i}{25} + \frac{(7 + i)(7 + i)}{49 + 1} + \frac{3}{25}(27 + 11i)$$

$$= \frac{-30 + 35i}{25} + \frac{49 + 14i - 1}{50} + \frac{3}{25}(27 + 11i)$$

$$= \frac{1}{25}(-30 + 35i + 24 + 7i + 81 + 33i)$$

$$= \frac{1}{25}(75 + 75i)$$

$$= 3 + 3i.$$

Exercise 144

1. Reduce the following expressions to the form $x + iy$.

(a) $\dfrac{2 + 2i}{1 - i}$

(b) $\dfrac{(a - bi)^2}{a + bi}$

(c) $\dfrac{10 + 20i}{3 - i}$

(d) $\dfrac{a}{1 + bi} + \dfrac{b}{1 + ai}$ (a, b, real)

(e) $\dfrac{2 + 11i}{2 + i} + \dfrac{17}{4 + i}$

2. If p is real, and if the complex number

$$\frac{1+i}{2+pi} + \frac{2+3i}{3+i}$$

is a real multiple of $1 + i$, show that $p = -5 \pm \sqrt{21}$.

13.4 ROOTS OF A POLYNOMIAL EQUATION WITH REAL COEFFICIENTS

Let $P(z)$ be a polynomial in z. Then by a root of the equation $P(z) = 0$ is meant any complex number say z_1, such that $P(z_1) = 0$. Some equations have all their roots real numbers; others have no real roots, such as the equation $x^2 + 1 = 0$ which initiated our discussion of complex numbers. Therefore, so long as we were restricted to real numbers, we had to accept that some polynomial equations had no roots. Now that we have extended our number system to the complex numbers we can use a theorem to the effect that every polynomial equation has at least one root. The proof of this theorem (the so-called *fundamental theorem of algebra*) is unfortunately beyond the scope of this book, but we *can* prove an easier result, which relates to polynomials in which all the coefficients are real numbers.

THEOREM 13.2.

If $P(z)$ is a polynomial whose coefficients are all real, then the roots of $P(z) = 0$ occur in conjugate pairs. That is, if $P(z_1) = 0$ then $P(\bar{z}_1) = 0$.

PROOF:
Let z_1 be a root; then $P(z_1) = 0$.
Taking conjugates we have

$$\overline{P(z_1)} = \bar{0} = 0.$$

(Clearly zero is its own conjugate.) But we have already shown that if the coefficients are real

$$\overline{P(z_1)} = P(\bar{z}_1).$$

Consequently $P(\bar{z}_1) = 0$, so that z_1 is also a root.

For a quadratic equation this follows immediately from the formula

$$\frac{-b \pm \sqrt{b^2 - 4ac}}{2a}$$

For if $b^2 - 4ac$ is negative, the roots are of the form $u \pm iv$.

13.5 THE ARGAND DIAGRAM

Just as it is sometimes convenient to think of real numbers as being "something like" points on a real line [for example by associating with the real number x the point $(x, 0)$ on the x-axis of the Cartesian plane] so it is convenient to set up a geometrical way of dealing with complex numbers. We do this by representing complex numbers as points of a plane, and the manner in which we do it is very simple; our original notation, in which the complex number $x + iy$ was written as (x, y), suggests the method. We make the point whose Cartesian coordinates are (x, y) represent the complex number $x + iy$. It is clear that every complex number is represented by some point of the plane, and that, conversely, every point of the Cartesian plane represents some complex number. Furthermore, real numbers will be represented by points on the x-axis; imaginary numbers by points on the y-axis. This method of representing complex numbers is known as the method of the *Argand diagram*, or *Argand plane*. Figure 13.1 shows some points of the Argand plane with the complex numbers that they represent.

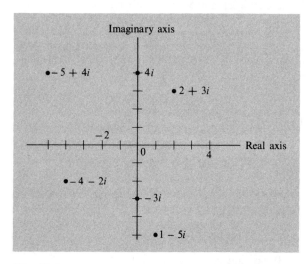

Fig. 13.1

The next thing we must do is to find geometrical interpretations of the various operations that we perform on complex numbers.

MODULUS

The modulus of $z(= x + iy$, say) is the real number $\sqrt{x^2 + y^2}$. This is precisely the distance of the point (x, y), representing z, from the origin (which represents the zero complex number). If P is the point which represents z, then $|z|$ is the distance OP in Fig. 13.2.

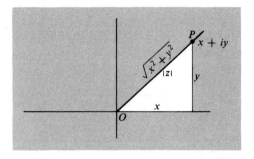

Fig. 13.2

ADDITION OF COMPLEX NUMBERS

If the complex numbers $z_1 = x_1 + iy_1$ and $z_2 = x_2 + iy_2$ are represented in the Argand diagram by the points P and Q, respectively, the point which represents $z_1 + z_2$ can be found by a simple construction, as follows. We construct a parallelogram which has OP and OQ as two of its sides; that is, we draw lines through P and Q, parallel to OQ and OP, respectively, to meet in a point R. The point R then represents $z_1 + z_2$.

The proof of this is straightforward. If L, M, N are the feet of the perpendiculars from P, Q, R to the x-axis, then the x coordinate of R is

$$ON = OM + MN$$
$$= OM + QK$$
$$= OM + OL$$

since it is easy to show that $MN = QK = OL$ (where QK is parallel to the x-axis). (See Fig. 13.3.)

Since $OL = x_1$ and $OM = x_2$, the x-coordinate of R is $x_1 + x_2$.

Similarly the y-coordinate of R is

$$RN = RK + KN$$
$$= PL + QM$$
$$= y_1 + y_2.$$

Hence the complex number represented by R is $(x_1 + x_2) + i(y_1 + y_2)$ $= z_1 + z_2$, as required.

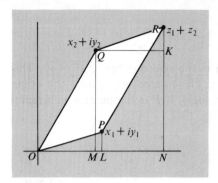

Fig. 13.3

SUBTRACTION

If P represents the complex number $z = x + iy$, then $-z$ is represented by the point P' whose coordinates are $(-x, -y)$. This point is the *reflection* of P in the origin, as shown in Fig. 13.4. Using this result we can obtain the construction for the point which represents the difference $z_1 - z_2$ of the complex numbers z_1 and z_2. We have merely to construct the point which represents the sum of z_1 and $(-z_2)$. This is shown in Fig. 13.5.

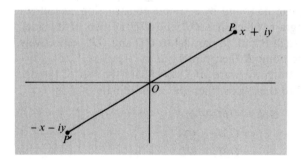

Fig. 13.4

Completing the parallelogram whose sides are OP and OQ' (P represents z_1 and Q' represents $-z_2$) we arrive at the point R, representing $z_1 - z_2$. It will be seen that we can construct R from P and Q without first constructing Q'. We have merely to construct a parallelogram which has OP as its diagonal, and OQ as one side. This we do by drawing a line through O parallel to QP, and one through P parallel to OQ. Where they meet is the point R.

The modulus of $z_1 - z_2$, by our first result, is the length OR, and from the geometry of Fig. 13.5 it is clear that this is the length PQ. Hence we have the result, very useful in interpreting complex numbers geometrically, that $|z_1 - z_2|$ is the distance between the points P and Q of the Argand diagram which represent z_1 and z_2.

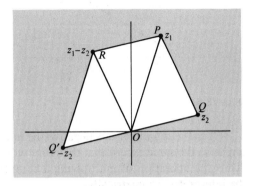

Fig. 13.5

Exercise 145

1. A, B and C are points on the Argand diagram representing the complex numbers $3 + 5i$, $2 - i$ and $5 + 2i$, respectively. Find the complex number represented by the foot of the perpendicular from A to BC.

2. Give a geometrical proof of the inequality

$$|z_1 + z_2| \leq |z_1| + |z_2|.$$

3. What is the complex number represented by the centroid of the triangle ABC defined in problem 1?

4. If z_0 is a fixed complex number, what can be said about points z for which

   (a) $|z - z_0| = 1$
   (b) $|z - z_0| = |z|$

13.6 MULTIPLICATION BY A REAL NUMBER

If a is a positive real number, and z is a complex number $x + iy$, then $az = ax + iay$. Thus if z is represented by the point $P = (x, y)$, az is represented by the point $Q = (ax, ay)$. These points are shown in Fig. 13.6. The point Q is on the line OP and its distance from O is a times the distance of P from O, i.e.,

$$\frac{OQ}{OP} = a.$$

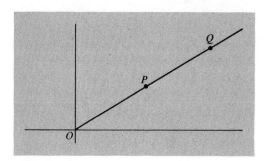

Fig. 13.6

If the real number a is negative the interpretation of the point representing az is the same except that the point will be on the other side of O from P. Thus the operation of multiplying a complex number by a real number can be interpreted geometrically.

The geometrical interpretation of multiplication of two complex numbers is a little more tricky and will be left until the end of the next section.

13.7 MODULUS—AMPLITUDE FORM OF A COMPLEX NUMBER

We shall now introduce a dodge similar to the one we introduced at the end of section 10.7. We can write a general nonzero complex number in the following form.

$$z = x + iy$$

$$(13.7.1) \qquad = \sqrt{x^2 + y^2} \left\{ \frac{x}{\sqrt{x^2 + y^2}} + \frac{iy}{\sqrt{x^2 + y^2}} \right\}.$$

Now $\dfrac{x}{\sqrt{x^2 + y^2}}$ and $\dfrac{y}{\sqrt{x^2 + y^2}}$ are respectively the cosine and sine of some angle. We shall denote this angle by θ, and let r denote the modulus $|z|$, which is, of course, $\sqrt{x^2 + y^2}$. We can then write (13.7.1) as

$$z = x + iy$$
$$= r\,(\cos\theta + i\sin\theta).$$

The relations between x, y, r and θ are easily seen from Fig. 13.7; they are

$$x = r\cos\theta$$
$$(13.7.2) \qquad y = r\sin\theta$$
$$r = \sqrt{x^2 + y^2}$$
$$\tan\theta = \frac{y}{x}.$$

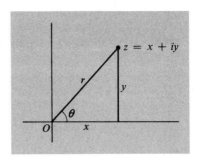

Fig. 13.7

The angle θ is called the *amplitude* or *argument* of z, and is often written as amp z or arg z. This form of writing a complex number is called the *modulus-amplitude form*.

EXAMPLE 1.

Find the modulus and amplitude of the complex number $5 + 12i$.
The modulus is $\sqrt{5^2 + 12^2} = \sqrt{169} = 13$.
The amplitude is $\tan^{-1} 12/5$.

EXAMPLE 2.

Express $1 + i$ in modulus-amplitude form

$$1 + i = \sqrt{2}\left(\frac{1}{\sqrt{2}} + \frac{1}{\sqrt{2}}i\right).$$

Now $\dfrac{1}{\sqrt{2}} = \cos\dfrac{\pi}{4} = \sin\dfrac{\pi}{4}$, so that

$$1 + i = \sqrt{2}\left(\cos\frac{\pi}{4} + i\sin\frac{\pi}{4}\right).$$

Let us now consider the product of two complex numbers which have been written in modulus-amplitude form. If

$$z_1 = r_1(\cos\alpha + i\sin\alpha)$$

and

$$z_2 = r_2(\cos\beta + i\sin\beta)$$

then $z_1z_2 = r_1r_2(\cos\alpha + i\sin\alpha)(\cos\beta + i\sin\beta)$.
Now the product r_1r_2 gives us no trouble, and we turn to consider the product

$$(\cos\alpha + i\sin\alpha)(\cos\beta + i\sin\beta).$$

Multiplying in the usual way we obtain

$$(\cos \alpha \cos \beta - \sin \alpha \sin \beta) + i(\sin \alpha \cos \beta + \cos \alpha \sin \beta).$$

Using the expressions for $\cos(\alpha + \beta)$ and $\sin(\alpha + \beta)$ from elementary trigonometry, viz.,

$$\cos(\alpha + \beta) = \cos \alpha \cos \beta - \sin \alpha \sin \beta$$
$$\sin(\alpha + \beta) = \sin \alpha \cos \beta + \cos \alpha \sin \beta$$

we find that

(13.7.3) $(\cos \alpha + i \sin \alpha)(\cos \beta + i \sin \beta)$
                          $= \cos(\alpha + \beta) + i \sin(\alpha + \beta).$

Hence $z_1 z_2 = r_1 r_2 \{\cos(\alpha + \beta) + i \sin (\alpha + \beta)\}$, a result which we can summarize as follows:

1. The modulus of the product of two complex numbers is the product of their moduli (this we already knew);

2. The amplitude of the product of two complex numbers is the sum of their amplitudes.

In symbols

$$|z_1 z_2| = |z_1| \cdot |z_2|$$
$$\operatorname{amp}(z_1 z_2) = \operatorname{amp} z_1 + \operatorname{amp} z_2.$$

A geometrical interpretation in the Argand diagram of the product $z_1 z_2$ is given in Fig. 13.8. If P and Q represent z_1 and z_2, and I represents the real number 1, then $z_1 z_2$ is represented by the point R which makes the tri-

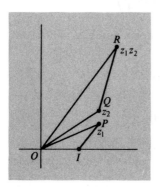

Fig. 13.8

angle OQR similar to triangle OIP, i.e., such that

$$\frac{OR}{OQ} = \frac{OP}{OI}$$

and angle ROQ = angle POI.

The reader should verify that this construction gives to the complex number represented by R the correct modulus and amplitude.

Exercise 146

1. Show that $\text{amp}(z_1 - z_2)$ is the angle that is made with the real axis (the x-axis) by the line joining the points representing z_1 and z_2 on the Argand diagram. Hence interpret geometrically the expression $\text{amp}\left(\dfrac{z_1 - z_2}{z_1 - z_3}\right)$.

2. Express the following complex numbers in modulus-amplitude form.

(a) $3 + 3i$ (c) $-\dfrac{1}{2} + i\dfrac{\sqrt{3}}{2}$ (e) $6 - 8i$

(b) $12 - 5i$ (d) $-7 + 9i$

13.8 DE MOIVRE'S THEOREM

From (13.7.3) we can fairly easily derive a very important theorem.

DE MOIVRE'S THEOREM

IF k IS A RATIONAL NUMBER, THEN ONE OF THE VALUES OF

$$(\cos\theta + i\sin\theta)^k$$

IS

$$\cos k\theta + i\sin k\theta.$$

PROOF:

The proof of the theorem proceeds by stages.

Stage 1. The theorem is true if k is a positive integer.

We have seen that the amplitude of the product of two numbers is the product of their amplitudes, and it is clear that this result will apply to the product of any number of complex numbers. Now consider $(\cos\theta + i\sin\theta)^k$ where k is an integer. It is the product of k numbers each equal to $\cos \theta + i \sin \theta$, having unit modulus. Hence the modulus of $(\cos \theta + i \sin \theta)^k$ is also unity, and its amplitude is the sum of the amplitudes, namely $k\theta$. Hence

$$(\cos \theta + i \sin \theta)^k = \cos k\theta + i \sin k\theta$$

when k is a positive integer.

Stage 2.   The theorem is true if k is a negative integer, say $k = -n$, where n is a positive integer.

We observe that

(13.8.1) $$(\cos \theta + i \sin \theta)(\cos \theta - i \sin \theta) = 1.$$

Hence

$$
\begin{aligned}
(\cos \theta + i \sin \theta)^k &= (\cos \theta + i \sin \theta)^{-n} \\
&= \{(\cos \theta + i \sin \theta)^{-1}\}^n \\
&= \{\cos \theta - i \sin \theta\}^n \quad \text{by (13.8.1)} \\
&= \{\cos (-\theta) + i \sin (-\theta)\}^n \\
&= \cos (-n\theta) + i \sin (-n\theta)
\end{aligned}
$$

by the result of stage 1. Note that $\cos \theta - i \sin \theta$ is the conjugate of $\cos \theta + i \sin \theta$, so that (13.8.1) is a special case of (13.3.1).

Hence

$$(\cos \theta + i \sin \theta)^k = \cos k\theta + i \sin k\theta$$

when k is a negative integer.

Stage 3.   The theorem is true for any rational number.

If $k = p/q$ where p and q are integers (we may assume $q > 0$) then $z^k = z^{p/q}$ will denote a complex number z_1 such that

$$z_1^q = z^p.$$

We shall see in a moment that there will be several such complex numbers. Let us write z_1 in modulus-amplitude form as

$$z_1 = \rho(\cos \alpha + i \sin \alpha).$$

Then we have

$$
\begin{aligned}
z_1^q &= \rho^q(\cos \alpha + i \sin \alpha)^q \\
&= \rho^q(\cos q\alpha + i \sin q\alpha),
\end{aligned}
$$

and this should be $(\cos\theta + i\sin\theta)^p = \cos p\theta + i\sin p\theta$, using stages 1 and 2. Thus we require that

(13.8.2) $\qquad\qquad p^q(\cos q\alpha + i\sin q\alpha) = \cos p\theta + i\sin p\theta.$

The modulus of the right-hand side of (13.8.2) is 1, so that the modulus of the left-hand side must also be 1. Hence $p^q = 1$ and hence $p = 1$, since p is real and positive. The amplitudes of the two sides of (13.8.2) must agree to within a multiple of 2π (since the values of $\sin x$ and $\cos x$ repeat at intervals of 2π). Hence $q\alpha$ and $p\theta$ differ by a multiple of 2π, i.e.,

$$q\alpha = p\theta + 2m\pi, \text{ where } m \text{ is some integer.}$$

This gives us the required amplitude α, namely

$$\alpha = \frac{p}{q}\theta + \frac{2m\pi}{q}.$$

Now *one* of these values of α is that for which $m = 0$, viz.,

$$\alpha = \frac{p}{q}\theta = k\theta.$$

Hence one possible number z_1 is the number,

$$\cos k\theta + i\sin k\theta,$$

and this is what the theorem stated.

Thus we have proved De Moivre's theorem in full.

EXAMPLE:
Find the value of $(1 + i)^6$

    (a) by De Moivre's Theorem
    (b) otherwise.

a. We have

$$1 + i = \sqrt{2}\left(\cos\frac{\pi}{4} + i\sin\frac{\pi}{4}\right).$$

Therefore

$$(1 + i)^6 = (\sqrt{2})^6\left(\cos\frac{\pi}{4} + i\sin\frac{\pi}{4}\right)^6$$
$$= 8\left(\cos\frac{6\pi}{4} + i\sin\frac{6\pi}{4}\right)$$

by De Moivre's theorem.

Hence

$$(1 + i)^6 = 8(0 + (-1)i)$$
$$= -8i.$$

b. By the binomial theorem

$$(1 + i)^6 = 1 + 6i + 15i^2 + 20i^3 + 15i^4 + 6i^5 + i^6$$
$$= 1 + 6i - 15 - 20i + 15 + 6i - 1$$
$$= -8i.$$

It might appear strange at first sight that there should be more than one value for $z^{p/q}$, and, in particular, for the qth root $z^{1/q}$ of a complex number. Yet this is just an extension of the familiar fact that there are two different values for the square root of a number. Thus $9^{1/2}$ or $\sqrt{9}$ has two possible values, $+3$ and -3. This ties up nicely with De Moivre's theorem, for

$$1 = \cos \theta + i \sin \theta$$

where $\theta = 0$, 2π, 4π, etc., or any multiple of 2π. Hence, by De Moivre's theorem

$$1^{1/2} = \cos \frac{\theta}{2} + i \sin \frac{\theta}{2}$$

(13.8.3) $$= \cos \phi + i \sin \phi$$

where ϕ is a multiple of π.

Since

$$\cos(-\pi) + i \sin(-\pi) = -1$$
$$\cos 0 + i \sin 0 = 1$$
$$\cos \pi + i \sin \pi = -1$$
$$\cos 2\pi + i \sin 2\pi = 1$$

etc.

there are only two distinct values of $1^{1/2}$, viz., $+1$ and -1.

Let us now consider the case $k = \frac{1}{3}$. The possible values for $1^{1/3}$ are $\cos \theta/3 + i \sin \theta/3$, namely:

$$\cos 0 + i \sin 0 = 1$$
$$\cos \frac{2\pi}{3} + i \sin \frac{2\pi}{3} = -\frac{1}{2} + i\frac{\sqrt{3}}{2}$$

and $$\cos \frac{4\pi}{3} + i \sin \frac{4\pi}{3} = \frac{-1}{2} - i\frac{\sqrt{3}}{2}$$

These are the only distinct values. They can be derived without using De Moivre's theorem; for if z is a cube root of 1, then

$$z^3 - 1 = 0$$

or $$(z - 1)(z^2 + z + 1) = 0.$$

Hence $z = 1$ is one root (obviously), and the others are given by

$$z^2 + z + 1 = 0.$$

Solving this as a quadratic in z we find that

$$z = -\frac{1}{2} \pm i\frac{\sqrt{3}}{2}.$$

Observe the complex roots occurring in conjugate pairs.

Consider $k = \frac{1}{4}$. We have

$$1 = \cos 2m\pi + i \sin 2m\pi \qquad m = 0, 1, 2, \ldots$$

$$1^{1/4} = \cos \frac{m\pi}{2} + i \sin \frac{m\pi}{2}.$$

The possible values are $1, i, -1, -i$, as is otherwise obvious.

In general the qth roots of 1 will be the complex numbers of the form

$$\cos \frac{2m\pi}{q} + i \sin \frac{2m\pi}{q}$$

for $m = 0, 1, 2, \ldots, q - 1$. Other values of m will also give the qth roots of 1, but they will be duplications of values already obtained.

INTERPRETATION ON THE ARGAND DIAGRAM
OF ROOTS OF UNITY

On the Argand diagram the points representing possible values of $1^{1/q}$ will be equally spaced round the circle, center at the origin, and radius 1 (the *unit circle* as it is called). Their moduli will be 1 and their amplitudes will be multiples of $2\pi/q$. This is shown, for $q = 5$, in Fig. 13.9.

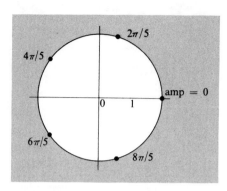

Fig. 13.9

Exercise 147

1. **(a)** Let us represent one of the complex cube roots of 1 by ω. Show that the other complex cube root is ω^2.

Note: The notation ω for a complex cube root of unity (either one) is standard.

(b) Prove that $1 + \omega + \omega^2 = 0$

(c) If ω_q is the complex qth root of unity given by

$$\omega_q = \cos\frac{2\pi}{q} + i\sin\frac{2\pi}{q},$$

show that the others are $\omega_q^2, \omega_q^3, \ldots, \omega_q^{q-1}$. Show also that

$$1 + \omega_q + \omega_q^2 \cdots + \omega_q^{q-1} = 0.$$

2. Prove that if n is a positive integer

$$(1 + i)^n + (1 - i)^n = 2^{n/2+1} \cos\frac{1}{4}n\pi.$$

From the binomial expansion of $(1 + x)^n$ deduce that

(a) $\binom{n}{0} - \binom{n}{2} + \binom{n}{4} - \cdots = 2^{n/2} \cos\frac{1}{4}n\pi$

(b) $\binom{n}{1} - \binom{n}{3} + \binom{n}{5} - \cdots = 2^{n/2} \sin\frac{1}{4}n\pi$

13.9 USES OF DE MOIVRE'S THEOREM

The uses of De Moivre's theorem are many and varied, and we shall meet the theorem again and again. But we might as well look at some of its uses right now.

Let us first find expressions for $\cos 3\theta$ and $\sin 3\theta$ in terms of $\cos \theta$ and $\sin \theta$.

By De Moivre's theorem, we have

$$\cos 3\theta + i \sin 3\theta = (\cos \theta + i \sin \theta)^3.$$

By the binomial theorem this is

$$\cos^3 \theta + 3i \cos^2 \theta \sin \theta + 3i^2 \cos \theta \sin^2 \theta + i^3 \sin^3 \theta$$
$$= (\cos^3 \theta - 3 \cos \theta \sin^2 \theta) + i(3 \cos^2 \theta \sin \theta - \sin^3 \theta).$$

Equating real and imaginary parts we obtain

$$\cos 3\theta = \cos^3 \theta - 3 \cos \theta \sin^2 \theta$$
$$\sin 3\theta = 3 \cos^2 \theta \sin \theta - \sin^3 \theta.$$

Using the fact that $\sin^2\theta = 1 - \cos^2\theta$, we can write the above equations as

$$\cos 3\theta = 4 \cos^3 \theta - 3 \cos \theta$$

and

$$\sin 3\theta = \sin \theta(4 \cos^2 \theta - 1).$$

This procedure can be generalized. To express $\cos n\theta$ and $\sin n\theta$ in terms of $\sin \theta$ and $\cos \theta$, we expand $(\cos \theta + i \sin \theta)^n$ by the binomial theorem. We get

$$\cos n\theta + i \sin n\theta = (\cos \theta + i \sin \theta)^n$$
$$= \cos^n \theta + \binom{n}{1} i \cos^{n-1} \theta \sin \theta + \cdots$$
$$+ i^n \sin^n \theta$$
$$= \sum_{r=0}^{n} \binom{n}{r} i^n \cos^{n-r} \theta \sin^n \theta.$$

Equating real and imaginary parts we obtain

(13.9.1) $\cos n\theta = \cos^n \theta - \binom{n}{2} \cos^{n-2} \theta \sin^2 \theta + \binom{n}{4} \cos^{n-4} \theta \sin^4 \theta \cdots$

(13.9.2) $\sin n\theta = \binom{n}{1} \cos^{n-1} \theta \sin \theta - \binom{n}{3} \cos^{n-3} \theta \sin^3 \theta + \cdots.$

Perhaps more useful still is the application of De Moivre's theorem to the reverse problem—that of expressing powers of $\cos \theta$ and $\sin \theta$ in terms of cosines and sines of multiples of θ. Let us put

$$z = \cos \theta + i \sin \theta$$
$$z^{-1} = \cos \theta - i \sin \theta.$$

Then $\qquad\qquad\qquad\qquad z + z^{-1} = 2 \cos \theta.$

Similarly

(13.9.3) $\qquad\qquad\qquad z^n + z^{-n} = 2 \cos n\theta.$

Now by the binomial theorem

(13.9.4) $\qquad (z + z^{-1})^n = z^n + \binom{n}{1} z^{n-1} \cdot z^{-1} + \binom{n}{2} z^{n-2} \cdot z^{-2} + \cdots + z^{-n}$

which can be rearranged as

$$(z + z^{-1})^n = (z^n + z^{-n}) + \binom{n}{1}(z^{n-2} + z^{-(n-2)})$$
$$+ \binom{n}{2}(z^{n-4} + z^{-(n-4)}) + \cdots$$

since $\binom{n}{r} = \binom{n}{n-r}$. Thus we have, from (13.9.4),

$$(2 \cos \theta)^n = 2 \cos n\theta + \binom{n}{1} 2 \cos(n-2)\theta + \cdots.$$

Dividing by 2 we find that

(13.9.5) $\qquad 2^{n-1} \cos^n \theta = \cos n\theta + \binom{n}{1} \cos(n-2)\theta$
$$+ \binom{n}{2} \cos(n-4)\theta + \cdots.$$

If n is an odd integer, then (13.9.5) will end with the term in $\cos \theta$, in which the coefficient will be $\binom{n}{\frac{1}{2}(n-1)}$. If n is even, however, there will be a constant term in (13.9.4), viz., $\binom{n}{\frac{1}{2}n}$ and this will be halved when we proceed to (13.9.5). Thus this final term will be half of a binomial coefficient. This is illustrated by the following expansions for $n = 3$, 4 and 5.

$$4 \cos^3 \theta = \cos 3\theta + 3 \cos \theta$$
(13.9.6) $\qquad 8 \cos^4 \theta = \cos 4\theta + 4 \cos 2\theta + 3$
$$16 \cos^5 \theta = \cos 5\theta + 5 \cos 3\theta + 10 \cos \theta.$$

Similar results can be obtained for powers of $\sin \theta$ in terms of sines and cosines of multiples of θ by use of the relation

$$z^n - z^{-n} = 2i \sin n\theta.$$

A complication arises, however, in that the required expression is in terms of sines if n is odd, and cosines if n is even. Alternatively, these results can be derived from those for $\cos^n \theta$ by using the relationship $\sin \theta = \cos(\pi/2 - \theta)$. Thus

$$16 \sin^5 \theta = 16 \cos^5\left(\frac{\pi}{2} - \theta\right)$$

$$= \cos 5\left(\frac{\pi}{2} - \theta\right) + 5 \cos 3\left(\frac{\pi}{2} - \theta\right) + 10 \cos\left(\frac{\pi}{2} - \theta\right)$$

$$= \sin 5\theta - 5 \sin 3\theta + 10 \sin \theta$$

$$8 \sin^4 \theta = 8 \cos^4\left(\frac{\pi}{2} - \theta\right)$$

$$= \cos 4\left(\frac{\pi}{2} - \theta\right) + 4 \cos 2\left(\frac{\pi}{2} - \theta\right) + 3$$

$$= \cos 4\theta - 4 \cos 2\theta + 3.$$

EXAMPLES:

We saw in Chapter 10 that the integrals of $\cos n\theta$ and $\sin n\theta$ are easily found. Expressions like (13.9.6) therefore enable us to evaluate the integrals of $\cos^n \theta$ and $\sin^n \theta$. Thus

$$\int \cos^5 \theta d\theta = \frac{1}{16} \int (\cos 5\theta + 5 \cos 3\theta + 10 \cos \theta) \, d\theta$$

$$= \frac{1}{16}\left[\frac{\sin 5\theta}{5} + \frac{5 \sin 3\theta}{3} + 10 \sin \theta\right] + C$$

and

$$\int \sin^4 \theta d\theta = \frac{1}{8} \int (\cos 4\theta - 4 \cos 2\theta + 3) d\theta$$

$$= \frac{1}{8}\left[\frac{\sin 4\theta}{4} - \frac{4 \sin 2\theta}{2} + 3\theta\right] + C.$$

Exercise 148

1. Find $\cos 5\theta$ and $\sin 5\theta$ in terms of $\cos \theta$ and $\sin \theta$.

2. Prove that for any positive integer n, $\cos n\theta$ and $\dfrac{\sin(n + 1)\theta}{\sin \theta}$ are polynomials in $\cos \theta$, of degree n, and find these polynomials when $n = 6$.

3. Find the values of

(a) $\displaystyle\int_0^{\pi/2} \sin^5 \theta \, d\theta$

(b) $\displaystyle\int_0^{\pi/2} \cos^4 \theta \, d\theta$

13.10 COMPLEX SERIES

In Chapter 8 we considered infinite series of real terms, and defined what we meant by the convergence of such a series and (when it converged) the sum of the series. We can do the same with series of complex terms, and the definitions and the general treatment are only slightly different.

Given a series

$$w_1 + w_2 + w_3 + \cdots + w_r + \cdots$$

of terms which are complex numbers we can define partial sums, as before, by

$$S_n = \sum_{r=1}^{n} w_r.$$

We now want to say that the infinite series converges if the sequence $\{S_n\}$ tends to a finite limit, but to do this we must define what we mean by the limit of a sequence $\{S_n\}$ of complex numbers. This definition, which is identical in written form to that for real sequences, is as follows:

Definition 13.3

The sequence $\{S_n\}$ of complex numbers is said to tend to the (complex) number a if, given any real number $\epsilon > 0$, there is an integer N such that

$$|S_n - a| < \epsilon \quad \text{for all } n > N.$$

Although this is virtually the same (except for minor notational differences) as Definition 8.2, there is an important conceptual difference in the interpretation of "$|S_n - a|$." For this is now the modulus of a complex number, not the absolute value of a real number. The interpretation on the Argand diagram is that the distance between the point $(P_n,$ say) representing the number S_n and the point $(A,$ say) representing the number a is less than ϵ, for $n > N$. Put another way, this means that all the points

$$P_{n+1}, P_{n+2}, P_{n+3}, \cdots$$

lie inside a circle, center A and radius ϵ.

EXAMPLE:

The sequence $\{S_n\}$, where

$$S_n = \frac{1}{1 + (\frac{1}{2}i)^n}$$

tends to the limit 1 as $n \to \infty$. For

(13.10.1)
$$|S_n - 1| = \left| 1 - \frac{1}{1 + (\frac{1}{2}i)^n} \right|$$
$$= \left| \frac{(\frac{1}{2}i)^n}{1 + (\frac{1}{2}i)^n} \right|$$
$$= \frac{(\frac{1}{2})^n}{|1 + (\frac{1}{2}i)^n|}, \quad \text{since } |\tfrac{1}{2}i| = \tfrac{1}{2}.$$

Now the point* $(\frac{1}{2}i)^n$ is never at a distance greater than $\frac{1}{2}$ from the origin, since $n \geqslant 1$, and hence never at a distance less than $\frac{1}{2}$ from the point 1. From this it is easily seen that

(13.10.2)
$$\left| 1 + \left(\frac{1}{2}i \right)^n \right| \geq \frac{1}{2} \quad \text{for } n \geq 1.$$

Hence from (13.10.1) and (13.10.2) we have

$$|S_n - 1| < \frac{(\frac{1}{2})^n}{\frac{1}{2}} = \left(\frac{1}{2} \right)^{n-1}.$$

It follows that since $(\frac{1}{2})^{n-1} \to 0$ as $n \to \infty$, so $|S_n - 1|$ can be made less than any given real number ϵ. The first few points of the Argand diagram for this sequence are shown in Fig. 13.10.

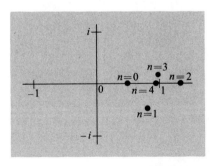

Fig. 13.10

* It will be convenient to use the phrase "the point z" when, strictly speaking, we should say "the point of the Argand diagram which represents the complex number z."

The handling of complex series is greatly simplified by the following theorems.

THEOREM 13.3.

IF $\sum w_r$ IS A SERIES OF COMPLEX TERMS, WHERE $w_r = u_r + iv_r$, THEN $\sum w_r$ CONVERGES IF, AND ONLY IF, $\sum u_r$ AND $\sum v_r$ BOTH CONVERGE.

PROOF:

If $\sum w_r$ converges then the sequence $\{S_n\}$ where

$$S_n = \sum_{r=1}^{n} w_r$$

$$= \sum_{r=1}^{n} (u_r + iv_r)$$

$$= \sum_{r=1}^{n} u_r + i \sum_{r=1}^{n} v_r$$

tends to some limit, say $\alpha + i\beta$. Hence, for any number $\epsilon > 0$, there is an integer N such that

(13.10.3) $|S_n - (\alpha + i\beta)| < \epsilon$ for all $n > N$,

or, equivalently, that

(13.10.4) $|(U_n - \alpha) + i(V_n - \beta)| < \epsilon$ for $n > N$

where $U_n = \sum_{r=1}^{n} u_r$ and $V_n = \sum_{r=1}^{n} v_r$.

From the modulus-amplitude form of a complex number, viz., from

$$x + iy = r \cos \theta + ir \sin \theta$$

we see that $|x| \leq r$ and $|y| \leq r$—the real and imaginary parts of a complex number are numerically not greater than the modulus. Hence from (13.10.4) we have

$$|U_n - \alpha| < \epsilon$$

and

$$|V_n - \beta| < \epsilon   \text{ for } n > N$$

since the modulus itself is $< \epsilon$. Since U_n and V_n are partial sums for $\sum u_r$ and $\sum v_r$ it follows that these two series converge.

Conversely, if $\sum u_r$ and $\sum v_r$ converge, to α and β say, then given any $\epsilon > 0$ we can find an integer N_1 such that

$$|U_n - \alpha| < \frac{1}{2}\epsilon \qquad \text{for } n > N_1,$$

and an integer N_2 such that

$$|V_n - \beta| < \frac{1}{2}\epsilon \qquad \text{for } n > N_2.$$

Thus if $n > \max(N_1, N_2)$ we have

$$\begin{aligned}
|S_n - (\alpha + i\beta)| &= |U_n + iV_n - (\alpha + i\beta)| \\
&= |(U_n - \alpha) + i(V_n - \beta)| \\
&\le |U_n - \alpha| + |V_n - \beta| \\
&< \frac{1}{2}\epsilon + \frac{1}{2}\epsilon \\
&= \epsilon.
\end{aligned}$$

Here we use the fact that $|x + iy| \le |x| + |y|$, which is a special case of Exercise 145.2. Hence the series $\sum w_n$ converges.

Another useful theorem asserts that if the series of moduli of the w_r converges (in which case the original series is said to be *absolutely convergent*, just as in Chapter 8), then the original series converges. In other words

THEOREM 13.4.

IF THE SERIES $\sum |w_r|$ OF REAL TERMS CONVERGES, THEN SO DOES THE COMPLEX SERIES $\sum w_r$.

PROOF:

Let $\sum |w_r|$ converge.

Then $0 \le |u_r| \le |w_r|$ as we saw above.

Hence the series $\sum |u_r|$ of non-negative terms is term for term less than the convergent series $\sum |w_r|$, and hence is convergent. Hence the series $\sum u_r$ is absolutely convergent, and hence is convergent. In exactly the same way we see that $\sum v_r$ is convergent. The previous theorem then shows that $\sum w_r$ is convergent.

These two theorems indicate that we can often handle the convergence of complex series by manipulating only *real* series, namely the series of real and imaginary parts, or the series of moduli. We shall now use these theorems to prove some results on complex power series.

Exercise 149

1. Prove that the sequence $\{z^n\}$ tends to zero if $|z| < 1$, and diverges if $|z| > 1$. What happens if $|z| = 1$?

2. Prove that if $\{u_n\}$ and $\{v_n\}$ are sequences of complex numbers, and $u_n \to a$ and $v_n \to b$, then $u_n + v_n \to a + b$ and $u_n v_n \to ab$.

3. Prove that the series $\sum \dfrac{z^n}{n!}$ is absolutely convergent, and hence is convergent, for all z.

13.11 COMPLEX POWER SERIES

A *complex power series* (or just *power series* if the presence of complex numbers is assumed) is a series of the form

$$(13.11.1) \qquad \sum a_n z^n = a_0 + a_1 z + a_2 z^2 + a_3 z^3 + \cdots + a_r z^r + \cdots$$

where z and the coefficients a_r are complex numbers.

The most important result concerning power series is the following:

THEOREM 13.5.

A POWER SERIES CONVERGES AT ALL POINTS (OF THE ARGAND PLANE) WITHIN A CERTAIN CIRCLE WITH ITS CENTER AT THE ORIGIN, AND DIVERGES FOR ALL POINTS OUTSIDE THAT CIRCLE.

A few observations are in order before we prove this result. The circle in question is called the *circle of convergence* and its radius R is the *radius of convergence*. Any power series converges when $z = 0$ (clearly). If this is the only point at which it converges we put $R = 0$ conventionally. Some series converge for *all* values of z. For these we write $R = \infty$. Note that the theorem says nothing about what happens *on* the circle of convergence—this is a much more difficult problem.

PROOF:

If $\sum a_n z^n$ converges only when $z = 0$ then $R = 0$ and the theorem is true. If not, then let z_1 be any nonzero complex value of z for which the series converges, i.e., $\sum a_n z_1^n$ converges. Let r_1 be the modulus of z_1.

Now if $\sum w_n$ converges then the individual terms w_n must become small (compare Theorem 8.1). For if the partial sums S_{n-1} and S_n tend to the same limit, their difference w_n must tend to 0.

Hence we know that $a_n z_1^n \to 0$ as $n \to \infty$, and we can therefore find a real number K such that

(13.11.2) $$|a_n z_1^n| < K \quad \text{for all } n.$$

Now consider any number z whose modulus r is $< r_1$. We have

$$|a_n z^n| = \left| a_n z_1^n \cdot \left(\frac{z}{z_1}\right)^n \right|$$

$$= |a_n z_1^n| \left(\frac{r}{r_1}\right)^n$$

$$< K\left(\frac{r}{r_1}\right)^n.$$

Thus $\sum |a_n z^n|$ is term for term less than the series $K \sum (r/r_1)^n$ which is a convergent geometric series (since $r < r_1$). Hence $\sum |a_n z^n|$ is convergent, (13.11.1) is absolutely convergent, and hence is convergent.

We have so far proved that if (13.11.1) is convergent at a point P it is convergent at any point nearer to the origin than P.

Now let $\mathscr{E}$ be the set of moduli of all values of z for which the series converges. $\mathscr{E}$ is a set of real numbers, and if $\mathscr{E}$ is unbounded above then whatever point Q we choose we can find a point P for which the modulus is larger and for which the series converges. Our result so far then shows that the series is convergent at Q. Hence it is convergent at every point of the plane, and $R = \infty$.

If $\mathscr{E}$ is bounded above, let R be its least upper bound. We have to show that if $|z| < R$ then $\sum a_n z^n$ is convergent. Since $|z|$ is not an upper bound of $\mathscr{E}$ there is an element of $\mathscr{E}$, say r_1, which exceeds $|z|$, and there is a value of z, having modulus r_1, for which the series converges (by the definition of $\mathscr{E}$). Hence our previous result shows that the series converges at the point z.

Hence the series converges at every point within the circle of radius R with center at the origin.

On the other hand if $|z| > R$ the series must diverge. For if it converged then $|z|$ would belong to $\mathscr{E}$ and R would not be an upper bound. This completes the proof of the theorem.

EXAMPLE 1:

The series

$$1 + \frac{z}{a} + \frac{z^2}{a^2} + \frac{z^3}{a^3} + \cdots$$

is absolutely convergent if $|z| < |a|$, and divergent if $|z| > |a|$. Hence the radius of convergence R is $|a|$.

EXAMPLE 2:

The series

$$z - \frac{1}{2}z^2 + \frac{1}{3}z^3 - \cdots.$$

Here we have

$$\left| \frac{w_{n+1}}{w_n} \right| = \left| -\frac{z^{n+1} \cdot n}{(n+1)z^n} \right|$$

$$= \left| \frac{n}{n+1} \right| \cdot |z|$$

$$< |z|.$$

Hence (by d'Alembert's test) the series is absolutely convergent, and hence convergent, for $|z| < 1$. If $|z| > 1$ then $\frac{n}{n+1}|z|$ will be > 1 for sufficiently large n, in which case the terms do not tend to zero and the series cannot converge. Hence $R = 1$.

EXAMPLE 3.

The series $\sum \frac{z^n}{n!}$ converges for all values of z. ($R = \infty$).

EXAMPLE 4:

The series $\sum n!z^n$ diverges except when $z = 0$. ($R = 0$).

The proofs of (3) and (4) will be left as an exercise to the reader.

Exercise 150

1. Show that the series

$$1 + \frac{\alpha \cdot \beta}{1!\gamma}z + \frac{\alpha(\alpha+1)\beta(\beta+1)}{2!\gamma(\gamma+1)}z^2 + \frac{\alpha(\alpha+1)(\alpha+2)\beta(\beta+1)(\beta+2)}{3!\gamma(\gamma+1)(\gamma+2)}z^3 + \cdots$$

where α, β, γ are constants, and γ is not a negative integer or zero, is absolutely convergent if $|z| < 1$, and divergent if $|z| > 1$.

This series, called the *hypergeometric series*, is a generalization of many well-known series.

2. What does the hypergeometric series of problem 1 become when
    (a) $\alpha = \beta = \gamma = 1$
    (b) $\alpha = 2$, $\beta = \gamma = 1$
    (c) $\alpha = -2$

3. Prove that the series

(a) $1 - \dfrac{z^2}{2!} + \dfrac{z^4}{4!} - \dfrac{z^6}{6!} + \cdots$

(b) $z - \dfrac{z^3}{3!} + \dfrac{z^5}{5!} - \dfrac{z^7}{7!} + \cdots$

are convergent for all values of z.

4. Find the radii of convergence of the following series:

(a) $\sum nz^n$

(d) $\sum \dfrac{(iz)^n}{n}$

(b) $\sum \dfrac{n^4 + i}{n - 1} z^{2n}$

(e) $\sum \dfrac{n!}{(2 + i)^n} z^n$

(c) $\sum (-1)^{n+1} \dfrac{z^n}{n}$

13.12 EXTENSIONS OF COMMON FUNCTIONS TO COMPLEX ARGUMENTS

We saw in Chapters 8 and 9 that the series

$$1 + x + \frac{x^2}{2!} + \frac{x^3}{3!} + \cdots$$

converges for all x, and defines a function which we now write as e^x. We shall extend the definition of this function to complex arguments as follows:

$$(13.12.1) \qquad e^z = 1 + z + \frac{z^2}{2!} + \frac{z^3}{3!} + \cdots + \frac{z^n}{n!} + \cdots$$

by definition. Since the series $\sum \dfrac{z^n}{n!}$ converges for all z, the function e^z is defined for all z.

If z is an imaginary number, say $z = i\theta$, then

$$e^{i\theta} = 1 + i\theta + \frac{(i\theta)^2}{2!} + \frac{(i\theta)^3}{3!} + \cdots$$
$$= \left(1 - \frac{\theta^2}{2!} + \frac{\theta^4}{4!} - \cdots\right) + i\left(\theta - \frac{\theta^3}{3!} + \frac{\theta^5}{5!} - \cdots\right)$$
$$= \cos\theta + i\sin\theta,$$

using the Taylor expansions of $\cos\theta$ and $\sin\theta$ given in Chapter 9. It follows from this, and the fact that any complex number can be written in modulus-amplitude form, that any complex number can be written as $re^{i\theta}$; r is the modulus, θ the amplitude.

For a general complex number $z = x + iy$ we define

$$e^{x+iy} = e^x \cdot e^{iy}$$
$$= r(\cos y + i \sin y)$$

where $r = e^x$. This gives e^z in modulus-amplitude form.

By way of justification for writing e^z in that form, i.e., as if it were "e to the power z", we may note that the first law of indices holds. If $z_1 = x_1 + iy_1$ and $z_2 = x_2 + iy_2$, then e^{z_1} and e^{z_2} are defined to be

$$e^{x_1}(\cos y_1 + i \sin y_1) \text{ and } e^{x_2}(\cos y_2 + i \sin y_2),$$

respectively, and their product is therefore

$$e^{x_1}e^{x_2} (\cos y_1 + i \sin y_1)(\cos y_2 + i \sin y_2)$$
$$= e^{x_1+x_2}[\cos (y_1 + y_2) + i \sin (y_1 + y_2)]$$

by (9.12.2) and De Moivre's Theorem. This is, by definition,

$$e^{x_1+x_2+i(y_1+y_2)} = e^{z_1+z_2}.$$

Hence

$$e^{z_1+z_2} = e^{z_1} \cdot e^{z_2}.$$

Note: We can use $\exp(z)$ to denote e^z where convenient (for example, when the argument of this function is a complicated expression).

This extension of the exponential function to complex numbers has several interesting consequences. For one thing De Moivre's theorem becomes

$$(e^{i\theta})^k = e^{i(k\theta)}$$

a special case of the second law of indices.

Another result concerns the solutions of the second order linear differential equation

$$ay'' + by' + cy = 0$$

when the roots of the auxiliary equation are not real. In Chapter 11 we tentatively examined the expressions $e^{(p+iq)x}$ and $e^{(p-iq)x}$. We now see that they are:

$$e^{px}(\cos qx + i \sin qx)$$

and

$$e^{px}(\cos qx - i \sin qx).$$

But if these functions are solutions of the equation, then so are their sum $2e^{px} \cos qx$ and difference $2ie^{px} \sin qx$. Thus $e^{px} \cos qx$ and $e^{px} \sin qx$ are solutions, and $e^{px}(A \cos qx + B \sin qx)$ is the general solution, just as before.

Yet another example of the use of the function e^z of a complex argument arises in the summation of (finite) trigonometric series. Let us consider the series

$$(13.12.2) \qquad C_n = 1 + \cos \theta + \cos 2\theta + \cos 3\theta + \cdots + \cos n\theta.$$

There are techniques for summing series of this kind without the use of complex numbers, but it is much easier with their help. We observe that since $\cos r\theta$ is the real part of $e^{ri\theta}$, therefore (13.12.2) is the real part of the series

$$(13.12.3) \qquad Z_n = 1 + e^{i\theta} + e^{2i\theta} + e^{3i\theta} + \cdots + e^{ni\theta}$$

a relation which we can write as $C_n = \mathscr{R}(Z_n)$, where $\mathscr{R}$ stands for "the real part of." But (13.12.3) is a geometric series with common ratio $e^{i\theta}$, and it is easily verified that the fact that the numbers in question are complex makes no difference to the method by which the sum is obtained. Hence we have

$$(13.12.4) \qquad Z_n = \frac{1 - e^{(n+1)i\theta}}{1 - e^{i\theta}}$$

We must now do some juggling to sort out the real and imaginary parts of this expression. We have

$$\frac{1 - e^{(n+1)i\theta}}{1 - e^{i\theta}} = \frac{1 - \cos(n+1)\theta - i \sin(n+1)\theta}{1 - \cos \theta - i \sin \theta}$$

$$= \frac{2 \sin^2\left(\frac{n+1}{2}\theta\right) - 2i \sin\left(\frac{n+1}{2}\theta\right) \cos\left(\frac{n+1}{2}\theta\right)}{2 \sin^2 \frac{\theta}{2} - 2i \sin \frac{\theta}{2} \cos \frac{\theta}{2}}$$

$$= \frac{\sin\left(\frac{n+1}{2}\theta\right)}{\sin \frac{\theta}{2}} \cdot \frac{\sin\left(\frac{n+1}{2}\theta\right) - i \cos\left(\frac{n+1}{2}\theta\right)}{\sin \frac{\theta}{2} - i \cos \frac{\theta}{2}}$$

$$= \frac{\sin\left(\frac{n+1}{2}\theta\right)}{\sin \frac{\theta}{2}} \cdot \frac{\cos\left(\frac{n+1}{2}\theta\right) + i \sin\left(\frac{n+1}{2}\theta\right)}{\cos \frac{\theta}{2} + i \sin \frac{\theta}{2}}$$

on multiplying top and bottom by i. Using De Moivre's theorem we find that Z_n reduces to

$$(13.12.5) \qquad \frac{\sin\left(\frac{n+1}{2}\theta\right)}{\sin \frac{\theta}{2}} \left[\cos \frac{n\theta}{2} + i \sin \frac{n\theta}{2} \right].$$

Hence the sum of the series (13.12.2) is

$$C_n = \sum_{r=0}^{n} \cos r\theta$$

$$= \frac{\sin\left(\dfrac{n+1}{2}\theta\right)}{\sin\dfrac{\theta}{2}} \cos \frac{n\theta}{2}.$$

Note that we can get another result with very little trouble. If S_n denotes the trigonometric series

$$\sin \theta + \sin 2\theta + \sin 3\theta + \cdots + \sin n\theta$$

then it is clear that S_n is the imaginary part of Z_n, which we can write as $S_n = \mathscr{I}(Z_n)$, where $\mathscr{I}$ means "the imaginary part of." Taking the imaginary part of (13.12.5) we find that

$$S_n = \frac{\sin\left(\dfrac{n+1}{2}\theta\right)}{\sin\dfrac{\theta}{2}} \sin \frac{n\theta}{2}.$$

Exercise 151

1. Show that $e^{i\pi} + 1 = 0$.
Note: This result, which links together the fundamental constants e and π, together with i and the integer 1, may well be regarded as one of the most beautiful equations in mathematics.
2. If $\alpha = 2\pi/n$, where n is a positive integer, show that

$$1 + 3 \cos \alpha + 5 \cos 2\alpha + \cdots + (2n - 1) \cos(n - 1)\alpha = n$$

and

$$3 \sin \alpha + 5 \sin 2\alpha + \cdots + (2n - 1) \sin(n - 1)\alpha = -n \cot \frac{\alpha}{2}.$$

3. Prove that

$$\sin \alpha - \sin(\alpha + \beta) + \sin(\alpha + 2\beta) - \cdots \pm \sin\{\alpha + (2n - 1)\beta\}$$
$$= -\sin n\beta \cos\left\{\alpha + \left(n - \frac{1}{2}\right)\beta\right\} \sec \frac{1}{2}\beta,$$

where α and β are real constants.

13.13 EXTENSIONS OF TRIGONOMETRIC AND OTHER FUNCTIONS TO COMPLEX ARGUMENTS

In the same way that we extended the definition of the function "exp" so as to apply to complex arguments, so we can similarly extend the definitions of other functions. Thus since we have

$$\cos x = 1 - \frac{x^2}{2!} + \frac{x^4}{4!} - \frac{x^6}{6!} + \cdots$$

it is natural to define $\cos z$, where z is a complex number, by the series

$$(13.13.1) \qquad \cos z = 1 - \frac{z^2}{2!} + \frac{z^4}{4!} - \frac{z^6}{6!} + \cdots$$

It is readily verified that this series converges for all values of z (thus $R = \infty$) and therefore defines the function for all z. If z is a real number then the definition coincides with our previous definition of $\cos x$. In the same way we define

$$(13.13.2) \qquad \sin z = z - \frac{z^3}{3!} + \frac{z^5}{5!} - \cdots.$$

We remarked in Chapter 10 on the similarity between the properties of the hyperbolic functions and of the trigonometric functions, but at that time there was no obvious reason why these two sets of functions—defined in quite different ways—should have anything much in common. We are now in a position to see what the connection is. If z is an imaginary number, say $z = iy$, then (13.13.1) becomes

$$\cos iy = 1 - \frac{i^2 y^2}{2!} + \frac{i^4 y^4}{4!} - \frac{i^6 y^6}{6!} + \cdots$$

$$= 1 + \frac{y^2}{2!} + \frac{y^4}{4!} + \frac{y^6}{6!} + \cdots$$

But this is exactly the series for $\cosh y$. For, by definition,

$$\cosh y = \frac{1}{2}(e^y + e^{-y})$$

$$= \frac{1}{2}\left(1 + y + \frac{y^2}{2!} + \frac{y^3}{3!} + \frac{y^4}{4!} + \cdots\right)$$

$$+ \frac{1}{2}\left(1 - y + \frac{y^2}{2!} - \frac{y^3}{3!} + \frac{y^4}{4!} \cdots\right)$$

$$= 1 + \frac{y^2}{2!} + \frac{y^4}{4!} + \cdots.$$

Hence we have the result

(13.13.3) $\cos iy = \cosh y.$

In the same way, since

$$\sin iy = iy - \frac{i^3 y^3}{3!} + \frac{i^5 y^5}{5!} - \frac{i^7 y^7}{7!} + \cdots$$

$$= iy + \frac{iy^3}{3!} + \frac{iy^5}{5!} + \frac{iy^7}{7!} + \cdots$$

and

$$\sinh y = \frac{1}{2}(e^y - e^{-y})$$

$$= \frac{1}{2}\left(1 + y + \frac{y^2}{2!} + \frac{y^3}{3!} + \frac{y^4}{4!} + \cdots\right)$$

$$- \frac{1}{2}\left(1 - y + \frac{y^2}{2!} - \frac{y^3}{3!} + \frac{y^4}{4!} - \cdots\right)$$

$$= y + \frac{y^3}{3!} + \frac{y^5}{5!} + \cdots$$

we have

(13.13.4) $\sin iy = i \sinh y.$

Using these two results we can obtain directly the corresponding results for trigonometric and hyperbolic functions, each from the other. Thus from

$$\cosh^2 y - \sinh^2 y = 1$$

we obtain

$$\cos^2 iy - \left(\frac{\sin iy}{i}\right)^2 = 1$$

which gives

$$\cos^2 iy + \sin^2 iy = 1$$

which shows that the result $\cos^2 \theta + \sin^2 \theta = 1$ holds for imaginary values of θ.

We can write

$$e^{iz} = 1 + iz + \frac{i^2 z^2}{2!} + \frac{i^3 z^3}{3!} + \cdots$$

and

$$e^{-iz} = 1 - iz + \frac{i^2 z^2}{2!} - \frac{i^3 z^3}{3!} + \cdots$$

whence, by addition, we obtain

$$e^{iz} + e^{-iz} = 2\left\{1 - \frac{z^2}{2!} + \frac{z^4}{4!} - \cdots\right\}$$

$$= 2\cos z,$$

and by subtraction

$$e^{iz} - e^{-iz} = 2i\left\{z - \frac{z^3}{3!} + \frac{z^5}{5!} - \cdots\right\}$$

$$= 2i\sin z.$$

Thus we see that

(13.13.5) $$\cos z = \frac{1}{2}(e^{iz} + e^{-iz})$$

(13.13.6) $$\sin z = \frac{1}{2i}(e^{iz} - e^{-iz})$$

and we have the functions $\cos z$ and $\sin z$ defined in terms of the exponential function.

THE FUNCTION "LOGARITHM" AGAIN

Suppose $z = re^{i\theta}$, $w = u + iv$ and $z = e^w$. Then

$$re^{i\theta} = e^{u+iv}$$

$$= e^u \cdot e^{iv}$$

$$= e^u(\cos v + i\sin v).$$

Hence $r = e^u$ and $v = \theta + 2k\pi$. Note that θ and v may not be equal; all we can say is that they differ by a multiple of 2π. Now, if we write $z = e^w$ then it is natural to write $w = \log z$. Hence

$$\log z = u + iv$$

$$= \log r + i(\theta + 2k\pi).$$

Thus we have extended the definition of the function *log* to complex arguments. Clearly it is a multivalued function. We can make it into a single valued function by taking that value whose amplitude lies between $-\pi$ and π; this is called the *principal value* of the logarithm function.* If the number z is real and positive then $y = 0$ and the principal value is $\log r$; that is, it coincides with the value of the "log" function as previously defined. If z is real, but negative, then the amplitude of z is π, and we have

$$\log(-a) = \log a + i\pi$$

where a is a positive real number. In the particular case $a = 1$, we have

$$\log(-1) = i\pi$$

which is an alternative form of the equation $e^{i\pi} = -1$.

Some examples will show how these functions, defined in this wider sense, can be manipulated.

EXAMPLE 1:

Show that

$$\sin(z_1 + z_2) = \sin z_1 \cos z_2 + \cos z_1 \sin z_2.$$

The right-hand side of this equation is, by definition,

$$\frac{1}{2i}(e^{iz_1} - e^{-iz_1}) \cdot \frac{1}{2}(e^{iz_2} + e^{-iz_2}) + \frac{1}{2}(e^{iz_1} + e^{-iz_1}) \cdot \frac{1}{2i}(e^{iz_2} - e^{-iz_2})$$

$$= \frac{1}{4i}\{e^{i(z_1+z_2)} - e^{-iz_1+iz_2} + e^{iz_1-iz_2} - e^{-i(z_1+z_2)}\}$$

$$+ \frac{1}{4i}\{e^{i(z_1+z_2)} + e^{-iz_1+iz_2} - e^{iz_1-iz_2} + e^{-i(z_1+z_2)}\}$$

$$= \frac{1}{2i}\{e^{i(z_1+z_2)} - e^{-i(z_1+z_2)}\}$$

which is the left-hand side. Hence the result.

In a similar way we can show that all the usual trigonometric identities hold just as well for complex numbers.

EXAMPLE 2:

Express $\sin(x + iy)$, where x and y are real, in the form $u + iv$, and show that

$$|\sin(x + iy)| = \{\cosh^2 y - \cos^2 x\}^{1/2}.$$

---

* Very often "Log z" is used for the multivalued function, "log z" being reserved for the single-valued function.

From the previous example

$$\sin(x + iy) = \sin x \cos iy + \cos x \sin iy$$
$$= \sin x \cosh y + i \cos x \sinh y.$$

Therefore

$$|\sin(x + iy)|^2 = \sin^2 x \cosh^2 y + \cos^2 x \sinh^2 y$$
$$= (1 - \cos^2 x) \cosh^2 y - \cos^2 x(\cosh^2 y - 1)$$
$$= \cosh^2 y - \cos^2 x$$

which gives the required result.

EXAMPLE 3:

Find the general solution of the equation $\sin z = 3i \cos z$.

From the definitions of $\sin z$ and $\cos z$ we have

$$\frac{1}{2i}(e^z - e^{-z}) = 3i \cdot \frac{1}{2}(e^z + e^{-z}).$$

Thus, multiplying both sides by $2ie^z$, we obtain

$$e^{2z} - 1 = -3(e^{2z} + 1)$$

which yields

$$4e^{2z} = -2$$

or

$$e^{2z} = -\frac{1}{2}.$$

Now $-\frac{1}{2} = \frac{1}{2}(\cos \theta + i \sin \theta)$, where $\theta = \pi + 2k\pi$, and therefore

$$2z = \log\left(-\frac{1}{2}\right)$$
$$= \log \frac{1}{2} + i(2k + 1)\pi.$$

Thus the general solution is

$$z = -\frac{1}{2} \log 2 + i(2k + 1)\frac{\pi}{2}.$$

Exercise 152

1. Express $\tan z$ and $\sec z$ in the form $u + iv$, where $z = x + iy$. [Define $\tan z = \sin z/\cos z$, and $\sec z = 1/\cos z$].

2. Express $\sinh z$, $\cosh z$ and $\tanh z$ in the form $u + iv$.

3. If $\sin(x + iy) = e^{i\pi/3}$, prove that

$$x = \cos^{-1}\left(\frac{3}{4}\right)^{1/4} \quad \text{and} \quad y = \sinh^{-1}\left(\frac{3}{4}\right)^{1/4}.$$

4. If $u + iv = \coth(x + iy)$, show that

$$v = \frac{-\sin 2y}{\cosh 2x - \cos 2y}$$

5. The definition of $\log z$ enables us to define what is meant by raising a complex number to a complex power. For we define

$$w^z \text{ to mean } e^{z \log w}.$$

Prove that if $z = x + iy$ and $w = r(\cos \theta + i \sin \theta)$ then one value of w^z is

$$\exp{(x \log r - y\theta)}\{\cos(y \log r + x\theta) + i \sin(y \log r + x\theta)\}.$$

6. Prove that $i^i = e^{-\pi/2}$ [a real number(!)].

7. Express the following complex numbers in the form $u + iv$:

   (a) $(1 + i)^{2i}$ (c) $\left(\dfrac{2 - i}{2 + i}\right)^{-i/2}$

   (b) $i^{(1+i)}$ (d) $\left(\dfrac{1 + i}{1 - i}\right)^{2-3i} + \left(\dfrac{1 - i}{1 + i}\right)^{2+3i}$

8. Prove the laws of indices in their most general form, viz., that if z_1, z_2 and z_3 are any three complex numbers, then

   (a) $z_1^{z_2} \cdot z_1^{z_3} = z_1^{z_2 + z_3}$

   (b) $(z_1^{z_2})^{z_3} = z_1^{z_2 z_3}$

Exercise 153 (Chapter Review)

1. Express the following complex numbers in the form $a + ib$.

   (a) $\dfrac{2 + 2i}{1 - i}$ (c) $\dfrac{10 + 20i}{3 - i}$

   (b) $\dfrac{2 + 11i}{2 + i} + \dfrac{17}{4 + i}$ (d) $\left(\dfrac{2 - 3i}{3 + i} + \dfrac{5}{1 - i}\right) \cdot \dfrac{2 - 3i}{3 - 2i}$

2. *ABC* is an isosceles triangle in the Argand plane, and the equal sides are *AB* and *AC*. The ratio of the lengths *BC* and *AB* is 2: 3. If *B* represents the complex number $3 + i$ and *C* represents $1 - 3i$, find the complex number which is represented by *A*, given that *A* and the origin are on opposite sides of the line *BC*.

3. If θ is real, and *n* is a positive integer, prove that

$$(\cos \theta + i \sin \theta)^n (\sin \theta + i \cos \theta)^n = e^{in\pi/2}.$$

Express the three values of $(1 + \sqrt{3}\, i)^{1/3}(\sqrt{3} + i)^{1/3}$ in the form $a + ib$, where *a* and *b* are real, and the positive value of $\sqrt{3}$ is taken.

4. Indicate on the Argand diagram the positions of the points $z = 1 + \sin \theta \pm i \cos \theta$ for a given value of θ.

Hence, or otherwise, prove that one value of

$$\left(\frac{1 + \sin \theta + i \cos \theta}{1 + \sin \theta - i \cos \theta}\right)^k$$

is equal to $\cos k(\pi/2 - \theta) + i \sin k(\pi/2 - \theta)$.

Find all the values of

$$\left(\frac{\sqrt{2} + 1 + i}{\sqrt{2} + 1 - i}\right)^{1/4}$$

in the form $a + ib$, where *a* and *b* are real.

5. The simultaneous differential equations

$$\frac{dx}{dt} + a \cos \alpha \cdot x + a(1 - \sin \alpha)y = 0$$

$$\frac{dy}{dt} - a(1 - \sin \alpha)x + a \cos \alpha \cdot y = 0,$$

where α and *a* are constants, are linear, and therefore could be solved by the methods of section 11.13. Show that by adding *i* times the second equation to the first, these two equations can be expressed by the single equation

$$\frac{dz}{dt} + [a(\cos \alpha + i \sin \alpha) - ia]\, z = 0$$

where $z = x + iy$. By solving this as a Type III equation (as in section 11.5) show that

$$z = (A + iB)e^{-a \cos \alpha t}[\cos \theta + i \sin \theta]$$

where $\theta = a(1 - \sin \alpha)t$. *Hint:* Remember that the arbitrary constant which appears must be a complex number, say $A + iB$.

Hence find the functions $x(t)$ and $y(t)$.

6. Second order linear equations with constant coefficients in which the function on the right-hand side involves a sine or cosine can often be solved by regarding the right-hand side as the real or imaginary part of a complex expression. Thus to solve the equation

$$\frac{d^2x}{dt^2} + 5\frac{dx}{dt} + 6x = \cos 2t$$

we can consider the equation

$$\frac{d^2x}{dt^2} + 5\frac{dx}{dt} + 6x = e^{2it}$$

and take the real part of the solution.

Find the complementary function for this equation. Then show that a particular integral is given by

$$\frac{1}{D^2 + 5D + 6}e^{2it}.$$

Use the rule for moving the e^{2it} over to the other side of the operator (see section 11.10) to obtain a particular integral. Hence find the general solution of the original equation.

In the same way find the general solution of the equation

$$\frac{d^2x}{dt^2} + 5\frac{dx}{dt} + 6x = \sin 2t.$$

7. Solve the equations

(a) $\dfrac{d^2y}{dx^2} + 4\dfrac{dy}{dx} + 4y = 3 \sin x + 5 \cos x$

(b) $\dfrac{d^2y}{dx^2} + 9y = 5 \cos 3x$

8. If ω is a cube root of 1, show that
$(x + \omega y + \omega^2 z)(x + \omega^2 y + \omega z) = x^2 + y^2 + z^2 - yz - zx - xy$.

9. If a, b, u, v, x and y are real, and

$$x + iy = a \cos(u + iv) + ib \sin(u + iv)$$

prove that

$$x^2 \sec^2 u - y^2 \operatorname{cosec}^2 u = a^2 - b^2.$$

10. Prove that $\int_0^{\pi/2} \cos^n \theta \, d\theta$ is zero if n is even, and that if n is odd it is

$$(-1)^{n-1/2}\left(\frac{1}{2}\right)^{n-1}\left\{\binom{n}{0}\frac{1}{n} - \binom{n}{1}\frac{1}{n-2} + \binom{n}{2}\frac{1}{n-4} - \cdots \pm \left(\tfrac{1}{2}(n-1)\right)\right\}.$$

11. Sum the infinite series

(a) $1 - \dfrac{1}{2!}\cos 2\theta + \dfrac{1}{4!}\cos 4\theta - \dfrac{1}{6!}\cos 6\theta + \cdots$

(b) $\dfrac{1}{2!}\sin 2\theta - \dfrac{1}{4!}\sin 4\theta + \dfrac{1}{6!}\sin 6\theta - \cdots$

12. Express the function $\cos\{\cos(x + iy)\}$ in the form $u + iv$, where x, y, u, v are real numbers.

13. If $\cos^{-1}(a + ib) = p + iq$ show that $\cos^2 p$ and $\cosh^2 q$ are the roots of the equation

$$x^2 - (1 + a^2 + b^2)x + a^2 = 0.$$

Show also that if

$$\sin^{-1}(a + ib) = P + iQ$$

then $\sin^2 p$ and $\cosh^2 Q$ are the roots of the same equation.

Suggestions for Further Reading

The subject of complex numbers is not one on which whole books tend to be written —at least not at the level of the present chapter. Therefore the reader who wishes to supplement his reading will have to rely on textbooks whose primary concern is some other branch of mathematics, and which deal with complex numbers as a means to an end, rather than a subject in its own right.

There are many such books. The books listed in the suggested reading for chapters 8 and 9 deal with complex numbers to a greater or less extent and can be consulted on the subject. In addition there is the book cited below, whose main concern is with the subject matter of the *next* chapter, but which contains a good treatment of complex numbers from a fairly elementary standpoint.

Archbold, J.W., *Algebra*. London: Sir Isaac Pitman & Sons, Ltd., 1964.

14   Linear Algebra

In Chapter 7 we met the idea of a *vector*. In this chapter we shall have further dealings with vectors and meet a new kind of mathematical object known as a *matrix*. First we must deal with some fundamental concepts and notations.

Suppose we have a vector (x_1, x_2, x_3) of dimension 3, and we define another vector (y_1, y_2, y_3) by means of the equations

$$y_1 = a_1 x_1 + b_1 x_2 + c_1 x_3$$
(14.1.1) $\qquad y_2 = a_2 x_1 + b_2 x_2 + c_2 x_3$
$$y_3 = a_3 x_1 + b_3 x_2 + c_3 x_3$$

where all the quantities in these equations are real numbers. Equations (14.1.1) determine a unique vector (y_1, y_2, y_3) corresponding to any given vector (x_1, x_2, x_3). In other words we have defined a function (in the sense of Chapter 3) whose domain is the set of three-dimensional vectors over the real numbers, and whose converse domain is the same set of vectors. Such a function is usually called a *linear transformation* ("linear" since y_1, y_2 and y_3

are linear functions of x_1, x_2 and x_3). By way of introducing some of the notation needed in this chapter we shall see how to write equations (14.1.1) in a more compact form.

Our first action is to introduce a convention whereby we write vectors in one or other of two different ways. We have been accustomed to writing a vector as, for example, (x_1, x_2, x_3), and when written in this way a vector will be called a *row vector*. From now on, however, we shall have at our disposal another way of writing a vector, when we want to, which is to display it as a column, thus:

$$\begin{pmatrix} x_1 \\ x_2 \\ x_3 \end{pmatrix}$$

When written in this way a vector will be known as a *column vector*. There is no essential difference between a row vector and a column vector—they are just alternative ways of denoting the same thing; but we shall see that by writing vectors sometimes one way, sometimes the other, according to their function in the expressions in which they occur—and according to certain rules which we shall lay down—the understanding and manipulation of vector expressions becomes much more straightforward.

Next we have a rule for multiplying a vector (either a row vector or a column vector) by a scalar. This is essentially the same as we had before. The product of a scalar λ and the column vector

$$\begin{pmatrix} x_1 \\ x_2 \\ x_3 \end{pmatrix} \quad \text{is the column vector} \quad \begin{pmatrix} \lambda x_1 \\ \lambda x_2 \\ \lambda x_3 \end{pmatrix}$$

The product of λ with the row vector (x_1, x_2, x_3) is $(\lambda x_1, \lambda x_2, \lambda x_3)$. The extension to vectors of any dimension is obvious.

Next comes the definition of the product of a row vector by a column vector (and the order is important). We define the product of the row-vector (a_1, b_1, c_1) by the column vector $\begin{pmatrix} x_1 \\ x_2 \\ x_3 \end{pmatrix}$ to be

$$(14.1.2) \qquad (a_1, b_1, c_1)\begin{pmatrix} x_1 \\ x_2 \\ x_3 \end{pmatrix} = a_1 x_1 + b_1 x_2 + c_1 x_3.$$

The result is thus a scalar quantity—a real number.

This definition of the product of a row vector and column vector (in that order) is the fundamental definition on which everything else in the first part of this chapter will be based. Using (14.1.2) we can rewrite (14.1.1) as

$$(14.1.3) \qquad \begin{aligned} y_1 &= (a_1, b_1, c_1)\begin{pmatrix} x_1 \\ x_2 \\ x_3 \end{pmatrix} \\[1em] y_2 &= (a_2, b_2, c_2)\begin{pmatrix} x_1 \\ x_2 \\ x_3 \end{pmatrix} \\[1em] y_3 &= (a_3, b_3, c_3)\begin{pmatrix} x_1 \\ x_2 \\ x_3 \end{pmatrix} \end{aligned}$$

Now the left-hand sides of the equations (14.1.3) are the components of a vector, and the temptation is strong to put them together as a vector, rather than leaving them as three separate items. If we do so, it is reasonable to write them as a column-vector $\begin{pmatrix} y_1 \\ y_2 \\ y_3 \end{pmatrix}$, since $\begin{pmatrix} x_1 \\ x_2 \\ x_3 \end{pmatrix}$ is also a column vector; but what

shall we do about the right-hand sides? On the right-hand sides of (14.1.3)

the column vector $\begin{pmatrix} x_1 \\ x_2 \\ x_3 \end{pmatrix}$ is common, so that with any luck we shall be able to

get away with writing it only once. The three row vectors that multiply the common column vector must, of course, all be shown, and by analogy with what we are going to do with the left-hand sides, it would seem reasonable to write these row vectors one under the other. In this way we are led to simplify (14.1.3) still further and to write

$$(14.1.4) \qquad \begin{pmatrix} y_1 \\ y_2 \\ y_3 \end{pmatrix} = \begin{pmatrix} a_1 & b_1 & c_1 \\ a_2 & b_2 & c_2 \\ a_3 & b_3 & c_3 \end{pmatrix} \begin{pmatrix} x_1 \\ x_2 \\ x_3 \end{pmatrix}.$$

The square array of numbers

$$\begin{pmatrix} a_1 & b_1 & c_1 \\ a_2 & b_2 & c_2 \\ a_3 & b_3 & c_3 \end{pmatrix}$$

is called a *matrix*. We shall shortly give some formal rules for manipulating matrices and vectors, but for the moment we shall simply regard (14.1.4) as being an abbreviated way of writing (14.1.1) and press on to consider what happens if we follow the linear transformation (14.1.1) by another.

If we transform the vector $\begin{pmatrix} y_1 \\ y_2 \\ y_3 \end{pmatrix}$ into another vector $\begin{pmatrix} z_1 \\ z_2 \\ z_3 \end{pmatrix}$ by means of the

equations

$$(14.1.5) \qquad \begin{aligned} z_1 &= \alpha_1 y_1 + \beta_1 y_2 + \gamma_1 y_3 \\ z_2 &= \alpha_2 y_1 + \beta_2 y_2 + \gamma_2 y_3 \\ z_3 &= \alpha_3 y_1 + \beta_3 y_2 + \gamma_3 y_3 \end{aligned}$$

then with our condensed matrix notation we can write

$$(14.1.6) \qquad \begin{pmatrix} z_1 \\ z_2 \\ z_3 \end{pmatrix} = \begin{pmatrix} \alpha_1 & \beta_1 & \gamma_1 \\ \alpha_2 & \beta_2 & \gamma_2 \\ \alpha_3 & \beta_3 & \gamma_3 \end{pmatrix} \begin{pmatrix} y_1 \\ y_2 \\ y_3 \end{pmatrix}$$

Suppose we start with the column vector $\begin{pmatrix} x_1 \\ x_2 \\ x_3 \end{pmatrix}$ and transform it into the

vector $\begin{pmatrix} y_1 \\ y_2 \\ y_3 \end{pmatrix}$ by (14.1.4); and then transform this vector into $\begin{pmatrix} z_1 \\ z_2 \\ z_3 \end{pmatrix}$ by (14.1.5).

Then we shall have obtained $\begin{pmatrix} z_1 \\ z_2 \\ z_3 \end{pmatrix}$ as a function of $\begin{pmatrix} x_1 \\ x_2 \\ x_3 \end{pmatrix}$. What is then the

relation between $\begin{pmatrix} z_1 \\ z_2 \\ z_3 \end{pmatrix}$ and $\begin{pmatrix} x_1 \\ x_2 \\ x_3 \end{pmatrix}$ and, in particular, what are the equations

which connect them? If we actually make the substitutions for the z_i in terms
of the y_i, and the y_i in terms of the x_i ($i = 1, 2, 3$) we get

$$
\begin{aligned}
z_1 &= \alpha_1(a_1 x_1 + b_1 x_2 + c_1 x_3) + \beta_1(a_2 x_1 + b_2 x_2 + c_2 x_3) \\
&\quad + \gamma_1(a_3 x_1 + b_3 x_2 + c_3 x_3) \\
(14.1.7) \quad z_2 &= \alpha_2(a_1 x_1 + b_1 x_2 + c_1 x_3) + \beta_2(a_2 x_1 + b_2 x_2 + c_2 x_3) \\
&\quad + \gamma_2(a_3 x_1 + b_3 x_2 + c_3 x_3) \\
z_3 &= \alpha_3(a_1 x_1 + b_1 x_2 + c_1 x_3) + \beta_3(a_2 x_1 + b_2 x_2 + c_2 x_3) \\
&\quad + \gamma_3(a_3 x_1 + b_3 x_2 + c_3 x_3)
\end{aligned}
$$

or

$$
\begin{aligned}
z_1 &= (\alpha_1 a_1 + \beta_1 a_2 + \gamma_1 a_3)x_1 + (\alpha_1 b_1 + \beta_1 b_2 + \gamma_1 b_3)x_2 \\
&\quad + (\alpha_1 c_1 + \beta_1 c_2 + \gamma_1 c_3)x_3 \\
(14.1.8) \quad z_2 &= (\alpha_2 a_1 + \beta_2 a_2 + \gamma_2 a_3)x_1 + (\alpha_2 b_1 + \beta_2 b_2 + \gamma_2 b_3)x_2 \\
&\quad + (\alpha_2 c_1 + \beta_2 c_2 + \gamma_2 c_3)x_3 \\
z_3 &= (\alpha_3 a_1 + \beta_3 a_2 + \gamma_3 a_3)x_1 + (\alpha_3 b_1 + \beta_3 b_2 + \gamma_3 b_3)x_2 \\
&\quad + (\alpha_3 c_1 + \beta_3 c_2 + \gamma_3 c_3)x_3.
\end{aligned}
$$

In other words we have a linear relation.

A specific example may be useful here and will be convenient to refer to
later on. If

$$
\begin{aligned}
(14.1.9) \quad y_1 &= 2x_1 + 3x_2 - x_3 \\
y_2 &= x_1 - 2x_2 + x_3 \\
y_3 &= 5x_1 - x_2 - 4x_3
\end{aligned}
$$

Linear Algebra

and

(14.1.10)
$$z_1 = y_1 - y_2 + 2y_3$$
$$z_2 = 3y_1 + y_2 + y_3$$
$$z_3 = -y_1 + 4y_2 - 3y_3$$

then

(14.1.11)
$$z_1 = 11x_1 + 3x_2 - 10x_3$$
$$z_2 = 12x_1 + 6x_2 - 6x_3$$
$$z_3 = -13x_1 - 8x_2 + 17x_3$$

as the reader should verify.

The notation of (14.1.4) suggests that we could describe the transformation given there as a sort of "multiplication" (whatever that may turn out to mean) of the matrix in (14.1.4) by the column vector $\begin{pmatrix} x_1 \\ x_2 \\ x_3 \end{pmatrix}$ to obtain the vector

$\begin{pmatrix} y_1 \\ y_2 \\ y_3 \end{pmatrix}$. Of course, we have not yet defined any such "multiplication" between matrices and vectors, and how we do it is entirely up to us. Since we want (14.1.1.) and (14.1.4) to represent the same set of equations, the sensible way to define the product "matrix times column vector" is by means of the equivalence of (14.1.1) and (14.1.4). In other words we define the right-hand side of (14.1.4) to mean the column vector composed of the right-hand sides of (14.1.1). Thus we can say that the product of a 3 × 3 matrix (i.e., one having 3 rows and 3 columns) and a column vector of dimension 3 (taken in that order) is a column vector whose ith element ($i = 1, 2, 3$) is the product of the ith row of the matrix and the column vector. Since we have already defined the product of a row vector and a column vector, this definition is a meaningful one.

To make life simpler, let us denote our matrices and vectors by single symbols when we do not wish to display their individual elements. It is customary to use boldface type for these symbols. Thus we write (14.1.4) as

$$\mathbf{y} = \mathbf{A}\mathbf{x}$$

where $\mathbf{y}$ and $\mathbf{x}$ stand for the column vectors in (14.1.4) and $\mathbf{A}$ stands for the matrix. The definition of the product of a matrix and a column vector can then be generalized as follows:

The product of an $n \times n$ matrix $\mathbf{A}$ and a column vector $\mathbf{x}$ of dimension n is a column vector $\mathbf{y}$ whose ith element $(i = 1, 2, \ldots, n)$ is the product of the ith row of $\mathbf{A}$ and the vector $\mathbf{x}$.

If we write equations (14.1.1) as

$$(14.1.12) \qquad\qquad \mathbf{y} = \mathbf{Ax}$$

then, by the same token, we can write equation (14.1.6) as

$$(14.1.13) \qquad\qquad \mathbf{z} = \mathbf{By}$$

where $\mathbf{B}$ is the matrix in (14.1.6). By analogy with the way that ordinary algebraic quantities behave we are naturally led to deduce from (14.1.12) and (14.1.13) that

$$(14.1.14) \qquad\qquad \mathbf{z} = \mathbf{BAx}$$

in which we have what looks like the product of two matrices. No such product has yet been defined, but (14.1.14) gives a natural way of defining it, since it suggests that the product, $\mathbf{BA}$, is the matrix which connects $\mathbf{z}$ and $\mathbf{x}$. But the equations (14.1.8) can be written as

$$\mathbf{z} = \mathbf{Cx}$$

where $\mathbf{C}$ is the matrix

$$
\begin{pmatrix}
\alpha_1 a_1 + \beta_1 a_2 + \gamma_1 a_3 & \alpha_1 b_1 + \beta_1 b_2 + \gamma_1 b_3 & \alpha_1 c_1 + \beta_1 c_2 + \gamma_1 c_3 \\
\alpha_2 a_1 + \beta_2 a_2 + \gamma_2 a_3 & \alpha_2 b_1 + \beta_2 b_2 + \gamma_2 b_3 & \alpha_2 c_1 + \beta_2 c_2 + \gamma_2 c_3 \\
\alpha_3 a_1 + \beta_3 a_2 + \gamma_3 a_3 & \alpha_3 b_1 + \beta_3 b_2 + \gamma_3 b_3 & \alpha_3 c_1 + \beta_3 c_2 + \gamma_3 c_3
\end{pmatrix}
$$

—the matrix of the coefficients in (14.1.8).

This shows that although the definition of the product of two matrices is at our disposal, the reasonable course is to define the product $\mathbf{BA}$ as being the matrix $\mathbf{C}$. In other words, the product of any two 3×3 matrices will be defined by the equation

$$
\begin{pmatrix}
\alpha_1 & \beta_1 & \gamma_1 \\
\alpha_2 & \beta_2 & \gamma_2 \\
\alpha_3 & \beta_3 & \gamma_3
\end{pmatrix}
\begin{pmatrix}
a_1 & b_1 & c_1 \\
a_2 & b_2 & c_2 \\
a_3 & b_3 & c_3
\end{pmatrix}
$$

$$
= \begin{pmatrix}
\alpha_1 a_1 + \beta_1 a_2 + \gamma_1 a_3 & \alpha_1 b_1 + \beta_1 b_2 + \gamma_1 b_3 & \alpha_1 c_1 + \beta_1 c_2 + \gamma_1 c_3 \\
\alpha_2 a_1 + \beta_2 a_2 + \gamma_2 a_3 & \alpha_2 b_1 + \beta_2 b_2 + \gamma_2 b_3 & \alpha_2 c_1 + \beta_2 c_2 + \gamma_2 c_3 \\
\alpha_3 a_1 + \beta_3 a_2 + \gamma_3 a_3 & \alpha_3 b_1 + \beta_3 b_2 + \gamma_3 b_3 & \alpha_3 c_1 + \beta_3 c_2 + \gamma_3 c_3
\end{pmatrix}
$$

The elements of the matrix on the right-hand side of (14.1.15) may seem to be rather complicated, but by an improvement in our notation we can make the formation of these elements readily apparent and extend our definition of "product" beyond what we have had up to now.

To avoid undue complication in the early stages we have used different letters of the alphabet for the columns of a matrix—as in (14.1.4)—and have used subscripts only to distinguish the different rows. But since a matrix is a two-dimensional array, a more logical procedure would be to use a *double* subscript notation—the first subscript to denote the row, the other to denote the column. This is what we shall usually do from now on. We shall denote the element in the ith row and jth column of a matrix by say, a_{ij} (any other symbol could be used in place of a, of course). Thus a typical 3×3 matrix would appear as

$$\begin{pmatrix} a_{11} & a_{12} & a_{13} \\ a_{21} & a_{22} & a_{23} \\ a_{31} & a_{32} & a_{33} \end{pmatrix}$$

We shall make no restriction on the numbers of rows and columns, not even requiring that there are the same number of rows as columns. A matrix with m rows and n columns will be called an $m \times n$ matrix. In particular, a $1 \times n$ matrix is just a row vector of dimension n; while an $m \times 1$ matrix is a column vector of dimension m. A matrix whose general element is a_{ij} can be denoted by (a_{ij}) or, as before, by a single letter (often the corresponding capital letter) in boldface type. Thus we can write

$$\mathbf{A} = (a_{ij}) = \begin{pmatrix} a_{11} & a_{12} & a_{13} & \cdots & a_{1n} \\ a_{21} & a_{22} & a_{23} & \cdots & a_{2n} \\ & & \cdot & \cdots & \cdot \\ a_{m1} & a_{m2} & a_{m3} & \cdots & a_{mn} \end{pmatrix}$$

We should define here what is meant by equality of matrices. Two matrices will be said to be equal if, and only if, they have equal elements in the corresponding positions: that is, if $\mathbf{A} = (a_{ij})$ and $\mathbf{B} = (b_{ij})$ are two $m \times n$ matrices, then $\mathbf{A} = \mathbf{B}$ means that for all i and j ($i = 1, 2, \ldots, m; j = 1, 2, \ldots, n$) $a_{ij} = b_{ij}$.

Let us now look at a typical element in the matrix on the right-hand side of (14.1.15), say the one in row 2 and column 3. It is $\alpha_2 c_1 + \beta_2 c_2 + \gamma_2 c_3$, which can be written as

$$(\alpha_2, \beta_2, \gamma_2) \begin{pmatrix} c_1 \\ c_2 \\ c_3 \end{pmatrix}$$

i.e., the product of row 2 of the first matrix on the left-hand side of (14.1.15) with column 3 of the second matrix. Inspection shows that the element in the ith row and jth column of the product matrix is the product of the ith row of the first matrix and the jth column of the second. We therefore take this as the basis of a definition of the product of two matrices (not necessarily both 3×3 matrices). We define the product matrix to be the matrix whose element in row i column j is the product of the ith row of the first matrix and the jth column of the second.

Now the definition of the product of row and column vectors that we had at the beginning of this chapter presupposed that the vectors being multiplied were of the same dimension, i.e., had the same number of elements. Hence the definition just given of the product of two matrices will not work unless the first matrix has as many columns as the second has rows, since otherwise the rows and columns cannot be multiplied in the required way. We therefore stipulate that this must be the case. If it is not then we do not define a product—multiplication of the matrices is then not possible.

We can therefore multiply an $m \times p$ matrix and a $p \times n$ matrix, and the result will be an $m \times n$ matrix. Thus, for example,

$$\begin{pmatrix} 1 & 0 & 5 & 3 \\ 2 & -1 & 0 & -3 \end{pmatrix} \begin{pmatrix} 4 & 1 & -2 \\ 0 & 3 & 2 \\ -1 & 3 & 1 \\ 6 & 0 & 5 \end{pmatrix} = \begin{pmatrix} 17 & 16 & 18 \\ -10 & -1 & -21 \end{pmatrix}$$

In this, $m = 2$, $p = 4$, $n = 3$, and the element -21 in row 2, column 3 on the right-hand side is obtained from

$$(2 \quad -1 \quad 0 \quad -3) \begin{pmatrix} -2 \\ 2 \\ 1 \\ 5 \end{pmatrix} = -21$$

—the product of row 2 of the first matrix and column 3 of the second. The other elements are obtained similarly.

Suppose $\mathbf{A} = (a_{ij})$ is an $m \times p$ matrix, and $\mathbf{B} = (b_{ij})$ is a $p \times n$ matrix. Then $\mathbf{AB}$ is an $m \times n$ matrix (c_{ij}), in which c_{ij} is the product of the ith row

$$(a_{i1}, a_{i2}, \ldots, a_{ip})$$

of $\mathbf{A}$, with the jth column

$$\begin{pmatrix} b_{1j} \\ b_{2j} \\ \ldots \\ b_{pj} \end{pmatrix}$$

of $\mathbf{B}$. Hence

$$c_{ij} = a_{i1}b_{1j} + a_{i2}b_{2j} + \cdots + a_{ip}b_{pj}$$

or

(14.1.16) $\qquad c_{ij} = \sum_{r=1}^{p} a_{ir}b_{rj} \quad (i = 1, 2, \ldots, m, j = 1, 2, \ldots, n).$

This puts in a concise form the definition of the product of any two matrices. Note that all the rules for multiplication that we have had so far are included in (14.1.16). The product of a row vector and a column vector is the special case where $m = n = 1$ (a 1×1 matrix is, of course, just a number). The product of a matrix and a column vector is the case $n = 1$, and so on.

Exercise 154

1. If

$$\mathbf{A} = \begin{pmatrix} 0 & 2 & -1 & 3 \\ 5 & 0 & 1 & 1 \\ 2 & -4 & 0 & 7 \end{pmatrix};$$

$$\mathbf{B} = \begin{pmatrix} 1 & 2 \\ 0 & 7 \\ 1 & 0 \\ -1 & -3 \end{pmatrix}; \quad \mathbf{C} = \begin{pmatrix} 1 & 0 & -1 & 0 \\ 0 & 2 & 3 & 1 \\ -1 & -1 & 4 & 2 \end{pmatrix}$$

$$\mathbf{x} = \begin{pmatrix} 3 \\ 0 \\ 1 \\ 4 \end{pmatrix}; \quad \mathbf{y} = \begin{pmatrix} 2 \\ -1 \end{pmatrix}$$

find the following vectors and matrices.

    (a) **Ax**   (d) **AB**   (f) **By**

    (b) **ABy**   (e) **CBy**   (g) **CB**

    (c) **Cx**

2. If, also, $\mathbf{u} = (3, -1, 1)$ and $\mathbf{v} = (2, 1, 0, -1)$ find the following vectors and matrices

    (a) **uA**   (c) **vB**   (e) **uC**

    (b) **vx**   (d) **uCBy**   (f) **vBy**

14.2 FURTHER REMARKS ON THE MULTIPLICATION OF MATRICES

Multiplication of matrices has much in common with multiplication in arithmetic or ordinary algebra, but there is one important difference. By the definition of multiplication of matrices we have

$$\begin{pmatrix} 1 & 2 \\ 3 & -1 \end{pmatrix} \begin{pmatrix} 2 & 0 \\ -1 & 4 \end{pmatrix} = \begin{pmatrix} 0 & 8 \\ 7 & -4 \end{pmatrix}$$

If we now multiply these matrices in the reverse order, we get

$$(14.2.2) \qquad \begin{pmatrix} 2 & 0 \\ -1 & 4 \end{pmatrix} \begin{pmatrix} 1 & 2 \\ 3 & -1 \end{pmatrix} = \begin{pmatrix} 2 & 4 \\ 11 & -6 \end{pmatrix}$$

Thus the product of two matrices will, in general, depend on the order in which they are taken. If **A** and **B** are two matrices, then in general

$$\mathbf{AB} \neq \mathbf{BA}.$$

Multiplication in arithmetic does not depend on the order; 2×3 is the same as 3×2 for example. We say that in arithmetic, multiplication is *commutative*. Matrix multiplication is *not* commutative, and this is something that must always be kept in mind. We must further remember that two matrices will not have a product unless the first has the same number of columns as the second has rows; and that even if the product is defined for two matrices in one order, it will not in general be defined if they are taken in the reverse order. In certain cases it may happen that two matrices commute, for example

$$\begin{pmatrix} 7 & 6 \\ 3 & 4 \end{pmatrix} \begin{pmatrix} 1 & 4 \\ 2 & -1 \end{pmatrix} = \begin{pmatrix} 19 & 22 \\ 11 & 8 \end{pmatrix} = \begin{pmatrix} 1 & 4 \\ 2 & -1 \end{pmatrix} \begin{pmatrix} 7 & 6 \\ 3 & 4 \end{pmatrix}$$

so that $\begin{pmatrix} 7 & 6 \\ 3 & 4 \end{pmatrix}$ and $\begin{pmatrix} 1 & 4 \\ 2 & -1 \end{pmatrix}$ commute. But this is the exception and not the rule.

Let us return briefly to the set of equations (14.1.1). We chose to write

$$a_1 x_1 + b_1 x_2 + c_1 x_3 = (a_1, b_1, c_1) \begin{pmatrix} x_1 \\ x_2 \\ x_3 \end{pmatrix}$$

but we could just as well have written

$$a_1 x_1 + b_1 x_2 + c_1 x_3 = (x_1, x_2, x_3) \begin{pmatrix} a_1 \\ b_1 \\ c_1 \end{pmatrix}.$$

Then (14.1.3) would become

$$y_1 = (x_1, x_2, x_3) \begin{pmatrix} a_1 \\ b_1 \\ c_1 \end{pmatrix}$$

(14.2.3)
$$y_2 = (x_1, x_2, x_3) \begin{pmatrix} a_2 \\ b_2 \\ c_2 \end{pmatrix}$$

$$y_3 = (x_1, x_2, x_3) \begin{pmatrix} a_3 \\ b_3 \\ c_3 \end{pmatrix}.$$

For much the same reasons that prompted the transition from (14.1.3) to (14.1.4) we write (14.2.3) as

(14.2.4)
$$(y_1, y_2, y_3) = (x_1, x_2, x_3) \begin{pmatrix} a_1 & a_2 & a_3 \\ b_1 & b_2 & b_3 \\ c_1 & c_2 & c_3 \end{pmatrix}.$$

Equation (14.2.4) expresses (14.1.1) as a relation between row vectors, instead of between column vectors as does (14.1.4). The matrix in (14.2.4) is the same as that in (14.1.4) except that we have interchanged the rows and columns, or *transposed* the matrix, to use the technical term. This term *transpose* we now define.

If $\mathbf{A}$ *is an* $m \times n$ *matrix whose general element is* a_{ij} $(i = 1, 2, \ldots, m;$ $j = 1, 2, \ldots, n)$, *the transpose of* $\mathbf{A}$ *is the* $n \times m$ *matrix whose element in the ith row and jth column is* a_{ji}. *The transpose of* $\mathbf{A}$ *will be denoted by* $\mathbf{A}^T$ (*though several other notations are in use*). *Thus if*

$$\mathbf{A} = \begin{pmatrix} a_{11} & a_{12} & \cdots & a_{1n} \\ a_{21} & a_{22} & \cdots & a_{2n} \\ & & \cdots & \\ a_{m1} & a_{m2} & \cdots & a_{mn} \end{pmatrix}, \text{ then } \mathbf{A}^T = \begin{pmatrix} a_{11} & a_{21} & \cdots & a_{m1} \\ a_{12} & a_{22} & \cdots & a_{m2} \\ & & \cdots & \\ a_{1n} & a_{2n} & \cdots & a_{mn} \end{pmatrix}$$

The transpose of a column vector of dimension n, i.e., of an $n \times 1$ matrix, is the corresponding row vector (a $1 \times n$ matrix), and vice versa. Hence if we write (14.1.4) as

$$\mathbf{y} = \mathbf{A}\mathbf{x}$$

then (14.2.4) will be written as

$$\mathbf{y}^T = \mathbf{x}^T \mathbf{A}^T$$

as is easily seen. This is a special application of a more general rule, namely, if

$$\mathbf{C} = \mathbf{A}\mathbf{B}$$

then

$$\mathbf{C}^T = \mathbf{B}^T \mathbf{A}^T$$

or what amounts to the same thing,

THEOREM 14.1.

$$(\mathbf{A}\mathbf{B})^T = \mathbf{B}^T \mathbf{A}^T$$

IN WORDS: THE TRANSPOSE OF A PRODUCT OF TWO MATRICES IS THE PRODUCT, IN THE REVERSE ORDER, OF THEIR TRANSPOSES.

PROOF:

If $\mathbf{A} = (a_{ij})$ is an $m \times p$ matrix, $\mathbf{B} = (b_{ij})$ is a $p \times n$ matrix, then $\mathbf{C} = \mathbf{A}\mathbf{B} = (c_{ij})$ is an $m \times n$ matrix where

$$c_{ij} = \sum_{r=1}^{p} a_{ir} b_{rj}$$

as we have already seen.

Let

$$\mathbf{A}^T = (\alpha_{ij}), \quad \mathbf{B}^T = (\beta_{ij})$$

so that

$$\alpha_{ij} = a_{ji}, \quad \beta_{ij} = b_{ji}.$$

Then $\mathbf{B}^T\mathbf{A}^T = (\gamma_{ij})$ where

$$\gamma_{ij} = \sum_{r=1}^{p} \beta_{ir}\alpha_{rj}$$

$$= \sum_{r=1}^{p} a_{jr}b_{ri}$$

$$= c_{ji}.$$

Hence $(\gamma_{ij}) = (c_{ji}) = \mathbf{C}^T$, from which the result follows.

It is hardly surprising that the order of the factors has to be reversed when we take transposes. For if the product $\mathbf{AB}$ is defined, then $\mathbf{B}^T\mathbf{A}^T$ will always be defined; but in general $\mathbf{A}^T\mathbf{B}^T$ will not be defined.

Exercise 155

1. By performing the multiplications and comparing the elements, show that the matrices

$$\begin{pmatrix} a & b \\ c & d \end{pmatrix} \quad \text{and} \quad \begin{pmatrix} \alpha & \beta \\ \gamma & \delta \end{pmatrix}$$

commute if, and only if,

$$\frac{c}{\gamma} = \frac{a-d}{\alpha-\delta} = \frac{b}{\beta}.$$

2. Find the matrix $\mathbf{AB} - \mathbf{BA}$ when

(a) $\mathbf{A} = \begin{pmatrix} 1 & 0 & -2 \\ 2 & 1 & 0 \\ 0 & -1 & -1 \end{pmatrix}; \quad \mathbf{B} = \begin{pmatrix} 2 & 0 & 3 \\ -1 & 0 & 5 \\ 0 & 1 & 6 \end{pmatrix}$

(b) $\mathbf{A} = \begin{pmatrix} 3 & -1 & -1 \\ -1 & 3 & -1 \\ -1 & -1 & 3 \end{pmatrix}; \quad \mathbf{B} = \begin{pmatrix} 2 & 1 & 1 \\ 1 & 0 & -2 \\ -1 & 2 & 1 \end{pmatrix}$

14.3 MULTIPLICATION OF A MATRIX BY A SCALAR

We next define the product of a scalar with a matrix, i.e., an expression of the form $\lambda\mathbf{A}$ where λ is a scalar, (i.e., a number as opposed to a vector or matrix), and $\mathbf{A}$ is a matrix. We would expect that if $\mathbf{y} = \mathbf{A}\mathbf{x}$ then $(\lambda\mathbf{A})\mathbf{x} = \lambda\mathbf{y}$, i.e., the matrix $\lambda\mathbf{A}$ gives a vector λ times the vector that $\mathbf{A}$ gives. To ensure this we must multiply every element in $\mathbf{A}$ by λ. Hence we shall define $\lambda\mathbf{A}$ to be the matrix (λa_{ij}). For example,

$$-6\begin{pmatrix} 1 & 0 & 3 \\ -2 & 1 & 8 \\ 5 & 9 & -7 \end{pmatrix} = \begin{pmatrix} -6 & 0 & -18 \\ 12 & -6 & -48 \\ -30 & -54 & 42 \end{pmatrix}.$$

14.4 ADDITION OF MATRICES

We now consider what is to be meant by an expression of the form $\mathbf{A} + \mathbf{B}$, where $\mathbf{A}$ and $\mathbf{B}$ are matrices of the same size (we shall not define addition of matrices of different sizes). If we put $\mathbf{S} = \mathbf{A} + \mathbf{B}$ then, by analogy with the familiar processes of arithmetic we might expect that

$$\mathbf{S}\mathbf{x} = (\mathbf{A} + \mathbf{B})\mathbf{x} = \mathbf{A}\mathbf{x} + \mathbf{B}\mathbf{x}.$$

That is, if $\mathbf{y}_1 = \mathbf{A}\mathbf{x}$ and $\mathbf{y}_2 = \mathbf{B}\mathbf{x}$ then $\mathbf{S}$ would be the matrix that transforms $\mathbf{x}$ into $\mathbf{y}_1 + \mathbf{y}_2$, where $\mathbf{x}$ is any column vector.

Now the ith element in the column vector $\mathbf{A}\mathbf{x}$ is

$$\sum_{r=1}^{n} a_{ir}x_r,$$

and that in $\mathbf{B}\mathbf{x}$ is $\sum_{r=1}^{n} b_{ir}x_r$. Hence the ith element in $\mathbf{A}\mathbf{x} + \mathbf{B}\mathbf{x}$ is

$$\sum_{r=1}^{n} (a_{ir} + b_{ir})x_r$$

from which it readily follows that $\mathbf{S} = (s_{ij})$ where

$$(14.4.1) \qquad\qquad s_{ij} = a_{ij} + b_{ij}.$$

Thus we are naturally led to define the sum $\mathbf{A} + \mathbf{B}$ of two matrices by equation (14.4.1). To add two matrices we add together corresponding elements.

Clearly this definition applies only when the matrices are of the same size, that is, have the same number of rows and the same number of columns.

14.5 SUBTRACTION OF MATRICES

Subtraction of one matrix from another can be defined in terms of definitions already given. The result of multiplying a matrix by the scalar -1 can be written as $(-1)\mathbf{B}$, or more briefly, as $-\mathbf{B}$. We call this the negative of $\mathbf{B}$. We now define $\mathbf{A} - \mathbf{B}$ to be $\mathbf{A} + (-\mathbf{B})$. This is the result of *subtracting* $\mathbf{B}$ from $\mathbf{A}$.

Exercise 156

1. If

$$\mathbf{A} = \begin{pmatrix} 2 & 0 & 1 & 4 \\ 0 & -1 & -1 & 3 \\ 1 & -2 & 0 & 0 \end{pmatrix}; \quad \mathbf{B} = \begin{pmatrix} 0 & 2 & -4 & 5 \\ -1 & 0 & 1 & 2 \\ 1 & 3 & 2 & 0 \end{pmatrix};$$

$$\mathbf{x} = \begin{pmatrix} 1 \\ 2 \\ 0 \\ -1 \end{pmatrix} \quad \text{and} \quad \mathbf{y} = \begin{pmatrix} 3 \\ 0 \\ 1 \\ 4 \end{pmatrix}, \quad \text{find}$$

   (a) $\mathbf{A} + \mathbf{B}$ (d) $\mathbf{A} + 5\mathbf{B}$

   (b) $(2\mathbf{A} - 3\mathbf{B})\mathbf{x}$ (e) $(4\mathbf{A} - 7\mathbf{B})\mathbf{y}$

   (c) $3\mathbf{A}\mathbf{x} - \mathbf{B}\mathbf{y}$ (f) $(4\mathbf{A} - 3\mathbf{B})(2\mathbf{x} + 3\mathbf{y})$

2. Using the same matrices and vectors as in problem 1, find

   (a) $\mathbf{B}^T\mathbf{A}\mathbf{y}$

   (b) $\mathbf{x}^T\mathbf{A}^T\mathbf{B}$

   (c) $\mathbf{x}^T\mathbf{y}$

3. Prove that $(\lambda\mathbf{A} + \mu\mathbf{B})^T = \lambda\mathbf{A}^T + \mu\mathbf{B}^T$.

14.6 THE RULES FOR MATRIX ALGEBRA.

It now begins to look as though the rules for addition and multiplication of matrices are turning out to be very similar to the corresponding rules in ordinary algebra or arithmetic, except for the non-commutativity of multiplication and the fact that these operations are defined only when the matrices have the correct numbers of rows and columns. To obviate this last source of difficulty let us look at the situation as it applies to square matrices having a fixed number n of rows, and the same number of columns. The similarity between matrix addition and multiplication and arithmetical addition and multiplication is then seen in the following theorem.

THEOREM 14.2.

IF A, B, C ARE ANY MATRICES FROM THE SET OF ALL $n \times n$ MATRICES THEN

1. $A + B = B + A$
2. $(A + B) + C = A + (B + C)$
3. $(AB)C = A(BC)$
4. $A(B + C) = AB + AC$
5. $(B + C)A = BA + CA$
6. There is a *zero matrix* 0 with the property that $A + 0 = A$ for any A in the set.
7. There is a *unit matrix* I with the property that $AI = IA = A$ for any A in the set.

PROOF:

The proofs of (1) through (5) come readily from the definitions of addition and subtraction and will be left as an exercise to the reader [the proof of (3) is a little lengthy, but presents no problems].

The zero matrix in (6) is the matrix all of whose elements are zero. It is readily verified that this has the required properties.

The unit matrix in (7) is the matrix

$$
I_n = \begin{pmatrix}
1 & 0 & 0 & 0 & \cdots & 0 \\
0 & 1 & 0 & 0 & \cdots & 0 \\
0 & 0 & 1 & 0 & \cdots & 0 \\
0 & 0 & 0 & 1 & \cdots & 0 \\
  &   &   & \cdots &   &   \\
0 & 0 & 0 & 0 & \cdots & 1
\end{pmatrix}
$$

in which all elements are zero except these in the *main diagonal*, which are 1. The main diagonal consists of those elements whose row and column numbers are the same. Thus the unit matrix I is defined by

$$
I_n = (\delta_{ij})
$$

where $\delta_{ij} = 0$ if $i \neq j$ and $\delta_{ii} = 1$ $(i, j = 1, 2, \ldots, n)$. The symbol δ_{ij} for the elements of the unit matrix is standard, and is often called the *Kronecker delta*. By the definition of multiplication $\mathbf{AI}_n$ is the matrix whose (i, j) element is

$$\sum_{r=1}^{n} a_{ir}\,\delta_{rj} = a_{ij}$$

since $\delta_{rj} = 0$ unless $r = j$. Hence $\mathbf{AI}_n = \mathbf{A}$. Similarly $\mathbf{I}_n\mathbf{A} = \mathbf{A}$, and $\mathbf{I}_n$ has the required property.

Note how, because matrix multiplication is not commutative we need to assert both $\mathbf{I}_n\mathbf{A} = \mathbf{A}$ and $\mathbf{AI}_n = \mathbf{A}$. It is conceivable (though not in fact true) that one of these might have been true while the other was not. For the same reason we have listed (4) and (5) as separate assertions.

All the fundamental operations of arithmetic now have their parallels in matrix algebra, with one exception. This is the operation of division or, what is really fundamental, the operation of taking the reciprocal. We are therefore led to ask the following question: "Given a matrix $\mathbf{A}$, is there another matrix, $\mathbf{A}^*$ say, which bears the same relation to $\mathbf{A}$ as a number does to its reciprocal in ordinary arithmetic?" The relation between a number x and its reciprocal $1/x$ is that their product is 1. Hence, by analogy, the matrix $\mathbf{A}^*$ that we look for is one such that

$$\mathbf{AA}^* = \mathbf{I}_n.$$

If we can find such a matrix $\mathbf{A}^*$, then we can reasonably call it the *reciprocal* or (more usually) the *inverse* of $\mathbf{A}$. Does such an inverse always exist? The answer is no. Although in general the inverse of a square matrix does exist, there are exceptional matrices that have no inverses. This is a question that we shall need to look at very carefully; but in order to do so we shall have to embark on a rather lengthy digression and consider some objects known as *determinants*. When we have defined determinants (which are of great importance) and established some of their properties, we shall return to our discussion of matrices.

Exercise 157

1. Prove the various rules for matrix algebra given in the last section.
2. By solving the 4 equations in a, b, c and d given by the matrix equation

$$\begin{pmatrix} 2 & 3 \\ -1 & -2 \end{pmatrix} \begin{pmatrix} a & b \\ c & d \end{pmatrix} = \begin{pmatrix} 1 & 0 \\ 0 & 1 \end{pmatrix}$$

show that the matrix $\begin{pmatrix} 2 & 3 \\ -1 & -2 \end{pmatrix}$ has an inverse, and find it.

3. In a similar manner, show that the matrix $\begin{pmatrix} 1 & 3 \\ -2 & -6 \end{pmatrix}$ does not have an inverse.

14.7 MORE ABOUT PERMUTATIONS

We have already met (in Chapter 4) the idea of a permutation of a set of objects, that is to say a rearrangement of these objects in some order. Suppose we have a set of n objects, distinguishable each from the other. Without any loss of generality we can call the objects "object 1," "object 2", and so on; or, what is even more simple, we can think of the objects as being the integers, $1, 2, \ldots, n$ themselves. A permutation of these objects is then simply an arrangement of these integers in some other order. We can specify a permutation by a notation of the following kind (with $n = 7$)

$$\begin{pmatrix} 1 & 2 & 3 & 4 & 5 & 6 & 7 \\ 7 & 1 & 3 & 2 & 4 & 6 & 5 \end{pmatrix}$$

which shows that the objects have been rearranged in such a way that for example the fourth object is in the second place after the rearrangement, the seventh moves into the fifth place, and so on.

We shall now show that there are two different kinds of permutations, to be called *even* and *odd*. Our first step towards this end is to define a special type of permutation known as an interchange. This is a permutation which merely interchanges two of the objects while leaving all the others in their original positions. Thus

$$\begin{pmatrix} 1 & 2 & 3 & 4 & 5 & 6 & 7 \\ 1 & 2 & 5 & 4 & 3 & 6 & 7 \end{pmatrix}$$

is an interchange, since it merely interchanges objects 3 and 5.

Our next step is to prove that any permutation can be expressed as the product of a number of interchanges, where by the product of two or more interchanges (or of two permutations of any kind for that matter) we mean the permutation resulting from performing the interchanges one after the other. This theorem is most easily proved diagrammatically.

Suppose we write the integers $1, 2, 3, \ldots, 7$ on a horizontal line, and repeat this on a lower line. We then join with a line each integer in the top line to the integer in the lower line to which it corresponds in the permutation. Thus, if the permutation is $\begin{pmatrix} 1 & 2 & 3 & 4 & 5 & 6 & 7 \\ 7 & 1 & 4 & 2 & 3 & 6 & 5 \end{pmatrix}$ we obtain Fig. 14.1, showing that in the permutation, 1 "becomes" 7, 2 "becomes" 1, and so on. The lines that we draw need not be straight, since their only purpose is to indicate which integers in the top line correspond to which integers in the bottom line. We can therefore always distort the lines where necessary so as to ensure that we do not have three or more lines all crossing at one point, and that no two of these crossing points are on the same "level." The way is now clear to express the permutation given above as the product of interchanges.

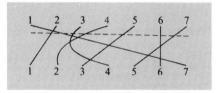

Fig. 14.1

Imagine that we draw a horizontal line just below the topmost crossing point in Fig. 14.1 (shown dotted in the figure). If we supply integers 1, 2, ..., 7 at the various points along the dotted line where it is crossed by the lines running down the page, and forget about what is below the dotted line, we get Fig. 14.2.

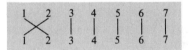

Fig. 14.2

This is a diagram representing the permutation

$$\begin{pmatrix} 1 & 2 & 3 & 4 & 5 & 6 & 7 \\ 2 & 1 & 3 & 4 & 5 & 6 & 7 \end{pmatrix}$$

which is an interchange—the interchange of 1 and 2. If we now draw another dotted line below the next crossing point and consider the portion of the diagram between the two dotted lines, we get a diagram for the interchange of 2 and 3. See Fig. 14.3.

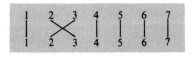

Fig. 14.3

Proceeding in this way we can divide the whole diagram up into sections, by horizontal dotted lines, one below each crossing point, such that each section represents an interchange. Distorting the original diagram slightly so as to keep the integers in corresponding positions, we can display the whole process for our given permutation in Fig. 14.4. On the right of each section is written the interchange which is represented by that section. Following what happens to the integer 1 as we pass through the sections, we see that $1 \to 2 \to 3 \to 4 \to 4 \to 5 \to 5 \to 6 \to 7 \to 7$. Thus ultimately 1 becomes 7 in the permutation. Similarly 2 eventually becomes 1, and so on. Thus the permutation has been expressed as a product of interchanges.

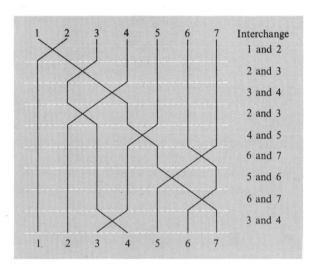

Fig. 14.4

Now although we have followed through this proof with a particular permutation, it is clear that it would apply equally well to any permutation of any number of objects. Note that we have in fact proved a somewhat stronger result, namely that any permutation can be expressed as the product of interchanges of *adjacent* objects.

If the number of interchanges (which is also the number of crossings in the diagram) is even, we say that the permutation is an *even permutation*. If the number is odd, then the permutation is an *odd permutation*. Thus we distinguish these two kinds of permutations. Before this definition is really acceptable we must show that it is consistent, that is, that a permutation cannot be both even and odd. The proof that this is the case is given in outline in Exercise 158.2 below, and the reader should fill in the details of this proof.

Exercise 158

1. Determine which of the following permutations are even, and which are odd:

(a) $\begin{pmatrix} 1 & 2 & 3 & 4 & 5 \\ 2 & 3 & 1 & 5 & 4 \end{pmatrix}$

(b) $\begin{pmatrix} 1 & 2 & 3 & 4 & 5 & 6 & 7 & 8 & 9 \\ 3 & 2 & 1 & 6 & 5 & 4 & 8 & 9 & 7 \end{pmatrix}$

(c) $\begin{pmatrix} 1 & 2 & 3 & 4 & 5 \\ 2 & 4 & 1 & 3 & 5 \end{pmatrix}$

(d) $\begin{pmatrix} 1 & 2 & 3 & 4 & 5 & 6 \\ 6 & 1 & 2 & 3 & 5 & 4 \end{pmatrix}$

(e) $\begin{pmatrix} 1 & 2 & 3 & 4 \\ 4 & 1 & 2 & 3 \end{pmatrix}$

(f) $\begin{pmatrix} 1 & 2 & 3 & 4 & 5 & 6 \\ 4 & 2 & 6 & 5 & 1 & 3 \end{pmatrix}$

2. Let $\alpha_1, \alpha_2, \ldots, \alpha_n$ be any n variables, and consider the expression

$$\Delta = (\alpha_1 - \alpha_2)(\alpha_1 - \alpha_3) \cdots (\alpha_1 - \alpha_n)(\alpha_2 - \alpha_3) \cdots$$
$$(\alpha_2 - \alpha_n) \cdots (\alpha_{n-1} - \alpha_n)$$

that is, the product of all factors of the form $(\alpha_r - \alpha_s)$ where $1 \leq r < s \leq n$. Let P be the permutation

$$\begin{pmatrix} 1 & 2 & 3 & 4 & \cdots & n \\ p_1 & p_2 & p_3 & p_4 & \cdots & p_n \end{pmatrix}.$$

Prove that
  (a) The effect of permuting the variables $\alpha_1, \alpha_2, \ldots, \alpha_n$ by the permutation P, that is, of replacing α_i by α_{p_i} is either to leave Δ unchanged, or to change the sign of Δ.
  (b) The interchange of two adjacent variables, say α_i and α_{i+1} changes the sign of Δ.
  Deduce that if P is expressed as a product of a number of interchanges of adjacent elements, then this number is either always odd (whatever the method of expression) or always even. In other words, it cannot be odd for some expressions and even for others.

Note: We could have used the function Δ to define even and odd permutations, as follows. A permutation P is odd if Δ is changed in sign when the variables are permuted by P. It is even if Δ is unchanged. However the method of drawing a diagram is a very handy one for finding, in an actual example, whether a given permutation is even or odd.

3. Prove that the interchange of two nonadjacent objects changes the sign of Δ.

Hint: To interchange i and j $(i < j)$ move i up next to j by interchanging adjacent objects; then move j back to where i was. Show that the number of interchanges required is odd.

14.8 PROPERTIES OF PRODUCTS OF PERMUTATIONS

The product of two permutations, as has already been mentioned, is the permutation obtained by applying the two given permutations one after the other. Thus if

$$A = \begin{pmatrix} 1 & 2 & 3 & 4 & 5 \\ 1 & 5 & 2 & 4 & 3 \end{pmatrix} \quad \text{and} \quad B = \begin{pmatrix} 1 & 2 & 3 & 4 & 5 \\ 5 & 1 & 4 & 3 & 2 \end{pmatrix}$$

then the product of these two permutations *in this order* is

$$AB = \begin{pmatrix} 1 & 2 & 3 & 4 & 5 \\ 5 & 2 & 1 & 3 & 4 \end{pmatrix}$$

since, for example, $3 \longrightarrow 2$ in A and $2 \longrightarrow 1$ in B, so that $3 \longrightarrow 1$ in AB, and so on. The order in which the permutations are taken is important, for if we take B first and then A we get

$$BA = \begin{pmatrix} 1 & 2 & 3 & 4 & 5 \\ 3 & 1 & 4 & 2 & 5 \end{pmatrix}$$

since, for example, $2 \longrightarrow 5$ in B and $5 \longrightarrow 1$ in A. Hence for permutations, as for matrices, it is not necessarily true that $AB = BA$.

If we have three permutations P, Q and R, we can form the product PQ and then take the product of this with R. The result would be written $(PQ)R$. On the other hand, we can multiply P by the permutation QR, and write the result as $P(QR)$. Each of these results stands for the result of applying the permutations P, Q and R in succession. Hence we see that permutations obey the associative law.

(14.8.1) $(PQ)R = P(QR)$.

Since the parentheses thus have no significance, we write PQR for either of these expressions.

One permutation of n objects is of particular importance. It is the one which, in a sense, does *not* permute the objects, but leaves each object in its original position, viz., the permutation

$$I = \begin{pmatrix} 1 & 2 & 3 & \cdots & n \\ 1 & 2 & 3 & \cdots & n \end{pmatrix}$$

It is called the *identity permutation*. It is readily verified that, for any permutation P, we have

(14.8.2) $$PI = IP = P$$

provided, of course, that P and I are permutations of the same set of objects (otherwise the product is not defined).

Corresponding to any permutation P, there is another permutation denoted by P^{-1}, called the *inverse of* P. It satisfies the following equations

$$PP^{-1} = P^{-1}P = I$$

which means that it, so to speak, puts back in their original order the objects permuted by P, so that the net effect of P and P^{-1} is I, namely to leave every object in its original place.

Thus if $P = \begin{pmatrix} 1 & 2 & 3 & 4 & 5 \\ 3 & 1 & 4 & 5 & 2 \end{pmatrix}$, so that P moves 1 to 3, 2 to 1 and so on, then P^{-1} moves 3 to 1, 1 to 2 and so on. Hence in this case

$$P^{-1} = \begin{pmatrix} 1 & 2 & 3 & 4 & 5 \\ 2 & 5 & 1 & 3 & 4 \end{pmatrix}$$

which is seen to be obtained from P by interchanging the two rows in the expression for P. [Actually this would give $\begin{pmatrix} 3 & 1 & 4 & 5 & 2 \\ 1 & 2 & 3 & 4 & 5 \end{pmatrix}$, but it is customary, though not necessary, to rearrange the columns so that the upper row is in order $(1, 2, \ldots, n)$.]

The diagram for the inverse permutation P^{-1} of P is obtained from that of P by merely turning it upside down. Thus, for the permutation considered, the diagram of P is

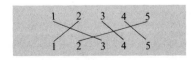

Hence that of P^{-1} is

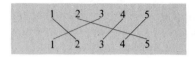

It follows that P and P^{-1} are either both even or both odd, since the number of crossings is the same in these two diagrams.

Exercise 159

1. Prove that
   (a) The product of two even permutations is an even permutation.
   (b) The product of two odd permutations is an even permutations.
   (c) The product of an odd and an even permutation (either way) is an odd permutation.
2. Find the following permutation products

   (a) $\begin{pmatrix} 1 & 2 & 3 & 4 \\ 3 & 1 & 4 & 2 \end{pmatrix} \begin{pmatrix} 1 & 2 & 3 & 4 \\ 3 & 1 & 2 & 4 \end{pmatrix}$

   (b) $\begin{pmatrix} 1 & 2 & 3 & 4 & 5 \\ 3 & 4 & 2 & 1 & 5 \end{pmatrix} \begin{pmatrix} 1 & 2 & 3 & 4 & 5 \\ 1 & 2 & 4 & 3 & 5 \end{pmatrix}$

   (c) $\begin{pmatrix} 1 & 2 & 3 & 4 & 5 & 6 & 7 & 8 & 9 \\ 2 & 8 & 7 & 9 & 1 & 5 & 4 & 6 & 3 \end{pmatrix} \begin{pmatrix} 1 & 2 & 3 & 4 & 5 & 6 & 7 & 8 & 9 \\ 9 & 8 & 4 & 5 & 1 & 2 & 3 & 6 & 7 \end{pmatrix}$

   (d) $\begin{pmatrix} 1 & 2 & 3 & 4 \\ 3 & 1 & 2 & 4 \end{pmatrix} \begin{pmatrix} 1 & 2 & 3 & 4 \\ 3 & 1 & 4 & 2 \end{pmatrix}$

   (e) $\begin{pmatrix} 1 & 2 & 3 & 4 & 5 & 6 \\ 6 & 5 & 2 & 1 & 4 & 3 \end{pmatrix} \begin{pmatrix} 1 & 2 & 3 & 4 & 5 & 6 \\ 5 & 6 & 2 & 1 & 4 & 3 \end{pmatrix}$

3. Find the inverses of the following permutations

   (a) $\begin{pmatrix} 1 & 2 & 3 & 4 \\ 2 & 3 & 1 & 4 \end{pmatrix}$

   (b) $\begin{pmatrix} 1 & 2 & 3 & 4 & 5 & 6 \\ 3 & 4 & 1 & 5 & 6 & 2 \end{pmatrix}$

   (c) $\begin{pmatrix} 1 & 2 & 3 & 4 & 5 \\ 4 & 5 & 2 & 3 & 1 \end{pmatrix}$

   (d) $\begin{pmatrix} 1 & 2 & 3 & 4 & 5 & 6 & 7 \\ 2 & 1 & 4 & 3 & 6 & 5 & 7 \end{pmatrix}$

14.9 GROUPS

We have seen various properties of permutations and of the operation "product" that holds between them (provided we restrict ourselves to permutations of the same set of objects). There are many systems of mathematical constructions that possess these properties, so let us summarize the properties without specific mention of permutations.

We have a set G of elements of some sort. We have also an operation (which we may call "product," or anything we choose) which associates some element of G with every pair of elements of G (just as the product of any two permutations is a permutation). We denote by $a, b, c, \ldots$ the elements of G, and the result of the operation by juxtaposition, so that the "product" of a and b is written ab. The following properties may hold:

1. The "product" of any two elements of G is an element of G. We have already said this, but we list it again for completeness. This is known as the property of *closure*.

2. The associative law holds, i.e., for any three elements a, b, c, of G we have

$$(ab)c = a(bc).$$

3. There is an identity element I with the property

$$aI = Ia = a \text{ for every } a \in G.$$

4. To every element $a \in G$ there corresponds an element $a^{-1} \in G$ (the inverse of a) with the property

$$aa^{-1} = a^{-1}a = I.$$

A set of elements, together with an operation satisfying these four properties is called a *group*. The operation is called the *group operation*. Note that a group is not just a set, but a set plus a group operation.

We see therefore that the set of permutations of a given collection of objects, together with the operation of taking successive permutations as described above, forms a group. It is called a *symmetric group*.

We shall not have to deal very much with groups in this book, but the concept of a group is of fundamental importance in modern mathematics, and the reader should be familiar with the definition, if nothing else.

Exercise 160

Below are given various sets together with operations on the elements of the set. State, in each case, whether the set forms a group with the given operation, and if not, why not.

(a)   The set of all nonzero rational numbers, with ordinary multiplication as the operation.

(b)   The set of all positive integers, under the operation of ordinary multiplication.

(c)   The set of all integers (negative, positive, zero) with addition as the operation.

(d)   The set of all real numbers with an operation (denoted by "$_0$") defined by $a_0 b = a + b + ab$.

(e)   The set of all even permutations of n objects, with products of permutations defined as above in the section.

(f)   The set of all odd permutations of n objects, with products defined as above.

14.10   DETERMINANTS

We now return to the main business of this chapter. With every square matrix we shall associate a number, called the *determinant of the matrix*. A typical $n \times n$ matrix can be written as

$$\begin{pmatrix} a_{11} & a_{12} & a_{13} & \cdots & a_{1n} \\ a_{21} & a_{22} & a_{23} & \cdots & a_{2n} \\ a_{31} & a_{32} & a_{33} & \cdots & a_{3n} \\ & & \cdots & & \\ a_{n1} & a_{n2} & & \cdots & a_{nn} \end{pmatrix}$$

What we shall do now is to choose, in all possible ways, a set of n elements of this matrix, in such a way that in each such set no two elements occur in the same row or the same column of the matrix. It follows that we must choose exactly one element from each row, for if we do not choose an element from one row, we must choose two at least from some other row to make up the total of n elements. Let the element chosen from row i be that in column p_i, namely the element $a_{i p_i}$. Then since no two elements may be chosen from the same column, it follows that no two of the integers $p_1, p_2, p_3, \cdots, p_n$ are the same, and hence that these must be the integers 1, 2, 3, ..., n, in some order. Hence with every choice of elements according to the above rules we can associate a permutation, viz.,

(14.10.1)
$$\begin{pmatrix} 1 & 2 & 3 & \cdots & n \\ p_1 & p_2 & p_3 & \cdots & p_n \end{pmatrix}$$

Conversely, every permutation of the integers $1, 2, \ldots, n$ gives a choice of elements. The number of choices is therefore $n!$—the number of permutations.

For each choice of elements we now write down the product of the elements, and put a $+$ or a $-$ sign in front of the term so obtained, according as the permutation (14.10.1) is even or odd. Finally we sum the $n!$ terms to obtain an expression of the form

$$\sum \pm a_{1p_1} a_{2p_2} \cdots a_{np_n}$$

where the summation is over all permutations like (14.10.1). The number that we get in this way is called the determinant of the matrix (a_{ij}). It is usually written

(14.10.2)
$$\begin{vmatrix} a_{11} & a_{12} & \cdots & a_{1n} \\ a_{21} & a_{22} & \cdots & a_{2n} \\ a_{31} & a_{32} & \cdots & a_{3n} \\ & \cdots & & \\ a_{n1} & a_{n2} & \cdots & a_{nn} \end{vmatrix}$$

or for short, $|a_{ij}|$.

We first note two simple special cases.

Case 1. When $n = 2$.

If $n = 2$, the determinant is $\begin{vmatrix} a_{11} & a_{12} \\ a_{21} & a_{22} \end{vmatrix}$ and it has two terms. The one corresponding to the identity permutation $\begin{pmatrix} 1 & 2 \\ 1 & 2 \end{pmatrix}$ is $+ a_{11} a_{22}$, while that corresponding to the (odd) permutation $\begin{pmatrix} 1 & 2 \\ 2 & 1 \end{pmatrix}$ is $- a_{12} a_{21}$. Hence

(14.10.3)
$$\begin{vmatrix} a_{11} & a_{12} \\ a_{21} & a_{22} \end{vmatrix} = a_{11} a_{22} - a_{12} a_{21}$$

Case 2. When $n = 3$.

There are six possible permutations, viz.,

$$\begin{pmatrix} 1 & 2 & 3 \\ 1 & 2 & 3 \end{pmatrix}, \quad \begin{pmatrix} 1 & 2 & 3 \\ 3 & 1 & 2 \end{pmatrix}, \quad \begin{pmatrix} 1 & 2 & 3 \\ 2 & 3 & 1 \end{pmatrix}$$

$$\begin{pmatrix} 1 & 2 & 3 \\ 2 & 1 & 3 \end{pmatrix}, \quad \begin{pmatrix} 1 & 2 & 3 \\ 1 & 3 & 2 \end{pmatrix}, \quad \begin{pmatrix} 1 & 2 & 3 \\ 3 & 2 & 1 \end{pmatrix}$$

Of these, those in the first line are even, those in the second line are odd. Hence, forming the terms as described above, we have

$$(14.10.4) \quad \begin{vmatrix} a_{11} & a_{12} & a_{13} \\ a_{21} & a_{22} & a_{23} \\ a_{31} & a_{32} & a_{33} \end{vmatrix} = \begin{aligned} &a_{11}a_{22}a_{33} + a_{13}a_{21}a_{32} + a_{12}a_{23}a_{31} \\ &- a_{12}a_{21}a_{33} - a_{11}a_{23}a_{32} - a_{13}a_{22}a_{31} \end{aligned}$$

THE RULE OF SARRUS

There are many interesting properties that are shared by all determinants. For 2×2 determinants these properties are usually trivial. On the other hand determinants of size 4×4 or larger are rather difficult to compute (even a 4×4 determinant has 24 terms, and the number increases rapidly with n). Thus it happens that 3×3 determinants are the most useful for illustrating the properties and applications of determinants. It is fortunate therefore that there is a simple rule for evaluating a 3×3 determinant. It is called the *rule of Sarrus*, and goes as follows:

Imagine that the determinant has been written down in the usual way and that the first two columns have been repeated to the right of the third column, as shown in Fig. 14.5; then the three diagonals that go downwards from left to right (the heavy arrows in Fig. 14.5), give the terms corresponding to even permutations, while the diagonals from right to left (the dotted arrows in the figure) give the terms corresponding to odd permutations.

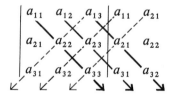

Fig. 14.5

Comparison of Fig. 14.5 with the result (14.10.4) will show what is meant.

The reader should get some practice in evaluating 3×3 determinants by this rule, and suitable examples are given in the exercise which follows.

It is important to remember that the Rule of Sarrus is a rule which happens to work for 3×3 determinants. It does *not* work for determinants of any other size. This is clear, because the number of terms that one would obtain by its use is $2n$, which is not the correct number of terms if $n \neq 3$.

Exercise 161

1. Evaluate the following determinants

(a) $\begin{vmatrix} 1 & 5 \\ 2 & 12 \end{vmatrix}$

(e) $\begin{vmatrix} 3 & -8 \\ 4 & -2 \end{vmatrix}$

(h) $\begin{vmatrix} 17 & 1 \\ -9 & -5 \end{vmatrix}$

(b) $\begin{vmatrix} -1.2 & 6.3 \\ -8.1 & -3.7 \end{vmatrix}$

(f) $\begin{vmatrix} 1 & 2 & -5 \\ 0 & 1 & 6 \\ 7 & 3 & 3 \end{vmatrix}$

(i) $\begin{vmatrix} -1 & 0 & 6 \\ 3 & 5 & 2 \\ 5 & 10 & 10 \end{vmatrix}$

(c) $\begin{vmatrix} x & 1+x & -x \\ 2x+5 & 0 & x-7 \\ 2x-4 & x & 5x \end{vmatrix}$

(g) $\begin{vmatrix} 1 & x & x^2 \\ 1 & y & y^2 \\ 1 & z & z^2 \end{vmatrix}$

(d) $\begin{vmatrix} x-y & x+y & 1 \\ x^2-y^2 & x^2+y^2 & 1 \\ 1 & x & y \end{vmatrix}$

2. By enumerating all permutations of 4 objects find the value of the 4×4 determinant

$$\begin{vmatrix} 1 & 0 & -3 & 7 \\ -2 & 1 & 5 & 3 \\ 7 & -9 & 0 & 6 \\ -3 & 1 & 5 & 2 \end{vmatrix}$$

14.11 MANIPULATION OF DETERMINANTS

We shall now consider some properties of determinants.

PROPERTY I

A matrix and its transpose have the same determinant, i.e., $|\mathbf{A}| = |\mathbf{A}^T|$.

PROOF.
Except possibly for their signs, the terms in $|\mathbf{A}|$ are the same as those in $|\mathbf{A}^T|$. For they are obtained by taking in all possible ways products of

n elements of the matrix such that no two elements are in the same row or the same column, and these choices of elements will not be affected by transposing the matrix. Hence we need only show that the signs agree. A typical term in $|\mathbf{A}|$ will be

(14.11.1) $$\pm a_{1p_1} a_{2p_2} \cdots a_{np_n}$$

corresponding to the permutation $\begin{pmatrix} 1 & 2 & 3 & \cdots & n \\ p_1 & p_2 & p_3 & \cdots & p_n \end{pmatrix}$. If we denote $\mathbf{A}^T$ by (b_{ij}), where $b_{ij} = a_{ji}$, then the term (14.11.1) can be written

(14.11.2) $$\pm b_{p_1 1} b_{p_2 2} \cdots b_{p_n n}$$

and this, with the appropriate sign, is a term in $|\mathbf{A}^T|$. It is the term which is obtained by choosing in row p_i the element in column i $(i = 1, 2, \ldots, n)$, and therefore corresponds to the permutation

$$\begin{pmatrix} p_1 & p_2 & p_3 & \cdots & p_n \\ 1 & 2 & 3 & \cdots & n \end{pmatrix}$$

Now this permutation is the inverse of the permutation which gave (14.11.1), and we have already seen that a permutation and its inverse have the same *parity*, that is, they are either both even or both odd. Thus not only do $|\mathbf{A}|$ and $|\mathbf{A}^T|$ have the same terms, but these terms occur with the same signs. Hence the two determinants have the same value.

Because of this result, properties of determinants which concern rows have a corresponding property for columns. The remaining properties of determinants will be stated for rows and columns, but we shall prove them only one way (either for rows or columns, whichever is more convenient), and rely on Property I for the proof the other way round.

PROPERTY II

The interchange of two rows (columns) in a determinant changes its sign. Thus interchanging the first two columns in a 3×3 determinant we have

$$\begin{vmatrix} b_1 & a_1 & c_1 \\ b_2 & a_2 & c_2 \\ b_3 & a_3 & c_3 \end{vmatrix} = - \begin{vmatrix} a_1 & b_1 & c_1 \\ a_2 & b_2 & c_2 \\ a_3 & b_3 & c_3 \end{vmatrix}.$$

PROOF.

We consider the interchange of columns 1 and 2. A typical term in $|A|$ is

$$\pm a_{1p_1} a_{2p_2} a_{3p_2} \cdots a_{np_n}$$

where the $+$ or the $-$ sign is taken according as the permutation

(14.11.3)
$$\begin{pmatrix} 1 & 2 & 3 & \cdots & n \\ p_1 & p_2 & p_3 & \cdots & p_n \end{pmatrix}$$

is even or odd. A typical term in the modified determinant is $\pm a_{1p_2} a_{2p_1} a_{3p_3} \cdots a_{np_n}$, the sign being $+$ or $-$ according as the permutation

(14.11.4)
$$\begin{pmatrix} 1 & 2 & 3 & \cdots & n \\ p_2 & p_1 & p_3 & \cdots & p_n \end{pmatrix}$$

is even or odd. But permutation (14.11.4) is the product of the interchange of 1 and 2 and the permutation (14.11.3). Hence if (14.11.3) is even (14.11.4) is odd and vice versa. Thus in the modified determinant we have the same terms, but they occur with reversed signs. Hence the modified determinant has the same numerical value but the opposite sign.

What we have said for the interchange of columns 1 and 2 applies equally well to the interchange of any two columns. Hence the property holds for columns. By Property I it holds also for rows.

PROPERTY III

If two rows (columns) of a determinant are identical, the determinant vanishes, i.e., its value is zero.

PROOF.

Let the value of the determinant be D. If we interchange the two identical rows (columns) we change the sign of the determinant (Property II). But since the rows (columns) are identical we have not altered the determinant. The only way that this can happen is that $D = -D$, i.e., $D = 0$. Hence the determinant vanishes.

PROPERTY IV

If one row (column) of a determinant is expressed as the sum of two rows (columns) (in the sense of addition of vectors) then the determinant can be expressed as the sum of two determinants, alike except for the row (column) in question. The rows (columns) that differ in the two determinants, and which occur in corresponding positions, sum to the corresponding row (column) in the original determinant.

EXAMPLE:

$$
\begin{vmatrix} a_1 + d_1 & b_1 & c_1 \\ a_2 + d_2 & b_2 & c_2 \\ a_3 + d_3 & b_3 & c_3 \end{vmatrix} = \begin{vmatrix} a_1 & b_1 & c_1 \\ a_2 & b_2 & c_2 \\ a_3 & b_3 & c_3 \end{vmatrix} + \begin{vmatrix} d_1 & b_1 & c_1 \\ d_2 & b_2 & c_2 \\ d_3 & b_3 & c_3 \end{vmatrix}
$$

Here the first column has been written as the sum of two column vectors, $\begin{pmatrix} a_1 \\ a_2 \\ a_3 \end{pmatrix}$ and $\begin{pmatrix} d_1 \\ d_2 \\ d_3 \end{pmatrix}$. The determinant is therefore the sum of the two determinants on the left, in which these two vectors appear as the first columns.

PROOF:

Suppose the ith row has been expressed as the sum of two row vectors in the following way:

$$
(a_{i1}, a_{i2}, \cdots, a_{in}) = (b_{i1}, b_{i2}, \cdots, b_{in}) + (c_{i1}, c_{i2}, \cdots, c_{in})
$$

Then

$$
|a_{ij}| = \sum \pm a_{1p_1} a_{2p_2} \cdots (b_{ip_i} + c_{ip_i}) \cdots a_{np_n}
$$
$$
= \sum \pm a_{1p_1} a_{2p_2} \cdots b_{ip_i} \cdots a_{np_n} +
$$
$$
\sum \pm a_{1p_1} a_{2p_2} \cdots c_{ip_i} \cdots a_{np_n}
$$

$$
= \begin{vmatrix} a_{11} & a_{12} & \cdots & a_{1n} \\ & & \cdots & \\ b_{i1} & b_{i2} & \cdots & b_{in} \\ & & \cdots & \\ a_{n1} & a_{n2} & \cdots & a_{nn} \end{vmatrix} + \begin{vmatrix} a_{11} & a_{12} & \cdots & a_{1n} \\ & & \cdots & \\ c_{i1} & c_{i2} & \cdots & c_{in} \\ & & \cdots & \\ a_{n1} & a_{n2} & \cdots & a_{nn} \end{vmatrix}
$$

which proves the property for rows. By Property I it holds also for columns.

PROPERTY V

If one row (column) of a determinant is multiplied by a scalar the value of the determinant is multiplied by the scalar.

EXAMPLE:

$$\begin{vmatrix} a_1 & b_1 & \lambda c_1 \\ a_2 & b_2 & \lambda c_2 \\ a_3 & b_3 & \lambda c_3 \end{vmatrix} = \lambda \begin{vmatrix} a_1 & b_1 & c_1 \\ a_2 & b_2 & c_2 \\ a_3 & b_3 & c_3 \end{vmatrix}$$

PROOF:

Suppose we replace the ith row in $|a_{ij}|$ by

$$\lambda a_{i1}, \lambda a_{i2}, \cdots, \lambda a_{in}$$

Then the determinant becomes

$$\sum \pm a_{1p_1} a_{2p_2} \cdots \lambda a_{ip_i} \cdots a_{np_n} = \lambda \sum \pm a_{1p_1} a_{2p_2} \cdots a_{ip_i} \cdots a_{np_n}$$

since we take outside the summation the factor λ, which is common to all the terms. Thus this factor multiplies the original determinant $|a_{ij}|$.

PROPERTY VI

The addition (in the vector sense) to a row (column) of a scalar multiple of another row (column) leaves the value of a determinant unchanged.

PROOF:

Let us write a determinant as

$$|C_1 \quad C_2 \quad \cdots \quad C_n|$$

where C_i denotes the ith column of the determinant. Suppose we add λ times column s to column r. Then this property asserts that the determinant that we get, viz.,

$$(14.11.5) \qquad |C_1, C_2, \cdots, C_{r-1}, C_r + \lambda C_s, C_{r+1}, \cdots, C_s, \cdots, C_n|$$

has the same value as the original determinant.

By Property IV, we can write (14.11.5) as

$$|C_1, C_2, \cdots, C_{r-1}, C_r, C_{r+1}, \cdots, C_s, \cdots, C_n|$$
$$+ |C_1, C_2, \cdots, C_{r-1}, \lambda C_s, C_{r+1}, \cdots, C_s, \cdots, C_n|$$

The first of these determinants is the original one. By Property V, the second is λ times the determinant.

$$|C_1, C_2 \cdots, C_{r-1}, C_s, C_{r+1}, \cdots, C_s, \cdots, C_n|$$

and this is zero (Property III) since it has two identical columns (C_s). This proves the result.

Property VI gives us a powerful method of evaluating determinants. With its aid we can manipulate a determinant, altering its appearance (i.e, altering it to the determinant of a different matrix) without altering its value. If we do this intelligently, we can get a determinant in a form which is more easily evaluated—usually by introducing zero elements. A few examples will show how this can be done. Since we already have in the Rule of Sarrus a convenient method for evaluating 3×3 determinants, let us look at some 4×4 examples.

EXAMPLE 1:

Evaluate

$$\begin{vmatrix} 1 & 2 & 2 & 8 \\ 8 & 2 & 2 & 2 \\ 2 & 1 & 1 & 1 \\ 4 & 9 & -2 & -1 \end{vmatrix}$$

Let us add -2 times row 3 to row 2. Row 2 then becomes 4, 0, 0, 0. Hence the determinant

$$\begin{vmatrix} 1 & 2 & 2 & 8 \\ 4 & 0 & 0 & 0 \\ 2 & 1 & 1 & 1 \\ 4 & 9 & -2 & -1 \end{vmatrix}$$

has the same value as the original.

We now subtract twice row 3 from row 1. We obtain

$$\begin{vmatrix} -3 & 0 & 0 & 6 \\ 4 & 0 & 0 & 0 \\ 2 & 1 & 1 & 1 \\ 4 & 9 & -2 & -1 \end{vmatrix}$$

This new determinant has several zeros, enough to make its evaluation easy. In choosing sets of 4 elements (no two in the same row or column) we can ignore any which do not have the 4 from row 2. For otherwise they must have a zero from row 2, and the corresponding term is zero. Similarly, for a nonzero term, we must choose a nonzero term, the 6, from row 1 (we cannot choose the -3 since it is in the same column as the 4 in row 2 which we have already chosen). The other two elements must be chosen from rows 3 and 4 and from columns 2 and 3. Thus there are only two nonvanishing terms in the expansion of this determinant, viz.,

$$-(6 \cdot 4 \cdot 1 \cdot -2) = 48$$

and

$$6 \cdot 4 \cdot 1 \cdot 9 = 216$$

The first of these has a minus sign, since it corresponds to the permutation $\begin{pmatrix} 1 & 2 & 3 & 4 \\ 4 & 1 & 2 & 3 \end{pmatrix}$, which is odd. The other corresponds to $\begin{pmatrix} 1 & 2 & 3 & 4 \\ 4 & 1 & 3 & 2 \end{pmatrix}$, which is even. Hence the value of the determinant is $48 + 216 = 264$.

EXAMPLE 2:
Evaluate

$$\begin{vmatrix} 1 & 2 & 3 & -6 \\ -1 & 0 & -2 & 10 \\ 4 & 10 & 12 & -19 \\ 2 & 4 & 6 & -11 \end{vmatrix}$$

We shall perform the following operations on the determinants, none of which will alter the value:
  (a) subtract 4 times row 1 from row 3
  (b) subtract 2 times row 1 from row 4
  (c) add row 1 to row 2
  (d) subtract row 2 from row 3.
We get, in succession

$$\begin{vmatrix} 1 & 2 & 3 & -6 \\ -1 & 0 & -2 & 10 \\ 0 & 2 & 0 & 5 \\ 2 & 4 & 6 & -11 \end{vmatrix}, \quad \begin{vmatrix} 1 & 2 & 3 & -6 \\ -1 & 0 & -2 & 10 \\ 0 & 2 & 0 & 5 \\ 0 & 0 & 0 & 1 \end{vmatrix}, \quad \begin{vmatrix} 1 & 2 & 3 & -6 \\ 0 & 2 & 1 & 4 \\ 0 & 2 & 0 & 5 \\ 0 & 0 & 0 & 1 \end{vmatrix}$$

and

$$\begin{vmatrix} 1 & 2 & 3 & -6 \\ 0 & 2 & 1 & 4 \\ 0 & 0 & -1 & 1 \\ 0 & 0 & 0 & 1 \end{vmatrix}.$$

This last determinant is in what is known as *upper-triangular form*, that is to say having all its nonzero elements on or above the main diagonal (from the top left to bottom right of the determinant) and only zeros below the diagonal. The value of such a determinant is easily found, since there is only one term that does not vanish, this being the product of the diagonal elements. Thus in the given determinant we must choose the 1 from row 4 (all other elements are zero), we must choose the -1 from row 3 (since column 4 is already taken), the 2 from row 2 and the 1 from row 1. Hence the value of this determinant is $1 \cdot 2.-1 \cdot 1 = -2$. This reduction to upper triangular form (or to *lower triangular form*, defined similarly) is a common method for evaluating determinants.

EXAMPLE 3:

Evaluate

$$\begin{vmatrix} -10 & 7 & 2 & 5 \\ 4 & 3 & 4 & -1 \\ -4 & 5 & -4 & 4 \\ -6 & 2 & 6 & 1 \end{vmatrix}$$

If we subtract row 4 from row 1, we get

$$\begin{vmatrix} -4 & 5 & -4 & 4 \\ 4 & 3 & 4 & -1 \\ -4 & 5 & -4 & 4 \\ -6 & 2 & 6 & 1 \end{vmatrix}$$

Here we have two identical rows (1 and 3); hence the value of the determinant is zero (Property III).

Exercise 162

1. Use the rules given in the last section to simplify, and hence evaluate, the following determinants.

(a) $\begin{vmatrix} 3 & 2 & -1 & 4 \\ 4 & 4 & -1 & 4 \\ 2 & 1 & 3 & -2 \\ 3 & 1 & 3 & 2 \end{vmatrix}$

(c) $\begin{vmatrix} 3 & 2 & 4 & 5 \\ 3 & 1 & 2 & 3 \\ 1 & 3 & 1 & 3 \\ 2 & 2 & 2 & 3 \end{vmatrix}$

(b) $\begin{vmatrix} 2007 & 198 & 108 \\ 1964 & 194 & 96 \\ 539 & 53 & 36 \end{vmatrix}$

(d) $\begin{vmatrix} 201 & 36 & 145 & 4 \\ 150 & 28 & 108 & 3 \\ 100 & 19 & 73 & 2 \\ 301 & 54 & 216 & 6 \end{vmatrix}$

2. Prove that the value of the determinant

$$\begin{vmatrix} \cos\theta & (1 + \tfrac{1}{2}\sec\theta)\sec^2\dfrac{\theta}{2} & \sin\theta \\ \cos^2\theta & 1 & \sin\theta\cos\theta \\ 2\sin\theta\cos^2\dfrac{\theta}{2} & k & \sin^2\theta - \cos\theta \end{vmatrix}$$

is $\tfrac{1}{2}\cos\theta\cos^2(\theta/2)$.

3. Find all the roots of the equations

(a) $\begin{vmatrix} 1 & 1 & 1 \\ x & 2 & 1 \\ x^3 & 8 & 1 \end{vmatrix} = 0.$

(b) $\begin{vmatrix} x & x^2 & a^3 - x^3 \\ b & b^2 & a^3 - b^3 \\ c & c^2 & a^3 - c^3 \end{vmatrix} = 0.$

4. Find the 3 values of λ for which

$$\begin{vmatrix} 13 - \lambda & -10 & 8 \\ -10 & 16 - \lambda & -2 \\ 8 & -2 & 7 - \lambda \end{vmatrix} = 0.$$

14.12 MINORS AND COFACTORS

In the expression for $|a_{ij}|$, what is the sum of the terms which contain a_{11}? The terms that contain a_{11} are of the form

$$\pm\, a_{11}\, a_{2 p_2}\, a_{3 p_3} \cdots a_{n p_n}$$

where $p_2, p_3, \ldots, p_n$ are the integers $2, 3, \ldots, n$ in some order. Hence the required sum is

(14.12.1) $$a_{11} \sum \pm\, a_{2 p_2}\, a_{3 p_3} \cdots a_{n p_n},$$

when we take outside the summation the common factor a_{11}. We must remember here that the choice between $\pm$ depends on the permutation

(14.12.2) $$\begin{pmatrix} 1 & 2 & 3 & \cdots & n \\ 1 & p_2 & p_3 & \cdots & p_n \end{pmatrix}$$

of n objects, rather than the permutation

(14.12.3)
$$\begin{pmatrix} 2 & 3 & \cdots & n \\ p_2 & p_3 & \cdots & p_n \end{pmatrix}$$

of $n - 1$ objects. However, it is easy to show by means of a diagram, that the parities of (14.12.2) and (14.12.3) are the same. Thus the summation in (14.12.1) is, by definition, a determinant, and we can therefore write (14.12.1) as

$$a_{11} \begin{vmatrix} a_{22} & a_{23} & \cdots & a_{2n} \\ a_{32} & a_{33} & \cdots & a_{3n} \\ \cdot & & \cdots & \cdot \\ \cdot & & \cdots & \cdot \\ a_{n2} & a_{n3} & \cdots & a_{nn} \end{vmatrix}$$

The determinant we have just written down can be thought of as having been obtained from the original determinant by deleting the row and column through the element a_{11}. This is an idea that can be generalized. If we consider a particular element a_{st} of an $n \times n$ determinant, and delete the row and column in which it stands (i.e., row s and column t) we get an $(n - 1) \times (n - 1)$ determinant, which we shall call the *minor* of a_{st}.

Let us now ask "What is the sum of the terms that contain a_{st}?" We could tackle this question from the beginning, but it is simpler to base our answer on what we have just done. If we interchange row s and the row above; then again interchange the original row s with the one above (which will be row $s - 2$) and so on, we can move row s successively upwards until it becomes the first row. It is readily seen that $s - 1$ interchanges will be required. By a similar succession of interchanges, we now arrange for column t to be moved to the left until it becomes the first column. In all, $t - 1$ interchange will be necessary to effect this latter move.

The result of these operations will be to bring the element a_{st} into the top left hand corner. The minor of a_{st} in this position is the same as its minor in its original position, since the order of the other rows and columns has not been changed. In performing the interchanges, however, we will have changed the sign of the determinant $(s - 1) + (t - 1)$ times, and will therefore have multiplied it by

$$(-1)^{s-1}(-1)^{t-1} = (-1)^{s+t-2} = (-1)^{s+t}.$$

From what went before we know that the sum of the terms containing a_{st} will therefore be $(-1)^{s+t}a_{st}$ times the minor. This sum, or rather the factor that goes with a_{st} in the sum, is of importance, and to avoid the variation in sign we introduce a new definition.

COFACTOR.

The *cofactor* of the element a_{st} in $|a_{ij}|$ is $(-1)^{s+t}$ times the minor of a_{st}. Thus to find the cofactor of an element in a determinant we merely delete the row and column in which the element stands and evaluate the resulting determinant. This gives us the minor. To get the cofactor we must attach the correct sign. We therefore multiply the minor by $(-1)^{s+t}$.

It is customary to denote elements of a determinant (or matrix) by small subscripted letters, and to use the corresponding capital letters to denote cofactors. Thus in the determinant $|a_{ij}|$, the cofactor of a_{st} would be denoted by A_{st}. It follows from what we have done that the terms in $|a_{ij}|$ which contain a_{st} will sum to $a_{st}A_{st}$.

EXAMPLE 1:

To find the cofactor of the 3 (row 2, column 3) in the determinant

$$\begin{vmatrix} 1 & 6 & 5 & 4 \\ 2 & 2 & 3 & -1 \\ -4 & 0 & 1 & -3 \\ 2 & -1 & 7 & 2 \end{vmatrix}$$

Deleting the row and column through this element, we obtain the determinant

$$\begin{vmatrix} 1 & 6 & 4 \\ -4 & 0 & -3 \\ 2 & -1 & 2 \end{vmatrix}$$

whose value is $0 - 36 + 16 - 0 - 3 + 48 = 25$ by the Rule of Sarrus. This is the minor of the element in question. To get the cofactor we must multiply by $(-1)^{2+3} = -1$. Thus the required cofactor is -25.

EXAMPLE 2.

To find the cofactor of the -4 (row 3, column 1). The minor is

$$\begin{vmatrix} 6 & 5 & 4 \\ 2 & 3 & -1 \\ -1 & 7 & 2 \end{vmatrix} = 131$$

Hence the cofactor is $(-1)^{3+1} 131 = 131$.

EXPANSION OF A DETERMINANT BY A ROW OR A COLUMN.

Let us consider a particular row of the determinant $|a_{ij}|$, say row i. From the definition of a determinant we know that every term in the expansion of $|a_{ij}|$ must contain exactly one element from row i. Now the terms that contain a_{i1} sum to $a_{i1} A_{i1}$, where A_{i1} is the cofactor of a_{i1}; those that contain a_{i2} sum to $a_{i2} A_{i2}$, and so on. It follows that the sum

$$a_{i1}A_{i1} + a_{i2}A_{i2} + \cdots + a_{in}A_{in}$$

contains every term in the expansion of the determinant, and therefore gives the value of the determinant. Thus we have an expansion of the determinant by the ith row, obtained by multiplying each element of this row by its cofactor and summing the products. We may write

$$|a_{ij}| = \sum_{t=1}^{n} a_{it}A_{it}.$$

In the same way, if we take each element of a particular column, multiply by its cofactor, and sum the products thus obtained, we again get an expansion of the determinant. If the column in question is the jth, we have

$$|a_{ij}| = \sum_{s=1}^{n} a_{sj}A_{sj}.$$

Thus we see that the problem of evaluating an $n \times n$ determinant can be made to depend on the evaluation of n determinants of size $(n - 1) \times (n - 1)$. These in turn, can be evaluated by expressing them in terms of $(n - 2) \times (n - 2)$ determinants and so on. Thus we can, in theory at any rate, evaluate determinants of any size, without having to enumerate the $n!$ permutations of the subscripts. What one does in practice is to use a combination of methods, including manipulations of rows and columns to product zero elements. Various other techniques are available which can be used as the occasion demands, but which we cannot take space to describe.

EXAMPLE 1:
Evaluate

$$\begin{vmatrix} 2 & 1 & 5 & -3 & 7 \\ 4 & 2 & 6 & 0 & 8 \\ 1 & 3 & 2 & -6 & 14 \\ 7 & 0 & 0 & 5 & 3 \\ 1 & 0 & 2 & -3 & 4 \end{vmatrix}$$

If we subtract twice row 1 from row 2, and also from row 3 we get

$$\begin{vmatrix} 2 & 1 & 5 & -3 & 7 \\ 0 & 0 & -4 & 6 & -6 \\ -3 & 1 & -8 & 0 & 0 \\ 7 & 0 & 0 & 5 & 3 \\ 1 & 0 & 2 & -3 & 4 \end{vmatrix}$$

If we now expand this by column 2 we get

$$-\begin{vmatrix} 0 & -4 & 6 & -6 \\ -3 & -8 & 0 & 0 \\ 7 & 0 & 5 & 3 \\ 1 & 2 & -3 & 4 \end{vmatrix} - \begin{vmatrix} 2 & 5 & -3 & 7 \\ 0 & -4 & 6 & -6 \\ 7 & 0 & 5 & 3 \\ 1 & 2 & -3 & 4 \end{vmatrix}$$

The first of these we can expand by row 2 to get

$$+(-3)\begin{vmatrix} -4 & 6 & -6 \\ 0 & 5 & 3 \\ 2 & -3 & 4 \end{vmatrix} - (-8)\begin{vmatrix} 0 & 6 & -6 \\ 7 & 5 & 3 \\ 1 & -3 & 4 \end{vmatrix}$$

to which we can apply the Rule of Sarrus, getting

$$-3(-80 + 36 + 60 - 36) + 8(18 + 126 + 30 - 168) = 108.$$

In the second determinant we can add column 3 to column 4 to obtain

$$-\begin{vmatrix} 2 & 5 & -3 & 4 \\ 0 & -4 & 6 & 0 \\ 7 & 0 & 5 & 8 \\ 1 & 2 & -3 & 1 \end{vmatrix} = 4\begin{vmatrix} 2 & -3 & 4 \\ 7 & 5 & 8 \\ 1 & -3 & 1 \end{vmatrix} + 6\begin{vmatrix} 2 & 5 & 4 \\ 7 & 0 & 8 \\ 1 & 2 & 1 \end{vmatrix}$$

on expansion by the second row. We then have

$$4(10 - 24 - 84 - 20 + 48 + 21) + 6(40 + 56 - 32 - 35)$$
$$= -196 + 174 = -22.$$

Hence the value of the determinant is $108 - 22 = 86$. As can be seen from these examples, the calculation is simpler if we expand by a row or column having zeros, and we normally choose that having the largest number of zeros.

Exercise 163

1. Find the cofactors of the elements a_{24} and a_{32} in each of the following determinants.

(a) $\begin{vmatrix} 2 & 1 & 7 & 4 \\ -1 & 3 & 0 & 6 \\ 5 & 1 & 1 & -2 \\ 3 & 9 & 0 & 5 \end{vmatrix}$ (c) $\begin{vmatrix} -1 & -3 & 4 & 1 \\ 3 & -1 & 2 & 2 \\ -4 & -2 & -1 & 5 \\ -1 & -2 & -5 & -1 \end{vmatrix}$

(b) $\begin{vmatrix} 1 & -1 & -2 & -3 \\ -1 & 2 & 0 & 2 \\ -1 & 0 & 3 & 3 \\ -3 & 2 & 3 & 4 \end{vmatrix}$ (d) $\begin{vmatrix} -2 & 1 & 0 & 1 \\ 1 & -3 & 0 & 1 \\ 0 & 0 & -2 & 1 \\ 1 & 1 & 1 & -3 \end{vmatrix}$

2. If A_{ij} is the cofactor of a_{ij} in the determinant $|a_{ij}|$, find the value of the determinant $|A_{ij}|$ when $|a_{ij}|$ is

(a) $\begin{vmatrix} 2 & 1 & -4 \\ 0 & -7 & 2 \\ 3 & 1 & 1 \end{vmatrix}$ (c) $\begin{vmatrix} 4 & 0 & -3 \\ 1 & 2 & 5 \\ -6 & 3 & 4 \end{vmatrix}$

(b) $\begin{vmatrix} 3 & 0 & 7 \\ -1 & 4 & 2 \\ 7 & 3 & -4 \end{vmatrix}$ (d) $\begin{vmatrix} 2 & -2 & -1 \\ 2 & 1 & 2 \\ 1 & 2 & -2 \end{vmatrix}$

Compare the determinants $|a_{ij}|$ and $|A_{ij}|$. What appears to be the relation between them?

3. Expand each of the determinants in problem 1 by the elements of the second column.

14.13 EXPANSION BY WRONG COFACTORS

What would happen if we took the elements of one row of a determinant and multiplied them not by their own cofactors but by the cofactors of the corresponding elements of some other row? To be more specific, suppose we multiply the elements of row r by the cofactors of the corresponding elements of row s ($r \neq s$). It is as if we wanted to expand the determinant by row r but inadvertently took the cofactors pertaining to row s. For this reason this process can be called *expansion by wrong cofactors* or by *alien cofactors*.

The expression we would get is

(14.13.1) $\qquad a_{r1}A_{s1} + a_{r2}A_{s2} + \cdots + a_{rn}A_{sn}.$

Compare this with the correct expansion by the elements of row s, viz.,

(14.13.2) $\qquad a_{s1}A_{s1} + a_{s2}A_{s2} + \cdots a_{sn}A_{sn}$

and we see that (14.13.1) is the expansion of the determinant which is obtained from $|a_{ij}|$ by replacing a_{s1} by a_{r1}, a_{s2} by a_{r2}, and so on. In other words (14.13.1) is the correct expansion of the determinant

(14.13.3)

$$
\begin{array}{c}
 \\
 \\
 \\
\text{row } r \\
 \\
\text{row } s \\
 \\
 \\
\end{array}
\begin{vmatrix}
a_{11} & a_{12} & a_{13} & \cdots & a_{1n} \\
a_{21} & a_{22} & a_{23} & \cdots & a_{2n} \\
\cdot & & \cdots & & \cdot \\
a_{r1} & a_{r2} & a_{r3} & \cdots & a_{rn} \\
\cdot & & \cdots & & \cdot \\
a_{r1} & a_{r2} & a_{s3} & \cdots & a_{rn} \\
\cdot & & \cdots & & \cdot \\
a_{n1} & a_{n2} & a_{n3} & \cdots & a_{nn}
\end{vmatrix}
$$

Clearly, the value of this determinant is zero, since it has two identical rows. A similar result applies to columns. Hence we have the following important result.

PROPERTY VII
Expansion by wrong cofactors always gives the value zero.

Exercise 164

Verify the result of this section (in several ways) using the following determinant

$$
\begin{vmatrix}
2 & 1 & 3 & 0 \\
-1 & 5 & 2 & 7 \\
6 & -1 & 0 & 3 \\
0 & 2 & -5 & 2
\end{vmatrix}
$$

14.14 SOLUTION OF EQUATIONS

We now apply the results we have just proved to the solution of a set of simultaneous linear equations. For convenience we shall consider a set of

three equations in three unknowns, but the treatment will be general, and applicable to any number of equations in an equal number of unknowns.
Let our equations be

(14.14.1) $$a_{11}x_1 + a_{12}x_2 + a_{13}x_3 = k_1$$

(14.14.2) $$a_{21}x_1 + a_{22}x_2 + a_{23}x_3 = k_2$$

(14.14.3) $$a_{31}x_1 + a_{32}x_2 + a_{33}x_3 = k_3.$$

This set of equations can be written in matrix notation as

$$\mathbf{Ax} = \mathbf{k}.$$

Now there are methods, with which the reader is doubtless familiar, for solving such a set of equations by successive elimination of unknowns, but we shall proceed to obtain the solution by a somewhat different route. We shall multiply equation (14.14.1) by A_{11} (the cofactor of a_{11} in the determinant of the matrix (a_{ij})); equation (14.14.2) by A_{21}, and equation (14.14.3) by A_{31}. If we now add corresponding sides of the resulting equations we obtain

(14.14.4)
$$(a_{11}A_{11} + a_{21}A_{21} + a_{31}A_{31})x_1 + (a_{12}A_{11} + a_{22}A_{21} + a_{32}A_{31})x_2$$
$$+ (a_{13}A_{11} + a_{23}A_{21} + a_{33}A_{31})x_3 = k_1A_{11} + k_2A_{21} + k_3A_{31}.$$

In equation (14.14.4) the coefficient of x_1 is

$$a_{11}A_{11} + a_{21}A_{21} + a_{31}A_{31}$$

and this is simply the expansion of the determinant $|a_{ij}|$ ($i, j = 1, 2, 3$) by the elements of its first column.
If we now look at the coefficient of x_2, viz.,

$$(a_{12}A_{11} + a_{22}A_{21} + a_{32}A_{31})$$

we find that this is an "expansion" of the same determinant by wrong cofactors (the elements are those of the second column, while the cofactors are those of the elements of the first column). Hence the coefficient of x_2 is zero. In exactly the same way the coefficient

$$a_{13}A_{11} + a_{23}A_{21} + a_{33}A_{31}$$

of x_3 is also an expansion by wrong cofactors, and is also zero.
On the right-hand side we have

$$k_1A_{11} + k_2A_{21} + k_3A_{31}$$

which is the expansion, by the elements of the first column, of the determinant

$$
\begin{vmatrix}
k_1 & a_{12} & a_{13} \\
k_2 & a_{22} & a_{23} \\
k_3 & a_{32} & a_{33}
\end{vmatrix}
$$

that is, the determinant obtained from $|a_{ij}|$ by replacing its first column by the column vector $\mathbf{k}$. Thus equation (14.14.4) reduces to

$$
(14.14.5) \qquad \Delta \cdot x_1 =
\begin{vmatrix}
k_1 & a_{12} & a_{13} \\
k_2 & a_{22} & a_{23} \\
k_3 & a_{32} & a_{33}
\end{vmatrix}
$$

where Δ stands for the determinant $|a_{ij}|$ of the coefficients on the left-hand sides of the equations.

Thus, at one blow, we have eliminated both x_2 and x_3 from the equations, leaving one simple equation from which x_1 can be found—or rather, can usually be found; for there is still a possible snag. Solving (14.14.5) for x_1 we get

$$
(14.14.6) \qquad x_1 = \frac{
\begin{vmatrix}
k_1 & a_{12} & a_{13} \\
k_2 & a_{22} & a_{23} \\
k_3 & a_{32} & a_{33}
\end{vmatrix}
}{\Delta}
$$

provided Δ is not zero.

Let us assume for the moment that $\Delta \neq 0$; we shall worry later about what to do if it is zero.

The other two variables can be found in the same sort of way. If we multiply (14.14.1) by A_{12}, (14.14.2) by A_{22} and (14.14.3) by A_{32} (using the cofactors for the second column—the coefficients of x_2) and add, we find that the coefficients of x_1 and x_3 both vanish, and we are left with

$$
(14.14.7) \qquad \Delta \cdot x_2 =
\begin{vmatrix}
a_{11} & k_1 & a_{13} \\
a_{21} & k_2 & a_{23} \\
a_{31} & k_3 & a_{33}
\end{vmatrix}
$$

where the determinant on the right is obtained by replacing the coefficients of x_2 by the constant terms. Hence, under our supposition that $\Delta \neq 0$, we have

(14.14.8)
$$x_2 = \frac{\begin{vmatrix} a_{11} & k_1 & a_{13} \\ a_{21} & k_2 & a_{23} \\ a_{31} & k_3 & a_{33} \end{vmatrix}}{\Delta}$$

By the same method (and the reader should check this for himself) we can eliminate x_1 and x_2 to obtain

(14.14.9)
$$x_3 = \frac{\begin{vmatrix} a_{11} & a_{12} & k_1 \\ a_{21} & a_{22} & k_2 \\ a_{31} & a_{33} & k_3 \end{vmatrix}}{\Delta}$$

Equations (14.14.6), (14.14.8) and (14.14.9) give us the complete solution of the set of equations.

For a general set of equations, n equations in n unknowns, the procedure is the same. For the equations $\mathbf{Ax} = \mathbf{k}$, the value of x_i is the quotient of two determinants. In the denominator we have $|\mathbf{A}| = |a_{ij}|$; in the numerator we have the determinant which is obtained from $\mathbf{A}$ by replacing its ith column by the column-vector $\mathbf{k}$.

EXAMPLE 1:

Solve the equations

$$x + y + z = 1$$
$$x + 2y + 3z = 4$$
$$x + 3y + 4z = 6.$$

The determinant Δ is

$$\begin{vmatrix} 1 & 1 & 1 \\ 1 & 2 & 3 \\ 1 & 3 & 4 \end{vmatrix} = 8 + 3 + 3 - 2 - 9 - 4$$
$$= -1$$

by the Rule of Sarrus.

To find x we evaluate the determinant

$$\begin{vmatrix} 1 & 1 & 1 \\ 4 & 2 & 3 \\ 6 & 3 & 4 \end{vmatrix} = 8 + 18 + 12 - 12 - 9 - 16$$

$$= 1$$

Hence $x = 1/-1 = -1$.

To find y we look at the determinant

$$\begin{vmatrix} 1 & 1 & 1 \\ 1 & 4 & 3 \\ 1 & 6 & 4 \end{vmatrix} = 16 + 3 + 6 - 4 - 18 - 4$$

$$= -1.$$

Hence $y = -1/-1 = 1$.

Finally, to find z we evaluate

$$\begin{vmatrix} 1 & 1 & 1 \\ 1 & 2 & 4 \\ 1 & 3 & 6 \end{vmatrix} = 12 + 4 + 3 - 2 - 12 - 6$$

$$= -1.$$

Hence $z = -1/-1 = 1$.

The solution is therefore $x = -1; y = z = 1$.

EXAMPLE 2:

Solve

$$2x + y - z = 0$$
$$x - 5y + 3z = 31$$
$$6x + 2y - 5z = 3.$$

The four determinants that we need to evaluate are:

$$\begin{vmatrix} 2 & 1 & -1 \\ 1 & -5 & 3 \\ 6 & 2 & -5 \end{vmatrix} = 50 + 18 - 2 - 30 - 12 + 5 = 29$$

$$\begin{vmatrix} 0 & 1 & -1 \\ 31 & -5 & 3 \\ 3 & 2 & -5 \end{vmatrix} = 9 - 62 - 15 + 155 = 87$$

$$\begin{vmatrix} 2 & 0 & -1 \\ 1 & 31 & 3 \\ 6 & 3 & -5 \end{vmatrix} = -310 - 3 + 186 - 18 = -145$$

and

$$\begin{vmatrix} 2 & 1 & 0 \\ 1 & -5 & 31 \\ 6 & 2 & 3 \end{vmatrix} = -30 + 186 - 124 - 3 = 29.$$

Hence

$$x = 87/29 = 3$$
$$y = -145/29 = -5$$

and $$z = 29/29 = 1.$$

The method just described for writing down the solution of a set of simultaneous equations is known as *Cramer's rule*.

Exercise 165

1. Solve the equations

$$4x + 3y + 5z = 11$$
$$9x + 4y + 15z = 13$$
$$12x + 10y - 3z = 4$$

by the methods of the last section.

2. If a, b, c are three numbers, no two of which are equal, solve the equations

$$x + y + z = 1$$
$$ax + by + cz = 1$$
$$a^2x + b^2y + c^2z = 1.$$

3. For what value of k do the equations

$$2x - 3y + 7z = a$$
$$5x + 4y - 2z = -3$$
$$x - 13y + kz = 9$$

not have a unique solution? Find the solution of these equations when k does not have this value.

14.15 EQUATIONS FOR WHICH $\Delta = 0$

If $\Delta = 0$ then we can carry out the manipulation of the equations up to the stage where we obtain equation (14.14.5), since we have not yet made any assumption about the value of Δ. But if $\Delta = 0$, then the left-hand side of (14.14.5) is zero, irrespective of what the value of x_1 may be. Hence the right-hand side must also be zero if equation (14.14.5) is to be satisfied. This condition can be written

(14.15.1) $\qquad k_1 A_{11} + k_2 A_{21} + k_3 A_{31} = 0.$

Now there is something funny here; for unless A_{11}, A_{21} and A_{31} are all zero (and this possibility we will leave for the moment) then we can certainly find values of k_1, k_2, and k_3 which will falsify equation (14.15.1). For example, if $A_{11} \neq 0$ then equation (14.15.1) will not be satisfied if $k_1 = 1, k_2 = k_3 = 0$. But the constants on the right-hand sides of our equation might well have just these values, in which case equation (14.15.1) contradicts equation (14.14.5). What has gone wrong?

Our discussion of the *reductio ad absurdum* form of argument (Chapter 1) gives us the answer. We must have made an assumption somewhere in the argument, an assumption that is not true. We have been discussing the set of equations (14.14.1), (14.14.2) and (14.14.3) and manipulating them, and in so doing we have tacitly assumed that it is possible for these equations to be simultaneously true. We have said, in effect, "If x_1, x_2, x_3 are numbers such that these equations are true, then such and such follows; in particular, equation (14.15.1) holds." But unless k_1, k_2, k_3 have rather special values, equation (14.15.1) does *not* hold. Hence it must be false that there exist numbers x_1, x_2, x_3 satisfying the original equations. In short, the equations have *no* solution.

In some simple examples this conclusion is very obvious. For instance, in the equations

(14.15.2) $\qquad\qquad x + y + z = 3$

(14.15.3) $\qquad\qquad x + y + 2z = 4$

(14.15.4) $\qquad\qquad 2x + 2y + 3z = 8$

it is easily seen that the left-hand side of (14.15.4) is the sum of the left-hand sides of (14.15.2) and (14.15.3). Since these latter left-hand sides have the values 3 and 4, respectively, their sum must be 7. But equation (14.15.4) says that it is 8. Hence if we take *any* values of x, y and z which satisfy (14.15.2) and (14.15.3), they will automatically *not* satisfy (14.15.4), since they will make its left-hand side into 7.

We therefore come to the conclusion that if the equations are such that $\Delta = 0$ and $k_1 A_{11} + k_2 A_{21} + k_3 A_{31} \neq 0$ then the equations have no solution. We say they are *incompatible*.

A completely different situation arises if the constants k_1, k_2, k_3 are such that

$$k_1 A_{11} + k_2 A_{21} + k_3 A_{31} = 0$$

for then equation (14.14.5) is automatically satisfied whatever the value of x_1. We are thus led to the possibility of a set of equations having more than one solution. That this can happen is easily demonstrated. We have already seen that if

$$x + y + z = 3$$

and

$$x + y + 2z = 4$$

then

$$2x + 2y + 3z = 7.$$

Hence in the set of equations

$$\begin{aligned}
x + y + z &= 3 \\
(14.15.5) \qquad x + y + 2z &= 4 \\
2x + 2y + 3z &= 7
\end{aligned}$$

any values of x, y and z which satisfy the first two equations will automatically satisfy the third. Hence we have, in effect, only two equations in the three unknowns since the third equation tells us nothing that is not implied by the first two. We could expect, in general, that two equations in three unknowns would have many solutions, since we could give one variable (say x) an arbitrary value and still have the right number (two) of equations for the remaining two unknowns. It is readily verified that $x = y = z = 1$ is a solution of (14.15.5) and so is

$$x = 0; \quad y = 2; \quad z = 1.$$

The reader can further verify that for equations (14.15.5) to be satisfied we must have $z = 1$, but that x and y can be any values satisfying $x + y = 2$. Thus here we have a whole lot of solutions—an infinity of solutions, in fact.

The question of whether a set of n equations in n unknowns has no solution, a unique solution, or an infinite number of solutions is more complex than we have indicated so far. For example, we have not shown whether the condition

$$k_1 A_{11} + k_2 A_{21} + k_3 A_{31} = 0$$

is sufficient to ensure that there are infinitely many solutions; nor have we dealt with the case $A_{11} = A_{21} = A_{31} = 0$ which arose earlier. One thing that is certain is that if $\Delta \neq 0$ then the equations have a unique solution, for which an explicit expression has been obtained. Just exactly what happens in the other cases has been left in some doubt, and we shall keep it that way for the time being.

EXAMPLE 1:

Show that the equations

$$5x + 2y - 7z = 7$$
$$3x + y - 3z = 5$$
$$2x + y - 4z = 10$$

are incompatible.

Here we have

$$\Delta = \begin{vmatrix} 5 & 2 & -7 \\ 3 & 1 & -3 \\ 2 & 1 & -4 \end{vmatrix} = 0$$

Substituting the constants into the first column of Δ we get

$$\begin{vmatrix} 7 & 2 & -7 \\ 5 & 1 & -3 \\ 10 & 1 & -4 \end{vmatrix} = 8 \neq 0.$$

Hence the equations are incompatible. We can see this more directly by subtracting the second equation from the first. We get $2x + y - 4z = 2$. If x, y, z satisfy the first two equations, then this equation too is satisfied. But it contradicts the third of the given equations.

EXAMPLE 2:

Find all the solutions of

$$x - 2y + z = -2$$
$$5x + y - 6z = 1$$
$$3x + 5y - 8z = 5.$$

Here

$$\Delta = \begin{vmatrix} 1 & -2 & 1 \\ 5 & 1 & -6 \\ 3 & 5 & -8 \end{vmatrix} = 0.$$

Putting the constants in column 1 gives

$$\begin{vmatrix} -2 & -2 & 1 \\ 1 & 1 & -6 \\ 5 & 5 & -8 \end{vmatrix} = 0$$

(clearly, since two columns are identical). Thus it looks as though x could take any value. Let us write

$$-2y + z = -2 - x$$
$$y - 6z = 1 - 5x$$
$$5y - 8z = 5 - 3x.$$

Adding twice the second equation to the first we get

$$-11z = -11x \quad \text{or} \quad z = x.$$

Adding 6 times the first to the second we get

$$-11y = -11 - 11x$$

or

$$y = 1 + x.$$

Hence it appears that we can give x any value, and that y and z are given by the equations

$$y = 1 + x, \, z = x.$$

Thus if we make $x = 1$, then $y = 2$ and $z = 1$. This is a solution. If we make $x = -2$, then $y = -1$ and $z = -2$; this is another solution. Any number of solutions can be constructed in this way.

We did not use the third equation in obtaining this result. The reason is that, given the first two equations, we get nothing new from the third. In fact, if we subtract twice the first equation from the second we arrive at the third, which is therefore a consequence of the other two.

Exercise 166

1. A set of equations like

$$a_1x + b_1y + c_1z = 0$$
$$a_2x + b_2y + c_2z = 0$$
$$a_3x + b_3y + c_3z = 0$$

in which the right-hand sides are all zeros, is called a set of *homogeneous equations*.

Show that these equations have the *trivial* solution $x = y = z = 0$, and deduce that they have other (*nontrivial*) solutions if, and only if,

$$\Delta = \begin{vmatrix} a_1 & b_1 & c_1 \\ a_2 & b_2 & c_2 \\ a_3 & b_3 & c_3 \end{vmatrix} = 0.$$

2. If the homogeneous equations

$$ax + by + cz = 0$$
$$bx + cy + az = 0$$
$$cx + ay + bz = 0.$$

have a nontrivial solution, show that

$$a^3 + b^3 + c^3 - 3abc = 0.$$

3. Show that the equations

$$3x - 4y + z = a$$
$$x + 3y - 2z = 9$$
$$3x - 17y + 8z = 49$$

do not have a unique solution. Find for what values of a they have

(a) no solution

(b) an infinity of solutions.

14.16 THE INVERSE OF A MATRIX

The reader may well have wondered why we left the topic of matrices to embark on this digression into determinants and their properties. In this section we return to our study of matrices, and it will be seen how we need determinants to settle the question that we left hanging a while back, namely, whether there is anything in matrix algebra corresponding to the concept of a reciprocal in ordinary algebra and arithmetic.

If $A = (a_{ij})$ is a square $n \times n$ matrix, then we can form the matrix (A_{ij}) where A_{ij} is the cofactor of a_{ij} in A. This matrix can be called the *matrix of cofactors* of A. This matrix is not very important, but its transpose is. The transpose of the matrix of cofactors is called the *adjoint*, or sometimes the *adjugate* of A, and is denoted by adj A. Thus

$$\text{adj } A = (A_{ij})^T$$
$$= (A_{ji}).$$

Let us consider the product of A with its adjoint. If B_{ij} denotes a typical element of adj A in the ith row and jth column, so that

$$B_{ij} = A_{ji}$$

then the element in the ith row and jth column of the product $A \cdot$ adj A is

(14.16.1) $$\sum_{r=1}^{n} a_{ir}B_{rj} = \sum_{r=1}^{n} a_{ir}A_{jr}$$
$$= a_{i1}A_{j1} + a_{i2}A_{j2} + \cdots + a_{in}A_{jn}$$

by the rule of multiplying two matrices. But the right-hand side of (14.16.1) is the "expansion" of the determinant $\Delta = |A|$ using the elements of the ith row and the cofactors of the elements of the jth row. If $i \neq j$ then this is an "expansion by wrong cofactors" and gives the value zero. If $i = j$ it is a proper expansion and gives the value Δ. Thus the (i, j) element in $A \cdot$ adj A is

Δ if $i = j$ and 0 otherwise, i.e.,

$$(14.16.2) \qquad A \cdot \text{adj } A = \begin{pmatrix} \Delta & 0 & 0 & \cdots & 0 \\ 0 & \Delta & 0 & \cdots & 0 \\ 0 & 0 & \Delta & \cdots & 0 \\ & & \cdot & \cdots & \cdot \\ 0 & 0 & 0 & \cdots & \Delta \end{pmatrix}$$

or

$$(14.16.3) \qquad A \cdot \text{adj } A = \Delta I$$

where I is the unit matrix. Remember that Δ is just a number.

So far we have made no assumptions about the matrix A (other than that it is square). Now we shall assume for the moment that $\Delta = |A| \neq 0$. If this is the case, then we can divide both sides of (14.16.3) by Δ and obtain

$$A \cdot \left(\frac{1}{\Delta} \text{adj } A \right) = I.$$

At last we have the matrix that we have been looking for, the one that behaves like a reciprocal. If we denote it by A^{-1} we have

$$A^{-1} = \frac{1}{\Delta} \text{adj } A$$

and its characteristic property is that

$$A \cdot A^{-1} = I.$$

The matrix A^{-1} is called the *inverse* of A.

SINGULAR AND NON-SINGULAR MATRICES

A matrix A for which $|A| = 0$ is said to be a *singular* matrix. One for which $|A| \neq 0$ is *nonsingular*. Hence we have the following result.

THEOREM 14.3

A NONSINGULAR SQUARE MATRIX HAS AN INVERSE; THAT IS, IF A IS NON-SINGULAR THEN THERE EXISTS A MATRIX A^{-1} SUCH THAT

$$(14.16.4) \qquad A \cdot A^{-1} = A^{-1} \cdot A = I.$$

We shall see that a singular matrix does not have an inverse. A rectangular matrix cannot have an inverse in the sense of (14.16.4) since if $\mathbf{A}$ is a $p \times q$ matrix ($p \neq q$), $\mathbf{A}^{-1}$ would have to be a $q \times p$ matrix for both products to exist; but then $\mathbf{A}\mathbf{A}^{-1}$ would be $p \times p$, and $\mathbf{A}^{-1}\mathbf{A}$ would be $q \times q$. Hence these two products could not be equal.

Let the columns of $\mathbf{A}^{-1}$ be denoted by $\mathbf{c}_1, \mathbf{c}_2, \ldots, \mathbf{c}_n$, these being column vectors. Then the equation

$$\mathbf{A}\cdot\mathbf{A}^{-1} = \mathbf{A}(\mathbf{c}_1, \mathbf{c}_2, \ldots, \mathbf{c}_n) = \mathbf{I}$$

can be split up into n sets of equations, viz.,

$$\mathbf{A}\,\mathbf{c}_1 = \begin{pmatrix} 1 \\ 0 \\ 0 \\ \cdots \\ 0 \end{pmatrix}$$

(14.16.5)
$$\mathbf{A}\,\mathbf{c}_2 = \begin{pmatrix} 0 \\ 1 \\ 0 \\ \cdots \\ 0 \end{pmatrix}$$

$$\cdots$$

$$\mathbf{A}\,\mathbf{c}_n = \begin{pmatrix} 0 \\ 0 \\ \cdots \\ 0 \\ 1 \end{pmatrix}$$

Hence the columns of the inverse of $\mathbf{A}$ (and hence $\mathbf{A}^{-1}$ itself) can be found by solving n sets of n equations in n unknowns. In each case the determinant of coefficients is $|\mathbf{A}|$ which we have assumed to be nonzero. Hence there is a unique solution for each column $\mathbf{c}_i$, and hence $\mathbf{A}^{-1}$ is found uniquely. This shows that the matrix $\mathbf{A}^{-1}$ in Theorem 14.3 is unique. If we set out to find $\mathbf{A}^{-1}$ by this method we might be tempted to use the method of determinants for the solution of the sets of equations. But if we did this, we would find that we were effectively tracing the same steps that we would take if we constructed the matrix of cofactors (the determinants in the numerators in the solutions would contain the cofactors). Hence there would be no advantage over using the original definition. The point is, however, that *any* method for solution of equations can be used to find the inverse, and some are better than the method of finding all the cofactors, especially for large matrices.

The reader's mind may boggle at the thought of having to solve n sets of equations, each of n equations in n unknowns, but it is not so bad. The sets of equations are such that they differ only in the column of constants, and it is therefore possible to write down all these columns at once and deal with them simultaneously. The method is described in detail in Example 3 below:

EXAMPLE 1:

Find the inverse of $\mathbf{A} = \begin{pmatrix} 1 & 7 \\ 2 & 9 \end{pmatrix}$.

The matrix of cofactors is $\begin{pmatrix} 9 & -2 \\ -7 & 1 \end{pmatrix}$, and the adjoint is thus $\begin{pmatrix} 9 & -7 \\ -2 & 1 \end{pmatrix}$. The determinant is -5. Hence

$$\mathbf{A}^{-1} = -\frac{1}{5} \begin{pmatrix} 9 & -7 \\ -2 & 1 \end{pmatrix} = \begin{pmatrix} -\dfrac{9}{5} & \dfrac{7}{5} \\ \dfrac{2}{5} & -\dfrac{1}{5} \end{pmatrix}.$$

Check:

$$\mathbf{A} \cdot \mathbf{A}^{-1} = \begin{pmatrix} 1 & 7 \\ 2 & 9 \end{pmatrix} \begin{pmatrix} -\dfrac{9}{5} & \dfrac{7}{5} \\ \dfrac{2}{5} & -\dfrac{1}{5} \end{pmatrix}$$

$$= \begin{pmatrix} -\dfrac{9}{5} + \dfrac{14}{5} & \dfrac{7}{4} - \dfrac{7}{5} \\ -\dfrac{18}{5} + \dfrac{18}{5} & \dfrac{14}{5} - \dfrac{9}{5} \end{pmatrix}$$

$$= \begin{pmatrix} 1 & 0 \\ 0 & 1 \end{pmatrix}.$$

EXAMPLE 2:

Find the inverse of the matrix

$$\mathbf{A} = \begin{pmatrix} 4 & -1 & -1 \\ -3 & 1 & 1 \\ 1 & 1 & -1 \end{pmatrix}$$

If we calculate all nine cofactors we obtain

$$\begin{pmatrix} -2 & -2 & -4 \\ -2 & -3 & -5 \\ 0 & -1 & 1 \end{pmatrix}$$

as the matrix of cofactors, and hence

$$\begin{pmatrix} -2 & -2 & -0 \\ -2 & -3 & -1 \\ -4 & -5 & 1 \end{pmatrix}$$

as the adjoint matrix. We easily calculate that $|\mathbf{A}| = -2$. Thus, dividing each element of adj $\mathbf{A}$ by -2 we find the inverse to be

$$\mathbf{A}^{-1} = \begin{pmatrix} 1 & 1 & 0 \\ 1 & \dfrac{3}{2} & \dfrac{1}{2} \\ 2 & \dfrac{5}{2} & -\dfrac{1}{2} \end{pmatrix}.$$

The reader should check that $\mathbf{A} \cdot \mathbf{A}^{-1} = \mathbf{A}^{-1} \cdot \mathbf{A} = \mathbf{I}$.

EXAMPLE 3.
Find the inverse of

$$\mathbf{A} = \begin{pmatrix} 50 & 5 & -5 \\ 25 & -17 & 10 \\ -5 & 10 & -5 \end{pmatrix}$$

We shall do this by solving the equations

$$\mathbf{A}\,\mathbf{x} = \mathbf{b}$$

where $\mathbf{b}$ is a vector one of whose elements is 1, the others being zero. Thus the equations to be solved are

$$\begin{matrix} 50x + 5y + 5z = 1 & 0 & 0 \\ 25x - 17y + 10z = 0 & \text{or} \quad 1 \quad \text{or} \quad 0 \\ -5x + 10y - 5z = 0 & 0 & 1 \end{matrix}$$

(14.16.6)

For convenience, we shall write down the 18 numbers that occur here in a double array, as follows:

(14.16.7)
$$\begin{pmatrix} 50 & 5 & 5 & : & 1 & 0 & 0 \\ 25 & -17 & 10 & : & 0 & 1 & 0 \\ -5 & 10 & -5 & : & 0 & 0 & 1 \end{pmatrix}$$

We shall now attempt to simplify the three equations by replacing them by other equations which are linear combinations of the existing ones. For example, we can add twice the third equation to the second. The effect on the double array (14.16.7) is to replace it by

$$\begin{pmatrix} 50 & 5 & 5 & : & 1 & 0 & 0 \\ 15 & 3 & 0 & : & 0 & 1 & 2 \\ -5 & 10 & -5 & : & 0 & 0 & 1 \end{pmatrix}$$

since whatever we do to the left-hand sides of the equations must also be done to the right-hand sides.

We can now add row 3 to row 1, and we get

$$\begin{pmatrix} 45 & 15 & 0 & : & 1 & 0 & 1 \\ 15 & 3 & 0 & : & 0 & 1 & 2 \\ -5 & 10 & -5 & : & 0 & 0 & 1 \end{pmatrix}.$$

Let us now subtract 5 times row 2 from row 1; this will give us

$$\begin{pmatrix} -30 & 0 & 0 & : & 1 & -5 & -9 \\ 15 & 3 & 0 & : & 0 & 1 & 2 \\ -5 & 10 & -5 & : & 0 & 0 & 1 \end{pmatrix}.$$

We now add $\frac{1}{2}$ times row 1 to row 2. This gives

$$\begin{pmatrix} -30 & 0 & 0 & : & 1 & -5 & -9 \\ 0 & 3 & 0 & : & \frac{1}{2} & -\frac{3}{2} & -\frac{5}{2} \\ -5 & 10 & -5 & : & 0 & 0 & 1 \end{pmatrix}$$

Now subtract $\frac{1}{6}$ times row 1 from row 3. We get

$$\begin{pmatrix} -30 & 0 & 0 & : & 1 & -5 & -9 \\ 0 & 3 & 0 & : & \frac{1}{2} & -\frac{3}{2} & -\frac{5}{2} \\ 0 & 10 & -5 & : & -\frac{1}{6} & \frac{5}{6} & \frac{5}{2} \end{pmatrix}.$$

Now subtract $\frac{10}{3}$ times row 2 from row 3. We get

$$\begin{pmatrix} -30 & 0 & 0 & : & 1 & -5 & -9 \\ 0 & 3 & 0 & : & \dfrac{1}{2} & -\dfrac{3}{2} & -\dfrac{5}{2} \\ 0 & 0 & -5 & : & -\dfrac{11}{6} & \dfrac{35}{6} & \dfrac{65}{6} \end{pmatrix}.$$

The matrix to the left of the dotted line is now a diagonal matrix. Since we can always multiply both sides of an equation by a constant, let us multiply each row of this double array by a constant in such a way as to make the left-hand matrix into the unit matrix. Multiplying row 1 by $-\frac{1}{30}$, row 2 by $\frac{1}{3}$ and row 3 by $-\frac{1}{5}$ we get

$$\begin{pmatrix} 1 & 0 & 0 & : & -\dfrac{1}{30} & \dfrac{1}{6} & \dfrac{3}{10} \\ 0 & 1 & 0 & : & \dfrac{1}{6} & -\dfrac{1}{2} & -\dfrac{5}{6} \\ 0 & 0 & 1 & : & \dfrac{11}{30} & -\dfrac{7}{6} & -\dfrac{13}{6} \end{pmatrix}.$$

The left-hand sides of the equations now read

$$\begin{aligned} x &= -\dfrac{1}{30} & \dfrac{1}{6} & & \dfrac{3}{10} \\ y &= \dfrac{1}{6} & -\dfrac{1}{2} & & -\dfrac{5}{6} \\ z &= \dfrac{11}{30} & -\dfrac{7}{6} & & -\dfrac{13}{6} \end{aligned}$$

where, for the right-hand sides, we take the three columns of the right-hand half of the double array. Hence these three columns are the solutions of the three sets of equations and hence, by what we said earlier, are the columns of the inverse matrix. Hence

$$\mathbf{A}^{-1} = \begin{pmatrix} -\dfrac{1}{30} & \dfrac{1}{6} & \dfrac{3}{10} \\ \dfrac{1}{6} & -\dfrac{1}{2} & -\dfrac{5}{6} \\ -\dfrac{11}{30} & -\dfrac{7}{6} & -\dfrac{13}{6} \end{pmatrix}$$

We can boil this procedure down to a very simple recipe. One way of finding the inverse of a matrix $\mathbf{A}$ is to write down $\mathbf{A}$ and the unit matrix $\mathbf{I}$ side by side in a double array

(14.16.8) $(\mathbf{A} : \mathbf{I})$.

Then, by adding multiples of one row to another, or by multiplying a row by a constant, in any convenient manner, we contrive to convert the left-hand matrix (originally $\mathbf{A}$) into a unit matrix. When this has been done, it will be found that the right-hand matrix (originally $\mathbf{I}$) will have been converted into the inverse of $\mathbf{A}$.

Looking at the problem from this point of view we can see how it works. Our manipulations of the rows are equivalent to multiplying the matrix (14.16.8) on the left by a matrix. For example, to add twice row 3 to row 2 is equivalent to multiplying $\mathbf{A}$ on the left by the matrix

$$\begin{pmatrix} 1 & 0 & 0 \\ 0 & 1 & 2 \\ 0 & 0 & 1 \end{pmatrix}.$$

Hence the final result of our manipulations will be that we shall have multiplied on the left by some matrix or other; call it $\mathbf{T}$. We therefore end up with the matrix

$$\mathbf{T}(\mathbf{A} : \mathbf{I}) = (\mathbf{TA} : \mathbf{T}).$$

Thus if the left-hand half is $\mathbf{I}$, we have $\mathbf{TA} = \mathbf{I}$; and hence, since the inverse is unique, we must have $\mathbf{T} = \mathbf{A}^{-1}$. The right-hand side is therefore the required inverse.

Exercise 167

1. Find, by any method, the inverses of the following matrices

(a) $\begin{pmatrix} 2 & 3 \\ -5 & 7 \end{pmatrix}$

(c) $\begin{pmatrix} 1 & 0 & 0 \\ 1 & 1 & 0 \\ 1 & 0 & 1 \end{pmatrix}$

(b) $\begin{pmatrix} 1 & 0 & 3 \\ 2 & -1 & -1 \\ 4 & 1 & -3 \end{pmatrix}$

(d) $\begin{pmatrix} 1 & 4 & 2 \\ 0 & 1 & 3 \\ -5 & 0 & -2 \end{pmatrix}$

2. By the method described at the end of the last section, find the inverse of the matrix

$$\begin{pmatrix} 1 & 0 & 2 & -1 \\ 2 & 1 & 0 & 1 \\ -1 & 2 & 1 & 0 \\ 1 & 1 & 1 & 2 \end{pmatrix}$$

3. Prove that

$$(\mathbf{AB})^{-1} = \mathbf{B}^{-1}\mathbf{A}^{-1}$$

where $\mathbf{A}$ and $\mathbf{B}$ are any two square, nonsingular matrices.

4. Show that the set of all nonsingular $n \times n$ matrices forms a group (in the sense of section 14.9) under the operation of matrix multiplication.

14.17 THE DETERMINANT OF THE PRODUCT OF TWO MATRICES

One important result in the theory of matrices that we have not yet had occasion to use, but that we shall need, is that which is expressed by the following theorem:

THEOREM 14.4

THE DETERMINANT OF THE PRODUCT OF TWO SQUARE MATRICES IS THE PRODUCT OF THEIR DETERMINANTS. IN SYMBOLS

$$|\mathbf{AB}| = |\mathbf{A}| \cdot |\mathbf{B}|.$$

PROOF:

There are many ways of proving this theorem. The one we choose has the advantage of being elementary but the disadvantage that it is difficult to describe for a pair of general $n \times n$ matrices. Accordingly we shall prove the theorem for the case of 3×3 matrices, and leave it to the reader to satisfy himself that a similar argument would apply to matrices of any size.

We take as our two matrices

$$\mathbf{A} = \begin{pmatrix} a_1 & b_1 & c_1 \\ a_2 & b_2 & c_2 \\ a_3 & b_3 & c_3 \end{pmatrix} \quad \text{and} \quad \mathbf{B} = \begin{pmatrix} \alpha_1 & \beta_1 & \gamma_1 \\ \alpha_2 & \beta_2 & \gamma_2 \\ \alpha_3 & \beta_3 & \gamma_3 \end{pmatrix}$$

We want to show that

$$|\mathbf{A}| \cdot |\mathbf{B}| = \begin{vmatrix} a_1 & b_1 & c_1 \\ a_2 & b_2 & c_2 \\ a_3 & b_3 & c_3 \end{vmatrix} \cdot \begin{vmatrix} \alpha_1 & \beta_1 & \gamma_1 \\ \alpha_2 & \beta_2 & \gamma_2 \\ \alpha_3 & \beta_3 & \gamma_3 \end{vmatrix}$$

is the same as

(14.17.1)

$$|\mathbf{AB}| = \begin{vmatrix} a_1\alpha_1 + b_1\alpha_2 + c_1\alpha_3 & a_1\beta_1 + b_1\beta_2 + c_1\beta_3 & a_1\gamma_1 + b_1\gamma_2 + c_1\gamma_3 \\ a_2\alpha_1 + b_2\alpha_2 + c_2\alpha_3 & a_2\beta_1 + b_2\beta_2 + c_2\beta_3 & a_2\gamma_1 + b_2\gamma_2 + c_2\gamma_3 \\ a_3\alpha_1 + b_3\alpha_2 + c_3\alpha_3 & a_3\beta_1 + b_3\beta_2 + c_3\beta_3 & a_3\gamma_1 + b_3\gamma_2 + c_3\gamma_3 \end{vmatrix}.$$

By a two-fold application of Rule II we can express the first column of (14.17.1) as the sum of three row vectors, and hence express (14.17.1) as the sum of 3 determinants, viz.,

$$\begin{vmatrix} a_1\alpha_1 & a_1\beta_1 + b_1\beta_2 + c_1\beta_3 & a_1\gamma_1 + b_1\gamma_2 + c_1\gamma_3 \\ a_2\alpha_1 & a_2\beta_1 + b_2\beta_2 + c_2\beta_3 & a_2\gamma_1 + b_2\gamma_2 + c_2\gamma_3 \\ a_3\alpha_1 & a_3\beta_1 + b_3\beta_2 + c_3\beta_3 & a_3\gamma_1 + b_3\gamma_2 + c_3\gamma_3 \end{vmatrix}$$

and two similar determinants whose first columns are

$$\begin{pmatrix} b_1\alpha_2 \\ b_2\alpha_2 \\ b_3\alpha_2 \end{pmatrix} \quad \text{and} \quad \begin{pmatrix} c_1\alpha_3 \\ c_2\alpha_3 \\ c_3\alpha_3 \end{pmatrix}$$

the other two columns being the same in all three cases.

Each of these 3 determinants can be similarly expressed as the sum of three determinants by splitting the second column. Furthermore each of these 9 determinants can also be expressed as the sum of three determinants by splitting up the third column. The result of all this is that the original determinant can be expressed as the sum of 27 determinants.

Several of these determinants, however, will be zero. For example, if we take the first part of the first column of $|\mathbf{AB}|$, and the first part of the second column, then the determinant will be

(14.17.2)

$$\begin{vmatrix} a_1\alpha_1 & a_1\beta_1 & \cdots \\ a_2\alpha_1 & a_2\beta_1 & \cdots \\ a_3\alpha_1 & a_3\beta_1 & \cdots \end{vmatrix}$$

where the third column has been left unspecified. Whatever the third column may be, determinant (14.17.2) is zero, because its first two columns are proportional. Taking out the common factors α_1 and β_1 from rows 2 and 3 we obtain a determinant with two identical columns. In the same way it can be shown that 21 of the 27 determinants have the value zero. Of these, 18 vanish because similar parts (first, second or third) have been chosen from two different columns, the other 3 vanish because similar parts were chosen from all three columns. The determinants which do not vanish are those where a different part has been chosen from each column of (14.17.1). Let us look at one of these determinants, say the one in which the third part has been chosen from column 1, the first part from column 2 and the second part from column 3, namely the determinant

$$
\begin{vmatrix}
c_1\alpha_3 & a_1\beta_1 & b_1\gamma_2 \\
c_2\alpha_3 & a_2\beta_1 & b_2\gamma_2 \\
c_3\alpha_3 & a_3\beta_1 & b_3\gamma_2
\end{vmatrix}.
$$

This is the same as

(14.17.3)
$$
\alpha_3\beta_1\gamma_2
\begin{vmatrix}
c_1 & a_1 & b_1 \\
c_2 & a_2 & b_2 \\
c_3 & a_3 & b_3
\end{vmatrix}.
$$

The determinant here is almost the determinant $|\mathbf{A}|$ except that the columns are out of order. To restore the order requires an even number (2) of interchanges, corresponding to the fact that the permutation

$$
\begin{pmatrix}
1 & 2 & 3 \\
3 & 1 & 2
\end{pmatrix}
$$

is an even permutation. Hence (14.17.3) is the same as

$$
\alpha_3\beta_1\gamma_2
\begin{vmatrix}
a_1 & b_1 & c_1 \\
a_2 & b_2 & c_2 \\
a_3 & b_3 & c_3
\end{vmatrix}.
$$

If, on the other hand, we choose the first part of column 1, the third part of column 2 and the second part of column 3 in (14.17.1), we obtain the determinant

$$
\begin{vmatrix}
a_1\alpha_1 & c_1\beta_3 & b_1\gamma_2 \\
a_2\alpha_1 & c_2\beta_3 & b_2\gamma_2 \\
a_3\alpha_1 & c_3\beta_3 & b_3\gamma_2
\end{vmatrix}
= \alpha_1\beta_3\gamma_2
\begin{vmatrix}
a_1 & c_1 & b_1 \\
a_2 & c_2 & b_2 \\
a_3 & c_3 & b_3
\end{vmatrix}
= -\alpha_1\beta_3\gamma_2
\begin{vmatrix}
a_1 & b_1 & c_1 \\
a_2 & b_2 & c_2 \\
a_3 & b_3 & c_3
\end{vmatrix}.
$$

The minus sign is required since one interchange of columns is required to restore the columns of the determinant above to their original order. This reflects the fact that the subscripts in $\alpha_1\beta_3\gamma_2$ form an *odd* permutation of 1, 2, 3.

In this way, each of the 6 nonvanishing determinants can be expressed as a multiple of the determinant

$$|\mathbf{A}| = \begin{vmatrix} a_1 & b_1 & c_1 \\ a_2 & b_2 & c_2 \\ a_3 & b_3 & c_3 \end{vmatrix}.$$

Taking this term out as a factor we find that the original determinant becomes

$$(14.17.4) \qquad \begin{vmatrix} a_1 & b_1 & c_1 \\ a_2 & b_2 & c_2 \\ a_3 & b_3 & c_3 \end{vmatrix} \{\alpha_1\beta_2\gamma_3 + \alpha_3\beta_1\gamma_2 + \alpha_2\beta_3\gamma_1 \\ - \alpha_1\beta_3\gamma_2 - \alpha_3\beta_2\gamma_1 - \alpha_2\beta_1\gamma_3\}.$$

The expression in braces in (14.17.4) is simply the expansion of

$$|\mathbf{B}| = \begin{vmatrix} \alpha_1 & \beta_1 & \gamma_1 \\ \alpha_2 & \beta_2 & \gamma_2 \\ \alpha_3 & \beta_3 & \gamma_3 \end{vmatrix}$$

since the subscripts run through all permutations of 1, 2, 3, (any other combination of subscripts gives a vanishing determinant), and the sign is $+$ if the permutation is even, and $-$ if it is odd. Hence the original determinant (14.17.1) is equal to

$$\begin{vmatrix} a_1 & b_1 & c_1 \\ a_2 & b_2 & c_2 \\ a_3 & b_3 & c_3 \end{vmatrix} \cdot \begin{vmatrix} \alpha_1 & \beta_1 & \gamma_1 \\ \alpha_2 & \beta_2 & \gamma_2 \\ \alpha_3 & \beta_3 & \gamma_3 \end{vmatrix}$$

as was to be proved.

For a general value of n it would be extremely tedious to write out an argument such as this in detail. However it is not difficult to see that the same approach would yield a proof of the theorem. The determinant $|\mathbf{AB}|$ can be split up into the sum of n^n determinants by splitting each of the n columns into n parts. Of these determinants only $n!$ will not automatically vanish; for if we choose the r_i-th part from the ith column, $i = 1, 2, \ldots, n$, then the resulting determinant will vanish unless $(r_1, r_2, \ldots, r_n)$ is a permutation of $(1, 2, \ldots, n)$. Each nonvanishing term will have $|\mathbf{A}|$ as a factor, and when this factor is taken out, what is left is the expansion of $|\mathbf{B}|$, each term having the correct sign. Thus the theorem follows.

Exercise 168

1. Verify the rule for the determinant of the product of two matrices by means of the following pairs of matrices:

(a) $\begin{pmatrix} 2 & 0 & 4 \\ 1 & 3 & 6 \\ 7 & -1 & 4 \end{pmatrix} \begin{pmatrix} 5 & 1 & -1 \\ 3 & 6 & 4 \\ -2 & 0 & 7 \end{pmatrix}$

(b) $\begin{pmatrix} 3 & -8 & 5 \\ 6 & 1 & 9 \\ 9 & 3 & -5 \end{pmatrix} \begin{pmatrix} 7 & 3 & -4 \\ -9 & 2 & 3 \\ 1 & -2 & 5 \end{pmatrix}$

2. If $\mathbf{A} = (a_{ij})$ is a 3×3 matrix, and $\mathbf{B} = (A_{ij})$ is its matrix of cofactors, prove that

$$|\mathbf{B}^T| \cdot |\mathbf{A}| = \begin{vmatrix} \Delta & 0 & 0 \\ 0 & \Delta & 0 \\ 0 & 0 & \Delta \end{vmatrix}$$
$$= \Delta^3$$

where $\Delta = |a_{ij}|$. Hence show that $|\mathbf{B}| = \Delta^2$ (compare Exercise 164.2). Generalize this result for $n \times n$ matrices.

14.18 FACTORS OF ZERO

Although there is much similarity between ordinary algebra and matrix algebra there are nevertheless many differences. In ordinary algebra we can deduce from the equation $ab = 0$ that either $a = 0$ or $b = 0$ (or possibly both). This is not so in matrix algebra, as the following example shows:

$$\begin{pmatrix} 1 & 2 \\ 1 & 2 \end{pmatrix} \begin{pmatrix} 2 & 2 \\ -1 & -1 \end{pmatrix} = \begin{pmatrix} 0 & 0 \\ 0 & 0 \end{pmatrix}.$$

Hence from $\mathbf{AB} = 0$ it does not follow that either $\mathbf{A}$ or $\mathbf{B}$ is a zero matrix. In matrix algebra, that is to say, a zero matrix can have nonzero factors. If $\mathbf{AB} = 0$ we say that $\mathbf{A}$ is a *left factor of zero* and that $\mathbf{B}$ is a *right factor of zero.* If $AB = 0$, then $|AB| = |A||B| = |0| = 0$, and since $|A|$ and $|B|$ are just numbers it *does* follow that either $|A| = 0$ or $|B| = 0$ (or both). Hence we have the following theorem.

THEOREM 14.5

IF THE PRODUCT OF TWO SQUARE MATRICES IS THE ZERO MATRIX, THEN AT LEAST ONE OF THE MATRICES IS SINGULAR.

A similar result holds if we consider the product of a square matrix and a column vector. Suppose $Ax = 0$ where A is a square matrix and x is a non-zero column vector. Then A must be singular. For if it were not, it would have an inverse A^{-1} and from $Ax = 0$ we could deduce that

$$(14.18.1) \qquad A^{-1}Ax = A^{-1}0 = 0.$$

But the left-hand side of (14.18.1) is just x, so we have $x = 0$, which contradicts the datum that x was nonzero. Hence our assumption that A was nonsingular is not true, and A is therefore singular.

Exercise 169

1. For each of the matrices given below assume a right factor of zero of the

form $\begin{pmatrix} x \\ y \\ z \end{pmatrix}$, and hence, by solving for x, y and z, find a right factor of zero.

Do any of these matrices have two *different* right factors of zero, that is, ones that are not simply multiples of each other? (Clearly if x is a right factor of zero of A, so is λx. where λ is any constant).

(a) $\begin{pmatrix} 1 & 0 & 3 \\ 2 & 2 & 1 \\ 7 & 6 & 6 \end{pmatrix}$ (c) $\begin{pmatrix} 2 & -1 & 7 \\ 4 & 5 & -18 \\ 8 & 3 & -4 \end{pmatrix}$

(b) $\begin{pmatrix} 3 & -9 & 6 \\ 2 & 1 & -3 \\ 5 & 0 & -5 \end{pmatrix}$ $\begin{pmatrix} 1 & -4 & 2 \\ -3 & 12 & -6 \\ 2 & -8 & 4 \end{pmatrix}$

2. Find left factors of zero of the matrices in Exercise 1.

3. Show that the vector $(A_{11}, A_{21}, \ldots, A_{n1})$ is a left factor of zero of the $n \times n$ matrix (a_{ij}) if, and only if, (a_{ij}) is singular. (A_{ij} denotes the cofactor of a_{ij} in $|a_{ij}|$.)

14.19 LINEAR DEPENDENCE

If $C_1, C_2, \ldots, C_k$ are k column vectors of order n, we know what is meant by

$$\lambda_1 C_1 + \lambda_2 C_2 \cdots + \lambda_k C_k$$

where $\lambda_1, \lambda_2, \ldots \lambda_k$ are scalars (see page 245). Such an expression is called a *linear combination* of the vectors $C_1, C_2, \ldots, C_k$, and the vector which it represents is said to be linearly dependent on $C_1, C_2, \ldots, C_k$. If there exist scalars $\lambda_1, \lambda_2, \ldots, \lambda_k$, *not all zero*, such that

$$\lambda_1 C_1 + \lambda_2 C_2 + \cdots + \lambda_k C_k = 0$$

then we say that the vectors $C_1, C_2, \ldots, C_k$ are *linearly dependent*. If there is no such set of scalars $\lambda_1, \lambda_2, \ldots, \lambda_k$, then we say that the k vectors are *linearly independent*.

EXAMPLES 1:

The 3 vectors

$$\begin{pmatrix} 1 \\ 2 \\ 0 \\ 5 \end{pmatrix}, \begin{pmatrix} 3 \\ 2 \\ 1 \\ 8 \end{pmatrix}, \begin{pmatrix} 3 \\ -2 \\ 2 \\ 1 \end{pmatrix}$$

are linearly dependent. For

$$3\begin{pmatrix} 1 \\ 2 \\ 0 \\ 5 \end{pmatrix} - 2\begin{pmatrix} 3 \\ 2 \\ 1 \\ 8 \end{pmatrix} + \begin{pmatrix} 3 \\ -2 \\ 2 \\ 1 \end{pmatrix} = \begin{pmatrix} 0 \\ 0 \\ 0 \\ 0 \end{pmatrix}$$

EXAMPLE 2:

The vectors

$$\begin{pmatrix} 1 \\ 0 \\ 0 \\ 0 \end{pmatrix}, \begin{pmatrix} 0 \\ 1 \\ 0 \\ 0 \end{pmatrix}, \begin{pmatrix} 0 \\ 0 \\ 1 \\ 0 \end{pmatrix}, \begin{pmatrix} 0 \\ 0 \\ 0 \\ 1 \end{pmatrix}$$

are linearly independent. For

$$\lambda_1 \begin{pmatrix} 1 \\ 0 \\ 0 \\ 0 \end{pmatrix} + \lambda_2 \begin{pmatrix} 0 \\ 1 \\ 0 \\ 0 \end{pmatrix} + \lambda_3 \begin{pmatrix} 0 \\ 0 \\ 1 \\ 0 \end{pmatrix} + \lambda_4 \begin{pmatrix} 0 \\ 0 \\ 0 \\ 1 \end{pmatrix} = \begin{pmatrix} \lambda_1 \\ \lambda_2 \\ \lambda_3 \\ \lambda_4 \end{pmatrix}.$$

This right-hand side is the zero-vector if, and only if, $\lambda_1 = \lambda_2 = \lambda_3 = \lambda_4 = 0$. Hence the result.

EXAMPLE 3:
Find a value of a such that the vectors

$$\begin{pmatrix} 3 \\ 1 \\ 2 \end{pmatrix}, \begin{pmatrix} a \\ 0 \\ 8 \end{pmatrix}, \begin{pmatrix} 6 \\ -2 \\ a \end{pmatrix}$$

are linearly dependent.
We have

$$\lambda_1 \begin{pmatrix} 3 \\ 1 \\ 2 \end{pmatrix} + \lambda_2 \begin{pmatrix} a \\ 0 \\ 8 \end{pmatrix} + \lambda_3 \begin{pmatrix} 6 \\ -2 \\ a \end{pmatrix} = \begin{pmatrix} 3\lambda_1 + a\lambda_2 + 6\lambda_3 \\ \lambda_1 \qquad\quad - 2\lambda_3 \\ 2\lambda_1 + 8\lambda_2 + a\lambda_3 \end{pmatrix}.$$

If this is the zero vector we have

$$3\lambda_1 + a\lambda_2 + 6\lambda_3 = 0$$
$$\lambda_1 \qquad\quad - 2\lambda_3 = 0$$
$$2\lambda_1 + 8\lambda_2 + a\lambda_3 = 0.$$

If we eliminate λ_1 from these equations (using the second one) we have

$$a\lambda_2 + 12\lambda_3 = 0$$

and

$$8\lambda_2 + (a + 4)\lambda_3 = 0$$

whence

$$\frac{\lambda_2}{\lambda_3} = \frac{12}{a} = -\frac{a + 4}{8}.$$

Thus $96 = a(a + 4)$ from which $a = 8$ or -12.

The vectors

$$\begin{pmatrix} 3 \\ 1 \\ 2 \end{pmatrix}, \begin{pmatrix} 8 \\ 0 \\ 8 \end{pmatrix} \text{ and } \begin{pmatrix} 6 \\ -2 \\ 8 \end{pmatrix}$$

are linearly dependent, as also are the vectors

$$\begin{pmatrix} 3 \\ 1 \\ 2 \end{pmatrix}, \begin{pmatrix} -12 \\ 0 \\ 8 \end{pmatrix} \text{ and } \begin{pmatrix} 6 \\ -2 \\ -12 \end{pmatrix}.$$

EXAMPLE 4:
Consider the four vectors

$$\mathbf{x}_1 = \begin{pmatrix} 1 \\ 2 \\ 9 \\ 7 \\ 5 \end{pmatrix}, \mathbf{x}_2 = \begin{pmatrix} 3 \\ 0 \\ 8 \\ -1 \\ -6 \end{pmatrix}, \mathbf{x}_3 = \begin{pmatrix} 2 \\ 1 \\ 8 \\ 3 \\ -1 \end{pmatrix}, \mathbf{x}_4 = \begin{pmatrix} -3 \\ 6 \\ 11 \\ 23 \\ 27 \end{pmatrix}.$$

It is easily verified that $3\mathbf{x}_1 - 2\mathbf{x}_2 - \mathbf{x}_4 = 0$. Hence these vectors are linearly dependent. This example shows that, although the λs are required to be "not all zero" in the definition of linear dependence, it is *not* required that they be "all nonzero." In this example the multiplier for $\mathbf{x}_3$ is zero. In an extreme case it may happen that there is only one nonzero λ. Thus any set of vectors $\{0, \mathbf{x}_1, \mathbf{x}_2, \mathbf{x}_3, \ldots\}$ which contains the zero vector is automatically dependent. For we have

$$\lambda 0 + 0\mathbf{x}_1 + 0\mathbf{x}_2 + 0\mathbf{x}_3 + \cdots = 0 \qquad (\lambda \neq 0)$$

which satisfies the definition of dependence.

We now prove some theorems on linear dependence.

THEOREM 14.6

IF k VECTORS $\mathbf{x}_1, \mathbf{x}_2, \ldots, \mathbf{x}_k$, EACH OF DIMENSION n, ARE LINEARLY DEPENDENT, THEN EITHER

    (a) $\mathbf{x}_1$ IS LINEARLY DEPENDENT ON $\mathbf{x}_2, \mathbf{x}_3, \ldots, \mathbf{x}_k$

OR (b) $\mathbf{x}_2, \mathbf{x}_3, \ldots, \mathbf{x}_k$ ARE LINEARLY DEPENDENT.

PROOF:
There exist scalars $\lambda_1, \lambda_2, \ldots, \lambda_k$, not all zero such that

(14.19.1) $\lambda_1\mathbf{x}_1 + \lambda_2\mathbf{x}_2 + \lambda_3\mathbf{x}_3 + \cdots + \lambda_k\mathbf{x}_k = 0.$

If $\lambda_1 \neq 0$ then (14.19.1) can be written as

$$\mathbf{x}_1 = -\frac{\lambda_2}{\lambda_1}\mathbf{x}_2 - \frac{\lambda_3}{\lambda_1}\mathbf{x}_3 \cdots -\frac{\lambda_k}{\lambda_1}\mathbf{x}_k,$$

which proves (a). On the other hand, if $\lambda_1 = 0$ then

$$\lambda_2\mathbf{x}_2 + \lambda_3\mathbf{x}_3 + \cdots + \lambda_k\mathbf{x}_k = 0$$

which proves (b).

THEOREM 14.7

IF n COLUMN VECTORS $\mathbf{x}_1, \mathbf{x}_2 \ldots, \mathbf{x}_n$ OF DIMENSION n ARE LINEARLY DEPENDENT, THEN $\Delta = 0$ AND CONVERSELY, WHERE Δ IS THE $n \times n$ DETERMINANT WHOSE COLUMNS ARE THE VECTORS $\mathbf{x}_1, \mathbf{x}_2, \ldots, \mathbf{x}_n$.

PROOF:

If the vectors are linearly dependent there exist numbers $\lambda_1, \lambda_2, \ldots, \lambda_n$ not all zero, such that

$$\lambda_1\mathbf{x}_1 + \lambda_2\mathbf{x}_2 + \cdots + \lambda_n\mathbf{x}_n = 0.$$

This can be written in matrix notation as

$$\mathbf{A}\begin{pmatrix} \lambda_1 \\ \lambda_2 \\ \cdot \\ \cdot \\ \cdot \\ \lambda_n \end{pmatrix} = 0$$

where $\mathbf{A} = (\mathbf{x}_1, \mathbf{x}_2, \ldots, \mathbf{x}_n)$ is the matrix whose columns are the vectors $\mathbf{x}_1, \mathbf{x}_2, \ldots, \mathbf{x}_n$. Since this matrix has a right factor of zero it must be singular, by the results of section 14.18, and its determinant therefore vanishes. Hence

$$\Delta = |\mathbf{A}| = 0.$$

The converse of the theorem is a little more difficult to prove. Let the column vectors $\mathbf{x}_1, \mathbf{x}_2, \ldots, \mathbf{x}_n$ be linearly independent. Now $\mathbf{x}_1$ has a nonzero element somewhere, say in row i_1 of Δ, and by adding suitable multiples of $\mathbf{x}_1$ to the other columns of Δ we can arrange for all elements of row i_1 except the first to become zeros. We next choose a nonzero element of the second column of Δ (which used to be $\mathbf{x}_2$ but, in general, will now be something else), say that in row i_2. By adding suitable multiples of column 2 to the other columns to its right we arrange that column 2 has a nonzero element in row i_2, but that no subsequent column has. We now repeat this procedure with column 3, and so on. Note that no column can be reduced to the zero vector by this procedure. For if this happened it would mean that some linear combination of the original columns was the zero vector, and this is impossible, since the columns were independent.

We end up with a determinant, equal in value to Δ, for which column 1 is the only column with a nonzero element in row i_1. Hence any nonvanishing term in the expansion of the determinant must contain this term (call it $a_{i_1 1}$). Similarly, all columns have a zero in row i_2 except column 2 (which has a nonzero element $a_{i_2 2}$) and column 1. But we are forced to choose element $a_{i_1 1}$ from column 1 if we are to get a nonzero contribution to the determinant. Hence it follows that the element $a_{i_2 2}$ must be included in any nonvanishing term of the determinant. Proceeding in this way we see that the only nonvanishing term in the determinant is

$$\pm a_{i_1 1}\, a_{i_2 2}\, a_{i_3 3}, \ldots a_{i_n n}$$

and this is nonzero, since none of these elements is zero. Hence $\Delta \neq 0$.

We shall give an example which should clarify this part of the proof of the theorem and also show how the procedure just explained, in addition to providing the required proof, gives a systematic method of evaluating a determinant and of testing whether a set of vectors is linearly independent.

EXAMPLE:
Show that the vectors

$$\begin{pmatrix} 2 \\ 1 \\ 4 \\ 3 \end{pmatrix}, \begin{pmatrix} 4 \\ 2 \\ 1 \\ 1 \end{pmatrix}, \begin{pmatrix} 3 \\ -1 \\ -1 \\ 2 \end{pmatrix} \text{ and } \begin{pmatrix} -1 \\ 1 \\ 5 \\ 0 \end{pmatrix}$$

are linearly independent.
We have to show that the determinant

$$\Delta = \begin{vmatrix} 2 & 4 & 3 & -1 \\ 1 & 2 & -1 & 1 \\ 4 & 1 & -1 & 5 \\ 3 & 1 & 2 & 0 \end{vmatrix}$$

is nonzero. Following the procedure in the proof of the theorem we choose a nonzero element in column 1, say the 2, and
  (a)   add -2 times column 1 to column 2;
  (b)   add $-\frac{3}{2}$ times column 1 to column 3;
  (c)   add $\frac{1}{2}$ times column 1 to column 4.
We obtain the determinant

$$\Delta = \begin{vmatrix} 2 & 0 & 0 & 0 \\ 1 & 0 & -\dfrac{5}{2} & \dfrac{3}{2} \\ 4 & -7 & -7 & 7 \\ 3 & -5 & -\dfrac{5}{2} & \dfrac{3}{2} \end{vmatrix}.$$

We now choose a nonzero element (-7) in column 2 and
  (a)   subtract column 2 from column 3;
  (b)   add column 2 to column 4.
We obtain

$$\begin{vmatrix} 2 & 0 & 0 & 0 \\ 1 & 0 & -\dfrac{5}{2} & \dfrac{3}{2} \\ 4 & -7 & 0 & 0 \\ 3 & -5 & \dfrac{5}{2} & -\dfrac{7}{2} \end{vmatrix}$$

We now choose a nonzero element (say $-\frac{5}{2}$) from column 3, and adding $\frac{3}{5}$ times column 3 to column 4 we obtain

$$\begin{vmatrix} 2 & 0 & 0 & 0 \\ 1 & 0 & -\dfrac{5}{2} & 0 \\ 4 & -7 & 0 & 0 \\ 3 & -5 & \dfrac{5}{2} & -2 \end{vmatrix}.$$

The only nonvanishing term in the expansion of this determinant is

$$-\left(2 \cdot -7 \cdot -\frac{5}{2} \cdot -2\right) = 70$$

which is therefore the value of the determinant.

Note: The reader should compare this evaluation with example 2 of section 14.11 where, without setting out any system, we reduced a determinant to upper triangular form. In the present example we have effectively reduced the determinant to lower triangular form, since on interchanging columns 2 and 3 we obtain

$$\begin{vmatrix} 2 & 0 & 0 & 0 \\ 1 & -\dfrac{5}{2} & 0 & 0 \\ 4 & 0 & -7 & 0 \\ 3 & \dfrac{5}{2} & -5 & -2 \end{vmatrix}.$$

Exercise 170

1. Determine whether the following sets of vectors are linearly dependent or not.

(a) $\begin{pmatrix} 1 \\ 0 \\ 3 \\ 2 \end{pmatrix}, \begin{pmatrix} 2 \\ 1 \\ 0 \\ 5 \end{pmatrix}, \begin{pmatrix} -11 \\ -2 \\ 21 \\ -31 \end{pmatrix}$

(b) $\begin{pmatrix} 12 \\ 0 \\ 1 \\ 4 \end{pmatrix}, \begin{pmatrix} 3 \\ 1 \\ 0 \\ 2 \end{pmatrix}, \begin{pmatrix} 6 \\ 6 \\ 1 \\ 7 \end{pmatrix}, \begin{pmatrix} 4 \\ 9 \\ 1 \\ -3 \end{pmatrix}, \begin{pmatrix} 2 \\ 0 \\ 0 \\ 7 \end{pmatrix}$

(c) $\begin{pmatrix} 1 \\ 3 \\ 0 \end{pmatrix}, \begin{pmatrix} 2 \\ 1 \\ 6 \end{pmatrix}, \begin{pmatrix} 1 \\ -1 \\ -3 \end{pmatrix}$

(d) $\begin{pmatrix} 2 \\ 0 \\ 1 \\ -3 \end{pmatrix}, \begin{pmatrix} 3 \\ 1 \\ 7 \\ -5 \end{pmatrix}, \begin{pmatrix} -4 \\ 0 \\ 3 \\ 2 \end{pmatrix}, \begin{pmatrix} -13 \\ -1 \\ 8 \\ 4 \end{pmatrix}$

14.20 THE RANK OF A MATRIX

An $m \times n$ matrix $\mathbf{A}$ can be regarded as being made up of m rows (each a row vector) or of n columns (each a column vector). We can ask "What is the greatest number of linearly independent rows we can choose from $\mathbf{A}$?"

We can ask a similar question for the columns of **A**. The next theorem links the answers to these two questions.

THEOREM 14.8

IN A MATRIX $\mathbf{A} = (a_{ij})$ $(i = 1, 2, \ldots, m; \ j = 1, 2, \ldots, n)$ THE MAXIMUM NUMBER OF LINEARLY INDEPENDENT ROWS IS THE SAME AS THE MAXIMUM NUMBER OF LINEARLY INDEPENDENT COLUMNS. THE NUMBER IN QUESTION IS CALLED THE RANK OF **A**.

PROOF:

Suppose the maximum number of independent rows is r. Then there are r linearly independent rows. Rearrange the rows of the matrix **A** so that these rows are the first r rows. It is easy to see that this will not affect the linear dependence, or otherwise, of the columns of **A**. The matrix **A** then takes the form

$$(14.20.1) \qquad \mathbf{A} = \begin{pmatrix} \mathbf{R}_1 \\ \mathbf{R}_2 \\ \cdots \\ \mathbf{R}_r \\ \lambda_1\mathbf{R}_1 + \lambda_2\mathbf{R}_2 + \cdots + \lambda_r\mathbf{R}_r \\ \mu_1\mathbf{R}_1 + \mu_2\mathbf{R}_2 + \cdots + \mu_r\mathbf{R}_r \\ \cdots \end{pmatrix}$$

Each of the last $m - r$ rows must be a linear combination of $\mathbf{R}_1, \mathbf{R}_2,$ $\ldots, \mathbf{R}_r$ since otherwise it would form, with them, a set of $r + 1$ independent vectors (from Theorem 14.6), and this is impossible. Equation (14.20.1) can be written as

$$(14.20.2) \qquad \mathbf{A} = \begin{pmatrix} 1 & 0 & 0 & \cdots & 0 \\ 0 & 1 & 0 & \cdots & 0 \\ & & \cdot & \cdots & \cdot \\ 0 & 0 & 0 & \cdots & 1 \\ \lambda_1 & \lambda_2 & \lambda_3 & \cdots & \lambda_r \\ \mu_1 & \mu_2 & \mu_3 & \cdots & \mu_r \\ & & \cdot & \cdots & \cdot \end{pmatrix} \begin{pmatrix} \mathbf{R}_1 \\ \mathbf{R}_2 \\ \mathbf{R}_3 \\ \cdots \\ \cdots \\ \mathbf{R}_r \end{pmatrix}$$

$$= \mathbf{B} \begin{pmatrix} \mathbf{R}_1 \\ \mathbf{R}_2 \\ \mathbf{R}_3 \\ \cdots \\ \mathbf{R}_r \end{pmatrix}$$

as the reader should verify. Here $\mathbf{B}$ is an $m \times r$ matrix. If we denote its columns by $\mathbf{C}_1, \mathbf{C}_2, \ldots, \mathbf{C}_r$, we can write (14.20.2) as

$$(14.20.3) \qquad \mathbf{A} = (\mathbf{C}_1, \mathbf{C}_2, \ldots, \mathbf{C}_r) \begin{pmatrix} \mathbf{R}_1 \\ \mathbf{R}_2 \\ \ldots \\ \mathbf{R}_r \end{pmatrix}$$

Now just as (14.20.2) shows that the rows of $\mathbf{A}$ all depend linearly on the rows $\mathbf{R}_1, \mathbf{R}_2 \ldots \mathbf{R}_r$, so (14.20.3) shows that the columns of $\mathbf{A}$ depend linearly on the r columns $\mathbf{C}_1, \mathbf{C}_2, \ldots, \mathbf{C}_r$. Since these columns may be linearly dependent among themselves, we cannot say that there are as many as r linearly independent columns in $\mathbf{A}$. What we *can* say is that there cannot be *more* than r linearly independent columns in $\mathbf{A}$.

If we now follow through this same line of reasoning starting with the columns of $\mathbf{A}$ instead of the rows, we can show similarly that the maximum number of linearly independent rows in $\mathbf{A}$ is not *more* than the maximum number of linearly independent columns.

Taking these two results together we see that the only possibility is that the two numbers are the same. This proves the theorem.

THEOREM 14.9

IF A MATRIX $\mathbf{A}$ HAS n COLUMNS, AND IS OF RANK r, THEN IT HAS $n - r$ LINEARLY INDEPENDENT RIGHT FACTORS OF ZERO.

PROOF:

Let $\mathbf{A} = (a_{ij})$ be an $m \times n$ matrix of rank r.

If $\mathbf{x} = \begin{pmatrix} x_1 \\ x_2 \\ \ldots \\ x_n \end{pmatrix}$ is a right factor of zero then

$$(14.20.4) \qquad \sum_{j=1}^{n} a_{ij} x_j = 0 \quad (i = 1, 2, \ldots, m).$$

Thus we can find right factors of zero by solving a system of m equations (one for each row of $\mathbf{A}$) in n unknowns. But since the rank is r there are only r linearly independent rows in $\mathbf{A}$, and hence only r of these equations are linearly independent. Hence we have, in effect, only r equations in n unknowns. Without loss of generality we may assume that these are the first r equations.

Let us allot arbitrary values to the $n - r$ variables $x_{r+1}, x_{r+2}, \ldots, x_n$.

We can then solve equations (14.20.4) as a set of r equations in r unknowns. Suppose, to be definite, that we give to the variables x_{r+1}, $x_{r+2}, \ldots, x_n$, the $n - r$ sets of values in which one variable is 1 and all the others are zero. The $n - r$ solutions of (14.20.4) are then of the form

$$
\begin{pmatrix} x_1 \\ x_2 \\ x_3 \\ \cdots \\ x_r \\ 1 \\ 0 \\ 0 \\ \cdots \\ 0 \end{pmatrix},
\begin{pmatrix} x_1 \\ x_2 \\ x_3 \\ \cdots \\ x_r \\ 0 \\ 1 \\ 0 \\ \cdots \\ 0 \end{pmatrix},
\begin{pmatrix} x_1 \\ x_2 \\ x_3 \\ \cdots \\ x_r \\ 0 \\ 0 \\ 1 \\ \cdots \\ 0 \end{pmatrix},
\cdots,
\begin{pmatrix} x_1 \\ x_2 \\ x_3 \\ \cdots \\ x_r \\ 0 \\ 0 \\ 0 \\ \cdots \\ 1 \end{pmatrix}.
$$

These vectors must be independent since each of them contains a 1 in a position where all the others have a zero. This completes the proof of Theorem 14.9.

We now make an observation which gives us another way of defining the rank of a matrix. For convenience we introduce the term *r-rowed minor* of a matrix $\mathbf{A}$, which is a determinant formed by the elements belonging to r chosen rows of $\mathbf{A}$ and r chosen columns. Thus if $\mathbf{A}$ is an $m \times n$ matrix, then $\mathbf{A}$ has $\binom{m}{r}\binom{n}{r}$ r-rowed minors, since there are $\binom{m}{r}$ ways of choosing r rows, and $\binom{n}{r}$ ways of choosing r columns.

If an $m \times n$ matrix $\mathbf{A}$ has rank r then there are, in $\mathbf{A}$, r linearly independent rows. Let us consider the $r \times n$ *submatrix* of $\mathbf{A}$ consisting of just these r rows; call it $\mathbf{B}$. The rank of $\mathbf{B}$ is r, since the rows are independent, and hence, by Theorem 14.8, $\mathbf{B}$ has at least one set of r linearly independent columns. These r rows and r columns determine an $r \times r$ minor of $\mathbf{A}$, which is nonzero, since its columns are linearly independent.

Hence an $m \times n$ matrix of rank r has at least one nonvanishing r-rowed minor.

Conversely, if a matrix $\mathbf{A}$ has a nonvanishing $r \times r$ minor, then its rank is at least r. For otherwise the r rows of $\mathbf{A}$ which define the minor would be linearly dependent, and the same linear relation holding between them would also hold for the rows of the minor. Hence the rows of the minor would be linearly dependent, which would mean that the minor was zero, which is not the case.

Further, if all $(r + 1) \times (r + 1)$ minors vanish then the rank of **A** cannot exceed r; for if the rank is $s > r$, then there is an s-rowed minor which does not vanish, and this is impossible if all $(r + 1)$-rowed minors vanish. For an $s \times s$ minor can be expanded by the cofactors of a row or column. These cofactors can be similarly expanded, and so on, until we arrive at an expansion of the $s \times s$ minor in terms of $(r + 1) \times (r + 1)$ minors. Since these all vanish, so does the original $s \times s$ minor.

Thus we have

THEOREM 14.10

A MATRIX **A** HAS RANK r IF, AND ONLY IF, IT HAS A NONVANISHING r-ROWED MINOR, BUT NO NONVANISHING $(r + 1)$-ROWED MINOR.

This result could have been taken (and sometimes is) as another (equivalent) definition of the term *rank*.

Exercise 171

1. Show that the rank of a matrix is not altered by the following operations:
   (a) multiplying a row or column by a non-zero scalar constant.
   (b) adding a multiple of one row (column) to another row (column).
   (c) interchanging two rows (columns).
2. By means of the operations described in problem 1 convert the matrices

(a)
$$\begin{pmatrix} 2 & 7 & 8 & 16 & -14 \\ 1 & 0 & -1 & -4 & 8 \\ 4 & 1 & 5 & -5 & 5 \\ -3 & 1 & 3 & 14 & -24 \\ 1 & 2 & 4 & 5 & -7 \end{pmatrix}$$

(b)
$$\begin{pmatrix} 3 & 6 & -2 & 4 \\ 0 & -3 & -1 & -1 \\ 2 & -1 & -3 & 1 \\ -1 & 10 & 4 & 2 \\ 7 & -17 & -15 & -1 \end{pmatrix}$$

into other matrices whose rank can be determined easily by inspection. (Try to obtain matrices with many zeros.) Hence find the ranks of these matrices.

3. Show that by means of the operations described in problem 1, a matrix can be converted to the following form

$$
\begin{pmatrix}
1 & 0 & 0 & \cdots & 0 & 0 & 0 & 0 & \cdots & 0 \\
0 & 1 & 0 & \cdots & 0 & 0 & 0 & 0 & \cdots & 0 \\
0 & 0 & 1 & \cdots & 0 & 0 & 0 & 0 & \cdots & 0 \\
  & & . & \cdots & & . & & \cdots & . \\
0 & 0 & 0 & \cdots & 1 & 0 & 0 & 0 & \cdots & 0 \\
0 & 0 & 0 & \cdots & 0 & 1 & 0 & 0 & \cdots & 0 \\
0 & 0 & 0 & \cdots & 0 & 0 & 0 & 0 & \cdots & 0 \\
  & & . & \cdots & & . & & \cdots & . \\
0 & 0 & 0 & \cdots & 0 & 0 & 0 & 0 & \cdots & 0
\end{pmatrix}
$$

in which the first r rows and columns constitute the unit matrix I_r, and the other elements are zero, where r is the rank of the original matrix.

4. List a sequence of operations (as in problem 1) that will convert each of the matrices of problem 2 into the form given in problem 3.

5. Find the ranks of the matrices

(a)
$$
\begin{pmatrix}
2 & 6 & 4 & 2 & 0 \\
-9 & -2 & 1 & -4 & -6 \\
17 & 1 & -4 & 7 & 12 \\
21 & -12 & -15 & 6 & 18
\end{pmatrix}
$$

(c)
$$
\begin{pmatrix}
1 & 1 & -3 & 0 \\
7 & 3 & -17 & -6 \\
-1 & 2 & 1 & 3
\end{pmatrix}
$$

(b)
$$
\begin{pmatrix}
1 & 3 & 5 & 3 \\
5 & 8 & 4 & 2 \\
14 & 18 & 7 & -1 \\
4 & 4 & -1 & 3
\end{pmatrix}
$$

6. Show that $n+1$ vectors, each of dimension n, must necessarily be linearly dependent. *Hint:* Consider the rank, defined by columns and by rows, of the $n \times n + 1$ matrix formed by those vectors.

14.21 ORTHOGONAL MATRICES

Linear transformations, with which we started this chapter, have appeared in this book before. In particular, the equations that we derived in Chapter 5 for a change of coordinate axes, keeping the same origin, viz.,

(14.21.1)
$$
\begin{aligned}
x &= x' \cos \theta - y' \sin \theta \\
y &= x' \sin \theta + y' \cos \theta
\end{aligned}
$$

exemplify a linear transformation, and if we write $\mathbf{x} = \begin{pmatrix} x \\ y \end{pmatrix}$ and $\boldsymbol{\xi} = \begin{pmatrix} x' \\ y' \end{pmatrix}$, equations (14.21.1) become

(14.21.2)
$$\mathbf{x} = \mathbf{P}\,\boldsymbol{\xi},$$

where $\mathbf{P}$ is the 2×2 matrix $\begin{pmatrix} \cos\theta & -\sin\theta \\ \sin\theta & \cos\theta \end{pmatrix}$. Clearly, $\mathbf{P}$ is far from being a general 2×2 matrix.

In three dimensions there is a similar result for the relation between the coordinates (x, y, z) of a point with respect to one set of mutually perpendicular (orthogonal) axes, and its coordinates (x', y', z') with respect to another set of orthogonal axes through the same origin. The equations are of the form

(14.21.3)
$$\begin{aligned} x &= l_1 x' + m_1 y' + n_1 z' \\ y &= l_2 x' + m_2 y' + n_2 z' \\ z &= l_3 x' + m_3 y' + n_3 z' \end{aligned}$$

that is, they specify a linear transformation. [The fact that they must be of this linear form can be deduced from the fact that a linear equation in x, y and z must, on substituting from (14.21.3) give a linear equation in x', y' and z'. Otherwise we could change a plane into some other surface merely by choosing a different coordinate system!] Equations (14.21.3) can be written in matrix notation as

(14.21.4)
$$\mathbf{x} = \mathbf{P}\,\boldsymbol{\xi}$$

where

$$\mathbf{x} = \begin{pmatrix} x \\ y \\ z \end{pmatrix}, \quad \boldsymbol{\xi} = \begin{pmatrix} x' \\ y' \\ z' \end{pmatrix} \quad \text{and} \quad \mathbf{P} = \begin{pmatrix} l_1 & m_1 & n_1 \\ l_2 & m_2 & n_2 \\ l_3 & m_3 & n_3 \end{pmatrix}$$

We shall now see that the matrix $\mathbf{P}$ is a rather special one.

The point $\boldsymbol{\xi}_1 = \begin{pmatrix} 1 \\ 0 \\ 0 \end{pmatrix}$ becomes the point $\mathbf{x}_1 = \begin{pmatrix} l_1 \\ l_2 \\ l_3 \end{pmatrix}$ by the transformation

(14.21.4). But since a change of coordinate system, keeping the origin fixed, will not alter the distance of a point from the origin, and since $\boldsymbol{\xi}_1$ is at distance 1 from the origin, it follows that $\mathbf{x}_1$ is also at distance 1 from the origin. Thus

(14.21.5)
$$l_1^2 + l_2^2 + l_3^2 = 1.$$

In the same way we see that

(14.21.6) $$m_1^2 + m_2^2 + m_3^2 = 1$$

and

(14.21.7) $$n_1^2 + n_2^2 + n_3^2 = 1.$$

If we call $\mathbf{x}$ the old coordinate vector of a point, and $\boldsymbol{\xi}$ the new coordinate vector, then $\boldsymbol{\xi}_1$ is a point on the "new" x-axis. Hence, from (14.21.5) we see that l_1, l_2 and l_3 are the direction cosines of the new x-axis relative to the old axes. Similarly m_1, m_2, m_3 are the direction cosines of the new y-axis relative to the old axes. Since both sets of axes are orthogonal we have

(14.21.8) $$l_1 m_1 + l_2 m_2 + l_3 m_3 = 0,$$

and there are similar results for the other pairs of columns of $\mathbf{P}$.

Let us now express $\boldsymbol{\xi}$ in terms of $\mathbf{x}$. This we do by the equation

$$\boldsymbol{\xi} = \mathbf{P}^{-1}\mathbf{x}$$

where $\mathbf{P}^{-1}$ is, shall we say, $\begin{pmatrix} l_1' & m_1' & n_1' \\ l_2' & m_2' & n_2' \\ l_3' & m_3' & n_3' \end{pmatrix}$. Now l_2' (for example) is the cosine of the angle which the old x-axis makes with the new y-axis, and hence is the same as m_1. Similarly $m_1' = l_2$, and so on. In short, the matrix $\mathbf{P}^{-1}$ is just the transpose of $\mathbf{P}$. Hence we have

(14.21.9) $$\mathbf{P}^{-1} = \mathbf{P}^T$$

or, better

(14.21.10) $$\mathbf{P}^T\mathbf{P} = \mathbf{P}\mathbf{P}^T = \mathbf{I}.$$

Our previous results are now readily deducible from (14.21.10). For if we multiply the first row of $\mathbf{P}^T$ by the first column of $\mathbf{P}$ we get $l_1^2 + l_2^2 + l_3^2$, and this must be 1 since it is the top left-hand element of $\mathbf{I}$. The other diagonal elements of the product give (14.21.6) and (14.21.7). By considering the off-diagonal elements we obtain (14.21.8) and its companion results. By taking the product $\mathbf{P}\mathbf{P}^T$ we also obtain

$$l_1^2 + m_1^2 + n_1^2 = 1$$

and

$$l_1 l_2 + m_1 m_2 + n_1 n_2 = 0$$

together with four similar results.

A square matrix $\mathbf{P}$ for which $\mathbf{P}^T\mathbf{P} = \mathbf{P}\mathbf{P}^T = \mathbf{I}$ is called an *orthogonal* matrix. In deriving this property we used the fact that a transformation of coordinates from one set of orthogonal axes to another could not alter the distance of a point from the origin. We can now deduce this result algebra-ically. If $\mathbf{x} = \begin{pmatrix} x \\ y \\ z \end{pmatrix}$ is the coordinate vector of a point P, we define the *norm* of $\mathbf{x}$ to be

$$\mathbf{x}^T\mathbf{x} = (x, y, z)\begin{pmatrix} x \\ y \\ z \end{pmatrix}$$

$$= x^2 + y^2 + z^2.$$

Clearly this is the square of the distance of P from the origin. We now state and prove

THEOREM 14.11

TRANSFORMATION BY AN ORTHOGONAL MATRIX LEAVES UNCHANGED THE NORM OF ANY VECTOR. THUS IF $\mathbf{x} = \mathbf{P}\boldsymbol{\xi}$ WHERE $\mathbf{P}$ IS ORTHOGONAL, THEN THE NORM OF $\mathbf{x}$ EQUALS THE NORM OF $\boldsymbol{\xi}$.

PROOF:
Since

$$\mathbf{x} = \mathbf{P}\boldsymbol{\xi},$$
$$\mathbf{x}^T = \boldsymbol{\xi}^T\mathbf{P}^T$$

Hence
$$\mathbf{x}^T\mathbf{x} = (\boldsymbol{\xi}^T\mathbf{P}^T)(\mathbf{P}\boldsymbol{\xi})$$
$$= \boldsymbol{\xi}^T(\mathbf{P}^T\mathbf{P})\boldsymbol{\xi}$$
$$= \boldsymbol{\xi}^T\boldsymbol{\xi}$$

since $\mathbf{P}^T\mathbf{P} = \mathbf{I}$. This proves the result.

In this treatment we have mostly concerned ourselves with orthogonal 3×3 matrices. Notice, however, that our definition of an orthogonal matrix does not specify any size, though it does specify that the matrix is square. Thus we can have orthogonal $n \times n$ matrices for any value of n. In particular, the matrix

$$\begin{pmatrix} \cos\theta & -\sin\theta \\ \sin\theta & \cos\theta \end{pmatrix}$$

for rotation of axes in the plane is easily seen to be an orthogonal matrix.

Exercise 172

1. Construct an orthogonal 3×3 matrix whose first row is $(\frac{2}{3}, \frac{1}{3}, \frac{2}{3})$.
2. A matrix $\mathbf{A}$ is said to be *symmetric* if $\mathbf{A}^T = \mathbf{A}$, i.e., $a_{ij} = a_{ji}$. Show that if $\mathbf{A}$ is symmetric and $\mathbf{P}$ is orthogonal, then $\mathbf{P}^{-1}\mathbf{AP}$ is symmetric.
3. Prove that the product of two orthogonal matrices is an orthogonal matrix and interpret this result geometrically.
4. Show that the determinant of an orthogonal matrix is either $+1$ or -1.

14.22 EIGENVALUES AND EIGENVECTORS

We have had a good deal to say about linear transformations of the form $\mathbf{y} = \mathbf{Ax}$. A fairly natural question to ask is whether such a transformation ever converts a non-zero vector into itself, i.e., whether we can have $\mathbf{x} = \mathbf{Ax}$. The answer to this is, in general, "No," so we ask instead a slightly more general question, namely, "Does this transformation ever give a vector proportional to the original vector?" Since the answer to this question is "Yes," we are on more interesting ground.

If $\mathbf{Ax}$ is indeed proportional to $\mathbf{x}$, then we have

(14.22.1) $$\mathbf{Ax} = \lambda\mathbf{x}$$

where λ is a scalar. The question then is, given a matrix $\mathbf{A}$, what sort of solutions are there to equation (14.22.1)? In this equation the elements of $\mathbf{A}$ are given, but the elements of $\mathbf{x}$ and the scalar λ are unknown. We can write (14.22.1) as

(14.22.2) $$(\mathbf{A} - \lambda\mathbf{I})\mathbf{x} = \mathbf{0}$$

and at once we narrow down our search. For equation (14.22.2) shows that the matrix $\mathbf{A} - \lambda\mathbf{I}$ is singular, since it has a right factor of zero. Hence the determinant of this matrix must vanish, and this gives us an equation, namely

(14.22.3) $$|\mathbf{A} - \lambda\mathbf{I}| = 0$$

which λ must satisfy.

EXAMPLE:

If

$$\mathbf{A} = \begin{pmatrix} 2 & 2 & 6 \\ -1 & 3 & 5 \\ 2 & 1 & 1 \end{pmatrix}$$

then

$$A - \lambda I = \begin{pmatrix} 2 - \lambda & 2 & 6 \\ -1 & 3 - \lambda & 5 \\ 2 & 1 & 1 - \lambda \end{pmatrix}.$$

Since this matrix has a right factor of zero it must be singular, and hence

$$\begin{vmatrix} 2 - \lambda & 2 & 6 \\ -1 & 3 - \lambda & 5 \\ 2 & 1 & 1 - \lambda \end{vmatrix} = 0.$$

By the Rule of Sarrus, this equation is

$$(2 - \lambda)(3 - \lambda)(1 - \lambda) + 20 - 6 - 12(3 - \lambda)$$
$$- 5(2 - \lambda) + 2(1 - \lambda) = 0$$

which reduces to

$$\lambda^3 - 6\lambda^2 - 4\lambda + 24 = 0$$

whence

$$\lambda = -2, 2 \text{ or } 6.$$

These, then, are the only possible values that λ can take. It follows almost at once that for each such value of λ we can find a corresponding vector x to satisfy (14.22.1). For if λ has one of these three values, the matrix $A - \lambda I$ has a right factor of zero, and this will be the required x. In our example, if $\lambda = -2$, then

$$A - \lambda I = \begin{pmatrix} 4 & 2 & 6 \\ -1 & 5 & 5 \\ 2 & 1 & 3 \end{pmatrix}.$$

A right factor of zero is readily found. It is $\begin{pmatrix} 10 \\ 13 \\ -11 \end{pmatrix}$, and can be obtained by finding the cofactors of the elements 4, 2 and 6 of the first row of $A - \lambda I$. The reader should verify that

$$\begin{pmatrix} 4 & 2 & 6 \\ -1 & 5 & 5 \\ 2 & 1 & 3 \end{pmatrix} \begin{pmatrix} 10 \\ 13 \\ -11 \end{pmatrix} = \begin{pmatrix} 0 \\ 0 \\ 0 \end{pmatrix}$$

If $\lambda = 2$, then

$$A - \lambda I = \begin{pmatrix} 0 & 2 & 6 \\ -1 & 1 & 5 \\ 2 & 1 & -1 \end{pmatrix}$$

which has the right factor of zero $\begin{pmatrix} 2 \\ -3 \\ 1 \end{pmatrix}$. Similarly if $\lambda = 6$ then

$$A - \lambda I = \begin{pmatrix} -4 & 2 & 6 \\ -1 & -3 & 5 \\ 2 & 1 & -5 \end{pmatrix}$$

for which $\begin{pmatrix} 2 \\ 1 \\ 1 \end{pmatrix}$ is a right factor of zero.

Thus there are three values of λ for which solutions of (14.22.1) are possible, and to each of them there corresponds a vector $\mathbf{x}$, which is unique to within a scalar multiple.

For the general case of an $n \times n$ matrix $\mathbf{A}$, the possible values of λ are given by the equation

(14.22.4) $$|\mathbf{A} - \lambda \mathbf{I}| = 0$$

which is of degree n in λ, since the product of the diagonal elements in the determinant gives a term in λ^n, while no other term will contain λ to such a high power. The equation (14.22.4) is called the *characteristic equation* of the matrix $\mathbf{A}$. This equation will have n roots, which are not necessarily distinct, though in general they will be. These roots are called the *eigenvalues* of $\mathbf{A}$.*

Suppose λ_1 is an eigenvalue of $\mathbf{A}$. Does it follow that there is a column vector $\mathbf{x}_1$ such that $\mathbf{Ax}_1 = \lambda_1 \mathbf{x}_1$? Yes, for $\mathbf{A} - \lambda_1 \mathbf{I}$ is a matrix of rank $r \leq n - 1$, so that the equation $(\mathbf{A} - \lambda_1 \mathbf{I})\mathbf{x} = 0$ has $n - r$ linearly independent solutions, by Theorem 14.9. These solutions can be found by the methods used in the proof of that theorem. A vector $\mathbf{x}$ such that $\mathbf{Ax} = \lambda \mathbf{x}$ for some scalar λ is called an *eigenvector* of $\mathbf{A}$. A very simple example will show that the roots of the characteristic equation may not be all distinct. If $\mathbf{A} = \begin{pmatrix} 1 & 1 \\ -1 & 3 \end{pmatrix}$ then

$$|\mathbf{A} - \lambda \mathbf{I}| = \begin{vmatrix} 1 - \lambda & 1 \\ -1 & 3 - \lambda \end{vmatrix} = (1 - \lambda)(3 - \lambda) + 1 = \lambda^2 - 4\lambda + 4.$$

* Other terms are *characteristic roots* or *latent roots*.

Hence the characteristic equation is $(\lambda - 2)^2 = 0$, which has the single root $\lambda = 2$, counted twice. For this value of λ

$$\mathbf{A} - \lambda\mathbf{I} = \begin{pmatrix} -1 & 1 \\ -1 & 1 \end{pmatrix}$$

which has the right factor of zero $\begin{pmatrix} 1 \\ 1 \end{pmatrix}$. This is the only eigenvector. When the roots of the characteristic equation, the eigenvalues, are all distinct we might expect that the corresponding eigenvectors would also be distinct. This is in fact the case. What is more, we can prove a stronger result, expressed in the following theorem.

THEOREM 14.12

IF THE n EIGENVALUES OF $\mathbf{A}$ ARE DISTINCT, THEN THE n CORRESPONDING EIGEN-
VECTORS ARE LINEARLY INDEPENDENT.

PROOF:

The proof is by a *reductio ad absurdum* argument.

Let $\lambda_1, \lambda_2, \ldots \lambda_n$, be the eigenvalues, and $\mathbf{x}_1, \mathbf{x}_2, \ldots, \mathbf{x}_n$ the corresponding eigenvectors. Suppose these eigenvectors are linearly dependent; then there exist scalars $k_1, k_2, \ldots, k_n$, not all zero, such that

$$(14.22.5) \qquad k_1\mathbf{x}_1 + k_2\mathbf{x}_2 + \cdots + k_n\mathbf{x}_n = \mathbf{0}.$$

We shall assume that $k_n \neq 0$. There is no loss of generality in this assumption; for the order in which we write the vectors is immaterial and we can always arrange that the last one is one for which the corresponding scalar coefficient is nonzero (we know that at least one of the ks is nonzero).

Let us multiply equation (14.22.5) on the left by the matrix $\mathbf{A} - \lambda_1\mathbf{I}$. This gives

$$(14.22.6) \qquad k_1(\mathbf{A} - \lambda_1\mathbf{I})\mathbf{x}_1 + k_2(\mathbf{A} - \lambda_1\mathbf{I})\mathbf{x}_2 + \cdots$$
$$+ k_n(\mathbf{A} - \lambda_1\mathbf{I})\mathbf{x}_n = \mathbf{0}$$

In (14.22.6) the first term is zero, since $(\mathbf{A} - \lambda_1\mathbf{I})\mathbf{x}_1 = \mathbf{0}$. The second term is

$$\begin{aligned} k_2(\mathbf{A} - \lambda_1\mathbf{I})\mathbf{x}_2 &= k_2\mathbf{A}\mathbf{x}_2 - k_2\lambda_1\mathbf{x}_2 \\ &= k_2\lambda_2\mathbf{x}_2 - k_2\lambda_1\mathbf{x}_2 \\ &= k_2(\lambda_2 - \lambda_1)\mathbf{x}_2. \end{aligned}$$

Proceeding similarly with the other terms in (14.22.6) we see that (14.22.6) becomes

(14.22.7) $k_2'\mathbf{x}_2 + k_3'\mathbf{x}_3 + \cdots + k_n'\mathbf{x}_n = 0$

where k_r' stands for $k_r(\lambda_r - \lambda_1)$. If we now multiply (14.22.7) by $\mathbf{A} - \lambda_2\mathbf{I}$ in a similar manner, the vector $\mathbf{x}_2$ will disappear and we shall obtain an equation of the form

(14.22.8) $k_3''\mathbf{x}_3 + k_4''\mathbf{x}_4 + \cdots + k_n''\mathbf{x}_n = 0$

where

$$k_r'' = k_r(\lambda_r - \lambda_1)(\lambda_r - \lambda_2).$$

Multiplying successively by $\mathbf{A} - \lambda_3\mathbf{I}$, $\mathbf{A} - \lambda_4\mathbf{I}$, etc., we progressively eliminate all the vectors except $\mathbf{x}_n$, and obtain an equation of the form

$$k_n(\lambda_n - \lambda_1)(\lambda_n - \lambda_2)\cdots(\lambda_n - \lambda_{n-1})\mathbf{x}_n = 0.$$

Now $\mathbf{x}_n$ is not the zero vector, and the factors $(\lambda_n - \lambda_1)$, $(\lambda_n - \lambda_2)$, etc., are not zero, since the λs are distinct. Hence we must have $k_n = 0$, which contradicts the previous statement that $k_n \neq 0$. This completes the *reductio ad absurdum* argument.

A consequence of this theorem is that an $n \times n$ matrix with n distinct eigenvalues has n distinct eigenvectors. The example of a 2×2 matrix just given shows that if the eigenvalues are not distinct the matrix may have fewer than n distinct eigenvectors. On the other hand, it may have its full complement of eigenvectors, even though the eigenvalues are not distinct, as the following example shows.

EXAMPLE:
Find the eigenvalues of the matrix

$$\mathbf{A} = \begin{pmatrix} 1 & -6 & -4 \\ 0 & 4 & 2 \\ 0 & -6 & -3 \end{pmatrix}$$

The characteristic equation is

$$\begin{vmatrix} 1 - \lambda & -6 & -4 \\ 0 & 4 - \lambda & 2 \\ 0 & -6 & -3 - \lambda \end{vmatrix} = 0$$

which reduces to $\lambda(\lambda - 1)^2 = 0$, whence the eigenvalues are $\lambda = 0$, and $\lambda = 1$ counted twice. Thus we have here a matrix for which the eigenvalues are not distinct. If $\lambda = 0$, $\mathbf{A} - \lambda\mathbf{I} = \begin{pmatrix} 1 & -6 & -4 \\ 0 & 4 & 2 \\ 0 & -6 & -3 \end{pmatrix}$, and it is easily verified that $\begin{pmatrix} 2 \\ -1 \\ 2 \end{pmatrix}$ is a right factor of zero. If $\lambda = 1$, $\mathbf{A} - \lambda\mathbf{I}$ becomes

$$\begin{pmatrix} 0 & -6 & -4 \\ 0 & 3 & 2 \\ 0 & -6 & -4 \end{pmatrix}$$

which has two right factors of zero (since the rank is only 1). They can be taken as

$$\begin{pmatrix} 1 \\ 0 \\ 0 \end{pmatrix} \quad \text{and} \quad \begin{pmatrix} 0 \\ 2 \\ -3 \end{pmatrix}$$

(though any two linear combinations of these vectors could do as well).

Exercise 173

1. If $\mathbf{P}$ is a nonsingular $n \times n$ matrix, and $\mathbf{A}$ is an $n \times n$ matrix, the matrix $\mathbf{P}^{-1}\mathbf{A}\mathbf{P}$ is called the *transform* of $\mathbf{A}$ by $\mathbf{P}$. Show that the relation $\sim$, where $\mathbf{A} \sim \mathbf{B}$ means "$\mathbf{B}$ is the transform of $\mathbf{A}$ by some nonsingular matrix $\mathbf{P}$," i.e., "there is a nonsingular matrix $\mathbf{P}$ such that $\mathbf{B} = \mathbf{P}^{-1}\mathbf{A}\mathbf{P}$," is an equivalence relation. (Matrices $\mathbf{A}$ and $\mathbf{B}$ related in this way are said to be equivalent matrices.)

2. Prove that equivalent matrices have the same characteristic equation.

3. Find the eigenvalues of the following matrices.

(a) $\begin{pmatrix} p & q \\ r & s \end{pmatrix}$

(c) $\begin{pmatrix} 4 & 3 & 2 \\ 3 & 4 & 2 \\ 2 & 2 & 0 \end{pmatrix}$

(b) $\begin{pmatrix} 2 & 1 & \dfrac{10}{3} \\ 0 & -1 & -10 \\ -3 & -1 & 1 \end{pmatrix}$

4. Prove that if (a_{ij}) is an $n \times n$ matrix then the coefficient of λ^{n-1} in the characteristic equation of $\mathbf{A}$ is $-\sum_{i=1}^{n} a_{ii}$. The number $\sum_{i=1}^{n} a_{ii}$, the sum of the diagonal elements of a matrix, is called the *trace* of the matrix. Prove that equivalent matrices have the same trace.

14.23 SOME APPLICATIONS

We shall now look at the analysis of a simple economic situation—an analysis which will use and illustrate some results of this chapter and of Chapter 11. Let us consider $m + 1$ goods, numbered $0, 1, 2, \ldots, m$, and let p_i be the price per unit of the ith good. We could take these prices as being in any arbitrary units, but there are some advantages in expressing them all in terms of the price of one particular good—the *numéraire*. We shall choose good number 0 to be the *numéraire*, so that $p_0 = 1$ by definition. The other prices are functions of the time t.

The demand for the ith good will depend on the prices of all the goods, as also will its supply. Thus if we denote by D_i and S_i the demand and supply for the ith good we can write

$$D_i = D_i(p_1, p_2, \cdots, p_m)$$
$$S_i = S_i(p_1, p_2, \cdots, p_m). \quad (i = 0, 1, 2, \ldots, m)$$

Note that the price p_0 of the *numéraire* need not be explicitly included, since it is constant. We can also define a function X_i, the *excess demand function*, which is the excess of demand over supply for the ith good. Thus

$$(14.23.1) \qquad X_i(p_1, p_2, \cdots, p_m) = D_i(p_1, p_2, \cdots, p_m) - S_i(p_1, p_2, \cdots, p_m)$$
$$(i = 0, 1, 2, \cdots, m).$$

If the supply and demand for each good are equal, we have a *general equilibrium* situation. In other words, all excess demands are zero. When this occurs we have

$$(14.23.2) \qquad X_i = 0 \qquad (i = 0, 1, 2, \cdots, m).$$

This is a set of $m + 1$ equations in the variables $p_1, p_2, \ldots, p_m$, which in general would be difficult to solve (they need not be linear equations); indeed there is no guarantee that *any* such equilibrium solution will exist. If there *is* a solution to equations (14.23.2) this determines an equilibrium price vector which we shall denote by $(p'_1, p'_2, \cdots, p'_m)$. There may quite possibly be more than one such vector.

In the more interesting case where supply and demand are not equal we need to investigate how the prices will change in order to counter the discrepancy between supply and demand. If demand exceeds supply for a particular good, one would expect the price to rise, and it is reasonably plausible to assume, at least as an approximation, that the rate at which the price rises is proportional to the excess of demand over supply. In short, that

$$(14.23.3) \qquad \frac{dp_i}{dt} = k_i X_i \qquad (i = 1, 2, \ldots, m)$$

where the k_i are positive constants.

This is about as far as we can go without making some more assumptions about the system we are studying and, in particular, being more specific about the functions X_i.

We shall assume now that the system is one for which an equilibrium is possible, and that $(p'_1, p'_2, \ldots, p'_m)$ is an equilibrium price vector. We shall further assume that the functions X_i can be expanded by Taylor's Theorem about the equilibrium value of the prices. We can then write

$$(14.23.4) \qquad X_i(p_1, p_2, \cdots, p_m) = X_i(p'_1, p'_2, \cdots, p'_m) + \sum_j \frac{\partial X_i}{\partial p_j}(p_j - p'_j)$$

$$+ \text{ terms of the second degree in } (p_i - p'_i), \text{ etc.,}$$

this being an extension to m variables of equation (9.23.4). By (14.23.2) we have $X_i(p'_1, p'_2, \ldots, p'_m) = 0$, and (14.23.4) can thus be written as

$$(14.23.5) \qquad X_i = \sum_j \frac{\partial X_i}{\partial p_j}(p_j - p'_j) + \text{ second degree terms, etc.}$$

The partial derivatives $\partial X_i/\partial p_j$ are, of course, evaluated at the equilibrium point. We shall now simplify (14.23.5) by writing it in matrix notation.

Let $\mathbf{A}$ be the matrix $(\partial X_i/\partial p_j)$, i.e., the matrix whose i, j-element is $a_{ij} = \partial X_i/\partial p_j$, evaluated at the equilibrium point, and let $\mathbf{P}$ denote the column vector whose ith element is $P_i = p_i - p'_i$. Then (14.23.5) becomes

$$(14.23.6) \qquad \mathbf{X} = \mathbf{AP}$$

where $\mathbf{X}$ is a column vector whose ith element is X_i. Here we have ignored the second degree terms in $p_j - p'_j$. This means that from now on the treatment is approximate and applies only for small departures from the equilibrium price vector.

Now if $\mathbf{U} = (u_i)$ is any column vector whose elements are functions of a single variable t we may conveniently write $d\mathbf{U}/dt$ to denote the column vector (du_i/dt). In other words the operator "d/dt" applied to a column vector will have the effect of differentiating every element of the vector. Using this notation we can write (14.23.3) as

$$\frac{d\mathbf{P}}{dt} = \begin{pmatrix} k_1 & 0 & 0 & \cdots & 0 \\ 0 & k_2 & 0 & \cdots & 0 \\ 0 & 0 & k_3 & \cdots & \\ & & & \cdots & \cdot \\ 0 & 0 & 0 & 0 & k_m \end{pmatrix} \begin{pmatrix} X_1 \\ X_2 \\ \cdot \\ \cdot \\ \cdot \\ X_m \end{pmatrix}$$

or

(14.23.7)
$$\frac{d\mathbf{P}}{dt} = \mathbf{KX}$$

where, on the right-hand side we have a *diagonal* matrix $\mathbf{K}$ multiplying the column vector $\mathbf{X}$, and this is easily seen to give the column vector whose general element is $k_i X_i$. On the left-hand side we have

$$\frac{dp_i}{dt} = \frac{d(p_i - p_i')}{dt} = \frac{dP_i}{dt},$$

since the p_i' are constants. Thus we end up with a system of differential equations of the form

(14.23.8)
$$\frac{dP_i}{dt} = \sum_{j=1}^{m} k_i a_{ij} P_j \quad (i = 1, 2, \ldots, m)$$

expressed in matrix notation as

(14.23.9)
$$\frac{d}{dt}\mathbf{P} = \mathbf{KAP},$$

since $\mathbf{X} = \mathbf{AP}$. We now find the form of the general solution of these equations.

Let $\lambda_1, \lambda_2, \ldots, \lambda_m$ be the eigenvalues of the matrix $\mathbf{A}$, and let $\mathbf{x}_1, \mathbf{x}_2, \ldots, \mathbf{x}_m$ be the corresponding eigenvectors. We have already shown (Theorem 14.12) that if the eigenvalues are distinct (and we shall assume them to be) then the eigenvectors are linearly dependent. We also showed (Exercise 171.6) that $m + 1$ vectors of dimension m are always linearly dependent. Hence the $m + 1$ vectors

$$\mathbf{P}, \mathbf{x}_1, \mathbf{x}_2, \ldots, \mathbf{x}_m$$

(where $\mathbf{P}$ is the solution vector we are looking for) are linearly dependent. Thus $\mathbf{P}$ is linearly dependent on $\mathbf{x}_1, \mathbf{x}_2, \ldots, \mathbf{x}_m$; for the alternative (by Theorem 14.6) is that $\mathbf{x}_1, \mathbf{x}_2, \ldots, \mathbf{x}_m$ are dependent, which is not so. Hence we can write

(14.23.10) $$\mathbf{P} = \sum_{j=1}^{m} c_j(t)\mathbf{x}_j.$$

Note that the $\mathbf{x}_j$ are *constant* vectors; only the multipliers c_j depend on t. Differentiating (14.23.10) with respect to t we have

$$\frac{d\mathbf{P}}{dt} = \sum_{j=1}^{m} c'_j(t)\mathbf{x}_j.$$

Making this substitution in the left-hand side of (14.23.9), and substituting for $\mathbf{P}$ in the right-hand side, from (14.23.10), we obtain

$$\sum_{j=1}^{m} c'_j(t)\mathbf{x}_j = \mathbf{KA} \sum_{j=1}^{m} c_j(t)\mathbf{x}_j$$

$$= \mathbf{K} \sum_{j=1}^{m} c_j(t)\mathbf{A}\mathbf{x}_j$$

$$= \mathbf{K} \sum_{j=1}^{m} \lambda_j c_j(t)\mathbf{x}_j$$

since $\mathbf{A}\mathbf{x}_j = \lambda_j \mathbf{x}_j$. Hence we have

$$\sum_{j=1}^{m} c'_j(t)\mathbf{x}_j = \sum_{j=1}^{m} k_j \lambda_j c_j(t)\mathbf{x}_j.$$

We deduce that

(14.23.11) $$\sum_{j=1}^{m} \{c'_j(t) - k_j \lambda_j c_j(t)\}\mathbf{x}_j = \mathbf{0}.$$

But the vectors $\mathbf{x}_j$ are linearly independent, and hence (14.23.11) can be true only if all the coefficients of the $\mathbf{x}_j$ are zero, i.e., only if

(14.23.12) $$c'_j(t) - k_j \lambda_j c_j(t) = 0 \quad (j = 1, 2, \ldots, m).$$

Thus we see that we have decomposed the system of equations into a set of m linear differential equations each in one variable $[c_j(t)]$ only. The general solutions of equations (14.23.12) are

$$c_j(t) = C_j e^{k_j \lambda_j t}$$

for $j = 1, 2, \ldots, m$, where C_j are arbitrary constants. Thus the required solution vector $\mathbf{P}$ is

(14.23.13) $$\mathbf{P} = \sum_{j=1}^{m} C_j e^{k_j \lambda_j t}\mathbf{x}_j$$

If the ith component of the vector $\mathbf{x}_j$ is α_{ij}, then we have, for the individual price differences $p_i - p'_i$

$$P_i = p_i - p'_i = \sum_{j=1}^{m} C_j \alpha_{ij} e^{k_j \lambda_j t}.$$

Thus in the general solution each P_i is a linear combination of terms of the form $e^{k_j \lambda_j t}$.

It is worth looking at the conditions under which the solution vector will tend to equilibrium, i.e., under which $\mathbf{P} \to \mathbf{0}$ as $t \to \infty$. If λ_j is real then $e^{k_j \lambda_j t} \to 0$ if, and only if $\lambda_j < 0$, since k_j is real and positive. Complex eigenvalues will occur in conjugate pairs (see Theorem 13.2) and will give rise to solutions of the form

$$e^{k_j u_j t} (A \cos k_j v_j t + B \sin k_j v_j t)$$

where $\lambda_j = u_j + iv_j$ (compare section 13.12). Solutions of this kind will tend to zero if u_j, the real part of λ_j, is negative.

It is easily verified that if these conditions do not obtain then the solution will not tend to zero. What is more, the solution not only does not tend to zero, it does not even remain small. Since our whole treatment was based on the supposition that we were considering *small* changes from the equilibrium price vector, the whole method breaks down if the above conditions are not satisfied, and we can make no statement about what happens to the system. Hence we have the result that for *stability*, that is, for the price vector to tend to the equilibrium vector, every eigenvalue of $\mathbf{A}$ must have negative real part. Note that we have assumed that all the latent roots are distinct. If they are not, then certain difficulties can arise, but it would take us too much out of our way to discuss these exceptional cases here.

INPUT-OUTPUT MODELS

As a further example of how matrices can arise in economic problems let us look at the basic concepts behind input-output economics. We shall merely brush the surface of this important branch of economic theory, but with this introduction the reader should not find it difficult to read further from the extensive literature on the subject.

Let us start with a simple example. Suppose that the national economy of a country is divided into four "sectors." Three of these need no explanation; they are Agriculture, Light Industry and Heavy Industry. The products of these sectors are for the most part intended to be used by the general body of consumers in the country, but some may be used by other sectors. Thus some products of Heavy and Light industries may be used in Agriculture (tractors, spades, etc.); some products of a sector may be used by that sector itself, and so on. The fourth sector we shall consider, which we can conveniently call *Households*, is a sort of "other" category, including all users of the products of the various sectors that are not accounted for under the headings of those sectors.

During a given interval of time, say one year, there will be a quantity associated with each ordered pair of sectors, namely the amount of the output of the first sector that is used as input by the second sector. Similar quantities will exist for the amount of output of each sector that is "ploughed back" as input to the same sector. These quantities can conveniently be displayed in the form of a two-way table, of which Table 14.1 is an example.

Table 14.1

From/To	Agr.	H.I.	L.I.	Households	Total output
Agriculture	200	80	120	200	600 units
Heavy Industry	15	25	35	15	90 units
Light Industry	120	150	60	90	420 units
Households	30	60	20	50	160 units

Note that we have added the total output of each sector on the right of the table.

The units in which the quantities in the table have been expressed have purposely been left vague. This is partly because this table is for illustration only and is not meant to be significant, and partly because the question of units is one which we must examine more closely. It is possible that for any given sector a unit could be found in terms of which the whole output of the sector could be expressed. But it would be most unlikely that a unit suitable for Heavy Industry, say (which would probably be "tons" of something or other) would be suitable for Agriculture, for which the unit would probably be "bushels" of this or that. In the absence of a common unit between the different sectors there is no significant way of comparing items in different rows of Table 14.1, which therefore consists of four quite separate rows rather than one two-way table.

This difficulty disappears if, instead of quantities of output, we use the monetary value of the output. Translating the quantities of Table 14.1 into their cash value we obtain a new table (Table 14.2) in which all the entries are directly comparable.

Because of the compatibility between units it now makes sense to add the columns as well as the rows, and this has been done. These column totals are the total inputs to the various sectors. A table such as Table 14.2 is called an *Input-Output Table*. Although the entries in Table 14.2 are all in the same monetary unit we have not specified what the unit is; there would be no

Table 14.2

From/To	Agriculture	Heavy Industry	Light Industry	Households	Total output
Agriculture	50	20	30	50	150
Heavy Industry	30	50	70	30	180
Light Industry	40	50	20	30	140
Households	30	60	20	50	160
Total input	150	180	140	160	

point in doing so for a purely illustrative example. In practice the units would probably be millions of dollars, or some other round number of dollars.

We see from the table that during the specified year 50 units worth of agricultural output (a third of the total output) was put back into Agriculture, and the same amount went to the ultimate users ("Households"). The total input to Households was 160 units worth of goods, which is therefore the annual demand by consumers. The output of Households will be mainly in the form of labor and amounts to 160 units per year also.

It is easy to go from this simple example to a description of a general input-output table. We have, say, n sectors, numbered "1, 2, ..., n," and for each ordered pair (i, j), where $i, j = 1, 2, ..., n$, we define a number x_{ij} which is the monetary value of the output of sector i which is used as input by sector j. These numbers can be displayed as a matrix $\mathbf{X} = (x_{ij})$ which together with the row and column totals and suitable captions and headings, makes up the input-output table.

It is a consequence of elementary bookkeeping that provided the n sectors cover the whole of the economy, the row and column totals should agree, that is, that the value of the total input to any sector should equal the total output from that sector. For any discrepancy that might otherwise arise between the total input to a sector and the total output from it will normally be allowed for by the inclusion of a heading "Profits," defined as the value of output minus the value of input. This profit is a payment for an input provided by Households, and by its very definition will ensure the agreement of the row and column totals.

We shall now make a slight modification to the above description, one which will emphasize the fact that the sector Households is on a rather different footing from the others, being an agglomeration of all the ultimate users of the products of the various sectors.

For our slightly modified table we shall assume n sectors *other* than Households, and denote by x_{ij}, as before, the value of output from sector i that is input to sector j. We denote by y_i the output from sector i that is input to Households, and we let x_j be the total output of sector j. Our input-output table is therefore in the form given by Table 14.3.

Table 14.3

From/To	Sector 1	2	3	$\cdots$	n	Households	Total output
S 1	x_{11}	x_{12}	x_{13}	$\cdots$	x_{1n}	y_1	x_1
e 2	x_{21}	x_{22}	x_{23}	$\cdots$	x_{2n}	y_2	x_2
c 3	x_{31}	x_{32}	x_{33}	$\cdots$	x_{3n}	y_3	x_3
t $\cdots$	.		$\cdots$	.		..	..
o r n	x_{n1}	x_{n2}	x_{n3}	$\cdots$	x_{nn}	y_n	x_n

For convenience, the output from Households has not been included. The last two columns of Table 14.3 are two column vectors: **y**, which we can call the *final demand vector*, and **x**, the *total output vector*. Clearly we have

$$(14.23.14) \qquad y_i = x_i - \sum_{j=1}^{n} x_{ij} \quad (i = 1, 2, \ldots, n)$$

directly from the definition of total output. The vector $(y_i + \sum_j x_{ij})$, or $\mathbf{y} + \sum_j \mathbf{x}_j$, where $\mathbf{x}_j = (x_{ij})$, is known as the *total demand vector*. Equation (14.23.14), in the form $\mathbf{x} = \mathbf{y} + \sum_j \mathbf{x}_j$ expresses the equality of supply and demand.

It is convenient to introduce numbers a_{ij} defined by

$$(14.23.15) \qquad a_{ij} = \frac{x_{ij}}{x_j}$$

called *input coefficients*. The input coefficient a_{ij} is the ratio of the value of the input of good i into sector j to the value of the output of sector j. We shall assume that the input coefficients are constant for all levels of output so that we can write $x_{ij} = a_{ij}x_j$ for all values of the variable x_j. The matrix $\mathbf{A} = (a_{ij})$ is known as the *structural matrix* of the economy. For the numerical example treated above it is

$$\begin{pmatrix} \dfrac{1}{3} & \dfrac{1}{9} & \dfrac{3}{14} \\[2mm] \dfrac{1}{5} & \dfrac{5}{18} & \dfrac{1}{2} \\[2mm] \dfrac{4}{15} & \dfrac{5}{18} & \dfrac{1}{7} \end{pmatrix}$$

as the reader should verify.

In terms of the input coefficients, equation (14.23.14) can be written as

(14.23.16) $\qquad y_i = x_i - \sum_{j=1}^{n} a_{ij}x_j \quad (i = 1, 2, \ldots, n)$

or in matrix notation as

$$\mathbf{y} = \mathbf{x} - \mathbf{Ax}$$

or

(14.23.17) $\qquad\qquad\qquad \mathbf{y} = (\mathbf{I} - \mathbf{A})\mathbf{x}.$

Let us write $\qquad\qquad \mathbf{B} = (\mathbf{I} - \mathbf{A})^{-1}; \quad$ then

(14.23.18) $\qquad\qquad\qquad \mathbf{x} = \mathbf{By}$

which gives us the various total outputs in terms of the demands. If $\mathbf{B} = (b_{ij})$ then b_{ij} is the increase in output of sector i if the demand for the products of sector j increases by one unit. This follows from the fact that

$$x_i = \sum_{j=1}^{n} b_{ij}y_j.$$

The matrix $\mathbf{B} = (\mathbf{I} - \mathbf{A})^{-1}$ which has just been introduced is called the *input-output inverse* or the *Leontief inverse*.

We shall now assume that the structural matrix $\mathbf{A}$ is characteristic of the economy and therefore that its elements do not change from year to year. On the other hand demand and output will not be the same each year. An important question is "Can the economy satisfy any demand?" In other words, if a demand vector $\mathbf{y}$ is given, does there exist a corresponding output vector $\mathbf{x}$ such that $\mathbf{x}$ and $\mathbf{y}$ are related by (14.23.17)? Since (14.23.18) gives $\mathbf{x}$ in terms of $\mathbf{y}$ the answer would at first sight appear to be "yes"; but we must remember that the elements of $\mathbf{x}$ must be non-negative to be economically meaningful. If $\mathbf{A}$ and $\mathbf{y}$ are such that $\mathbf{By}$ has a negative element, then the answer to the question is "No". Thus the economy can satisfy any demand if, and only if, $\mathbf{By}$ has no negative elements whatever $\mathbf{y}$ may be. This is a condition on the matrix $\mathbf{B}$ and hence, ultimately, on $\mathbf{A}$.

If $\mathbf{B}$ has the above property, then if $\mathbf{y}$ is the column vector $\begin{pmatrix} 1 \\ 0 \\ 0 \\ \cdots \\ 0 \end{pmatrix}$

we see that $\mathbf{x}$ is the first column of $\mathbf{B}$; hence this column has no negative elements. In a similar way we see that every element of $\mathbf{B}$ must be non-negative.

Hence for an economy to be able to satisfy any demand, the matrix **B** must have no negative elements.

This condition on the matrix **B** was easily obtained. It is less easy to derive an equivalent condition on the structural matrix **A**, and we shall not go into this question.

The brief outline given above touches only on the simplest concepts of input-output models of an economy. There are many complications that could have been introduced, not the least of which is the question of how best to set about inverting a matrix which may have several hundred rows and columns, as many structural matrices have. Fortunately the advent of the electronic computer has made such computations a practical possibility. The input-output model here considered was essentially a static one; more interesting and more meaningful results are obtained when one considers how the system changes with time, though the analysis is, naturally, more involved. But for these and other matters the reader is referred to the references at the end of the chapter.

Exercise 174

1. Show that if the excess-demand functions X_i ($i = 1,2,3$) are given by

$$X_1 = 7p_1 \qquad + 2p_3 - 24$$
$$X_2 = \qquad 5p_2 - 2p_3 + 5$$
$$X_3 = 2p_1 - 2p_2 + 6p_3 - 32$$

then there is an equilibrium point at which

$$p_1 = 2, p_2 = 1 \text{ and } p_3 = 5.$$

Use the matrix method of the last section to determine whether the system is stable about this equilibrium point, given that $k_1 = k_2 = k_3 = 1$.
2. Verify that the matrix

$$\mathbf{A} = \begin{pmatrix} 0.3 & 0.2 & 0 & 0.4 \\ 0.2 & 0 & 0.3 & 0.3 \\ 0.4 & 0.2 & 0.4 & 0 \\ 0 & 0.3 & 0.3 & 0.3 \end{pmatrix}$$

satisfies the condition mentioned in the latter part of section 14.23, namely that $(\mathbf{I} - \mathbf{A})^{-1}$ contains no negative elements.

Exercise 175 (Chapter Review)

1. Given that

$$A = \begin{pmatrix} 1 & -1 & 1 \\ 2 & 0 & 3 \\ 5 & 1 & 4 \end{pmatrix}; \qquad B = \begin{pmatrix} 12 & 0 & 5 \\ 0 & 3 & 7 \\ 1 & 1 & 0 \end{pmatrix}$$

$$C = \begin{pmatrix} 3 & 3 & 1 \\ 3 & 1 & 3 \\ 1 & 3 & 3 \end{pmatrix}; \qquad x = \begin{pmatrix} 1 \\ -1 \\ 2 \end{pmatrix}$$

$$u = (1, 0, 5) \quad \text{and} \quad v = (0, 2, -1)$$

evaluate

(a) **Ax**   (c) **uB**   (e) **vCx**   (g) **AC**   (i) **vx**

(b) **uCB**   (d) **vABx**   (f) **xu**   (h) **vxu**

2. Prove that if **A, B, C** are matrices such that

$$AC = CA \quad \text{and} \quad BC = CB$$

then

$$(\lambda AB + \mu BA)C = C(\lambda AB + \mu BA)$$

where λ, μ are any two scalars.

3. Find a general expression for A^n when

(a) $A = \begin{pmatrix} \lambda & 1 \\ 0 & \lambda \end{pmatrix}$

(b) $A = \begin{pmatrix} \lambda & 1 & 0 \\ 0 & \lambda & 1 \\ 0 & 0 & \lambda \end{pmatrix}$

4. If P, Q, R, S are permutations (of 6 objects) defined by

$$P = \begin{pmatrix} 1 & 2 & 3 & 4 & 5 & 6 \\ 3 & 2 & 4 & 1 & 6 & 5 \end{pmatrix}; \qquad Q = \begin{pmatrix} 1 & 2 & 3 & 4 & 5 & 6 \\ 3 & 4 & 5 & 1 & 2 & 6 \end{pmatrix}$$

$$R = \begin{pmatrix} 1 & 2 & 3 & 4 & 5 & 6 \\ 1 & 3 & 6 & 2 & 4 & 5 \end{pmatrix}; \qquad S = \begin{pmatrix} 1 & 2 & 3 & 4 & 5 & 6 \\ 3 & 4 & 2 & 1 & 6 & 5 \end{pmatrix}$$

find the following permutations

    (a) PQ

    (b) RS

    (c) PS^{-1}

    (d) $SRQ,$

    (e) $P^{-1}QP$

    (f) $PQSRQ$

    Determine which of these permutations are even and which are odd.

5. Evaluate the following determinants

(a) $\begin{vmatrix} 11 & 3 \\ 15 & 4 \end{vmatrix}$
    (c) $\begin{vmatrix} 4 & 2 & 0 \\ 3 & 2 & 5 \\ 5 & 3 & 7 \end{vmatrix}$
    (e) $\begin{vmatrix} 3 & -5 & 3 \\ 5 & -5 & 5 \\ 2 & -5 & 1 \end{vmatrix}$

(b) $\begin{vmatrix} 5 & -1 & -2 \\ -1 & -3 & 8 \\ 6 & -1 & 1 \end{vmatrix}$
  (d) $\begin{vmatrix} 1 & -1 & 1 & -4 \\ 4 & 2 & 1 & 2 \\ 1 & 1 & 1 & 4 \\ 4 & -2 & 1 & -2 \end{vmatrix}$
  (f) $\begin{vmatrix} 3 & 7 & 5 & 2 \\ 4 & 10 & 6 & 6 \\ -1 & 3 & 1 & 4 \\ 1 & 2 & 3 & 4 \end{vmatrix}$

6. If A is the $n \times n$ matrix whose diagonal elements are all a, and whose off-diagonal elements are all b, where a and b are any two real numbers, find the value of $|A|$.

7. Solve the following sets of equations

    (a) $3x + 2y + 5z = 1$

            $x + 3y + 2z = 2$

          $2x + y - z = 3.$

    (b) $x + y + z = 4$

          $3x - 8y + 7z = 8$

      $11x - 44y + 31z = 24.$

    (c) $x + 2y + 3z + 4t = 34$

        $4x + 7y + 3z + 4t = 39$

        $4x + 8y + 5z + 6t = 69$

        $x - 3y - 7z - 2t = 36.$

8. Find, by any method, the inverses of the following matrices

    (a) $\begin{pmatrix} -4 & 3 \\ 1 & -2 \end{pmatrix}$

(b) $\begin{pmatrix} 3 & 6 & 5 \\ 1 & 3 & 2 \\ 2 & 5 & 4 \end{pmatrix}$

(c) $\begin{pmatrix} -1 & 1 & 2 \\ 2 & -1 & 1 \\ 1 & 2 & -1 \end{pmatrix}$

(d) $\begin{pmatrix} 3 & 3 & 2 & 4 \\ 0 & 12 & 4 & 8 \\ 2 & 5 & 2 & 3 \\ 3 & 6 & 3 & 2 \end{pmatrix}$

9. Prove that the n equations

$$\mathbf{Ax} = \mathbf{b}$$

where $\mathbf{A}$ is an $n \times n$ matrix have a solution if, and only if, the rank of $(\mathbf{A} : \mathbf{b})$, i.e., the $n \times (n + 1)$ matrix obtained from $\mathbf{A}$ by adding $\mathbf{b}$ as an extra column, is the same as the rank of $\mathbf{A}$.

10. Find the ranks of the following matrices

(a) $\begin{pmatrix} 1 & 2 & 0 \\ 3 & 0 & -1 \\ 5 & 1 & 4 \end{pmatrix}$

(b) $\begin{pmatrix} 1 & 9 & 7 & 6 \\ 0 & 6 & 6 & 6 \\ 3 & 5 & -1 & -4 \\ 2 & 8 & 4 & 2 \end{pmatrix}$

(c) $\begin{pmatrix} 1 & 0 & 3 & 2 \\ 4 & -1 & 0 & 7 \\ 6 & -1 & -3 & 2 \\ 5 & 0 & 6 & 1 \end{pmatrix}$

(d) $\begin{pmatrix} 1 & 0 & 2 & 1 & 6 \\ 0 & 1 & -4 & -2 & 0 \\ 4 & -1 & 3 & 0 & 2 \\ 10 & 0 & 6 & 0 & 12 \\ 5 & 1 & 1 & -1 & 4 \end{pmatrix}$

11. Prove that if the eigenvalues of matrix **A** are all distinct, then **A** satisfies its characteristic equation; that is, that if

$$\lambda^n + a_1\lambda^{n-1} + a_2\lambda^{n-2} + \cdots + a_{n-1}\lambda + a_n = 0$$

is the characteristic equation of **A**, then

$$\mathbf{A}^n + a_1\mathbf{A}^{n-1} + a_2\mathbf{A}^{n-2} + \cdots + a_{n-1}\mathbf{A} + a_n\mathbf{I} = \mathbf{0}.$$

Hint: Consider factors of the form $\mathbf{A} - \lambda_i\mathbf{I}$.

12. Deduce from problem 11 that if **A** is nonsingular

$$\mathbf{A}^{-1} = -\frac{a_{n-1}}{a_n}\mathbf{I} - \frac{a_{n-2}}{a_n}\mathbf{A} - \cdots - \mathbf{A}^{n-1}$$

and use this result to find the inverse of the matrix $\begin{pmatrix} 1 & 4 & 2 \\ 1 & 3 & 2 \\ 1 & 3 & 3 \end{pmatrix}$.

13. A real matrix **N** for which

$$\mathbf{N}^T = -\mathbf{N}$$

is said to be *skew-symmetric*. Prove that the diagonal elements of **N** are all zero.

If **N** is skew-symmetric, show that the matrix

$$(\mathbf{I} + \mathbf{N})^{-1}(\mathbf{I} - \mathbf{N})$$

is orthogonal.

14. Find an example of a matrix whose elements are real numbers but whose eigenvalues are complex.

15. Prove that the eigenvalues of a real symmetric matrix **A** are real.
Hint: From $\mathbf{A}\mathbf{x}_1 = \lambda_1\mathbf{x}_1$, by taking complex conjugates, deduce that $\bar{\mathbf{x}}_1^T\mathbf{A} = \bar{\lambda}_1\bar{\mathbf{x}}_1$. Now consider the expression $\bar{\mathbf{x}}_1^T\mathbf{A}\mathbf{x}_1$.

16. By the same method as in problem 15 prove that

(a) the eigenvalues of a skew-symmetric matrix are imaginary or zero

(b) the eigenvalues of an orthogonal matrix have unit modulus.

17. The eigenvalues $\lambda_1, \lambda_2, \ldots \lambda_n$, of a symmetric matrix **A** are distinct, and $\mathbf{x}_1, \mathbf{x}_2, \ldots, \mathbf{x}_n$ are the corresponding eigenvectors. If **T** is the matrix

$$\mathbf{T} = (x_1, x_2, \ldots, x_n)$$

show that $\mathbf{T}^{-1}\mathbf{A}\mathbf{T}$ is a diagonal matrix.

Show that by multiplying each eigenvector by a suitable constant the matrix **T** can be made into an orthogonal matrix.

To what extent and for what reason are the above results not true if A is not symmetric?

Suggestions for Further Reading

There are many books available that deal with matrices, either as a subject by itself or in conjunction with other branches of algebra. The following three books represent different approaches to the subject.

Archbold, J. W., *Algebra*. London: Sir Isaac Pitman & Sons, Ltd., 1964.

This book gives a good treatment of matrices among the many branches of algebra that it covers. It deals also with summation of series and many other topics discussed already in this book, and it was recommended at the end of Chapter 13 for its treatment of complex numbers.

Ayres, F., *Matrices*. New York: Schaum Publishing Company, 1962.

A strictly practical book containing a great many worked and unworked examples.

Littlewood, D. E., *A University Algebra*. London: Heinemann Ltd., Second Edition, 1958.

This is pitched at a more advanced level than the other two books and contains much that the reader will not need. Nevertheless the chapters dealing with matrices, as well as some other chapters, will repay study.

The following book is recommended for a better understanding of input-output matrices.

Leontief, W., *Input-Output Economics*. New York: Oxford University Press, 1966.

This consists of eleven essays written by a pioneer in this branch of mathematical economics.

15 Iterative Processes,

Difference Equations

and Markov Chains

Life is just one damned thing after another.

E. HUBBARD.
A Thousand and One Epigrams.

15.1 ITERATION

An iterative process is one which is performed over and over again in essentially the same manner. In a sense, any sort of repetitive task could be called an iterative process, from the soul-destroying occupation of the factory worker endlessly screwing nut A to bolt B, to a finance minister's annual job of preparing a national budget; but in mathematics the term *iteration* is applied to the repetition of a calculation with some particular aim in view, in general that of approaching some desired result to within a given degree of accuracy. But the best method of illustrating what is meant by an iterative process is by means of some examples.

We shall start with a very simple example which was used as an exercise on limits in Chapter 8. Let x_0 be any real number, and let the sequence $\{x_n\}$ be defined for $n = 1, 2, 3, \ldots$ by the equation

$$(15.1.1) \qquad\qquad x_{n+1} = \frac{1}{2}\left(x_n + \frac{2}{x_n}\right).$$

An equation like (15.1.1) is often called a *recursive equation;* it expresses x_{n+1} in terms of previous terms in the sequence—in this case just the previous one. The reader should verify that the first few terms in this sequence, if we take $x_0 = 1$, are

$$1, \quad 1.5, \quad 1.417, \quad 1.414, \quad \ldots$$

to three decimal places whereas, if we start by taking $x_0 = 3$ we get

$$3, \quad 1.833, \quad 1.462, \quad 1.415, \ldots$$

It rather looks as though each of these series tends to a finite limit and, what is more, that they tend to the *same* limit. Let us see whether this conjecture is true. We shall assume, for convenience, that x_0 is positive.

The number to which these sequences seem to be tending is about 1.414 or so, which the reader will recognize as being approximately $\sqrt{2}$. Can it be that the sequence tends to this value? To test this idea, let us write

$$x_n = \sqrt{2} + y_n.$$

Then (15.1.1) becomes

$$\sqrt{2} + y_{n+1} = \frac{1}{2}\left(\sqrt{2} + y_n + \frac{2}{\sqrt{2} + y_n}\right)$$

or

$$y_{n+1} = \frac{(\sqrt{2} + y_n)^2 + 2}{2(\sqrt{2} + y_n)} - \sqrt{2}.$$

Hence

(15.1.2)
$$y_{n+1} = \frac{y_n^2}{2(\sqrt{2} + y_n)}.$$

Since we are assuming that $x_0 = \sqrt{2} + y_0$ is positive, it follows from (15.1.2) that y_1 and all the y_ns are positive.

Further

$$y_{n+1} = \frac{1}{2}\left(\frac{y_n}{\sqrt{2} + y_n}\right) y_n < \frac{1}{2} y_n$$

so that $\{y_n\}$ is a *decreasing* sequence of positive numbers. From Theorem 8.11 it follows that y_n tends to a finite limit, and hence that x_n does too. Moreover since $y_{n+1} < \frac{1}{2}y_n$ it follows that $y_n < (\frac{1}{2})^n y_0$, and hence that $y_n \to 0$. The limit of the sequence $\{x_n\}$ is therefore $\sqrt{2}$. Note that this is true whatever the value x_0 that we start with.

This then is an iterative method of calculating $\sqrt{2}$, and we see at once that it has two great advantages over other methods. In the first place we note that it is proof against errors. We need not worry whether we have made any mistakes—it doesn't matter. If, for example, we start with $x_0 = 1$, as we did above, and having calculated the value of x_2 as 1.417 we write it down wrongly as 1.427 and carry on from there, we shall still get to the right answer in the end. For from then on it is just the same as if we had decided to take $x_0 = 1.427$; since the sequence converges to the same limit irrespective of the value of x_0, the answer we get when we have reached the required degree of accuracy will be the same. The worst that can happen if we make an error is that it takes us longer to reach the desired degree of accuracy. But if our mistakes are lucky we might even get there *more* quickly!

The other advantage of iterative methods is that the computation consists not of a very complicated series of operations but of one series of operations repeated many times. This is important if the calculation is to be done on an electronic computer (where the possibility of error can usually be ignored) since it means that once arrangements have been made for the performance of this sequence of operations, all that remains to be done is to repeat it many times—a comparatively simple task.

Exercise 176

1. Calculate the first few terms of the sequence $\{x_n\}$ where $x_0 = 1$ and $x_{n+1} = 1 + 1/x_n$.
Show that $\{x_{2k}\}$, the sequence of even-numbered terms, is an increasing sequence, and $\{x_{2k+1}\}$ is a decreasing sequence. Hence show that $x_n \to$ a finite limit as $n \to \infty$. What is this limit?
2. Find, by experiment, what happens if you calculate the terms of a sequence $\{x_n\}$ from the relation

$$x_{n+1} = 1 - \frac{1}{x_n}$$

starting with any value of $x_0 \neq 0$ or 1. Make a conjecture as to the behavior of $\{x_n\}$, and prove it.

15.2 DIFFERENCE EQUATIONS—DIFFERENCES

In the example just given the value of y_{n+1} depended only on y_n. In more general recursive relations the new term will depend on several of the previous terms in the sequence and will be of the form

$$y_{n+k} = f(y_{n+k-1}, y_{n+k-2}, \ldots, y_n)$$

or as

$$(15.2.1) \qquad F(y_{n+k}, y_{n+k-1}, \ldots, y_n) = 0.$$

Equation (15.2.1) is extremely general, and it would be too much to expect that there would be any all-purpose method for solving it. We shall therefore consider a very special case of (15.2.1) namely

$$(15.2.2) \qquad a_0 y_{n+k} + a_1 y_{n+k-1} + a_2 y_{n+k-2} + \cdots + a_k y_n = \phi(n)$$

where $a_0, a_1, a_2, \ldots, a_k$ are constants. An equation such as (15.2.2) is called a *linear difference equation with constant coefficients*. The adjective "constant" here refers only to the coefficients on the left-hand side of (15.2.2); the function $\phi(n)$ can be quite general.

The reason why it should have this name is worth mentioning. We saw in Chapter 9, how the derivative at $x = x_0$ of a function $f(x)$ of a real variable depended on the quotient

(15.2.3)
$$\frac{f(x_0 + h) - f(x_0)}{h}.$$

If we have a function y_n of an integral variable n, then the increment h that we can make to the dependent variable n cannot be less than 1. Hence the nearest we can get to (15.2.3) is

$$\frac{y_{n+1} - y_n}{1} = y_{n+1} - y_n.$$

If we define Δy_n to mean $y_{n+1} - y_n$, then it turns out that the operator Δ, thus defined, has many properties in common with the operator D introduced in Chapter 11. In particular we can repeat the operator, for

$$\begin{aligned}
\Delta^2 y_n &= \Delta(\Delta y_n) = \Delta(y_{n+1} - y_n) \\
&= y_{n+2} - y_{n+1} - (y_{n+1} - y_n) \\
&= y_{n+2} - 2y_{n+1} + y_n.
\end{aligned}$$

Similarly

$$\begin{aligned}
\Delta^3 y_n &= \Delta(\Delta^2 y_n) \\
&= \Delta(y_{n+2} - 2y_{n+1} + y_n) \\
&= y_{n+3} - 3y_{n+2} + 3y_{n+1} - y_n
\end{aligned}$$

and so on. The operator Δ is called the *difference operator* and Δy_n, $\Delta^2 y_n$, $\Delta^3 y_n$, etc., are called the first, second, third, etc., differences of y_n.

Just as the derivative of a polynomial is a polynomial of one less degree so

$$\begin{aligned}
\Delta n^p &= (n + 1)^p - n^p \\
&= \text{a polynomial of degree } p - 1
\end{aligned}$$

so that a similar result holds for differences.

It is convenient to introduce, at this stage, another operator E defined by

$$E y_n = y_{n+1}.$$

Thus E is the operator which increases n by 1. This operator can also be applied many times, and we have $E^k y_n = y_{n+k}$. We can immediately write down a symbolic equation between E and Δ. For

(15.2.4)
$$\begin{aligned}
\Delta y_n &= y_{n+1} - y_n \\
&= E y_n - y_n \\
&= (E - 1)y_n
\end{aligned}$$

where 1 denotes the identity operator, as in Chapter 11. We can write (15.2.4) symbolically as

$$(15.2.5) \qquad\qquad \Delta = E - 1$$

or as

$$(15.2.6) \qquad\qquad E = \Delta + 1.$$

Thus equation (15.2.2) can be written as

$$a_0 E^k y_n + a_1 E^{k-1} y_n + a_2 E^{k-2} y_n + \cdots + a_{k-1} E y_n + a_k y_n = \phi(n)$$

or as

$$(15.2.7) \qquad (E^k + a_1 E^{k-1} + a_2 E^{k-2} \cdots + a_k) y_n = \phi(n).$$

If we now make the substitution $E = \Delta + 1$, then (15.2.7) becomes

$$(15.2.8) \qquad\qquad P(\Delta) y_n = \phi(n)$$

where P is some polynomial. We see here a close connection, at least in form, with differential equations. Allowing for the fact that in (15.2.8) the independent variable is n and the dependent variable is y_n, we observe that (15.2.8) is formally the same as the differential equation

$$P(D)y = \phi(x).$$

It is for this reason the equations like (15.2.1) or (15.2.8) are called difference equations.

Although the analogy with differential equations is most easily seen when the equation is expressed in terms of the operator Δ, we shall nevertheless find it more convenient to carry on most of the discussion of the next section with our equations in the form (15.2.7). Later on we shall have occasion to use the Δ notation again.

Exercise 177

1. We saw in the last section that

$$\Delta y_n = y_{n+1} - y_n$$
$$\Delta^2 y_n = y_{n+2} - 2y_{n+1} + y_n.$$

Prove by mathematical induction that

$$\Delta^k y_n = \sum_{i=0}^{k} \binom{k}{i} (-1)^i y_{n+k-i}.$$

2. Show that the result of problem 1 follows immediately from the relation

$$\Delta = E - 1$$

and the binomial theorem.

15.3 SOLUTION OF REDUCED DIFFERENCE EQUATIONS

We shall take our difference equation in the form

$$a_0 y_{n+k} + a_1 y_{n+k-1} + a_2 y_{n+k-2} + \cdots + a_k y_n = \phi(n)$$

or, symbolically, as

$$(E^k + a_1 E^{k-1} + a_2 E^{k-2} + \cdots + a_{k-1} E + a_k) y_n = \phi(n)$$

or, even more symbolically, as

$$(15.3.1) \qquad\qquad F(E) y_n = \phi(n),$$

where F is a polynomial in E of degree k. We first give a theorem analogous to Theorem 11.3.

THEOREM 15.1

IF $u_1(n)$, $u_2(n) \ldots , u_k(n)$ ARE DIFFERENT SOLUTIONS OF

$$F(E) y_n = 0$$

AND $\psi(n)$ IS ANY SOLUTION OF $(15.3.1)$, THEN

$$(15.3.2) \qquad y_n = A_1 u_1(n) + A_2 u_2(n) + \ldots + A_k u_k(n) + \psi(n)$$

IS THE GENERAL SOLUTION OF $(15.3.1)$, WHERE A_1, A_2, ETC., ARE ARBITRARY CONSTANTS.

We shall not take space to give a full proof of this theorem, but we can very easily show that $(15.3.2)$ is a solution of $(15.3.1)$. For

$$F(E)(A_1 u_1(n) + A_2 u_2(n) + \ldots + A_k u_k(n) + \psi(n))$$
$$= A_1 F(E) u_1(n) + A_2 F(E) u_2(n) + \ldots + A_k F(E) u_k(n) + F(E)\psi(n)$$
$$= 0 + 0 + \ldots + 0 + F(E)\psi(n)$$
$$= 0,$$

by the conditions of the theorem.

This shows that, as with differential equations, the general solution of (15.3.1) is the sum of the *complementary function*, which is a solution of the *reduced* equation

$$(15.3.3) \qquad F(E)y_n = 0$$

and a *particular integral* (any solution) of the original equation. Considerations similar to those for differential equations lead us to expect that the general solution of a difference equation of order k, i.e., one for which F is a polynomial of degree k, will have k arbitrary constants, as the theorem says it has.

To find solutions of the reduced equation we proceed as with differential equations, by trying a suitable function of n. This time the one that pays off is λ^n, where λ is a constant. For if $y_n = \lambda^n$ then

$$F(E)y_n = a_0\lambda^{n+k} + a_1\lambda^{n+k-1} + \cdots + a_k\lambda^n = 0$$

whence

$$(15.3.4) \qquad a_0\lambda^k + a_1\lambda^{k-1} + \cdots + a_k = 0.$$

Thus, if λ_1 is a root of the equation (15.3.4) then λ_1^n is a solution of (15.3.3). Equation (15.3.4) is called the *auxiliary equation*, just as with differential equations.

If (15.3.4) has k distinct roots $\lambda_1, \lambda_2 \ldots, \lambda_k$, then the complementary function we are looking for will be

$$A_1\lambda_1^n + A_2\lambda_2^n + \cdots + A_k\lambda_k^n.$$

However, if some of the roots coincide the situation is more complicated.

At this stage we shall leave more general results and concentrate on a specific type of difference equations, namely those of order 2. The general form for the reduced equation is

$$(15.3.5) \qquad ay_{n+2} + by_{n+1} + cy_n = 0.$$

Letting $y_n = \lambda^n$ we have

$$a\lambda^{n+2} + b\lambda^{n+1} + c\lambda^n = 0$$

or

$$(15.3.6) \qquad a\lambda^2 + b\lambda + c = 0,$$

as the auxiliary equation. If the auxiliary equation has distinct real roots

λ_1 and λ_2 then the general solution of (15.3.5) is

$$y_n = A_1 \lambda_1^n + A_2 \lambda_2^n.$$

If y_0 and y_1 are known, then A_1 and A_2 can be found.

EXAMPLE:

Solve the difference equation $y_{n+2} - 5y_{n+1} + 6y_n = 0$ given that $y_0 = 1$ and $y_1 = 3$.
The auxiliary equation in this case is

$$\lambda^2 - 5\lambda + 6 = 0$$

which has roots $\lambda = 2, 3$. Hence the general solution is

$$y_n = A_1 2^n + A_2 3^n.$$

Since $y_0 = 1$, $y_1 = 3$, then, with $n = 0$

$$1 = A_1 + A_2$$

and with $n = 1$

$$3 = 2A_1 + 3A_2.$$

Solving for A_1 and A_2 we have $A_1 = 0$, $A_2 = 1$. Hence in this particular case

$$y_n = 3^n.$$

Thus when (15.3.6) has distinct real roots there is no difficulty in obtaining the general solution of the reduced difference equation.

Exercise 178

1. Find the general solution of each of the following difference equations
   (a) $y_{n+2} - 5y_{n+1} + 4y_n = 0$
   (b) $y_{n+2} - 5y_{n+1} - 66y_n = 0$
   (c) $y_{n+3} - 6y_{n+2} + 11y_{n+1} - 6y_n = 0$
   (d) $y_{n+2} - 25y_n = 0$
   (e) $y_{n+2} + 3y_{n+1} - 20y_n = 0$
2. Find the solutions of the following difference equations that satisfy the given conditions.
   (a) $y_{n+2} - 3y_{n+1} + 2y_n = 0$
       where $y_0 = 1, y_1 = -2$

(b) $y_{n+2} + 4y_{n+1} + 3y_n = 0$

where $y_0 = 2, \quad y_1 = 0$

(c) $y_{n+2} = y_{n+1} + y_n$

where $y_0 = y_1 = 1$ (The sequence $\{y_n\}$ generated by this equation is known as the *Fibonacci sequence*.)

15.4 COINCIDENT ROOTS OF THE AUXILIARY EQUATION

If (15.3.6) has coincident roots (equal to λ_1 say) we have

$$a\lambda^2 + b\lambda + c = a(\lambda - \lambda_1)^2,$$

and the difference equation is of the form

$$(15.4.1) \qquad y_{n+2} - 2\lambda_1 y_{n+1} + \lambda_1^2 y_n = 0.$$

Clearly λ_1^n is one solution, by the same reasoning as before, but we are lacking a second solution to go with it. Let the second solution be $\lambda_1^n v_n$ where v_n is some function of n. Then (15.4.1) becomes

$$\lambda_1^{n+2} v_{n+2} - 2\lambda_1^{n+2} v_{n+1} + \lambda_1^{n+2} v_n = 0$$

whence

$$(15.4.2) \qquad v_{n+2} - 2v_{n+1} + v_n = 0.$$

Equation (15.4.2) is satisfied by $v_n = 1$ (which gives us the solution λ_1^n that we already have), and is also satisfied by $v_n = n$ which gives us a new solution. It is $n\lambda_1^n$. Hence the general solution of (15.4.1) is

$$A\lambda_1^n + Bn\lambda_1^n = \lambda_1^n(A + Bn).$$

EXAMPLE:

To solve $y_{n+2} - 6y_{n+1} + 9y_n = 0$.

Here the roots of the quadratic coincide at $\lambda = 3$. Hence the general solution is $(A + Bn) \cdot 3^n$. The reader should verify by substitution that this function satisfies the equation whatever the values of A and B.

Exercise 179

1. Find the general solutions of the following difference equations

    (a) $y_{n+2} + 4y_{n+1} + 4y_n = 0$

(b) $y_{n+2} - 10y_{n+1} + 25y_n = 0$
(c) $y_{n+3} + 3y_{n+2} - 4y_n = 0$

2. Find the solution of the difference equation

$$y_{n+2} - 4y_{n+1} + 4y_n = 0$$

for which $y_0 = -5$, $y_1 = -1$.

15.5 COMPLEX ROOTS

If the quadratic equation $a\lambda^2 + b\lambda + c = 0$ has complex roots, then these roots are conjugate complex numbers and we can take them either as $p \pm iq$, or as $r(\cos\theta \pm i\sin\theta)$. From our previous discussion it follows that

(15.5.1) $(p + iq)^n = r^n(\cos n\theta + i\sin n\theta)$

and

(15.5.2) $(p - iq)^n = r^n(\cos n\theta - i\sin n\theta)$

are solutions. By Theorem 15.1 any linear combination of solutions is a solution. Hence adding (15.5.1) and (15.5.2) and multiplying by $\frac{1}{2}$, we see that $r^n \cos n\theta$ is a solution. Subtracting this solution from (15.5.1) we see that $ir^n \sin n\theta$, and hence $r^n \sin n\theta$, is a solution. Clearly these solutions are distinct, and the general solution of the equation is therefore

(15.5.3) $\begin{aligned} y_n &= Ar^n \cos n\theta + Br^n \sin n\theta \\ &= r^n(A \cos n\theta + B \sin n\theta). \end{aligned}$

We could, of course, have written the general solution as

$$y_n = A(p + iq)^n + B(p - iq)^n$$

but if A and B are real this expression is necessarily complex. Usually we are interested in real solutions only. Although it is possible to choose *complex* values for A and B in the above expression in such a way that y_n is always real, this would be a tedious method of procedure. It is easier to use (15.5.3).

EXAMPLE:
Solve the difference equation

$$y_{n+2} = y_{n+1} - y_n$$

or

$$y_{n+2} - y_{n+1} + y_n = 0.$$

The auxiliary equation in λ is

$$\lambda^2 - \lambda + 1 = 0$$

whence

$$\lambda = \frac{1}{2}(1 \pm \sqrt{-3})$$

$$= \cos\frac{\pi}{3} \pm i\sin\frac{\pi}{3}.$$

Hence the general solution is

$$y_n = A\cos\frac{n\pi}{3} + B\sin\frac{n\pi}{3}.$$

Exercise 180

1. Show that a reduced second order difference equation will have a trigonometric solution, arising from complex roots of the auxiliary equation, if, and only if, it is of the form

$$y_{n+2} - 2py_{n+1} + (p^2 + q^2)y_n = 0.$$

2. Solve the following difference equations
   (a) $y_{n+2} - 2y_{n+1} + 5y_n = 0$
   (b) $y_{n+2} + 3y_{n+1} + 10y_n = 0$
3. Solve the difference equation

$$y_{n+2} - \sqrt{3}y_{n+1} + y_n = 0$$

given that $y_0 = y_1 = 1$.

15.6 FINDING PARTICULAR INTEGRALS

We now turn to methods of finding particular integrals of the difference equation.

(15.6.1) $$F(E)y_n = \phi(n).$$

The methods that we shall develop are very similar to those that we obtained for differential equations in Chapter 11, using the "D" operator. As in Chapter 11 we shall not give methods that will work for *any* function $\phi(n)$ but only enough to cover the commonest cases. First we must look closer at the difference operator Δ and the operator E.

The methods given in Chapter 11 for finding particular integrals depended very heavily on the fact that the operator D obeys the usual rules of algebra, subject to certain provisos. We must first prove that the same is true of Δ and E. This is not difficult, and the proof of results such as the following

$$(a\Delta + b)(c\Delta + d) = ac\Delta^2 + (bc + ad)\Delta + bd$$
$$\Delta E = E\Delta$$

etc.

will be left as exercises to the reader. We have also remarked that if Δ acts on a polynomial in n of degree p, then the result is a polynomial of degree $p - 1$. It follows that if Δ^p acts on a polynomial of degree p the result is a polynomial of degree 0, i.e., a constant. A further consequence is that the effect of Δ^k on such a polynomial is to give zero if $k > p$. Hence the successive differences of a polynomial eventually become, and stay, zero. This is exemplified by the following example.

If

$$\phi(n) = n^4 + 2n^3 + 7$$

then

$$\begin{aligned}
\Delta\phi(n) &= (n + 1)^4 + 2(n + 1)^3 + 7 - n^4 - 2n^3 - 7 \\
&= 4n^3 + 6n^2 + 4n + 1 + 2(3n^2 + 3n + 1) \\
&= 4n^3 + 12n^2 + 10n + 3.
\end{aligned}$$

Hence

$$\begin{aligned}
\Delta^2\phi(n) &= 4(n + 1)^3 + 12(n + 1)^2 + 10(n + 1) + 3 \\
&\qquad\qquad - (4n^3 + 12n^2 + 10n + 3) \\
&= 4(3n^2 + 3n + 1) + 12(2n + 1) + 10 \\
&= 12n^2 + 36n + 26; \\
\Delta^3\phi(n) &= 12(n + 1)^2 + 36(n + 1) + 26 - (12n^2 + 36n + 26) \\
&= 12(2n + 1) + 36 \\
&= 24n + 48; \\
\Delta^4\phi(n) &= 24(n + 1) + 48 - (24n + 48) = 24 \\
\Delta^5\phi(n) &= 24 - 24 = 0.
\end{aligned}$$

Finally, before deriving our various methods, we note that given a polynomial $P(\Delta)$ with a nonzero constant term we can find a formal series $Q(\Delta) = a_0 + a_1\Delta + a_2\Delta^2 + \ldots$ such that

$$(a_0 + a_1\Delta + a_2\Delta^2 + \ldots) \cdot P(\Delta) = 1$$

i.e., the series $Q(\Delta)$ is the formal reciprocal of the polynomial $P(\Delta)$. We now apply these results.

Case I. When $\phi(n)$ is a polynomial of degree p, say.
By means of the relationship $E = 1 + \Delta$ we can express the difference equation in the form

(15.6.2) $$P(\Delta)y_n = \phi(n)$$

and we shall suppose this done. If we operate on (15.6.2) with the operator $Q(\Delta)$ we get

$$Q(\Delta) \cdot P(\Delta)y_n = Q(\Delta)\phi(n)$$

in which the left-hand side reduces to $1 \cdot y_n = y_n$. Since $\Delta^k\phi(n) = 0$ if $k > p$, only a finite number of terms of the series $Q(\Delta)$ are relevant to the calculation of the right-hand side, and hence this can be evaluated. In point of fact we could avoid any suggestion that $Q(\Delta)$ was an infinite series of operations by defining $Q(\Delta)$ in this particular example to be any *polynomial* such that

$$Q(\Delta) \cdot P(\Delta) = 1 + \text{terms in } \Delta^{p+1} \text{ or higher.}$$

This is similar to what we did in Chapter 11.
The final outcome of this is that from (15.6.2) we can deduce

(15.6.3) $$y_n = Q(\Delta)\phi(n)$$

which gives a particular integral.

Example:
Find a particular integral of

$$y_{n+2} - 5y_{n+1} + 6y_n = n^2.$$

We write this as

$$(E^2 - 5E + 6)y_n = n^2$$

from which, by the substitution $E = 1 + \Delta$, we get

$$(\Delta^2 - 3\Delta + 2)y_n = n^2.$$

Now

$$\frac{1}{2 - 3\Delta + \Delta^2} = \frac{1}{2\left\{1 - \dfrac{3\Delta - \Delta^2}{2}\right\}}$$

$$= \frac{1}{2}\left\{1 + \frac{3\Delta - \Delta^2}{2} + \left(\frac{3\Delta - \Delta^2}{2}\right)^2 + \cdots\right\}$$

$$= \frac{1}{2} + \frac{3}{4}\Delta + \frac{7}{8}\Delta^2 + \cdots.$$

Hence

$$\left(\frac{1}{2} + \frac{3}{4}\Delta + \frac{7}{8}\Delta^2\right)(\Delta^2 - 3\Delta + 2) = 1 + \text{terms in } \Delta^3, \text{ etc.}$$

Thus we have

$$y_n = \left(\frac{1}{2} + \frac{3}{4}\Delta + \frac{7}{8}\Delta^2\right)n^2$$

$$= \frac{1}{2}n^2 + \frac{3}{4}(2n + 1) + \frac{7}{8}\cdot 2$$

$$= \frac{1}{2}n^2 + \frac{3}{2}n + \frac{5}{2}$$

which is the required integral.

Note that we have so far avoided writing $Q(\Delta)$ as $1/P(\Delta)$. There is no harm in doing this however provided it is borne in mind that $1/P(\Delta)$ stands for a formal reciprocal power series, and that (for the case at present being considered) only a finite number of terms of this power series are required. We can then say, very concisely, that if $\phi(n)$ is a polynomial, a particular integral of

$$P(\Delta)y_n = \phi(n)$$

is

$$y_n = \frac{1}{P(\Delta)}\phi(n).$$

Exercise 181

1. Prove that the operators Δ and E satisfy the following relations.
   (a) $\Delta E = E\Delta$
   (b) $(a\Delta + b)(c\Delta + d) = ac\Delta^2 + (bc + ad)\Delta + bd$

(c) $(\Delta + a)(E + b) = \Delta E + aE + b\Delta + ab$

(d) $E^m \Delta^n = \Delta^n E^m$,

where we observe the usual convention that a constant (such as a) stands for a constant multiple of the identity operator.

2. Obtain particular integrals for the following difference equations, and hence solve them completely.

(a) $y_{n+2} - 7y_{n+1} + 10y_n = n^2$

(b) $y_{n+2} - 12y_{n+1} + 36y_n = 3n^2 + 5n + 2$

(c) $y_{n+2} - 7y_{n+1} + 12y_n = n^3 - n + 1$

(d) $y_{n+2} - y_{n+1} - 6y_n = n^2 - 6n + 7$

(e) $y_{n+2} + 4y_n = n^4$

15.7 FURTHER PARTICULAR INTEGRALS

Case II. $\phi(n)$ is of the form $a^n \psi(n)$ where $\psi(n)$ is a polynomial, and a is a constant.

If $\phi(n)$ is of this form we can reduce the problem to a Case I problem by means of a useful property of the E operator. We have

$$E^r(a^n \psi(n)) = a^{n+r} \psi(n + r)$$
$$= a^{n+r} E^r \psi(n)$$
$$= a^n (aE)^r \psi(n).$$

If we now consider a polynomial $F(E) = \sum_{r=0}^{m} a_r E^r$, then

$$F(E)(a^n \psi(n)) = \sum_{r=0}^{m} a_r E^r (a^n \psi(n))$$
$$= \sum_{r=0}^{m} a_r \cdot a^n (aE)^r \psi(n)$$
$$= a^n \left\{ \sum_{r=0}^{m} a_r (aE)^r \right\} \psi(n).$$

Hence

(15.7.1) $$F(E)(a^n \psi(n)) = a^n F(aE) \psi(n).$$

Hence we have the rule that we can bring the a^n over to the other side of the operator, provided we replace E by aE. An example will show how this result is used.

EXAMPLE:

Find a particular integral of

$$y_{n+1} - 2y_n = 3^n n^2$$

i.e., of

$$(E - 2)y_n = 3^n n^2.$$

Symbolically a particular integral is given by

$$y_n = \frac{1}{E - 2} 3^n n^2$$

$$= 3^n \frac{1}{3E - 2} n^2$$

by the rule just given. Hence

$$y_n = 3^n \frac{1}{3\Delta + 1} n^2 \qquad \text{(putting } E = 1 + \Delta\text{)}$$

$$= 3^n (1 - 3\Delta + 9\Delta^2 + \ldots) n^2$$

$$= 3^n (n^2 - 3(2n + 1) + 9 \cdot 2)$$

$$= 3^n (n^2 - 6n + 15).$$

Note that we applied the rule to the function $1/(E - 2)$, even though this is not a polynomial. This is justified by the fact that although, for convenience, we write $1/(E - 2)$, we really mean a polynomial $Q(\Delta)$ in Δ whose product with $P(\Delta)$ is 1 + terms of sufficiently high degree in Δ. In short, the fact that our operands are eventually polynomials, means that the operators are effectively polynomials, too, though the way that we write them may not always give this impression.

Case III. Operators $P(\Delta)$ with no constant term.

The preceding arguments cannot be applied if $P(\Delta)$ has no constant term (or, more correctly, no term in the identity operator 1), for it is not then possible to expand $1/P(\Delta)$ as a power series in Δ. As an example let us consider the difference equation

(15.7.2)
$$y_{n+1} - 2y_n = 2^n n^2$$
$$(E - 2)y_n = 2^n n^2.$$

A particular integral is given by

$$y_n = \frac{1}{E-2} 2^n n^2$$

$$= 2^n \frac{1}{2E-2} n^2 \qquad \text{by (15.7.1)}$$

$$= 2^n \frac{1}{2\Delta} n^2.$$

We must now interpret an expression of the form

$$(15.7.3) \qquad\qquad u_n = \frac{1}{\Delta} \phi(n).$$

This should be the same as $\Delta u_n = \phi(n)$, so that the function u_n, in (15.7. 3) is one whose first difference is the given function $\phi(n)$. We see here an analogy with integration, and the way that we interpreted $1/D$ in Chapter 11.

In our example we want a function whose first difference is n^2. Remembering that taking the first difference of a polynomial lowers the degree by 1, we see that the required function must be a cubic polynomial. Let it be

$$an^3 + bn^2 + cn + d.$$

Its first difference is then

$$a(3n^2 + 3n + 1) + b(2n + 1) + c.$$

Hence we have

$$a(3n^2 + 3n + 1) + b(2n + 1) + c = n^2.$$

Equating coefficients of n^2, n and 1 we have

$$3a = 1$$

$$3a + 2b = 0$$

$$a + b + c = 0$$

whence $a = \frac{1}{3}$, $b = -\frac{1}{2}$, $c = \frac{1}{6}$. Hence the required particular integral for (15.7.2) is

$$y_n = 2^n \cdot \frac{1}{2} \left(\frac{1}{3} n^3 - \frac{1}{2} n^2 + \frac{1}{6} n \right)$$

$$= 2^n \cdot \frac{1}{6} n(n-1)(2n-1).$$

The interpretation just given of $1/\Delta$ will cover all cases where $P(\Delta)$ has no constant term (assuming still that the operand is a polynomial). It is worth looking at a more elegant method of applying it, a method which brings out even more forcibly the analogy between differences and derivatives. Let us define an expression $n^{(k)}$ by

(15.7.4) $n^{(k)} = n(n - 1)(n - 2) \ldots (n - k + 1).$

Thus $n^{(k)}$ is the number of permutations of n objects k at a time, a number that we have met before. If we form the first difference of this function we find that it is

$$\begin{aligned}
\Delta n^{(k)} &= (n + 1)n(n - 1) \ldots (n - k + 2) \\
&\quad - n(n - 1)(n - 2) \ldots (n - k + 1) \\
&= n(n - 1)(n - 2) \ldots (n - k + 2)\{(n + 1) - (n - k + 1)\} \\
&= kn^{(k-1)}.
\end{aligned}$$

Thus functions of the form $n^{(k)}$ behave with respect to the difference operator in the same way as functions of the form x^k behave with respect to the differential operator D. Just as $Dx^k = kx^{k-1}$, so $\Delta n^{(k)} = kn^{(k-1)}$. From the result $\Delta n^{(k+1)} = (k + 1)n^{(k)}$ we conclude that

(15.7.5) $$\frac{1}{\Delta} n^{(k)} = \frac{1}{k + 1} n^{(k+1)}$$

which is the analog of $\dfrac{1}{D} x^k = \dfrac{1}{k + 1} x^{k+1}$. This gives us a method of applying the operator $1/\Delta$ when $\phi(n)$ is a polynomial of the form $n^{(k)}$. Even when this is not so, the polynomial $\phi(n)$ is expressible in terms of such polynomials, and this can be done in a systematic manner. This is best shown by an example.

Suppose we wish to express $\phi(n) = 4n^3 - 2n^2 + 4n - 7$ in terms of expressions of the form $n^{(k)}$. The highest value of k will be 3, and since the coefficient of n^3 in $\phi(n)$ is 4 this is the coefficient of $n^{(3)}$. Hence $\phi(n) = 4n^{(3)} +$ other terms, and since

$$4n^{(3)} = 4n^3 - 12n^2 + 8n$$

we have

$$\phi(n) = 4n^{(3)} + 10n^2 - 4n - 7.$$

If we express $10n^2 - 4n - 7$ in terms of these expressions, we must have a term $10n^{(2)}$, which accounts for $10n^2 - 10n$, leaving us a further

$6n - 7$ to deal with. This is clearly $6n^{(1)} - 7$, and our task is complete. We have

$$4n^3 - 2n^2 + 4n - 7 = 4n^{(3)} + 10n^{(2)} + 6n^{(1)} - 7n^{(0)}$$

if, conventionally, we put $n^{(0)} = 1$.

If we now wish to operate on this polynomial with the operator $1/\Delta$, this is very easy by use of (15.7.5). We have

$$\frac{1}{\Delta}(4n^3 - 2n^2 + 4n - 7) = 4 \cdot \frac{1}{\Delta} n^{(3)} + 10 \cdot \frac{1}{\Delta} n^{(2)} + 6 \cdot \frac{1}{\Delta} n^{(1)}$$

$$- 7 \cdot \frac{1}{\Delta} n^{(0)}$$

$$= 4 \cdot \frac{1}{4} n^{(4)} + 10 \cdot \frac{1}{3} n^{(3)} + 6 \cdot \frac{1}{2} n^{(2)} - 7n^{(1)}$$

$$= n^{(4)} + \frac{10}{3} n^{(3)} + 3n^{(2)} - 7n$$

which can then be expressed as a straight polynomial if required. For our previous example we have

$$\frac{1}{\Delta} n^2 = \frac{1}{\Delta} (n^{(2)} + n^{(1)})$$

$$= \frac{1}{3} n^{(3)} + \frac{1}{2} n^{(2)}$$

$$= \frac{1}{6} \{2n(n-1)(n-2) + 3n(n-1)\}$$

$$= \frac{1}{6} n(n-1)\{2n - 4 + 3\}$$

$$= \frac{1}{6} n(n-1)(2n-1)$$

as before.

Exercise 182

1. Obtain particular integrals for the following difference equations, and hence solve them.

   (a) $y_{n+2} - 3y_{n+1} + 2y_n = 5^n n^2$
   (b) $y_{n+2} + y_{n+1} + y_n = 2^n n^3$
   (c) $y_{n+2} - 9y_n = (-2)^n n^2 + (-4)^n (n+1)$

2. Solve the following difference equations

   (a) $y_{n+2} - 3y_{n+1} + 2y_n = n^2 + 3n - 1$

(b) $y_{n+2} - 5y_{n+1} + 4y_n = 4^n n$

(c) $y_{n+2} + 6y_{n+1} + 9y_n = (-3)^n(n + 1)(n + 2)$

15.8 THE COBWEB MODEL OF SUPPLY AND DEMAND

One very well-known application in economics of an iterative process such as those that have been discussed is the so-called cobweb model of supply and demand. Let x be the price of a certain article, and let the two functions $S(x)$ and $D(x)$ be the supply and demand schedules—the amounts of the article that will be supplied and demanded if the price is x.

One might expect that the price of the article would be determined by the criterion that the market be exactly cleared, which happens when supply exactly equals demand. This condition of *market equilibrium* is expressed by the equation

$$(15.8.1) \qquad\qquad S(x) = D(x).$$

If we draw the graphs of the functions $S(x)$ and $D(x)$ as in Fig. 15.1 then the solution of (15.8.1) is given by the point of intersection of the two

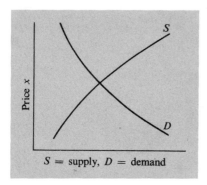

$S = $ supply, $D = $ demand

Fig. 15.1

curves marked S and D.* This point determines the price at which there will be equilibrium between supply and demand. At this price consumers are willing to purchase exactly the amount that firms are willing to supply. If the price were higher, there would be excess supply—firms would supply more than consumers would demand; if lower, there would be excess demand.

---

* In deference to current economic practice the independent variable, *price*, has been assigned to the *vertical* axis.

In practice this is not what usually happens. Instead of a static situation there is a dynamic one, with the supply, demand and price all varying with time. In order to express this sort of process in mathematical terms we must make some plausible assumptions. We shall first assume that all changes occur at discrete moments of time, say, for example, at the beginning of each month, or of each year. Thus we shall think of time as being, for this purpose, an integer variable, and write $t = 0, 1, 2, \ldots$, etc.

We next observe that there is bound to be some delay in the response of firms to a change in the price of the article. When the manufacturer decides how many articles to make in the next time period he has no information on what the price will be then, only on what it has been in the past. Our first assumption concerning this delay will be that the manufacturer bases his supply at time $t = n$ on the price prevailing at time $t = n - 1$. Thus there is a delay of one time unit between the supply and the price on which it is estimated.

If x_n denotes the price at time $t = n$, then the amount supplied at time $t = n$ will be $S(x_{n-1})$. The price will adjust so that demand equals supply, and the new price x_n will therefore be determined by

(15.8.2) $$D(x_n) = S(x_{n-1}).$$

Let us see diagrammatically what the effect of all this is. If the price is x_0 at time $t = 0$, then at time $t = 1$ the supply will be $S(x_0)$, because of the delay, and the price during the period beginning at $t = 1$ will be that for which

$$D(x_1) = S(x_0).$$

The supply at time $t = 2$ will be $S(x_1)$ and the price x_2 at time $t = 2$ is determined by

$$D(x_2) = S(x_1)$$

and so on. In the diagram of Fig. 15.2 the horizontal line at x_0 meets the curve S in P_0 giving the supply $S(x_0)$ at time $t = 1$. Hence x_1, the value of x for which $D(x_1) = S(x_0)$ is given by the point on the curve D "below" P_0, viz., the point Q_1 in the figure. The supply for $t = 2$ is then given by the point P_1 on the S curve (corresponding to the same price x_1) and this in turn gives rise to the point Q_2 on the curve D, which gives us the price and the demand at time $t = 2$. Carrying on in this way we see thất we get a sequence of prices $x_0, x_1, x_2, \ldots$, which have every appearance (in the figure, at least) of tending to some limiting value. Whether this is so in general remains to be seen; the figure might be misleading.

Because of the cobweb-like appearance of Fig. 15.2 this particular model

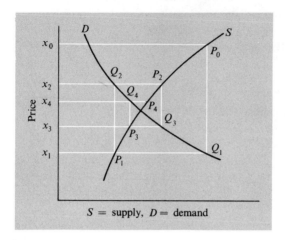

Fig. 15.2

of the interplay between supply and demand is usually called the *cobweb model.*

The behavior of the sequence of prices will clearly depend on the functions S and D and could possibly be quite complicated. The simplest assumption that we can make about these functions is that they are linear. This may not be a reasonable assumption, but it will do for a start. To make it as realistic as possible we shall assume throughout that the slope of the supply curve is positive, while that of the demand curve is negative. If the equilibrium point X has coordinates (α, β), we can take $S(x)$ and $D(x)$ to be

$$S(x) = a(x - \alpha) + \beta$$

and

$$D(x) = -b(x - \alpha) + \beta$$

where a and b are positive constants.

From the equation $D(x_n) = S(x_{n-1})$ we have

$$-b(x_n - \alpha) + \beta = a(x_{n-1} - \alpha) + \beta$$

which yields the difference equation

$$(15.8.3) \qquad\qquad bx_n + ax_{n-1} = (a + b)\alpha$$

to which we shall apply the methods established earlier in this chapter.

We consider first the reduced equation

$$bx_n + ax_{n-1} = 0$$

for which we can straightaway write the solution

$$(15.8.4) \qquad x_n = \left(-\frac{a}{b}\right)^n A$$

where A is a constant. We now need a particular integral of (15.8.3). This is easily found by inspection. If we try the simplest function of all, namely $x_n = k$, a constant, then (15.8.3) is satisfied if

$$bk + ak = (a + b)\alpha$$

i.e., if $k = \alpha$. Hence the general solution of (15.8.3) is

$$(15.8.5) \qquad x_n = \left(-\frac{a}{b}\right)^n A + \alpha.$$

We can specify the constant A in terms of the initial value x_0 of x; for if $n = 0$ then $x_0 = A + \alpha$. Thus the general solution can be written as

$$(15.8.6) \qquad x_n = \left(-\frac{a}{b}\right)^n (x_0 - \alpha) + \alpha.$$

It is clear from (15.8.6) that as $n \to \infty$ so $x_n \to \alpha$, provided $a/b < 1$. Hence if the angle that the supply curve makes with the x-axis is less than the angle that the demand curve makes, then the situation will tend to equilibrium at $x = \alpha$. In the contrary case the oscillations about the equilibrium point will get larger and larger, as can be seen in Fig. 15.3. There is an intermediate possibility, namely that the two angles are equal and hence that $a = b$. We then have

$$x_n = (-1)^n (x_0 - \alpha) + \alpha$$

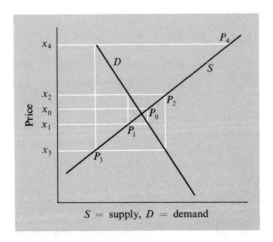

$$S = \text{supply}, \quad D = \text{demand}$$

Fig. 15.3

and x_n takes alternately the values $-x_0 + 2\alpha$ and x_0. This possibility is such an exceptional one that we can ignore it for practical purposes. When $a/b < 1$ the situation is said to be *stable* or *damped;* when $a/b > 1$ it is *unstable* or *explosive.*

What we have here considered is a very artificial model of an economic situation. The forms of the functions $S(x)$ and $D(x)$ would be difficult to determine in a given application at the best of times, but one is reasonably safe in assuming that they will be continuous, differentiable functions whose shapes are something like those shown in Fig. 15.1. It may not be a reasonable assumption to assume that they are linear functions, as we have done; but the assumption of linearity is not so outrageous if we confine our attentions to the neighborhood of the equilibrium point, for in this neighborhood the curves S and D will be fairly closely approximated by their tangents at that point. Hence we would expect that the linear model for which we have just found the solution will give a close approximation to what happens in a more realistic model provided that the price does not fluctuate too widely about the equilibrium value.

In the model just considered we assumed that the supply at time $t = n$ was determined by the price for time $t = n - 1$. This would be a reasonable assumption for a supplier to make at the beginning of the operation, but after a few cycles of this model the supplier would certainly notice the fact that prices were alternately increasing and decreasing, and would therefore be inclined to change his method of calculating the supply. He might well decide to base his prediction on the average of the prices for the last two times, and assume that the new price would be $\frac{1}{2}(x_{n-1} + x_{n-2})$. More generally he might take a *weighted* average of these two prices, that is, a convex combination of them, in the sense of Chapter 7. He would then predict that the price would be $\lambda x_{n-1} + (1 - \lambda)x_{n-2}$, where λ is a number between 0 and 1. In this case the equation analogous to (15.8.2) becomes

$$(15.8.7) \qquad D(x_n) = S(\lambda x_{n-1} + (1 - \lambda)x_{n-2}).$$

If we now make the assumption, as before, that $S(x)$ and $D(x)$ are linear then equation (15.8.7) becomes

$$-b(x_n - \alpha) + \beta = a[\lambda x_{n-1} + (1 - \lambda)x_{n-2} - \alpha] + \beta,$$

which can be rearranged as

$$(15.8.8) \qquad bx_n + a\lambda x_{n-1} + a(1 - \lambda)x_{n-2} = (a + b)\alpha.$$

Note, as a check, that if $\lambda = 1$ equation (15.8.8) reduces to (15.8.3).

We can simplify (15.8.8) somewhat by writing it as

$$(15.8.9) \qquad b(x_n - \alpha) + a\lambda(x_{n-1} - \alpha) + a(1 - \lambda)(x_{n-2} - \alpha) = 0.$$

If we write k for a/b and y_n for $x_n - \alpha$, the equation simplifies even further, and we have

$$(15.8.10) \qquad y_n + k\lambda y_{n-1} + k(1 - \lambda)y_{n-2} = 0.$$

This difference equation can be solved by the method given earlier in the chapter, and we find that the solution is

$$(15.8.11) \qquad y_n = Au^n + Bv^n$$

where u and v are the roots of the quadratic equation

$$(15.8.12) \qquad x^2 + k\lambda x + k(1 - \lambda) = 0,$$

and are therefore given by

$$\tfrac{1}{2}\{-k\lambda \pm \sqrt{k^2\lambda^2 - 4k(1 - \lambda)}\}.$$

We first consider for what values of k and λ these roots are real. For this we must have the expression under the root sign non-negative, i.e.,

$$k\lambda^2 + 4\lambda - 4 \geq 0$$

(taking out a factor k). It is easily seen (say from graphical considerations) that this will be true if λ does not lie between the two roots of the equation $k\lambda^2 + 4\lambda - 4 = 0$, i.e., between the two values

$$\frac{-2 \pm \sqrt{4 + 4k}}{k}.$$

Since the smaller of these values is negative, it follows that λ must be greater than the larger root, i.e.,

$$(15.8.13) \qquad \lambda > \frac{\sqrt{4 + 4k} - 2}{k}.$$

We shall first consider what happens when (15.8.13) is satisfied and the roots are real. By a well-known result of elementary algebra the product of the roots of (15.8.12) is the constant term in (15.8.12)—the positive number $k(1 - \lambda)$—and the sum of the roots is minus the coefficient of x in (15.8.12), viz., $-k\lambda$. From this it follows that both roots are negative. The original price variable x_n can then be written as

$$(15.8.14) \qquad x_n = A(-p)^n + B(-q)^n + \alpha$$

where $p(= -u)$ and $q(= -v)$ are positive numbers. From what we said above we have $p + q = k\lambda$. Provided neither p nor q is greater than 1, x_n is a constant plus two terms, each of which alternates in sign and tends to zero. Hence x_n tends to the equilibrium value α, and the situation is a stable one. Moreover, the size of the oscillations about the equilibrium point is determined by the numbers p and q, whose sum is less than $k = a/b$, the number which determined the size of the oscillations in the original model [see equation (15.8.5)]. If we remember also that the two terms in (15.8.14) may tend to cancel each other (if A and B have opposite signs) then we would expect the price to tend more quickly towards the equilibrium value than before.

The numbers p and q are the roots of the equation

$$(15.8.15) \qquad x^2 - k\lambda x + k(1 - \lambda) = 0$$

obtained from (15.8.12) by changing the sign of x. If both roots are less than 1 the left-hand side of (15.8.15) is positive for $x = 1$, and

$$1 - k\lambda + k(1 - \lambda) > 0$$

which reduces to

$$(15.8.16) \qquad \lambda < \frac{1 + k}{2k}.$$

If this condition is not satisfied, then one root, say q, exceeds 1. The term $B(-q)^n$ in (15.8.14) then produces ever-increasing oscillations, and the situation represented by the model is unstable; x_n does not tend to a limit as $n \to \infty$.

We now turn to the possibility $k\lambda < \sqrt{4 + 4k} - 2$, for which the roots of (15.8.12) are complex numbers, conjugate to each other. The product of the roots is $k(1 - \lambda)$ as before, and this is the square of their modulus [compare equation (13.3.1)]. Hence if we take the roots in the form

$$r(\cos \theta \pm i \sin \theta)$$

we have $r = +\sqrt{k(1 - \lambda)}$. The solution of the difference equation assumes the form

$$(15.8.17) \qquad y_n = r^n(A \cos n\theta + B \sin n\theta).$$

Now the factor $A \cos n\theta + B \sin n\theta$ is numerically less than $|A| + |B|$ whatever θ or n may be. Hence the behavior of y_n depends on that of r^n. If $r < 1$ then $y_n \to 0$ as $n \to \infty$; otherwise the sequence $\{y_n\}$ diverges. Hence the system is stable, and approaches equilibrium if $k(1 - \lambda) < 1$.

It is instructive to summarize these results in a graphical form, and this has been done in Fig. 15.4, in which λ has been plotted against k. We assume

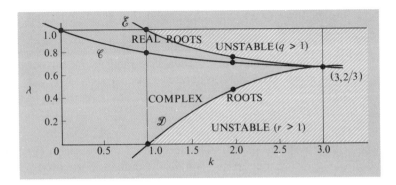

Fig. 15.4

that k is positive in order that the model should be economically realistic, and λ is necessarily between 0 and 1. The curve $\mathscr{C}$ given by the equation $k\lambda = \sqrt{4 + 4k} - 2$ marks the boundary between those values of λ and k which give real roots to equation (15.8.12) and those which give complex roots. The curve $\mathscr{D}$ with equation $k(1 - \lambda) = 1$ is the boundary between points for which $r < 1$ and those for which $r > 1$, and is therefore a boundary between stability and instability. The curve $\mathscr{E}$ with equation $2k\lambda = 1 + k$ is also a boundary between stability (with p and q both less than 1) and instability (with one root greater than 1).

Note that in the original model [equation (15.8.5)] stability was possible only for $k < 1$. This is no longer true in the present model. For k between 1 and 3 stability is still possible provided λ is not too near 0 or 1. If $\lambda = 1$ we naturally get the same result as before—stability if $k < 1$. The same applies if $\lambda = 0$, and the reason is not hard to understand. If $\lambda = 0$, the supplier is basing his calculation on the price established two time units in the past, and the effect is that we have exactly the same model except that the time intervals are twice as long. If the time intervals were months, then it would mean that the prices in odd months were quite independent of prices in even months, and taken separately would follow the pattern of our original model.

It is interesting to note that all three curves in Fig. 15.4 intersect at the point $k = 3$, $\lambda = 2/3$. For values of k beyond 3 all points are unstable, and no equilibrium will be reached.

Exercise 183

1. An economic model of the relations between income, consumption and investment can be set up in the following way. We start with the equation

$$\text{Income} = \text{Consumption} + \text{Investment}$$

representing the two sources of demand that generate output and, hence, income. We assume that consumption at time $t = n$ depends on the expected income, and that this is x_{n-1}, the income at time $t = n - 1$; that is, consumption is planned on a basis of no change in income. We further assume that investment grows exponentially, and is given by Ak^n where k is a constant (> 1). Then

$$x_n = \phi(x_{n-1}) + Ak^n$$

where ϕ is the function that consumption is of income. In the linear case, when $\phi(x) = \alpha x$ (α a constant), obtain the difference equation

$$x_n - \alpha x_{n-1} = Ak^n.$$

Find the general solution of this equation in terms of x_0, the income at time $t = 0$.

2. Analyze carefully the conditions under which the solution of the difference equation

$$x_n = ax_{n-1} + b(x_{n-1} - x_{n-2})$$

is   **(a)**   damped
     (b)   explosive.

15.9 MARKOV CHAINS

Let us suppose that the population of a certain country can be divided into two classes. What distinguishes the members of the two classes is not at present important; it might be the level of income, level of education, marital status, or any sort of dichotomy. Suffice it that there is some characteristic which every individual either has or does not have, which determines to which of the two classes he belongs. We shall need to have names for the two classes, and although the terms are far from neutral we shall call them the *Uppers* and the *Lowers*.

We shall assume that there is some social mobility; that Uppers do not necessarily remain Uppers, and that Lowers need not remain Lowers, but that individuals can change from one class to the other. What we want to do is to set up a mathematical model of what happens to this community in the course of time.

We need to know two things: the number in each class at a given time, and information concerning the transfer of population between the classes. Typical data might be as follows: in 1968 there were 200,000 Uppers and 800,000 Lowers; each year 2 percent of the Uppers become Lowers, while

5 percent of the Lowers become Uppers. Let us see what can be done with these figures. For convenience we shall work in thousands as the unit of population.

Let U_0, L_0 be the numbers of Uppers and Lowers in 1968, so that $U_0 = 200$, $L_0 = 800$. Let U_1, L_1 be the numbers at the end of the first year, i.e., by 1969. During that year, the number of Uppers who became Lowers was $\frac{2}{100} \times 200 = 4$ thousand, while $\frac{5}{100} \times 800 = 40$ thousand Lowers became Uppers. Hence at the end of the year we have

$$U_1 = U_0 - 4 + 40 = 236.$$
$$L_1 = L_0 + 4 - 40 = 764.$$

We can now repeat the same sort of calculation for the next year, during which $\frac{2}{100} \times 236 = 4.72$ thousand Uppers became Lowers, and $\frac{5}{100} \times 764 = 38.2$ thousand Lowers became Uppers, a net gain of 33.48 for the Uppers. Hence the new numbers are

$$U = 236 + 33.48 = 269.48$$

and

$$L = 764 - 33.48 = 730.52.$$

Clearly we could repeat this calculation for any number of years. The calculation is simple, requiring nothing more than elementary arithmetic.

Now it is a curious fact that in mathematics it frequently happens that one gets a deeper insight into a problem by making it more general, even though this may well also make it more difficult. The present problem is a case in point. Instead of taking given numerical data, let us be quite general. Let U_0, L_0 be the numbers of Uppers and Lowers at a given instant of time, and suppose that in a year a fraction u of the Uppers become Lowers, and that a fraction l of the Lowers become Uppers. The numbers u and l can be interpreted as probabilities; u is the probability that any given Upper will become a Lower, and l is the probability that a Lower will become an Upper in the course of a year.

At the end of the first year, uU_0 Uppers will have become Lowers, and lL_0 Lowers will have become Uppers, Hence

$$U_1 = U_0 - uU_0 + lL_0$$
$$L_1 = L_0 + uU_0 - lL_0$$

which we may write as

(15.9.1)
$$U_1 = (1 - u)U_0 + lL_0$$
$$L_1 = uU_0 + (1 - l)L_0.$$

We can now recognize something that was completely obscured in the numerical example, namely that the connection between U_1, L_1, and U_0, L_0 is that of a linear transformation between vectors. For we can write (15.9.1) as

$$(15.9.2) \qquad (U_1, L_1) = (U_0, L_0)\begin{pmatrix} 1-u & u \\ l & 1-l \end{pmatrix}.$$

We can further simplify this by writing x_r for (U_r, L_r), and A for the matrix

$$\begin{pmatrix} 1-u & u \\ l & 1-l \end{pmatrix}.$$

Then (15.9.2) becomes

$$(15.9.3) \qquad\qquad\qquad x_1 = x_0 A.$$

If we now turn to what happens in the second year, we see that the equations connecting U_2, L_2 with U_1, L_1 are exactly similar to (15.9.1), being in fact

$$U_2 = (1-u)U_1 + lL_1$$
$$L_2 = uU_1 + (1-l)L_1.$$

All that has happened is that the subscripts have increased by 1. Using the matrix notation we have

$$(15.9.4) \qquad\qquad\qquad x_2 = x_1 A.$$

Hence from (15.9.3) and (15.9.4) we have

$$x_2 = (x_0 A)A = x_0 A^2.$$

The vector x_3 which gives the distribution of population at the end of the third year will likewise be given by

$$x_3 = x_2 A$$

from which it follows that

$$x_3 = x_0 A^3$$

and so on. In general, after r years, we have

$$(15.9.5) \qquad\qquad\qquad \begin{aligned} x_r &= x_{r-1} A \\ x_r &= x_0 A^r. \end{aligned}$$

An important question which arises in problems of social mobility is whether the distribution of population is tending towards some kind of equilibrium and, if so, what the distribution of population would be in such an equilibrium situation. In our present example, which is a particularly simple one, we can determine this with no trouble. If the numbers in the two classes remain constant then the number uU of Uppers that become Lowers must equal the number lL of Lowers that become Uppers. Hence

$$uU = lL \quad \text{or} \quad \frac{U}{L} = \frac{l}{u},$$

and equilibrium exists if the ratio of the numbers in the two classes has the value l/u. Note that this equilibrium does not imply an absence of social mobility; individuals still cross from one class to the other but do so in such numbers that the gains and losses in each class cancel, and each class remains the same size.

If $x = (U, L)$ gives the distribution in this steady state, we must have

$$x = xA$$

since xA, the distribution after a lapse of one year, must be the same as before. Hence x must be an eigenvector of A corresponding to the eigenvalue 1. Since, as we shall see later, 1 is always an eigenvalue of the sort of matrix we are now considering, it follows that an equilibrium situation always exists. Whether it can be reached starting with a given population distribution is another matter.

The example just considered is a little too simple to take us any further, so let us now consider a country with three social classes, Upper, Middle and Lower. Let p_{uu} be the probability that an Upper remains an Upper in the course of a year, p_{um} the probability that an Upper becomes a Middle, and let $p_{ul}, p_{mu}, p_{mm}, p_{ml}, p_{lu}, p_{lm}, p_{ll}$ be similarly defined (i.e., p_{xy} means the probability that an x becomes a y). Then it is easily verified that

$$U_r = p_{uu} U_{r-1} + p_{mu} M_{r-1} + p_{lu} L_{r-1}$$
$$M_r = p_{um} U_{r-1} + p_{mm} M_{r-1} + p_{lm} L_{r-1}$$
$$L_r = p_{ul} U_{r-1} + p_{ml} M_{r-1} + p_{ll} L_{r-1}$$

or, in matrix notation,

$$(15.9.6) \qquad (U_r, M_r, L_r) = (U_{r-1}, M_{r-1}, L_{r-1}) \begin{pmatrix} p_{uu} & p_{um} & p_{ul} \\ p_{mu} & p_{mm} & p_{ml} \\ p_{lu} & p_{lm} & p_{ll} \end{pmatrix}$$

which can be abbreviated to

(15.9.7) $$x_r = x_{r-1}P.$$

It follows that

$$x_r = x_0 P^r.$$

Note that the matrix P is a rather special one. The sum of the elements in its first row is

$$p_{uu} + p_{um} + p_{ul}$$

which can be interpreted as the probability that during the year an Upper remains an Upper, or becomes a Middle, or becomes a Lower. But these are the only possibilities so that the probability of one or another of them happening is 1. Hence

$$p_{uu} + p_{um} + p_{ul} = 1.$$

Similarly the sum of the elements in each of the other rows is also 1. Thus P is a matrix whose entries are numbers in the range $0 \leq x \leq 1$ (since they are probabilities) and every row total is 1. Such a matrix (of any size) is called a *stochastic matrix*.

It is often convenient to consider not the actual numbers of individuals in the various population classes but the fractions of the total population. Thus instead of specifying, for example, that on a certain date the distribution of population in our latest example was

$$\text{Uppers} = 200, \quad \text{Middles} = 500, \quad \text{Lowers} = 300$$

we say that it is

$$\text{Uppers} = 0.2, \quad \text{Middles} = 0.5, \quad \text{Lowers} = 0.3.$$

Provided we know the total population this gives us as much information as the actual numbers; and even if we do not know the total population it gives us the most important information, viz., the relative sizes of the different classes. If the information is given in this way, then the initial distribution vector

$$x_0 = (200, 500, 300)$$

is replaced by an initial *probability vector*

$$x_0 = (0.2, 0.5, 0.3)$$

whose entries give the probability that an individual member of the population, chosen at random, belongs to a particular class. Since every individual belongs to one of the classes, the sum of the elements in a probability vector must be 1.

Exercise 184

1. Which of the following matrices are stochastic?

(a) $\begin{pmatrix} 0.3 & 0.4 & 0.3 \\ 0.5 & 0.0 & 0.5 \\ 0.6 & 0.1 & 0.3 \end{pmatrix}$

(c) $\begin{pmatrix} 0.7 & 0.3 \\ 0.4 & 0.6 \end{pmatrix}$

(b) $\begin{pmatrix} 0.8 & 0.1 & 0.1 \\ 1.3 & -0.2 & -0.1 \\ 0.7 & 0.5 & -0.2 \end{pmatrix}$

(d) $\begin{pmatrix} 0.25 & 0.50 & 0.25 \\ 0.50 & 0.30 & 0.25 \\ 0.25 & 0.20 & 0.50 \end{pmatrix}$

2. The matrix

$$\begin{pmatrix} 0.7 & 0.1 & 0.2 \\ 0.1 & 0.8 & 0.1 \\ 0.7 & 0.0 & 0.3 \end{pmatrix}$$

is a social transition matrix for a certain country, showing the (constant) probabilities of moving from one social group to another in the course of a year. If the initial distribution into social groups is given by the vector

$$(7000, \quad 2000, \quad 1000)$$

calculate the evolution of the expected distribution into social classes over the first five years.

15.10 GENERAL TREATMENT

We shall now generalize what we have done so far. We consider a population, of a completely unspecified nature, in which each individual member can at any one time belong to one of a certain number, say n, of classes. These classes are usually referred to as *states*. Thus any given individual will be in one or another of the n states at any given moment. However, it may happen that an individual will change from one state to another. For simplicity we consider the changes in the population as taking place at discrete moments of time, for example every year, as in the problem discussed above.

Let p_{ij} be the probability that in one such time-interval an individual in state i "moves" into state $j (i, j = 1, 2, \ldots, n)$. These *transition probabilities*, as they are called, can be displayed as a matrix

$$P = (p_{ij})$$

called the *transition matrix*. It gives all the information concerning the probabilities of transitions from one state to another. Since an individual in state i must be in one of the n states at the end of the time interval (possibly the same one as before) the sum

$$p_{i1} + p_{i2} + \ldots + p_{in}$$

must be 1. Hence every row sum in P is 1 and P is a stochastic matrix.

The distribution of the population at any given time can be specified by a row vector

$$(U_1, U_2, \ldots, U_n)$$

where U_i is the number of individuals in state i. It is more convenient, however, to specify a probability vector

$$(u_1, u_2, \ldots, u_n)$$

where u_i is the fraction of the population in state i, or, what amounts to the same thing, the probability that a given individual is in state i.

The time interval between the moments when we scrutinize the population can conveniently be called a *step*. Let us see how many individuals will be in state j after the first step. Of the U_i individuals originally in state i, a fraction p_{ij} will migrate to state j during the step, contributing $U_i p_{ij}$ to the number in state j at the end of the step. Since all individuals ending up in state j must previously have been in one of the n states we have

$$U'_j = \sum_{i=1}^{n} U_i p_{ij} \qquad (j = 1, 2, \ldots, n)$$

where U'_j denotes the number in state j after the first step. Dividing by the total population, we have

(15.10.1) $$u'_j = \sum_{i=1}^{n} u_i p_{ij} \qquad (j = 1, 2, \ldots, n).$$

If we now let $u_0 = (u_1, u_2, \ldots, u_n)$ and $u_1 = (u'_1, u'_2, \ldots, u'_n)$ then (15.10.1) can be written as

(15.10.2) $$u_1 = u_0 P$$

by the rule for multiplication of a row vector by a matrix. Thus u_1 is the probability vector which gives the probability of an individual being in any given state after the first step.

Similarly, the change in the population during the second step is summarized by the matrix equation

$$u_2 = u_1 P$$

where u_2 is the probability vector describing the probability of being in the various states after step 2. As before we see that

$$u_r = u_{r-1} P$$

and hence that

(15.10.3) $$u_r = u_0 P^r$$

showing how the distribution of the population after r steps depends on the initial distribution. In this way the "future history" of the population can be computed for as many steps as are desired.

It has been convenient to think in terms of populations of many individuals, since this is the sort of context in which these problems usually arise. However, it is better, for theoretical purposes, to imagine that we are following the progress of one individual who starts off with a certain probability (possibly zero) of being in each of the possible states. Step by step, these probabilities change in accordance with equation (15.10.3). In very general problems the transition matrix may be different for different steps, and we have what is called a *Markov process*; if, as here, the transition probabilities are constant, the process is called a *Markov chain*.

We have envisaged the possibility that starting with a given probability vector x_0 we may find that the sequence of vectors $x_0, x_0 P, x_0 P^2, \ldots, x_0 P^n$, ... might tend to some limiting vector, x say. We shall need to consider under what circumstances this can happen, for if it does happen in, say, a Markov chain dealing with social mobility, it is of considerable demographic importance.

It is easy to see that it will not always happen. For consider the rather unlikely situation in a population of two classes where each year every Upper becomes a Lower, and vice versa. The transition matrix is

$$P = \begin{pmatrix} 0 & 1 \\ 1 & 0 \end{pmatrix}$$

and if $x_0 = (u, l)$, the vectors in the sequence are

$$(u, l), \quad (l, u), \quad (u, l), \quad (l, u), \ldots$$

and this sequence does not tend to a limit if $u \neq l$. This is of course a trivial example, but many nontrivial examples could be given. One naturally wonders whether they have anything in common.

If $p_{ij}^{(r)}$ is a typical element in the matrix $\boldsymbol{P}^r$, then $p_{ij}^{(r)}$ is the probability that, in r steps, an individual originally in state i will end up in state j. $\boldsymbol{P}^r$ is therefore a stochastic matrix, and is in fact the transition matrix for the Markov chain which results if we treat r steps of the original chain as a single step of the new chain. If $p_{ij}^{(r)} = 0$ then it is not possible for an individual in state i to change to state j in r steps; state j is inaccessible in r steps from state i. However, as r increases one would expect that the number of states that are accessible to an individual would increase, the most favorable situation being reached when after a certain number, say N, of steps all states are accessible to the individual irrespective of his initial state. This means that whatever state i the individual was in to start with, he has a nonzero probability of being in any other specified state j after N steps. This means that $p_{ij}^{(N)} > 0$ for all i and j, i.e., every element of $\boldsymbol{P}^N$ is nonzero. A transition matrix $\boldsymbol{P}$, some power $\boldsymbol{P}^N$ of which contains no zero elements is known as a *regular transition matrix*, and the corresponding Markov chain is called a *regular Markov chain*.

An example of a transition matrix which is not regular is the matrix

$$\begin{pmatrix} 0 & 1 \\ 1 & 0 \end{pmatrix}$$

which we just considered. Its powers are alternately

$$\begin{pmatrix} 0 & 1 \\ 1 & 0 \end{pmatrix} \quad \text{and} \quad \begin{pmatrix} 1 & 0 \\ 0 & 1 \end{pmatrix}$$

and therefore no power has all nonzero elements. The matrix

$$\begin{pmatrix} \dfrac{1}{2} & \dfrac{1}{2} & 0 & 0 \\[2mm] \dfrac{1}{3} & \dfrac{2}{3} & 0 & 0 \\[2mm] \dfrac{1}{4} & \dfrac{1}{4} & \dfrac{1}{4} & \dfrac{1}{4} \\[2mm] \dfrac{1}{5} & \dfrac{2}{5} & \dfrac{1}{5} & \dfrac{1}{5} \end{pmatrix}$$

is also not regular; all its powers will have an array of 4 zeros in the top right hand quarter. On the other hand, although it has zero elements, the matrix

$$\begin{pmatrix} \dfrac{1}{2} & \dfrac{1}{2} & 0 \\ \dfrac{1}{3} & \dfrac{1}{3} & \dfrac{1}{3} \\ \dfrac{1}{2} & 0 & \dfrac{1}{2} \end{pmatrix}$$

is regular, since its square has no zero elements.

Exercise 185

1. If A is a stochastic matrix, show that the vector sum of the columns of the matrix $A - I$ is the zero column vector. Hence show that 1 is an eigenvalue of A.
2. Determine whether the following stochastic matrices are regular or not.

(a) $\begin{pmatrix} 0.1 & 0.0 & 0.9 \\ 0.5 & 0.3 & 0.2 \\ 0.0 & 0.4 & 0.6 \end{pmatrix}$

(c) $\begin{pmatrix} 0.7 & 0.0 & 0.3 & 0.0 \\ 0.2 & 0.3 & 0.1 & 0.4 \\ 0.5 & 0.0 & 0.5 & 0.0 \\ 0.4 & 0.3 & 0.0 & 0.3 \end{pmatrix}$

(b) $\begin{pmatrix} 0.4 & 0.6 & 0.0 \\ 0.0 & 1.0 & 0.0 \\ 0.3 & 0.2 & 0.5 \end{pmatrix}$

(d) $\begin{pmatrix} 0.1 & 0.2 & 0.5 & 0.2 \\ 0.6 & 0.2 & 0.2 & 0.0 \\ 0.1 & 0.5 & 0.3 & 0.1 \\ 0.3 & 0.0 & 0.0 & 0.7 \end{pmatrix}$

15.11 A FUNDAMENTAL THEOREM

We now prove a fundamental result in the study of Markov chains.

THEOREM 15.2

IF P IS A REGULAR $n \times n$ STOCHASTIC MATRIX, THEN AS $r \to \infty$, P^n TENDS TO A MATRIX OF n IDENTICAL ROWS. EACH ROW IS A PROBABILITY VECTOR OF POSITIVE ELEMENTS.

PROOF:

Let y be any column vector, and let M_0 and m_0 be its largest and smallest elements. Let M_r and m_r be the largest and smallest elements of $P^r y$. Let ϵ be the smallest element of P. Clearly $0 \leq \epsilon \leq \frac{1}{2}$.

The ith element in Py is

(15.11.1) $\qquad z_i = \sum p_{ij} y_j \leq am_0 + (1 - a)M_0$

since some element a of P multiplies m_0, while the sum of the remaining terms is less than M_0 times the remaining elements of row i of P, and their sum is $1 - a$. Since $\epsilon \leq a$ we have

$$z_i \leq M_0 - a(M_0 - m_0)$$
$$\leq M_0 - \epsilon(M_0 - m_0)$$

and since this holds for all z_i, we have, in particular

(15.11.2) $\qquad M_1 \leq M_0 - \epsilon\,(M_0 - m_0).$

If we repeat this argument with the vector $-y$ whose largest and smallest elements are $-m_0$ and $-M_0$, we obtain

(15.11.3) $\qquad -m_1 \leq -m_0 - \epsilon\,(-m_0 + M_0).$

Adding (15.11.2) and (15.11.3) we find that

$$M_1 - m_1 \leq M_0 - m_0 - 2\,\epsilon\,(M_0 - m_0)$$
(15.11.4) $\qquad = (1 - 2\,\epsilon)(M_0 - m_0).$

Similarly, using $P^{r-1}y$ in place of y we have

(15.11.5) $\qquad M_r \leq M_{r-1} - \epsilon(M_{r-1} - m_{r-1})$
(15.11.6) $\qquad -m_r \leq -m_{r-1} - \epsilon(-m_{r-1} + M_{r-1})$

and

(15.11.7) $\qquad M_r - m_r \leq (1 - 2\epsilon)(M_{r-1} - m_{r-1}).$

From (15.11.5) and (15.11.6) it follows that $M_r \leq M_{r-1}$ and $m_r \geq m_{r-1}$, so that $\{M_r\}$ is a decreasing sequence and $\{m_r\}$ an increasing sequence. We must now show that $M_r - m_r \to 0$. We cannot do this from (15.11.7) since ϵ may be zero. But if P is a regular transition matrix, then some power, say P^N, has no zero elements. Let η be the smallest element in P^N. Then

(15.11.8) $\qquad M_N - m_N = (1 - 2\eta)(M_0 - m_0)$

since this is the result (15.11.4) applied to the matrix P^N. Similarly

$$M_{2N} - m_{2N} = (1 - 2\eta) \ (M_N - m_N)$$
$$= (1 - 2\eta)^2 (M_0 - m_0)$$

and, in general

$$M_{kN} - m_{kN} = (1 - 2\eta)^k (M_0 - m_0)$$

which tends to zero as $k \to \infty$ since $0 < 1 - 2\eta < 1$. Hence $M_r - m_r$, which is monotonic decreasing, can be made as near to zero as we please, and M_r and m_r tend to the same limit (compare Theorem 8.11). Hence $P^r y$ tends to a vector all of whose elements are equal.
We now write

$$P^r = P^r I = P^r(e_1, e_2, \ldots, e_n)$$
$$= (P^r e_1, P^r e_2, \ldots, P^r e_n)$$

where $e_1, e_2, \ldots, e_n$ are the columns of the unit matrix. By what we have already proved, each $P^r e_j$ tends to a vector of the form

$$\begin{pmatrix} a_j \\ a_j \\ \cdots \\ a_j \end{pmatrix}$$

and therefore

(15.11.9)
$$P^r \to \begin{pmatrix} a_1 a_2 a_3 \ldots a_n \\ a_1 a_2 a_3 \ldots a_n \\ \cdot \quad \cdots \quad \cdot \\ a_1 a_2 a_3 \ldots a_n \end{pmatrix}.$$

Since P is stochastic so is P^r and so is the limit (15.11.9). Thus each row of (15.11.9) is a probability vector. Moreover $m_N \neq 0$ and $M_N \neq 1$. Hence $0 < m_N < a_j < M_N < 1$, so that the elements of (15.11.9) are all positive. This completes the proof of the theorem.

We are now able to answer the question raised on page 774, namely under what conditions does the sequence of vectors $x_0, x_0 P, x_0 P^2, x_0 P^3, \ldots$ tend to some limiting vector. If we denote the matrix on the right-hand side of (15.11.9) by A, then as $r \to \infty$ so $x_0 P^r \to x_0 A$. Now if $x_0 = (\xi_1, \xi_2, \ldots, \xi_n)$ then the jth element in $x_0 A$ is

$$\xi_1 a_j + \xi_2 a_j + \cdots + \xi_n a_j = (\xi_1 + \xi_2 + \cdots + \xi_n) a_j$$
$$= a_j$$

since x_0 is a probability vector, and hence $\xi_1 + \xi_2 + \ldots + \xi_n = 1$. Thus we see that the vector $x_0 A$ is just $(a_1, a_2, \ldots, a_n)$ irrespective of what the initial vector x_0 may be.

Hence, provided the matrix P is regular, the sequence in question *will* tend to a limiting vector. This vector is the same vector which constitutes every row of the matrix $\lim_{r \to \infty} P^r$.

Exercise 186

1. At the end of the proof of Theorem 15.2 it was stated that since P is stochastic so is P^r and so is the limit of P^r as $r \to \infty$. Give a rigorous proof of these statements.

2. The datum that P is a regular matrix was used at only one point in the proof of Theorem 15.2. The reader should check that the proof does not hold if P is not regular.

15.12 TRANSIENT AND ERGODIC SETS

We have already remarked that for any two states s_i and s_j of a Markov chain it may or may not be the case that s_j is accessible from s_i, that is, that there is a nonzero probability that we can move from state s_i to state s_j in some number of steps. Let us write this as $s_i R s_j$. Clearly this relation R is transitive. We can now define an equivalence relation E on the set of all states by the definition

$$s_i E s_j = s_i R s_j \ \& \ s_j R s_i.$$

That is to say, $s_i E s_j$ means that s_i and s_j are mutually accessible. (We allow $s_i E s_i$ on the basis that we can go from a state to itself in zero steps.) This equivalence relation divides the states into equivalence classes in the usual way. Let us denote these equivalence classes by $C_1, C_2, C_3, \ldots$.

Consider two of these classes, say C_α and C_β, and let s_i, s_j be any two states in C_α and C_β, respectively. The states s_i and s_j cannot be mutually accessible, since they are in different equivalence classes, but one may be accessible from the other, say $s_i R s_j$. If so then every state y in C_β is accessible from every state x in C_α. For $x R s_i$, $s_i R s_j$ and $s_j R y$, so that $x R y$ follows since R is transitive. We shall then write $C_\alpha < C_\beta$. On the other hand it could happen that states in the two classes are not accessible either way, in which case the relation $<$ just defined does not apply to the two classes. Thus we can see that the relation $<$ is a partial ordering of the equivalence classes. We can now state an important theorem.

THEOREM 15.3

IN ANY MARKOV CHAIN WITH A FINITE NUMBER OF STATES THERE IS AT LEAST ONE EQUIVALENCE CLASS OF STATES WHICH HAS THE PROPERTY THAT ONCE IT IS ENTERED IT CANNOT BE LEFT; THAT IS, AN EQUIVALENCE CLASS C SUCH THAT THERE IS NO EQUIVALENCE CLASS C_α FOR WHICH THE STATEMENT $C < C_\alpha$ IS TRUE.

PROOF:

We first observe that if we once leave an equivalence class C_α, we cannot return to it. For if we go from a state $x \in C_\alpha$ to a state $y \notin C_\alpha$, and subsequently return to a state in C_α, it would then be possible to go on to the state x (since the states in C_α are mutually accessible) and this would make states x and y mutually accessible. This contradicts the fact that they are in different equivalence classes.

Hence if we are in one equivalence class, either it is possible to go from there to another equivalence class, or it is not. If it is not, then the equivalence class we are in satisfies the requirements of the theorem. Otherwise we can move to another equivalence class, thence (possibly) to another, and so on. Since, as we have seen, we can never return to an equivalence class having once left it, and since there is only a finite number of equivalence classes (since there is only a finite number of states) this process of moving from one equivalence class to another cannot continue indefinitely. But the only way in which it can end is by reaching an equivalence class from which no further progress is possible, i.e., an equivalence class satisfying the requirements of the theorem. Thus the existence of at least one such equivalence class has been demonstrated.

These equivalence classes which, once entered, cannot be left are known as *ergodic sets*. The others are known as *transient sets*.

EXAMPLE 1:

Regular Markov chains

If a Markov chain is regular then some power of the stochastic matrix P, say P^n, has no zero elements. It follows that every state is accessible from every other state in n steps. Hence the states are all mutually accessible, there is only one equivalence class, and hence just the one ergodic set, containing all the states. There are no transient sets.

EXAMPLE 2:

In a straight street there are 5 buildings A, B, C, D and E, in order. B, C and D are bars, while A and E are police stations. A man starts at the bar C and visits other buildings in turn. His moves are determined by chance, and when he emerges from a bar there is probability $\frac{1}{2}$ that he will turn and visit the next building to the right, and hence probability $\frac{1}{2}$ that he will go to the next building on the left; but if he once enters either of the police stations he will be kept there.

This is a simple example of what is known as a *random walk* problem and provides a good example of a Markov chain. If we number the buildings 1 through 5, we see that the transition matrix is

$$P = \begin{pmatrix} 1 & 0 & 0 & 0 & 0 \\ \frac{1}{2} & 0 & \frac{1}{2} & 0 & 0 \\ 0 & \frac{1}{2} & 0 & \frac{1}{2} & 0 \\ 0 & 0 & \frac{1}{2} & 0 & \frac{1}{2} \\ 0 & 0 & 0 & 0 & 1 \end{pmatrix}$$

Here $p_{11} = 1$ since if the man is at A it is certain (probability $= 1$) that after the next step he will still be at A. On the other hand p_{23}, for example, is $\frac{1}{2}$, since this is the probability that he will move from state 2 to state 3 (bar B to bar C) in one "step." Without calculating any powers of P we can see that the three bars B, C and D are mutually accessible in at most two steps, but that the two police stations are not *mutually* accessible to the bars or to each other. Hence the equivalence classes are three in number, viz.,

$$\{A\}, \quad \{B, C, D\} \quad \text{and} \quad \{E\}.$$

Of these $\{A\}$ and $\{E\}$ are ergodic sets, while $\{B, C, D\}$ is a transient set. This situation could be described as a random walk on a line with two *absorbing barriers*. The absorbing barriers are the police stations, which "absorb" the man, preventing him from continuing the walk any further.

EXAMPLE 3:

If we replace the police station E by another bar—the last one on the road—and assume that if the man goes from D to E at one step, he will go from E to D at the next, then P becomes

$$\begin{pmatrix} 1 & 0 & 0 & 0 & 0 \\ \frac{1}{2} & 0 & \frac{1}{2} & 0 & 0 \\ 0 & \frac{1}{2} & 0 & \frac{1}{2} & 0 \\ 0 & 0 & \frac{1}{2} & 0 & \frac{1}{2} \\ 0 & 0 & 0 & 1 & 0 \end{pmatrix}$$

The problem is now a random walk with one absorbing barrier (A) and one *reflecting barrier* (E). It is easily verified that there is just one ergodic set, containing A alone, and one transient set.

EXAMPLE 4:

"Gamblers ruin" problems can be expressed as Markov chains. Thus let two players A and B start with n dollars between them, and repeatedly play a game in which A has probability p of winning \$1 from B, and B has probability $1 - p$ of winning \$1 from A. Then if s_i $(i = 0, 1, \ldots, n)$ denotes the state characterized by the fact that A has i dollars, the transition matrix is (p_{ij}) where

$$p_{i,\,i+1} = p$$
$$p_{i,\,i-1} = 1 - p \qquad (i = 1, \ldots, n - 1)$$
$$p_{00} = p_{nn} = 1$$

and all other elements are zero. The element p_{00} is 1 because state s_0 is that for which A has been forced out of the game, and A cannot therefore leave this state. Similarly $p_{nn} = 1$ since state s_n is that for which B has been forced out of the game.

This Markov chain, like that of example (2) has two ergodic sets, $\{s_0\}$ and $\{s_n\}$, and one transient set $\{s_1, s_2, \ldots, s_{n-1}\}$. In fact, apart from the descriptive background of the problem (which is not important mathematically) example (2) is the special case of example (4) which arises when $p = \frac{1}{2}$ and $n = 4$.

Exercise 187

1. The following diagram is a map of six towns and the roads that connect them. A travelling salesman travels from town to town along these roads. On reaching a town he takes with equal probability the other roads leading from it; he does not go back to the town from which he has just come. Construct the transition matrix for this random walk.

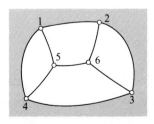

Fig. 15.5

Can you tell, without performing any algebraic manipulations, whether this matrix is regular or not?

2. A chess knight moves at random on a chess board with equal probability at each step of taking each of the moves available to it (including that which takes it back to the square from whence it previously came). It starts on one of the center 16 squares of the board and is removed from the board if it lands on a square other than these center 16 squares.

Analyze this problem as a random walk with an absorbing barrier, having 17 states. Obtain the relevant transition matrix and describe the transient and ergodic sets.

Note: A knight in chess is allowed to move from one corner square to the opposite corner of any 2×3 rectangle of squares, as follows:

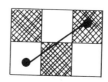

15.13 A THEOREM CONCERNING ERGODIC SETS

We conclude this chapter by proving a theorem concerning the probability that an ergodic set will have been reached after a certain number of steps. It is clear that there are random walks which never reach an ergodic set. For example, the man in example (2) of Section 15.12 could alternate between bars B and C indefinitely. The probability of this happening for n steps is $(\frac{1}{2})^n$ which becomes small as $n \to \infty$. The same will be true of other walks that do not enter an ergodic set.

These remarks are made precise by the following theorem:

THEOREM 15.4

IF, IN A MARKOV CHAIN, $p(N)$ IS THE PROBABILITY THAT AFTER N STEPS THE STATE REACHED IS IN AN ERGODIC SET, THEN $p(N) \to 1$ AS $N \to \infty$.

PROOF:

For each state s_i at which we start there is a number k_i such that the probability p_i that a state in an ergodic set is reached after k_i steps is nonzero. In other words it is possible in k_i steps to go from state s_i and reach an ergodic set. Let $k = \max(k_i)$, and $p = \min(p_i)$. Then whatever state s_i we start from there is probability p or more of reaching an ergodic set in k steps. Hence the probability of not reaching an ergodic set in k steps is α, where $\alpha \le 1 - p$.

The probability of not reaching an ergodic set in $2k$ steps is $\leq \alpha^2$. (This is a simple exercise in conditional probability.) More generally, the probability of not reaching an ergodic set in mk steps is $\leq \alpha^m$. Let us now consider the function $p(N)$ in the statement of the theorem. If we have reached an ergodic set in N steps, then either

(1) we reached an ergodic set in $N - 1$ steps and stayed in the set; or
(2) we reached, in $N - 1$ steps, a state which was one step away from a state in an ergodic set, and moved to the ergodic set on the Nth step. Hence

$$p(N) = p(N - 1) + \text{the probability of possibility (2) above,}$$

from which we deduce that $p(N) \geq p(N - 1)$, i.e., that $p(N)$ is a nondecreasing function of N.

Now $p(mk) \geq 1 - \alpha^m$, which tends to 1 as $m \to \infty$. Hence there are values of $p(N)$ as near to 1 as we wish. This together with the fact that $p(N)$ is a nondecreasing function of N, shows that $p(N) \to 1$ as $N \to \infty$, and completes the proof of the theorem.

Exercise 188

1. Closely allied to the question of the probability of reaching an ergodic set is that of the expected number of steps that will be taken before an ergodic set is reached. Show that if E_i is the expected number of steps required, starting from the state s_i then

$$E_i = 0 \quad \text{if } s_i \text{ is in an ergodic set}$$

and

$$E_i = 1 + \sum_{j=1}^{n} p_{ij} E_j \text{ otherwise.}$$

2. Use the result of problem 1 to find the expected number of steps before the man in example (2) of Section 15.12 ends up at a police station (either one). By symmetry, there will be two answers according as the man starts at bar C or either of bars B or D.
3. Repeat problem 2 with example (3) of Section 15.12.

Exercise 189 (Chapter Review)

1. Prove that in the iterative approximation given by

$$x_{n+1} = \frac{1}{2A} x_n (3A - x_n^2) \quad (A > 0)$$

x_n tends to a limit, and that the limit is $\sqrt{A}$, provided x_0 is sufficiently close to $\sqrt{A}$.

Note: This method of successively approximating to a square root converges less rapidly than that given at the beginning of the chapter, but it has the advantage that no division is required in each iteration. The only division is in the calculation of $1/2A$ and this is done once and for all before the iterative process starts.

What happens if A is < 0?

2. Find the general solution of the following difference equations.
   (a) $y_{n+2} - 14y_{n+1} + 49y_n = 0$
   (b) $y_{n+2} - 10y_{n+1} + 21y_n = 0$
   (c) $y_{n+2} - 2y_{n+1} + 10y_n = 0$
   (d) $2y_{n+2} - y_{n+1} - 2y_n = 0$
   (e) $y_{n+2} + 9y_{n+1} + 20y_n = 0$
   (f) $y_{n+2} + 4y_{n+1} + 29y_n = 0$

3. Solve completely the following difference equations.
   (a) $y_{n+2} - 9y_{n+1} + 18y_n = 3^n n^2$
   (b) $6y_{n+2} - 11y_{n+1} + 4y_n = n^2 + 3n$
   (c) $4y_{n+2} - 4y_{n+1} + y_n = 8^n + 5$
   (d) $3y_{n+2} + 10y_{n+1} + 3y_n = n^3$
   (e) $5y_{n+2} + 11y_{n+1} + 2y_n = 3n(n-1)$

4. Find the solution of each of the following difference equations for which $y_0 = y_1 = 1$.
   (a) $3y_{n+2} - 19y_{n+1} - 14y_n = 0$
   (b) $y_{n+2} - y_{n+1} + y_n = n$
   (c) $y_{n+2} - 5y_{n+1} + 6y_n = 5 \cdot 2^n$
   (d) $y_{n+2} - 2y_{n+1} + 2y_n = \frac{1}{2}n^2$
   (e) $y_{n+2} - 3y_{n+1} + 2y_n = 15$

5. The following difference equations are of order greater than 2 but can be solved by the methods of this chapter. Find the general solution in each case.
   (a) $y_{n+3} - 7y_{n+2} + 16y_{n+1} - 12y_n = 0$
   (b) $y_{n+3} - y_{n+2} - y_{n+1} + y_n = n$
   (c) $y_{n+4} - 16y_n = 2^n$

6. The following equations are not linear but can be brought to linear form by a substitution of the form $y_n = f(\eta_n)$, where f is a suitable function. By finding the appropriate function in each case, solve the equations completely.
   (a) $y_n y_{n+2} = 16y_{n+1}^2$
   (b) $3y_{n+2}^2 + 10y_{n+1}^2 + 3y_n^2 = 6$
   (c) $y_n y_{n+1} - 2y_{n+2} y_n + 10y_{n+1} y_{n+2} = 0$

7. In a study of American meat consumption during the years 1919–41 it was found that if the consumption X_n in year n could be described by a third

order difference equation, then this equation would be of the form

$$X_{n+3} = 0.219X_{n+2} + 0.281X_{n+1} + 0.231X_n + 0.430.$$

By drawing a graph, or otherwise, estimate roughly the roots of the equation

$$\lambda^3 - 0.219\lambda^2 - 0.281\lambda - 0.231 = 0$$

and hence obtain (also roughly) the solution of the above difference equation. (Do not attempt to work to more than two decimal places.)

Comment on the form of the solution.

8. In a four-class social structure the annual migration from one class to another is given by the following transition matrix.

$$\begin{pmatrix} 0.85 & 0.10 & 0.05 & 0.00 \\ 0.25 & 0.65 & 0.10 & 0.00 \\ 0.05 & 0.10 & 0.70 & 0.15 \\ 0.00 & 0.03 & 0.07 & 0.90 \end{pmatrix}$$

If the distribution of the population into the four classes is given initially by the probability vector $(0.60, 0.25, 0.10, 0.05)$ find the distribution in the three subsequent years.

9. Determine which of the following stochastic matrices are regular.

(a) $\begin{pmatrix} 0.3 & 0.0 & 0.0 & 0.7 \\ 0.5 & 0.0 & 0.0 & 0.5 \\ 0.0 & 0.6 & 0.0 & 0.4 \\ 0.0 & 1.0 & 0.0 & 0.0 \end{pmatrix}$ (c) $\begin{pmatrix} 0.2 & 0.3 & 0.1 & 0.4 \\ 0.7 & 0.1 & 0.1 & 0.1 \\ 0.2 & 0.6 & 0.2 & 0.0 \\ 0.4 & 0.3 & 0.3 & 0.0 \end{pmatrix}$

(b) $\begin{pmatrix} 0.0 & 0.3 & 0.0 & 0.0 & 0.7 \\ 0.4 & 0.2 & 0.0 & 0.0 & 0.4 \\ 0.0 & 0.6 & 0.0 & 0.1 & 0.3 \\ 0.4 & 0.0 & 0.0 & 0.0 & 0.6 \\ 0.1 & 0.0 & 0.4 & 0.0 & 0.5 \end{pmatrix}$ (d) $\begin{pmatrix} 0.0 & 0.4 & 0.0 & 0.6 \\ 0.5 & 0.0 & 0.5 & 0.0 \\ 0.0 & 0.1 & 0.0 & 0.9 \\ 0.7 & 0.0 & 0.3 & 0.0 \end{pmatrix}$

10. A typical example of a Markov process is that of shifts of patronage by customers from one store to another, or from one brand of produce to another. If the process is, in fact, a Markov chain, then these shifts of patronage are given by a transition matrix (p_{ij}) in the usual way, with p_{ij} being the probability that in one time unit a customer will transfer his patronage from store i to store j. A natural question to ask is whether a store could lose all its customers in this way. That is to say, can the equilibrium probability vector have a zero element?

If x is the equilibrium row vector, and P the transition matrix, then

$$xP = x.$$

Show that if x has just one zero element, x_i, then every element of the ith column of P is zero, except possibly for the element in the diagonal. More generally, show that if x has several zero elements, and S is the set of their subscripts, then for every pair of subscripts i, j for which $i \notin S$ and $j \in S$, we have $a_{ij} = 0$. Interpret this result.

Suggestions for Further Reading

A full treatment of difference equations is contained in the following book

Levy, H., and Lessman, F. *Finite Difference Equations*. London. Sir Isaac Pitman & Sons, Ltd., 1959.

For a treatment of difference equations together with differential equations, the reader can consult

Chorlton, F. *Ordinary Differential and Difference Equations*. Princeton, N.J.: D. Van Nostrand Co., Inc., 1965.

For further information on the cobweb model, as well as many other economic topics discussed in this book, the reader can refer to

Allen, R. G. D., *Mathematical Economics*. New York: the MacMillan Company, 1957.

The following book gives a much deeper treatment of the subject of Markov chains than has been given in this chapter, but the reader may profitably dip into it.

Kemeny, J. G., and Snell J. L., *Finite Markov Chains*. Princeton, N.J.: D. Van Nostrand Co., Inc., 1960.

In addition, the three "Finite Mathematics" books cited on page 27 each have a discussion of Markov chains and related topics.

16   Linear Programming

Consider the following four problems.

PROBLEM 1:
Given that

$$x + y \leq 9$$
$$x + 3y \leq 21$$
$$x - 5y \leq 3$$
$$4x + y \leq 24$$
$$-x + y \leq 5$$

and
$$9x + 8y \leq 72$$

where x and y are non-negative numbers; what is the maximum value that the function $2x + 3y$ can take?

A chemical manufacturing company produces three patent medicines. Each medicine is a mixture of 5 drugs, the proportions being different for the different medicines. If we denote the 5 drugs by A, B, C, D and E, the relevant facts concerning these medicines are as follows:

	Price	Grams per bottle				
		A	B	C	D	E
Medicine No. 1	$6 per bottle	3	11	3	7	15
Medicine No. 2	$1 per bottle	6	6	10	14	12
Medicine No. 3	$5 per bottle	4	12	6	4	15

The company's ability to produce these medicines depends on the rate at which the drugs can be manufactured. The maximum amounts of drugs, in grams, that can be manufactured each day are:

A	B	C	D	E
48	136	70	55	150

789

Assuming that there is a ready market for the 3 medicines in any amounts, how much of each medicine should be produced per day in order that the company should realize the largest sales?

PROBLEM 3:

Mr. Smith wishes to buy his son a collection of building bricks. These bricks come in different sizes, and for reasons best known to himself, Mr. Smith wishes to assemble a collection containing at least

194 bricks, size 0

152 bricks, size 1

98 bricks, size 2

34 bricks, size 3.

He discovers that the shops do not sell these bricks separately; they come only in kits, of which there are three kinds. The prices of the kits, and the numbers of different bricks they contain, are as follows:

	Price	Size 0	Size 1	Size 2	Size 3
Kit A	$8	29	14	11	4
Kit B	$3	2	8	2	1
Kit C	$5	13	10	7	2

Mr. Smith makes up the collection he requires by buying a certain number of each kit. What is the most economical way in which he can do this?

PROBLEM 4:

The manager of a chain store wishes to supply each of his four branches with quantities of a certain commodity. This commodity is available from any of three warehouses (A, B and C) which can supply, respectively, 150, 100 and 70 tons of the commodity per month. The quantities required by the four branches (1, 2, 3 and 4) are, respectively, 30, 40, 80 and 90 tons per month. The cost of transportation, in dollars per ton, from each warehouse to each branch is given by the following table:

Transportation Costs

From	To Branch 1	2	3	4
Warehouse A	30	30	50	90
B	30	40	40	70
C	40	20	30	60

What is the most economical way of supplying the branches; that is, how much of the commodity should be transported from each warehouse to each branch in order that the total cost of transportation should be a minimum?

These four problems may not seem, at first sight, to have much in common. Nevertheless, they are typical examples of a type of problem that has attracted a great deal of attention during the last twenty years or so, and that has given rise to a new branch of mathematics called *linear programming*. Linear programming has proved itself to be a useful tool in tackling many problems of great importance, especially in what has come to be known as "Operations Research", and is playing an ever-increasing part in the handling and the discussion of many topics in economics. Let us first see why these four problems present essentially the same kind of question to the mathematician.

Some resemblances between the problems are obvious. Each problem calls for the determination of some sort of optimum result—requiring that something should have a maximum or minimum value. In problem 1 we ask for the largest value of $2x + 3y$; in problem 2 we want the sales to be a maximum; while in problems 3 and 4 the expenditure is required to be a minimum. Further resemblances appear if we translate the problem into mathematical terms. Since problem 1 is already formulated mathematically, let us start by analyzing problem 2.

In this problem, as is not unusual, there are certain unknowns—quantities that we do not know but wish to find. Here they are the numbers of bottles of the three kinds of medicine produced by the manufacturing company per day. Let us call them x, y and z respectively. These unknowns are not arbitrary; there are certain restrictions on them because of the limitations on the amounts of the various drugs that are available each day. Thus to produce x bottles of Medicine 1 requires $3x$ grams of drug A; y bottles of Medicine 2 require $6y$ grams of the same drug; $4z$ grams of it are required to produce the z bottles of Medicine 3. Since only 48 grams of this drug are available, we must have

$$3x + 6y + 4z \leq 48.$$

Reasoning in exactly the same way for all the drugs, we obtain the following system of inequalities:

(16.1.1)
$$\begin{cases} 3x + 6y + 4z \leq 48 \\ 11x + 6y + 12z \leq 136 \\ 3x + 10y + 6z \leq 70 \\ 7x + 14y + 4z \leq 55 \\ 15x + 12y + 15z \leq 150 \end{cases}$$

which x, y and z must satisfy.

The money realized by the sale of these bottles of medicine is, in dollars,

(16.1.2) $6x + y + 5z.$

The problem is therefore to choose values x, y and z which satisfy the restrictions (or *constraints* as they are usually called) given by (16.1.1) and which make the linear expression (16.1.2) a maximum.

We now look at problem 3. Here the unknown quantities are the numbers of each kind of kit that Mr. Smith buys. Call them x, y and z, respectively. The purchase of these numbers of kits will result in the acquisition of

$$29x + 2y + 13z$$

bricks of size 0. Since Mr. Smith wishes to have at least 194 bricks of this size, we have the inequality

$$29x + 2y + 13z \geq 194.$$

We proceed in the same way with the requirements for the other sizes of bricks, and we find that x, y and z must satisfy the following system of linear inequalities

(16.1.3)
$$\begin{cases} 29x + 2y + 13z \geq 194 \\ 14x + 8y + 10z \geq 152 \\ 11x + 2y + 7z \geq 98 \\ 4x + y + 2z \geq 34. \end{cases}$$

The cost of buying these numbers of kits is

(16.1.4) $C = 8x + 3y + 5z$

and we are therefore required to choose values of x, y and z consistent with the constraints (16.1.3) in such a way that the cost (16.1.4) is a minimum.

In each of these problems the physical setup implies that x, y and z are not only non-negative but are integers.

In problem 4 the unknowns are the amounts transported from the warehouses to the branches. There are 12 unknowns, since any of the warehouses can deliver to any of the branches. Let us denote the number of tons per month delivered by each warehouse to each branch by the symbols in the following table.

	Branch			
	1	2	3	4
Warehouse A	x_1	x_2	x_3	x_4
Warehouse B	y_1	y_2	y_3	y_4
Warehouse C	z_1	z_2	z_3	z_4

We now look to see what restrictions there are on these variables. First, the amounts delivered to each branch must be correct. Hence, since the requirements of the four branches are 30, 40, 80 and 90 tons per month, respectively, we have the constraints

(16.1.5)
$$x_1 + y_1 + z_1 = 30$$
$$x_2 + y_2 + z_2 = 40$$
$$x_3 + y_3 + z_3 = 80$$
$$x_4 + y_4 + z_4 = 90.$$

Next we make sure that no warehouse is called upon to supply more of the commodity than it has in store. Note that we have not made any objection to a warehouse having some of the commodity left over; indeed this is bound to happen since the amount stored is more than the amount demanded. Warehouse A can supply 150 tons per month, and therefore

$$x_1 + x_2 + x_3 + x_4 \le 150.$$

We obtain similar inequalities for the other two warehouses, and thus add

(16.1.6)
$$x_1 + x_2 + x_3 + x_4 \le 150$$
$$y_1 + y_2 + y_3 + y_4 \le 100$$
$$z_1 + z_2 + z_3 + z_4 \le 70$$

to the system of constraints required to be satisfied by the unknowns.

Any set of values of the xs, ys and zs satisfying (16.1.5) and (16.1.6) will give a possible method of supplying the branches from the warehouses, but we want to find the cheapest method. From the table of costs we see that the cost of supplying the branches is given by

(16.1.7) $30x_1 + 30x_2 + 50x_3 + 90x_4 + 30y_1 + 40y_2 + 40y_3 + 70y_4$
$$+ 40z_1 + 20z_2 + 30z_3 + 60z_4.$$

Thus the values of the unknowns that we are looking for are those that are consistent with (16.1.5) and (16.1.6), and that make (16.1.7) a minimum. In this problem there is no need for the unknowns to be integers, but they must be non-negative.

The resemblance between the above four problems is now easy to see, and we can formulate a general type of problem of which these are special cases. First we have a set of linear inequalities (or equations, or both) in several variables, the *constraints*. Secondly we have a linear function of the same variables, called the *objective function*. The problem is to find non-negative values of the variables which satisfy the constraints and make the objective function a maximum (or, in some problems, a minimum).

The requirement that the variables are not negative is a most important one. Instead of stating it in words we can express it just as well, if not better, by adding to the set of constraints further constraints, namely, the inequalities stating that each variable is non-negative. For example, we could replace the system (16.1.3) by

(16.1.8)
$$\begin{cases} 29x + 2y + 13z \geq 194 \\ 14x + 8y + 10z \geq 152 \\ 11x + 2y + 7z \geq 98 \\ 4x + y + 2z \geq 34 \\ x \geq 0 \\ y \geq 0 \\ z \geq 0 \end{cases}$$

and drop the verbal statement that x, y and z are non-negative. Often we will not actually do this; we will, however, keep in mind the fact that all the variables that we are working with take only zero or positive values.

The planning of an operation, or of some economic activity, in order to achieve some optimum result has been given the name of *programming* (not to be confused with computer programming, which we shall discuss later). In the four problems above, and in all other problems that will be considered in this chapter, the left-hand sides of the inequalities or equations that form the constraints are linear expressions in the variables, as is also the objective function. For this reason these are all problems in *linear programming*. The general linear programming problem is therefore that of maximizing or minimizing a linear function of non-negative variables, subject to a number of linear constraints on these variables.

Programming problems that are not of this type do arise, of course— problems in which the constraints and the objective function are not linear. However, the consideration of such problems is beyond the scope of this book.

Before discussing how to set about solving linear programming problems let us summarize their nature. We have a set of inequalities, say k of them, involving a set of, say, n variables or unknowns, $x_1, x_2, \ldots, x_n$. They can be written as

(16.1.9)
$$\begin{cases} a_{11}x_1 + a_{12}x_2 + \cdots + a_{1n}x_n \leq b_1 \\ a_{21}x_1 + a_{22}x_2 + \cdots + a_{2n}x_n \leq b_2 \\ \cdots \qquad \cdots \qquad \cdots \qquad \cdots \\ a_{k1}x_1 + a_{k2}x_2 + \cdots + a_{kn}x_n \leq b_k \end{cases}$$

or, in matrix notation

(16.1.10) $$\mathbf{Ax} \leq \mathbf{b}$$

where $\mathbf{A}$ is an $k \times n$ matrix, $\mathbf{x}$ the column vector

$$\begin{pmatrix} x_1 \\ x_2 \\ \cdots \\ x_k \end{pmatrix} \quad \text{and } \mathbf{b} \text{ the column vector} \quad \begin{pmatrix} b_1 \\ b_2 \\ \cdots \\ b_k \end{pmatrix}.$$

These inequalities, or constraints, are to be satisfied in such a way that a given linear function, the objective function, of the form

(16.1.11) $$c_1 x_1 + c_2 x_2 + \cdots + c_n x_n$$

or $\mathbf{c} \cdot \mathbf{x}$ where $\mathbf{c}$ is the row vector $(c_1, c_2, \ldots, c_n)$, assumes a maximum value.

It may seem that this description is not quite general enough insofar as in some problems (a) some of the constraints may be equations, or (b) we may want to minimize the objective function rather than maximize it. The second objection is easily met. Minimizing the function $\mathbf{c} \cdot \mathbf{x}$ is the same as maximizing the function $-\mathbf{c} \cdot \mathbf{x}$. Hence we need only change the sign of the objective function to convert a minimizing problem into a maximizing one. For the same reason there is no loss of generality in assuming that the inequality signs in (16.1.9) are all the same way round, viz., $\leq$.

To show that constraints that are equations can be included in the general scheme described above we need only observe that we can replace an equation by two inequalities. Thus, for example, we could replace the equation

$$x_1 + 3x_2 + 4x_3 + x_4 = 10$$

by the two inequalities

$$x_1 + 3x_2 + 4x_3 + x_4 \leq 10$$
$$x_1 + 3x_2 + 4x_3 + x_4 \geq 10$$

or, to keep the $\leq$ sign,

$$-x_1 - 3x_2 - 4x_3 - x_4 \leq 10.$$

When we discuss how to solve linear programming problems we shall see that equations need not be treated in this way. This argument merely shows that the presence of equations among the constraints can be allowed for in the general formulation of the problem and would not be expected to give rise to any particular difficulty.

Exercise 190

1. A manufacturer has available 45 units of wood, 30 units of plastic and 20 units of screws. He can make three kinds of articles: ashtrays, bookshelves and candlesticks. Ashtrays use 2 units of wood, 1 unit of plastic and 2 of screws: bookshelves use 5 units of wood, 2 of plastic and 1 of screws; candlesticks use 4 units of wood, 3 of plastic and 1 of screws. Ashtrays sell for $7 each; bookshelves for $12 each; and candlesticks for $9 each.

The manufacturer wishes to choose the number of articles of each kind to be made from the available material in such a way that the total value of the articles produced is a maximum.

Show that this is a linear programming problem, and express it in purely mathematical terms.

2. Products A, B and C are manufactured in each of two factories. Factory 1 produces 5, 15 and 20 of these products, respectively, per day; factory 2 produces 10, 5 and 15 of them. The total cost per day of running each factory is the same. An order is placed for 180 of article A, 240 of article B, and 520 of article C.

Show that the problem of determining how many days each factory should be run in order to fulfill the order as cheaply as possible is a linear programming problem, and derive the inequalities and the function to be minimized.

16.2 GRAPHICAL SOLUTION OF LINEAR PROGRAMMING PROBLEMS

Some simple linear programming problems, especially those in which there are only two variables in the inequalities (i.e., $n = 2$), can be solved by graphical methods. It is rather unlikely that problems like this would arise very often in practice, but we can learn much from these simple problems that will help us to tackle the more difficult ones. Let us therefore look at the first of our four problems, which is of this particularly simple form, having only two variables.

The solution of this problem is a pair of values of x and y, and can therefore be represented by some point in the Cartesian plane, viz., the point (x, y). Since x and y must be non-negative this point must lie in the first quadrant $(x \geq 0, y \geq 0)$. Let us now look at the first constraint, $x + y \leq 9$. By what was said on page 202 the line whose equation is

$$x + y = 9$$

divides the plane into two halves. On one side of the line are those points for which $x + y < 9$; on the other are those for which $x + y > 9$. Points satisfying the constraint $x + y \leq 9$ therefore lie on this line or on the same side of it as the origin. In Fig. 16.1 this line is shown and the side of the line on which

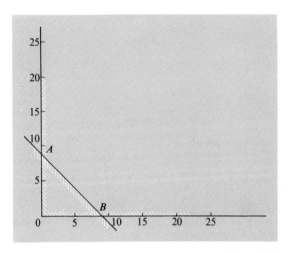

Fig. 16.1

(x, y) must lie in order to satisfy the constraint is indicated by the shading. The fact that x and y must be non-negative has also been indicated in Fig. 16.1 by shading the appropriate sides of the x-and y-axes.

It follows then that the points which satisfy the three constraints

$$x \geq 0; \qquad y \geq 0; \qquad x + y \leq 9$$

are those which lie inside or on the boundary of the triangle OAB in Fig. 16.1.

Each of the other constraints in problem 1 can be interpreted in a similar way, requiring that points representing candidates for the solution of the problem should lie on one side or the other of some specified line. If we add these lines to Fig. 16.1 and indicate, by shading, which side of each line is the one on which the points must lie to satisfy the constraint, we obtain Fig. 16.2.

The solution that we are looking for must be represented by a point which lies on the shaded side of each one of these lines. The set of points for which this is true defines the shaded area in Fig. 16.3. Pairs of values of x and y which satisfy all the constraints are said to form *feasible solutions* to the problem. Thus Fig. 16.3 shows graphically all the feasible solutions to this problem.

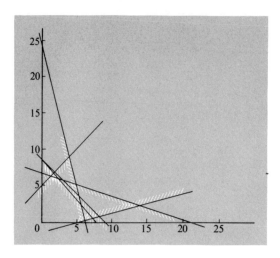

Fig. 16.2

Our problem is therefore reduced to that of finding which point of the shaded area in Fig. 16.3 is the one for which the objective function $2x + 3y$ has the greatest value.

Suppose we ask the question "Are there any points of the shaded area, i.e., any feasiblé solutions, for which $2x + 3y$ has the value 36?" This question is easily answered, for any such point must be on the line

$$2x + 3y = 36$$

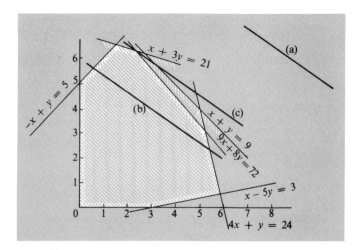

Fig. 16.3

which is the line a in Fig. 16.3. By just looking at the figure we see that no point of this line belongs to the shaded area. Are there points of the shaded area for which $2x + 3y = 18$? If there are then they lie on the line which has this equation, the line b in Fig. 16.3. This line is parallel to the line a and *does* intersect the shaded area. However, since there are points of the shaded area which are on the far side from the origin of line b, i.e., points for which $2x + 3y \geq 18$, it is clear that 18 is not the maximum value that the objective function can take.

It is now fairly evident what we must do. We need to find a line of the form

$$2x + 3y = \text{constant}$$

that is, a line parallel to those we have already drawn, as far as possible from the origin, provided that it intersects the shaded area. Moreover, the line that we seek must be such that the whole of the shaded area lies on the origin side of it; for if any points of the shaded area were on the far side of the line, the objective function would be more for those points than for points of the line. To find such a line we start with any line which, like line a, is too far from the origin; we now "move" it parallel to itself nearer to the origin until it just touches some part of the shaded area. Thus we obtain the line c in Fig. 16.3. The point P is the point representing the required solution.

Thus the solution of this problem is given by the point of intersection of the two lines

$$x + 3y = 21$$

and

$$9x + 8y = 72$$

viz., the point $\left(\dfrac{48}{19}, \dfrac{117}{19}\right).$

Exercise 191

1. By graphical means find the solution of the following linear programming problem.

$$\text{Minimize} \qquad 4x + 5y$$

subject to the constraints

$$x + 2y \geq 3$$
$$3x + y \geq 7$$
$$x - y \leq 5.$$

2. Solve the following linear programming problem.

$$\text{Maximize} \quad 3x + 5y$$

subject to the constraints

$$2x + y \le 30$$
$$x + y \le 20$$
$$x + 2y \le 36.$$

3. Consider the following linear programming problem.

$$\text{Maximize} \quad 2x + 3y$$

subject to the constraints

$$2x + 7y \ge 14$$
$$x + y \ge 5$$
$$8x + 3y \ge 24.$$

Show that this problem does not have a solution, in the sense that the objective function $2x + 3y$ can be made as large as we please by suitable choice of x and y satisfying the constraints. A problem of this type is said to be *unbounded*.

16.3 CONVEX SETS AGAIN

The graphical method discussed in section 16.2 can be used whenever there are only two variables in the problem. If there are three variables in a linear programming problem we could, in theory, make a three-dimensional model in order to solve it. Instead of lines we would have planes corresponding to each constraint, and the feasible solutions would be given by those points which were on the "correct" side of each of the planes. These would form a certain region of space, and so on. Nevertheless, as was remarked at the beginning of Chapter 6, three-dimensional models are tiresome to make and are inconvenient if depicted in perspective on a plane. Thus the graphical method is not really practical when there are three variables and, of course, if there are more than three variables the construction of a model which would enable one to "see" intuitively the solution of a linear programming problem is out of the question.

Despite this, the discussion of the graphical method is worthwhile since it enables one to understand certain important features that are common to all linear programming problems and to form some sort of mental picture of what is going on, by analogy with the two- or the three-dimensional problems. We shall use the two-dimensional problem that we have just solved to illustrate some of these important features.

We note first that the shaded area in Fig. 16.3 is a convex set of points. This is true in general, as we saw on page 260. We next note that there will, in general, be just one solution to a problem, i.e., just one optimal point. For the line that represents the value of the objective function, like line c in Fig. 16.3, when moved parallel to itself, will usually meet the shaded area at a corner—a single point. It could happen, however, that one edge of the shaded area is parallel to the line, and that the line meets the shaded area first by coinciding with this edge. In that case every point of the edge would have an equal right to be considered an optimal solution; there would then be a set of solutions to the problem and not just one. This is illustrated by the segment PQ in Fig. 16.4.

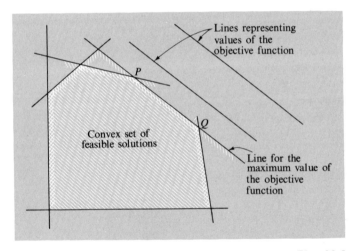

Fig. 16.4

Another thing worth noting is that the given constraints may not all contribute effectively to the definition of the convex set of points representing feasible solutions. This is the case with the constraint

$$x + y \leq 9$$

in our worked example (section 16.2). The convex set defined by the constraints would be exactly the same if this constraint were removed. In Fig. 16.2 we see that the line representing the constraint lies outside the convex set defined by the other constraints, and the feasible solutions automatically satisfy this constraint simply by virtue of the fact that they satisfy the others. In short, this constraint does not really "constrain" anything. It is redundant, and could be left out. Redundant constraints often appear in linear programming problems and, if spotted, they can be deleted. On the other hand, they may not be easily detected, and they give no trouble if allowed to remain (other than increasing the size of the problem).

From what has been said so far it might be thought that any linear programming problem has feasible solutions from which the desired optimal solution must be chosen. But this is not the case. It can easily happen that no feasible solutions exist; that, for example, one constraint is automatically *not* satisfied by the points that satisfy the others. (This is just the opposite of what happens with a redundant constraint.) In short, the constraints may be *inconsistent*. This is illustrated by the system of constraints

$$4x + y \leq 8$$
$$x + 2y \leq 6$$
$$6x + 5y \geq 30$$

for which a diagram is given in Fig. 16.5. It is easily seen from this diagram that there can be no values of x and y which satisfy all three of these constraints. For the first two (together with $x \geq 0$, $y \geq 0$) define the interior and boundary of the quadrilateral $OABC$, and no point of this region lies on the far side of the line PQ, as required by the third constraint.

Again there is no very easy way of checking, in more complicated problems, whether the constraints are consistent or not, but we shall not worry

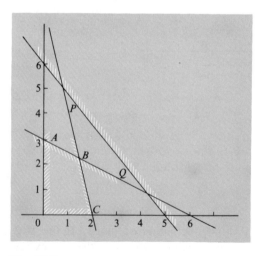

Fig. 16.5

about this either. It will be seen later that the technique is simply to press on as if we were confident that there were feasible solutions. If there are not, the mathematical procedure that we shall be using will ultimately tell us.

Our diagrammatical method requires us to "move" the line "objective function = constant" parallel to itself until it just touches the convex area of feasible solutions. This is all very well in the two-dimensional case, but another procedure is required when there are more variables. Let us therefore interpret this procedure in a different way.

One thing is certain—a solution of the problem will always be a point on the boundary of the convex region, since we could clearly improve on any point inside the region. If the point X in Fig. 16.6 is inside the region, we can

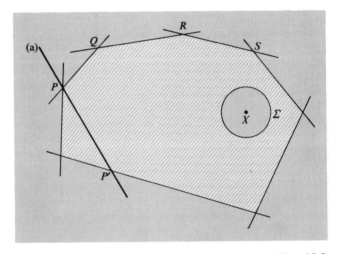

Fig. 16.6

surround it by a small circle Σ, as shown, also lying entirely in the region, and some points of Σ will be "better" than X, since they are further from the origin. We can therefore confine our investigation to the boundary, which, in the two-dimensional case, will be a polygon. A line such as a in Fig. 16.6 meets this polygon in two points P and P'. Concentrate on one of them, say P. As we move the line a away from the origin in search of the maximum value of the objective function, the point P moves around the polygon, going from vertex to vertex, from P to Q, R, S and so on.

At each vertex one of three things can happen; the next vertex, i.e., not the one just passed, may correspond to a smaller value of the objective function, or the objective function may be the same there, or it may have a larger value. In the first case the maximum value of the objective function has been found. In the second case any point of the line joining the present

vertex to the next is an optimum point. In the third case there is further possibility for improvement in the value of the objective function. If either of
the first two conditions holds, the problem is solved. Otherwise we go to the
next vertex and try again.

We cannot get the third possibility indefinitely, since there is only a
finite number of vertices and, since the value of the objective function increases every time, we can never get back to any vertex that we have visited
before. Hence this procedure must end some time.

To use this method in a two-dimensional problem would be a waste
of time, since the answer can be "seen" so much more easily. The reason for
describing this step-by-step procedure is that this description gives us the
essential details of a method, the *simplex method*, used to solve linear programming problems in any number of variables. We shall describe this
method in the next section and use it to solve the second of the problems
given at the beginning of this chapter.

Exercise 192

1. Show that in the linear programming problem

$$\text{Minimize} \quad 3x + 4y$$

subject to the constraints

$$x + 2y \geq 6$$
$$2x + 3y \geq 9$$
$$3x - y \geq 7$$
$$x - 2y \geq 1$$
$$2x + 4y \geq 11$$

three of the constraints are redundant. Solve this problem.

2. Show that there are no solutions to the linear programming problem

$$\text{Maximize} \quad 4x + y$$

subject to the constraints

$$3x + 5y \leq 15$$
$$5x + 3y \leq 15$$
$$-x + y \leq 2$$
$$4x + 5y \geq 20.$$

Note: Such a problem is said to be *infeasible*. Note that infeasibility is a property of the set of constraints alone; the objective function makes no difference.

3. Show that the linear programming problem

$$\text{Minimize} \quad 15x + 6y$$

subject to the constraints

$$5x + 2y \geq 10$$
$$-x + y \leq 7$$
$$x - y \leq 4$$
$$4x + 3y \leq 60$$
$$3x + 4y \geq 12$$

has many solutions, and find them.

16.4 THE SIMPLEX METHOD

The general linear programming problem is to maximize (or minimize) the value of a given linear function of, say, n variables (the objective function) subject to the restrictions that the variables are non-negative and that they satisfy a set of, say, k linear inequalities (the constraints) of the form

$$\mathbf{Ax} \leq \mathbf{b}$$

in the notation used before (page 795). Let us interpret the problem geometrically.

An equation of the form

$$a_{i1}x_1 + a_{i2}x_2 + \cdots + a_{in}x_n = b_i$$

is the equation of a hyperplane in space of n dimensions, and will divide the space into two halves. The inequality

$$a_{i1}x_1 + a_{i2}x_2 + \cdots + a_{in}x_n \leq b_i$$

then expresses the fact that the point

$$\mathbf{x} = (x_1, x_2, \ldots, x_n)$$

lies in one of these halves. In this way the set of constraints specifies a region

P of the space in which (or on the boundary of which) any point corresponding to a solution of the problem must lie. We know from section 7.8 (page 260) that P is convex. It will be bounded by portions of hyperplanes, just as in the two-dimensional case it was bounded by portions of lines forming a polygon.

This P is an n-dimensional convex polytope in the sense defined in Chapter 7. At each vertex of this polytope n hyperplanes will meet, i.e., for the coordinates of a vertex n of the inequalities become equalities. In exceptional circumstances it may happen that more than n hyperplanes meet at a vertex, just as in two dimensions it was possible for more than two lines to pass through a given vertex. We shall ignore this possibility for the time being.*

Points at which $n - 1$ inequalities become equalities will lie on a line— an *edge* of the polytope (or rather, a line of which the edge of the polytope is a portion). Similarly, by taking the equal sign in $n - 2$ of the constraints we determine a plane which will contain a *face* of the polytope.

We might remark here that the solution of the linear programming problem is what is known as a *finite problem*, that is, one that could (at least in theory) be solved by sheer brute force, by investigating all of a finite number of possibilities, and seeing which one was the correct solution. The argument would be as follows.

In general, the solution of a linear programming problem is given by a vertex of the polytope, i.e., by a point at which exactly n of the constraints become equalities. There are, properly speaking, $k + n$ constraints altogether (since we must include the n inequalities $x_1 \geq 0$, $x_2 \geq 0$, etc.), and therefore only a finite number $\binom{k + n}{n}$ of points at which n hyperplanes intersect. We can find these by choosing all combinations of hyperplanes n at a time. Suppose we do this, and for each such set solve the set of n equations in n unknowns of which it consists. We shall then have found the coordinates of all points that *might* be vertices of the polytope. Not all of them will be vertices, however, since the coordinates of many of them will not satisfy the remaining inequalities, and the points will therefore be outside the region of feasible solutions. We therefore test each of these points to see if it satisfies all the constraints and we reject those that do not. When we have done this we shall then have found the coordinates of all the vertices of the polytope. We now have merely to calculate the value of the objective function at each vertex and pick that vertex for which it is a maximum (or minimum, as the case may be). This solves the problem. Note that this argument still applies if we have not one but many solutions. All that will happen is that several vertices will give the same value to the objective function, and in this case, any point of the edge, face, etc., of the polytope determined by the 2, 3, etc., vertices will be a solution.

This method is valid in theory but is not of any use in practice because of

* See Exercise 200.6 and 200.7.

the inordinate amount of work it would entail. It is not that there are so many vertices of the polytope (there is not much that we can do about that anyway) but that the majority of the points calculated by this method would have to be rejected since they would not satisfy all the constraints. In other words, the set of vertices of the polytope is only a small subset of the set of all intersections of sets of n hyperplanes.

What is needed is a method that deals only with the vertices of the convex polytope—a method that requires no information concerning any other intersections of the hyperplanes. The simplex method, which we now consider, does just this, and the procedure is much the same as we described before for the two-dimensional case. We start with one vertex of the polytope and consider the edges which lead from that vertex to other vertices. If none of these other vertices is "better" than the one being considered, that is, gives a larger value to the objective function we are trying to maximize, then our search is over and we have the solution. Otherwise we go on to one of the better verties, and repeat the procdure, Since there is only a finite number of vertices, and we cannot arrive at any vertex more than once (since the value of the objective function is increasing all the time) this procedure must eventually end in the solution to the problem.

One difficulty that can arise is that in a certain type of problem (the so-called *degenerate case*) the method may not take us to a better vertex but merely to another set of n hyperplanes intersecting in the *same* vertex. There is then the possibility that the process might not terminate, but "cycle," i.e., go round and round the same succession of sets of hyperplanes. Although this is theoretically possible it never seems to arise in practice.*

We shall first illustrate the simplex method by solving the second of the problems given at the beginning of this chapter. Then we shall consider the general case and devise an algorithm† for its solution. Certain queries and tricky points will arise as we go along. Some of these we shall deal with as they arise; discussion of others will be postponed until later.

In problem 2 we had the following set of constraints

(16.4.1)
$$\begin{cases} 3x + 6y + 4z \leq 48 \\ 11x + 6y + 12z \leq 136 \\ 3x + 10y + 6z \leq 70 \\ 7x + 14y + 4z \leq 55 \\ 5x + 4y + 5z \leq 50 \end{cases}$$

in which we have removed a factor from the last inequality. It will be convenient to convert these inequalities into equations, which we can do by intro-

---

* Well, hardly ever! See Exercise 200.6.
† An algorithm is a set of rules for the solution of a particular type of problem.

ducing extra variables which will "take up the slack" between the left- and right-hand sides in (16.4.1). These variables, known as *slack variables*, are introduced in such a way that they are non-negative. Thus we replace (16.4.1) by

(16.4.2)
$$\begin{cases} 3x + 6y + 4z + s & = 48 \\ 11x + 6y + 12z & + t & = 136 \\ 3x + 10y + 6z & + u & = 70 \\ 7x + 14y + 4z & + v & = 55 \\ 5x + 4y + 5z & + w = 50. \end{cases}$$

Of course, we have not really eliminated the inequalities from the system by this means. In order that the inequalities in (16.4.2) should be the right way round, we have to stipulate that the newly-introduced slack variables are all non-negative. Thus we have converted our inequalities to equations at the expense of adding five new inequalities to the system, viz.,

$$s \geq 0, \quad t \geq 0, \quad u \geq 0, \quad v \geq 0, \quad w \geq 0.$$

However, the inequalities in the system now are all of the same kind, stating that the variables (both the original, or *main variables*, and the slack variables) are non-negative. Thus we now have a set of 5 equations in 8 variables, and 8 inequalities (stating that the variables are all non-negative).

Now we need somewhere to start—a feasible solution. In point of fact we want more than that; we want what is called a *basic feasible solution*, i.e., one in which exactly three of the variables are zero, the remaining five being nonzero. If the three zero variables are all slack variables then three of the inequalities are satisfied with the $=$ sign, and the point in question is a vertex of the polytope. If one of the main variables, say y, is zero then the inequality $y \geq 0$ becomes an equality, and we still have a vertex of the polytope, this time one lying on one of the coordinate planes. Thus a basic feasible solution corresponds to a vertex of the polytope. The variables that take nonzero values at the vertex are called *basic variables;* the others are *nonbasic*. In the general case, where we have n main variables and k inequalities, a basic solution will be determined by the fact that n of the inequalities are satisfied with equality.

The first thing to do, then, is to find a basic feasible solution to the equations and inequalities. There are ways and means of doing this which, by rights, we ought to discuss before going any further; but we shall not do this. Instead we shall postpone this question, and for the moment assume that somehow, whether by low cunning or divine inspiration, we have found a basic feasible solution. Actually, in our example, this is very easily accomplished; we make the slack variables basic and the main variables nonbasic, and allocate the following values:

Nonbasic			Basic				
x	y	z	s	t	u	v	w
0	0	0	48	136	70	55	50

It is clear that, with these values, all the equations are satisfied and all the variables are non-negative. However, the value of the objective function

$$6x + y + 5z$$

is 0, which is not the maximum value; we can certainly do better than that! At each stage during the solution of the problem by the simplex method we shall do two things. We express the basic variables in terms of the nonbasic variables, and we see if we can get any improvement in the objective function by making one of the nonbasic variables into a basic variable. In the present example, if we express the basic variables in terms of the nonbasic variables we obtain

$$s = 48 - 3x - 6y - 4z$$
$$t = 136 - 11x - 6y - 12z$$
(16.4.3)
$$u = 70 - 3x - 10y - 6z$$
$$v = 55 - 7x - 14y - 4z$$
$$w = 50 - 5x - 4y - 5z.$$

We now look at the objective function and see if we can do anything to improve it. The variables x, y and z are all zero, and must be non-negative, so that *any* change in these variables will increase the objective function, and will therefore be to the good. Let us suppose that we increase the value of z, making suitable adjustments to the values of the other variables. Then we shall no longer have a basic feasible solution unless the value we give to z is such that some other variable, previously nonzero, becomes zero. Speaking geometrically, we are at a certain vertex of the polytope, a vertex at which three planes meet, one of these being the plane $z = 0$. If we alter the value of z we obtain points of the line determine by the other *two* planes. If z is increased until another variable becomes zero, this means that we have progressed along the line just enough to reach another plane. In short, we have travelled along an edge of the polytope and have reached another vertex.

In this way we can go from one vertex to another, always in such a way as to improve the objective function, insofar as this is possible. When it is no longer possible the required solution will, in general, have been reached.

At each stage one nonbasic variable becomes basic, and a basic variable becomes nonbasic; we choose the nonbasic variable in such a way that the change causes an improvement in the objective function. It may well turn out that we have a choice of which nonbasic variable to increase, as in our example where increases in x, y or z will all improve the objective function. (Geometrically, there may be more than one edge from the present vertex to a "better" vertex.) If so, it usually matters little in the long run which one we take, and z will do as well as any here.

By how much can we increase z? Remembering that x and y are not going to alter and will continue to be zero, we see that s will become negative if $48 - 4z < 0$, i.e., if $z > 12$. Similarly, t, u, v and w will be negative if z exceeds $\frac{136}{12}$, $\frac{70}{6}$, $\frac{55}{4}$ and $\frac{50}{5}$, respectively. Hence the new value of z must not exceed any of these numbers. We can however make it equal to the smallest, viz., $\frac{50}{5} = 10$, without breaking any of the rules. Thus if we put $z = 10$, w will become zero, and the other basic variables will still be positive. This is exactly what we wanted; z becomes the new basic variable, changing places with w, which becomes nonbasic.

The next thing is to express the basic variables and the objective function in terms of the nonbasic variables, as before. This is easy. We turn round the equation for w in terms of x, y, z so as to get an equation for z in terms of x, y, w.

Thus from

$$w = 50 - 5x - 4y - 5z$$

we get

(16.4.4)
$$z = 10 - x - \frac{4}{5}y - \frac{1}{5}w.$$

We now use equation (16.4.4) to eliminate z from the other equations in (16.4.3), and we obtain

$$s = 8 + x - \frac{14}{5}y + \frac{4}{5}w$$

$$t = 16 + x + \frac{18}{5}y + \frac{12}{5}w$$

(16.4.5)
$$u = 10 + 3x - \frac{26}{5}y + \frac{6}{5}w$$

$$v = 15 - 3x - \frac{54}{5}y + \frac{4}{5}w$$

$$z = 10 - x - \frac{4}{5}y - \frac{1}{5}w$$

$$C = 50 + x - 3y - w.$$

We have now arrived at the vertex $(0, 0, 10)$ and the value of the objective function has increased to 50.

Now that the pattern of what has to be done at each step has been set up, we can carry on more quickly with our solution. Looking at the new expression for the objective function C we see that any increase in y or w will reduce it. Since these variables cannot be decreased (being already zero) no advantage can be gained by altering them. The only way to increase C is to increase x.

By how much can we increase x? Only as much as will make one of the present basic variables become zero. Looking at (16.4.5) we see that only v and z will eventually become zero if we increase x (since in the other equation x has a positive coefficient); and v becomes zero if $x = 5$ while z becomes zero if $x = 10$. Hence we can increase x to 5, thereby reducing v to zero while z remains positive. Thus x becomes the new basic variable and v takes its place as a nonbasic variable.

The reader should verify that if we now express the basic variables in terms of the nonbasic variables we get the following equations:

$$s = 13 - \frac{1}{3}v - \frac{32}{5}y + \frac{16}{5}w$$

$$t = 21 - \frac{1}{3}v + \frac{8}{3}w$$

(16.4.6) $\qquad u = 25 - v - 16y + 2w$

$$x = 5 - \frac{1}{3}v - \frac{18}{5}y + \frac{4}{15}w$$

$$z = 5 - \frac{1}{3}v + \frac{14}{5}y - \frac{7}{15}w$$

$$C = 55 - \frac{1}{3}v - \frac{33}{5}y - \frac{19}{15}w.$$

Remembering that all nonbasic variables have value zero, we see that we have now reached the vertex with coordinate $(x, y, z) = (5, 0, 5)$, and the value of the objective function C is 55.

If we now try to repeat the procedure used up to now, we find that we cannot. Any change in the variables v, y and w can now only *decrease* C, since those variables have negative coefficients in the expression for C. In geometrical terms, we have reached a vertex of the polytope from which every edge leads in a direction which decreases the value of the objective function. Thus we have arrived at the optimum point, and the problem is solved.

We will do well to make quite sure that we are not mistaken about this. Let M be the vertex we have now reached, so that along any edge from

M (in our case there are three of them) the objective function decreases. Then for any point P in the convex polytope, and near M, the objective function is less than it is at M; for its value at P is a convex combination of its values at points Q, R and S on the edges from M, and coplanar with P, and these values are all less than the value at M. Hence, certainly, M is a *local* maximum point. What we must show is that it is also the *overall* maximum, and this will certainly follow if we can show that M is the *only* maximum.

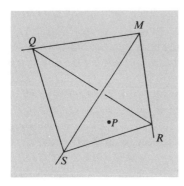

Fig. 16.7

That M is in general the only maximum follows from the convexity of the polytope. For if N is another local maximum, then by the definition of "convex" every point on the line MN lies in the polytope. Now the coordinates of the points of a line can be expressed as linear functions of a parameter t (see page 252), and by substituting these coordinates in the objective function C we obtain a linear expression for C in terms of t, say

$$C = a + bt,$$

for points of the line MN. We now see that whatever the signs of a and b may be, C either increases steadily from M to N or else decreases steadily. In either case one of the points M, N cannot be a maximum point. Hence the supposition of two maximum points leads to a contradiction, and we deduce that there can be only one.

The one possible exception to the above conclusion occurs if it turns out that $b = 0$. In this case M and N are maxima of equal standing, so to speak. It can therefore happen that there are other points at which the objective function is *as large as* it is at M, but in this case at every point of the edge joining the two points (or the triangle formed by three points, if there are three, and so on) the objective function has the same value. We thus have a whole set of points, all equally good.

Hence we see that the simplex method will lead us to the unique optimum point, if there is one. Failing that, it will lead us to one of a set of points, all of which are solutions to the problem. For many purposes this is just as good.

Since we have just found the solution to our second problem, expressed in mathematical form, we might as well interpret this solution in terms of the original formulation of the problem. It means that the best plan for the manufacturer is to produce 5 bottles each of Medicine No. 1 and Medicine No. 3, and none of Medicine No. 2. The amounts, in grams, of drugs used per day to make the medicines are shown in the following table.

Drugs	A	B	C	D	E
5 bottles of No. 1	15	55	15	35	75
5 bottles of No. 3	20	60	30	20	75
Totals	35	115	45	55	150

Notice that the amounts required of drugs A, B and C are less than the amounts available. This is only to be expected. The amounts by which they fall short are 13, 21 and 25 grams respectively, and these are the amounts of slack given by the slack variables, s, t and u. These also can be read off from the final equations, as were the values of x and z. For drugs D and E there is no slack; the corresponding slack variables v and w are therefore zero.

Exercise 193

1. Use the simplex method to solve the linear programming problem that was described verbally in Exercise 190.1.

2. Solve the following linear programming problem.

$$\text{Maximize} \quad 100x + 200y + 50z$$

subject to the constraints

$$5x + 5y + 10z \leq 1000$$
$$10x + 8y + 5z \leq 2000$$
$$10x + 5y \quad\quad \leq 500.$$

3. Find the solutions in non-negative numbers of the constraints

$$3x + 4y - z + 2t \leq 10$$
$$8x + 2y + 5z - t \leq 30$$
$$2x + y + z - 3t = 8$$

which (a) minimize (b) maximize the function $2x + 3y - z + t$.

Note: When one of the constraints is an equation, no slack variable is needed for that restraint.

Hint: To find a basic feasible solution, try putting $x = t = 0$ and substituting for y from the third constraint.

4. The unbounded problem given in Exercise 191.3 was to maximize $2x + 3y$ subject to the constraints $2x + 7y \geq 14$, $x + y \geq 5$, $8x + 3y \geq 24$. Starting with the feasible solution $x = \frac{21}{5}$, $y = \frac{4}{5}$ attempt to solve this problem by the simplex method. Where does the method break down, thus enabling you to deduce that the problem is unbounded?

16.5 THE SIMPLEX ALGORITHM

In this section and the next we shall consider ways of carrying out the simplex algorithm, informally described in the last section, in a concise and efficient manner. We should note at this stage that it is unlikely that the reader will ever need to solve a linear programming problem of any consequence by hand. The linear programming problems that occur in real life usually contain a large number, hundreds possibly, of inequalities, and are far beyond the range of convenient hand computation. These problems are nowadays entrusted to electronic computers, and every self-respecting computing center has programs for handling linear programming problems.

It can be argued, therefore, that there is no need to be acquainted with the more intimate details of the simplex algorithm; that it is sufficient to have a general idea of what the algorithm does, and to know that having formulated a linear programming problem in mathematical terms, one can turn it over to the nearest computing center, and wait for the answer to come back. This is all perfectly true; but it is also true that the working of a few concrete examples can give one an insight into a mathematical procedure—a "feel" for it, so to speak—which no amount of theoretical discussion is likely to give. Moreover, in the context of this book, we really need to know rather more about the simplex algorithm if we are to follow the proof of the fundamental theorem of linear programming, the *Duality Theorem*, which is discussed in section 16.9.

Therefore if the reader is quite happy to accept that linear programming problems have solutions (usually), and that there are ways and means of finding them, but is not greatly concerned with how it is done, he can omit this and the next section. He will also have to omit section 16.8 and the proof of the Duality Theorem in 16.9. He should not omit the other sections, including the rest of 16.9.

For the reader who wants to have a stab at understanding the inner workings of the simplex algorithm, we shall start by considering again the problem solved in the last section, namely to maximize $6x + y + 5z$ subject to the constraints

$$3x + 6y + 4z \leq 48$$
$$11x + 6y + 12z \leq 136$$
$$3x + 10y + 6z \leq 70$$
$$7x + 14y + 4z \leq 55$$
$$5x + 4y + 5z \leq 50.$$

In solving this problem we interspersed the various mathematical steps with a good deal of explanation which does not need to be repeated every time we tackle a linear programming problem. It is therefore useful to reduce the simplex method to an algorithm, that is, a set of rules which can be followed automatically, so that one does not need to explain (or, ideally, even think about) what is being done at each stage of the solution. We shall therefore run through the problem, deriving the required algorithm, and see how to set out the work in such a way that we do not have to write down any more than is necessary.

Our starting point will be a set of equations (16.4.2). Instead of writing down these equations in full, as we did before, we shall simply display the coefficients and constraints in the form of a tableau, as in (16.5.1).

(16.5.1)

Constant	x	y	z	s	t	u	v	w
48	3	6	4	1	0	0	0	0
136	11	6	12	0	1	0	0	0
70	3	10	6	0	0	1	0	0
55	7	14	4	0	0	0	1	0
50	5	4	5	0	0	0	0	1
C 0	6	1	5	0	0	0	0	0

The order of the columns is: constants, main variables, slack variables.

We next need a basic feasible solution to start with. As before, we obtain one by equating each of the slack variables with the corresponding constant. Thus these slack variables become the basic variables, and the main variables become the nonbasic variables. We can write these values above the tableau, thus obtaining (16.5.2).

(16.5.2)

	0	0	0	48	136	70	96	50
Constant	x	y	z	s	t	u	v	w
48	3	6	4	1	0	0	0	0
136	11	6	12	0	1	0	0	0
70	3	10	6	0	0	1	0	0
55	7	14	4	0	0	0	1	0
50	5	4	⑤	0	0	1	0	1
C 0	6	1	5	0	0	0	0	0

Having completed these preliminaries we can now start to apply the method.

Step 1.   We first look for a nonbasic variable whose increase (from zero) would result in an increase of the objective function. In other words we want a nonbasic variable with positive coefficient in C. Here we have the choice of any of them; let us choose z.

Step 2.   We must now see by how much we can increase z. If in each row of (16.5.2) we divide the constant by the entry under z we obtain the numbers

$$12, \ 11\tfrac{1}{3}, \ 11\tfrac{2}{3}, \ 13\tfrac{3}{4}, \ 10$$

which are the amounts by which we can increase z without making the corresponding basic variable become negative. If none of the basic variables is to become negative we must pick out the smallest of these numbers, viz., 10. This will then be the new value of z. In the tableau we circle the corresponding entry in the column under z, as shown in (16.5.2). We call this entry the *pivot*.

The new nonbasic variable will be w and we must again express the basic variables in terms of the nonbasic variables which are now x, y and w.

Step 3.   To express z in this way, we have to go from the equation

(16.5.3) $50 = 5x + 4y + 5z \qquad + w$

to

(16.5.4) $10 = x + \dfrac{4}{5}y + z \qquad + \dfrac{1}{5}w.$

Hence, in the tableau, we merely divide the row in question, the *pivot row*, by the coefficient of z.

To express the other variables in terms of the new nonbasic variables we must eliminate z between (16.5.4) and the equation for the variable. Thus from (16.5.4) and

(16.5.5) $48 = \ 3x + \ 6y + 4z + s$

we obtain

(16.5.6) $8 = -x + \dfrac{14}{5}y \qquad + s \qquad - \dfrac{4}{5}w$

which is obtained by subtracting 4 times (16.5.4) from (16.5.5). Thus, in the tableau, for each nonpivot row, we subtract a multiple of the new pivot row in such a way as to make the coefficient of z in the nonpivot row zero. If we carry out this procedure with each row of (16.5.2) we obtain (16.5.7).

(16.5.7)

Constant	x	y	z	s	t	u	v	w
8	-1	$\frac{14}{5}$	0	1	0	0	0	$-\frac{4}{5}$
16	-1	$-\frac{18}{5}$	0	0	1	0	0	$-\frac{12}{5}$
10	-3	$\frac{26}{5}$	0	0	0	1	0	$-\frac{6}{5}$
15	3	$\frac{54}{5}$	0	0	0	0	1	$-\frac{4}{5}$
10	1	$\frac{4}{5}$	1	0	0	0	0	$\frac{1}{5}$
$C-50$	1	-3	0	0	0	0	0	-1

Note that the z column now has only a single "1" in it, just as have the other columns for basic variables, while the w column is full of nonzero elements, like those for the other nonbasic variables.

Step 4. This step is not really necessary, but it emphasizes the fact that we have now arrived at a situation similar to that with which we started. We interchange the columns for z and w, thus putting all the nonbasic variables to the left of the basic variables, as at first. We thus obtain (16.5.8).

(16.5.8)

Constant	x	y	w	s	t	u	v	z
8	-1	$\frac{14}{5}$	$-\frac{4}{5}$	1	0	0	0	0
16	-1	$-\frac{18}{5}$	$-\frac{12}{5}$	0	1	0	0	0
10	-3	$\frac{26}{5}$	$-\frac{6}{5}$	0	0	1	0	0
15	③	$\frac{54}{5}$	$-\frac{4}{5}$	0	0	0	1	0
10	1	$\frac{4}{5}$	$\frac{1}{5}$	0	0	0	0	1
$C-50$	1	-3	-1	0	0	0	0	0

Note that the entry in the C row under "Constant" is -50. This number is minus the new value of C (the old one was zero). The reason for the minus sign is that the objective function is now

$$C = x - 3y - w + 50$$

in which the variables and constant are on the same side of the equation. In the rows above, however, the constant is assumed to be on the opposite side of the equation from the variables. We could have written equations (16.4.2), etc., with the constants on the same side as the variables, in which case this anomalous sign would not have appeared; but it is not usual to do this, and having to change the sign of the constant in C is at most a very minor annoyance!

Note also that the elements in the column for the new nonbasic variable are those of the old nonbasic variable multiplied by minus the reciprocal of the pivot entry; except for the element in the pivot row, which is the reciprocal of the pivot entry.

We now repeat the above four steps.

Step 1.   There is now no choice. The only candidate for the new basic variable is x, since this is the only one with positive coefficient in C. Hence the x column is the new pivot column.

Step 2.   The limits on the increase in x for the rows are now just 5 and 10, for the last two rows. Any row with a negative coefficient in the pivot column can be ignored, since an increase in x will *increase* the basic variable for that row. Hence we ignore rows 1, 2 and 3, and make $x = 5$. The new nonbasic variable is thus v, and the pivot entry is the one circled in (16.5.8).

Step 3.   Dividing the pivot row by the pivot gives

$$5 \quad 1 \quad \frac{18}{5} \quad -\frac{4}{15} \quad 0 \quad 0 \quad 0 \quad \frac{1}{3} \quad 0$$

and by subtracting suitable multiples of this from the other rows we obtain (16.5.9).

(16.5.9)

Constant	x	y	w	s	t	u	v	z
13	0	$\frac{32}{5}$	$-\frac{16}{15}$	1	0	0	$\frac{1}{3}$	0
21	0	0	$-\frac{8}{3}$	0	1	0	$\frac{1}{3}$	0
25	0	16	-2	0	0	1	1	0
5	1	$\frac{18}{5}$	$-\frac{4}{15}$	0	0	0	$\frac{1}{3}$	0
5	0	$-\frac{14}{5}$	$\frac{7}{15}$	0	0	0	$-\frac{1}{3}$	1
$C - 55$	0	$-\frac{33}{5}$	$-\frac{11}{15}$	0	0	0	$-\frac{1}{3}$	0

Step 4. Interchanging the columns for x and v we have (16.5.10)

(16.5.10)

Constant	v	y	w	s	t	u	x	z
13	$\frac{1}{3}$	$\frac{32}{5}$	$-\frac{16}{15}$	1	0	0	0	0
21	$\frac{1}{3}$	0	$-\frac{8}{3}$	0	1	0	0	0
25	1	16	-2	0	0	1	0	0
5	$\frac{1}{3}$	$\frac{18}{5}$	$-\frac{4}{15}$	0	0	0	1	0
5	$-\frac{1}{3}$	$-\frac{14}{15}$	$\frac{7}{15}$	0	0	0	0	1
$C-55$	$-\frac{1}{3}$	$-\frac{33}{5}$	$-\frac{11}{15}$	0	0	0	0	0

Hence the objective function now has the value 55.

Since all the coefficients in C are now negative, no change in any of the nonbasic variables will increase C, and we have come to the end of the computation. The values which give the optimum value 55 to C are

$$x = 5, \qquad y = 0, \qquad z = 5.$$

For y is a nonbasic variable, and hence is zero, and we can read off the values of x and z from the fourth and fifth equations, remembering that v, y and w are zero. The arrows in tableau (16.5.10) show how this is done.

Exercise 194

1.   Minimize the function $2x + y + 4z$

subject to the constraints

$$x + 2y - z \leq 3$$
$$-2x + 3y + z = 5.$$

2.   Solve the problem of maximizing

$$3x + 2y + 3z + 4u + v$$

subject to the constraints

$$3x + 2y + z + 2u + 5v \leq 20$$
$$2x + 3y + z + 3u + v \leq 14$$
$$2x + 4y + z + 6u + v \leq 25.$$

3.   Maximize $9x + 2y + 5z$

subject to the constraints

$$2x + 3y - 5z \leq 12$$
$$2x - y + 3z \leq 3$$
$$3x + y - 2z \leq 2.$$

16.6 FURTHER EXAMPLES AND TECHNIQUES

By way of further illustration of the simplex method we shall solve problem 3, which was to minimize

$$C = 8x + 3y + 5z$$

subject to the constraints

(16.6.1)
$$29x + 2y + 13z \geq 194$$
$$14x + 8y + 10z \geq 152$$
$$11x + 2y + 7z \geq 98$$
$$4x + y + 2z \geq 34.$$

Although there are some slight differences in the method required for this problem (resulting from the fact that the inequality signs are the other way round, and the objective function is to be minimized instead of maximized) the general method is the same as before.

First we must find a basic feasible solution. We cannot do this the same way as before, by making all the main variables zero, but we can instead make one of the main variables (say z) large enough to satisfy the inequalities of (16.6.1) with zero values for the other main variables (x and y). Furthermore we can choose the value of z just big enough to make the left-hand side of one of the inequalities equal to the right-hand side. The reader can readily verify that the values

$$x = 0, \quad y = 0, \quad z = 17$$

will do the trick. The last inequality of (16.6.1) then becomes an equation. Introducing slack variables as before we obtain the equations

(16.6.2)

$$
\begin{aligned}
29x + 2y + 13z - t \qquad\qquad &= 194 \\
14x + 8y + 10z \quad - u \qquad\qquad &= 152 \\
11x + 2y + 7z \qquad\quad - v \quad &= 98 \\
4x + y + 2z \qquad\qquad - w &= 34
\end{aligned}
$$

which have the feasible solution

$$
\begin{array}{ccccccc}
x & y & z & t & u & v & w \\
0 & 0 & 17 & 27 & 18 & 21 & 0.
\end{array}
$$

Hence z, t, u and v are the basic variables; x, y and w are the nonbasic variables. To start the ball rolling we must express the basic variables in terms of the nonbasic ones, which calls for a little bit of manipulation of the equations. We find that z, t, u and v are given by

(16.6.3)

$$
\begin{aligned}
17 &= \quad 2x + \frac{1}{2}y + z \qquad\qquad\qquad - \frac{1}{2}w \\
27 &= -3x + \frac{9}{2}y \qquad + t \qquad\qquad - \frac{13}{2}w \\
18 &= \quad 6x - 3y \qquad\qquad + u \quad\; - 5w \\
21 &= \quad 3x + \frac{3}{2}y \qquad\qquad\qquad + v - \frac{7}{2}w \\[2mm]
C &= 85 - 2x + \frac{1}{2}y \qquad\qquad\qquad + \frac{5}{2}w
\end{aligned}
$$

Note: If the reader wishes, he can skip the derivation of equations (16.6.3), for it will turn out later that even the elementary algebra required for the transition from (16.6.2) to (16.6.3) can be dispensed with.

We now write equations (16.6.3) in the form of a tableau as in (16.6.4).

(16.6.4)

Constant	x	y	z	t	u	v	w
17	2	$\frac{1}{2}$	1	0	0	0	$-\frac{1}{2}$
27	-3	$\frac{9}{2}$	0	1	0	0	$-\frac{13}{2}$
18	⑥	-3	0	0	1	0	-5
21	3	$\frac{3}{2}$	0	0	0	1	$-\frac{7}{2}$
$C\ -85$	-2	$\frac{1}{2}$	0	0	0	0	$\frac{5}{2}$

Note that we have written -85 instead of 85 in the C row. This is because, in conformity with the last problem, we shall assume that the constant is on the other side of the equation.

We are now ready to start the algorithm. Since we wish to minimize C we must choose, as a variable to be increased, a variable that has *negative* coefficient in C. There is only one, x, so we now see by how much we can increase it.

As before we can ignore the negative entries in the x column. The positive entries we divide into the corresponding constant terms and take the smallest quotient. The smallest of $\frac{17}{2}$, $\frac{18}{6}$ and $\frac{21}{3}$ is the middle one, and hence the pivot element will be the one circled in (16.6.4).

Now we perform the row manipulations necessary to make the pivot become 1, and to make the other elements in its column become zero. The new pivot row is

$$3 \quad 1 \quad -\frac{1}{2} \quad -\frac{5}{6} \quad 0 \quad 0 \quad \frac{1}{6} \quad 0$$

and by subtracting suitable multiples (2, -3, 3 and -2 to be precise) of this from the other rows we eventually end up with (16.6.5).

(16.6.5)

Constant	x	y	z	t	u	v	w
11	0	$\frac{3}{2}$	1	0	$-\frac{1}{3}$	0	$\frac{7}{6}$
36	0	3	0	1	$\frac{1}{2}$	0	-9
3	1	$-\frac{1}{2}$	0	0	$\frac{1}{6}$	0	$-\frac{5}{6}$
12	0	3	0	0	$-\frac{1}{2}$	1	-1
C -79	0	$-\frac{1}{2}$	0	0	$\frac{1}{3}$	0	$\frac{5}{6}$

On rearranging the columns we get (16.6.6)

(16.6.6)

Constant	u	y	w	z	t	x	v
11	$-\frac{1}{3}$	$\frac{3}{2}$	$\frac{7}{6}$	1	0	0	0
36	$\frac{1}{2}$	3	-9	0	1	0	0
3	$\frac{1}{6}$	$-\frac{1}{2}$	$-\frac{5}{6}$	0	0	1	0
12	$-\frac{1}{2}$	③	-1	0	0	0	1
C -79	$\frac{1}{3}$	$-\frac{1}{2}$	$\frac{5}{6}$	0	0	0	0

We again look for a negative coefficient in the C row and find the pivot element. It is the "3," circled in (16.6.6). Applying the rules once again, we obtain (16.6.7).

(16.6.7)

Constant	u	v	w	z	t	x	y
5	$-\dfrac{1}{12}$	$-\dfrac{1}{2}$	$\dfrac{5}{3}$	1	0	0	0
24	1	-1	-8	0	1	0	0
5	$\dfrac{1}{12}$	$\dfrac{1}{6}$	-1	0	0	1	0
4	$-\dfrac{1}{6}$	$\dfrac{1}{3}$	$-\dfrac{1}{3}$	0	0	0	1
$C\ -77$	$\dfrac{1}{4}$	$\dfrac{1}{6}$	$\dfrac{2}{3}$	0	0	0	0

Since there are no more negative entries in the C row, the problem is solved. The solution is

$$x = 5, \quad y = 4, \quad z = 5.$$

These values can be read off from the tableau (16.6.7). Interpreting these values in terms of the original problem, we see that the best way for Mr. Smith to buy his kits of bricks is in accordance with the following table.

	Size 0	Size 1	Size 2	Size 3
5 of Kit A	145	70	55	20
4 of Kit B	8	32	8	4
5 of Kit C	65	50	35	10
	218	152	98	34

We have 24 too many bricks of size 0. We would therefore expect that the slack variable t would have the value 24, as is indeed the case.

Now it was all very well for us to start the solution of a linear programming problem by "somehow" finding a basic feasible solution. For some problems, like those considered up to now, this is not difficult. However, it would be comforting if we could have at our disposal a sure-fire method for obtaining a basic feasible solution (if one exists) no matter what the form of the problem. Fortunately, there *is* such a method. Of course, no method will give a basic feasible solution if none exists; but if no solution exists, the method that we are about to establish will tell us so.

It is most easy to spot a basic feasible solution when the constraints are, as in problem 2, all of the form

(16.6.8) $$a_1x_1 + a_2x_2 + \cdots + a_nx_n \le b$$

where b is positive. For then, when we introduce the slack variables, the constraints become equations of the form

(16.6.9) $$a_1x_1 + a_2x_2 + \cdots + a_nx_n + t = b$$

where t is the slack variable. We can then obtain a basic feasible solution at once by making all the main variables zero ($x_1 = x_2 = \ldots = x_n = 0$) and giving to each slack variable the value of the corresponding constant ($t = b$, etc.).

It would be very pleasant if the equations were all like (16.6.9), with each slack variable having $+1$ as its coefficient. Unfortunately, if the inequality in (16.6.8) were $\geq$ then (16.6.9) would become

$$a_1x_1 + a_2x_2 + \cdots + a_nx_n - t = b$$

and the method of obtaining a basic feasible solution would no longer apply, since we cannot put $t = -b$. However, there is nothing to stop our putting into the equation another variable having a coefficient of the right sign, in addition to the slack variable. We shall in fact do this and the extra variables thus added are called *artificial variables*. We need do this only for those constraints where the inequality sign goes the "wrong" way.

An example will make clear the general method. Let us solve the problem of minimizing

$$5x_1 + 4x_2 - 7x_3$$

subject to the constraints

(16.6.10)
$$
\begin{aligned}
2x_1 + 5x_2 - 6x_3 &\leq 7 \\
3x_1 - 2x_2 + 4x_3 &\geq 4 \\
6x_1 + x_2 + x_3 &\leq 5 \\
5x_1 - 3x_2 - 4x_3 &\geq 2.
\end{aligned}
$$

Adding slack variables in the normal way, we obtain

(16.6.11)
$$
\begin{aligned}
2x_1 + 5x_2 - 6x_3 + t_1 &= 7 \\
3x_1 - 2x_2 + 4x_3 \quad - t_2 &= 4 \\
6x_1 + x_2 + x_3 \quad + t_3 &= 5 \\
5x_1 - 3x_2 - 4x_3 \quad - t_4 &= 2.
\end{aligned}
$$

In the second and fourth equations in (16.6.11) the sign of the slack variables is the wrong way round for the application of the simple method of obtaining a basic feasible solution. We therefore add extra variables in these two equations, to obtain

$$(16.6.12) \quad \begin{aligned} 2x_1 + 5x_2 - 6x_3 + t_1 &= 7 \\ 3x_1 - 2x_2 + 4x_3 \quad - t_2 \quad + u_1 &= 4 \\ 6x_1 + x_2 + x_3 \quad + t_3 &= 5 \\ 5x_1 - 3x_2 - 4x_3 \quad - t_4 \quad + u_2 &= 2. \end{aligned}$$

Clearly a basic feasible solution of (16.6.12) is obtained at once by making $t_1 = 7$, $u_1 = 1$, $t_3 = 5$, $u_2 = 2$ and all other variables zero.

Suppose then that we apply the simplex method to the solution of (16.6.12) and that when we have finished we find that the artificial variables u_1 and u_2 are both zero, i.e., they are nonbasic variables. Then we will have obtained the solution of the original problem. For if $u_1 = u_2 = 0$ equations (16.6.12) are the same as equations (16.6.11). On the other hand, if u_1 and u_2 are not both zero then the solution will not be a solution of the original problem at all but of a somewhat different problem.

How then are we to ensure that u_1 and u_2 will eventually be given the values zero? We can do this by tinkering with the objective function, which at the moment is

$$5x + 4x_2 - 7x_3.$$

Let us add to this two extra terms, multiples of the artificial variables, chosen in such a way that nonzero values of u_1 or u_2 will preclude the possibility of the objective function having reached a minimum value. We take as the new objective function

$$(16.6.13) \quad C = 5x_1 + 4x_2 - 7x_3 + Mu_1 + Mu_2$$

where M is some large positive number, sufficiently large that the contributions from the terms Mu_1 and Mu_2 to the value of the objective function will swamp the contributions from the other three terms if u_1 and u_2 have any values other than zero. We do not need to know what M is; it suffices that it is sufficiently large that a minimum value of C can occur only when u_1 and u_2 are zero.

If the problem calls for us to *maximize* the objective function, then we add terms of the form

$$-Mu_1 - Mu_2$$

to the objective function C, thereby ensuring that any sets of values of the variables in which u_1 and u_2 were *not* zero could not give a maximum for C because of the presence of these two large negative terms.

If we now proceed in the normal manner with equations (16.6.12), using the modified objective function, sooner or later the two artificial variables will take the value zero and become nonbasic variables. When this happens we have a basic feasible solution of the original set of equations (16.6.11), and we can then forget about the artificial variables, drop them from the tableaux, and carry on as if they had never existed. Let us see how it works out. The top part of our first tableau will look like (16.6.14)

(16.6.14)

Constant	x_1	x_2	x_3	t_1	t_2	t_3	t_4	u_1	u_2
7	2	5	−6	1	0	0	0	0	0
4	3	−2	4	0	−1	0	0	1	0
5	6	1	1	0	0	1	0	0	0
2	5	−3	−4	0	0	0	−1	0	1

in which t_1, t_3, u_1 and u_2 are the basic variables. For the objective function we might be tempted to write

$$C = 5x_1 + 4x_2 - 7x_3 + M(u_1 + u_2).$$

This however, does not express C in terms of the nonbasic variables. We must therefore substitute for u_1 and u_2, which we do from the second and fourth rows of (16.6.12). We have

$$C = 5x_1 + 4x_2 - 7x_3 + M(4 - 3x_1 + 2x_2 - 4x_3 + t_2$$
$$+ 2 - 5x_1 + 3x_2 + 4x_3 + t_4)$$
$$= 5x_1 + 4x_2 - 7x_3 + M(6 - 8x_1 + 5x_2 + t_2 + t_4).$$

Since C consists of some terms that contain M and some that do not, it is convenient to use two rows for C, one for each kind of term. Thus the full starting tableau becomes

(16.6.15)

Constant	x_1	x_2	x_3	t_1	t_2	t_3	t_4	u_1	u_2
7	2	5	−6	1	0	0	0	0	0
4	3	−2	④	0	−1	0	0	1	0
5	6	1	1	0	0	1	0	0	0
2	5	−3	−4	0	0	0	−1	0	1
C	5	4	−7	0	0	0	0	0	0
× M −6	−8	5	0	0	1	0	1	0	0

As before we have changed the sign of the constant, and it will be seen that, except for the coefficients of u_1 and u_2, the $\times M$ row is the sum of the second and fourth rows of (16.6.14), with the signs reversed. Thus the full tableau is easily written down from the original equations.

Since we are trying to minimize the objective function, we look for a negative coefficient in it. There are two: the coefficient of x_1 (namely $5 - 8M$, which is certainly negative since M is large) and that of x_3. It does not make much difference which one we take; let us choose the column for x_3. The pivot element then becomes the "4" in row 2 of this column, circled in (16.6.15).

Applying the rules for the simplex method we arrive at the new tableau (16.6.16) in which we observe that one of the artificial variables has already become a nonbasic variable.

(16.6.16)

Constant		x_1	x_2	x_3	t_1	t_2	t_3	t_4	u_1	u_2
	13	$\frac{13}{2}$	2	0	1	$-\frac{3}{2}$	0	0	$\frac{3}{2}$	0
	1	$\frac{3}{4}$	$-\frac{1}{2}$	1	0	$-\frac{1}{4}$	0	0	$\frac{1}{4}$	0
	4	$\frac{21}{4}$	$\frac{3}{2}$	0	0	$\frac{1}{4}$	1	0	$-\frac{1}{4}$	0
	6	⑧	-5	0	0	-1	0	-1	1	1
C	7	$\frac{41}{4}$	$\frac{1}{2}$	0	0	$-\frac{7}{4}$	0	0	$\frac{7}{4}$	0
$\times$ M	-6	-8	5	0	0	1	0	1	0	0

The coefficient of x_1 in C is still negative, and we easily find the new pivot to be the 8 circled in (16.6.16). We again apply the rules and obtain the new tableau (16.6.17).

(16.6.17)

Constant		x_1	x_2	x_3	t_1	t_2	t_3	t_4	u_1	u_2
	$\frac{65}{8}$	0	$\frac{97}{16}$	0	1	$-\frac{11}{16}$	0	$\frac{13}{16}$	$\frac{11}{16}$	$-\frac{13}{16}$
	$\frac{7}{16}$	0	$-\frac{1}{32}$	1	0	$-\frac{5}{32}$	0	$\frac{3}{32}$	$\frac{5}{32}$	$-\frac{3}{32}$
	$\frac{1}{16}$	0	$\frac{153}{32}$	0	0	$\frac{29}{32}$	1	$\frac{21}{32}$	$\frac{29}{32}$	$-\frac{21}{32}$
	$\frac{3}{4}$	1	$-\frac{5}{8}$	0	0	$-\frac{1}{8}$	0	$-\frac{1}{8}$	$\frac{1}{8}$	$\frac{1}{8}$
C	$-\frac{11}{16}$	0	$\frac{221}{32}$	0	0	$-\frac{15}{32}$	0	$\frac{41}{32}$	$\frac{15}{32}$	$-\frac{41}{32}$
$\times$ M	0	0	0	0	0	0	0	0	1	1

Both the artificial variables have now become nonbasic variables and therefore are zero. Hence we have a basic feasible solution of the original problem, and we can forget the variables u_1 and u_2. We can omit them from the tableau, which we now write as (16.6.18).

	Constant	x_2	t_2	t_4	t_1	x_3	t_3	x_1
	$\dfrac{65}{8}$	$\dfrac{97}{16}$	$-\dfrac{11}{16}$	$\dfrac{13}{16}$	1	0	0	0
	$\dfrac{7}{16}$	$-\dfrac{1}{32}$	$-\dfrac{5}{32}$	$\dfrac{3}{32}$	0	1	0	0
	$\dfrac{1}{16}$	$\dfrac{153}{32}$	$\dfrac{29}{32}$	$\dfrac{21}{32}$	0	0	1	0
	$\dfrac{3}{4}$	$-\dfrac{5}{8}$	$-\dfrac{1}{8}$	$-\dfrac{1}{8}$	0	0	0	1
C	$-\dfrac{11}{16}$	$\dfrac{221}{32}$	$-\dfrac{15}{32}$	$\dfrac{41}{32}$	0	0	0	0

(16.6.18)

We can now continue to solve this problem in exactly the same way as if we had arrived at the basic feasible solution shown in (16.6.18) by any other method. We shall not continue the solution now but leave it as an exercise to the reader.

There is one exception to the statement that our tinkering with the objective function will "force out" the artificial variables. It is not difficult to see what it is. If the problem *has* a solution then this solution will certainly be better than any feasible solution containing nonzero artificial variables, because of the extra terms in the objective function. Hence the artificial variables will certainly be forced out if the original problem has a solution. If the problem has no solution, however, one or other of the artificial variables will remain in the tableau. For otherwise we would have found a basic feasible solution of the original problem, whereas none in fact exists. Hence if the simplex algorithm terminates with one or more artificial variables still "in", i.e., still as basic variables taking nonzero values, this indicates that the original problem has no solution—the constraints are incompatible.

To sum up: the problem as augmented by the artificial variables and with the modified objective function certainly has a solution, since at least one basic feasible solution can be written down by inspection. If the original problem has a solution, this is also the solution of the augmented problem (since it makes the artificial variables zero, as is required by the modification of the objective function). Hence the artificial variables will become nonbasic as the algorithm proceeds. If the original problem has no solution, then in the solution of the augmented problem one or more artificial variables will have nonzero values.

Exercise 195

1. By taking a suitably large value for either x or y, or otherwise, find a basic feasible solution to the pair of constraints

$$9x + 2y + 4z \geq 6$$
$$6x + 3y - 2z \geq 8$$

and hence solve, by the simplex algorithm, the linear programming problem which seeks to minimize $2x + 5y + z$, subject to these constraints.

2. In a similar way solve the following linear programming problem

$$\text{Minimize} \quad 4x + y + 5z$$

subject to the constraints

$$x + y + z \geq 4$$
$$2x - y + 2z \geq 1$$
$$x \phantom{{}+y} + 2z \geq 3.$$

3. During the course of the solution of a linear programming problem it may happen that in the objective function, i.e., in the bottom row of the tableau, the coefficient of one of the nonbasic variables is zero. Show that this happens with the problem

$$\text{Minimize} \quad 5x + 2y + 3z$$

subject to

$$29x + 2y + 13z \geq 194$$
$$14x + 8y + 10z \geq 152$$
$$11x + 2y + 7z \geq 98$$
$$4x + y + 2z \geq 34.$$

Note: This is problem 16.6.1 with a different objective function; solve it in the same way.

Show that if we regard the zero coefficients as being negative (so that we can continue the algorithm for another cycle using the column of that coefficient as pivot column) then the continuation of the algorithm results in the interchange of a basic and nonbasic variable (as always) but that the objective function retains the same value. By completing the algorithm, obtain two independent solutions of this problem, giving the same value of the objective function. Prove that any convex combinations of the two solutions is also a solution.

Note: The above result is quite general. The appearance of such a zero coefficient signals that the problem has more than one solution. This possibility was discussed on page 801.

4. Complete the exercise discussed in section 16.6 from the point where it was left at tableau (16.6.18).

5. By introducing one artificial variable and using the simplex algorithm, show that the linear programming problem of maximizing $100x + 200y + 50z$

subject to the constraints

$$5x + 5y + 10z \leq 1000$$
$$10x + 8y + 5z \geq 2000$$
$$10x + 5y \qquad \leq 500$$

has no solution.

6. Show that to solve the problem of minimizing

$$2x + 4y + z$$

subject to the constraints

$$x + 2y - z \leq 5$$
$$2x - y + 2z \geq 2$$
$$-x + 2y + 2z \geq 1$$

by the method given in section 16.6, two artificial variables are needed. Hence, solve this problem.

7. Use the method of artificial variables to solve the problem that was described verbally in Exercise 190.2.

16.7 SOME APPLICATIONS

In this section we shall consider some typical linear programming problems of various kinds, examine the way they are formulated in mathematical terms, and introduce some useful concepts and nomenclature.

Problem 3 of section 16.1 (Mr. Smith and the bricks) may not seem to be a problem of any great economic significance, but it is, in fact, a typical, though simple, example of a type of linear programming problem known as a *diet problem*. The first linear programming problems ever to receive serious attention were of this type, so it is fitting that we should consider them first.

The background of the diet problem is the information that there are certain chemical substances, *nutrients* (such as carbohydrates, proteins, vitamins, etc.), that the body must have in sufficient amounts if it is to be fit, and that there are certain minimum amounts of these various nutrients that should be consumed per day by a healthy man if he is to stay that way. In general it is not harmful to consume *more* than the minimum amounts. The source of these nutrients is, of course, the food that we eat, and different foodstuffs contain different amounts of them.

Now it is not difficult to find a daily diet, consisting of certain amounts

of the available foodstuffs, that will supply the minimum nutritional requirements; there will be many of them. To take an extreme example: if there is
a foodstuff which contains something of each nutrient, then we could obtain
all the necessary minimum requirements by simply eating enough of this one
foodstuff. This would be monotonous, and quite possibly expensive, for the
other fact that we must consider is that these foodstuffs have to be bought.
The problem is whether, among the many diets that yield the minimum nutritional requirements, there is one that is less costly than the others. If there is,
we wish to find it.

Let us express this problem in mathematical terms. Let the various foodstuffs be numbered $1, 2, 3, \ldots, m$, and let $x_j (j = 1, 2, \ldots, m)$ be the amount
of foodstuff j which occurs in a particular diet under consideration. Thus a
diet can be considered as a vector $(x_1, x_2, \ldots, x_m)$ whose elements are the
quantities x_j. If p_j is the cost per unit of the jth foodstuff, then the cost of the
diet is

$$(16.7.1) \qquad C = p_1 x_1 + p_2 x_2 + \cdots + p_m x_m.$$

Each foodstuff will contain a certain amount (possibly zero) of each
nutrient, and perhaps other things besides. Let us number the nutrients
$1, 2, \ldots, n$, and denote by a_{ij} the amount of nutrient i that is contained in
one unit of foodstuff j. Finally let us denote by b_i the minimum daily requirement of nutrient i. These various numbers, p_j, a_{ij} and b_i are assumed to be
known, through knowledge of current prices, chemical analysis and nutri-
tion theory, respectively, though it is hardly necessary to add that in practice
there will be differences of opinion about what are the "correct" values,
especially for the b_i; but that is the dietitian's headache, not the mathematician's!

We can now formulate the diet problem mathematically. If we consume
x_j units of foodstuff j we obtain $a_{ij} x_j$ units of nutrient i. Hence the diet
$(x_1, x_2, \ldots, x_m)$ will yield

$$\sum_{j=1}^{m} a_{ij} x_j = a_{i1} x_1 + a_{i2} x_2 + \cdots + a_{im} x_m$$

units altogether of nutrient i. If we are to receive the minimum requirement
b_i of this nutrient we must have

$$(16.7.2) \qquad \sum_{j=1}^{m} a_{ij} x_j \geq b_i.$$

This argument applies for any nutrient, so we have an equation like (16.7.2)
for every value $i = 1, 2, \ldots, n$. The inequalities (16.7.2) are therefore the
constraints in a linear programming problem in which the objective is to

minimize the *cost function* (16.7.1). In other words we have to find the most economical of those diets that satisfy the nutritional requirements. We see, by comparison with (16.1.3) and (16.1.4), that the problem of Mr. Smith and the bricks is precisely of this type; the three kits correspond to foodstuffs, and the four kinds of bricks to nutrients. Mr. Smith's problem is to minimize the cost of providing at least the required numbers of bricks.

At this point we can use the diet problem to introduce and illustrate some terminology. Confronted with a problem which we suspect to be amenable to linear programming techniques we must first analyze it carefully to see exactly what is going on. In any such problem there will be certain *activities*, actions which, in effect, convert something into something else. In the diet problem there is just one kind of activity, namely, the purchase and consumption of some quantity of a foodstuff. This activity results in the conversion of money into certain amounts of various nutrients, and there are m activities of this kind.

As just remarked, activities convert something into something else. The "somethings" that are thus treated by the activities, either by being "consumed" or "produced," are called *items*. The items that are consumed are *input items* for the activity; those that are produced are the *output items*. Thus the activity "purchasing and consuming foodstuff 3" has one input item, viz., money, and up to n output items, the various nutrients contained in foodstuff 3.

With each activity we can associate an *activity level* which is a measure of the extent to which the activity does whatever it does. This level can be measured in any suitable units. Thus if we speak of buying and eating one pound of foodstuff 3, we not only specify an activity ("buying and eating foodstuff 3") but also a level at which this activity proceeds, viz., to the extent of one pound.

To maintain an activity at a given level, certain amounts of the input items are required, and the activity will then result in the production of certain amounts of the output items. The amounts of input or output items that are consumed or produced by a unit level of the activity are known as *input-output coefficients*. Thus in the diet problem the unit level of the activity "buying and eating foodstuff j" requires p_j units of money to buy the foodstuff j, and produces, on consumption, amounts a_{ij} of the nutrients ($i = 1, 2, \ldots, n$). The p_j are therefore input coefficients, and the a_{ij} are output coefficients.

We now turn to the various constraints imposed on this system. These will each relate to a particular item, and in the present example, which is a rather simple one, there are only two kinds of items, "money" and "nutrients." Money is an input from outside to the system as a whole, and the nutrients are output from the system as a whole; we do not have any outputs from one activity that are inputs to another, though in more complicated problems this can happen.

The constraints on the items "nutrients" are that the amounts of nutrients should be at least the stated amounts, the b_j. The restriction on the item "money" is that the total amount of it, i.e., the cost of the diet, should be a minimum. All these constraints depend on the levels of the various activities, which are at present unknown. So we assign variables $x, y, z \ldots$, etc., to these activites, express the constraints in terms of these unknowns, and in this way formulate the mathematical statement of the problem.

Let us now analyze in a similar way the second problem of section 16.1, concerning the production of three medicines. Here there are three activities "producing and selling medicine i" where $i = 1, 2$ or 3. The input items to an activity are the 5 drugs; the output item is, in each case, money. Thus for the activity "producing and selling Medicine 1" the input coefficients are the amounts of the five drugs required for the production of one bottle (this being taken as the unit activity level), and these are 3, 11, 3, 7 and 15 grams, respectively. The output coefficient is the amount of money produced by the unit level of this activity, in other words the price per bottle, $6.

In this problem the limitations are on the input items; they must not exceed the specified amounts. Moreover, it is the output item, money, which is to be optimized, but since it represents a profit and not a cost, the objective is to maximize it, instead of (as in the last example) to minimize it. Thus this problem is in many ways exactly opposite to the preceding problem.

Rather more elaborate is problem 4 of section 16.1, which requires more careful analysis. A typical activity is that of transporting material from, say, Warehouse A to Branch 1, and the unit level of this activity can be taken as the transportation of one ton of material. The inputs to this activity at this unit level will be $30 and 1 ton of material *at Warehouse A*; the output will be 1 ton of material *at Branch* 1. Note that although we have what is physically the same material featuring both as an input and as an output, nevertheless from our viewpoint we have two *different* items, since the material is in a different place. After all, the whole aim of the activity is to achieve this change of locale. Thus the input items are money and "material-at-a-warehouse." The output items are "material-at-a-branch." The input coefficients are the transportation costs per ton given in the table on page 790.

We expect a constraint of some kind to be associated with each item. If we allocate unknowns $x_1, x_2, \ldots, z_4$ as the levels of the several activities (as we did on page 792) then for the items material-at-a-warehouse we have the inequalities (16.1.6). For the items material-at-a-branch we obtain the equations (16.1.5), since the branches must be supplied with exactly the right amounts. Finally, the item "money" gives rise to the cost function (16.1.7), which has to be minimized.

This straightforward transportation problem is capable of considerable elaboration. We already know from the treatment of the simplex algorithm that in solving a linear programming problem we first introduce slack variables to convert the inequalities into equations. Thus we might replace (16.1.6) by

$$x_1 + x_2 + x_3 + x_4 + u = 150$$

(16.7.3) $\qquad y_1 + y_2 + y_3 + y_4 + v = 100$

$$z_1 + z_2 + z_3 + z_4 + w = 70.$$

The slack variables u, v and w correspond to the amounts left behind in the warehouses. In practice it would normally cost money to keep this excess of the commodity stored in the warehouse, and if this is taken into account, extra terms (multiples of u, v and w) will occur in the cost function representing these storage changes. By the introduction of slack variables, which will have an interpretation as the *excess* of something, or the *deficit* of something, the constraints in a linear programming problem can be taken as equations (they are often called *material balance equations*).

Further elaboration is possible. It could well happen that it would be profitable to transfer goods from a warehouse where storage charges are high to one where they are low. If this were allowed for, we would have to introduce a new kind of activity, viz., warehouse-to-warehouse transfer, together with the appropriate input-output coefficients, and extra terms in the cost function. We might also have to consider how long the material remained in the warehouse, cater for regular deliveries (not just one) to the various branches, and so on. Clearly such a system could become very complex; but a systematic analysis along the terms just given will enable one to keep track of what is happening and formulate the correct mathematical model.

Let us now consider a linear programming problem that arises in yet another context. A factory manager has to plan the production of his factory for the next n months. From knowledge of the state of the market, seasonal fluctuations in demand, and so on, he arrives at estimates of the amount that he can sell in each of these months. He now has to decide how he is going to meet those demands. Here we have a fairly complicated setup. To cater for those months when there is a very heavy demand he can use overtime production; but this will cost more in wages, and in any case there is a limit to how much can be produced in this way. An alternative is to produce more than enough when the demand is low, and store it for the times of heavy demand; but this means that money must be spent on storage. Despite these complexities, the problem is nevertheless fairly easy to formulate in mathematical terms.

Let us denote by a_i the amount in tons (say) of the product that is estimated will be demanded during month i. Let x_i be the amount of regular production in month i, and y_i the amount of overtime production in month i; let u be the cost of overtime production per ton, and v the monthly cost of storage per ton of the product. The last two are assumed to be constant, not subject to seasonal fluctuation, and hence not dependent on i.

We first look at the cost of a given schedule, or rather, at the excess cost over that for regular working conditions. This excess will be from two sources, overtime pay and storage. In month i the overtime payments will amount to uy_i. The extra product produced in month k will be $x_k + y_k - a_k$, and this will have to be stored; but in storage during that month will also be the extra produced during preceding months. Thus in month k the factory will be storing an amount

$$(16.7.4) \qquad X_k = \sum_{j=1}^{k} (x_j + y_j - a_j)$$

tons at a cost of v per ton.

The total cost in overtime and storage over the n months will therefore be

$$(16.7.5) \qquad C = \sum_{i=1}^{n} uy_i + v \sum_{k=1}^{n} \sum_{j=1}^{k} (x_j + y_j - a_j).$$

In the second (double) summation of (16.7.5) we see that $(x_1 + y_1 - a_1)$ occurs n times, once for each month; that $(x_2 + y_2 - a_2)$ occurs $(n - 1)$ times, and so on. In general, $x_i + y_i - a_i$ occurs $(n - i + 1)$ times. Hence we can write (16.7.5) as

$$(16.7.6) \qquad C = \sum_{i=1}^{n} uy_i + v \sum_{i=1}^{n} (n - i + 1)(x_i + y_i - a_i)$$

or

$$(16.7.7) \qquad C = v \sum_{i=1}^{n} (n - i + 1)x_i + \sum_{i=1}^{n} \{u + v(n - i + 1)\} y_i$$
$$- v \sum_{i=1}^{n} (n - i + 1)a_i.$$

The last term in (16.7.7) is a constant (not depending on the x_i or y_i) and hence can be ignored by the production manager whose purpose, naturally, is to choose the activity levels x_i and y_i so as to minimize the cost C. For if C is a minimum, so is

$$(16.7.8) \qquad C' = v \sum_{i=1}^{n} (n - i + 1)x_i + \sum_{i=1}^{n} \{u + v(n - i + 1)\} y_i.$$

Now we look at the restraints governing the choice of the x_i and y_i. First we observe that there are limits to the amounts of regular and overtime production (unless the factory employs more staff). Thus we have inequalities of the form

(16.7.9)
$$x_i \leq M_i$$
$$y_i \leq m_i$$

where M_i and m_i are the maximum possible amounts of regular and overtime production in month i. Next we note that one cannot take more from stores than has been put there; we cannot have a negative inventory. Thus the inventory in month k, viz.,

$$\sum_{i=1}^{k} (x_i + y_i - a_i)$$

must be positive. Hence we have, for $k = 1, 2, \ldots, n$,

(16.7.10)
$$\sum_{i=1}^{k} x_i + \sum_{i=1}^{k} y_i \geq \sum_{i=1}^{k} a_i.$$

Finally, the activity levels x_i and y_i must be positive, i.e.,

(16.7.11)
$$x_i \geq 0; \quad y_i \geq 0$$

since we cannot (in the usual sense of the word) produce a negative amount of a product.

Note: It is worth mentioning here, however, that if we *did* need to consider some kind of "negative production" we would normally define this as a different activity, having its own activity level. This negative production would be given a different name ("sabotage," for example) and be assigned an activity level z_i, say. The net regular production—normal production minus sabotage—would then appear in the cost function C as $x_i - z_i$ in place of x_i.* This trick of expressing a variable which is allowed to assume both positive and negative values as the difference of two non-negative variables enables us to maintain the standard linear programming restriction that all the variables are non-negative. We shall have more to say about this later.

We have now expressed the production manager's task in the form of a linear programming problem, namely, to minimize the cost function C', given by (16.7.8), subject to the linear restrictions (16.7.9), (16.7.10) and (16.7.11).

It might be as well to say a word or two here about how one would cope with a linear programming problem in practice. If the problem is a small one then it can be tackled by a hand computation, using the simplex method as

---

* This would be the only change; the idea of "overtime sabotage" is too bizarre to contemplate!

already given in this chapter. But, as was remarked earlier, if the problem is of any appreciable size, the use of an electronic computer is indicated. The data for the problem—the inequalities and equations that make up the constraints, the coefficients in the objective function, and so on—will have to be presented to the computer somehow, on punched cards, magnetic tape, etc. The precise details of how this is done will vary according to the particular program being used and do not concern us here. Once this is done, the data can be fed to the computer, which will then proceed to grind out the answer.

Now one might think that it would be enough to get back a straightforward answer to one's problem—the values of the variables (both main and slack) and the optimum value of the objective function—but one usually gets much more than this from the computer program. The reason is easy to see. When a factory manager, economist or other person with a linear programming problem goes to have the problem solved on a computer, it is very rarely that he goes away, happily clutching the solution to his problem, never to darken the computer center's doors again. On the contrary; what is more likely to happen is that the optimum solution to the problem will be far from satisfactory, and he will want to come back and put the problem to the computer again in a modified form in an attempt to find a more acceptable solution.

To take a specific example: the factory manager, whose problem we have been discussing, might well find that the optimum solution required the payment of an inordinate amount of costly overtime. If he is not willing to consider changing any of the conditions of the problem then there is, of course, nothing he can do about this; the solution is optimal for the problem *as set*. If he modifies the problem, however, he may get something better. He may, for instance, be willing to consider employing extra staff (we assumed above that he did not do this). To find out whether there would be any advantage in doing this, he could reformulate the problem on the basis of some increase in staff numbers, and see what the new solution looked like. If there was no improvement, he could try to think of some other way in which the situation might be advantageously modified.

Thus the usual pattern is that the attempt to solve a managerial or similar problem of the linear programming type requires many trips to the computer in order to try out different modifications to the problem in the hopes of finding a satisfactory solution. For this reason, the output that one gets from a computer will not usually be a bald statement of the solution to the problem as set, but will give also a great deal of other information, the purpose of which is to provide indications of which modifications to the problem are likely to be advantageous, and which are not. Here again it will depend on the particular computer program what will be included in this additional output, but the following are some of the items of information which might well be supplied.

REDUCED COST.

In the solution of a linear programming problem certain variables (the non-basic variables) have zero values, while others (basic variables) have nonzero values. The basic variables are the ones that really matter in the solution, and the simplex method is essentially a process of successively replacing a variable in the *basis* (the set of basic variables) by one outside it, each time getting a better value for the objective function.

Now if a change is made to one of the coefficients in the objective function, this will surely alter the solution, but if the change is not too large, it may not alter the *nature* of the solution. In other words, it may not change the basis—the same variables will be basic in the new solution as in the old. A large change may well cause a shift to a new basis.

To be more specific let us look again at problem 2, the manufacture of drugs. We saw, on page 813, that the optimum solution for this called for no bottles of Medicine No. 2 to be manufactured. In everyday terms, Medicine No. 2 was not worth making at the price for which it could be sold, namely $1 per bottle. Suppose, however, that the price can be increased without invalidating the assumption that the quantity produced can all be sold. What will happen to the optimum solution? Clearly if the price can be made high enough, Medicine No. 2 will appear in the basis. If it could be sold for its weight in plutonium there would be no point in producing the other two medicines, and it would certainly be in the basis then. Therefore, between the original price and this admittedly extreme price, there must be some critical price at which Medicine No. 2 enters into the basis, displacing some other medicine, which the increased profit from Medicine No. 2 now renders uneconomical to make.

Consider, on the other hand, a typical diet problem. If there are more foodstuffs than nutrients then some of the foodstuffs will not appear in the optimum solution. These nonbasic foodstuffs have costs that are too high to be economical. Suppose now that the cost of one of these foodstuffs were reduced lower and lower. There would, in general, come a point when this foodstuff would be included in the optimum solution. To take an extreme case again; if one of the foodstuffs contained a little of every nutrient and were being given away free, there would be no point in buying any other foodstuff. There will thus be some critical cost at which this foodstuff will appear in the solution. Note, however, that this will not always happen. For example, a foodstuff which contains none of the nutrients will not appear in the optimum solution, however cheap it is.

This critical value of a coefficient in the objective function is called the *reduced cost*. For a minimizing problem (like the diet problem) it is the level below which the cost of one of the nonbasic items must be reduced in order to bring it into the basis. Knowledge of the reduced cost of an item is very useful. There would be no advantage, for example, in taking steps to obtain a certain nonbasic foodstuff at a lower price (by changing one's butcher, say) unless the new price were below the reduced cost. For unless this happens, the foodstuff in question will still not appear in the solution, and, if the solution is followed, none of that foodstuff will be bought.

For a maximizing problem this critical value is the value above which the coefficient in the objective function must be increased if the variable in question is to enter into the basis. This is useful information for the same reason as before. There would be no sense in embarking on an advertising campaign to persuade people to pay a higher price for Medicine No. 2 unless the price that one expects them to pay is above the critical value.

COST RANGE.

If we look at those coefficients in the objective function which relate to the basic variables, we can ask a similar question. What will happen if we change one of these coefficients? As before, if we change one of these coefficients too much we will alter the nature of the solution. In general, there will be a range of values for a given coefficient, given by a lower and an upper limit, such that the nature of the solution will not be affected provided the coefficient lies between these two limits. This is called the *cost range* for that coefficient, and again is a useful thing to know. There is no point repeating the solution of a linear programming problem with one of these coefficients altered unless it has been made to fall outside the cost range, for otherwise the solution will be much the same as before.

If one of the coefficients goes outside its cost range then its variable will come out of the basis and another will go in. A computer program for linear programming, in addition to giving the cost range for each basic variable, will normally indicate which variable will come in when that variable goes out of the basis as a result of its coefficients going beyond one or other of the limits of the cost range.

MARGINAL COSTS.

We have considered variations in the coefficients in the objective function; now let us consider variations in the right-hand sides of the constraints. The problem of Mr. Smith and the bricks will serve us here. Suppose Mr. Smith decides that he would like to have more bricks of Size 3. In general he will have to buy some extra kits, and will consequently pay more for the bricks that he gets. If the numbers of the other bricks are the same (and in all this we are considering changing only one thing at a time), then this additional amount is paid solely in order to acquire the extra bricks. This can be expressed as so much per extra brick, and is called a *marginal cost*. Thus the marginal cost of a Size 3 brick is the cost of acquiring one more brick of this size. In general the marginal cost associated with a constraint is defined as c/h, where h is a small change in the right-hand side of the constraint, and c is the resulting change in the optimum value of the objective function. (We say "small change" to indicate that we do not want to change the constraint so much that we alter the basis.)

We shall discuss marginal costs again in section 16.10 and shall therefore say no more about them here. Suffice it to say that the various marginal costs, along with details of cost ranges, reduced costs, and probably many other pieces of useful information will be included in the computer output to the problem. This may well amount to a sizeable stack of paper, but on careful perusal this information can be of immense value in the search for a satisfactory solution to an optimization problem of the kind that we have been considering.

Exercise 196

1. The following is a problem faced by the production manager of a factory in which table lamps are assembled from three components made elsewhere, viz., pedestals, lampshades and bulbholders (together with the necessary wiring).

The available supply of these components varies from month to month, and so does the demand for the final product. If more table lamps are produced than are required, or if more components are bought than are needed, these must be stored in the factory's warehouse and will incur appropriate storage costs. Given that the available supplies of components and the demand for table lamps for each month of a 12-month period are known, and that other facts relevant to the problem are given (see below), what amounts of the various components should be bought each month and how many lamps should be produced in each month in order to minimize the storage costs for the 12-month period?

Components arrive at the beginning of each month, either from the suppliers or from the factory's warehouse, and those that are going to be used during the month (but only those) are kept on the factory floor at no cost. The rest (if any) are stored in the warehouse. Pedestals, lampshades and bulbholders cost, respectively, S_p, S_l and S_b dollars each per month to store. The completed lamps in demand are collected at the end of each month, and any produced in excess of the demand are stored at a cost of S_c dollars per month. If fewer lamps are made than are demanded, the deficit is made up from those in storage. At the beginning of the year there are p_0, l_0, b_0 and c_0 pedestals, lampshades, bulbholders and completed lamps already in store. Pedestals, lampshades, bulbholders and completed lamps occupy, respectively, v_p, v_l, v_b and v_c cubic feet each, and storage space in the warehouse is restricted to a maximum of V cubic feet.

Analyze this problem, identifying the items and activities. Derive the constraints and the objective function and formulate the problem as a linear programming problem. Use the following notation:

M = the maximum number of lamps that can be produced per month.

d_n = the number of lamps demanded in month $n(n = 1, 2, \ldots, 12)$.

c_n = the number of lamps produced in month n.

P_n, L_n, B_n = the numbers of pedestals, lampshades and bulbholders available at the beginning of month n.

p_n, l_n, b_n = the numbers of pedestals, lampshades and bulbholders bought at the beginning of month n.

Note: the problem as stated is concerned only with the cost of storage. No other costs are relevant.

What modifications would need to be made to this model if

(a) the manager takes into account a penalty of $\$A$ for every lamp that was not supplied when demanded? (The problem as stated above assumed that the demand was always met.) Assume that any shortfall in the number of lamps for one month is added to the demand for the following month.

(b) there is a charge for transporting components or completed lamps from the factory to the warehouse and back?

Suppose it were required to minimize the cost of the whole operation of the factory throughout the 12-month period. Discuss what further information would be needed in order to tackle this more comprehensive problem.

2. A company owns two oil wells. Well A produces 500 barrels of high-grade oil per day, 200 barrels of medium-grade and 100 barrels of low-grade oil. Well B produces per day 100, 200 and 400 barrels respectively of the three grades.

The company finds that it requires at least 1,000 barrels of high-grade oil, 1,200 barrels of medium-grade oil, and 1,200 barrels of low-grade oil. Well A costs $300 per day to operate, and Well B costs $200 per day. It is required to determine the number of days that each well should be operated in order to meet these requirements with the minimum expenditure.

Formulate this problem in mathematical terms, and solve it by the graphical method. Show that an excess of low-grade oil will be produced.

If the cost of operating Well B increases, at what stage will the optimum solution entail the production of an excess of high-grade oil?

3. In problem 2 of section 16.1 show that if Medicine No. 2 can be sold for $7.60 or more per bottle, then this medicine will appear in the optimum solution to the problem. Which variable will disappear from the solution if this happens?

Hint: Let the price of Medicine No. 2 be p dollars and, by following the simplex method as in sections 16.4 or 16.5, find what p must be in order to bring y into the basis.

4. Find the marginal costs of the bricks of sizes 1, 2, 3 in problem 3 of section 16.1.

Hint: Remember that the optimal solution is given by the last three constraints in (16.6.1) being satisfied with equality. The first constraint in (16.6.1) can be ignored.

What is the marginal cost of a Size 0 brick?

5. Find the cost range of the variable x in problem 3 of section 16.1.

Hint: Follow the simplex procedure, replacing the coefficient of x by a parameter. See at what stage the value of this parameter determines whether x is in the basis or not.

Do the same for the variables y and z.

16.8 THE CONDENSED NOTATION

The reader may have noticed that all the tableaux in this chapter contain a unit matrix. Thus in tableau (16.5.2) we have a 5×5 unit matrix at the right-hand end. This unit matrix serves two purposes; it enables one to tell which slack variable occurs in which equation, and because of its presence we are able to manipulate the rows of the tableau by simple operations of the kind that are familiar in matrix algebra and the solution of equations. On the other hand, this unit matrix takes up a good deal of space, and the question arises whether its presence is really necessary; that is, whether its function could not be performed in some other way. It turns out that we *can* manage without this unit matrix and thus work with a smaller array of numbers in a tableau.

If we omit this matrix, then the purposes it has served must be fulfilled in some other way. The first purpose is easily taken care of. In tableau (16.5.2), for example, we can indicate that s is the slack variable for the first equation by writing s to the left of the first row. Similarly we write t against the second row, and so on. We obtain the *condensed* tableau that follows.

(16.8.1)

	Constant	x	y	z
		0	0	0
s	48	3	6	4
t	136	11	6	12
u	70	3	10	6
v	55	7	14	4
w	50	5	4	⑤
C	0	6	1	5

The second purpose of the unit matrix was to make the rules for manipulating the tableau the familiar rules of matrix algebra. If we dispense with it we must not be surprised if the rules for going from one tableau to the next becomes slightly more complicated. Let us follow the steps of problem 2 yet once more, this time working with the condensed tableau.

The first two steps are not affected and are exactly the same as before. In step 3 we divided the pivot row by the pivot element. If we do this in (16.8.1) we get, for the fifth row

$$(16.8.2) \qquad\qquad w \quad 10 \quad 1 \quad \frac{4}{5} \quad \underline{1}$$

Now the 1 which is now in the pivot position (underlined above) will eventually be an element of the next unit matrix, which we are not going to write down. In the present unit matrix, which has not been written down, the 1 in the fifth row becomes 1/5 (compare the original manipulation of the tableau) and this will eventually be an element in the column which is to replace the z-column. Let us therefore put it in its final place now, that is, in the pivot position. We get tableau (16.8.3).

(16.8.3)

	Constant	x	y	z
s	48	3	6	4
t	136	11	6	12
u	70	3	10	6
v	55	7	14	4
w	10	1	$\frac{4}{5}$	$\frac{1}{5}$
C	0	6	1	5

Our next step, call it step $3A$, is to subtract suitable multiples of the row given in (16.8.2) so as to make the remaining elements in the z-column become zero. Take the first row. To eliminate the 4 in the z-column we must subtract 4 times row (16.8.2) from the first row. Thus from the 48 at the beginning of the row, we subtract 4 times 10, getting 8. From the 3 we subtract 4 times 1 to get -1, and so on. The rule is easy to see. Pick on any element, say the "11" in row 2. This element and the pivot element can be thought of as being at opposite corners of a rectangle. The rule then is that from the element we subtract the product of the elements at the other two corners. Figure 16.8 below should make this clear.

Fig. 16.8

The rule indicates that 11 must be replaced by $11 - 12 \times 1 = -1$. This rule applies to all elements except those in the pivot row (which have already been dealt with) and those in the pivot column, which we must now look at.

The effect of step 3 on the pivot column is to reduce all elements, except the pivot, to zero; but these zeros will end up in a column of the new unit matrix, and we therefore ignore them, since they will not appear in the new condensed tableau. However, while this reduction to zero has been going on, one column in the old unit matrix (the one with a "1" in the pivot row) has been changed by step 3 into something less simple. If we write it down (including, in its proper place, the number which has, in anticipation, already been put in place of the pivot) then before step 3A is performed we have (16.8.4)

(16.8.4)

z	w
4	0
12	0
6	0
4	0
1/5	1/5
5	0

Applying the rectangle rule we see that the old w-column becomes

$$
\begin{array}{c}
-4/5 \\
-12/5 \\
-6/5 \\
-4/5 \\
1/5 \\
\hline
-1
\end{array}
$$

When we perform our original step 4, interchanging the z- and w-columns, this will replace the present z-column. Hence for the next step—call it step 3B—we have the following rule. To go from the present pivot column to the same column for the new tableau, multiply every element (except the pivot element) by the pivot element, and change the sign.

Step 4 is as before and is mere tidying up. The new column which replaces the old first column is for the new nonbasic variable w, instead of z. Hence we interchange z and w in the tableau. The new tableau then becomes

	Constant	x	y	w
s	8	−1	$\dfrac{14}{5}$	$-\dfrac{4}{5}$
t	16	−1	$-\dfrac{18}{5}$	$-\dfrac{12}{5}$
u	10	−3	$\dfrac{26}{5}$	$-\dfrac{6}{5}$
v	15	3	$\dfrac{54}{5}$	$-\dfrac{4}{5}$
z	10	1	$\dfrac{4}{5}$	$\dfrac{1}{5}$
C	−50	1	−3	−1

(16.8.5)

It will be observed that the pivot element was replaced by its reciprocal at the earliest possible moment. There are two good reasons for this. One is that by doing this we do not lose track of what the original pivot element is. Had we divided the pivot row by the pivot element, getting a 1 in the pivot position, there would be no indication *in the tableau* of what the old pivot element was. Since we need this information, we would either have to record it outside the tableau or postpone the division until later. This latter alternative is possible, but the computation of the other elements then requires division and is more complicated. As it is (and this is the other reason) the replacing of the pivot element by its reciprocal is the only occasion during the whole process when division is required; the rest is all multiplication and subtraction. This will be clear if we summarize the process of going from one condensed tableau to the next.

Step 1.   Look for a nonbasic variable with positive coefficient in *C*, as before.

Step 2.   Chose the pivot element, as before.

Step 3.   Replace the pivot by its reciprocal. Multiply the other elements in the pivot row by the present (new) pivot.

Step 3*A*.   Alter all other elements except those in the pivot column by the rectangle rule, as described.

Step 3*B*.   Multiply every element in the pivot column (except the pivot) by minus the pivot.

Step 4.   Interchange the letters associated with the pivot row and column.

The reader should now verify that these rules when used to continue the problem just started, give the following condensed tableaux.

After step 3 we have

	Constant	x	y	w
s	8	-1	$\frac{14}{5}$	$-\frac{4}{5}$
t	16	-1	$-\frac{18}{5}$	$-\frac{12}{5}$
u	10	-3	$\frac{26}{5}$	$-\frac{6}{5}$
v	5	$\frac{1}{3}$	$\frac{18}{5}$	$-\frac{4}{15}$
z	10	1	$\frac{4}{5}$	$\frac{1}{5}$
C	-50	1	-3	-1

After step 3A we have

	Constant	x	y	w
s	13	-1	$\frac{32}{5}$	$-\frac{16}{15}$
t	21	-1	0	$-\frac{8}{3}$
u	25	-3	-6	-2
v	5	$\frac{1}{3}$	$\frac{8}{5}$	$-\frac{4}{15}$
z	5	1	$-\frac{14}{5}$	$-\frac{11}{15}$
C	-55	1	$\frac{33}{5}$	$-\frac{11}{15}$

Finally, after steps 3B and 4 we have

	Constant	x	y	w
s	13	$\frac{1}{3}$	$\frac{32}{5}$	$-\frac{16}{15}$
t	21	$\frac{1}{3}$	0	$-\frac{8}{3}$
u	25	1	16	-2
v	5	$\frac{1}{3}$	$\frac{18}{5}$	$-\frac{4}{15}$
z	5	$-\frac{1}{3}$	$-\frac{14}{5}$	$\frac{7}{15}$
C	-55	$-\frac{1}{3}$	$-\frac{33}{5}$	$-\frac{11}{15}$

The nonbasic variables (along the top) are zero. The values of the basic variables on the left are found in the constant column.

Although the rules for manipulating these condensed tableaux may seem confusing at first, once they are mastered they provide a very quick method of solving linear programming problems with the minimum of arithmetic and the minimum amount of writing down.

As a further example of the use of the condensed notation, and to illustrate some further general points, let us solve the following linear programming problem.

(16.8.6)

$$\text{Minimize}$$

$$48s + 136t + 70u + 55v + 50w$$

subject to the constraints

$$3s + 11t + 3u + 7v + 5w \geq 6$$
$$6s + 6t + 10u + 14v + 4w \geq 1$$
$$4s + 12t + 6u + 4v + 5w \geq 5.$$

If we introduce slack variables, calling them x, y and z, to make the inequalities into equations, we obtain

$$3s + 11t + 3u + 7v + 5w - x = 6$$
$$6s + 6t + 10u + 14v + 4w - y = 1$$
$$4s + 12t + 6u + 4v + 5w - z = 5.$$

Since there is no obvious basic feasible solution to these equations let us also introduce artificial variables p, q and r as well (as described in section 16.6). We must then add the terms $Mp + Mq + Mr$ to the objective function in order that the artificial variables will eventually be forced out of the solution. Our problem then appears as follows:

(16.8.7)

$$\text{Minimize}$$

$$48s + 136t + 70u + 55v + 50w + Mp + Mp + Mr$$

subject to the constraints

$$3s + 11t + 3u + 7v + 5w - x + p = 6$$
$$6s + 6t + 10u + 14v + 4w - y + q = 1$$
$$4s + 12t + 6u + 4v + 5w - z + r = 5.$$

There is now an obvious basic feasible solution, viz.,

$$p = 6, \quad q = 1, \quad r = 5.$$

Before we can set up the starting tableau we must express the objective function in terms of the nonbasic variables. It becomes

$$48s + 136t + 70u + 55v + 50w$$
$$+ M(6 + x - 3s - 11t - 3u - 7v - 5w)$$
$$+ M(1 + y - 6s - 6t - 10u - 14v - 4w)$$
$$+ M(5 + z - 4s - 12t - 6u - 4v - 5w)$$
$$= 12M + 48s + 136t + 70u + 55v + 50w$$
$$+ M(-13s - 29t - 19u - 25v - 14w + x + y + z).$$

Remembering that the constant in the objective function will occur in a tableau with the opposite sign, we set up the starting tableau as in (16.8.8).

(16.8.8)

	Constant	s	t	u	v	w	x	y	z
p	6	3	11	3	7	5	-1	0	0
q	1	6	⑥	10	14	4	0	-1	0
r	5	4	12	6	4	5	0	0	-1
	0	48	136	70	55	50	0	0	0
$\times M$	-12	-13	-29	-19	-25	-14	1	1	1

Our problem is a minimizing one, so we look for a negative coefficient in the bottom row. The usual rules determine the pivot element, circled in (16.8.8). Applying rules 3, 3A, 3B and 4 of this section we obtain tableau (16.8.9)

(16.8.9)

		s	q	u	v	w	x	y	z
p	$\frac{25}{6}$	-8	$-\frac{11}{6}$	$-\frac{46}{3}$	$-\frac{56}{3}$	$-\frac{7}{3}$	-1	$\frac{11}{6}$	0
t	$\frac{6}{1}$	1	$\frac{1}{6}$	$\frac{5}{3}$	$\frac{7}{3}$	$\frac{2}{3}$	0	$-\frac{1}{6}$	0
r	3	-8	-2	-14	-24	-3	0	②	-1
	$-\frac{68}{3}$	-88	$-\frac{68}{3}$	$-\frac{470}{3}$	$-\frac{787}{3}$	$-\frac{122}{3}$	0	$\frac{68}{3}$	0
$\times M$	$-\frac{43}{6}$	16	$\frac{29}{6}$	$\frac{88}{3}$	$\frac{128}{3}$	$\frac{16}{3}$	1	$-\frac{23}{6}$	1

The artificial variable q has now become nonbasic and takes the value zero. This is the value it must have in the final solution, and we can therefore ignore it from now on. Accordingly we shall delete the q column in (16.8.9)—the one enclosed in a rectangle. This means, incidentally, that we could have saved ourselves the trouble of calculating its elements. No matter; we will remember next time.

With only one negative entry in the bottom row we have no choice for the next pivot element, which is circled in (16.8.9). Applying the rules we obtain (16.8.10).

			s	u	v	w	x	r	z
(16.8.10)	p	$\dfrac{17}{12}$	$-\dfrac{2}{3}$	$-\dfrac{5}{2}$	$\left(\dfrac{10}{3}\right)$	$\dfrac{5}{12}$	-1	?	$\dfrac{11}{12}$
	t	$\dfrac{5}{12}$	$\dfrac{1}{3}$	$\dfrac{1}{2}$	$\dfrac{1}{3}$	$\dfrac{5}{12}$	0	?	$-\dfrac{1}{12}$
	y	$\dfrac{3}{2}$	-4	-7	-12	$-\dfrac{3}{2}$	0	$\dfrac{1}{2}$	$\dfrac{1}{2}$
		$-\dfrac{170}{3}$	$\dfrac{8}{3}$	2	$\dfrac{29}{3}$	$-\dfrac{20}{3}$	0	?	$\dfrac{34}{3}$
	$\times M$	$-\dfrac{17}{12}$	$\dfrac{2}{3}$	$\dfrac{5}{2}$	$-\dfrac{10}{3}$	$-\dfrac{5}{12}$	1	?	$-\dfrac{11}{12}$

Since r, another artificial variable, has now become zero, we can delete it from the problem. We have therefore not bothered to calculate the "r" column in (16.8.10).

At this stage there are three negative entries in the bottom row to choose from. When there is a choice like this it is often recommended to choose the numerically largest of the numbers, since it is likely, though not certain, that this will lead more quickly to the final solution. These considerations lead to the choice of the "10/3," circled in (16.8.10), as pivot element. The next tableau is then (16.8.11).

			s	u	p	w	x	z
(16.8.11)	v	$\dfrac{17}{40}$	$-\dfrac{1}{5}$	$-\dfrac{3}{4}$	$\dfrac{3}{10}$	$\dfrac{1}{8}$	$\dfrac{3}{10}$	$\dfrac{11}{40}$
	t	$\dfrac{11}{40}$	$\dfrac{2}{5}$	$\dfrac{3}{4}$	?	$\left(\dfrac{3}{8}\right)$	$\dfrac{1}{10}$	$-\dfrac{7}{40}$
	y	$\dfrac{33}{5}$	$-\dfrac{32}{5}$	-16	?	0	$-\dfrac{18}{5}$	$\dfrac{14}{5}$
		$-\dfrac{2431}{40}$	$\dfrac{23}{5}$	$\dfrac{37}{4}$	?	$-\dfrac{63}{8}$	$\dfrac{29}{10}$	$\dfrac{347}{40}$
	$\times M$	0	0	0	?	0	0	0

The one remaining artificial variable, p, has become nonbasic and can be left out. We now have a basic feasible solution for the original problem. It is

$$v = \frac{17}{40}; \quad t = \frac{11}{40}; \quad y = \frac{33}{5}$$

with all other variables zero. Notice that the terms in M in the objective function have now all become zero, as, of course, they should—a useful check on the accuracy of the computation so far.

One more iteration gives us the final tableau. There is one possibility only for the pivot element, and we obtain (16.8.12).

(16.8.12)

		s	u	t	x	z
v	$\frac{1}{3}$	$-\frac{1}{3}$	-1	-3	$-\frac{1}{3}$	$\frac{1}{3}$
w	$\frac{11}{15}$	$\frac{16}{15}$	2	$\frac{8}{3}$	$\frac{4}{15}$	$-\frac{7}{15}$
y	$\frac{33}{5}$	$-\frac{32}{5}$	-16	0	$-\frac{18}{5}$	$\frac{14}{5}$
	-55	13	25	21	5	5

The absence of negative coefficients in the bottom line signals the end of the calculation. The minimum value of the objective function is 55 (remember the change of sign), and is achieved for the values

$$v = \frac{1}{3}; \quad w = \frac{11}{15}; \quad s = t = u = 0$$

of the main variables. The slack variable y is nonzero, showing that the second inequality in (16.8.6) remains an inequality; there is a difference of 33/5 between the two sides. Since $x = z = 0$ the other two inequalities are satisfied with equality. The problem is therefore completely solved.

In this section we have solved two problems. The first was a maximizing problem which we had solved before. The second was a minimizing problem. The numbers of equations and variables was different in the two problems and the technique of solution was different in that for the second problem we needed artificial variables and rather more iterations. Despite these several differences the reader has probably spotted many similarities between the coefficients in the two problems, and the fact that the answer is 55 to each of the problems is a further point in common.

These resemblances are not coincidental. Besides providing examples of two common types of linear programming problems, these two problems are related in an important manner which will be the subject of the next section.

Exercise 197

1. Rework some of the problems given in the exercises to this chapter, using the condensed notation. Compare the successive tableaux with those obtained previously, thus verifying that the new method achieves the same result and by the same route (given the same choices where choices are available) as does the old method.

2. Solve the linear programming problem of maximizing

$$5x + 6y - 3z$$

subject to the constraints

$$3x + 2y - 2z \leq 8$$
$$4x - y + 2z \geq 9$$
$$-4x + 2y + z \leq 10$$
$$-2x + y + z \leq 4$$
$$3x - 2y + z \geq 5$$
$$x \geq 3$$
$$y \leq 6$$
$$z \leq 1.$$

16.9 DUALITY

The two problems that we solved in the previous section are related in a particular, important way. Let us look at these two problems once again, and see exactly what this relationship is. The first problem was

(16.9.1)

$$\text{Maximize} \quad 6x + y + 5z$$

subject to the constraints

$$3x + 6y + 4z \leq 48$$
$$11x + 6y + 12z \leq 136$$
$$3x + 10y + 6z \leq 70$$
$$7x + 14y + 4z \leq 55$$
$$5x + 4y + 5z \leq 50.$$

The second was

(16.9.2)

$$\text{Minimize} \quad 48x + 136t + 70u + 55v + 50w$$

subject to the constraints

$$3s + 11t + 3u + 7v + 5w \geq 6$$
$$6s + 6t + 10u + 14v + 4w \geq 1$$
$$4s + 12t + 6u + 4v + 5w \geq 5.$$

Notice first that the matrices of coefficients in these two problems, namely

$$
\begin{pmatrix} 3 & 11 & 3 & 7 & 5 \\ 6 & 6 & 10 & 14 & 4 \\ 4 & 12 & 6 & 4 & 5 \end{pmatrix} \quad \text{and} \quad \begin{pmatrix} 3 & 6 & 4 \\ 11 & 6 & 12 \\ 3 & 10 & 6 \\ 7 & 14 & 4 \\ 5 & 4 & 5 \end{pmatrix}
$$

are transposes each of the other. Further, the constants on the right-hand sides of (16.9.1) are the coefficients in the objective function of (16.9.2), and vice versa. Note finally that the inequalities in the two problems are the opposite way, and that one problem is a minimizing one while the other is a maximizing one.

Two linear programming problems related in this way are said to be *dual* to each other. To be quite general let us define this concept of *duality* for problems of any size and with general coefficients, using the matrix notation.

Duality.

The dual of the linear programming problem

(16.9.3)
$$
\begin{cases} \text{Maximize} \quad \mathbf{bx} \\ \text{subject to the constraints} \\ \mathbf{Ax} \le \mathbf{c} \end{cases}
$$

is the linear programming problem

(16.9.4)
$$
\begin{cases} \text{Minimize} \quad \mathbf{c}^T\mathbf{u} \\ \text{subject to the constraints} \\ \mathbf{A}^T\mathbf{u} \ge \mathbf{b}^T. \end{cases}
$$

We can call either of these problems the *primal* problem; the other will then be called the *dual* problem. Duality is a symmetric relation.

This relationship between two linear programming problems is all very pretty but would not be worth mentioning if there were not some deeper connection between the problems. If we look at the solutions of problems (16.9.1) and (16.9.2) we can see the optimum values of the objective functions agree. For one problem this optimum is a minimum, for the other it is a maximum, but the numerical value is the same, namely 55. This result is quite general—a linear programming problem and its dual always have the same optimum solution. This statement, that the maximum for problem (16.9.3) equals the minimum for problem (16.9.4), is known as the *Fundamental Duality Theorem* in linear programming.

In showing that this is so we shall demonstrate other connections between dual problems, but first we must stress an important point, one which although implicit in what has gone before has not been explicitly mentioned.

In carrying out a single stage of the simplex method, all that we do is to manipulate the set of equations resulting from the introduction of the slack variables and, if required, the artificial variables. We replace the equations by an equivalent set of equations, having the same solutions, and express a new set of nonbasic variables in terms of the remaining, basic variables. This process is continued until (for a maximizing problem) all the coefficients in the objective function are negative, at which stage the solution has been obtained. All the other paraphernalia of the simplex method—the choice of pivot element, the use of artificial variables and so on—serve one purpose only, namely, to tell us which manipulations to perform at each stage in order to achieve the required result. The simplex method is just a guidebook to the maze of edges of the complex polytope of feasible solutions; it tells us which turning to take at each vertex of the polytope in order to reach the optimum vertex in a reasonably efficient manner.

It follows that any sequence of manipulations of the equations that reaches the optimum vertex is as good as any other; there is nothing sacrosanct about the route laid out by the simplex method. It would not even matter if, in the course of these manipulations, we allow some of the variables to become negative, though they must eventually become positive when the final solution is reached. If we allowed this to happen we would visit (in the course of our search for the optimum vertex) certain intersection points of hyperplanes that did not belong to the convex set of feasible solutions. In the application of the simplex method we were careful to keep all variables non-negative at all times, for the very good reason that had we ever left the convex set of feasible solutions, there would have been no obvious method of getting back on to it. But if we can provide such a method then we need not observe the prohibition on negative values during the course of the solution.

Bearing these remarks in mind we shall now see that the solution of a problem and the solution of its dual can go hand in hand, the manipulations performed in one tableau being copied exactly in the other. We shall illustrate this with solution of problems (16.9.1) and (16.9.2), but it will be seen that the argument is quite general.

If we introduce slack variables into problem (16.9.2) in the usual way, but change the signs right through the resulting equations, we get

$$-3s - 11t - 3u - 7v - 5w + x = -6$$

(16.9.5)
$$-6s - 6t - 10u - 14v - 4w + y = -1$$

$$-4s - 12t - 6u - 4v - 5w + z = -5.$$

With the inequalities the way they are in this problem we cannot easily find a basic feasible solution to equations (16.10.5); it was for this reason that we needed to introduce artificial variables. But if we are willing to relax the prohibition on negative variables we can take

$$x = -6, \quad y = -1, \quad z = -5$$

as a solution (though not a *feasible* solution). If we now set up a condensed tableau in the usual way we obtain the following.

(16.9.6)

		s	t	u	v	w
x	-6	-3	-11	-3	-7	-5
y	-1	-6	-6	-10	-14	-4
z	-5	-4	-12	-6	-4	-5
C	0	48	136	70	55	50

Note the similarities between tableau (16.9.6) and tableau (16.8.1); if we transpose the matrix of tableau (16.8.1) and interchange the left-hand column with the bottom row, we obtain tableau (16.9.6) except that all signs, except those of the coefficients in the objective function, which appear in the bottom row of the tableau, have been changed. Note also that the letters used for the main and slack variables have been made to correspond. This is not essential, but it makes the exposition clearer.

We know that the sequence of manipulations to a tableau described by steps 3, 3A, 3B and 4 in section 16.8 will convert the tableau into an equivalent one, i.e., one corresponding to an equivalent set of equations, and will cause one basic variable and one nonbasic variable to change roles. We shall now apply these rules also to tableau (16.9.6); but what shall we use for pivot element? One penalty of allowing variables to become negative is that this plays havoc with our rule for choosing the pivot element. We therefore turn to the dual tableau (16.8.1) as a guide and use the same pivot that we used there. This was the "5" in the bottom right-hand corner of the matrix, and accordingly we choose the corresponding element of tableau (16.9.6) as our pivot element. It happens also to be the element in the bottom right-hand corner and is circled in (16.9.6).

We now apply the rules of section 16.8 to tableau (16.9.6). It will be left as an exercise to the reader to show that the result is

(16.9.7)

		s	t	u	v	z
x	-1	1	1	3	-3	-1
y	3	$-\dfrac{14}{5}$	$-\dfrac{18}{5}$	$-\dfrac{26}{5}$	$-\dfrac{54}{5}$	$\dfrac{4}{5}$
w	1	$\dfrac{4}{5}$	$\dfrac{12}{5}$	$\dfrac{6}{5}$	$\dfrac{4}{5}$	$-\dfrac{1}{5}$
C	-50	8	16	10	15	10

We now notice that tableau (16.9.7) is related to tableau (16.8.5) in exactly the same way as tableau (16.9.6) is related to (16.8.1). Moreover, this is not just a coincidence; they are so related because exactly the same computations have been performed in the two tableaux, as the reader should verify. Thus the tableaux for dual problems start out related in this way— transposed except for the signs of all except the bottom row—and when one iteration has been performed, using corresponding pivot elements, the resulting tableaux are related in exactly the same way.

Suppose that we complete the simplex method in the first (or primal) tableau. The final tableau that we get has the following properties.

(a) The numbers in the bottom row are all negative (or zero).
(b) The numbers in the left-hand column, not including that in the bottom row, are all positive. This is because they are the values of the basic variables, and has been true all along.

Therefore, if we carry out the same iterations in the dual tableau we shall end up with a tableau in which

(c) the numbers in the left-hand column (except for the bottom row) are all non-negative. This follows from (a) above and the relation between the two tableaux.
(d) The numbers in the bottom row, except for the first column, are all non-negative. This follows from (b) and the relation between the tableaux.

The interpretation of (c) is that in the final dual tableau the basic variables (and hence all the variables) are non-negative, whatever may have been the case during the intermediate stages. Thus the final tableau is consistent with the prohibition on negative values. The interpretation of (d) is that since the dual problem is a minimizing one, the simplex method cannot be carried any further, for since the coefficients are positive the increase of any nonbasic variable will cause an increase in the objective function.

Thus when we carry out the analogous iterations in the primal and dual tableaux, the simplex method terminates simultaneously in both. Moreover the value of the objective function (minus the bottom left-hand element) is the same in both tableaux after each iteration, and hence also at the termination of the whole procedure. This shows that the maximum for the primal problem equals the minimum for the dual problem, and proves the Duality Theorem.

Looking at the final stage of the dual tableau for the specific problem that we have been studying we see that it will be as follows.

(16.9.8)

		s	t	u	x	z
v	$\dfrac{1}{3}$	$-\dfrac{1}{3}$	$-\dfrac{1}{3}$	-1	$-\dfrac{1}{3}$	$\dfrac{1}{3}$
y	$\dfrac{33}{5}$	$-\dfrac{32}{5}$	0	-16	$\dfrac{18}{5}$	$\dfrac{14}{5}$
w	$\dfrac{11}{15}$	$\dfrac{16}{15}$	$\dfrac{8}{3}$	2	$\dfrac{4}{15}$	$-\dfrac{7}{15}$
	-55	13	21	25	5	5

This shows that the solution of the dual problem is given by

$$v = \frac{1}{3}; \quad y = \frac{33}{5}; \quad w = \frac{11}{15}$$

with the other variables zero. The reader should verify that these values satisfy the inequalities in (16.9.2), equality holding for the first and third; that the left-hand side of the second inequality is 38/5, which exceeds the right-hand side by 33/5, the value of the slack variable y; and that these values give the objective function the value 55. This is the same result that we obtained at the end of section 16.8.

It follows from all this that the routine for solving a linear programming problem is essentially the same as that for solving its dual, and hence it is not necessary to do both. Thus if we have a linear programming problem to solve, we can, if we wish and if there is any advantage, formulate the dual problem and solve that instead. Often there *is* some advantage. We have seen that problem (16.9.1) is easier to solve than problem (16.9.2) since a basic feasible solution is easily found, and hence artificial variables are not needed. If we were asked to solve problem (16.9.2) it would be prudent, therefore, to solve the dual problem (16.9.1) instead. Sometimes there is no obvious advantage. If the primal problem has inequalities going both ways, then since we reverse all inequalities in forming the dual problem, the dual problem will also have inequalities going both ways, and the best procedure may not be so obvious.

Exercise 198

1. Write down the linear programming problems dual to the following problems.

(a)   Maximize $7x - 2y - 3z$

subject to the constraints

$$x + y + z \leq 5$$
$$2x + 2y + z \leq 7$$
$$2x - y \qquad \leq 1$$

(b)   Minimize $6x + 6y + z$

subject to the constraints

$$x + 4y + 2z \geq 5$$
$$3x + 2y + 2z \geq 5$$
$$4x - 3y + 4z \geq 1$$

2. For each of the two problems in problem 1 solve either the primary problem or the dual problem (whichever you consider easier) and hence obtain the solutions of both problems and their duals.

3. Since linear programming problems having only two variables can be solved graphically, it follows that, by taking the dual, problems having only two constraints can be solved graphically. In this way, find graphically the solution of the problem of maximizing $5x - 2y + 3x$ subject to the constraints

$$3x - 4y + 2z \leq 9$$
$$2x + 7y + 5z \leq 11.$$

16.10 THE INTERPRETATION OF DUAL PROBLEMS

We now turn to the question of the interpretation of the dual of a linear programming problem. There is no difficulty about interpreting the primal problem; its interpretation is the original economic structure from which the problem was derived. It is not at all obvious that the dual problem, which, after all, was derived from the primal by a purely algebraic process—mainly that of transposition—will have a significant interpretation in the context of the original economic situation. In fact it does, though the interpretation is not always easy to formulate. Let us again consider the problem of Mr. Smith and the bricks.

In this problem the constraints (16.1.3) result from the fact that certain minimum requirements of bricks are required. Thus, for example,

$$29x + 2y + 13z \geq 194$$

must be satisfied since the left-hand side is the number of Size 0 bricks, and this is required to be at least 194. The optimization required for this problem is that the cost

$$C = 8x + 3y + 5z$$

should be a minimum.

The dual of this problem is the following:

Maximize

(16.10.1) $\qquad\qquad 194t + 152u + 98v + 34w$

subject to the constraints

$$29t + 14u + 11v + 4w \leq 8$$
$$2t + 8u + 2v + w \leq 3$$
$$13t + 10u + 7v + 2w \leq 5.$$

Whereas in the primal problem we had a variable—an activity level—associated with each kit, here we have one associated with each brick, according to size. If we assume that these variables, s, t, u and v stand for the amounts of something or other possessed by bricks of Size 0, 1, 2 and 3, respectively, then

$$29t + 14u + 11v + 4w$$

for example, is the total amount of this "something" possessed by Kit A, and the first restraint says that this should not be too much. The same applies to the other constraints. The objective in the dual problem is to maximize the total amount of this "something" in the kits eventually purchased.

What is this "something"? It is clearly in the nature of a price, since the quantities on the right-hand sides of the restraints in (16.10.1) are the prices of the kits. It is, in fact, a valuation. We remarked at the beginning of the chapter that the bricks could not be bought separately (if they could there would be no problem), but this need not deter Mr. Smith from having ideas about what the prices of the bricks might be if they *were* sold separately. He can attach an *imputed value* or *shadow price* to each brick, these values being the same for bricks of the same size. The objective of the dual problem now becomes clear; it is to maximize the imputed value of all the bricks bought. It now remains to interpret the constraints.

Let us look at the first restraint of (16.10.1) in a rather negative way, and consider what happens if it is *not* satisfied, i.e., if

(16.10.2) $29t + 14u + 11v + 4w > 8.$

Since this possibility is precluded, we might well expect it to correspond to some economically unrealistic situation, as is indeed the case. For the left-hand side is Mr. Smith's valuation of Kit A, and (16.10.2) states that this exceeds the actual price of the kit. One or other of two things is wrong here: if Mr. Smith's valuation of the individual bricks is reasonable, then the seller of the kits is behaving foolishly, since he is selling this kit for less than the total value of its component bricks; on the other hand, if the seller's price is reasonable, then Mr. Smith is over-valuing the bricks, since he asserts that he get more value from the kit than he paid for it. Whichever way one looks at it (16.10.2) represents an unrealistic situation, and it is therefore reasonable to include its contradictory, the first inequality of (16.10.1), as one of the constraints. The other constraints are interpreted in exactly the same way.

The duality theorem asserts that the optimal feasible solution to the primal problem is also the optimum feasible solution to the dual; that the minimum cost in the primal problem equals the maximum value in the dual. If we give the variables in the dual problem the same symbols that were given to the slack variables when we solved the primal problem in section 16.6, we shall perceive more easily some aspects of the primal-dual relationship that are of importance. For example, if we write tableau (16.6.7) in the condensed form, it becomes the following.

		u	v	w
z	5	$-\dfrac{1}{12}$	$-\dfrac{1}{2}$	$\dfrac{5}{3}$
t	24	1	-1	-8
x	5	$\dfrac{1}{12}$	$\dfrac{1}{6}$	-1
y	4	$-\dfrac{1}{6}$	$\dfrac{1}{3}$	$-\dfrac{1}{3}$
	-77	$\dfrac{1}{4}$	$\dfrac{1}{6}$	$\dfrac{2}{3}$

We see from this that the solution of the dual problem is

$$t = 0, \quad u = \frac{1}{4}, \quad v = \frac{1}{6}, \quad w = \frac{2}{3}$$

and that all three inequalities are satisfied with equality. In particular the variable t, which was basic in the solution of the primal problem, taking the value 24, is nonbasic in the solution of the dual problem. This means that the imputed value of bricks of Size 0 is zero. This may seem strange, until we consider that Mr. Smith's valuation of the bricks is dependent on the way in which he is compelled to buy them, that is, in kits. In the optimum program for buying the bricks he gets more than enough of the Size 0 bricks, as observed at the end of section 16.6. Since he gets them anyway, as a by-product, so to speak, of his efforts to get appropriate numbers of the other bricks, these bricks are of no value to him. To impute a value to them would imply some measure of incentive to obtain them, which, in this example, does not exist.

Another way of interpreting an imputed value or shadow price is as a *marginal cost*, in the sense used in other branches of economics. For example, the marginal cost of a Size 1 brick in the above problem will be the extra cost incurred by increasing the number of Size 1 bricks required from 152 to 153. At least, it will be, provided this increase in the number of bricks does not affect the *nature* of the optimum solution. If we alter one of the requirements we alter one of the hyperplanes which defines the convex set of feasible solutions. In other words we shall alter some of the vertices of this convex set. If we alter them too much, then the new convex set may give an optimum at some different vertex; but provided the alteration is small, the optimum will be achieved at the same vertex, though this will have been slightly displaced. This means that the division of the variables into basic and nonbasic will be the same. To be on the safe side let us suppose that Mr. Smith increases his requirement of Size 1 bricks by an amount h, small enough not to change the vertex of the polytope giving the optimum solution.

We already know that at the optimum solution of this problem the first inequality is automatically satisfied, and the other three are equations. Thus we have

$$14x + 8y + 10z = 152$$
(16.10.3)
$$11x + 2y + 7z = 98$$
$$4x + y + 2z = 34.$$

If we replace 152 by $152 + h$ and solve the problem again, we shall have a new solution with x, y, z replaced by $x + \delta x, y + \delta y, z + \delta z$, say, and the following equations satisfied

$$14(x + \delta x) + 8(y + \delta y) + 10(z + \delta z) = 152 + h$$
(16.10.4)
$$11(x + \delta x) + 2(y + \delta y) + 7(z + \delta z) = 98$$
$$4(x + \delta x) + (y + \delta y) + 2(z + \delta z) = 34.$$

If we subtract the equations in (16.10.3) from the corresponding equations in (16.10.4) we then obtain

$$14\delta x + 8\delta y + 10\delta z - h = 0$$
(16.10.5)
$$11\delta x + 2\delta y + 7\delta z = 0$$
$$4\delta x + \delta y + 2\delta z = 0.$$

The increase in cost δC is $8\delta x + 3\delta y + 5\delta z$, and the marginal cost u of Size 1 bricks is given by $u = \delta C/h$. Hence we have the equation

(16.10.6) $\qquad\qquad 8\delta x + 3\delta y + 5\delta z - uh = 0.$

In equations (16.10.5) and (16.10.6) we have a set of homogeneous equations in the variables δx, δy, δz and h. Since these have a nontrivial solution, we see, by the reasoning of Exercise 167.1 extended in an obvious way to four variables, that

(16.10.7)
$$\begin{vmatrix} 14 & 8 & 10 & -1 \\ 11 & 2 & 7 & 0 \\ 4 & 1 & 2 & 0 \\ 8 & 3 & 5 & -u \end{vmatrix} = 0.$$

Expanding the determinant by the last column we have

(16.10.8)
$$-u\begin{vmatrix} 14 & 8 & 10 \\ 11 & 2 & 7 \\ 4 & 1 & 2 \end{vmatrix} + \begin{vmatrix} 11 & 2 & 7 \\ 4 & 1 & 2 \\ 8 & 3 & 5 \end{vmatrix} = 0$$

which gives us $u = \frac{1}{4}$. This agrees with the solution of the dual problem in which the shadow price of Size 1 bricks is indeed $\frac{1}{4}$. That this is no coincidence can be seen from equations (16.10.1). We know that $t = 0$ (this corresponds to our omission of the first equation of the original problem) so that the solution of the dual problem is given by

(16.10.9)
$$14u + 11v + 4w = 8$$
$$8u + 2v + w = 3$$
$$10u + 7v + 2w = 5.$$

We now see that, with a little juggling of rows and columns, equation (16.10.8) is precisely what we get by solving (16.10.9) for u by Cramer's rule (see page 682). Although we have illustrated the identity of shadow prices and marginal costs by a specific example, it is easily seen that the argument is quite general.

Note that the cost of requiring an extra Size 0 brick is zero, since Mr. Smith has a surplus of these bricks anyway. Thus the marginal cost of a Size 0 brick is zero, in agreement with the imputed value that we remarked on earlier.

Let us now look at the transportation problem (problem 4 of the beginning of the chapter) but with a slight alteration. In this problem, as originally stated, each branch is to be supplied with *exactly* its monthly requirement. For the moment let us be satisfied with supplying each branch with *at least* its monthly requirement. The equations (16.1.5) will then become inequalities. Since this problem is a minimizing problem, we have to write the inequalities (16.1.6) with the sign $\geq$ in order to bring them in line with (16.9.2).

$$\text{Minimize}$$

$$30x_1 + 30x_2 + 50x_3 + 90x_4 + 30y_1 + 40y_2 + 40y_3 + 70y_4 \\ + 40z_1 + 20z_2 + 30z_3 + 60z_4$$

subject to the constraints

$$(16.10.10) \begin{cases}
x_1 + y_1 + z_1 & & & \geq & 30 \\
& x_2 + y_2 + z_2 & & \geq & 40 \\
& & x_3 + y_3 + z_3 & \geq & 80 \\
& & & x_4 + y_4 + z_4 \geq & 90 \\
-x_1 & -x_2 & -x_3 & -x_4 & \geq -150 \\
-y_1 & -y_2 & -y_3 & -y_4 & \geq -100 \\
-z_1 & -z_2 & -z_3 & -z_4 \geq & - 70.
\end{cases}$$

We now form the dual of this problem. Allocating variables, say p, q, r, s, t, u, v to correspond to the seven constraints, we find that the dual problem is

$$\text{Maximize}$$

$$30p + 40q + 80r + 90s - 150t - 100u - 70v$$

subject to the constraints

$$(16.10.11) \begin{cases}
p - t \leq 30 \\
p - u \leq 30 \\
p - v \leq 50 \\
q - t \leq 90 \\
q - u \leq 30 \\
q - v \leq 40 \\
r - t \leq 40 \\
r - u \leq 70 \\
r - v \leq 40 \\
s - t \leq 20 \\
s - u \leq 30 \\
s - v \leq 60.
\end{cases}$$

It remains to interpret this dual problem in an economically significant manner. Consider the variable p. It corresponds to the inequality in the primal which states that Branch 1 receives at least the required amount. We shall interpret it as the value per ton of the commodity on arrival at Branch 1. Similarly the variables q, r and s will be interpreted as values per ton of the commodity at Branches 2, 3 and 4, respectively. The variables t, u and v will be interpreted as the values per ton of the commodity at each of the three warehouses. The meaning of the dual problem now becomes clear. The first constraint can be written as

$$p \leq 30 + t$$

and simply states that the value of the commodity at Branch 1 cannot exceed the sum of its value at Warehouse A and the cost of transportation from there to the branch. The other inequalities in (16.10.11) bear similar interpretations for the eleven other warehouse-to-branch transportation routes.

The function to be maximized in the dual problem contains four positive terms, giving the total value of the commodity at the branches, and three negative terms whose positive sum is the total value at the warehouses. Hence what is being maximized is the excess of value at the branches over value at the warehouses.

In the original statement of this problem, the branches were to be supplied with exactly the right amounts of the commodity, and we had equations (16.1.5) instead of the first four inequalities of (16.10.3). Now it is not immediately obvious what the dual problem is for a primal which contains equations as well as inequalities. To find out, let us consider the simple example.

$$\text{Maximize} \quad 2x + 3y - z + t$$

Subject to the constraints

$$3x + 4y - z + 2t \leq 10$$
$$8x + 2y + 5z - t \leq 30$$
$$2x + y + z - 3t = 8.$$

We shall use the artifice mentioned on page 795 of replacing a single equation by two inequalities, stating that the left-hand side is both $\geq$ and $\leq$ the right-hand side. Thus we replace the constraints in (16.10.12) by

$$3x + 4y - z + 2t \leq 10$$
$$8x + 2y + 5z - t \leq 30$$
$$2x + y + z - 3t \leq 8$$
$$-2x - y - z + 3t \leq -8.$$

We now write down the dual program, in variables p, q, r and s, say. It is

$$(16.10.13) \quad \begin{cases} \text{Minimize} & 10p + 30q + 8r - 8s \\ \text{subject to} & 3p + 8q + 2r - 2s \geq 2 \\ & 4p + 2q + r - s \geq 3 \\ & -p + 5q + r - s \geq -1 \\ & 2p - q - 3r + 3s \geq 1. \end{cases}$$

What has appeared here is an example of another artifice, introduced on page 836, for making provision for a variable which may assume negative as well as non-negative values, namely that of expressing it as the difference of two non-negative variables, here r and s. Thus (16.10.13) could be reformulated as

$$\begin{cases} \text{Minimize} & 10p + 30q + 8w \\ \text{subject to} & 3p + 8q + 2w \geq 2 \\ & 4p + 2q + w \geq 3 \\ & -p + 5q + w \geq -1 \\ & 2p - q - 3w \geq 1 \end{cases}$$

with $w = r - s$, where it is understood that w is not restricted to being non-negative. Thus an equation in the primal problem gives rise to an unrestricted variable in the dual problem, and vice versa.

If we use this dodge with the transportation problem as originally stated, we find that the dual problem is formally the same as (16.10.11), but that we have to add the statement that the variables p, q, r and s are not restricted to non-negative values; or alternatively, we could replace each of these variables by the difference of two non-negative variables. Thus if we are to supply each branch with exactly the required amount of the commodity, we must allow the possibility that the value of the commodity at some branches may turn out to be negative.

The whole subject of dual linear programming problems is one that can be taken very much further than has been done in these last two sections, which have merely given an introduction to the subject. We shall meet the duality theorem again in the next chapter, in a rather different context, but that is all. The reader who wishes to delve still further into this important branch of mathematics will find ample material among the books referred to at the end of this chapter, which include fuller statements and more rigorous proofs of the duality theorem than have been given here.

Exercise 199

1. Formulate the dual of problem 2 of section 16.1 (the manufacture of medicines), and interpret it.

2. Do the same with the other verbal problems that have occurred as exercises in this chapter.

Exercise 200 (Chapter Review)

1. Solve graphically the following linear programming problems

   (a)   Maximize $2x + y$ subject to the constraints

$$x + y \geq 5$$
$$2x + 3y \leq 20$$
$$4x + 3y \leq 25.$$

   (b)   Maximize $x + y$ subject to

$$2x + y \leq 4$$
$$x + 3y \geq 1$$
$$3x + 2y \leq 10$$
$$2x + 3y \geq 3$$
$$7x + 10y \leq 35.$$

   (c)   As for (b), but minimize $x + y$.

   (d)   Maximize $3x + 4y$ subject to

$$2x + 2y \leq 9$$
$$3x + 4y \geq 12$$
$$x + y \leq 8$$
$$2x + 3y \leq 12$$
$$x + 4y \geq 3$$
$$5x + 7y \geq 23.$$

   (e)   Maximize $x + y$ subject to

$$-2x + 3y \leq 9$$
$$2x + 3y \leq 15$$
$$3x + y \leq 12.$$

In (b), (c) and (d) above, some of the constraints are redundant. Which ones?

2. **(a)** Maximize $5x + 2y - z$ subject to

$$x - 3y + 4z \leq 8$$
$$3x + 2y + 6z \leq 5$$
$$2x + 7y - z \leq 9.$$

(b) Maximize $2x - 3y + z$ subject to

$$3x + 6y + z \leq 6$$
$$4x + 2y + z \leq 4$$
$$x - y + z \leq 3.$$

3. Show that the constraints

$$3x + 2y - z \leq 14$$
$$x - 3y - 4z = 19$$
$$2x + 4y - z = 10$$
$$3x + 4y + 7z \leq 30$$

are inconsistent and, hence, that any linear programming problem which includes these constraints will be infeasible.

4. The transportation problem (problem 4 in the first section of this chapter), although fairly simple, gives rise to a linear programming problem in quite a few variables. Nevertheless it is amenable to the methods developed in this chapter, and the reader may like to try his hand at solving it. It should be remarked, however, that this type of problem is rather special and that, in consequence, there are simpler methods for solution, methods which, unfortunately, are not applicable to linear programming problems in general. Details of these methods can be found in the suggested reading given at the end of the chapter.

5. Minimize the function $5x + 4y$ subject to the constraints

$$2x + y \geq 1$$
$$x + 3y \geq 2$$
$$-x + y \geq 2$$

where x and y are *not* restricted to being non-negative.

Hint: This problem can be tackled graphically, and the lack of restriction to non-negative values of x and y then causes no difficulty.

6. Mention has already been made (Exercise 195.5) of degeneracy. When degeneracy occurs the algorithm may "cycle," that is, return to a previous basic feasible solution. This is possible since the objective function remains the same during these iterations, and the argument against cycling, namely that the objective function always increases (or decreases, as the case may be) no longer applies. Although this never seems to cause trouble in practice, examples have been deliberately constructed for which cycling can take place.

More for interest than for exercise we give one such example, which the reader may like to play with. It is the problem

$$\text{Minimize} \qquad z = \frac{3}{4}x + 150y - \frac{1}{50}z + 3u$$

subject to the constraints

$$\frac{1}{4}x - 60y - \frac{1}{25}z + 9u \leq 0$$

$$\frac{1}{2}x - 90y - \frac{1}{50}z + 3u \leq 0$$

$$z \qquad \leq 1.$$

7. Show that the geometrical interpretation of degeneracy is that the point in question is common to more than n hyperplanes, where n is the number of variables. (Refer to the discussion on page 807.)

8. Find, by any method, the maximum value of

$$2x + 4y + 3z + t$$

subject to the constraints

$$3x + 2y + 4z + 5t \geq 6$$
$$2x - 2y - 3z + 2t \geq 8$$
$$-x + y + 2z - 4t \geq 5.$$

9. Minimize $x + 3y + 4z + t + 2u$ subject to the constraints

$$2x + 5y + 3z - 2t + 6u \geq 5$$
$$-x - 2y - 4z + t + 2u \geq 6.$$

10. Solve the following linear programming problems by considering the corresponding dual problems:

(a)   Maximize $3x + 5y + 2z$

subject to

$$2x - y + 3z \leq 6$$
$$x + 2y + 4z \leq 8.$$

(b)   Minimize $4x + 5y + z$

subject to

$$18x - 3y + 5z \geq 42$$
$$x - \frac{1}{3}y - \frac{1}{4}z \leq 1.$$

11. Consider the following, which is not a linear programming problem. Find the point (x, y) whose coordinates satisfy the inequalities

$$4x + y \geq 8$$
$$5x + 3y \geq 15$$
$$x + 7y \geq 7$$

and whose distance from the origin is a minimum.

This is a very simple example of *quadratic programming*, namely minimizing a quadratic function (a polynomial of degree 2 in the variables, in this case $x^2 + y^2$, since it is enough to minimize the *square* of the distance) subject to a set of linear constraints.

This particular problem is easily solved graphically. Solve it, and show that unlike what happens with a linear programming problem, the unique minimum value does not occur at a vertex of the convex polytope of feasible solutions.

Suggestions for Further Reading

One of the most comprehensive books on linear programming is

Dantzig, G. B., *Linear Programming and Extensions*. Princeton, N.J.: Princeton University Press, 1963.

This is a high-powered book containing far more than the nonspecialist student is likely to need; but it is well worth consulting. Much the same applies to

Saaty, T.L., *Mathematical Methods of Operations Research*. New York: McGraw-Hill Book Co., 1959,

which deals with linear programming as a part of the general subject of operations research.

Of special interest to the economist is

Dorfmann, R., Samuelson, P., and Solow, R., *Linear Programming and Economic Analysis*. New York: McGraw-Hill Book Co., 1958,

which treats the subject from an economist's point of view, linking it up thoroughly with other methods of economic analysis.

Other useful books are:

Charnes, A., Cooper, W. W., and Henderson, A., *An Introduction to Linear Programming* New York: John Wiley & Sons, Inc., 1963.

Gass, S. I., *Linear Programming and Applications*. New York: McGraw-Hill Book Co., 1958.

Vajda, S., *Theory of Games and Linear Programming*. London: Methuen & Co., Ltd.,; New York: John Wiley & Sons, Inc., 1956.

Two of the "finite mathematics" books deal with linear programming, viz.,

Kemeny, J. G., Mirkil, H., Snell, J.L. and Thompson, G. L., *Finite Mathematical Structures*. Englewood Cliffs, N.J.: Prentice-Hall, Inc., 1959.

Lipschutz, S., *Theory and Problems of Finite Mathematics*. New York: Schaum Publishing Co., 1966.

17 Theory of Games

A Man with a Shotgun said to a Bird:
  "It is all nonsense, you know, about shooting being a cruel sport. I put my skill against your cunning—that is all there is to it. It is a fair game."
  "True," said the Bird, "but I don't wish to play."
  "Why not?" inquired the Man with a Shotgun.
  "The game," the Bird replied, "is fair as you say; the chances are about even; but consider the stake. I am in it for you, but what is there in it for me?"
  Not being prepared with an answer to the question, the Man with a Shotgun sagaciously removed the propounder.

AMBROSE BIERCE.

17.1 INTRODUCTION

It might seem strange, almost irreverent perhaps, to suggest that many economic problems are best handled by treating them as games. One might well imagine, for example, that the sort of competition which exists between three firms vying for supremacy in marketing a certain commodity is in a different category altogether from the sort of competition existing between three men playing a friendly, or not so friendly, game of cards. Yet since the publication in 1944 of the book *Theory of Games and Economic Behaviour* by Von Neumann and Morgenstern (London: John Wiley & Sons, Inc.), it has become increasingly obvious that the sort of mathematics which is used in analyzing games of skill and strategy of the cardtable or chessboard variety is very useful in the discussion and analysis of a wide range of problems relating to all sorts of economic activity. A knowledge of at least the more elementary aspects of the mathematical theory of games is therefore an important item in the tool-kit of any mathematical economist. It will appear later that there is a very close connection between the theory of games and linear programming; each subject supplements and assists the other.

It is a matter of common observation that there are many different kinds of games of strategy, and our first task will be to attempt some classification. One obvious means of classification is by the number of players needed to play the game. For many games, like poker, the number of players is not fixed; for others the number is always the same, as in solo whist for which the number must be four. However, for any given play of a game, a specific number of persons take part. ("A play" is a useful generic term for what is variously expressed as a "hand," of whist, poker, etc.; a "game," of chess; a "rubber," of bridge, and so on.) Very numerous are those games which only two can play, such as chess, tic-tac-toe, and the many two-handed card games. We call these games "two-person games," and in a similar way we define "three-person games," "four-person games," "n-person games," and so on. Note that "person" is really a technical term used to denote one of the conflicting interests in the game, and may or may not correspond to an individual human being. Thus, despite appearances, contract bridge is a two-person game, since the two conflicting interests are the two partnerships, forming two "composite persons," so to speak, between whom the game takes place. It is not a four-person game, since the players are not playing each against the others.

Games for more than two persons admit of a complication which does not arise in two-person games. With more than two persons it is possible for two or more players to gang up on the others and form a coalition (permanent or temporary) to their mutual advantage. For this reason, and others, the theory of games for more than two persons is extremely difficult and not by any means fully investigated, although much research has been done, and is being done, into the many problems to which it gives rise. Certainly the detailed study of such games is beyond the scope of this book. With two-person games no such coalitions are possible, and (with one further restriction) the topic becomes amenable to a comparatively simple mathematical treatment.

When two persons play a game, the usual procedure at the end of each "play" is that one player (the loser) pays a certain sum of money to the other (the winner). Thus what the loser loses is what the winner wins and, in consequence, the total amount of money possessed by the two players together remains constant. No money enters or leaves the "system" consisting of the two players. Such games are called zero-sum games, since the amount of money entering or leaving the system is zero.

Two-person games for which this is not true are uncommon in the gambling world, but there are many economic situations which approximate to this type of game. Consider two firms which are in competition in selling a certain product. There are various strategic moves open to each for improving their sales—increased advertising, purchase of more modern factory equipment, and so on. The firm which makes better use of the possibilities will probably attract customers from the other firm and hence increase its profits at the expense of its rival. But it could well happen that inept management (choice of strategy) might result in loss of profits by *both* firms so that money would leave the system; on the other hand good management might well increase the demand for the product, with resultant gain to both sides. This is therefore not a zero-sum game. In many ways this situation resembles a game for three persons, the third, rather passive, player being in this case a composite of the consumer public, the manufacturers of factory equipment, advertising agencies, and so on. The analysis of two-person, nonzero-sum games is therefore beset with much the same difficulties that beset three-person games (certainly it is not unknown for two firms, ostensibly in competition, to form a coalition against the consumer!). The discussion of games of this type is also beyond the scope of this book.

17.2 TWO-PERSON ZERO-SUM GAMES

We shall confine ourselves to the theory of two-person, zero-sum games. This may seem a severe restriction, but in point of fact this theory covers a great deal of important ground. Here are some simple examples of games of this kind:

GAME 1.

Two players A and B simultaneously place a dime on a table. If the coins match, i.e., both show heads or both tails, then player A takes both coins; if they do not match, i.e., one shows heads the other tails, then player B takes both.

GAME 2.

Dimes are placed on a table as in the first game. Payments are then made as follows

(a) If both coins are heads, A pays B \$1
(b) If both coins are tails, B pays A \$3
(c) If the coins do not match, B pays A \$2.

Here the dimes merely serve as counters and are retrieved by the players after each play.

GAME 3.

Alec and Bertram are both anxious to gain the affections of Diana, a charming brunette. Alec is an accomplished ice skater but an indifferent dancer; Bertram is clumsy on skates but an expert on the dance floor. On a certain evening it is known that unless she decides to stay at home, Diana will be either at the ice rink or at the local charity ball. Both Alec and Bertram wonder whether it would be best to go to the ice rink or to the ball or to stay at home. The possible advantage in staying at home is that if Diana also stays at home she may, on receiving a phone call, be willing to go out to some less elaborate form of entertainment. Unfortunately one cannot phone from the ice rink or from the ballroom, and Diana's phone, at present out of order, will not be repaired until after she goes out (if she does). The "game" for Alec and Bertram consists in making a choice of what to do for their evening's entertainment. To analyze this latter game we clearly need a good deal of quantitative data which will be supplied in due course.

Let us look at Game 1, the game of matching dimes. What strikes us at once about this game is that it is very much the same for both players. The players are betting on a 50—50 chance (the probability of the coins matching is $\frac{1}{2}$), and each receives the same amount if he wins. Each player has a choice between two alternatives: to put down the coin head-up or tail-up. For each combination of choices (4 in all) some payment is made by one player to the other. Let us consider the amounts paid *to A*, using the obvious convention that a negative amount means that *A* pays *B*. These amounts are called the *payoffs* and can be displayed in the following 2 × 2 array:

Player B

		H	T
Player A	H	1	−1
	T	−1	1

which is called the *payoff matrix*. This indicates for example, that if the coins match then player *A* receives 1 dime from player *B*, since one of the coins he picks up is his own. This payment is indicated by the top left and bottom right elements of the array. For the other two possibilities (the coins not matching) *A*'s payoff is −1 dime, i.e., he pays *B* 1 dime, which is what in effect happens.

Intuitively one can see that it makes no difference what *A* does; whatever choice he makes he stands an equal chance of winning or losing. Similarly there is no reason for *B* to favor one choice rather than the other. Of course, if *B* had some prior information about how *A* was going to play then this would not be so. If, for example, *B* knew that *A* intended to play heads, he, *B*, would naturally play tails. Even if he knew only that in the course of several plays of this game *A* would tend to play heads more often than tails, then *B* could win more often by playing tails more often than heads. Similarly *A* could take advantage of any preference by *B* for one choice rather than the other. All in all it would appear that the best bet for both players is to play heads and tails at random and in approximately equal numbers over a series of plays of the game. We shall see later that this intuitive conclusion is borne out by the mathematical theory.

The second game is similar to the first, but the payoff matrix is different. It is as follows:

Player B

		H	T
Player A	H	−1	2
	T	2	3

as the reader can verify, remembering that the entries are the payoffs made *to* player *A* according to the four possible outcomes of the game.

Player *A*, looking at this game, will observe that there is no advantage at all in playing heads. For if he does he may win $2 or lose $1; whereas if he plays tails he is certain to win at least $2 and may win $3. Hence player *A* will (if he is sensible) always play tails.

Player *B* is, we shall suppose, just as able as player *A* to analyze the situation, and knowing that player *A* will play tails, he will play heads, thus ensuring that he pays *A* only $2 and not $3. He would do this anyway, even if he did not consider *A*'s position, for by playing heads he pays out at most $2 (and gains $1 if *A* is foolish enough to play heads also) whereas playing tails brings an automatic loss of at least $2 and possibly $3.

Hence, unlike the first game, there is in this game a definite method of play for each player which, if departed from, brings immediate loss; *A* must play tails and *B* must play heads. The important element in the payoff matrix is thus the "2" in the bottom left-hand corner.

There is another way of seeing why this element is so important. It has the property that it is the smallest element in its row, but the largest element in its column. The first of these statements implies that, by playing tails, *A* can be sure of gaining at least $2 (the element in question), while the second implies that by playing heads *B* can ensure that he need never pay out *more* than $2. If *A* can be certain of always getting $2 or more, and *B* can arrange never to pay more than $2, then the given choices, for which *B* pays *A* exactly $2, are optimum for both players. Any other choice carries the risk of losing more (or gaining less).

This game has now ceased to be of any interest. The players will play the same way each time, and the result each time is that *B* pays *A* $2. Clearly the game is biased in favor of *A* who is $2 better off each time. This amount is called the *value* of the game to *A*. We shall give a precise definition of this term later but its meaning is sufficiently clear in the present context. One might wonder whether anyone would be foolish enough to take the part of player *B* in a game which is so manifestly "unfair." However, it often happens that a game is unfair in the sense that one player has a permanent advantage over the other so that there is a certain sum of money (the value of the game) which, on the average, he expects to make on each play of the game. Such a bias in favor of one player can be overcome in two ways. One is that the players interchange roles at intervals so that each has his share of the advantage, such as when the "bank" changes hands in games like poker, craps, blackjack and baccarat. The other is for the player who has the advantage to make a down payment (an *ante*) to the other player for the privilege of playing the game, this amount being chosen so as to offset exactly the bias of the game and thus make it "fair."

In the game we are considering the natural thing to do would be for *A* to pay B \$2, before each play, for the privilege of playing. Certainly this would be fair, since after the play A would have won back his \$2 and both players would be just as well off as before. Equally as certain, it makes the whole procedure quite pointless, but the game will serve as a stepping-stone to more significant examples.

Note that the effect of playing with an ante can be allowed for by subtracting the ante from every element in the payoff matrix. In our Game 2, this results in the following matrix:

		Player B	
		H	T
Player A	H	−3	0
	T	0	1

Since the relative sizes of the entries are unchanged, the reasoning is the same as before. The key element is now the zero in the bottom left-hand corner. Hence the value of this game is zero, and the game is said to be *fair*.

The game might seem to be rather less pointless if we alter the rules so that the payoff matrix reads

		Player B	
		H	T
Player A	H	−1	4
	T	2	3

Here it looks as though there would be some temptation for A to play heads in that he *might* then win \$4. But the situation for B is virtually unchanged; he still has no reason to play tails, and hence the possibility of A's winning \$4 will never arise. The alteration to the payoff matrix has not altered the vital thing about it, namely the peculiar property of the element "2," that it is the smallest in its row and the largest in its column. Thus, as we have seen, A can be sure of getting at least \$2, while B can be certain of limiting A's winnings to just this amount. Hence the choice which A should make (we shall call it his best *strategy*) is to play tails, while B's best strategy is to play heads. A departure from the best strategy by one player will inevitably result in a loss. A departure by *both* players from their best strategies might well be advantageous to one player, but this will not happen if both players are sensible. For example, if our payoff matrix were

		Player A	
		H	T
Player B	H	−1	−10
	T	2	3

then B might, by playing tails instead of heads, win $10. But this can only happen if A is obliging enough to depart from his best strategy also, and B can hardly expect A to help him to win when it is in his (A's) power to stop him.

We see from this that we are making a fundamental assumption in the theory of two-person zero-sum games that the players, in addition to trying to win, are both capable of analyzing the situation to its fullest extent. If, however, B knew that A was too stupid to realize that his only reasonable strategy was to play tails, he could better his position (using the last payoff matrix) by occasionally playing tails himself, and winning an occasional $10. This in no way invalidates the theory, because under these circumstances A and B are in fact playing a *slightly different game*. The assumption of sophistication on the part of both players virtually forces A to play tails, and it would be all the same if the rules of the game required him to do this. In much the same way, if it is known that A will sometimes play heads, then it would be the same as if the rules *required A* to choose this strategy from time to time. In short, any assumption that the players are *not* capable of analyzing the game effectively converts the game into a slightly different game.*

Exercise 201

1. In the children's game of "stone-scissors-paper" the two players simultaneously thrust forth (a) a clenched fist (stone), (b) two fingers (scissors), or (c) a flat palm (paper). If both players present the same object, the play is drawn; otherwise the winner is determined on the basis that "stone blunts scissors, scissors cut paper, and paper wraps stone."

Construct the payoff matrix for this game, assuming that the loser pays the winner $1.

On purely commonsense grounds, how would you rate the merits of the three possible moves available to each player?

2. In the game of "morra" the two contestants simultaneously hold out 1, 2 or 3 fingers, and each guesses the number of fingers his opponent is holding out (or, in some schools, guesses the total number of fingers extended by both players between them, which amounts to much the same thing). If both guess correctly, or both incorrectly, the play is drawn; otherwise the one who guesses correctly collects from the other a dollar for every finger extended.

The possible strategies for each player are the pairs (i, j), where $i, j = 1, 2, 3$, and (i, j) means "Hold out i fingers and guess j." Construct the 9×9 payoff matrix for this game.

---

* There seems to be a theorem here to the effect that any game between unsophisticated players is equivalent to another game between sophisticated players. I do not know to what extent this statement can be made precise.

17.3 STRICTLY DETERMINED GAMES AND NON-STRICTLY DETERMINED GAMES

We turn now to our third example concerning the rivalry between Alec and Bertram. First we must specify suitable data for the problem. We need to know for a start how likely Diana is to go to the ice rink, to go to the ball or to stay at home. We shall be quite general, and let x, y, z be the probabilities, respectively, for these three possibilities. Naturally $x + y + z = 1$.

Alec and Bertram estimate what their chances of success are in the various circumstances as follows, using units called *points*—doubtless rather ill-defined. Alec reckons that his ice-skating is worth 100 points but his dancing only 30 points; Bertram has the same estimate of his powers of dancing and ice-skating in that order. Hence if both decide to go to the ice rink Alec has an advantage of 70 points over Bertram, but this advantage is to no avail unless Diana is present (probability x). Hence Alec's *expected* advantage if both go to the rink is $70x$. If Alec goes to the rink and Bertram to the ball, then there is a probability x that Alec will have a 100 point advantage over Bertram, and probability y that Bertram will have a 100 point advantage over Alec. Hence Alec's expectation in this case is $100x - 100y$ points.

If Bertram stays at home instead there is a chance that if Diana also stays at home they may arrange an evening together. This is reckoned to be worth 20 points (assuming that Diana is at home) and hence staying at home has an expected advantage of $20z$ points. Thus if Alec goes to the rink and Bertram stays at home, Alec's expected advantage is $100x - 20z$ points.

These considerations determine the elements in the first row of the payoff matrix, and the rest of this matrix can be filled in in the same way, except for the element corresponding to the situation where Alec and Bertram both stay at home. If Diana stays at home then the advantage goes to the one who first contacts her after her phone has been repaired. Bertram, an only son, can phone at frequent intervals, but Alec has three teenage sisters who habitually hog the phone all evening. Under these circumstances Alec assesses his chances of phoning Diana first as virtually nil; hence if both stay at home there is an expected advantage of $20z$ to Bertram.

We now have the complete payoff matrix, which is as follows:

			Bertram	
		Rink	Ball	Home
Alec	Rink	$70x$	$100x - 100y$	$100x - 20z$
	Ball	$-30x + 30y$	$-70y$	$30y - 20z$
	Home	$-30x + 20z$	$-100y + 20z$	$-20z$

Let us suppose that x, y, z are 0.4, 0.4 and 0.2, respectively. The payoff matrix is then

		Bertram		
		R	B	H
	R	28	⓪	36
Alec	B	0	−28	8
	H	−8	−36	−4

(17.3.1)

As we learned from the previous example, one way of determining the best strategies for Alec and Bertram is to see if there is an element which is the smallest in its row and the largest in its column. There is such an element here, namely, the circled zero corresponding to Alec going to the rink and Bertram to the ball. If they do this then Alec is certain that whatever happens he will do no worse than break even, while Bertram is assured that Alec will do no better than this. Thus neither "player" gains advantage himself, but each prevents the other from gaining advantage.

Another way of seeing that this is the only reasonable strategy is to observe that every element in the second row of (17.3.1) is greater than the corresponding element in the third row. We say that row 2 *dominates* row 3. In general, we say that a row vector $(x_1, x_2, \ldots, x_n)$ *dominates* a row vector $(y_1, y_2, \ldots, y_n)$ if $x_i \geq y_i$ $(i = 1, 2, \ldots, n)$ and for at least one i we have $x_i > y_i$, i.e., the vectors are not equal. Clearly Alec is not well-advised to stay at home. Whatever Bertram does, Alec is better off at the ball than at home. In general the first player in a game (the one whose advantages are listed in the payoff matrix) should never choose as his strategy one whose row in the matrix is dominated by another row, and any dominated row can therefore be eliminated from the payoff matrix without materially affecting the game.

We see also in (17.3.1) that row 1 dominates row 2. Hence row 1 is the only row that really matters, and Alec's only sensible strategy is to go to the rink. This being the case Bertram will of course go to the ball so as to make his losses a minimum, zero in fact. The value of the game is therefore zero.

If we consider Bertram's strategy first we see that, since the payoff matrix lists Bertram's *losses*, and since column 2 is dominated by the other two columns (with an obvious extension to column vectors of the term "dominate"), column 2 is the only reasonable choice for Bertram. Alec then makes his advantage as much as possible by choosing the first row. In general we see that the second player should never choose a column which dominates another, and therefore that such a column can be removed from a payoff matrix without materially affecting the game.

Suppose now that $x = 0.3$, $y = 0.1$, $z = 0.6$. The payoff matrix then reads

(17.3.2)   Alec

		Bertram		
		R	B	H
	R	21	20	⑱
	B	−6	−7	− 9
	H	3	2	−12

We again have an element which is the minimum in its row and the maximum in its column; it is the circled "18". If there is such an element in a payoff matrix the element is called a *saddle point* and the game is said to be *strictly determined*. Hence all the games considered so far have been strictly determined except for the first game (matching dimes). From (17.3.2) we see that the best strategies are for Alec to go to the rink and for Bertram to remain at home. Intuitively, the fact that Diana is now much more likely to be at home means that it is worth while for Bertram to exchange his advantage on the dance floor for his advantage on the phone.

In this example as with the preceding, we can obtain the same result by deleting dominated rows from the matrix. In fact the first row dominates each of the others. However this is just coincidental and bears no relation to the existence (or otherwise) of a saddle point. If the payoff matrix had been

(17.3.3)   Alec

		Bertram		
		R	B	H
	R	21	20	18
	B	30	−7	−9
	H	−30	22	−12

then the same element "18" would still be a saddle point although now no row dominates another row, and no column dominates another. The technique of deleting dominated rows and dominating columns from a payoff matrix is often useful to simplify such a matrix, but only in trivial examples will it eliminate all but one of the rows or columns and hence determine the game.

When a game is strictly determined the row and column in which the saddle point occurs give the optimum strategies for the first and second players, respectively. The saddle point (element) is then the value of the game. Strictly determined games are less interesting than those which are not strictly determined, but they are simpler, which is why we have considered them first. We shall now consider a game which is not strictly determined, and to this end we consider the Alec-Bertram-Diana situation when $x = 0.1$, $y = 0.2$,

$z = 0.7$. The payoff matrix is then

(17.3.4)

		Bertram		
		R	B	H
Alec	R	7	-10	-4
	B	3	-14	-8
	H	11	-6	-14

The row minima are -10, -14, -14, respectively, and none of these is the maximum in its column. Hence there is no saddle point. Our previous method will therefore not work, and we must consider afresh what we should do.

We first observe that we can simplify the payoff matrix, using the idea of dominance, for row 1 dominates row 2. As far as any reasonable course of action is concerned the game is no different if we delete row 2 and make the payoff matrix

		Bertram		
		R	B	H
Alec	R	7	-10	-4
	H	11	-6	-14

Similarly since column 1 dominates column 2, it can never correspond to a sensible strategy for Bertram, and hence can be deleted. We end up with the 2×2 payoff matrix:

		Bertram	
		B	H
Alec	R	-10	-4
	H	-6	-14

which does not have a saddle point.

Suppose now that this "game" were to be played not once but many times. Then it would not be to Alec's advantage to make the same choice each time. If he always went to the rink, Bertram would counter by going to the ball; if Alec always stayed at home, Bertram would do so, too. But if Alec does sometimes one thing, sometimes the other, then Bertram's choice will often turn out to be the worse of the two possibilities, and his gain will be less. For the same reason Bertram would be ill-advised to play either strategy to the exclusion of the other.

Clearly then each player should adopt what is known as *mixed strategy*, whereby he sometimes makes one choice, sometimes the other. With what frequencies, or probabilities, should they make these choices? At present we do not know, so let us assume that Alec goes to the rink with probability p (and hence stays at home with probability $1 - p$), while Bertram goes to the ball with probability q, and stays at home with probability $1 - q$. We now analyze Alec's expectation.

If Alec goes to the rink, and Bertram to the ball (the probability of this double occurrence being pq) then Alec's gain is -10. Thus his expected gain from this source is $-10pq$. If Alec goes to the rink and Bertram stays at home, then Alec has a probability $p(1 - q)$ of gaining -4 points, or an expected gain of $-4p(1 - q)$. The other two possibilities yield expectations of $-6(1 - p)q$ and $-14(1 - p)(1 - q)$. Hence Alec's total expectation is

$$E = -10pq - 4p(1 - q) - 6(1 - p)q - 14(1 - p)(1 - q)$$
$$= -14 + 10p + 8q - 14pq.$$

Since this game is not strictly determined, and hence no *pure strategy* (making always the same choice) is advantageous for either player, the game now consists in Alec and Bertram choosing values for p and q, respectively, to their best advantage. It is Alec's aim to maximize E, and Bertram's aim to minimize it.

The reader can verify that E can be written in the following form:

$$E = -14(p - \tfrac{4}{7})(q - \tfrac{5}{7}) - \tfrac{58}{7}.$$

Suppose that Alec chooses to make $p = \tfrac{4}{7}$, then $E = -\tfrac{58}{7}$ irrespective of what q is. On the other hand if he chooses some other value, then Bertram can choose q (either $> \tfrac{5}{7}$ or $< \tfrac{5}{7}$ as necessary) so as to make

$$(p - \tfrac{4}{7})(q - \tfrac{5}{7})$$

a positive number, thus decreasing Alec's expectation. For example, if Alec makes $p = \tfrac{6}{7}$, then

$$E = -14 \cdot \tfrac{2}{7}(q - \tfrac{5}{7}) - \tfrac{58}{7} = -4(q - \tfrac{5}{7}) - \tfrac{58}{7}$$

and by choosing $q = 1$, say, Bertram can decrease Alec's expectation to $-\tfrac{66}{7}$.

In the same way, if Bertram makes $q = \tfrac{5}{7}$, this fixes Alec's expectation at $-\tfrac{58}{7}$ while if he chooses any other value for q, he runs the risk of allowing Alec to increase his expectation. Hence the best values for p and q are $\tfrac{4}{7}$ and $\tfrac{5}{7}$, respectively, enabling Alec to ensure that he gets at least $-\tfrac{58}{7}$ points, while enabling Bertram to ensure that Alec gets no more than this amount.

Thus Alec should go to the rink with probability $\tfrac{4}{7}$, and Bertram should go to the ball with probability $\frac{5}{7}$. The value of the game is $-\frac{58}{7}$ to Alec. The game is not a fair game, being biased in favor of Bertram, but it is unlikely that Bertram would offer any compensation to Alec for the privilege of playing this particular game! All's fair in love and war!

Let us now return to our first game, which also was not strictly determined. We shall assume that A plays heads with probability p, while B plays heads with probability q. From the payoff matrix

B

		q	$1-q$
A	p	1	-1
	$1-p$	-1	1

we find that A's expectation is

$$E = pq - p(1-q) - (1-p)q + (1-p)(1-q)$$
$$= 4(p - \tfrac{1}{2})(q - \tfrac{1}{2}).$$

We have much the same situation as before. By making $p = \frac{1}{2}$, A can ensure a gain of 0, whatever B does. On the other hand, by choosing $q = \frac{1}{2}$, B can also ensure a gain of 0. Both players then break even. A choice by either player of some other probability introduces the possibility of that player losing money. Hence over many plays of the game each player should choose heads and tails with equal likelihood. This is the conclusion we came to earlier from commonsense considerations. The value of the game is zero, i.e., it is a fair game.

We can reiterate here a point already made on page 877, namely, that we shall assume that both players can analyze the game fully. Thus the *best strategy* for A will be one which enables him to win the most that he can be sure of winning whatever B does. The best strategy for B will be one which enables him to keep his losses to the minimum possible against any play by A. Thus any departure from such a best strategy allows the possibility of gaining less or losing more.

This assumption, and this definition of "best strategy," will be followed throughout this chapter.

Exercise 102

1. Show that the games with the following payoff matrices are strictly determined, and hence find the optimum pure strategies for the players, and the value of the game.

(a)

5	8
3	−1

(b)

40	59
79	83

(c)

−1	3
2	10

(d)

2	5
3	4

(e)

3	−1	0	4
2	−2	−1	3

(f)

−2	0
2	4
6	8

(g)

−3	−2	−1	0
4	3	2	1
3	2	1	0
−4	−3	−2	−1

2. It is quite possible for a game to have more than one saddle point. Show that this is so with the following matrices, and find all the saddle points.

(a)

5	5	3	4	3
4	3	2	4	1
4	6	2	7	2
7	5	3	4	3

(b)

1	0	0	3
1	0	0	6
1	−2	−2	1
6	−4	−2	1

3. Prove that if a payoff matrix has more than one saddle point, then the payoffs at these saddle points must be equal.

17.4 GENERAL MATRIX GAMES

We are now in a position to formulate the definition of a *matrix game*, which generalizes the type of game we have been considering. There are two players, A and B; A has to choose between m possibilities, and B chooses between n possibilities. For each combination of choices there is a corresponding payoff—the amount of money (or something to which some quantitative valuation can be attached) which B pays to A. (If A pays B the payoff is regarded as negative.) These payoffs can be exhibited as $m \times n$ *payoff matrix*.

$$M = (a_{ij})$$

where a_{ij} is the payoff when A chooses strategy i and B chooses strategy j.

If M has an element $a_{\alpha\beta}$ which is the smallest element in its row and the largest in its column the game is strictly determined, and its value is $a_{\alpha\beta}$. The optimum method of play is for A to play the pure strategy α, and for B to play the pure strategy β. The justification for this is a straightforward generalization of what has been said about the strictly determined games earlier in the chapter.

If there is no such element in M, then the game is not strictly determined, and it is in neither player's interest to play a pure strategy. Let us suppose therefore that A plays the m strategies with probabilities $x_1, x_2, \ldots, x_m$ and B plays his n possible strategies with probabilities $y_1, y_2, \ldots, y_n$. The payoff a_{ij} will occur with probability $x_i y_j$ and hence A's expectation will be

$$(17.4.1) \qquad \sum_{i,j} a_{ij} x_i y_j = (x_1, x_2, \ldots, x_m) \begin{pmatrix} a_{11} & a_{12} & \cdots & a_{1n} \\ a_{21} & a_{22} & \cdots & a_{2n} \\ & & \cdots & \\ a_{m1} & a_{m2} & \cdots & a_{mn} \end{pmatrix} \begin{pmatrix} y_1 \\ y_2 \\ \cdots \\ y_n \end{pmatrix}$$

or

$$(17.4.2) \qquad E(x, y) = x^T M y$$

where x, y are the probability column vectors of the x_is and the y_is.

Suppose now that A chooses a particular strategy given by the vector $(x'_1, x'_2, \ldots, x'_m)$. If B knew what this strategy was he would plan his own strategy accordingly and choose $y_1, y_2, \ldots, y_n$ so as to make A's expectation

$$(17.4.3) \qquad \sum_{i,j} a_{ij} x'_i y_j$$

a minimum. But (17.4.3) can be written as

$$(17.4.4) \qquad \sum_j q_j y_j$$

where $q_j = \sum_i a_{ij} x'_i$, and we see that B's best strategy (assuming still a knowledge of A's strategy) will be to play a pure strategy. For if q_β is the smallest of the q_j, then

$$\sum_j q_j y_j \geq \sum_j q_\beta y_j$$
$$= q_\beta,$$

since

$$\sum_j y_j = 1.$$

Hence the choice of the pure strategy β ($y_\beta = 1$, all other $y_j = 0$) is at least as good as any other for B.

It follows that if A plays the strategy $(x'_1, x'_2, \ldots, x'_m)$ he will win as little as q_β, the smallest of the q_js, if B plays correctly, i.e., he can hope to gain only

$$(17.4.5) \qquad\qquad \min_j \sum_i a_{ij} x'_i.$$

Clearly then A will choose whatever vector makes (17.4.5) a maximum, and the most that he can expect from the game is therefore

$$(17.4.6) \qquad\qquad v_A = \max_x \min_j \sum_i a_{ij} x_i.$$

If we now consider B's point of view, and suppose he has chosen a strategy $(y'_1, y'_2, \ldots, y'_n)$, we see that the best that A can do is to choose the strategy x which maximizes

$$(17.4.7) \qquad\qquad \sum_i x_i p_i$$

where $p_i = \sum_j a_{ij} y'_j$. It is in B's interests to ensure that this maximum amount (17.4.7) is as small as possible. He therefore chooses the y'_j with this end in view, and the most he can expect from the game, unless A plays badly, is that he will have to pay to A the amount

$$(17.4.8) \qquad\qquad v_B = \min_y \max_i \sum_j a_{ij} y_j.$$

We therefore have the situation that A can ensure that he gets at least v_A, the largest of the amounts that he will win even against the best defence by B, with a possibility of a larger win if B plays badly. On the other hand B can restrict his losses to v_B, even against best defence by A. Now if it so happens that $v_A = v_B (= v$, say) then there is no more to be said. A cannot win more than v (since B can ensure that he does not lose more than v), and B cannot lose less than v (since A can ensure a win of at least v). Hence the strategies determined by the above reasoning are optimum for both players, and the value of the game is v.

Two questions therefore arise. Under what circumstances are v_A and v_B

equal, and when they are, how do we find the value of the game and discover the associated strategies? The first question is answered by the *Minimax theorem*—the fundamental theorem of the theory of games—which states that the *maximin* v_A is *always* equal to the *minimax* v_B. A discussion and proof of this important theorem will be given later in the chapter and will contain within it the answer to the second question.

Exercise 203

1. If A is an $m \times n$ payoff matrix, and J is the $m \times n$ matrix all of whose elements are 1, show that the games whose payoff matrices are A and $B = \lambda A + \mu J$, where λ and μ are real numbers, and $\lambda > 0$, have the same optimum strategies.

2. If the payoff matrix B is derived from a payoff matrix A by deleting dominated rows and dominating columns, prove that B is strictly determined if, and only if, A is strictly determined.

17.5 SOLUTION OF 2 × 2 GAMES

Before considering the general problem of solving matrix games we shall consider the solution of the general 2×2 games, since this turns out to be particularly simple. The method has been indicated by particular examples in section 17.3; it merely remains for us to follow the method in the general case.

Suppose the payoff matrix is

(17.5.1)

		B	
		q	$1-q$
A	p	a	b
	$1-p$	c	d

and A plays his two strategies with probabilities p and $1 - p$, while B plays his with probabilities q and $1 - q$, as shown in (17.5.1). A's expectation is

$$E(p, q) = pqa + p(1 - q)b + q(1 - p)c + (1 - p)(1 - q)d$$

which reduces to

(17.5.2) $\quad E(p, q) = (a - b - c + d) \times$
$$\left\{ pq - \frac{d - b}{a - b - c + d}p - \frac{d - c}{a - b - c + d}q + \frac{d}{a - b - c + d} \right\}.$$

The expression within braces in (17.5.2) can be expressed in the form $(p - \alpha)(q - \beta) + k$, and then reads

$$(17.5.3) \quad \left(p - \frac{d - c}{a - b - c + d}\right)\left(q - \frac{d - b}{a - b - c + d}\right) + \frac{ad - bc}{(a - b - c + d)^2}$$

as the reader should verify.

If the game is strictly determined, and a is the element giving the required pure strategies, then a must be the smallest element in its row, and the largest in its column. Hence $a < b$ and $a > c$, or $c < a < b$. Taking the other three elements in the same way we see that the game is strictly determined if any of the following hold:

$$(17.5.4) \quad \begin{array}{c} c < a < b \\ b < d < c \\ d < b < a \\ a < c < d. \end{array}$$

Hence the game is not strictly determined if, and only if, *none* of these statements (17.5.4) is correct. This means that a and d do not lie between b and c, and that b and c do not lie between a and d. In other words, either b and c are each less than a and d, or a and d are each less than b and c. It follows from this that

$$\alpha = \frac{d - c}{a - b - c + d} = \frac{(c - d)}{(b - a) + (c - d)}$$

and

$$\beta = \frac{d - b}{a - b - c + d} = \frac{(b - d)}{(c - a) + (b - d)}$$

are both positive and less than 1, since the bracketed expressions are either all positive or all negative.

We thus have

$$(17.5.5) \quad E(p, q) = (a - b - c + d)(p - \alpha)(q - \beta) + \frac{ad - bc}{a - b - c + d}.$$

It follows that A can play so as to guarantee winning

$$(17.5.6) \quad \frac{ad - bc}{a - b - c + d}$$

(by choosing $p = \alpha$), while B, by choosing $q = \beta$, can guarantee that A

gets no more than this amount. If A were to choose a probability p other than α, then B could choose q so as to make $(a - b - c + d)(p - \alpha)(q - \beta)$ in (17.5.5) negative, thereby decreasing A's expectation. Only by choosing $p = \alpha$ can A prevent B from doing this. Similarly, only by choosing $q = \beta$ can B prevent A from profiting at B's expense. Thus (17.5.6) is the value of the game, and the winning (mixed) strategies for A and B are given by

$$(17.5.7) \qquad p = \frac{d - c}{a - b - c + d}; \quad q = \frac{d - b}{a - b - c + d}.$$

This solves the problem of 2×2 games that are not strictly determined.

Exercise 204

1. Solve the 2×2 non-strictly determined games having the following pay-off matrices:

(a)
1	−2
−2	1

(c)
3	−5
−6	4

(e)
1	−3
−1	3

(b)
7	4
−3	21

(d)
1	−4
−4	3

(f)
3	0
1	3

2. By deleting dominated rows and dominating columns from the following matrices, solve the corresponding games.

(a)
2	1	1	3
3	1	3	4
−2	5	6	0
−6	3	5	0

(b)
2	5	2
19	−3	24
1	6	2

17.6 SOLUTION BY ITERATION

We now turn to a method of solution of games larger than 2×2, a method that has some advantages over the more "standard" method which we shall consider later. It is an *iterative* method, somewhat like those treated in Chapter 15, which enables us, by the repeated application of a simple calculation, to approximate to the solution of the game as closely as we wish.

To demonstrate this method we shall consider a particular 3×3 game, but it will be clear that the method will apply to games of any size. We take the game whose payoff matrix is

		B		
		p	q	r
A	a	8	2	2
	b	4	1	6
	c	6	5	0

In this game there is no saddle point. We shall imagine that it is played many times over by two players who are not capable of analyzing the game completely but who *are* able to decide what is their best pure strategy given foreknowledge of what their opponent's strategy will be. Each player plays as follows: he assumes that his opponent will play the mixed strategy obtained by choosing the three pure strategies with probabilities proportional to the number of times the strategies have been chosen in the past. That is, at each play, A will assume that B's strategy is going to be an *average* of B's previous plays, and having made this assumption about his opponent's strategy, chooses his best pure strategy on this assumption. B does the same.

This explanation is perhaps rather too general to be immediately lucid but should become clear if we consider an actual run of plays of the game by A and B.

The first time that A and B play, neither has any information concerning his opponent, so each will choose a strategy at random. Suppose A chooses strategy c while B chooses strategy q. We shall denote this by

			B		
			p	q	r
			0	1	0
A	a	0	8	2	2
	b	0	4	1	6
	c	1	6	5	0

The numbers above and to the left of the payoff matrix are the numbers of times each pure strategy has been used.

In the second play of the game A will assume that B will play as in the past, i.e., will choose q. On this assumption A's best strategy is c again, since this maximizes the payoff. B will assume that A, as before, will play c, and hence will choose to play strategy r, since this gives the least payoff to A.

The diagram now becomes

			B		
			p	q	r
			0	1	1
A	a	0	8	2	2
	b	0	4	1	6
	c	2	6	5	0

For the third play, A assumes that, since B has chosen q and r equally often in the past, he will now choose them with equal probabilities. His expectations for his three choices are therefore:

1. If he chooses a: $\frac{1}{2} \cdot 2 + \frac{1}{2} \cdot 2 = 2$
2. If he chooses b: $\frac{1}{2} \cdot 1 + \frac{1}{2} \cdot 6 = \frac{7}{2}$
3. If he chooses c: $\frac{1}{2} \cdot 5 + \frac{1}{2} \cdot 0 = \frac{5}{2}$.

His best strategy is therefore b. B will assume the pure strategy c on A's part, and will play r, as before. The diagram is now

			B		
			p	q	r
		3	0	1	2
A	a	0	8	2	2
	b	1	4	1	6
	c	2	6	5	0

where in the top left-hand corner we have put the total number of plays so far. One more iteration will put us on the track of a general procedure. A now assumes that B will play strategies q and r with probabilities $\frac{1}{3}$ and $\frac{2}{3}$, respectively. His expectations are

$$\frac{1}{3} \cdot 2 + \frac{2}{3} \cdot 2 = 2 \quad \text{if he plays } a,$$

$$\frac{1}{3} \cdot 1 + \frac{2}{3} \cdot 6 = \frac{13}{3} \quad \text{if he plays } b,$$

and $$\frac{1}{3} \cdot 5 + 0 = \frac{5}{3} \quad \text{if he plays } c.$$

He will therefore choose b. B assumes that A will play b and c with probabilities $\frac{1}{3}$ and $\frac{2}{3}$, respectively, so he expects the following losses:

$$\frac{1}{3}\cdot 4 + \frac{2}{3}\cdot 6 = \frac{16}{3} \quad \text{if he plays } p,$$

$$\frac{1}{3}\cdot 1 + \frac{2}{3}\cdot 5 = \frac{11}{3} \quad \text{if he plays } q,$$

and

$$\frac{1}{3}\cdot 6 + \frac{2}{3}\cdot 0 = 2 \quad \text{if he plays } r.$$

He therefore chooses r again, and the diagram becomes

(17.6.1)

			B		
			p	q	r
		4	0	1	3
A	a	0	8	2	2
	b	2	4	1	6
	c	2	6	5	0

Suppose now that we have reached the situation

			B		
			p	q	r
		n	y_1	y_2	y_3
A	a	x_1	8	2	2
	b	x_2	4	1	6
	c	x_3	6	5	0

where the x_is and y_js are integers. Clearly

$$x_1 + x_2 + x_3 = y_1 + y_2 + y_3 = n.$$

Now A, assuming that B will play his three strategies with probabilities y_1/n, y_2/n, y_3/n, will choose the strategy corresponding to the largest of

$$8y_1 + 2y_2 + 2y_3$$
$$4y_1 + y_2 + 6y_3$$

and

$$6y_1 + 5y_2.$$

Note that we can forget the denominator n in the probabilities, since by doing so we merely multiply A's expectations by a constant factor, which does not alter their relative magnitudes. In a similar way, B will choose the strategy corresponding to the smallest of

$$8x_1 + 4x_2 + 6x_3$$
$$2x_1 + x_2 + 15x_3$$

and
$$2x_1 + 6x_2.$$

We can summarize this as follows. A chooses the row which has the largest scalar product with the vector (y_1, y_2, y_3) of B's plays; B chooses the column which has the smallest scalar product with the vector $\begin{pmatrix} x_1 \\ x_2 \\ x_3 \end{pmatrix}$ of A's plays. To complete the iteration, 1 is added to the appropriate row and column counts.

Thus the computation required for each iteration is seen to be very simple. The question remains whether the successive probability vectors $\left(\frac{x_1}{n}, \frac{x_2}{n}, \frac{x_3}{n}\right)$ and $\left(\frac{y_1}{n}, \frac{y_2}{n}, \frac{y_3}{n}\right)$ that we get will converge to the optimal mixed strategies for A and B. It can be proved that they do, though the proof is beyond the scope of this book. It is also true that this convergence does not depend on the strategies chosen in the first play of the game.

This last remark prompts an observation of interest concerning the payoff matrix that has been chosen as our example. Column 1 of this matrix dominates column 2, and hence it can never be in B's interests to choose strategy p; B's optimum mixed strategy will thus choose this column with probability zero. But what if B, in ignorance of this, starts by playing strategy p? Will it matter? No, because the dominance will ensure that the scalar product of the (x_1, x_2, x_3) with the first column is never the smallest, so that strategy p will never be chosen again. Hence after n plays there will be a probability $1/n$ associated with the choice of strategy p. This will never become zero (its theoretical value) but can be made as close to zero as we like by taking sufficiently large n. This emphasizes the fact that the iterative method may not yield the *exact* solution of the game but can give it to any desired degree of accuracy.

A shortcut in the above method—one which the reader will find useful in the exercises which follow—enables us to avoid recalculating the scalar products completely at each stage. Before starting the iterations, write a row vector of zeros below the table, and a column vector of zeros to the right of it. We can call these the *key row* and *key column*. Now each iteration starts by choosing a row and a column of the matrix. When these have been chosen, add the chosen row (vectorially) to the key row, and add the chosen column to the key column. Thus the key row and the key column will change as the iterations progress. It will be left as an exercise to the reader to show that at each stage the elements of these key vectors are the required scalar products, and that to find the required row and column for the next iteration we look for the largest element in the key column and the smallest element in the key row.

In this way we keep track of the scalar products using addition only and no multiplication.

To illustrate this method let us consider the game whose payoff matrix is

3	5	2
4	1	2
1	2	4

The first few iterations yield the following results:

(1)

	0	0	0	
0	3	5	2	0
0	4	1	2	0
0	1	2	4	0

Key row→ 0   0   0   ↑
                      Key column

(2)

	1	0	1	0	
0		3	5	2	⑤
1		4	1	2	1
0		1	2	4	2
		4	①	2	

(3)

	2	0	2	0	
1		3	5	2	⑩
1		4	1	2	2
0		1	2	4	4
		7	6	④	

(4)

	3	0	2	1	
2		3	5	2	⑫
1		4	1	2	4
0		1	2	4	8
		10	11	⑥	

The circled element in the key row and key column are the smallest and largest, respectively. They determine the row and column for the *next* iteration. The reader should continue this example for several more iterations to get the general idea of the method.

Exercise 105

1. In the iterative method we choose A's next play by finding the row whose scalar product with B's vector of plays is a maximum (see page 893). Show that this scalar product, divided by the number of iterations, is the amount which A can be sure of winning, on the average, if B plays according to his past overall strategies. Deduce that this scalar product exceeds the value of the game.

Show similarly that the scalar product which determines B's next choice of strategy is less than the value of the game.

These results enable one to assign bounds to the value of the game, bounds which approach each other as the iterations proceed. This in turn enables one to assess the accuracy of the current (approximate) solution.

2. Carry on the solution of the following games by the iterative method for at least 10 iterations.

(a)

0	2	1
2	0	2
1	2	0

(b)

1	4	−1
6	1	−2
0	−2	0

Obtain bounds for the value of these games (as in problem 1).

3. The example worked in section 17.6 was a 3×3 game, but the method is clearly applicable to games of any size. Use the iterative method on the following game, taking at least 10 iterations, and obtain an estimate of its value.

3	3	−1	7	0
6	−2	6	4	5
−1	4	1	1	−1
2	1	1	2	6
7	3	0	−1	3

17.7 GENERAL METHOD OF SOLUTION

We now consider a general method for solving a two-person, zero-sum game. The method consists in reducing the problem to a problem in linear programming, which we can solve by the methods of Chapter 16. Insofar as these methods are iterative, the process we are about to describe is also an iterative process; but unlike the method of section 17.6, it is a process which must terminate after a finite number of steps. In the course of describing this process we shall prove the result that was left hanging at the end of section 17.4 as to whether the two values v_A and v_B there defined (the *maximin* and the *minimax*) are always equal. In other words, we shall prove the *Minimax theorem*, which is fundamental to the theory of two-person zero-sum games.

Referring back to section 17.4 we see that in the general matrix game the most that A can expect to gain is

$$(17.7.1) \qquad\qquad v_A = \max_x \min_j \sum_i a_{ij} x_i.$$

Let $g = \min_j \sum_i a_{ij} x_i$. Assuming, for the moment, that $g > 0$, we have

$$(17.7.2) \qquad\qquad \sum_i a_{ij} x_i \geq g > 0 \text{ for all } j.$$

If we write $\xi_i = x_i/g$, then (17.7.2) can be written

$$(17.7.3) \qquad\qquad \sum_i a_{ij} \xi_i \geq 1 \text{ for all } j.$$

We know also that $\sum_i x_i = 1$, since the x_is are the probabilities of the various pure strategies available to A. This can now be written as

$$(17.7.4) \qquad\qquad \sum_i \xi_i = \frac{1}{g}.$$

Now A's object is to choose the vector $x = (x_1, x_2, \dots)$ so as to maximize g [see (17.7.1)]. This is the same as minimizing $k = 1/g$. Thus we can describe A's problem as that of choosing $\xi_1, \xi_2, \dots$, so that

$$\sum_i \xi_i = k$$

is a minimum, subject to the linear inequalities (17.7.3). This is a straightforward linear programming problem, which we can summarize as

$$(17.7.5) \qquad \begin{cases} \text{Minimize} \quad \sum_i \xi_i \\ \text{subject to the constraints} \\ \sum_i a_{ij} \xi_i \geq 1 \quad \text{(all } j). \end{cases}$$

Note that provided $g > 0$ all the variables ξ_i will be non-negative.

Let us now look at B's problem. He wishes to choose probabilities $y_1, y_2, \ldots,$ such that

$$\max_i \sum_j a_{ij} y_j$$

is a minimum [see (17.4.8)]. Let

$$h = \max_i \sum_j a_{ij} y_j. \text{ We then have}$$

(17.7.6)
$$\begin{cases} \sum_j a_{ij} y_j \leq h & \text{for all } i \\ \sum_j y_j = 1. \end{cases}$$

If we write $\eta_j = y_j / h$ and put $h = 1/l$, then

$$\sum_j a_{ij} \eta_j \leq 1 \qquad \text{for all } i,$$

and

$$\sum_j \eta_j = l.$$

Now B's aim is to minimize h, that is, to maximize l. Hence B's problem can also be summarized as a linear programming problem, namely,

$$\text{Maximize} \quad \sum_j \eta_j$$

(17.7.7)   subject to the constraints

$$\sum_j a_{ij} \eta_j \leq 1 \qquad \text{for all } i.$$

If $h > 0$ all the η_i will be non-negative.

Thus provided the conditions $g > 0$ and $h > 0$ are met, the problem of A's strategy and that of B's strategy reduce to linear programming problems. But the fulfilling of these conditions is no trouble, for we saw on page 887 that if we add a constant amount to every element of a payoff matrix we do not materially affect the game; we merely introduce an *ante* to be paid by one player to the other for each game played. The strategies will be the same, and the value of the game (assuming that it has one—we have not yet proved that it must have) will be increased by the same constant amount. Hence, by adding a suitable constant amount to every element of a payoff matrix we can arrange that every element of it is positive. It then follows at once that g and h (and hence k and l) are positive.

This reduction of the solution of a game to two linear programming problems is elegant enough in itself. What is even more elegant is that the two problems (17.7.5) and (17.7.7) are *dual* problems, in the sense defined in section 16.10, as the reader should verify. That being the case we can immediately invoke the fundamental theorem of linear programming, that the optimum values for a problem and its dual are equal, and deduce that the minimum value of

$$k = \sum_i \xi_i$$

is the same as the maximum value of

$$l = \sum_j \eta_j.$$

Since $v_A = \max g$, we have $1/v_A = \min k$, and similarly we have $1/v_B = \max l$. Therefore, from the properties of dual linear programming we deduce that $v_A = v_B$.

Hence we have now settled the point which we raised at the end of section 17.4; v_A and v_B are always equal. This means that, by suitable strategy, A can be sure of winning an amount $v(= v_A = v_B)$, and that B, by playing correctly, can limit A's winnings to just this amount. These are therefore the best strategies, and the amount v is the value of the game.

Since the problems associated with the two players are dual to each other, it does not matter which one we solve. B's problem is a little simpler, since a basic feasible solution is ready to hand, and the initial condensed tableau will be

		η_1	η_2	η_3	$\cdots$	η_n
ξ_1	1	a_{11}	a_{12}	a_{13}	$\cdots$	a_{1n}
ξ_2	1	a_{21}	a_{22}	a_{23}	$\cdots$	a_{2n}
ξ_3	1	a_{31}	a_{32}	a_{33}	$\cdots$	a_{3n}
$\cdot$	$\cdot$	$\cdot$	$\cdot$	$\cdot$	$\cdots$	$\cdot$
ξ_m	1	a_{m1}	a_{m2}	a_{m3}	$\cdots$	a_{mn}
	0	1	1	1	$\cdots$	1

The basic feasible solution is $\xi_1 = \xi_2 = \ldots = \xi_m = 1$. This tableau is now manipulated according to the rules already given until no further iterations can be performed. The element in the bottom left-hand corner is then the *reciprocal* of the value of the game. The final values of the ξs and the ηs will be proportional to the probabilities of the various pure strategies and thus the probabilities themselves are easily found.

All this is best illustrated by example, and for our first example we shall take the problem that we used in the last section to illustrate the iterative method of solution. The payoff matrix was

$$\begin{pmatrix} 8 & 2 & 2 \\ 3 & 1 & 6 \\ 6 & 5 & 0 \end{pmatrix}$$

and we therefore set up the following initial tableau.

(17.7.9)

		η_1	η_2	η_3
ξ_1	1	⑧	2	2
ξ_2	1	4	1	6
ξ_3	1	6	5	0
	0	1	1	1

Using the pivot element shown we apply the usual procedure once to obtain tableau (17.7.10)

(17.7.10)

		ξ_1	η_2	η_3
η_1	$\frac{1}{8}$	$\frac{1}{8}$	$\frac{1}{4}$	$\frac{1}{4}$
ξ_2	$\frac{1}{2}$	$-\frac{1}{2}$	0	5
ξ_3	$\frac{1}{4}$	$-\frac{3}{4}$	$\left(\frac{7}{2}\right)$	$-\frac{3}{2}$
	$-\frac{1}{8}$	$-\frac{1}{8}$	$\frac{3}{4}$	$\frac{3}{4}$

With the indicated pivot element, the next tableau is

(17.7.11)

		ξ_1	ξ_3	η_3
η_1	$\frac{3}{28}$	$\frac{5}{28}$	$-\frac{1}{14}$	$\frac{5}{14}$
ξ_2	$\frac{1}{2}$	$-\frac{1}{2}$	0	⑤
η_2	$\frac{1}{14}$	$-\frac{3}{14}$	$\frac{2}{7}$	$-\frac{3}{7}$
	$-\frac{5}{28}$	$\frac{1}{28}$	$-\frac{3}{14}$	$\frac{15}{14}$

The next is

(17.7.12)

		ξ_1	ξ_3	ξ_2
η_1	$\frac{1}{14}$	$\left(\frac{3}{14}\right)$	$-\frac{1}{14}$	$-\frac{1}{14}$
η_3	$\frac{1}{10}$	$-\frac{1}{10}$	0	$\frac{1}{5}$
η_2	$\frac{4}{35}$	$-\frac{9}{35}$	$\frac{2}{7}$	$\frac{3}{35}$
	$-\frac{2}{7}$	$\frac{1}{7}$	$-\frac{3}{14}$	$-\frac{3}{14}$

and the final tableau is

(17.7.13)

		η_1	ζ_3	ζ_2
ξ_1	$\frac{1}{3}$	$\frac{14}{3}$	$-\frac{1}{3}$	$-\frac{1}{3}$
η_3	$\frac{2}{15}$	$\frac{7}{15}$	$\frac{1}{30}$	$\frac{1}{6}$
η_2	$\frac{1}{5}$	$\frac{6}{5}$	$\frac{1}{5}$	0
	$-\frac{1}{3}$	$-\frac{2}{3}$	$-\frac{1}{6}$	$-\frac{1}{6}$

Now in this problem η_1, η_2, η_3, are the main variables and ξ_1, ξ_2, ξ_3 are the slack variables. Hence the solution is given by

$$\eta_1 = 0; \qquad \eta_2 = \tfrac{1}{5}; \qquad \eta_3 = \tfrac{2}{15}.$$

Thus

$$y_1 = 0 \qquad y_2 = \tfrac{3}{5} \qquad y_3 = \tfrac{2}{5}.$$

The solution to the dual problem, for which ξ_1, ξ_2, ξ_3 are the main variables is

$$\xi_1 = 0; \qquad \xi_2 = \tfrac{1}{6}; \qquad \xi_3 = \tfrac{1}{6}$$

since ξ_1 is a nonbasic variable in the dual program. Hence A's best strategy is given by

$$x_1 = 0; \qquad x_2 = \tfrac{1}{2}; \qquad x_3 = \tfrac{1}{2}.$$

Thus the game is completely solved in a finite number of iterations. Note that any variables which end up in the same part of the tableau as they started (i.e., along the top, or the left-hand side) are zero, since they are slack variables in the appropriate problem. This is why ξ_1 and η_1 do *not* have the values 1/3 and 2/3, as one might mistakenly think.

One further example, somewhat larger than the problems considered hitherto, should make the method quite clear. We shall consider the 4 × 5 game whose payoff matrix is

$$\begin{pmatrix} 2 & 3 & ① & 4 & 2 \\ ① & 2 & 5 & 4 & 4 \\ 2 & 3 & 4 & 1 & 4 \\ 4 & ② & 2 & 2 & ② \end{pmatrix}$$

Now before we embark on any elaborate procedure it is only prudent to check that there is no obvious solution. We therefore look for a saddle point. The smallest elements in each row have been circled (in the last row there is a choice between four elements) but none of these is the largest element in its column. Hence there is no saddle point. It is as well to check also whether any row or column dominates another row or column, since this would enable us to reduce the size of the matrix. Again we draw a blank, and resign ourselves to having to use our most powerful weapons.

The initial tableau of the linear programming problem that we have to solve is (17.7.14) in which a suitable pivot element for the first iteration has been circled.

(17.7.14)

		p	q	r	s	t
a	1	2	3	1	4	2
b	1	1	2	⑤	4	4
c	1	2	3	4	1	4
d	1	4	2	2	2	2
	0	1	1	1	1	1

We now follow the rules for the simplex algorithm, exactly as before, and we obtain, in turn, the following tableaux:

(17.7.15)

		p	q	b	s	t
a	$\frac{4}{5}$	$\frac{9}{5}$	$\frac{13}{5}$	$-\frac{1}{5}$	$\frac{16}{5}$	$\frac{6}{5}$
r	$\frac{1}{5}$	$\frac{1}{5}$	$\frac{2}{5}$	$\frac{1}{5}$	$\frac{4}{5}$	$\frac{4}{5}$
c	$\frac{1}{5}$	$\textcircled{$\frac{6}{5}$}$	$\frac{7}{5}$	$-\frac{4}{5}$	$-\frac{11}{5}$	$\frac{4}{5}$
d	$\frac{3}{5}$	$\frac{18}{5}$	$\frac{6}{5}$	$-\frac{2}{5}$	$\frac{2}{5}$	$\frac{2}{5}$
	$-\frac{1}{5}$	$\frac{4}{5}$	$\frac{3}{5}$	$-\frac{1}{5}$	$\frac{1}{5}$	$\frac{1}{5}$

(17.7.16)

		c	q	b	s	t
a	$\frac{1}{2}$	$-\frac{3}{2}$	$\frac{1}{2}$	1	$\frac{13}{2}$	0
r	$\frac{1}{6}$	$-\frac{1}{6}$	$\frac{1}{6}$	$\frac{1}{3}$	$\frac{7}{6}$	$\frac{2}{3}$
p	$\frac{1}{6}$	$\frac{5}{6}$	$\frac{7}{6}$	$-\frac{2}{3}$	$\frac{11}{6}$	$\frac{2}{3}$
d	0	-3	-3	2	$\textcircled{7}$	-2
	$-\frac{1}{3}$	$-\frac{2}{3}$	$-\frac{1}{3}$	$\frac{1}{3}$	$\frac{5}{3}$	$-\frac{1}{3}$

(17.7.17)

		c	q	b	d	t
a	$\frac{1}{2}$	$\frac{9}{7}$	$\textcircled{$\frac{23}{7}$}$	$-\frac{6}{7}$	$\frac{13}{14}$	$\frac{13}{7}$
r	$\frac{1}{6}$	$\frac{1}{3}$	$\frac{2}{3}$	0	$-\frac{1}{6}$	1
p	$\frac{1}{6}$	$\frac{1}{21}$	$\frac{8}{21}$	$-\frac{1}{7}$	$\frac{11}{42}$	$\frac{1}{7}$
s	0	$-\frac{3}{7}$	$-\frac{3}{7}$	$\frac{2}{7}$	$\frac{1}{7}$	$-\frac{2}{7}$
	$-\frac{1}{3}$	$\frac{1}{21}$	$\frac{8}{21}$	$-\frac{1}{7}$	$-\frac{5}{21}$	$\frac{1}{7}$

(17.7.18)

		c	a	b	d	t
q	$\frac{7}{46}$	$\frac{9}{23}$	$\frac{7}{23}$	$-\frac{6}{23}$	$-\frac{13}{46}$	$\frac{13}{23}$
r	$\frac{3}{46}$	$\frac{5}{69}$	$-\frac{14}{69}$	$\frac{4}{23}$	$\frac{1}{46}$	$\frac{43}{69}$
p	$\frac{5}{46}$	$-\frac{7}{69}$	$-\frac{8}{69}$	$-\frac{1}{23}$	$\frac{17}{46}$	$-\frac{5}{69}$
s	$\frac{3}{46}$	$-\frac{6}{23}$	$\frac{3}{23}$	$\frac{4}{23}$	$\frac{1}{46}$	$-\frac{1}{23}$
	$-\frac{9}{23}$	$-\frac{7}{69}$	$-\frac{8}{69}$	$-\frac{1}{23}$	$-\frac{3}{23}$	$-\frac{5}{69}$

From the final tableau (17.7.18) we see that the variable t has ended up in the top line, where it began (in fact, it never moved); its value is therefore zero. All other variables have changed their places. Hence a, b, c and d are proportional to the corresponding elements in the bottom line (with a positive sign), and p, q, r and s are proportional to the numbers to their right. Thus

$$a : b : c : d = \tfrac{8}{69} : \tfrac{1}{23} : \tfrac{7}{69} : \tfrac{3}{23}$$
$$= 8 : 3 : 7 : 9$$

and

$$p : q : r : s : t = \tfrac{5}{46} : \tfrac{7}{46} : \tfrac{3}{46} : \tfrac{3}{46} : 0$$
$$= 5 : 7 : 3 : 3 : 0.$$

Since the probabilities for each player must sum to 1, we have, for the final solution

$$a = \tfrac{8}{27}, \qquad b = \tfrac{1}{9}, \qquad c = \tfrac{7}{27}, \qquad d = \tfrac{1}{3}$$

and

$$p = \tfrac{5}{18}, \qquad q = \tfrac{7}{18}, \qquad r = \tfrac{1}{6}, \qquad s = \tfrac{1}{6}, \qquad t = 0.$$

The common optimum value of the two linear programming problems is 9/23 (remember the change of sign), and this is the reciprocal of the value of the game. Thus the value of this game is 23/9. The reader may care to verify that the given probabilities will result in the payment to the first player, A, of exactly this amount.

It is worth asking why the variable t has the value zero. The interpretation is, of course, that the corresponding pure strategy (the fifth for player B) should never be used; there is always a better alternative. This better alternative cannot be another pure strategy since, if so, its column would be dominated by the t column, and we checked on this before beginning the simplex algorithm. What has happened here is that although the t column does not dominate any other one column, it does dominate a convex combination of other columns. In fact, a "50–50" mixture of the second and third columns of the payoff matrix, viz., $\begin{pmatrix} 2 \\ 3\frac{1}{2} \\ 3\frac{1}{2} \\ 2 \end{pmatrix}$, is dominated by the last column. Hence it

is always better for B to play these two columns with equal probabilities than it is for him to play the fifth pure strategy. Had we spotted this earlier we could have eliminated the final column from the matrix; but it is only rarely that this sort of dominance will be apparent on inspection.

Exercise 206

1. Two players each think of one of the numbers 1, 2, 3, 4. If both choose the same number, player A pays player B that number of dollars. If they choose different numbers, then B pays A the number of dollars that A chose.

Construct the payoff matrix for this game and solve it.

2. Solve the game whose payoff matrix is

−2	2	2	4
6	0	4	2
2	6	5	−2

3. Since linear programming problems having matrices whose sizes are $2 \times n$ or $n \times 2$ can be solved graphically, it follows that games of these sizes also admit a graphical solution. Solve graphically the following games.

(a)

7	0	1	3
−1	7	3	0

(b)

2	−1
−3	2
−1	3
−2	3

4. Solve the game whose payoff matrix is

− 7	22	7	13	5
−17	− 5	37	4	−11
32	35	2	−19	57
15	− 5	24	4	4

17.8 MORE COMPLEX GAMES—INFINITE GAMES

We began this chapter by discussing games in general but soon confined our scope to two-person zero-sum games. We then defined matrix games and devoted the rest of the chapter to various methods for their solution. We have not (apparently) considered any games more complicated than matrix games, which, by all appearances, seem to be quite unsophisticated. The reader may well be asking "When do we get on to *real* games of skill, like chess, poker, business competition or international politics?"

The surprising answer to this very natural question is that although matrix games may appear to be very simple and quite specialized, it is nevertheless true (at least in theory) that *any* two-person zero-sum game can be formulated as a matrix game! The way in which this is done is far from obvious, however, and will require a certain amount of discussion.

Let us consider a game in which the players do not play simultaneously, and in which a player uses information given by the other's play. Two players, A and B, each have 3 coins. Player A takes a certain number of his coins in his hand, and without showing them to player B calls a number between 0 and 6. Player B, not knowing how many coins A is holding, but knowing what A has called, then takes a certain number of coins in his hand and also calls a number, which must be different from that called by A. The coins are then displayed. If one of the players has called correctly the total number of coins held by both players together, that player wins. Otherwise the play results in a draw. The winner receives a sum of money from the loser, this sum of money depending on what the players held and what they called. Since we are not going to analyze this game in detail we shall not specify any particular rules for determining the sum of money. (The simplest rule would be that the loser pays the winner a fixed amount, but in this case the game is a little too simple for our purpose.)

Now A's possible strategies are all of the form

"Hold m coins, and call n,"

where $0 \leq m \leq 3$ and $0 \leq n \leq 6$. It may happen (according to how the payment to the winner is determined) that for some pairs (m, n) the corresponding strategy is clearly disadvantageous. In practice any such strategies would be ignored, but we shall not assume that this has been done. So far everything is straightforward.

We turn to consider B's play which will be partly based on what A has already called. Let us imagine that B is a lazy person who wishes to spare himself the agonizing decision required by every play of the game; he therefore prepares for himself a list of instructions concerning what to do under all possible circumstances. This list could be, for example, as in (17.8.1).

If A calls 0, hold 2 and call 2.

If A calls 1, hold 2 and call 3.

If A calls 2, hold 1 and call 3.

(17.8.1)   If A calls 3, hold 2 and call 4.

If A calls 4, hold 2 and call 5.

If A calls 5, hold 3 and call 6.

If A calls 6, hold 2 and call 5.

The instructions in the list might not all be sensible. We might have, for example

"If A calls 0, hold 0 and call 6;"

but as with A's strategies, we shall not assume that obviously disadvantageous instructions have necessarily been excluded from the list.

Now given such a list we can predict exactly how B will reply to any "opening move" that A can make. Since this determines the outcome of the game we can compute the payoff to A for each of his opening moves, assuming that B obeys the instructions in the list. There are, however, many possible lists that B might choose to follow; some sensible, some not so. Let us imagine that B has constructed *all possible lists* of instructions. There will be a very large number of them, but this does not matter. If we are now told what A's opening move is, and which list B is going to follow, we can predict the outcome of the game, and compute the payoff. Thus if A opens by holding 1 and calling 3, then B, using list (17.8.1) will hold 2 and call 4. The total is thus 3, and A wins. Had B used some other list the outcome would, in general, have been different.

We can therefore construct a payoff matrix in which the pure strategies for A are the pairs (m, n) as described above, and in which the pure strategies for B are all the many "lists" which B can use to dictate his move. We now have an equivalent matrix game. A chooses an opening move, and B simultaneously chooses one of the lists. This double choice determines what the outcome of the game will be when it is played according to the original rules, and hence determines the payoff. Because of the large number of choices available (especially to B) this new formulation of the game is much more cumbersome than the original description of it, but theoretically is simpler, being a straightforward matrix game.

It is clear that for a game of any complexity, requiring successive "moves" by the two players, the payoff matrix for the equivalent matrix game will be astronomically large. If, for example, we augment the above game by allowing A an extra "move" in which he can revise his original estimate of the total, then each of A's possible strategies will be a list of instructions, each instruction being of the form

"If I hold m and call n, and he then calls p, I will revise my call to q,"

each list having enough instructions to cover all possible eventualities. The set of strategies for A is then the set of all such lists! Clearly it would be tedious, if not impracticable, to prepare all such lists, but it is theoretically possible.

The same is true even of a game like chess. A pure strategy for a player would consist of a *book of rules* (and a very large book it would be, too!) which told the player exactly how to reply to any given situation (the replies not necessarily being the best ones). We can imagine an enormous library of such books. A game of chess then consists of each player choosing one book from this library and then playing out the game in accordance with the instructions therein. Once the double choice has been made the outcome of the game is determined. A beginner will probably choose a book in which most of the instructions specify a reply which is not the best; the expert is characterized by his ability to choose, fairly reliably, books in which most of the instructions dictate good replies to the opponent's move. It can be proved that somewhere in this library there is either a book which, if followed, will enable one player to win against any opposition, or else a book which will tell each player how to force a draw. Despite this, the fact that the "library" is so enormous ensures that the game of chess will be worth playing for many years to come, and that even master players will occasionally lose games, when they hit a bad instruction in the particular book that they choose!

INFINITE GAMES

Let us consider the following, very simple, matrix "game." In a small town there are only two grocery stores, the "Apex" and the "Buywell". The owners of these stores are each about to embark on extensions to their property; the owner of the Apex store has $5,000 to spend, while the owner of the Buywell store has $4,000 to spend. Both owners can choose whether to spend the whole sum on improvements in building, or on advertising and customer amenities. Whatever choices they make they cannot bring in more customers between them, since all customers in the town shop at one store or the other anwyay; but as a result of the improvements customers may transfer their loyalty from one store to the other. (This remark ensures that the game is zero-sum.)

This situation can be set up as a 2×2 matrix game once the payoffs are known. We shall assume these to be as in table (17.8.2)

		Buywell	
		Building	Advertising
Apex	Building	1000	−1000
	Advertising	0	2000

(17.8.2)

The payoffs are the numbers of customers that will transfer from Buywell to Apex for each of the four possible combinations of choices.

To solve this game we first look for a saddle point and find that there is none. We therefore use the method of section 17.5, and we find that the best (mixed) strategy for Apex is to choose building or advertising with equal probabilities, while for Buywell it is to choose building with probability $\frac{3}{4}$ and advertising with probability $\frac{1}{4}$. The value of the game is 500 to Apex. Thus the owner of the Apex store should toss a coin and make his choice accordingly; the owner of the Buywell store should toss two coins and sink his money into advertising if they both land heads, and into building otherwise. In this way they will make their choices with the correct probabilities. Notice however that since this particular "game" will (presumably) be played only once, each player will in fact play a pure strategy. It is not easy, therefore, to see in what way the tossing of coins (or the employment of any other chance device) is relevant. Since only one strategy can be chosen on the one and only occasion on which the game is played, can it make any difference how the choice was arrived at?

The rough answer to this question (and we shall not attempt here to provide a better one) is that although a game such as the above is played only once, nevertheless, in the lifetime of a grocery store owner (business man, politician, etc.) a large number of similar game-theoretical decisions will have to be made. The contention is that if each decision, as the occasion arises, is made by choosing a strategy with the probability required by the game-theoretical analysis, as above, then the total result will be better than if the decisions are made on any other basis. That is to say, one will reap much the same sort of benefit from following the calculated probabilities over a large number of one-shot plays of different games as one does over many plays of the same game.

In some games there is no way out of this difficulty, but in others, and in particular the one we have just analyzed, there is a loophole which is worth investigating. Rather than choosing to put all his money in either building or advertising, with equal probabilities for the two possibilities, suppose the owner of the Apex store invests equal amounts ($2,500) in each. Would this not be much the same as the "overall" result of playing the original game many times over? In the same way, would it not be best for the owner of the Buywell store to put $3,000 into building and $1,000 into advertising? It would seem that this would, for example, prevent the possibility of Buywell losing 2,000 customers, for which there is at present a 1 in 8 chance. On the whole this "compromise" solution is not unreasonable, but there are difficulties.

The first difficulty is that we are tacitly assuming that there is no disadvantage in dividing the available capital between two projects. This may or may not be true. It could happen, for example, that no significant building could be effected for $2,500, or that, because of initial costs, $1,000 worth of new advertising was hardly better than none at all. If so, this would materially complicate the situation. But even if this difficulty does not arise, the fact remains that we are now considering a different game. In the present game Apex has not just two, but a large number of strategies, each of the form

"Invest x in building, and $(5,000 - x)$ in advertising."

Now since dollars are not infinitely divisible, there is still only a finite number (500,001 to be precise) of strategies for Apex. But it is easier to assume that x is a real variable and give Apex an infinite number of strategies. In the same way we allow Buywell an infinite number of strategies of the form

"Invest y in building, and $(4,000 - y)$ in advertising"

where y is a real number. This results in a game which is fundamentally different from any that we have considered so far, and we require a new theory; but it is easier to develop a new theory than to work with a $500,001 \times 400,001$ payoff matrix!*

The type of game that we are now considering is known as an *infinite game*. By suitable scaling of the numbers chosen by the two players we can formulate a general infinite game as follows:

Two players A and B each choose a real number in the closed interval $[0, 1]$. If A chooses x and B chooses y then B pays A an amount $f(x, y)$, where $f(x, y)$ is a function defined over the square $0 \leq x \leq 1, 0 \leq y \leq 1$.

The function $f(x, y)$ is called the *payoff function*; it is the continuous analogue of the payoff matrix.

The discussion of infinite games is beyond the scope of this book, being more tricky than that of finite games. Nevertheless we can close with a few remarks on what we might expect to happen with infinite games. If we have a finite game in which the strategies are such that we can attach a physical meaning to a mixture of strategies (as we can in the example of this section) then we would expect to be able to construct a corresponding infinite game. If the strategies for one player are $S_1, S_2, \ldots, S_n$ in the finite game, then the strategies in the infinite game are all the convex combinations of the form

(17.8.3) $\lambda_1 S_1 + \lambda_2 S_2 + \cdots + \lambda_n S_n$

---

* Compare the observation on page 418 concerning the approximation of economic variables (which are essentially discrete) by real numbers.

where the λ_i are non-negative and $\lambda_1 + \lambda_2 + \cdots + \lambda_n = 1$. (We are assuming that such a combination has a meaning.) Constructing the payoff function might not be so easy (compare what was said about possible disadvantages of dividing the available capital) but one would expect that, under suitable assumptions, a mixed strategy given by probabilities λ_1, $\lambda_2, \ldots, \lambda_n$ in the finite game would correspond to the "pure" strategy (17.8.3) in the infinite game. Thus the infinite game would necessarily have a saddle point. One might be tempted to conjecture that every infinite game has a saddle point, but unfortunately things are not quite that simple!

Exercise 207 (Chapter Review)

1. Two players, A and B, each place two coins on a table. They can choose whether to place them both head up, both tail up, or one head and one tail up. The coins are compared, and B pays A a number of dollars (0, 1 or 2) equal to the number of his coins that show the same face as one of A's coins.

Find the payoff matrix for this game, determine the optimum strategies for A and B and find the value of the game.

2. Solve the following 2×2 games.

(a)

2	1
4	9

(b)

2	3
8	9

(c)

9	18
20	6

(d)

−1	1
3	4

(e)

2	8
9	5

(f)

8	12
14	6

(g)

−5	−2
1	4

(h)

3	20
10	7

(i)

−9	−2
0	−5

3. Consider the $2 \times n$ game whose payoff matrix is

		\multicolumn{6}{c}{B}
		1	2	3	4	$\cdots$	n
A	1	u_1	u_2	u_3	u_4	$\cdots$	u_n
	2	v_1	v_2	v_3	v_4	$\cdots$	v_n

If B chooses his pure strategy i and A mixes his two strategies in the ratio $1 - x : x$, show that A will win an expected amount

$$y = (1 - x)u_i + xv_i.$$

Hence show that if $1 - \xi : \xi$ is A's optimum strategy, and η is the value of this game, then the point (ξ, η) lies "below" the line segment joining the points $(0, u_i)$ and $(1, v_i)$.

If S is the convex set of points lying "below" all the n line segments corresponding in the above way to B's n strategies (see Fig. 17.1 below), then the solution of the game is given by the "highest" point (ξ, η) of S, i.e., the one with the largest ordinate.

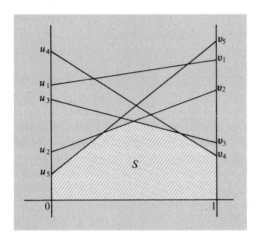

Fig. 17.1

4. Using the method of problem 3, solve graphically the following $2 \times n$ games

(a)

5	−1	3
3	3	5

(b)

1	−1	3
−5	−3	−1

(c)

4	1	2	5
3	5	7	8

(d)

9	7	14	16	8
8	10	7	6	9

(e)

11	3	85	29	53	55	65
27	63	43	21	11	21	5

Hint: To find B's optimum strategy remember that, by following it, **B** can realize the value of the game no matter what strategy A uses.

5. Solve by graphical means the following $m \times 2$ games
Hint: Interchange the roles of A and B.

(a)

4	1
2	4
1	7

(b)

4	5
5	3
1	6

(c)

9	7
15	11
11	18
14	9

(d)

−15	30
−1	13
17	−5
5	−45

(e)

39.	25
81	−27
185	−41
29	77
−35	133
33	61

6. By considering the linear programming method for solving an $m \times n$ game show that if $m \le n$ then at most m of B's strategies are *active* in his optimum mixed strategy, i.e., are chosen with nonzero probability.

Solve the following game.

0	2	−6	−1	2
−8	−4	−2	−3	−2
−2	0	0	1	1

7. Prove that if the payoff matrix of a game is skew-symmetric then the optimum strategies for A and B are the same, and the value of the game is zero.

8. Use the iterative method of section 17.6 to find approximate solutions to the following games; obtain estimates of their values and the accuracy of the solution.

(a)

5	4	2
4	1	3
−2	5	1

(b)

2	0	5
0	4	1
4	1	6
0	5	2

	6	3	3	5	2
	5	6	2	6	3
(c)	7	10	2	5	4
	5	7	0	7	2
	4	5	-1	10	-1

9. In the game discussed in the first part of section 17.8 the details of the transfer of money at the end of a play were left unspecified. If the payment is of a fixed amount, say $1, by the loser to the winner, show that the game assumes a more simple form, in that it is never to A's advantage to call anything but 3. Hence, if he is wise, he will always do this, and this call then gives no information to B. (It could, therefore, be omitted.) Show that the game thus reduces itself to a matrix game without our assistance.

There will be 4 strategies available to A (according to what he holds) and 24 strategies available to B (depending on what he holds and what he calls—remember that he cannot call the same number as A). Construct the 4×24 payoff matrix for this game and solve it by any suitable method.

10. The discussion in the first part of section 17.8 was intended to show how games in which players take turns in making moves can be forced into the general category of matrix games. Another common feature of some games, one which might seem to preclude their inclusion under the general heading of matrix games, is the presence of a *chance element*—something in the game which is (or should be!) beyond the control of the players, such as the deal in bridge or the draw in poker. This problem gives an indication of how, by using the idea of expectation, those two-person games that have a chance element can also be subsumed in the class of matrix games.

Player A draws a card at random from a standard deck of cards, and calls either "black" or "red." What he calls may or may not be the color of the card he has drawn. Player B draws a card at random from another standard deck and can then either "fold"—in which case the cards are returned to their respective decks and B pays A $8—or he can "see"—in which case the two cards are displayed. If the cards are different colors the holder of the red card pays $8 to the holder of the black card; if both cards have the same color B pays A $12.

Note: Although A is holding a red card, he may yet call "black"; thus this game contains an elementary form of bluffing.

Show that A has 4 pure strategies, each of the form (x, y) where x and y each stand for either "black" or "red," and (x, y) means

"If I draw a black card I will call x;

If I draw a red card I will call y."

Show that B has 16 pure strategies, each of the form (p, q, r, s) where

p, q, r, s each stand for either "fold" or "see," and (p, q, r, s) means

"If A calls "black" and my card is "black", do p;

If A calls "black" and my card is "red", do q;

If A calls "red" and my card is "black", do r;

If A calls "red" and my card is "red", do s."

There are 4 equiprobable holdings (black or red cards) for the two players, and given the pure strategy used by each player the payoff can be calculated for each holding, and an expected value of the payoff obtained. Show thus that for the pure strategies

(Black, Red) by A, and (See, See, See, Fold) by B

the expected payoff to A is $5; while for the strategies

(Black, Black) by A and (Fold, See, See, See) by B

the expected payoff to A is $9. In this way construct the whole 4 × 16 payoff matrix for this game.

By eliminating dominating columns, reduce this matrix to a 4 × 4 matrix and show that it has two saddle points. Hence show that A has two equally good pure strategies, namely, to call "black" regardless of this holding, or to call "red," regardless of his holding; while B should "see" if he holds a black card, and "fold" if he holds a red card. Find the value of the game to A.

Suggestions for Further Reading

For a light-hearted, but instructive, introduction to the theory of games the reader cannot do better than to read

Williams, J.D., *The Compleat Strategyst*. New York: McGraw-Hill Book Co., 1966.

The following two books give a more advanced treatment of the subject:

McKinsey, J.C.C., *Introduction to the Theory of Games*. New York: McGraw-Hill Book Co., 1952.

Vajda, S., *Theory of Games and Linear Programming*. London: Methuen & Co., Ltd.; New York: John Wiley & Sons, Inc., 1956.

For obvious reasons the theory of games is often taken with linear programming, as if the two formed a single subject. Thus some of the books in the suggested reading for Chapter 16 will serve also for this chapter. In particular, the two "Finite Mathematics" books which deal with linear programming (see page 869) also deal with the theory of games and can also be profitably consulted.

18 Graph Theory

The proteiform graph itself is a polyhedron of scripture.

JAMES JOYCE.
Finnegan's Wake.

18.1 WHAT IS A GRAPH?

It is truly remarkable that many complicated mathematical structures can be built on very modest foundations and that theorem upon theorem can be derived from just a few simple basic concepts. For example, the theory of sets (to which Chapter 2 gives a brief introduction) is founded upon such simple notions as that of a set itself, membership of a set, and so on. The theory of groups stems entirely from the four group axioms which were given on page 659. One should not be altogether surprised, then, to learn that an important branch of mathematics known as the *Theory of graphs* has its origins in what is essentially the childish pursuit of drawing dots on a piece of paper and joining some of them with lines.

Before the reader is completely misled it must be pointed out that the term *graph theory* or *theory of graphs* is a misnomer. The "graphs" referred to are nothing like the graphs we dealt with in Chapter 3 of this book, which, these days, are common enough objects in many walks of life. The graphs we shall consider in this chapter are what one would be prone to call *networks*, a term that is sometimes used, though we shall use it only in a rather restricted sense. The word *graph* (or its equivalent in other languages) is unfortunately the one which is almost universally employed, however; and confusing though it may be to use this word (because of its alternative meaning) it would be even more confusing to use some other, less generally accepted, word. The reader should be warned, however, that he may meet other terms; in fact the nomenclature of graph theory is at present in a state of chaos from which it is only slowly emerging. Hence in the definitions which follow shortly there may be several names associated with a concept. These names are given so that the reader will be able to recognize them if he meets them in his reading. The first name given will be the one that we shall use in this book.

After this apology for the general confusion of terminology we naturally ask first "What is a graph, in the sense in which the word is now going to be used?" If we make some dots on a piece of paper and then join some of them by line segments, we shall obtain a figure looking something like that in Fig. 18.1. This figure is a graph of a sort, but before we can know exactly

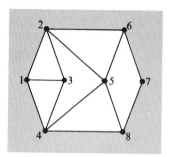

Fig. 18.1

what is meant by the term we must decide which parts of Fig. 18.1 are *essential* and which parts are *incidental*. First of all, the location of the dots on the paper will not be regarded as important. Secondly, although we have joined the dots by straight line segments we shall not consider it necessary to join them in this way; we could use instead any sort of curve. The purpose of the curve is simply to indicate that two particular dots are joined, as opposed

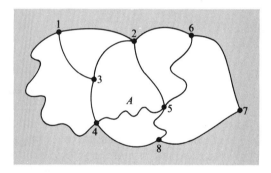

Fig. 18.2

to other pairs of dots that are not joined. Hence Fig. 18.2, in which the locations of the dots have been slightly altered from their positions in Fig. 18.1, and in which the line segments have been replaced by curves, is in fact the "same" graph as that of Fig. 18.1. For convenience the dots in the two figures have been labelled 1 through 8 to indicate the correspondence between them.

What we have so far called "dots" we shall now refer to by their technical name of *nodes*. Other names for nodes are *points*, or *vertices*. What we have called "curves" we shall henceforth refer to as *edges* (other terms are *lines*, *branches* or *arcs*). But what exactly is an edge? The size and the shape of an edge are irrelevant, as its only purpose is to indicate that a certain two nodes are joined by it. When we have specified a pair of nodes we have specified all that is relevant about the edge which joins them. Hence, for logical economy we cannot do better than to define an edge simply as an unordered pair of nodes.

This raises the question "What is a node?" We have thought of nodes as dots on a piece of paper, but this is just a convenience and does not reflect any essential property of nodes. In fact we are not going to postulate *any* properties that nodes must have; consequently our nodes will be considered to be completely undefined. We are now able to give a definition of an *abstract graph*.

Definition 18.1

An abstract graph G is a non-empty set $\mathcal{N}$ of undefined objects, called the nodes of G, together with a set $\mathcal{E}$ of unordered pairs of nodes. The pairs of nodes (elements of $\mathcal{E}$) are called the edges of G.

Thus the graph represented by Fig. 18.1 has eight nodes denoted by the digits 1, 2, ..., 8. Since nodes are undefined it would be equally correct to say that the nodes of this graph *are* the digits 1 through 8. The edges of the graph are the following *unordered* pairs (1, 2), (1, 3), (1, 4), (2, 3), (2, 5), (2, 6), (3, 4), (4, 5), (4, 8), (5, 6), (5, 8), (6, 7), (7, 8).

Exactly the same information is imparted by Fig. 18.2; hence these two graphs are, in some sense, the same, as already remarked. Even if we had labelled the nodes differently in the two figures the graphs would still be essentially the same, since the labels were put in merely for convenience and form no part of the graphs as such. This concept of two graphs being the "same" or, to use the technical word, *isomorphic* is made precise by the following definition.

Definition 18.2

Two graphs G and G' are isomorphic if there exists a one-to-one correspondence between their nodes such that two nodes of G form an edge of G if, and only if, the corresponding two nodes in G' form an edge of G'.

Although abstract graphs, as just defined, are what we shall be working with, it is inevitable that one tends to think of graphs in terms of more tangible objects. Any set of actual objects (as opposed to nebulous undefined objects) which are linked in pairs according to the specifications of the edges of a graph will be called a *realization* of that graph. Very often, though not always, the realization one forms of a graph is a geometrical one, such as those of Figs. 18.1 and 18.2, in which actual physical dots on a piece of paper are the nodes, and drawn curves indicate which pairs of nodes are edges. Such a realization will be called a *geometrical realization* of the graph. Any abstract graph has a geometrical realization, for we· can choose arbitrary points in three-dimensional space to represent the nodes of the graph and construct a curve (of arbitrary shape) between any two points which represent nodes forming an edge of the graph.

It will be noticed that this method for constructing a geometrical realization of an arbitrary graph specifically mentions *three*-dimensional space. There is a good reason for this, since we can get into difficulties if we try to represent a graph geometrically in only two dimensions. For example, since the positions of the nodes in Fig. 18.2 are not important it would be all the same if we moved node 8 of that figure to inside the region marked *A*. But if we do this we find that we cannot then draw curves between the nodes to represent the edges without making some of the curves intersect, as in Fig. 18.3.

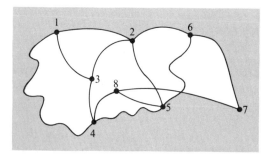

Fig. 18.3

The fact that some of the edges cross in this way need not cause us any trouble so long as we make sure that we do not confuse the "accidental" crossing points with the nodes. This is best done by drawing the nodes in some unmistakable way, and for this reason we shall in the future draw the nodes of our graphs as small circles.

There are many graphs for which no geometrical realization can be drawn in the plane with no edge crossing over another. The two simplest graphs of this kind are shown in Figs. 18.4 and 18.5. These figures have been drawn so as to have no more crossings than are absolutely necessary—one crossing in each case. The graph of Fig. 18.4 is associated with a well-known puzzle, namely, to lay (if possible) water, gas and electricity mains from the three utilities, denoted by W, G and E (water, gas, electricity) to each of three houses (A, B and C), in such a way that no main crosses over another. As we have implied, this is in fact not possible. The graph of Fig. 18.5 is an example of a *complete* graph, i.e., one in which every node is joined to every other node. If there are n nodes there are $\binom{n}{2}$ pairs of distinct nodes, and this is therefore the number of edges in the complete graph. For $n = 5$, it is 10 as we see from Fig. 18.5.

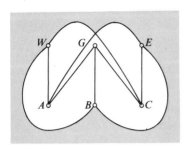

Fig. 18.4

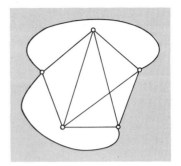

Fig. 18.5

A graph that has a geometrical realization in a plane with no edges crossing is known as a *planar* graph. Note that the only question is whether the abstract graph has *some* plane realization with no crossings. Thus the graph of Fig. 18.3 is still a planar graph even though that particular representation of it contains edges which cross. As we see from Figs. 18.1 and 18.2, this graph *can* be drawn in the plane with no crossings, although we are under no compulsion to draw it so; hence it is planar. A graph that is not planar is *nonplanar*. Figures 18.4 and 18.5 show the nonplanar graphs with the fewest edges and nodes, respectively.

If we construct a geometrical realization of a graph in three-dimensional space we do not run into any difficulty over edges intersecting. There is, so to speak, sufficient room in the space for an edge to be "drawn" from any node to any other node without intersecting any edges already drawn. Although there are interesting problems associated with the differences between planar and nonplanar graphs, we shall not be concerned with them.

Since graphs are usually pictured geometrically in this way, the nomenclature of graph theory is very strongly geometrical. Suppose that two nodes, say A and B, form an edge which we call e. Then we say that the nodes A and B are *joined* by e (we have already used this term). We can also say that A and B are each *incident* with the edge e. If two nodes are joined by an edge we say that they are *adjacent* to each other.

Exercise 208

1. Prove that the complete graph on n nodes has a geometrical realization in three-dimensional space in which no two edges intersect. Deduce that every graph with a finite number of nodes has such a realization.

2. A graph G has 6 nodes labelled 1, 2, ..., 6, and its edges are given by the pairs

$$(1, 2), \quad (1, 3), \quad (1, 5) \quad (1, 6), \quad (2, 3), \quad (2, 4), \quad (2, 5),$$
$$(3, 4), \quad (3, 5), \quad (3, 6), \quad (4, 5), \quad (5, 6).$$

Draw a geometrical realization of this graph for which no two edges cross.

3. If the nodes of a graph G are labelled $1, 2, \ldots, n$, we can define a matrix (a_{ij}) by

$$a_{ij} = 1, \text{ if node } i \text{ is joined to node } j$$

and $a_{ij} = 0,$ otherwise.

This matrix is known as the *adjacency matrix* of G.

Draw the graphs given by the following adjacency matrices.

(a)
$$\begin{pmatrix} 0 & 1 & 1 & 1 & 1 & 1 \\ 1 & 0 & 1 & 1 & 0 & 1 \\ 1 & 1 & 0 & 1 & 1 & 0 \\ 1 & 1 & 1 & 0 & 1 & 1 \\ 1 & 0 & 1 & 1 & 0 & 1 \\ 1 & 1 & 0 & 1 & 1 & 0 \end{pmatrix}$$

(c)
$$\begin{pmatrix} 0 & 1 & 1 & 1 & 1 \\ 1 & 0 & 1 & 1 & 1 \\ 1 & 1 & 0 & 1 & 0 \\ 1 & 1 & 1 & 0 & 1 \\ 1 & 1 & 0 & 1 & 0 \end{pmatrix}$$

(b)
$$\begin{pmatrix} 0 & 1 & 0 & 0 & 0 \\ 1 & 0 & 1 & 0 & 0 \\ 0 & 1 & 0 & 0 & 0 \\ 0 & 0 & 0 & 0 & 1 \\ 0 & 0 & 0 & 1 & 0 \end{pmatrix}$$

(d)
$$\begin{pmatrix} 0 & 1 & 0 & 1 & 1 & 0 & 1 \\ 1 & 0 & 1 & 0 & 1 & 1 & 0 \\ 0 & 1 & 0 & 1 & 1 & 1 & 1 \\ 1 & 0 & 1 & 0 & 0 & 0 & 0 \\ 1 & 1 & 1 & 0 & 0 & 0 & 1 \\ 0 & 1 & 1 & 0 & 0 & 0 & 1 \\ 1 & 0 & 1 & 0 & 1 & 1 & 0 \end{pmatrix}$$

Why are these matrices all symmetric?

18.2 GENERAL TERMINOLOGY.

Although the basic terms *node, edge, incident* and *adjacent* have already been introduced, they are still not quite precise. In our definition of a graph we defined edges as pairs of nodes, for example (A, B), (P, Q), etc., but should we include pairs like (A, A), or not? In geometrical terms, do we allow an edge to join a node to itself? This is a matter of definition, and although such edges (called *loops* or *slings*) were not explicitly excluded by Definition 18.1, we shall assume from now on that they are not allowed unless the contrary is stated.

Another point on which there might be some doubt is whether it is possible for two or more edges to join a node to another. Strictly speaking, Definition 18.1 excludes this possibility; a pair of nodes either forms an edge or it does not. As we saw on page 34 an element either belongs to a set or it does not; there is no question of an element appearing several times in a set. Nevertheless for some purposes it is convenient to allow *multiple edges*, as they are called, and the edges can then be regarded as being a selection, with repetitions allowed, from the set of all possible pairs of nodes.

If loops and multiple edges are both allowed, we shall call the graph a *general graph*; if loops are allowed but multiple edges are not, we shall call the graph a *graph with loops*. A graph in which multiple edges are allowed but loops are not is called a *multigraph*. When the term *graph* alone is used it will be assumed that neither loops nor multiple edges are allowed. Figures 18.6, 18.7 and 18.8 illustrate these terms.

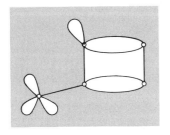

Fig. 18.6

A General Graph

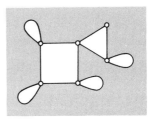

Fig. 18.7

A Graph with Loops

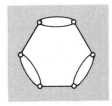

Fig. 18.8

A Multigraph

The number of edges that are incident with a given node is called the *valency* (or *valence* or *degree*) of that node. A node which is not incident with any edge, i.e., one whose valency is zero, is called an *isolated node*. Consider a graph G having n nodes, and let the valencies of the nodes (in some order) be $v_1, v_2, \ldots, v_n$. Then

THEOREM 18.1

$\sum_{i=1}^{n} v_i = 2E$, WHERE E IS THE NUMBER OF EDGES.

PROOF:

Each edge is incident with two nodes, and hence contributes 1 to the valency of each of those two nodes. Hence each edge contributes 2 to the sum of the valencies, from which the theorem follows. Looking at it another way, we can imagine each edge to be broken in the middle.

The ith node will then be attached to v_i half-edges. Hence $\sum_{i=1}^{n} v_i$ is the total number of half-edges, that is, $2E$.

An alternative way of stating the result of Theorem 18.1 is as follows. If we let n_i denote the number of nodes of valency i, then the sum of the valencies is $\sum_i in_i$, since each of the n_i nodes of valency i contributes i to the total. Hence Theorem 18.1 can be written as

$$\sum_i in_i = 2E.$$

18.3 SUBGRAPHS

Let G be a graph whose node set is $\mathcal{N}$ and whose edge set is $\mathcal{E}$, as in our original definition. Let $\mathcal{M}$ be some subset of $\mathcal{N}$, and $\mathcal{D}$ the subset of $\mathcal{E}$ consisting of those edges which join nodes in $\mathcal{M}$. Finally let $\mathcal{C}$ be a subset of $\mathcal{D}$. Then the graph H whose node set is $\mathcal{M}$ and whose edge set is $\mathcal{C}$ is a *subgraph* of G. Thus a subgraph of G can be formed by choosing some, possibly all, of the nodes of G, and then choosing some, possibly all, of the edges of G which join nodes chosen for the subset. Thus in Fig. 18.9 the graph indicated by the heavy lines is a subgraph of the whole graph depicted in the figure. If a subgraph H of G contains all the nodes of G ($\mathcal{M} = \mathcal{N}$) so that the only difference is that some edges of G are not edges of H, then we call H a *spanning subgraph* of G.

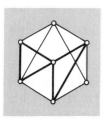

Fig. 18.9

Paths and Circuits

An example of something from everyday life which is at any rate "graph-like" is a system of railroad stations and the interconnecting railway lines; the stations are the nodes and the lines are the edges. Thinking of graphs as being something like railroad networks we are led to ask questions like "Can we get from one given node to another by going along edges of the graph?" To make this question more precise we first define a *path* in a graph.

Definition 18.3

A path in a graph G is a sequence of distinct nodes $A_1, A_2, \ldots, A_k$ of G such that any two successive nodes in the sequence are adjacent, together with the edges (A_1, A_2), (A_2, A_3), etc., joining successive nodes. The length of a path is the number of edges it contains.

The concept of a path is an obvious one intuitively, and is probably more easily seen from a diagram than from the definition. Figure 18.10 shows a path between the nodes A and B in a graph, the edges joining the successive nodes of the path being drawn heavily. Note that the nodes in a path are distinct so that there is no possibility of a path crossing itself in the sense of "going through" a node more than once, or of ever reaching a node that it has reached before.

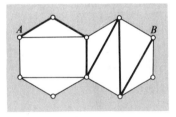

Fig. 18.10

We now define a *circuit* in a graph. To say that a circuit is a closed path, i.e., one that returns to the node from which it started, will probably convey the concept quite clearly, but a formal definition is as follows:

Definition 18.4

A circuit (or cycle) in a graph G is a sequence of nodes of G, any two successive nodes of which are adjacent, together with the edges joining successive nodes (as for a path), the edges being all distinct, and the nodes being distinct except that the first is the same as the last. The length of a circuit is the number of edges it contains.

In the definition of a path the distinctness of the nodes guarantees that the edges are also distinct. This is not so for circuits, and we must explicitly state that the edges are distinct. For otherwise we could have circuits of length 2, (e.g., A_1, A_2, A_1) in a graph, and this we wish to avoid. Note, however, that although a graph cannot have circuits of length 2, a multigraph may have them; if there are two (or more) edges joining nodes A_1 and A_2, then there is a circuit of length 2 containing nodes A_1 and A_2 and two *different* edges joining them.

Figure 18.11 shows a graph with two circuits marked on it (by the same device of drawing heavily the edges joining successive nodes). It is clear that the only reason why one node occurs twice in Definition 18.4 is that to specify a circuit one has to start somewhere, and whatever node one writes down first must be repeated at the end of the sequence in order to indicate that the circuit closes up. The node in question is not, for this reason, any different from the others. In the same way we see that Definitions 18.3 and 18.4 do not ascribe any *sense* to the paths and circuits, although they may appear to do so. Thus the path defined in Definition 18.3 is not *from* A_1 to A_k, nor yet from A_k to A_1. It is simply *between* these two nodes, with no implication of direction one way or the other.

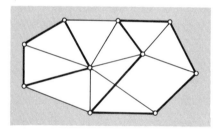

Fig. 18.11

If, in a graph G, there is a path between nodes A and B, or if $A = B$, we say that A and B are *accessible* each to the other. Let us write "$A \sim B$" to mean that A and B are accessible to each other. It is easily verified that the relation $\sim$ of accessibility is an equivalence relation. Hence it will divide the nodes of G into a number of equivalence classes (see page 53) where the nodes in each class are mutually accessible, but nodes from different classes are not accessible to each other. If we take all the nodes of one such equivalence class, and the edges of G which join them, we obtain a subgraph of G. There is one such subgraph for every equivalence class, and they are called the *connected components* or just the *components* of G. Fig. 18.12 shows a graph which has three components. Colloquially we would say that it consists of three completely separate pieces. If a graph has only one component then its nodes are all mutually accessible—there is a path between

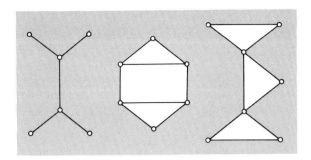

Fig. 18.12

any pair of nodes. Such a graph is said to be *connected*. The graphs in Figs. 18.1 through 18.11 are all connected graphs.

Exercise 209

1. We usually talk about the *components* of a graph, although the full term is *connected component*, as given above. The full term is apposite since each such component, considered as a graph by itself, is a connected graph. Prove this.

2. Let the nodes of a graph G be labelled $1, 2, \ldots, n$, and let the edges be labelled $1, 2, \ldots, k$. Define a matrix **B** as follows

$$b_{ij} = 1, \text{ if node } i \text{ is incident with edge } j$$

$$b_{ij} = 0, \text{ otherwise.}$$

The matrix $\mathbf{B} = (b_{ij})$ is called the *incidence matrix of G*. Prove that

(a) The sum of the elements in row i is the valency of node i.

(b) The sum of the elements in any column is 2.

(c) $\mathbf{BB}^T = \mathbf{A} + \Delta$, where **A** is the adjacency matrix (see Exercise 208.3) and Δ is the diagonal matrix whose diagonal elements are the valencies of the nodes (all other, nondiagonal, elements of Δ are zero).

3. If the stipulation that the nodes (and hence the edges) of a path are distinct is dropped from Definition 18.3, we obtain the definition of a *walk*. Thus a walk, unlike a path, can pass through a node, or traverse an edge, more than once.

From the definition of matrix multiplication show that if $\mathbf{P} = \mathbf{A}^2$, where **A** is the adjacency matrix of a graph G, then a general element p_{ij} of **P** is the number of walks of length 2 in G between nodes i and j. Obtain a similar result for the matrix $\mathbf{A}^m$, where m is a positive integer.

4. Prove that if there is a walk of length k between nodes A_i and A_j of a graph G then there is a path of length $\leq k$ between these two nodes.

5. Show that a general element q_{ij} of the matrix $Q = (\mathbf{I} + \mathbf{A})^m$ is nonzero if, and only if, there is a path between node i and node j of length m *or less*. Deduce that if the graph is connected then $(\mathbf{I} + \mathbf{A})^{n-1}$ contains no zero elements, and conversely. (n is the number of nodes.)

6. Find the number of distinct paths between A and B in the graph of Fig. 18.10.

18.4   SOME EXAMPLES

This will be a good point at which to consider some examples of systems which are capable of being represented by graphs. We have already observed that a railroad network has much in common with the sort of structure that we are considering. More generally any kind of communication network—of shipping routes between ports, highways between cities, and so on—can be represented by a graph. These graphs will show which nodes (stations, ports, cities) are directly joined by edges (lines, shipping lanes, roads), but clearly this will not tell us everything that we might wish to know. Usually we want to know more about the way the nodes are joined, such as the distance from one node to another, or the cost of transporting a unit amount of some commodity from one node to another, or some other number associated with an edge. For this reason we frequently have occasion to consider graphs in which a number is allocated to each edge. Such a graph is called a *network*. As a common example of a network we may take the kind of map that is often seen in air travel agencies. This shows the airports in a particular territory and indicates by a line joining them those pairs of airports between which a direct flight is possible. If we ignore any coastlines, towns, rivers, etc., that the map may also depict (these are irrelevant) the map is a graph, with the airports as its nodes. If, in addition, the map indicates the distances between adjacent airports, then it is more than a graph, it is a *network*. If we had another map, identical to the first except that each edge was marked with the fare from the one airport to the other, then the two *graphs* would be the same (more strictly, isomorphic) but the two *networks* would be different.

Another example of a graphical structure is provided by any symmetric relation R defined over a set of objects. We take the objects to be the nodes of the graph, and we say that two objects (nodes) form an edge of the graph if, and only if, the relation holds between them. Since the relation R is assumed to be a symmetric one, it either does not hold at all between two objects A and B, or else it holds both between A and B *and* between B and A. The order of the objects does not matter. This is essential, since the edges of a graph are *unordered* pairs of nodes. It is readily seen from examples of this sort that the set of nodes of a graph need not be a finite set. We could, for example, take the set of integers as our node set, and the relation to be the one defined by

$$pRq \equiv (p = 2q) \lor (2p = q)$$

that is, the relations R holds between two integers p and q if, and only if, one is twice the other (in one order or the other). The graph corresponding to this relation (or rather part of it) is shown in Fig. 18.13.

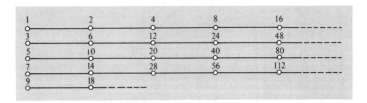

Fig. 18.13

There are many complications that can arise in the study of infinite graphs (graphs with an infinite number of nodes) which are absent from the study of finite graphs. Furthermore, infinite graphs are unlikely to occur in the analysis of economic problems. Consequently we shall assume from now on that our graphs are all finite.

Let us consider the graph obtained by taking as nodes the various countries of the world (or some part of it) and stating that two nodes are joined by an edge if, and only if, the two countries in question have a common frontier. This is illustrated in Fig. 18.14 for the countries of South America. Here the nodes have been located at the capitals of the corresponding countries.

Such a graph is really a particular case of a graph associated with a symmetric relation. For the relation of having a common frontier is a symmetric relation between countries. One can think of many other such relations—for example, the existence of a nonaggression pact, a trade agreement, common currency, or any kind of mutual agreement. It is true that the graph in Fig. 18.14 contains less information than does the map, since from the map we can get information about the shapes and the sizes of the various countries, and so on. But in a context where this latter information was irrelevant, the graph would be more useful precisely because it excluded what was not important. This advantage of the representation of symmetrical relations by graphs can be seen even better in other examples. Thus, if in Fig. 18.15 the edges mean that a nonaggression pact has been signed by the two countries represented by joined nodes, then the existence of three well-defined "blocs" (in the political sense) is at once apparent. A mere list of signatories

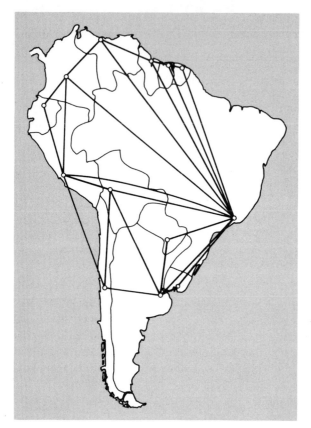

Fig. 18.14

to the various pacts, though it contains the same information, would bring this out far less clearly.

Similar remarks would apply if the edges in Fig. 18.15 stood for trade agreements between business firms, the relation of friendship between

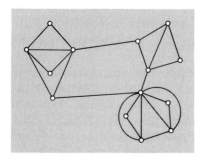

Fig. 18.15

individuals in a small group, or for any of the many symmetrical relations that occur in economics and sociology.

Exercise 210

1. If a graph W were drawn like Fig. 18.14, but included all the countries of the world, what interpretation could you give to
  (a) the components of W
  (b) the isolated nodes of W
  Would this graph be planar?
2. Find in an atlas four countries, every two of which have a common frontier. How could you describe briefly the subgraph of W (the graph of problem 1) which is determined by these four countries?
3. If the nodes of a graph G are countries, and the edges of G join countries between which there exists a nonaggression pact, show, by means of a hypothetical example, that G need not be planar.

18.5 TREES

Definition 18.5
A connected graph having no circuits is called a tree.

Figure 18.16 gives an example of a tree and we can readily see how it gets its name.

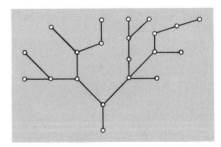

Fig. 18.16

We have already met trees in connection with the more complicated questions in probability, as for example in section 4.15. Trees are graphs of a very special kind, and for this reason they play an important part in many applications of graph theory. For the time being we content ourselves with proving just two results concerning trees.

THEOREM 18.2

IN A TREE THE NUMBER OF NODES IS ONE MORE THAN THE NUMBER OF EDGES.

PROOF:

This is most easily proved by seeing what happens if we set out to draw a tree. We start by drawing a node. Then we draw an incident edge and its other node. We now have 2 nodes and 1 edge. We pick an edge incident with a node already drawn, draw it and its other node. We now have 3 nodes and 2 edges. We continue in this way. Thus if we wanted to draw Fig. 18.16, starting with the node nearest the bottom of the page, we might do it as follows:

and so on. At each stage both the number of nodes and the number of edges are increased by 1, and hence the number of nodes remains one more than the number of edges. Hence this is true of the tree finally obtained.

Note that this is not true if the graph has a circuit. For then, at some stage, the addition of an edge will complete a circuit, in which case both nodes incident with the new edge have already been drawn, and the operation adds 1 to the number of edges, but the number of nodes remains the same.

Our second theorem about trees is the following:

THEOREM 18.3

EVERY CONNECTED GRAPH HAS A SPANNING TREE, I.E., A SPANNING SUB-GRAPH WHICH IS A TREE.

PROOF:

We shall not only prove the existence of a spanning tree but also give a method of obtaining one.

Let G be any connected graph. We choose any edge of G, then choose any other edge, then any other edge not forming a triangle (circuit of length 3) with the first two, then a fourth edge, a fifth, and so on. The only restriction in our choices is that when we choose an edge, we do not choose one which, taken with some or all of the others, forms a circuit; otherwise the choice is unrestricted. This process must terminate, since there is only a finite number of edges in G, and we thus arrive at a collection of edges of G. These edges, together with their end-nodes, form a subgraph T of G. We now show that T is a spanning tree of G.

Clearly T contains no circuits, since the formation of a circuit was disallowed. We must show that T is connected. Suppose it is not, and let A be a node in one component α of T, and B a node in another, different, component β. Now there is a path, in G, from A to B, since G is connected, and this path must contain at least one edge e which is not in T (otherwise A and B would be in the same component of T). Further, the addition of the edge e to T could not create a circuit, for the most that can happen is that, as a result of its inclusion, some node of α is joined, by e, to some node of β. But if e were part of a circuit we could go from α to β (via e) and then, by the rest of the circuit, back to α again, and this implies the existence of a path from β to α not containing e, which is impossible.

Thus if T is not connected we have not finished the process of choosing edges; consequently, when the process *has* terminated, T must be connected.

In the same way we see that T spans G. For if any node of G were not in T we could treat it as a 1–node component of T, and the argument given above shows that some more edges could have been added to T. This completes the proof of the theorem.

Exercise 211

1. Prove the converse of Theorem 18.2, namely, that if a connected graph has one more node than it has edges then it is a tree.

2. A graph (not necessarily connected) having no circuits is called a *forest*. Show that a forest is an assemblage of trees (as one might expect!). If a forest has n nodes, k edges and m components prove that

$$n = k + m.$$

18.6 DIGRAPHS

In the definition of a graph, an edge was defined as an unordered pair of nodes. For this reason our method of representing relations by graphs was confined to symmetric relations only. Insofar as there are many important relations that are not symmetric it is profitable to consider what happens if we replace the word *unordered* by *ordered* in the definition. If we do this, then since the pair (A, B) is now not the same as the pair (B, A), we must allow for the possibility of *two* edges being associated with two nodes; we may have the edge (A, B) which goes *from A to B*, or the edge (B, A) which goes *from B to A*, or neither, or both of these. Thus an edge in the new sense will have a *direction* associated with it, and the structure given by the modified definition is therefore called a *directed graph*. This term is now usually abbreviated to *digraph*, and we shall follow this usage. The formal definition of an (abstract) digraph is as follows:

Definition 18.6

A digraph D is a set $\mathcal{N}$ of undefined objects called the nodes of D, together with a set $\mathcal{E}$ of ordered pairs of nodes. The pairs of nodes are called the edges of D.

All that was said about geometrical realizations of graphs will apply equally to digraphs, provided that we have some means of indicating the directions of the edges. This is easy. If (A, B) is an edge of a digraph we construct a curve of some sort joining A and B (as for a graph) and indicate that it is from A to B, and not the other way round, by an arrow-head pointing away from A and towards B. If the edges (A, B) and (B, A) are both present in the digraph we can either draw two curves between the nodes, with arrows in different directions, as with Fig. 18.25(b) on page 938, or we can make one curve suffice and put two arrowheads on it. Thus Fig. 18.17 represents the digraph with nodes identified by the digits 1, 2, 3, 4 and 5, and whose (directed) edges are the ordered pairs (1, 2), (1, 5), (2, 3), (2, 4), (3, 4), (4, 1), (4, 2), (4, 5) and (5, 1).

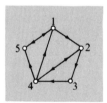

Fig. 18.17

The terms used in talking about digraphs are usually simple extensions of the terms used for graphs. If (A, B) is an edge of a digraph, we say that A is *adjacent to B*, but that B is *adjacent from A*. As with graphs one may wish to exclude edges which join a node to itself (directed loops), though often it is convenient to allow them. The possibility of having more than one edge joining two nodes in the same direction (multiple directed edges) is also one which may be allowed or disallowed according to circumstances. For many purposes it is convenient to regard a graph as a special kind of digraph, namely one in which any two nodes are either joined by one edge in each direction or are not joined at all. Such a digraph is called a *symmetrical digraph* and can, for practical purposes, be identified with the graph obtained by simply ignoring all the arrowheads. The symmetrical digraph and graph of Fig. 18.18 illustrate this way of regarding a graph as a special sort of digraph.

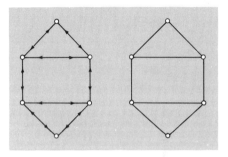

Fig. 18.18

The number of edges that go *from* a given node of a digraph to other nodes is known as the *out-valency* (or *out-degree*) of the node; the number of edges that come *to* a node from the other nodes is called the *in-valency* (or *in-degree*) of the node. If $u_1, u_2, \ldots, u_n$ are the out-valencies of the nodes of a graph, and $v_1, v_2, \ldots, v_n$, are the in-valencies, we have the following theorem:

THEOREM 18.4

$$\sum_{i=1}^{n} u_i = \sum_{i=1}^{n} v_i = E, \text{ THE NUMBER OF DIRECTED EDGES.}$$

PROOF:

Each edge contributes 1 towards the out-valency of the node *from* which it is incident (i.e., the first node of the ordered pair) and contributes 1 towards the in-valency of the node *to* which it is incident. Hence the result follows.

The definitions of *path* and *circuit* in a digraph are similar to those for graphs.

Definition 18.7

A path in a digraph D is a sequence of distinct nodes $A_1, A_2, \ldots, A_k$, such that any two successive nodes are adjacent in D in the order in which they occur in the sequence, together with the directed edges joining successive nodes, that is, each pair (A_r, A_{r+1}), for $r = 1, 2, \ldots, k - 1$, is an edge of D, and of the path.

Thus the successive edges associated with a path must be directed in a consistent manner; each edge goes *to* a node *from* which its successor goes. Thus in Fig. 18.19 a path is shown from node *A* to node *D*, but in Fig. 18.20 the nodes joined by heavy lines do not form a path from *C* to *D* since one edge (*E, F*) is directed in the "wrong" direction.

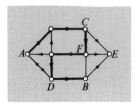

Fig. 18.19   **Fig. 18.20**

Circuits. If we amend the definition of a path in a digraph to stipulate that the final node of the sequence is the same as the first (the nodes otherwise being distinct)—just as we did for graphs—we obtain the definition of a *circuit* in a digraph. Figure 18.21 shows a circuit in a digraph, and the concept is a sufficiently obvious modification of that of a path that we shall not give a formal definition. The important thing, as with a path, is that the successive edges determined by the nodes of the circuit should be consistently oriented. Colloquially, as we go along the path, or around the circuit we should always be going *with* the arrows.

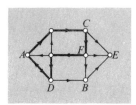

Fig. 18.21

Note that circuits of length 2 are possible in digraphs. If node *A* is joined to node *B*, and *B* is joined to *A*, then these two nodes, together with the edges (*A,B*) and (*B,A*) form a circuit of length 2.

Connectedness. Let us write $A\pi B$ to indicate that (in a given digraph) either $A = B$ or there is a path from node *A* to node *B*. This is analogous to the relation $A \sim B$ which we considered for graphs, but unfortunately the relation π, unlike $\sim$, is not an equivalence relation. For example, in Fig. 18.19 we have $B\pi E$ but not $E\pi B$. Hence the relation is not symmetrical, one of the requirements of an equivalence relation. In consequence, the idea of connectedness in digraphs is more complex and less intuitively obvious than it is for graphs. We can in fact define three types of connectedness for digraphs by the following definitions:

Definition 18.8

A digraph D is said to be strongly connected if for any two nodes A and B of D there is a path from A to B and a path from B to A; that is, for any two nodes A and B we have $(A\pi B)$ and $(B\pi A)$.

Definition 18.9

A digraph D is said to be semistrongly connected if for any two nodes A and B of D there is a path connecting them in one direction or the other; that is $(A\pi B) \vee (B\pi A)$.

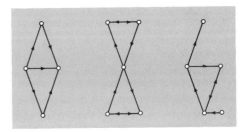

Fig. 18.22

The absence of one or other of these two kinds of connectedness does not show itself in a digraph in any very spectacular manner; certainly nothing as obvious as lack of connectedness in a graph, which means that the graph breaks up into several quite distinct pieces. Yet we can clearly have digraphs that are in several pieces, such as that of Fig. 18.22. To allow for digraphs of this kind (or rather, those not of this kind) we have one further type of connectedness.

Definition 18.10.

A digraph D is said to be weakly connected if the graph G obtained from D by ignoring all the arrows is a connected graph. More strictly, G is defined as the graph whose nodes are the nodes of D and in which two nodes A and B are joined in G if, and only if, either (A, B) or (B, A) is an edge of D (or both are).

Digraphs that are strongly connected, semi-strongly connected or weakly connected are usually referred to as *strong* digraphs, *semi-strong* digraphs and *weak* digraphs, respectively. Figure 18.23 shows digraphs of all three kinds. Figure 18.22 gives an example of a digraph that is not even weakly connected. It consists of three quite separate pieces, each of which is a weak digraph. These are known as its *weak components*.

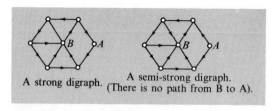

A strong digraph.

A semi-strong digraph.
(There is no path from B to A).

A weak digraph.
(No path from A to B or from B to A).

Fig. 18.23

Exercise 212

1. Prove that every strong digraph is a semi-strong digraph, and that every semi-strong digraph is a weak digraph.

2. A digraph can be specified by an adjacency matrix **A** just as a graph can. We define

$$a_{ij} = 1, \text{ if node } i \text{ is adjacent } to \text{ node } j$$

$$a_{ij} = 0, \text{ otherwise.}$$

Construct the digraphs whose adjacency matrices are

(a)
$$\begin{pmatrix} 0 & 1 & 1 & 0 & 1 & 0 \\ 1 & 0 & 1 & 1 & 0 & 0 \\ 0 & 1 & 0 & 0 & 1 & 1 \\ 1 & 0 & 1 & 0 & 0 & 1 \\ 1 & 1 & 1 & 1 & 0 & 1 \\ 1 & 0 & 0 & 0 & 1 & 0 \end{pmatrix}$$

(c)
$$\begin{pmatrix} 0 & 1 & 1 & 1 & 0 \\ 0 & 0 & 1 & 1 & 1 \\ 1 & 0 & 0 & 1 & 1 \\ 1 & 1 & 0 & 0 & 1 \\ 0 & 0 & 0 & 0 & 0 \end{pmatrix}$$

(b)
$$\begin{pmatrix} 0 & 1 & 0 & 0 & 1 \\ 1 & 0 & 0 & 0 & 1 \\ 0 & 0 & 0 & 1 & 0 \\ 0 & 0 & 1 & 0 & 0 \\ 1 & 1 & 0 & 0 & 0 \end{pmatrix}$$

(d)
$$\begin{pmatrix} 0 & 1 & 1 & 1 & 1 & 1 \\ 0 & 0 & 1 & 1 & 0 & 0 \\ 0 & 0 & 0 & 1 & 1 & 0 \\ 0 & 0 & 0 & 0 & 0 & 0 \\ 0 & 0 & 1 & 1 & 0 & 0 \\ 0 & 0 & 0 & 1 & 1 & 0 \end{pmatrix}$$

3. Determine what kind of connectedness is possessed by the digraphs of problem 2.

4. If $\mathbf{A} = (a_{ij})$ is an $n \times n$ technology matrix (i.e., a matrix associated with an input-output table, as in section 14.23), we can define a digraph D as follows: D has n nodes, numbered $1, 2, \ldots, n$, and node i is joined to node j if, and only if, $a_{ij} > 0$.

Show that each industry purchases, either directly or indirectly, from every other industry if, and only if, D is strongly connected.

18.7 SOME APPLICATIONS

One-choice structures. Digraphs are frequently useful for depicting the various relationships that may hold between elements of some set, for example individuals in a group of people, nations, business firms, and so on. We have already mentioned how graphs can be useful in this way; with digraphs there are even greater potentialities since we are not restricted to symmetrical relations. Hence we can consider such relationships as love, hatred, consumer preference for one article over another, supremacy of one team over another on the sports field, or of one nation over another on the battlefield, and so on, which are not symmetrical. The method of constructing such digraphs is obvious. The nodes of the digraph are the elements of the set in question, and two elements (nodes) A and B form a directed edge (A,B) if, and only if, A has the specified relation to B. To begin with we shall consider a special kind of digraph called a *one-choice structure*, or a *functional digraph*.

Let us suppose that we have a collection of people, and we ask each person to choose one person in the collection according to some criterion. We can summarize the choices in the form of a digraph by taking each person to be a node of the digraph and by drawing a directed edge from each node (person) to the node (person) whom he chooses. To be specific, suppose we have a room full of mathematicians, and we ask each one the question "Who is the best mathematician in this room?" Displaying the answers in the form of a digraph, as just indicated, we will get a digraph looking, in all probability, like Fig. 18.24!

Fig. 18.24

So we rephrase our question and ask "Who, apart from yourself, is the best mathematician in this room?" This time we get a digraph that might look like one of those in Fig. 18.25.

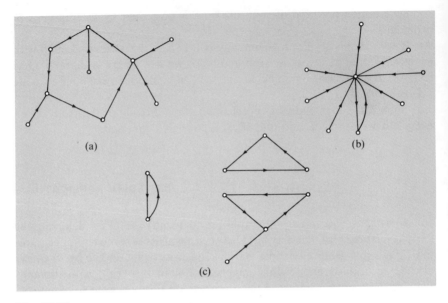

Fig. 18.25

Now we can learn a lot just by looking at such a digraph. Figure 18.25 (*a*) is a fairly typical example, with quite a variety of persons being chosen by someone or other. Figure 18.25(*b*), however, is clearly an entirely different situation; there is one person whom everyone else chooses. Both these digraphs are weakly connected, indicating a certain amount of cohesion among the persons represented by the nodes. Figure 18.25(*c*) is different again. Here the assembly has been broken up into three sets, or factions, such that no person in one faction chooses a person in another faction. These factions are the weak components of the digraph.

A digraph obtained in this manner is often called a *one-choice structure*. It is also called a *functional digraph* since it is the digraph corresponding to a functional relation, as defined in Chapter 3. We recall that *x R y* is a functional relation if there is exactly one *y* corresponding to a given *x*. This is the case here, since we have specified that each person makes just one choice; the relation "*x* chooses *y*" is thus a functional relation.

Let us now see what a functional digraph looks like. We know already from Fig. 18.25(*c*) that it may be disconnected. If it is, then each of its weak components will itself be a functional digraph. Thus we can confine ourselves to asking "What does a weakly connected functional digraph look like?"

Let *F* be any weakly connected functional digraph. If we ignore the directions of the edges of *F*, then we shall, in general, obtain a graph. Exceptionally we may get a multigraph; if *F* has a symmetric edge, that is, an edge directed in both senses—as, for example, in Fig. 18.25(*b*)—this will give rise to a double edge when we ignore all the arrows. This will make no difference to our argument, and we shall call the graph (or multigraph) *G*.

We notice first that *G* cannot be a tree. For corresponding to each node there is exactly one edge, namely, that joining the node to the corresponding "chosen" node. Hence the numbers of nodes and edges are equal, whereas for a tree they differ by 1. However, by the reasoning of Theorem 18.2, we see that if in the course of drawing a graph we complete just one circuit, then this makes the numbers of nodes and edges equal. Thus we see that the graph *G* has just one circuit, for if we were to complete *another* circuit we would then have more edges than nodes. This circuit may contain only two edges, in the special case already mentioned when *F* has a symmetric edge. The circuit then consists of the two edges of the double edge which *G* then has.

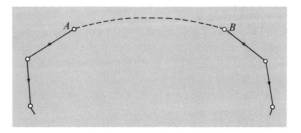

Fig. 18.26

Let us consider whether this circuit of *G* is also a circuit (in the digraph sense) of *F*. If it is not, then we shall have a situation like that depicted in Fig. 18.26, where the directions of the arrows from *A* and *B* are inconsistent with the definition of a circuit of a digraph. But this cannot happen, since clearly some node between *A* and *B* would have to have *two* arrows leading from it, and this contradicts the definition of a functional digraph. Hence a weakly connected functional digraph consists of a circuit together with some other edges. It is easy to see that since no more circuits must be completed the other edges (ignoring their directions for the moment) must form trees that are attached by one of their nodes to a node of the circuit. Figure 18.27 is a typical example.

What can we say about the directions assigned to the edges of the trees? First it is clear that any edge joining a node to a node of the circuit must be directed towards the circuit, for otherwise the node of the circuit would have *two* arrows going from it, which is not allowed. In the same way, if *A* is a node joined to a node of the circuit, then all the other edges incident at *A* must be directed *towards A*. Only one edge in a functional digraph can be

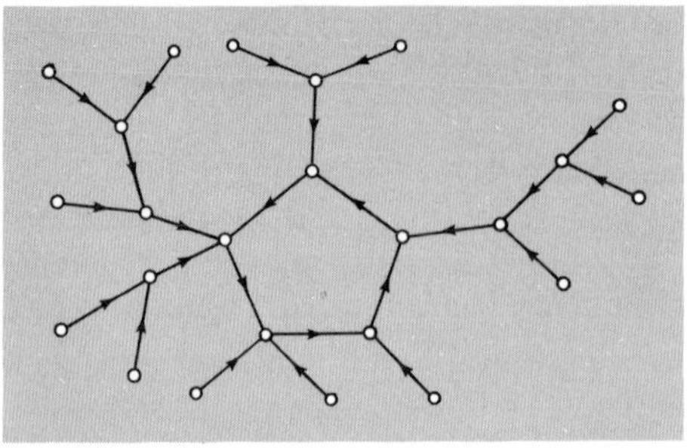

Fig. 18.27

directed *away* from a given node. Thus we see that the edges of the (directed) trees that are attached to the nodes of the circuit must all be directed *towards* the node of the circuit. Putting this another way, there is a (directed) path from any node to a node of the circuit.

This characterization of function digraphs, or one-choice structures, leads to interpretations that are not at all intuitively obvious. It means for example that there will always be a nucleus of people who choose each other cyclically. This nucleus could consist of two persons only who choose each other, but cannot consist of just one person (since he must choose someone else). Furthermore, everyone else in the group of people either chooses someone in this nucleus, or chooses someone who chooses someone in the nucleus and so on. Thus if the relation in question implies some sort of respect, then there is at least one group of persons whom everyone respects either directly or indirectly. If the digraph is disconnected there will be several such nuclei, each holding sway over a separate section of the group of people.

Exercise 213

1. Draw all possible one-choice structures on four nodes.

2. A one-choice structure S has exactly two weak components A and B. An individual x in A changes his choice and chooses an individual in B. Show that the resulting one-choice structure S' is weakly connected if, and only if, some individual in A chose x.

18.8 SIGNED GRAPHS

We shall now consider graphs of a different kind, called *signed graphs*, which are of some sociological interest and significance. In these graphs there are two kinds of edges, which in our diagrams we shall denote by unbroken lines on the one hand, and dotted lines on the other. We shall call them *positive* and *negative* edges, respectively. Two nodes may be joined by a positive edge, or by a negative edge, but not by both. In the commonest sociological interpretation of signed graphs, the nodes represent human beings; a positive edge between two nodes indicates friendship between the persons they represent, and a negative edge indicates enmity. Until further notice we shall insist that one or the other of the relations must hold between any two persons (nodes) under consideration, and we shall call such a signed graph a *complete* signed graph.

The simplest complete signed graph has one node only and no edges (we do not allow loops). This represents a "Robinson Crusoe," with no friends or enemies. If there are two nodes we have two possible signed graphs; one in which the nodes are joined by a positive edge, and one in which they are joined by a negative edge. These represent a pair of friends and a pair of enemies, respectively. So far we have met nothing of any great significance.

There are four signed graphs on three nodes, and they are shown in Fig. 18.28. Here we can make some pertinent observations.

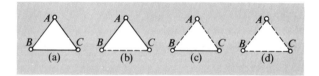

Fig. 18.28

In the graph (*a*) we have a happy situation—three mutual friends—which gives rise to no problems. Graph (*b*) is different, however; *A* is friendly with both *B* and *C*, but *B* and *C* are enemies. Such a situation creates problems for all concerned. For example, *B* might seek *A*'s company but is very likely to find his enemy *C* with him, and so on. Anticipating a later definition we shall say that graph (*b*) represents an *unbalanced* situation, whereas that represented by graph (*a*) is *balanced*. The same adjective can be applied to the graphs themselves.

This lack of "balance" is not due merely to the presence of both positive and negative edges in the same graph, as can be seen from graph (*c*). Here *B* and *C* are friends, and both dislike *A*. This is a balanced situation, giving rise to no conflicts; *B* and *C* will keep to themselves and shun *A*, who is quite content with this state of affairs. In graph (*d*), on the other hand, we have only one kind of edge, but the situation is "unbalanced." *B* is an enemy of *A*, and so is *C*. This ought to bring *B* and *C* together, but instead they are enemies. Hence there are problems inherent in this arrangement also.

Thus of the four possible complete signed graphs on three nodes (*signed triangles*, as we can call them) two are balanced and two unbalanced. Note that this information can be summed up in the following definition.

Definition 18.11

A signed triangle is balanced (unbalanced) if the number of negative edges in it is even (odd).

We now want to extend the concept of balance to groups of any number of people in a manner that will be sociologically significant. Taking what is perhaps a rather strict attitude we could say that the interplay of love and hate between the persons comprising a group cannot be said to be satisfactory, or balanced, if there are, anywhere in the group, three people between whom an unbalanced situation exists. Thus if balance for a complete signed graph on n nodes $(n > 3)$ is to mean a complete absence of any conflict, we are almost inevitably led to the following definition.

Definition 18.12

A complete signed graph is balanced if, and only if, every (signed) triangle that it contains is balanced.

Thus, according to this definition, the graph in Fig. 18.29(a) is balanced, while that in Fig 18.29(b) is not, since, for example, the triangle ACE is not balanced.

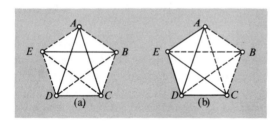

Fig. 18.29

We now prove a useful theorem.

THEOREM 18.5

If S is a balanced complete signed graph, then the nodes of S can be divided into two disjoint sets, say U and V, such that

(a) Any two nodes of U are joined by a positive edge,
(b) Any two nodes of V are joined by a positive edge,
(c) Any node of U and any node of V are joined by a negative edge.

Conversely, if the nodes of a complete signed graph S can be split into disjoint sets in this way, then S is balanced.

PROOF:

The converse is easy to prove. Any triangle in S either has all three nodes in the same set (U or V) in which case all three edges are positive and the triangle is balanced; or else it has two nodes in one set and one in the other, in which case two edges are negative and one is positive. Thus every triangle is balanced; hence S is balanced.

The proof of the first part of the theorem is rather less obvious, but not at all difficult. Choose any node A, and let V be the set of nodes of S that are joined to A by negative edges. Let all the other nodes (of which A will be one) constitute the set U. We must now prove that U and V have the required properties.

(a) Any node in U, except A, is joined to A by a positive edge, since otherwise it would be joined by a negative edge and would be in V, by the definition of V. If A, X, Y are distinct nodes of U, then A is joined by positive edges to X and to Y, and therefore (since the triangle AXY is balanced) the third edge, joining X and Y must be positive.

Thus, since X and Y were arbitrary nodes of U, it follows that any two nodes of U are joined by a positive edge.

(b) If P and Q are any two nodes of V, then A is joined by negative edges to P and to Q. Since S is assumed to be balanced the remaining edge of the triangle APQ must be a positive edge. Thus P and Q, and hence any two nodes of V, are joined by a positive edge.

(c) Let X be any node in U, and P any node in V. If X is the node A then it is joined to P by a negative edge. If not, then A is joined to X by a positive edge, and to P by a negative edge. Hence the edge joining X and P must be negative in order to balance the triangle XAP. Hence a node in U and a node in V are joined by a negative edge.

This completes the proof of the theorem. Note that the proof remains valid if one of the sets U, V is empty.

Thus in order to have a balanced situation, the group of persons must split into two rival factions, each faction consisting of persons who are friends each with the other, and with every person in one faction hating everyone in the other—a "Montague and Capulet" situation. Any departure from this setup—a relation of love where there should be hate (as with Romeo and Juliet)—unbalances the situation and causes problems.

By means of Theorem 18.5 we can give an alternative definition of balance of a complete signed graph, viz.,

Definition 18.13

A complete signed graph is balanced if, and only if, every circuit has an even number of negative edges. (We can say that every circuit is "balanced.")

We must, of course, prove that Definition 18.12 is equivalent to Definition 18.13; i.e., that we are not giving two different meanings to the word *balanced*. First we notice that if a signed graph S is balanced in the sense of Definition 18.13, it is balanced in the sense of Definition 18.12. For all the triangles of S are circuits, and hence have an even number (0 or 2) of negative edges. This means that they are all balanced.

We therefore have to show that if all the triangles of S are balanced, then all the circuits are balanced. We shall make use of the sets U and V of Theorem 18.5. Consider any circuit of S. If all its nodes lie in the set U, or are all in the set V, then all its edges are positive and it is balanced. If some nodes are in U and some in V, let us start with a node in U and traverse the circuit, returning ultimately to our starting node. Whenever we traverse a negative edge (and only then), we go from one of the sets U, V to the other. Hence, to return to out starting node, we must traverse an even number of such edges; for if we traversed an odd number of negative edges we would end up in the other set to the one we started from. Hence any cycle has an even number of negative edges. (See Fig. 18.30.)

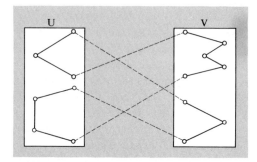

Fig. 18.30

This shows that Definitions 18.12 and 18.13 are equivalent. They define the same kind of graph, and we can use whichever one we like.

18.9 SIGNED GRAPHS THAT ARE NOT COMPLETE

Of the two definitions of balance given in the preceding section, the first (Definition 18.12) is the easier to apply since it requires us to examine only the triangles in a graph and not all the circuits. There would have been no point in introducing the second, less practical, definition therefore, unless we had some further use for it. We shall see what this is.

Suppose we allow the possibility of two persons in a sociological group neither loving nor hating each other, but having a relationship which we can describe as *indifference*. This relation can be indicated in a signed graph by leaving the corresponding nodes unjoined by either a positive or negative edge. Thus we are led to generalize the sort of situation we are considering and to study signed graphs defined exactly as before; they are, however, not complete, as hitherto assumed.

If we choose three nodes of a signed graph there will not in general be a triangle of edges joining them. In fact the graph may not have any triangles at all. Thus we can forsee difficulties in applying Definition 18.12 to signed graphs that are not complete. Definition 18.13, however, applies just as well as before, and we can state

Definition 18.14

A signed graph S is balanced if, and only if, every circuit in S is balanced, i.e., has an even number of negative edges.

Our object in this section will be to prove a theorem analogous to Theorem 18.5, namely,

THEOREM 18.6

A SIGNED GRAPH S IS BALANCED IF, AND ONLY IF, ITS SET OF NODES CAN BE DIVIDED INTO TWO DISJOINT SUBSETS, U AND V, SUCH THAT

(a)   every positive edge of S joins nodes in the same set
(b)   every negative edge of S joins nodes in different sets.

PROOF:

If such a division of the node set is possible, let us join by a positive edge any pair of nodes, both in U, that are not already joined. Similarly, join by a positive edge any pair of nodes, both in V, that are not already joined. Finally, wherever a node of U is not joined to a node of V, join them by a negative edge. The result will be a *complete* signed graph S', which will be balanced (it satisfies the conditions of Theorem 18.5).

Hence every circuit in S' is balanced (Definition 18.13). But any circuit of S is also a circuit of S', and hence is balanced. Thus S is balanced.

We must now prove the converse of this, namely, that if S is balanced, then sets U and V exist having the specified properties. Before doing this we prove a lemma, and for convenience we define the *sign* of a path in a signed graph.

Definition 18.15

The sign of a path in a signed graph S is defined to be

    (1)   positive, if the path has an even number of negative edges, and

    (2)   negative if it has an odd number of negative edges.

Our lemma is as follows.

LEMMA:

    IF S IS A BALANCED SIGNED GRAPH, AND X AND Y ARE ANY TWO NODES OF S, THEN ALL PATHS IN S BETWEEN X AND Y HAVE THE SAME SIGN.

    PROOF:

Consider any two paths between X and Y in S. These paths may have certain sequences of edges in common, and the general form of the two paths will be as in Fig. 18.31 (in which no distinction of positive and negative edges has been made). Note that in going *from X to Y*, some common edges may be traversed in opposite directions in the two paths, as is possible with edge AB for example. This makes no difference to the lemma or to its proof.

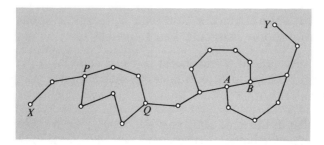

Fig. 18.31

Clearly the common edges will have the same effect on the sign of both paths. Let us start from X and continue till we come to a node (P) at which the paths diverge. We continue along one of the paths until we next come to a node (Q) common to the two paths. This will certainly happen when we arrive at Y, if not before. Between P and Q we have two paths with no common edges, and the numbers of negative edges in these two paths must be both odd or both even, for otherwise the circuit which they form would have an odd number of negative edges, which is impossible since S is balanced. Thus the two paths from X to Q have the same sign. We now proceed further along the two paths, repeating the above argument, until we arrive at Y. Hence the lemma is proved.

We can now prove the "only if" part of Theorem 18.6. If S is connected, let A be any node of S, let V be the set of nodes that are accessible from A by some negative path, and U the set of nodes (including A) that are accessible from A by a positive path. Since S is connected every node is accessible from A, and hence

$$U \cup V = \text{the node set of } S.$$

Moreover, from the lemma, there cannot be both a positive *and* a negative path from A to any node. Hence

$$U \cap V = \emptyset.$$

If two nodes X and Y of U were joined by a negative edge, then the circuit from X to A, A to Y, Y to X would have an odd number of negative edges, which is not possible since S is balanced. The same would happen for two nodes of V. Hence the negative edges (if any) in S must each join a node of U to a node of V. If $X \in U$ and $Y \in V$ were joined by a positive edge then the circuit X to A, A to Y, Y to X would be unbalanced. Hence any positive edges in S must join nodes from the same set (U or V). This proves the theorem for connected signed graphs.

If S is not connected, and has k components, we form sets U_i and V_i for each component ($i = 1, 2, \ldots, k$). We then define U to be the union of the U_i, and V to be the union of the V_i. It is readily verified that U and V satisfy the required conditions.

Exercise 214

1. In the preceding two sections we have made a distinction only between balanced and unbalanced signed graphs. We did not consider in what way some signed graphs might be "more unbalanced" than others. Several methods have been suggested for expressing the *degree of unbalance* of a signed graph when it is not balanced.

One such definition, that of Harary and Cartwright (see the "Suggestions for Further Reading" for this chapter), is

$$\text{Degree of unbalance} = 1 - \frac{\text{Number of positive circuits in } S}{\text{Total number of circuits in } S}$$

Show that this degree of unbalance must lie between 0 and 1. Use this definition to calculate the degree of unbalance of the following signed graphs.

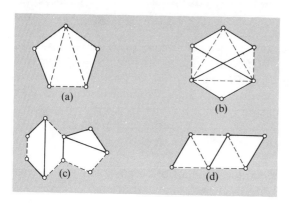

Fig. 18.32

2. Another method for expressing the degree of unbalance is to define it
as the minimum number of edges whose change from positive to negative,
or vice versa, will convert the signed graph into a balanced graph. Clearly
this is zero if the graph is already balanced.

Using this definition, find the degree of unbalance of the signed graphs
in problem 1.

18.10 DIGRAPHS AND MARKOV CHAINS

The treatment of Markov chains that was given in Chapter 15, and especially that of ergodic and transient sets, can be clarified further by the use of
a representation of the chain by means of a diagram. In a Markov chain we
have a certain number of states, $s_1, s_2, \ldots, s_n$, and a matrix (p_{ij}) of probabilities, where p_{ij} is the probability of going from state s_i to state s_j in one
step. Let us take the n states as the nodes of a digraph and associate with
the directed edge from s_i to s_j the number p_{ij}. Since the numbers p_{ii} are not
necessarily zero, we must allow loops in this digraph. This digraph will strictly
speaking be a *directed network* since we have associated a number with
each directed edge. Clearly this directed network contains all the information
that is given by the matrix (p_{ij}).

We can simplify this network by leaving out those directed edges which
have a zero probability associated with them. Thus for the first of the two
random walk problems discussed in Chapter 15 (page 779) we obtain the
diagram of Fig. 18.33, while for the second random walk problem we have
that of Fig. 18.34.

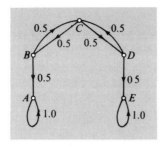

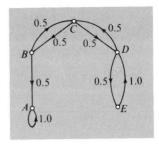

Fig. 18.33   **Fig. 18.34**

Another example of this kind of representation is given by the problems in Chapter 4 for which we drew tree diagrams. These trees (strictly speaking, they were *directed trees*) are rather special for digraphs representing Markov chains since they have the property that there is a unique path from the root to any node. On the other hand, these trees have an unnecessarily large number of nodes (states) since, on inspection, it will be seen that many nodes in different parts of the tree represent exactly the same state of affairs. Thus, for example, the various nodes in Fig. 4.3 which represent a successful conclusion to the experiment could be identified to form a single node, since they denote the same state. Similarly the nodes representing an unsuccessful conclusion could be lumped together in one node. The same is true of two other nodes, namely, those labelled s and v in Fig. 4.3. The reader should verify that in this way we can replace the tree of Fig. 4.3 by the digraph of Fig. 18.35.

This gives a simpler diagram, with no unnecessary nodes, but we lose the advantage of having a tree. To calculate the probability of a success in the experiment we must now consider all possible (directed) paths from the root to the node representing a successful conclusion (node Q), multiply the probabilities associated with the edges of each path (this gives the probability of a success via that path) and add the probabilities thus obtained. Using Fig. 18.35 we have to make sure that we do not overlook any paths between the root and Q. When we used the tree there was no danger of doing this since each path had a distinct end-node.

In the discussion of ergodic and transient sets in Chapter 15 we took no interest in the precise probabilities associated with the transition from one state to another; we had interest only in whether the transition was possible (probability > 0) or impossible (probability 0). In terms of our graphical representation of the Markov chain we were interested in the digraph rather than in the directed network. Let us see, therefore, what ergodic and transient sets become when interpreted in terms of the digraph.

We said that state s_j was accessible from state s_i if it was possible to

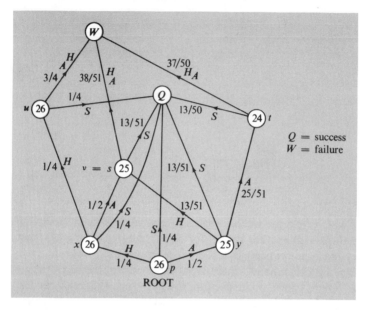

Fig. 18.35

get from s_i to s_j in a certain number of steps. This means that there must be a path from node s_i to node s_j in the digraph. The product of the probabilities associated with the edges of the path will be the probability of going from s_i to s_j by this path (there may well be other routes), but this does not interest us. We require only that the probability is nonzero and this is ensured by the fact that only directed edges with nonzero probabilities are included in the digraph. If states s_i and s_j are mutually accessible then there is a path from each node to the other. We saw in section 18.6 that if this relation (which is an equivalence relation) holds between each pair of nodes of a digraph D then D is said to be strongly connected. Hence a reasonable name for the equivalence classes of nodes which this relation defines is *strong component*, by analogy with the concept of components of an undirected graph. Thus the digraph of Fig. 18.36 has 4 strong components, viz., those containing nodes (A, B, C, D, E), (F), (G, H) and (I, J, K).

Corresponding to such a digraph D we can construct another digraph D' in the following way. The nodes of D' are the strong components of D, and there is a directed edge from a node a' of D' to a node b' of D' if, and only if, there is, in D, a directed edge from some node belonging to a' to some node of b'. Thus the digraph D' corresponding to the digraph D of Fig. 18.36 is that of Fig. 18.37. The digraph D' is essentially the result of ignoring transitions between nodes in the same equivalence class and concentrating only on those transitions which go from one equivalence class to another.

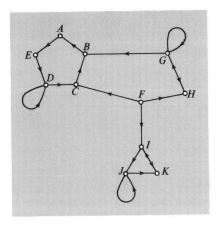

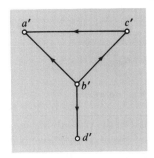

Fig. 18.36

Fig. 18.37

The essential property of D' is that it has no (directed) circuits. This is easily seen; for if D' had a circuit, say $a'_1, a'_2, \ldots, a'_k, a'_1$ then all nodes of all these equivalence classes would be mutually accessible. This contradicts the fact that $a'_1, a'_2, \ldots, a'_k$ are different equivalence classes.

We next show that D' has at least one node with zero out-valency. Starting at any node of D' we proceed to another adjacent node. If this node has out-valency > 0 we can go on to yet another node, and so on. If there were no node with out-valency zero we could continue this process indefinitely. But this is impossible; for there is only a finite number of nodes, and we cannot at any stage come back to a node already encountered (since D' has no circuits). This treatment covers much the same ground as section 15.12 but from a graphical point of view, which many find easier to grasp. The nodes of D' that have zero out-valency are the ergodic sets, which, once entered, cannot be left. The others are the transient sets. In Fig. 18.37 the nodes a' and d' represent ergodic sets.

One observation that we did not make in section 15.12 is that the digraph D' must have at least one node with zero in-valency. This follows by reversing all the arrows on the digraph D' and repeating the proof for zero out-valency. These nodes of D' represent transient sets with the property that they cannot be reached from any other set. In other words, the states in one of these sets will not occur in the Markov chain at all unless the chain starts in the set. In Fig. 18.37 the node b' represents such a set.

Exercise 215

1. The digraph that represents a Markov chain can often be used to tell whether the transition matrix of the chain is regular or not. Prove that if the transition matrix is regular then the digraph representing the chain must be strongly connected.

Consider the following digraph:

Fig. 18.38

Here the probability $\frac{1}{2}$ is associated with each edge. Show that the converse of the above statement is not true—that the matrix may fail to be regular, even though the digraph is strongly connected.

2. Draw the digraphs corresponding to the transition matrices of Exericse 185.2 thereby checking your answers to that exercise.

3. Draw a diagram, as in the last section, to replace the tree diagram for the craps problem in section 4.15.

4. Draw similar diagrams for the problems of Exercise 37.1 and 37.4.

Exercise 216 (Chapter Review)

1. The *complement* $\bar{G}$ of a graph G is defined as follows: The node set of $\bar{G}$ is the same as for G, and a pair of nodes is an edge of $\bar{G}$ if, and only if, it is not an edge of G. Thus $\bar{G}$ has all the edges, and only those, that are not in G.

Draw the complements of the following graphs.

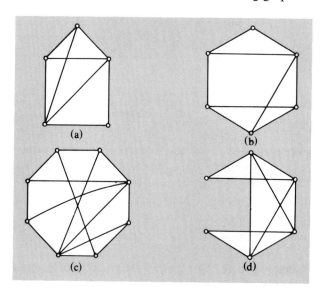

Fig. 18.39

2. Show that each of the following graphs is self-complementary, that is to say, it is isomorphic to its complement. (The definition of *isomorphism* was given on page 917.)

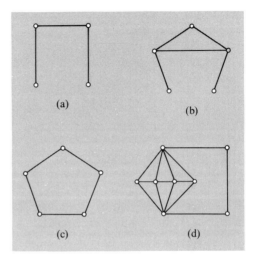

Fig. 18.40

3. By considering the maximum number of edges that can be present in a graph having m components, show that if a graph G is not connected then its complement $\bar{G}$ must be connected.

4. Show that if a graph G has m components then, by numbering the nodes in a suitable way, the adjacency matrix of G can be written in *block diagonal form*, that is, in the form

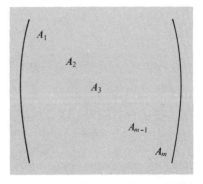

where the submatrices $\mathbf{A}_1, \mathbf{A}_2, \ldots, \mathbf{A}_m$ are square, and the shaded portions of the matrix consist entirely of zeros.

5. Let D_1' be a digraph having no (directed) circuits. Let B_1 be the set of nodes of D_1' having zero out-valency (the existence of such nodes was proved in this chapter).

Let D_2' be the digraph obtained from D_1' by deleting the nodes of B_1 and any edges incident with them. Let B_2 be the set of nodes of D_2' which have out-valency zero.

Define similarly sets $B_3, B_4, \ldots$, up to B_k, say, until no more nodes of D_1' remain.

Prove that

(a) $B_1, B_2, \cdots, B_k$ are disjoint.

(b) $B_1 \cup B_2 \cup \cdots \cup B_k =$ the node set of D_1'.

Hence show that if the nodes of D_1' are numbered in order, the nodes in B_1 being numbered first, then those in B_2, and so on, then the adjacency matrix of D_1' has lower triangular form.

By considering the circuit-less digraph D' associated with a general digraph D in the manner described in section 18.10, and using the method of problem 4, show that by a suitable numbering of its nodes the adjacency matrix of D can be put in *block lower triangular form*, viz., in the form

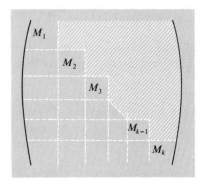

where the shaded portion consists entirely of zeros.

Show that the matrices $M_1, M_2, \ldots, M_k$ on the diagonal are the adjacency matrices of the strong components of D.

6. Using problem 5 show that the transition matrix of any Markov chain can be written in a standard form which is block lower triangular.

7. A circuit which contains each of the nodes of a graph G exactly once is called a *Hamiltonian circuit* of G. Which of the following graphs have Hamiltonian circuits? Can you *prove* that no Hamiltonian circuit is possible for two of these graphs?

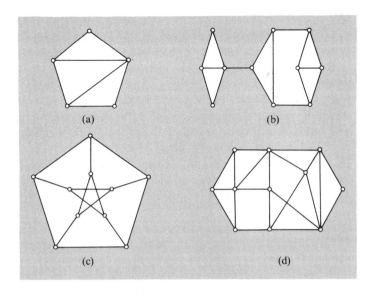

8. Which of the following signed graphs are balanced, and which are unbalanced?

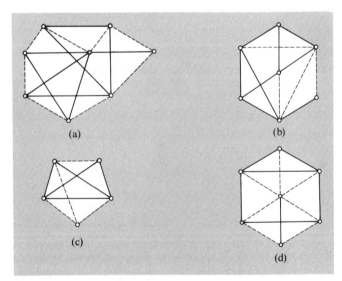

9. Prove that if any two nonadjacent nodes of a tree are joined by a new edge, the resulting graph contains one, and only one, circuit.

10.   A deck of 24 cards consists of the cards 1 through 6 in each of the four
suits. A card is drawn at random. If it is a spade, then the experiment is
at an end. If it is a heart or a diamond then the card is replaced; if it is a
club, the card is discarded. Another card is then drawn and the process
repeated. Draw a transition diagram (digraph with probabilities) for this
process, as a Markov chain with 8 states (the end state and the states for
which the deck contains 6, 5, . . . , 0 clubs). Construct the transition matrix,
and hence, or otherwise, calculate the probabilities of reaching the end state
after 1, 2 and 3 drawings of a card.

11.   The nodes of a digraph correspond to the integers 1, 2, 3, . . . , 10.
Node i is joined to node j if, and only if, the remainder when i is divided by 2
equals the remainder when j is divided by 3. Draw this digraph. How many
strong components does it have? Draw the digraph D' which corresponds
to this digraph in the manner of section 18.10.

12.   A digraph is defined on the same nodes as in problem 11 by the statement that node i is joined to node j if j is the largest integer (other than i
itself) that divides i. Draw this digraph, and show that it is not a functional
digraph but that it becomes one if the node 1 is joined to another node.

Suggestions for Further Reading

Graph theory is a subject that has quite recently made very rapid strides from humble
beginnings. Not many years ago there was only one textbook of graph theory (in
German) in existence; now there are many to choose from. The suggestions below
are only a few of the many books that the reader can profitably consult.

   For an easy-going, but careful, survey of the elements of graph theory the following book is recommended:

Ore, O., *Graphs and Their Uses*. New York: Random House, Inc., 1963.

For a much more detailed coverage of the whole subject (going far beyond the scope
of this chapter) there is

Berge, C., *The Theory of Graphs and Its Applications*. London: Methuen & Co.,
   Ltd.; New York: John Wiley & Sons, Inc., 1964.

The following book

Busacker, R., and Saaty, T., *Finite Graphs and Networks*. New York: McGraw-
   Hill Book Co., 1965,

contains a wealth of applications of graph theory in many different fields, while

Avondo-Bondino, G., *Economic Applications of the Theory of Graphs*. Glasgow:
   Blackie and Son Ltd., 1962,
Flament, C., *Applications of Graph Theory to Group Structure*. Englewood Cliffs,
   N. J.: Prentice-Hall, Inc., 1963,

treat the subject specifically in terms of applications in economics and sociology.
   The standard book on digraphs is

Harary, F., Norman, R.Z., and Cartwright, D., *Structural Models*. New York: John Wiley & Sons, Inc., 1956,

which also contains a wealth of applications to problems in economics, sociology and psychology.

A more recent book that is recommended is

Harary, F., *Graph Theory*. Reading, Mass.: Addison-Wesley Publishing Co., 1969.

19 Computers and

Computer Programming

19.1 INTRODUCTION

Among the various lines of technological progress which are apparent today, the truly phenomenal advances in computer techniques are outstanding. In its rapid development, its use of the latest discoveries in the electronic and engineering sciences, in the notice that it has attracted not only among scientists but among the general public, and in the constant publicity that it has received in the popular press, the field of computer technology is second only to the even more spectacular advances in space research.

Just how rapid has been this advance is seen if we consider that the first electronic computer was built as recently as 1944. For several years after that the computers in the whole world could be numbered on the fingers of one hand; but during the 1950s there was an "explosion" in the development and the use of computers so that at the end of that decade there were computers in almost every country, totalling many thousands. At the same time that the demand for and the production of computers expanded, so did the evolution of the computers themselves. Recent discoveries in physics were eagerly seized and pressed into service; a single small ring of magnetic material took the place of numerous radio tubes and electronic components. When transistors became commercially available in the late fifties computers using tubes were rendered obsolete. Computers became increasingly "bigger" in the sense of being able to handle more and more complicated problems, though because of the increasing use of miniaturized components (largely a by-product of space research) they were often physically smaller than their forerunners.

All this tends to give the impression of a modern computer as being something quite fantastically complicated, difficult to understand and rather remote from everyday life. Many popular accounts of computers tend to give the same idea. Now there is a certain amount of truth in this attitude; but insofar as it also tends to suggest that dealing with a computer is purely a matter for the highly trained specialist, and not a province into which someone with only an ordinary knowledge of mathematics can intrude, this impression is misleading. It is true that if you are interested in computers in the sense of wanting to take them apart and put them together again then you will indeed have to become highly trained in computer technology; but if, as is more likely to be the case, you are interested in them because you have some difficult problems that you want to solve, then the fact of the matter is that with comparatively little study you can learn enough about the subject to achieve that end. Computer programming, which is the art of persuading a computer to do what you want it to do, is more easily learned than most people imagine, at least for the general run of mathematical problems. The object of this chapter is to deal with the fundamentals of this new and fascinating branch of mathematics, and to take the reader some way towards learning a particular programming language in which many well-known types of computers can be programmed.

Computers can be regarded as machines of an extraordinary degree of complexity, but this, as we have just said, is the point of view of the person who wants to take one apart to find out how it ticks. There is an analogy here between a computer and a television receiver. The latter is also an extremely complicated device, which uses all sorts of up-to-date inventions. Manufacture and repair of television receivers is a highly technical business. Yet all this is of little consequence to the person who merely wishes to watch the shows. To him a television receiver is one of the simplest devices imaginable—you have only to switch on, turn to the proper channel, adjust the volume and brilliance, and that is that. The only time that the average viewer is aware of the complexity of the device which he normally takes so much for granted is if something goes wrong, and then it is only to the extent of realizing that he must call the repair man and not try to fix the trouble himself.

Much the same is true of a computer. The user of a computer does not need to know how it performs the various operations in its repertoire; he need know only how to make it do the particular operations that he requires. Agreed, this is more difficult than switching on a television receiver, but it is something which anyone with some background in mathematics can reasonably expect to be able to manage without too much trouble.

19.2 COMPUTER OPERATIONS

Now that we have put computers in their place, let us see the sort of things a computer does, not bothering ourselves with how it does them. A computer is capable of performing a specific, fairly limited, number of operations. These are of several kinds. First there are the arithmetical operations; the computer can add or multiply two numbers together; it can subtract one number from another, and divide one number by another. So far there is nothing very special— an ordinary desk calculating machine will do as much.

In order to be able to perform these arithmetical operations the computer must be able to retain the numbers that it is dealing with. A desk calculating machine does this too; when you have set up a number on the keyboard of such a machine, the number is retained there in the configuration of keys that have been depressed. One can say that the number is "stored" in the machine, or even "remembered" by the machine (though this does rather stretch the meaning of the word "remember"). For reasons that will become apparent, an electronic computer is designed to be able to store very many numbers at the same time, unlike a calculating machine which can usually store only three or four numbers simultaneously. The part of the computer in which numbers are stored is called the *storage*, or sometimes the *memory* of the computer.

For greater flexibility, a computer is given the ability not only to store numbers within itself but also to transfer numbers from one part of its storage to another. This, then, is another kind of operation that a computer can perform, in addition to the arithmetical operations already mentioned.

It is convenient to think of a computer as consisting (among other things) of a large number of boxes or pigeonholes, into each of which can be put a single number. The "transmit" instructions, which move data from one part of storage to another, can then be likened to the process of taking a number out of one box and putting it in another. More strictly, it is a question of putting into the second box a copy of the number that is in the first box. Depending on the type of computer, the number in the first box may be erased in the process, or it may remain there, in which case there will be two exemplars of that number, one in each box. Some desk calculating machines have the ability to transfer a number from one register to another, but only to a very limited extent.

Two other kinds of operations that a computer must be able to perform are those of *input* and *output*. It must be possible to put numbers into the computer and to get the answers out of it. Here again we shall not ask how this is done, but merely note that provision is made for doing it. On a desk calculator input is usually effected by depressing keys or knobs; output is effected by displaying the answer in the windows of the appropriate register.

We have not yet come across anything in an electronic computer which is fundamentally different from what is found in a desk calculating machine, but we must now discuss two characteristics which make an electronic computer something quite different. These are its speed, and the concept of a *program*. Since the second is an outcome of the first, we shall first consider the implications of the very high speeds at which electronic computers operate.

19.3 THE CONCEPT OF A PROGRAM

A desk calculating machine will take perhaps two or three seconds to multiply two five-digit numbers (not counting the time taken to set them up), and seven or eight seconds to divide such a number by another. An electronic computer, on the other hand, will perform these operations in times of the order of a thousandth of a second or less. In other operations, besides arithmetic ones, the electronic computer works correspondingly faster than a desk calculator. This might seem to be a purely quantitative difference between the two types of instruments rather than a fundamental distinction; the electronic device works much faster than the mechanical device, but otherwise does much the same operations. Nevertheless, the difference *is* fundamental, and the reason lies in the way in which the machine (mechanical or electronic) is used by its human operator.

Imagine a mathematician starting to perform a complicated calculation using a desk calculating machine. He knows, shall we say, that he must first multiply two numbers together, so he looks at the two numbers, sets them on the machine, and presses the button which causes them to be multiplied. He then knows that he must add a third number, and therefore sets up this number and causes it to be added to the previous product, and so it goes on. The mathematician supplies the data for the problem (by setting up the numbers on the keyboard of the machine) and initiates the correct operations (by pressing the appropriate buttons). The machine's contribution is to do the actual arithmetic. In a sense the mathematician and the machine work together as a team, and the speed of the calculation is determined by the speed of the slowest participant—the mathematician or the machine, as the case may be.

With an electronic computer the situation is quite different. With the computer working thousands, maybe millions of times as fast as the hand machine, this sort of collaboration between operator and machine is just not possible. It is no use having a computer that can multiply two numbers in a ten-thousandth of a second if it is going to take the operator a second or more to specify which two numbers the computer must multiply. In short, if computers were operated in the same way as desk machines, their high speed of operation would be completely wasted. For this reason a completely different approach to computation is necessitated. The operator must be taken out of the picture as far as possible, and the computer must be left to carry on at its own speed.

The upshot of all this is that in order to make efficient use of the very high speed of the electronic computer, *all* the instructions concerning the manner of performing a particular calculation must be set out in advance and must be presented (in some way) to the computer so that, for example, as soon as the computer has finished multiplying two numbers together it immediately gets on with the next operation of, say, adding a given number to the result, without having to wait while an unbelievably slow human being tells it what the number is! Thus the high speed of the modern computer necessitates a *program*, that is, a sequence of instructions that is presented to the machine before any calculation begins, and which tells the computer exactly what operations it must perform at each stage of the computation in order to arrive at the final result.

At this stage an example will probably help clarify matters. Suppose we have five numbers x_1, x_2, x_3, x_4 and x_5, and we wish to calculate their sum $x_1 + x_2 + x_3 + x_4 + x_5$. This presents no difficulty when performed with pencil and paper, or on a desk calculating machine, and we would probably carry out this computation without giving much thought to what we were doing. To perform this calculation on an electronic computer we would have to give a list of instructions specifying exactly what steps are to be taken, and in what order, to arrive at the answer. A moment's consideration will show that the following is the sort of list to be produced:

$$
\boxed{
\begin{array}{l}
\text{Store } x_1 \\
\text{Add } x_2 \text{ to it} \\
\text{Add } x_3 \text{ to the total} \\
\text{Add } x_4 \text{ to the total} \\
\text{Add } x_5 \text{ to the total}
\end{array}
}
$$

Fig. 19.1

The reader will probably object, quite rightly, that whatever the method by which the operator conveys the instructions to the computer, short of telepathy, the time taken to do this would be longer than the time required to perform the calculation by hand. Even if the instructions and numbers were presented in the form of punched cards (a common method), which can be read very quickly by the computer, the fact remains that someone would have to punch the cards first, and this, again, would take as long as the calculation by hand. All this is perfectly true; if we had to specify individually every operation the computer had to perform in the course of a calculation there would be no advantage in using the computer at all. The point here is that it is not necessary to specify separately every occurrence of every instruction that the computer has to perform. Most involved calculations consist of smaller calculations which are repeated over and over again. If we have therefore given a computer the instructions for one of these smaller calculations and we find that we need to perform that same calculation again, we need only refer the computer to the instructions already given; we do not have to repeat them. This supposes that there is some method of making a computer refer back to instructions that it has previously been given. Fortunately modern computers do have this ability; it is one of their essential characteristics.

There is an analogy here with mathematical notation. If we wanted to denote the sum of five numbers x_1, x_2, x_3, x_4 and x_5, we would usually write it as

$$x_1 + x_2 + x_3 + x_4 + x_5$$

as we did above. But if we wanted to represent the sum of a thousand numbers, $x_1, x_2, \ldots, x_{1000}$, we would hardly write it out in full in this way. We would write instead

$$\sum_{i=1}^{1000} x_i.$$

This expresses in a concise form what would otherwise take a great deal of time and paper to write out. In the same way, if we want the computer to add together a thousand numbers, we do not need to write out a thousand addition instructions. Let us see what we would do instead.

Let us suppose that the thousand numbers which are to be added are stored in the computer in such a way that if we specify an integer i, the computer can pick out the corresponding number x_i. This can easily be done. If we think of the storage of the computer as being like a long row of boxes, each of which contains one number, we can imagine that these boxes have been labelled "Box 1," "Box 2," "Box 3," and so on. We can then instruct the computer to deal with a particular number by telling it the label of the box in which the number to be dealt with is to be found. If we store our thousand numbers in the first thousand boxes, then the subscript i will simply be the label of the box, and for the computer to pick out the number x_i, given the integer i is an easy task. It is an operation that is built into the computer, part of its circuitry, or *hardware*, as it is called. In order that the computer can keep track of the current value of i at the various stages of the computation, this number i must be available to the computer, and will be stored somewhere, in some other "box."

We can now think up some instructions that will at least start the computer on its task of adding the thousand numbers, as follows:

> Store x_1
> Put $i = 2$
> Add x_i to the stored number
> Add 1 to i
> Go back to the third instruction.

Fig. 19.2

We have assumed so far that the computer carries out the instructions, one after the other, in the order that we have written them; but the last instruction of Fig. 19.2 is one which instructs the computer to depart from this normal procedure. It tells the computer that instead of executing the *next* instruction (which in this case does not exist), it must execute some other instruction. Instructions that cause, or may cause, the computer to depart from executing the instructions in the given order are called *branch instructions*. This particular one is an *unconditional branch instruction*.

There is one very obvious fault with the set of instructions in Fig. 19.2; it does not contain any provision for stopping the additions. The computer would begin by storing x_1, that is by copying into some convenient "box" the number x_1 (which is in Box 1). This box into which x_1 is put could be anywhere in storage; let us say it is Box 1500 for the sake of referring to it. Then having put $i = 2$, i.e., having put a 2 into a certain box, it adds x_2 to the number in Box 1500. It then increases the "2" to a "3", and adds the appropriate x_i, namely x_3, to the total accumulating in Box 1500. Again it increases i by 1 and adds x_4 to the total, and so on. Eventually i will be 1000, and the computer will add x_{1000} to the total. This is where we want the computer to stop, or to carry on with something else, but with the instructions as they are in Fig. 19.2 this will not happen. The computer will again add 1 to i and try to add x_{1001}, i.e., the contents of Box 1001 to the total. What happens then will depend on what is in that box; but whatever it is, it is not what we want to happen.

19.4 LOOPS AND FLOW CHARTS

We must therefore arrange somehow that when x_{1000} has been added to the total the computer terminates the string of additions and goes on with something else (or perhaps stops). This is achieved by what is known as a *conditional branch instruction*, a branch instruction which tells the computer to carry on with the next instruction in the normal manner under some circumstances, but under other circumstances to branch to some other instruction. By inserting an instruction of this kind we can complete the set of instructions for adding a thousand numbers as follows:

> Store x_1
> Put $i = 2$
> Add x_i to the stored total
> Increase i by 1
> If $i \leq 1000$ go back to the third instruction
> Stop.

Fig. 19.3

With this set of instructions the computer will do what we want it to do. Every time it adds 1 to i it will test whether i is now greater than 1000 (we need not worry about how it does this). If it is not, then a branch is made to the third instruction. If i is in fact 1001, then the branch is not made and the next instruction is taken in the normal way. This causes the computer to stop.

A portion of a program that is executed several times until the satisfying of some criterion (here the fact that $i > 1000$) causes the program to carry on with something else is called a *loop*. The concept of a loop is perhaps the most important concept in the whole subject of computer programming, since if we did not use loops we would have to specify every occurrence of every instruction, which, as we have seen, would render the whole operation pointless.

Figure 19.3 represents what is essentially a portion of a computer program. It would probably be presented to the computer in the form of a deck of punched cards, each containing an instruction. These instructions would be in some standard coded form, rather than as the English sentences of Fig. 19.3, but they would express the same commands to the computer.

Machines for punching cards, and the typewriters that are attached to some computers for the input of instructions, are usually fairly limited in the number of different characters they can punch or type. A common type of keyboard has the alphabet in capital letters only, the digits 0 to 9, and a dozen or so special characters like $+$, $-$, $/$, $=$ and various punctuation marks. Thus it would not be possible to punch or type "x_i", and an alternative way of expressing the very useful idea of a variable with a subscript must be devised. A convenient way of writing a subscripted variable is "X(I)", and this is what we shall use in future.

Since we are making some changes in the notation of Fig. 19.3, we might as well introduce a few other improvements. We have seen that in Fig. 19.3 we have occasion to refer to one of the instructions (the third) when we instruct the computer to go back to it. A more convenient method of doing this is to give the instruction a number and to refer to it by that number.

Another innovation that we can make is in our method of referring to the numbers we are dealing with. We have assumed, for the sake of argument, that the total is accumulated in Box 1500, but it does not much matter whereabouts the total is stored, and quite probably we would neither know nor care where it was stored. In order to be able to refer to this box, let us give it a name or symbol—TOTAL would be an appropriate one. Remember that strictly speaking, this name refers to the box, or storage location, rather than to the contents of that box, but the two being so closely connected there is little harm in thinking of TOTAL as representing the number, too. The same is true of the symbols we have already used, namely "I" and "X(I)", which really represent storage locations. Thus the instruction "Put I = 2" means, in effect, "put a 2 in location I." However, we shall continue to write it this way.

With these changes the program of Fig. 19.3 becomes

	Put X(1) in TOTAL
	Put I = 2
7	Add X(I) to TOTAL
	Add 1 to I
	If I ≤ 1000 go back to instruction 7
	STOP

Fig. 19.4

Notice that the number that we have put against the third instruction is arbitrary; we could have used any number we liked, since its purpose is solely for identification. Notice also that there is no need to number any instructions that we do not wish to refer to.

Figure 19.4 is not the only program that will perform this particular computation. The following would do just as well.

```
         Put TOTAL = 0.0
         Put I = 1
  193    Add X(I) to TOTAL
         Add 1 to I
         If I ≤ 1000 go to 193
         STOP
```

Fig. 19.5

In this example we have avoided giving the number X(1) any special treatment, and it is now added in to the total just like all the other numbers. This requires that we start the program by making sure that the location TOTAL is set to zero (there is no knowing what junk the last user of the computer may have left in there!). After this is done the numbers X(1), X(2), ..., are added in succession until the fact that I = 1001 causes the computer to stop.

When a computer executes a conditional branch instruction, like the next-to-last one of Fig. 19.5 it is in a sense making a decision. The use of this phrase may seem out of place applied to a machine, but what the computer is doing is carrying out one or the other of two possible courses of action on the basis of what has happened in the past, and this, after all, is the essential part of decision making. The application to computers of anthropomorphic terms, like "memory," "making a decision," "looking at a number" and so on, is often decried, especially by those who do not have much to do with computers. In practice it is difficult to avoid this manner of speaking, and it would be cumbersome to do so. A computer programmer might well say of a computer at a certain stage of a calculation that "it is looking to see whether it has got to the end of the list" rather than say that "a comparison is being made between the number in location I and the number 1000, according to the result of which the instruction next to be executed will be determined." In doing this he is well aware that he is speaking metaphorically and is unlikely to be misled by this sort of personification of the computer. For this reason we shall use this manner of speaking without apology whenever it will help the exposition of the subject.

The program that we have been considering is an extremely simple one. In a more involved calculation there may be literally hundreds of places at which the computer has to make a decision of one sort or another, and it then becomes a matter of some difficulty for the programmer to keep track of all the possible outcomes of these decisions in their various combinations. To ease the programmer's burden a diagrammatical method of portraying a program is often employed. This is known as a *flow chart*. In the programs given so far we have set out the instructions one under the other in what is in fact a one-dimensional ordered array. A flow chart makes the "flow" of the program from one instruction to another more apparent by spreading the program out in two dimensions. We can see what this means by looking at the flow chart for Fig. 19.5. It is as follows:

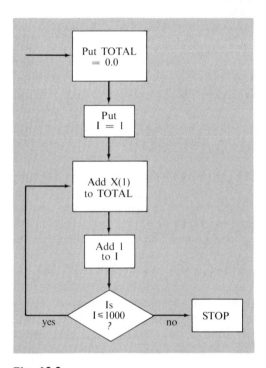

Fig. 19.6

The majority of instructions in a program will be of one of three kinds, namely, *input-output* instructions, *executive* instructions and *branch* instructions. In a flow chart the descriptions of executive instructions, which simply tell the computer to perform some calculations, are enclosed in a rectangle; descriptions of branch instructions, at which the computer "makes a decision," are written inside a diamond. Input-output instructions (to be considered later) are written either in a rectangle with rounded ends, or inside the following shape.

These shapes, representing portions of the program, are connected by arrows which indicate the flow of the program, that is, indicate which instructions come after which. After a branch instruction there may be two or more possible ways in which the program can continue. The arrows emanating from such an instruction are therefore labelled so as to indicate their connection with the outcome of the decision. All this can be seen in Fig. 19.6, in which there are four executive instructions and one branch instruction. The instruction STOP, which ends the program is usually written inside a square.

We shall now consider a somewhat more involved problem, that of finding the average of a set of numbers. This time we shall not assume, as before, that the numbers $X(1)$, $X(2)$, . . . , have somehow been stored in the computer. We shall make provision for getting them into storage. A common method of doing this is to punch the numbers in question on to punched cards and to instruct the computer to "read" each card and to transfer the number punched on it to a suitable location in storage. The details of this process will vary from one computer to another, but the general idea is much the same. In our flow chart we shall simply write "read a card," and leave it at that.

An important point arises here. We must either instruct the computer beforehand how many cards are to be read in, or else we must make some provision whereby the computer will "know" (i.e., can detect) that it has read in all the numbers. If we follow the first alternative, we must first make the computer read a card containing the number of numbers to expect. The computer must then be made to read in just that number of cards. We could, if we wished, read in all the cards, store all the numbers somewhere in storage, and then, when they were all stored, proceed with the calculation of their average. But this would be wasteful in storage space; it would unnecessarily use up a lot of storage locations (we shall drop the term "boxes" now). The main part of the calculation is to add up the input numbers, and if we add each number to a total immediately after it has been read in, we can forget about it and use the location into which it was read for other purposes, such as holding the next number to be read in. Thus each number can be dealt with as soon as it is read in, and then forgotten.

To keep track of how many numbers have been read in we shall set up a *counter* in the computer, that is a number to which we shall add 1 every time a card is read in. If we set this number to zero before the reading of cards starts, then this counter will always contain a record of how many cards have so far been read in. Let us refer to the location in which this counter is stored as I; let us call the location into which we read the numbers X, and that in which we accumulate the total, TOTAL, as before. The first card (containing the number of cards that are coming) will be read into a location N. After the total has been found we must, of course, divide by the number of numbers to get the average, but this presents no problems. Bearing these points in mind the reader should have no difficulty in understanding the following flow chart.

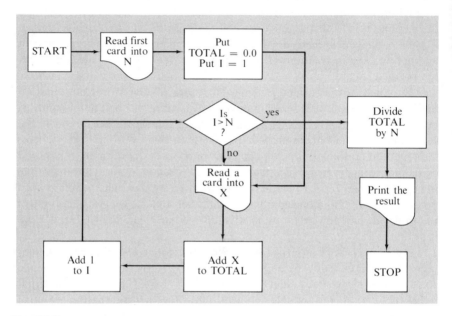

Fig. 19.7

One instruction has not been explained, the instruction "print the result." The meaning of this will depend on the particular computer being used. What is required is for the result of the calculation to be displayed in some visible form. This may be achieved by having the computer type the answer on a typewriter, or print it on a high-speed printer, or by many other methods. The particular method employed is, however, immaterial to our purpose, which is merely to grasp the fundamentals of computer programming. Notice the word START which has been put in to indicate where the program begins. Like STOP it is usually put inside a square.

Suppose we did not wish to specify in advance the number of numbers to be read in. What other course of action could we take? In many computers the card-reading device is equipped with an indicator which shows whether the card just read in was the last card of the batch or not. This indicator can be tested by the computer in the course of executing the program. When

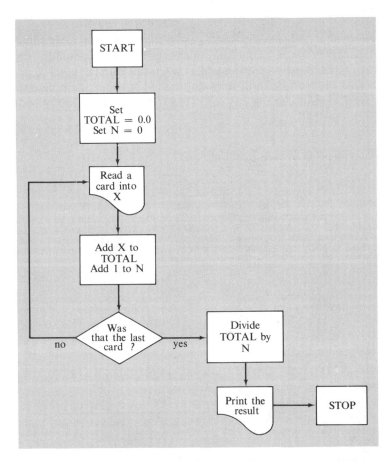

Fig. 19.8

the computer has such an indicator the programming of this problem is greatly simplified, as shown in Fig. 19.8.

We have caused a zero to be put in the location N before reading any cards, and by adding 1 to this number each time a card is read we keep track of their number. When the last card has been read, the location N contains the number by which the total must be divided. Notice that some instructions which were written in their own separate rectangle in Fig. 19.7 have now been put together, thus reducing the size of the flow chart. There is no harm in doing this, and it saves space and often clarifies the flow chart.

What if our computer has no mechanism for detecting the last card, or if we want several batches of cards to be loaded into the card-reading unit? How is the computer to tell when it has come to the end of a batch of cards? Clearly we need to put in some sort of marker, and this is most easily done in the form of a *trailer* card which comes at the end of the batch and which is punched in some distinctive way. Thus if we knew that none of the numbers punched on the data cards was greater than, say, 2000.0, we could follow the data cards by a trailer card on which was punched the number 9999.9. All we then have to do is to arrange that when this number is read, the computer behaves in the same way as it did in the last example on detecting that the card just read was the last card. The following flow chart gives the outline of this method.

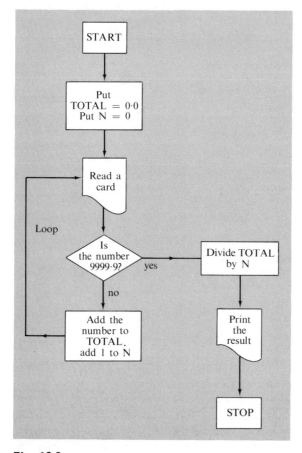

Fig. 19.9

Exercise 217

1. The following is a sequence of instructions to a computer. What calculation will the computer perform?

       1   Read X
           Square X

Multiply the result by 3
Add 1 to the result
Store this result
Add 2 to X
Divide this result by the result previously stored
Print the answer
Go back to 1.

2. What would be the result of executing the following instructions?

1   Read A and R
    Put I $= 0$
    Put TERM $=$ A
23 Print the value of TERM
    Multiply TERM by R
    Add 1 to I
    If I $\leq$ 50 go to statement 23
    STOP.

3. Draw a flow chart for a program that will read in four numbers, and print out the largest and smallest of them.

4. Draw a flow chart for a program to read in 20 numbers, and to print their total if this is positive, but to print the number zero if the total is negative.

19.5 LOOPS WITHIN LOOPS

These programs for computing an average are very simple programs containing one *loop*, that is, a sequence of instructions that the computer executes over and over again until the satisfying of some criterion causes it to come out of the loop and carry on with another part of the program. We shall now consider a more complicated example in which there are two loops, one inside the other. We shall draw a flow chart for a program to calculate values of e^x from the formula

(19.5.1) $$e^x = 1 + x + \frac{x^2}{2!} + \frac{x^3}{3!} + \cdots$$

which we obtained in Chapter 9. We shall do this for values of x from 1 to 2, in steps of .05, and produce a table of results.

First a few remarks. Equation (19.5.1) defines e^x as an infinite series, which cannot (in general) be summed exactly, but can be summed to any specified degree of accuracy. Suppose we stop summing the series when we have reached the term $\frac{x^n}{n!}$. Then our result will be in error by

$$\frac{x^{n+1}}{(n+1)!} + \frac{x^{n+2}}{(n+2)!} \cdots = \frac{x^{n+1}}{(n+1)!}\left\{1 + \frac{x}{n+2} + \frac{x^2}{(n+2)(n+3)} + \cdots\right\}$$

which, if $n > 2$, is certainly less than

$$\frac{x^{n+1}}{(n+1)!}\left(1 + \frac{x}{4} + \frac{x^2}{4^2} + \cdots\right)$$

$$\leq \frac{x^{n+1}}{(n+1)!}\left(1 + \frac{1}{2} + \frac{1}{2^2} + \cdots\right), \text{ since } x \leq 2$$

$$= 2\,\frac{x^{n+1}}{(n+1)!}.$$

Thus the error cannot exceed twice the first term to be omitted. If we want our results to be correct to 6 places of decimals, we must not have an error of more than .0000005, and we can achieve this by stopping the summation of the series as soon as we get to a term which is $< .00000025$.

Our program for performing this calculation will contain two loops. There will be an *outer* loop, in which the calculation of e^x is repeated for all the various values of x; and an *inner* loop, in which the repeated calculation is that of calculating the successive terms of the series. By not worrying about how to calculate e^x for the moment, we can draw a flow chart for the outer loop. It is as follows.

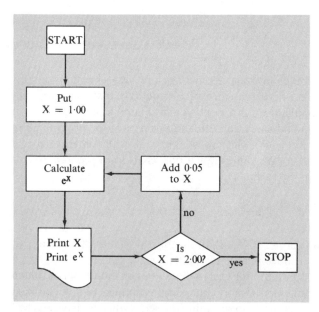

Fig. 19.10

This simply repeats the calculation of e^x for $x = 1, 1.05, 1.10, \ldots, 2.00$. When x becomes 2.00 the program comes out of the loop and goes to the next part of the program which is the stop instruction.

We turn now to the calculation of e^x. We need to compute the successive terms $1, x, \dfrac{x^2}{2}$, etc., and the simplest way is to compute each one from the previous one. We have, in fact,

$$\text{Term in } x^r = \frac{x}{r} \text{ times the term in } x^{r-1}.$$

This enables us to go from one term to the next by means of a multiplication by x and a division by r. Of course r will have to be increased each time we calculate a new term. The following flow chart will carry out this part of the calculation. It will also perform the required addition of terms, and includes the arrangement for coming out of the loop when the term just calculated is $< .00000025$.

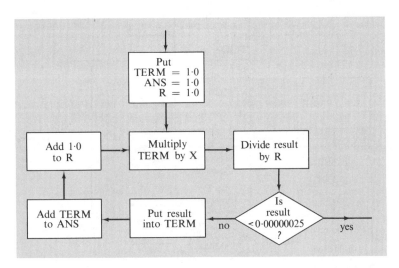

Fig. 19.11

We start with ANS (which will eventually contain the answer, e^x), TERM and R all equal to 1. Multiplying by X and dividing by R gives the next term (it will of course be just X) and since this is not $< .00000025$ the program enters the loop. The result is put into TERM, replacing the old term, and is added to ANS. ANS now has the sum of the first two terms of the series. R is increased by 1 ready for the next go round the loop. Multiplication by X and division by R gives the next term, which is made to replace the old term, and is added to ANS, which now contains the sum of the first three terms.

This process continues until the calculation of the next term results in a number $<$ .00000025. The answer (ANS) has now been calculated to the required degree of accuracy and the program leaves the loop.

All this must go in place of the rectangle which said "calculate e^x" in Fig. 19.10. If we do this we get a detailed flow chart for the whole calculation, which is given in Fig. 19.12. The significance of the numbers by the side of the rectangles will be seen shortly.

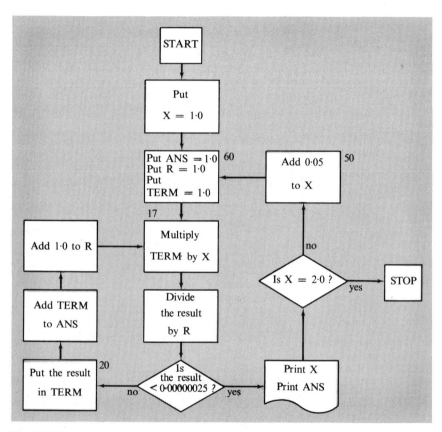

Fig. 19.12

19.6 FROM FLOW CHART TO PROGRAM

Flow charts are very useful, but when it comes to presenting a program to a computer the instructions have to be given in a linear sequential order as in Figs. 19.1 through 19.6, and a flow chart will not do. But once a flow chart has been drawn for a program it is fairly easy to make out the corresponding list of instructions. Loops in the program, such as those in Fig. 19.12, are indicated by conditional branch instructions which cause the computer to go to some instruction other than the one next in sequence, and to repeat the execution of a loop if it has not already been executed the correct number of times. In order to identify the instruction to which the program must branch we shall number it, as we did before, with some arbitrary integer. In this way, from the flow chart of Fig. 19.12 we obtain the following program

	Put X = 1.00
60	Put ANS = 1.0
	Put R = 1.0
	Put TERM = 1.0
17	Multiply TERM by X
	Divide the product by R
	If the result is ≥ .00000025 go to 20
	Print X
	Print ANS
	If X < 2.0 go to 50
	STOP
20	Put the result in TERM
	Add TERM to ANS
	Add 1.0 to R
	Go to 17
50	Add 0.05 to X
	Go to 60

Fig. 19.13

To facilitate the appreciation of the correspondence between this program and the flow chart, the numbers of those instructions in Fig. 19.13 which we labelled were written in the appropriate places in the flow chart in Fig. 19.12.

Notice that the "main stream" of the program is from the first instruction to the STOP instruction. The instructions in the two loops form two completely separate sections of the program, namely the four instructions beginning with instruction 20, and the two beginning with instruction 50. Note that in this program we have two examples of an *unconditional* branch instruction, namely "Go to 17" and "Go to 60," at which the statement to which the branch is made does not depend on what has happened in the past.

The reader should now be able to draw flow charts of the sort we have been considering for quite a variety of problems. The wording which is put in the rectangles or diamonds need not be in any standard or conventional form; any sort of description of what the computer should be doing at that stage in the computation will do. One further example should make the various points clear, and the reader can then go on to try the exercises which follow.

EXAMPLE.

Draw a flow chart and write a program for the computer to read in three numbers a, b, c, and compute and print the roots of the quadratic equation

$$ax^2 + bx + c = 0.$$

If the roots are complex, the computer should print the word COMPLEX and then the real and imaginary parts; if the roots coincide, it should print EQUAL and the value of the common root; if the roots are real it should print REAL and the two roots.

We first make a few observations. The formula required is that the roots are given by

$$\frac{-b \pm \sqrt{b^2 - 4ac}}{2a}.$$

We have three courses of action, according as $b^2 - 4ac$ is negative, zero, or positive. If it is negative we must calculate the two quantities

$$-\frac{b}{2a} \text{ and } \frac{\sqrt{4ac - b^2}}{2a};$$

if $b^2 - 4ac = 0$, only $\dfrac{b}{2a}$ need be calculated; while if $b^2 - 4ac > 0$ we calculate

$$\frac{-b + \sqrt{b^2 - 4ac}}{2a} \text{ and } \frac{-b - \sqrt{b^2 - 4ac}}{2a}.$$

We therefore draw a flow chart which, at the first decision instruction, branches into three streams, according to these three possibilities. Each stream is then a straightforward sequence of instructions, and we can specify comparatively easily what the computer must do. We arrive at the flow chart of Fig. 19.14 as a first attempt. We shall see that it can be improved. After the calculation of $B^2 - 4AC$ (which we have put in one rectangle although there would be many steps in the calculation) the program goes its three separate ways. We have denoted by D the quantity $\sqrt{B^2 - 4AC}$ or $\sqrt{4AC - B^2}$, as the case may be, and the two roots by ALPHA and BETA (when the roots are not complex).

This flow chart can be improved. It is not wrong, but it is inefficient. For example, we have to calculate $-B/2A$ at some stage or other, whatever happens. At present this calculation is performed in each of the three branches, but we could just as well calculate it at the beginning, before the program splits. We need to give the quantity $-B/2A$ (or rather its location) a symbol, so that we can refer to it later; let us call it

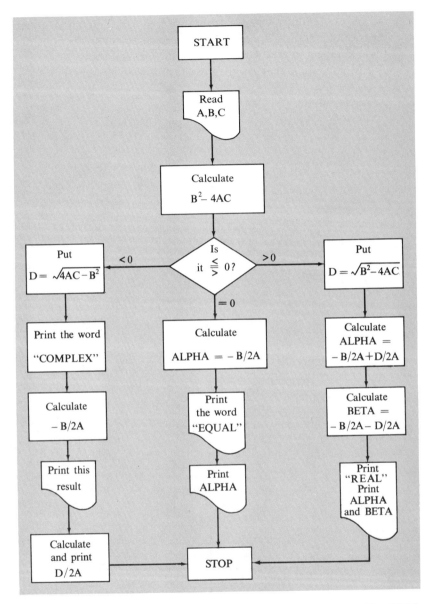

Fig. 19.14

P. It will be convenient also to use Q to stand for D/2A, where D is $\sqrt{B^2 - 4AC}$ or $\sqrt{4AC - B^2}$ as the case may be. Moreover, except when $B^2 = 4AC$, we end the program by printing two numbers, and it is better to have these numbers always in the same locations, ALPHA and

BETA, even though, when the roots are complex, these numbers are not the roots themselves. We can then use the same print instructions to print these numbers in the two possiblities for unequal roots.

Making these changes gives us the following flow chart.

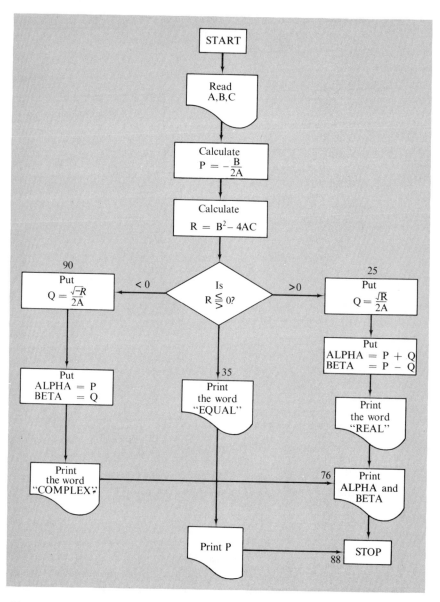

Fig. 19.15

Having constructed a flow chart for the problem, we complete this example by arranging our instructions in sequence. This is not difficult if we work from the flow chart, making the appropriate branches where necessary, and we obtain the sequence of instructions given in Fig. 19.16.

	Read in A, B, C
	Calculate P $= -$B/2A
	Calculate B$^2 - $4AC $=$ R
	If R is $<$ 0, go to 90
	If R $=$ 0, go to 35
25	Put Q $= \sqrt{R}$/2A
	Put ALPHA $=$ P $+$ Q
	Put BETA $=$ P $-$ Q
	Print the word 'REAL'
	Go to 76
35	Print the word 'EQUAL'
	Print P
88	STOP
90	Put Q $= \sqrt{-R}$/2A
	Put ALPHA $=$ P
	Put BETA $=$ Q
	Print the word 'COMPLEX'
76	Print ALPHA
	Print BETA
	Go to 88

Fig. 19.16

The reader should follow this sequence of instructions in conjunction with Fig. 19.15 and satisfy himself that it expresses the same pattern of operations and calculations as the flow chart.

Exercise 218

1. Draw a flow chart for a program that will read a number x and print out the value of log x, computed to 6 decimal places from the series

$$\log x = 2\left\{\frac{1-x}{1+x} + \frac{1}{3}\left(\frac{1-x}{1+x}\right)^3 + \frac{1}{5}\left(\frac{1-x}{1+x}\right)^5 + \cdots\right\}$$

Write a series of instructions (after the manner of the previous section) forming a program for this calculation.

2. Write a program to read in two numbers u_0, u_1, and print out the first 50 terms of the sequence $\{u_n\}$ given by the recurrence relation

$$u_{n+2} = 5u_{n+1} - 3u_n - 2n.$$

3. Write a program to produce a table of values of the polynomial

$$y = x^5 + 3x^4 - 2x^3 + 5x^2 - x + 7$$

for $x = -1.0, -0.9, -0.8, \ldots, 0.8, 0.9, 1.0$.

Hint: the number of additions and multiplications required is reduced if the polynomial is taken in the form

$$y = (\{[(x + 3)x - 2]x + 5\}x - 1)x + 7.$$

4. Write a program to produce a table of sin x for $x = 0, 0.01, 0.02, \ldots,$ 0.50 to 5 figure accuracy, using the series

$$\sin x = x - \frac{x^3}{3!} + \frac{x^5}{5!} - \cdots$$

to as many terms as are required.

19.7 WHAT IS A PROGRAMMING LANGUAGE?

In the previous sections we have considered the problem of how to devise a sequence of instructions that, if suitably communicated to a computer, would cause the computer to carry out a desired calculation. We made no attempt to answer the question of how, or in what form, these instructions were to be fed to the computer. In writing our instructions we were content to express them in any convenient way, though we made a few rules about the use of capital letters, numbering instructions, and so on. In consequence, our instructions have not been in a form which a computer will be able to digest. What must we do, then, to produce a program that a computer can accept?

There are three possible courses of action. The first is to learn the computer's own language, *machine language* as it is often called. This entails learning how to write the various instructions that the computer can perform, and these are the simple, basic instructions of addition, subtraction, etc., the instructions for moving data around in the storage of the computer, the various branch instructions, and the input-output operations. Each such instruction causes the computer to perform one simple operation. Consequently, writing in machine language is a tedious business. One cannot just write "Calculate $\sqrt{B^2 - 4AC}$"; the whole calculation has to be spelled out in the finest detail, including the sequence of additions, subtractions, etc., that the computer must follow in order to calculate the square root. Professional programmers, whose occupation it is to communicate with computers, may sometimes have to use machine language, though even they avoid it whenever possible. For the scientist who merely wants to solve a problem it is definitely not the way to program the computer.

The second possibility is to make use of these programmers and their intimate knowledge of the computer. If we were to give an experienced programmer the list of instructions in Fig. 19.16, for example, could he not produce a suitable machine-language program for us? Indeed he could. What is more it would probably not be a very difficult task for him. Many of the instructions in the program would correspond to straight machine language instructions; others, such as taking the square root, would be represented by lists of instructions (called *subroutines*) which the programmer would certainly have had occasion to use before, and he could merely look up what instructions were used on the previous occasion and use them again. This would then seem to be a good method. We break our problem down to a sequence of standard calculations, branches and input-output specifications, and then hand it to the programmer to translate the various parts of it into the appropriate batches of machine-language instructions.

This method would certainly work (though the programmers might not like it!), but there is a better way. The translation which the programmer would make would be a tedious, possibly complicated, task, but essentially it would be just routine; certain instructions that we have specified would correspond to certain sequences of machine-language instructions, in accordance with "rules" of translation depending on the machine. Here we come to the crucial point. Tedious, complicated, but purely routine tasks are exactly the sort of things that computers excel in. Why not get the computer itself to convert our list of instructions into the appropriate list of machine-language instructions? This can certainly be done. It does mean, of course, that someone at some stage has to write a program for this translation process, specifying what the computer must do for every kind of input instruction we are likely to give it, and this is indeed a formidable task. But once it has been done, it is done once and for all, and thereafter anyone can use the translation program.

Now it would be too much to expect any team of programmers, however competent, to produce a program that would accept any English sentence expressing a mathematical instruction, sort out what it meant, and produce the corresponding machine-language instructions. Therefore the person who wishes to write a program for a computer must learn what sorts of instructions the computer (or, more correctly, the translation program) can deal with, and what it cannot deal with. He must, in fact, learn the language that the translation program has been designed to accept. This is what is known as a *programming language*. Learning it requires a certain amount of time and trouble on the part of the would-be programmer, but this is very much easier than learning to write in machine-language.

Let us just recapitulate what has been said so far, and introduce a few terms. In the first half of this chapter we gained some experience at breaking down a calculation into simple steps. Many of these steps, however, would still correspond to a large number of basic operations that a computer is designed to perform. To break the program further into these basic operations would be extremely tedious for a human being—it is so alien to the normal manner of mathematical thought in which, for example, taking a square root is a *simple* operation, whereas to most computers it is not. Fortunately it has been possible to write programs for computers which will take the essential steps of a program (such as those we wrote earlier in the chapter), written in some conventional way (ours were not), and translate them into the sequences of machine-language instructions. Such a conventional way of writing programs is called a programming language, and the program which carries out the translation from the programming language to the machine language is called the *processor* or the *compiler* for that programming language.

What we must do, therefore, is to learn the conventional way of expressing our wishes to the computer. This will depend on the particular programming language which we decide to learn, for there are many. The one we shall look at now is known as FORTRAN, which stands for *FO Rmula TRANs*-lator. It is a programming language which can be used to write programs for a large number of different computers, and is quite easy to learn. The particular version of it ("dialect," if you like) that we shall study is the very simplest version of all—FORTRAN without FORMAT—which can be used on, for example, the IBM 1620 computer. There are more versatile versions for larger computers, but with a few minor exceptions, a program written in FORTRAN for a small machine will always work if compiled on a larger machine, though the opposite is not generally true.

19.8 THE FORTRAN LANGUAGE

THE ALPHABET

As with learning any language we must first learn the alphabet; that is, we must learn which characters may and may not be used in making statements in the language. (We shall use the word *statement* when referring to FORTRAN, rather than *instruction*.) The FORTRAN alphabet is quite familiar, and consists of three sets of characters, as follows:

Letters: A B C D E F G H I J K L M N O P Q R S T U V W X Y Z
Digits: 0 1 2 3 4 5 6 7 8 9
Special characters: $+ - * / ( ) = , .$

No other characters may be used in the dialect of FORTRAN that we are considering. In particular, lower case letters, subscripts and superscripts (written below or above the line) are not permitted.

CONSTANTS

In program statements we shall frequently need to specify constants, that is, numbers whose values are known. These must specified in the proper ways. There are two ways of doing this.

(1) Fixed point constants. These are essentially integers. They are written in normal notation without a decimal point, and must not exceed 4 digits. If negative they *must* be preceded by a minus sign; if positive they *may* be preceded by a plus sign, but this is not necessary. Examples of valid fixed point numbers are the following:

$$23 \quad +9721 \quad -53 \quad -8863$$

The following are not valid fixed point numbers:

$$21374 \quad -23.596 \quad 12D7 \quad 789226$$

(2) Floating point constants. These are the numbers on which calculations will usually be performed. They can be indicated by writing the number in a normal fashion, including a decimal point (this is essential). Examples of floating point constants are:

$$214.3 \quad +1. \quad 0.0 \quad -.000173298654$$

The sign of a floating point number is indicated by a preceding plus or minus sign, but the plus sign may be omitted.

Another way of indicating a floating point number reflects the common method, used by scientists especially, for representing very large or very small numbers. Instead of writing

$$.00000000000000015986$$

a scientist would write 1.5986×10^{-16}, and instead of

$$-56683927000000000000$$

he would write $-5.6683927 \times 10^{18}$. The same system is used in FORTRAN, but since superscripts are not allowed, the letter E is used to separate the number containing the significant digits (the *mantissa*) from the power of ten (the *characteristic*). The above numbers would be written as

$$1.5986E-16 \qquad \text{and} \qquad -5.6683927E18$$
mantissa characteristic   mantissa characteristic

Note that they could also be written as

	15986.E–20	and	−56683927.E11
or as	15.986E–17	and	−566.83927E16

and so on. The mantissa must contain a decimal point, and the characteristic must not.

The difference between fixed and floating point constants rests solely on how they are stored in the computer. Fixed point constants are stored as four-digit numbers with an indication of their signs. A floating point number is stored in two parts, as the mantissa and the characteristic. The mantissa is an eight-digit number, the characteristic a two-digit number, and when the constant is stored these two numbers are automatically adjusted so that, in storage, the mantissa is less than 1, and not less than .1 (so that the first digit after the decimal point is not a zero). The floating point zero itself is the only exception; its mantissa is all zeros. Thus the mantissas of the constants given just now would be stored as

$$.15986000 \quad \text{and} \quad .56683927$$

and the characteristics would be adjusted accordingly.

The characteristic can be any integer in the range −50 to +49. More than 8 significant figures can be specified in a FORTRAN constant, but since the mantissa is restricted to 8 digits, any extra digits will not be stored, and will be ignored. Thus a constant specified as 12793.52976 would enter the machine as if it had been specified as 12793.529.

Exercise 219

Determine whether the following are valid FORTRAN constants, and if valid, whether they are floating point or fixed point.

(a)	223	(e)	−.039E26	(i)	1.26585	(m)	.0251967
(b)	5931E16	(f)	72.61583	(j)	+7081	(n)	25108
(c)	−276.39E–5	(g)	−392	(k)	279,715.8	(o)	−27.39E1.0
(d)	−1970P6	(h)	27.916E1	(l)	55567.2981	(p)	1

19.9 SYMBOLS

While some numbers which occur in a calculation will be known in advance, most of the numbers that we must instruct the computer to handle will be the results of the various computations that the computer performs, and will not be known to us when we write the program. We must have some way of referring to these numbers. This is done in FORTRAN by giving the numbers a symbol, much as in ordinary algebra; but since the FORTRAN alphabet is limited we are allowed to use a sequence of up to 5 letters or digits to designate a number. Strictly speaking the symbol designates the location in the storage of the machine where the number is stored, but this amounts to more or less the same thing. There are two classifications of symbols, according to whether the numbers that they represent are fixed point numbers or floating point. They are defined as follows.

(1) Fixed point symbol. Any sequence of up to 5 letters or digits of which the first is one of the six letters—I, J, K, L, M, or N.

(2) Floating point symbol. Any sequence of up to 5 letters or digits of which the first is a letter *other than* I, J, K, L, M or N.

Thus all symbols must begin with a letter. Notice that the symbols TOTAL, ROOT, etc., that we used in the first part of this chapter were chosen to conform with the above definitions.

As examples of valid fixed point symbols we have

ITEM   J3P   MAX   KAPPA   NOTE

As examples of valid floating point symbols we have

X   ALPHA   SUM   SQ75   P1970   QUEEN

The following are examples of invalid symbols

ABCDEF   2PORQ   FU*S   )PICK

Subscripts. If we want to indicate a subscripted variable we can use a sub-scripted symbol, and the subscript is indicated, as in the first part of the chapter, by putting it in parentheses. Any valid symbol can be subscripted, but the subscript itself *must* denote a fixed point number. Thus we can have

MONEY(K)   TABLE(14)   MSQ(INDEX)   VECTR(J)

Two-dimensional arrays (matrices, two-way tables, etc.) can be designated by the use of two subscripts, separated by a comma inside the parentheses. Thus a matrix (a_{ij}) could be denoted by "A(I,J)" and so on. Further examples are:

ARRAY(ITEM,INVCE)   TABLE(INCOM, LEXP)

Exercise 220

State whether the following symbols (some of which are subscripted) are valid, and if they are, state whether they are floating or fixed point symbols.

POSN(INK)	34GT5	FLOAT
LILY	ANNE	TIM(S)
NOI(E)	KING(JOHN)	PQ7.0
FIXED	VELOC	TEMP
COL(NU)	SIGMA	SET(NU,PI)
EPSILON	ACC(MU,NU)	QUOT

19.10 ARITHMETICAL EXPRESSIONS

There are five arithmetical operations, which are denoted by the following characters—

+ plus, addition
− minus, subtraction
* multiplication
/ division
** exponentiation.

These are used in a fairly obvious way to form what are known as *arithmetical expressions*. Thus "A + B" denotes the expression which, in ordinary algebra could be written as $a + b$. Similarly

P − Q denotes $p - q$
VELOC*TIME denotes $v \times t$ (for example)
WORK/HOURS denotes, say w/t
X**Y denotes x^y
R**2 denotes r^2 and so on.

More complicated expressions can be built up. Thus corresponding to the expression $b^2 - 4ac$ we would have

(19.10.1)   B**2 − 4.0*A*C

In forming an arithmetical expression we must ensure that *either* all symbols are fixed-point *or* all symbols are floating-point. An expression in which fixed- and floating-point symbols are mixed (called a *mixed-mode* expression), such as

$$INCOM - EXPND$$

is invalid. For this reason we had to use 4.0 and not just 4 in expression (19.10.1). There is one exception to this rule. An exponent may be fixed-point in a floating-point expression. Thus

$$B**2 \quad \text{and} \quad B**2.0$$

are both valid. They represent different ways of computing b^2; "B**2" causes the computer to multiply B by itself, but "B**2.0" causes it to take the logarithm of B, multiply it by 2.0 and take the antilog. Clearly the first method is preferable here, but if we had to compute b^{176}, it would be better (and certainly quicker) to write "B**176.0" than "B**176," while if we wished to compute $b^{17.6}$ then "B**17.6" is the only possibility.

There are certain priorities between the arithmetical operations, exactly as in ordinary algebra. Multiplication and division take priority over addition and subtraction and are performed first. Thus

$$A + B*C \quad \text{means } a + bc \text{ and not } (a + b)c$$

$$P/Q + R \quad \text{means } \frac{p}{q} + r, \text{ not } \frac{p}{q + r}.$$

Similarly exponentiation takes precedence over the other operations, so that

$$A*B**3 \quad \text{means } ab^3, \text{ not } (ab)^3$$

$$P**Q/R \quad \text{means } p^q/r, \text{ not } p^{q/r}.$$

These priorities can be overridden, however, by the use of parentheses, again exactly as in algebra. Thus to represent $(a + b)c$ we would write "(A + B)*C," and the expressions

$$\frac{p}{q + r}, (ab)^3 \text{ and } p^{q/r}$$

would be written

$$P/(Q + R) \quad (A*B)**3 \quad \text{and} \quad P**(Q/R)$$

Pairs of parentheses can be used within other pairs of parentheses to any depth, as for example

$$T/(2.0 + ((A + B)**2 + (P - Q)**3)/(3.1416*(R + S)**2))$$

which would stand for

$$\cfrac{t}{2 + \cfrac{(a + b)^2 + (p - q)^3}{3.1416(r + s)^2}}$$

In the absence of any priorities arithmetical expressions are processed from left to right. This again is as in ordinary algebra, but there is one type of expression which may cause confusion. An expression of the form

$$A*B/C*D$$

might be thought to mean $\dfrac{ab}{cd}$. But in FORTRAN there are no priorities between multiplication and division, so the expression is processed from left to right. A is multiplied by B, the result is divided by C and the result multiplied by D. Thus this expression stands for $\dfrac{abd}{c}$, just as if it had been written "(A*B/C)*D" If we want $\dfrac{ab}{cd}$ we must write "A*B/(C*D)"

Examples:

The following are some algebraic expressions and the corresponding FORTRAN expressions.

$$\left(\frac{a + b}{c + d}\right)^3 \qquad\qquad ((A + B)/(C + D))**3.$$

$$1 + \frac{2}{c} + \frac{6}{1 + c^3} \qquad 1.0 + 2.0/C + 6.0/(1.0 + C**3)$$

(Note that the constants must have a decimal point.)

$$\frac{ax^2 + bx + c}{px^2 + qx + r} \qquad (A*X**2 + B*X + C)/(P*X**2 + Q*X + R)$$

$$\left(\frac{T}{P}\right)^{x-1} \qquad\qquad (T/P)**(X - 1.0)$$

Exercise 221

In the following lines supply the missing expression—either the algebraic expression on the left or the FORTRAN expression on the right.

(a) $\dfrac{a + bc}{d + 7}$

(b) $\qquad\qquad\qquad\qquad\qquad\qquad\qquad$ A + B/C + D

(c) $\qquad\qquad\qquad\qquad\qquad\qquad\qquad$ (I + K)*5*M

(d) $\left[1 + \dfrac{x^3 - 36}{x^2 + 7}\right]^{3/2} \div \left(\dfrac{1}{m}\right)^{z-8}$

(e) $\qquad\qquad\qquad\qquad\qquad$ 1.6*H/(T + U/V − 5.0)

(f) $\dfrac{28cf}{1 + h^2} + \dfrac{9p^2}{14 + 26}R^{7/3}$

(g) $\qquad\qquad\qquad\qquad\qquad$ 2.0*P/X*(Y − 2.0)

(h) $-\dfrac{2}{3} \cdot \dfrac{(1 - e)\,eD}{(1 + e)\,2\pi r(1 + e)}$

19.11 ARITHMETICAL STATEMENTS

So far we have not produced anything in FORTRAN which really tells the computer to do anything. We have, so to speak, considered the words and phrases of the language without as yet having put these together to form sentences. We now consider what are known as *arithmetical statements*, which instruct the computer to perform a numerical calculation.

An arithmetical statement consists of three parts; first a symbol, then the character =, and thirdly an arithmetical expression. Thus

$$\text{AREA} = 3.1416*R**2$$

is an arithmetical statement. An arithmetical statement gives the following instructions to the computer:

Evaluate the arithmetical expression to the right of the =, store the result, and henceforth refer to this by the symbol on the left of =.

Thus in the example just given, the computer would be instructed to calculate the right hand side (πr^2) from the value of R (which would have to be in storage already), store the result, and designate the result by the symbol AREA. That is, if later in the program the symbol AREA were used, it would refer to the result of this calculation.

It is important to realize that an arithmetical statement, despite its appearance, is not an equation in the usual sense. It is really a definition of the left-hand symbol in terms of the symbols and constants on the right-hand side, and for this reason the left-hand side must be a single symbol, possibly subscripted, but no more. The difference is readily seen from the following example, which is of common occurence:

$$I = I + 1$$

This instructs the computer to take the value of I, or more precisely, to take the number stored in the location designated by I; to add 1 to it; and to call the result I, that is, to store this number in the same location from which I was originally taken. In other words, this statement is equivalent to

"increase I by 1".

Note that as an algebraic equation $I = I + 1$ is impossible, or at least rather unlikely!

It is not necessary for the two sides of an arithmetical statement to be in the same mode. One may be fixed-point, the other floating-point, and it is of interest to see what happens if this is so. The statement

$$R = I$$

in which the symbol being defined is floating-point, will cause the computer to take the fixed-point variable I, convert it to floating-point form, and store it as a floating-point variable R.

The statement

$$I = R$$

in which the modes are the other way round, causes the computer to convert the floating-point variable to a fixed-point number, i.e., an integer. To do this it simply cuts off the decimal part of the floating-point number R and stores the integral part in fixed-point form. Thus if R were 235.7938, the above statement would result in I being equivalent to 235.

A similar procedure is adopted with division of fixed-point numbers. The result of a fixed-point calculation is always fixed-point, so that the result of, say, the division 307/25 is not 12.28 (which is not an integer) but just 12. The decimal part is eliminated.

We shall postpone considering any examples of arithmetical statements until we have considered functions.

Functions. The calculation of several of the commoner functions can be indicated to the computer by the use of function names. These are as follows:

SQRT	(square root)
SIN	(sine—argument in radians)
COS	(cosine—argument in radians)
ATAN	(arctangent, or $\tan^{-1}$. In radians)
LOG	(logarithm to base e)
EXP	(antilogarithm to the base e).

The argument of the function is placed in parentheses after the function name. These functions are included in arithmetical expressions in an obvious way. Thus the arithmetical statement

$$Y = SIN(X)$$

causes the computer to compute the sine of X and to refer to the result as Y.

Functions of functions can be taken to any level, and the arithmetical operations can be used with functions just as if the latter were symbols for variables. In this way very complicated calculations can be concisely specified. Thus

$$Y = EXP(1.0 + ATAN(3.0*SIN(X) + 2.0*COS(X)))*SQRT(LOG(X**2 + 59.0))$$

specifies the calculation of the function

$$\exp\{1 + \tan^{-1}(3 \sin x + 2 \cos x)\} \cdot \sqrt{\log(x^2 + 59)}$$

The function SQRT(X) is equivalent to X**0.5 as far as the result is concerned, but when the function SQRT is used the calculation is performed using a subroutine which does the job much quicker than the alternative method (indicated by X**0.5) of taking the logarithm, halving it and taking the antilogarithm.

We can now look at some examples of arithmetical statements, including functions. The following are typical examples.

$$Q = SQRT (B**2 - 4.0*A*C)/(A + A)$$
$$AREA = .33333333*H*(Y(1) + 4.0*Y(2) + Y(3))$$
$$SUM = SUM + A(I)*B(I)$$
$$P = -Z/(LOG(1.0 - Z))$$
$$OMEGA = 7.915*EXP(-P*T)*(2.78*COS(G*T) - 5.19*SIN(G*T))$$

Exercise 222

1. Write arithmetical statements to evaluate the following functions, giving the result the symbol ANS.

(a) $\dfrac{x^3 + 1}{1 - 2x^5}$

(b) $\dfrac{1}{\sqrt{2\pi x}} e^{-x^2/2}$

(c) $\tan^{-1}\left(\dfrac{y}{x}\right)$

(d) $\dfrac{1}{2} \log\left(\dfrac{1 + x}{1 - x}\right)$

(e) $\sqrt{x^2 + y^2}$

(f) $\left(\dfrac{p + q}{p - q}\right)^{3t/2} \cdot \tan\dfrac{x^2 - 1}{2x}$

2.   If A is 3.62, B is 5.09, I is 10 and J is 3, state what M will be as a result of the following statements

$$M = A + B \qquad M = I + J$$
$$M = A/B \qquad M = I/J$$
$$M = B**2 \qquad M = I/J*J$$

19.12   STATEMENTS FOR TRANSFER OF CONTROL

We are now in a position where we could write in FORTRAN many of the instructions in the examples in the first half of this chapter, but not by any means all of them. If we write a sequence of FORTRAN statements one after the other, then when these statements have been translated into machine language, and the program is being run, the calculations equivalent to these statements will be performed in the order in which the statements occur. This may not always be what we want to happen, but as yet we have no means of instructing the computer to depart from this sequential performance. Methods for this *transfer of control*, as it is called, are provided by three types of FORTRAN statement, two of which we shall consider now.

First we remark that in order to tell the computer to execute next some statement other than the one which follows it in the program we must have some method of referring to the statement which we want to come next. This we do by attaching a number to the statement we wish to refer to, exactly as in the first part of the chapter. We need to know what restrictions there are in FORTRAN on these *statement numbers*, but this is settled by the following definition.

A statement number can be any unsigned fixed point integer between 1 and 9999.

It is unwise, though not wrong, to attach statement numbers to statements that you will not have occasion to refer to.

The first statement for transfer of control is an unconditional branch. It is called a GO TO statement and takes the following form

GO TO 26

This simply instructs the computer that the next statement to be executed is the one which is numbered 26 (or whatever the statement number following the words GO TO may be). We used exactly this wording earlier in the chapter so that this type of statement is already familiar.

The second kind of statement for transfer of control is the IF statement. A typical IF statement is the following:

$$IF \quad (B*B - 4.0*A*C) \ 27, 91, 5196$$

This statement will cause the computer to perform the following sequence of operations. It will first compute the value of the arithmetical expression which is enclosed in parentheses after the word IF. It then continues the program with statement 27 if the result of the computation is negative. If the result is zero it goes on to execute statement 91; if the result is positive it goes to statement 5196. Thus an IF statement is a three-way branch in which the particular branch chosen is dependent on the result of evaluating a specified arithmetical expression.

The general form of an IF statement is

1. The word IF
2. A valid arithmetical expression in parentheses
3. Three statement numbers, separated by commas.

Note that there must not be a comma after the parentheses, before the first statement number, and there must be exactly three statement numbers. FORTRAN is extremely fussy about punctuation; a misplaced comma will render a statement invalid. It is permissible to insert blanks in a FORTRAN statement where desired, however, so as to space the component parts for easier reading or greater clarity. Blanks in a FORTRAN statement are ignored. Thus the statement just written could just as well have been written

$$IF(B*B-4.0*A*C)27,91,5196$$

or

$$IF \quad (B*B - 4.0* A*C) \quad 27,91 ,5196$$

Exercise 223

Examine the following control transfer statements and say which are valid and which are invalid.
   (a) IF $(I - J + MAX)$ 26, 519, 78
   (b) GOTO7 6
   (c) IF (T45(ITEM)), 1, 2, 3
   (d) IF $((2.0 + X)/(P + LOG(1.25 + COS(A*X)))$ 7,918,56
   (e) GO TO, 70
   (f) IF (27 6.39*X − Y) 209, 1, 17
   (g) IF $(A + B − C**2)$ 378,2941

19.13 INPUT AND OUTPUT

With the GO TO and IF statements at our command we could start to write something like a genuine computer program. But we shall be able to do much better when we can also handle input and output. We shall therefore take these next.

In the version of FORTRAN that we are studying, input to the machine is by an ACCEPT statement. We shall assume that data will be entered into the computer via an electric typewriter which is *on-line* to the computer, that is, permanently attached to it. To get the computer to accept data in this way, a statement of the following form is used

ACCEPT, A, B, C

The symbols following the word ACCEPT, separated from it, and from each other, by commas, designate the variables whose values are to be read in. When such an instruction is executed the keyboard of the typewriter is released (it is locked during computation) and it is up to the operator to type the value of A. When this is done, a special key is depressed, signifying the termination of typewriter input and the number is entered into storage in the location designated by the symbol A. The carriage of the typewriter is then returned automatically, the next number is typed, and so on. When the last number in the list following ACCEPT has been typed and entered the computer carries on with the next statement in the program.

The details of input from a typewriter will vary from one make of computer to another, and there are, of course, other ways of entering data, such as from punched cards, magnetic tape, and so on. These details are not very germane to our purpose, and we shall take the above type of statement as sufficiently typical for our needs.

We shall similarly simplify our consideration of output, and look at only one output statement—that for typewriter output. A statement of the form

PRINT, VAR, VALUE

will cause the values of the variables denoted by VAR and VALUE to be printed on the computer's typewriter. The list following the word PRINT, like that following ACCEPT, can contain any number of symbols (subject to whatever limitations there may be on the number of characters permitted in a statement).

The symbols which occur in an ACCEPT statement need not have appeared previously in the program. If the symbols A, B, C have not previously been used, the statement

ACCEPT, A, B, C

will cause the computer to select suitable locations and to store therein the numbers being entered from the typewriter. The symbols A, B, C, will thenceforth refer to these locations. If the symbols have been used before, the numbers typed in will be put into the storage locations which these symbols already designate. In a PRINT statement, on the other hand, the symbols must have been defined; that is they must have occurred previously either in an ACCEPT statement (which gives them each a value, as we have just seen) or on the left-hand side of an arithmetical statement, in which case they are defined in terms of the variables on the right-hand side of that statement. This is all quite reasonable; the computer cannot be expected to print the value of a variable that it has never met!

One other statement is required before we look at a few typical FORTRAN programs, and that is the one that causes the computer to stop. This one is easy; it consists simply of the word STOP. When this statement is executed the computer stops, and the computation is at an end.

SOME PROGRAMS

In the first part of this chapter we drafted out some programs for simple calculations, writing the instructions without much reference to any conventional format (though, as the reader will have noticed, we introduced, even then, one or two concepts and notations from the FORTRAN language). In order to convert these programs into genuine FORTRAN programs it is necessary only to rewrite the instructions in the correct FORTRAN manner. We do not need to consider again the mathematics and the logic of the program. Let us see how these programs look in their new clothes.
The program of Fig. 19.4 now becomes

```
        SUM = X(1)
        I = 2
   7    SUM = SUM + X(I)
        I = I + 1
        IF (I − 1000) 7,7,8
   8    STOP
```

The program of Fig. 19.5 now becomes

```
          TOTAL = 0.0
          I = 1
   193    TOTAL = TOTAL + X(I)
          I = I + 1
          IF (I − 1000) 193, 193, 205
   205    STOP
```

The program of Fig. 19.6 now becomes

```
              ACCEPT, A,B,C
              P = −B/(A + A)
              Z = B*B − 4.0*A*C
              IF(Z) 90, 35, 25
      25      Q = SQRT(Z)/(A + A)
              ALPHA = P + Q
              BETA = P − Q
              ITEM = 1
              PRINT, ITEM
              GO TO 76
      35      ITEM = 2
              PRINT, ITEM
              PRINT, P
      88      STOP
      90      Q = SQRT(−Z)/(A + A)
              ALPHA = P
              BETA = Q
              ITEM = 3
              PRINT, ITEM
      76      PRINT, ALPHA, BETA
              GO TO 88
```

Fig. 19.17

Notice that since the version of FORTRAN that we have studied has no facilities for printing *words* (higher versions of FORTRAN have this facility) we have caused the computer to print a number (ITEM) which is, respectively, 1, 2 and 3, for real, equal and complex roots. Clearly the information content of the output is the same.

Exercise 224

1. Write a series of FORTRAN statements for the calculation of Exercise 217.4.
2. Write the FORTRAN statements necessary to read in a specified number of numbers and pick out which are the largest and smallest of them. (Compare Exercise 217.3.)
3. Write a series of FORTRAN statements for the problem given in Exercise 218.1.

19.14   THE DO STATEMENT

As already remarked, one extremely important feature of computer programs is the presence of loops, as in Figs. 19.9 and 19.12, for example.

A glance at these examples will show that there are in general three parts
to a loop.

1. An instruction which sets up an index or counter to count the number
   of times the loop is performed;
2. A sequence of instructions which form the body of the loop and are
   executed each time the program performs the loop.
3. An instruction which determines whether the loop has been performed
   the correct number of times, and hence whether it should be performed
   again or whether the program should continue with something else.

In some programs, such as that of Fig. 19.8, no index is set up. The loop is
performed over and over again until the satisfying of some criterion (probably
not directly related to the number of times the loop has been performed)
is satisfied. In such programs part (1) will not be present, but otherwise
the three parts given above will be present.

So important are loops in the makeup of computer programs that the
FORTRAN language has been made to include provision for setting up everything connected with a loop by means of a single statement. This is done by
what is known as a "DO" statement. A typical DO statement is the following

$$\text{DO } 24 \text{ I} = 1, 60$$

The interpretation of this is "Execute the statements which follow this,
up to and including the statement numbered 24, giving I the value of 1;
execute them again with I = 2, then with I = 3 and so on, until these statements have been executed with I = 60. Then carry on with the program by
executing the next statement, that is, the one following statement number 24."

The DO instruction therefore defines an index (in this case I) and sets
it up at its initial value (here I starts at 1). The set of instructions to be included in the loop are specified; they are all those from the one following the
DO statement to the one whose statement number is specified. The exit from
the loop is likewise determined; it occurs when the second specified value for
I is reached. The general form for a DO statement is

1. The word DO
2. A statement number
3. A *fixed-point* symbol, followed by =
4. Two or three fixed-point constants or symbols separated by commas.

Thus the following are examples of DO statements

DO 127 INDEX = 1,507
DO 100 N = 1, 10, 3

DO 2 MIN = 2, N
DO 17 KOOL = NP, NQ, J
DO 5199 LINE = 3, 13, 2

The third number after the = sign specifies the increment in the index on each performance of the loop. Thus the last of the above examples means "Execute all the statements up to and including statement 5199, first with LINE = 3, then with LINE = 5, and so on in steps of 2, until they have been executed with LINE = 13. Then carry on with the program." If no third number is specified it is assumed to be 1, as in our first example. If the difference between the first and second numbers specified after the = sign is not a multiple of the third, then the loop finishes when a further addition of the increment would cause the index to *exceed* the second number. Thus

DO 50 KING = 7,33,4

would cause the execution of the loop to cease when KING was 31, since a further addition of 4 would make it exceed 33. This loop would be performed 7 times, with KING taking successively the values 7, 11, 15, 19, 23, 27, and 31.

Some examples should make this clear. Suppose we wish to enter into the computer a list of numbers, which in algebra might be denoted by $x_1, x_2, \ldots, x_{100}$, but which we shall denote by X(1), X(2), X(3), ..., X(100). This can be done using only two statements, as follows:

DO 3 N = 1, 100
3   ACCEPT, X(N)

Here the body of the loop (the *range* of the DO statement) consists of just one instruction, the ACCEPT instruction. This instruction is therefore executed, first with N = 1, then with N = 2 and so on up to N = 100. Hence the computer will accept the numbers X(1), X(2), etc., one after the other.

Suppose now we want to find the sum of these numbers. This can be done by the following three statements.

SUM = 0.0
DO 19 J = 1, 100
19   SUM = SUM + X(J)

Statement 19 has the effect of adding X(J) to SUM. Since this statement is in a DO loop, it will be repeated with J = 1, 2, 3, ..., 100, and hence SUM, which was previously set to zero will have in it the sum of the X(J) when the DO loop is satisfied and the program continues.

Notice that a single DO loop would serve both to accept these 100 numbers and to find their sum. We could write

$$SUM = 0.0$$
$$DO\ 19\ J = 1,\ 100$$
$$ACCEPT,\ X(J)$$
$$19 \qquad SUM = SUM + X(J)$$

If we wished to find the sum of the squares of the X(J) as well, we could do this easily by including another statement in the DO loop.

$$SUM = 0.0$$
$$SQR = 0.0$$
$$DO\ 19\ J = 1,\ 100$$
$$ACCEPT,\ X(J)$$
$$SUM = SUM + X(J)$$
$$19 \qquad SQR = SQR + X(J)**2$$

The reader who knows some statistics will appreciate that with $\sum\limits_{j=1}^{100} x_j$ and $\sum\limits_{j=1}^{100} x_j^2$ calculated we are well on the way towards the evaluation of the standard deviation of these 100 numbers. The formula is

$$\sigma^2 = \frac{1}{n} \sum_{j=1}^{n} x_j^2 - \left(\frac{1}{n} \sum_{j=1}^{n} x_j \right)^2$$

where, in our case, $n = 100$. Let us complete a program for the evaluation of σ, but let us make it so that it can apply for differing numbers of X(J). To do this, we shall first type in how many X(J)s there are to come; after that we shall type the X(J)s one by one. The reader should verify that the following program does all that is required, including printing the answer.

$$ACCEPT,\ N$$
$$SUM = 0.0$$
$$SQR = 0.0$$
$$DO\ 19\ J = 1,\ N$$
$$ACCEPT,\ X(J)$$
$$SUM = SUM + X(J)$$
$$19 \qquad SQR = SQR + X(J)**2$$
$$EN = N$$
$$SUM = SUM/EN$$
$$SIGMA = SQRT(SQR/EN - SUM**2)$$
$$PRINT,\ SIGMA$$
$$STOP$$

Note how we have to include the statement EN = N to convert the *fixed-point* N into the *floating*-point EN, in order to avoid a mixed-mode expression in the following statement.

At the conclusion of the above portion of a program the numbers X(1), X(2), ... , X (N)would all be stored in the computer, ready for later use. If it so happened that these numbers were not required again, there would be no point in keeping them. This could be avoided by rewriting the program in the following way

```
         ACCEPT, N
         SUM = 0.0
         SQR = 0.0
         DO 19 J = 1, N
         ACCEPT, X
         SUM = SUM + X
   19    SQR = SQR + X**2
         EN = N
         SUM = SUM/EN
         SIGMA = SQRT(SQR/EN − SUM**2)
         PRINT, SIGMA
         STOP
```

Here, each number is read into the location denoted by X, thus replacing the previous number read in. It is added to SUM, and its square is added to SQR. Then the loop is repeated for the next number, until all have been read.

Let us now apply our technique of using a DO loop to the problem of calculating $\exp(x)$ for $x = 1.00$ up to 2.00, as in Fig. 19.13. Of course, since FORTRAN contains allowance for calculating EXP(X) in one statement, one would never want to implement the program of Fig. 19.13 in the FORTRAN language; but it will serve as an example of the method of summing an infinite series to a given degree of accuracy. The outer loop of this program is performed a specified number of times, viz., 21 times, and therefore is very amenable to treatment by a DO statement. The number of times the inner loop is performed depends in a not very obvious way on X. Hence it is probably best to leave that as it is, the criterion for terminating the loop being dealt with by an IF statement. We obtain in this way the following FORTRAN program.

```
            X = 1.00
            DO 100 K = 1, 21
   60       ANS = 1.0
            R = 1.0
            TERM = 1.0
   17       RESLT = TERM*X/R
            IF(RESLT − 2.5E − 7) 10,20, 20
   20       TERM = RESLT
            ANS = ANS + TERM
            R = R + 1.0
            GO TO 17
   10       PRINT, X, ANS
   100      X = X +0.05
            STOP
```

Exercise 225

Write FORTRAN programs for the calculations in Exercise 218.2, 218.3, 218.4.

19.15 FURTHER STATEMENTS

The statements given so far are probably sufficient to give the reader the "feel" of computer programming. However, in order that the reader can be able to write *complete* programs for at least one type of computer, we briefly describe below some additional statements that are possible (and in some cases required) in the FORTRAN without FORMAT language for the IBM 1620 computer.

COMPUTED GO TO

The GO TO statement is a sort of one-way branch. There is also provision for a many-way branch; this is the *computed* GO TO statement. In this statement, a list of statement numbers is specified, and the program branches to a statement bearing one of these numbers. Which one it branches to depends on the value of a fixed-point variable specified in the GO TO statement. A typical example is

$$\text{GO TO } (5, 276, 1, 93, 9, 6), \text{ MU}$$

The effect of this statement is that if $MU = 1$ the program branches to the *first* statement specified, viz., statement number 5; if $MU = 2$ the program branches to the second statement specified, viz., statement 276; similarly if $MU = 3$ then the program branches to statement 1; if $MU = 4$ it goes to statement 93, and so on.

DIMENSION

When a subscripted variable is used in a program its values, $X(1)$, $X(2)$, $X(3)$, ..., for example, are stored sequentially in the computer. In order for this to be done it is necessary to tell the computer what is the maximum number of numbers that are represented by this subscripted variable. This information is imparted by a "DIMENSION" statement, in which the subscripted variables used in a program are presented together with the maximum values that the subscripts will take. If in a program we use a subscripted variable $X(I)$ and I goes from 1 to 100, and we use also a doubly-subscripted variable ARRAY (M, N) in which M can go from 1 to 50, and N from 1 to 15, we set out these specifications as follows.

$$\text{DIMENSION } X(100), \text{ ARRAY}(50, 15)$$

This statement causes the computer to set aside 100 locations to receive the numbers X(I), and $50 \times 15 = 750$ locations to receive the numbers ARRAY(M, N). Dimension statements should be at the beginning of the program. Clearly if no subscripted variables are used then no DIMENSION statement needs to be specified.

PAUSE

A statement consisting of the word "PAUSE" causes the computer to stop computing; but whereas the STOP statement implies the end of all computing for the program, PAUSE indicates only a temporary cessation of computing. On the console of the computer is a button marked START. If this is pressed after the execution of a PAUSE statement the program will continue, beginning at the statement following the PAUSE. The PAUSE statement is often used to interrupt a program in order, for example, to change the paper in the typewriter, or to load a deck of cards, and for many other purposes.

CONTINUE

For technical reasons, which we need not go into, the final statement of a DO loop (the one that is referenced in the DO statement) must not be an IF statement or a GO TO (ordinary or computed). To cover cases where one would want to have such a statement ending a DO loop a dummy statement "CONTINUE" is supplied. A CONTINUE statement gives rise to no computation of any sort; it contributes nothing towards the final program. It simply serves as a convenient peg on which to hang the end of the DO loop which otherwise would end in an IF or a GO TO statement.

CONTINUE is also useful in other respects. If a program splits into several branches which subsequently reunite, a CONTINUE statement often serves as a convenient meeting place for the several branches. Because of the manner in which it is used, a CONTINUE statement *must* be numbered.

As an example of the use of a CONTINUE statement, consider the following program, which adds 25 positive numbers X(1), X(2), ..., X(25) and, in the ordinary course of events, carries on with statement 500, but which goes to statement 600 if, at any stage, the sum of the numbers so far added exceeds 1000.

```
          SUM = 0.0
          DO 88 M = 1, 25
          SUM = SUM + X(M)
          IF (SUM — 1000.) 88, 88, 600
   88     CONTINUE
   500    . . .
```

The CONTINUE statement is necessary here to avoid ending the DO loop on an IF statement.

END

As we have already seen, a FORTRAN program, once written, must be translated into machine language before it can be run on the computer. This is done by a special program, the compiler which is loaded into the computer. The FORTRAN statements are then entered into the computer one by one, and the corresponding batches of machine instructions are punched on cards. Since a FORTRAN program can be of any length, and will not necessarily end with a STOP statement (see Fig. 19.13), there is no way for the computer to "know" when the operator has entered the last statement of his program. There might be more to come, and the computer will await the entry of further statements unless the operator gives an indication that there are no more. This is done by entering the word "END." This statement *must* therefore be the last statement of any FORTRAN program. Note that END signifies the end of the FORTRAN program; it has nothing to do with how the program will end (if it does) when it is being executed. When the END statement is encountered, the compiler winds up the business of translating from FORTRAN into machine language, and the deck of cards that has been produced is then ready for use whenever the program is required to be run. When the program is to be run, the operator simply loads this deck of cards into the machine and the computation is then carried out.

This concludes our account of the FORTRAN language. Since we have discussed only the simplest version of FORTRAN, the reader who has occasion to write a program for a particular computer will probably find that he has more statements and more rules to learn. But what has been done in this chapter will serve as a foundation for what the reader will need; if he has grasped the basic principles with which this chapter has been mainly concerned, he will have little difficulty in adapting what he has learnt to the particular requirements of the computer to which he has access. And even if it should turn out that the computer is not one on which FORTRAN can be used, the contents of this chapter should still help towards learning whatever programming language does apply.

Exercise 226 (Chapter Review)

1. Consider how you would arrange for a computer to read in an unordered list of numbers (say from cards—one number per card) and store them in *ascending order of magnitude* inside the computer. Draw a flow chart for the method which you consider the most practical. Write such a program, either in the informal language used in the first part of this chapter or, if you wish, in FORTRAN.

2. Write a FORTRAN program to print a table of values of the function defined by

$$J(x) = 1 - \frac{x^2}{(2!)^2} + \frac{x^4}{(4!)^2} - \frac{x^6}{(6!)^2} + \cdots$$

for $x = 0$ (0.1) 2.0, i.e., x from 0 to 2.0 in steps of 0.1, to an accuracy of 6 decimal places.

3. Write a FORTRAN program to read in and evaluate a 4×4 determinant.

4. Write a FORTRAN program to solve a set of 3 linear equations in 3 variables.

5. Draw a flow chart for a program to read in the payoff matrix of a game and to determine whether the game is strictly determined or not. Realize this flow chart by a FORTRAN program.

   Note: Make the output a printed "0" if the game is not strictly determined, and a "1" if it is.

5. Write a program that will compute the successive iterations leading to the solution of the game given in section 17.6. What sort of criterion could be used to determine when the iterations should stop?

7. Write a program to compute the successive probability vectors in a Markov chain with 5 states. The program should read in the stochastic matrix, check that it is indeed stochastic (typing out, shall we say, the number "9999" if it is not); then read in the initial probability vector and check that its elements sum to 1. Then it should type out successive probability vectors. Make some provision for stopping the program at a suitable stage.

Suggestions for Further Reading

The main difference, indeed the only *essential* difference, between the very simple dialect of FORTRAN described in this chapter, and the "higher" versions of FORTRAN used on the biggest and most modern computers is that the latter make provision for control of the *format* of the input and output data of a program so that it becomes possible to exercise control over the way in which the data are printed, their positioning on the page, the number of decimal places retained, and so on. In addition, it is possible to print out messages, table headings, captions, etc., along with the output data. (The reader may have noticed that, although we allowed statements like "Print the word COMPLEX" in the first half of the chapter, the language developed in the second half made no provision for doing this.) Even programmers of the IBM 1620 (on which the programs in this chapter *can* be run as they stand) will normally use a more elaborate version of the FORTRAN language.

The reader who wishes to write and run his own programs will therefore almost certainly find that he will need to learn about control of format, and the *format statements* by which this is effected. A discussion of these statements was deliberately omitted from this chapter since, though they are undoubtedly important, they do not contribute much to the general understanding of the basic techniques of programming with which this chapter has been chiefly concerned. Moreover, the rules governing their use are fairly elaborate and would have constituted a major digression from the overall theme of the chapter.

These rules are not difficult to learn, however, and if the reader cares to curl up for an hour or so with one of the books recommended below, he can learn enough about format for most of his programming needs. With the addition of suitable format statements where appropriate, the FORTRAN programs given in this chapter could be run on almost any modern computer.

In addition to providing for format control the higher FORTRAN languages have many additional statements and various tricks of the trade to enable programmers to write quicker and more efficient programs with less time and trouble. Generally speaking these additional instructions perform calculations which *could* be specified in the simple FORTRAN of this chapter, but would be cumbersome if done in this way. These extra facilities in the language, too, are not difficult to pick up, and to a large extent can be learned piecemeal, as the reader finds he has need of them.

The following books will help the reader to amplify the material of this chapter, and to expand his knowledge of FORTRAN to the stage where he can look any modern computer straight in the eye.

Harvill, J.B., *Basic Fortran Programming*. Englewood Cliffs, N.J.: Prentice-Hall, Inc., 1966.

McCracken, D.D., *Introduction to FORTRAN Programming*. New York: John Wiley & Sons, Inc., 1961.

Smith, R.E., *FORTRAN Autotester*. New York: John Wiley, Sons, Inc., 1962.
    This latter book is an amusingly written "programmed learning" textbook, which is very easy to learn from.

The student who has access to a particular computer on which to run his programs should also consult the FORTRAN manual issued by the manufacturers of the computer. This will give information on the idiosyncracies of the particular computer. Thus (to take but one example) on many computers you can have fixed-point numbers with more than four digits. Details of this kind are often not given in the more general textbooks.

Considering the rapid improvements in computer technology, and the corresponding rapid obsolescence of computers themselves, the reader may well wonder whether FORTRAN might itself become out-of-date in the near future. He can be reassured on this point. It may not be *quite* true that

"Computers may come and computers may go,
But FORTRAN goes on forever."

but since FORTRAN is not tied to any one particular computer, it will certainly be used for many years to come. FORTRAN progams can be run on any computer for which a FORTRAN compiler has been written—and this, nowadays, means nearly all of them.

Nevertheless one should not overlook the fact that FORTRAN is only one of many programming languages in common use. On the severely practical side, used for accounting and other routine business calculations, is the language COBOL (*CO*mmon *B*usiness *O*riented *L*anguage), to which the following book gives a good introduction.

McCracken, D. D., *A Guide to COBOL Programming*. New York: John Wiley & Sons, Inc., 1963.

On the more theoretical side there is the language ALGOL (*Algo*rithm *l*anguage) which is becoming extensively used not only for computer programming but also for discussion of computational procedures between human beings. The reader who wishes to learn about this language can consult

Baumann, R., Feliciano, M., Bauer, F.L., and Samuelson, K., *Introduction to ALGOL*. Englewood Cliffs, N.J.: Prentice-Hall, Inc., 1964.

Index

(Where a whole section is devoted to an entry in the index, the section number has been given in parentheses after the page number.)